Advanced Nanomaterials in Biomedical Application

Advanced Nanomaterials in Biomedical Application

Editors

Goran Kaluđerović
Nebojša Pantelić

MDPI

Basel • Beijing • Wuhan • Barcelona • Belgrade • Novi Sad • Cluj • Manchester

Editors
Goran Kaluđerović
Department of Engineering and Natural Sciences
University of Applied Sciences Merseburg
Merseburg
Germany

Nebojša Pantelić
Faculty of Agriculture
University of Belgrade
Belgrade
Serbia

Editorial Office
MDPI
St. Alban-Anlage 66
4052 Basel, Switzerland

This is a reprint of articles from the Special Issue published online in the open access journal *Nanomaterials* (ISSN 2079-4991) (available at: www.mdpi.com/journal/nanomaterials/special_issues/adv_nano_biomedical).

For citation purposes, cite each article independently as indicated on the article page online and as indicated below:

Lastname, A.A.; Lastname, B.B. Article Title. *Journal Name* **Year**, *Volume Number*, Page Range.

ISBN 978-3-0365-8613-7 (Hbk)
ISBN 978-3-0365-8612-0 (PDF)
doi.org/10.3390/books978-3-0365-8612-0

Contents

nanomaterials

Editorial

Advanced Nanomaterials in Biomedical Application

Goran N. Kaluđerović [1,*] and Nebojša Đ. Pantelić [1,2,*]

[1] Department of Engineering and Natural Sciences, University of Applied Sciences Merseburg, Eberhard-Leibnitz-Str. 2, 06217 Merseburg, Germany
[2] Department of Chemistry and Biochemistry, Faculty of Agriculture, University of Serbia, 11080 Belgrade, Serbia
* Correspondence: goran.kaluderovic@hs-merseburg.de (G.N.K.); pantelic@agrif.bg.ac.rs (N.Đ.P.)

Citation: Kaluđerović, G.N.; Pantelić, N.Đ. Advanced Nanomaterials in Biomedical Application. *Nanomaterials* **2023**, *13*, 1625. https://doi.org/10.3390/nano13101625

Received: 18 April 2023
Accepted: 10 May 2023
Published: 12 May 2023

Over the last few decades, great efforts have been dedicated to the discovery of various nanomaterials. Due to their unique optical, magnetic, and electrical properties (among others), they have found applications in medicine (drug delivery), agriculture, electronics, catalysis, etc. Thus, due to the increasing possibilities, the need to design and fabricate novel nanoparticles is rapidly increasing. In this Special Issue "Advanced Nanomaterials in Biomedical Application", a total of seventeen articles—including five reviews—have been published, addressing the most recent advances in nanomaterials in terms of both synthesis and characterization as well as technological applications. In the following, we provide a brief overview of the key findings presented in this Special Issue.

In recent decades, platinum-based drugs have been widely used for the treatment of many types of cancer. However, their clinical utility is limited by severe side effects, such as neuro-, nephro-, and ototoxicity, and the development of resistance. One of the strategies for overcoming these issues is the development of nanoparticles that can enhance cellular accumulation in target cells and reduce the associated toxicity of drugs in normal cells. In this regard, Giusto et al. [1] prepared a 2D graphene-oxide-based nanoplatform functionalized with highly branched, eight-arm polyethylene-glycol, which enhanced the efficiency and loading capacity of platinum-based drugs, achieving a high-performance and stable nanodelivery system. The results have shown that the fabricated nanocarrier enables the application of lower amounts of Pt-drugs than a Pt-free complex to attain similar outcomes. Furthermore, the nanoplatform achieves excellent cellular proliferation inhibition in osteosarcoma, which is also observed in glioblastoma but in a less pronounced manner. Moreover, the presented nanoplatform also shows promise for inhibiting migration, especially in highly invasive breast carcinoma (e.g., MDA-MB-231 cells). Accordingly, the prepared nanoplatform represents an interesting tool for the treatment of various cancers. In addition, Predarska and co-workers [2] reported the synthesis of three novel platinum(IV) conjugates of cisplatin, containing derivatives of caffeic and ferulic acid in their axial positions, and their immobilization into SBA-15 particles for the preparation of corresponding mesoporous silica nanoparticles (MSNs). The prepared complexes showed higher or comparable antiproliferative activity with respect to cisplatin against four human breast cancer cells (BT-474, MCF-7, MDA-MB-468, and HCC1937). This activity increased significantly after immobilization in SBA-15, and the IC_{50} values were more than 1000 times lower compared to cisplatin. Furthermore, the derivative with the highest activity cisplatin–diacetyl caffeate conjugate and its MSNs induced apoptotic cell death by causing potent caspase activation. Moreover, in vivo studies conducted using BALB/c mouse models with breast tumors showed that the same compound and its MSNs exhibit tumor growth inhibition with a reduced necrotic area and lowered mitotic rate. The review paper by Spoială [3] focused on smart magnetic drug delivery systems for cancer treatment. The authors comprehensively describe the effectiveness of using nanotechnology and magnetic nanoparticles to facilitate the early detection and selective destruction of cancer cells. This review will help new researchers to obtain comprehensive information in the field of drug delivery systems for cancer diagnosis and treatment. Peserico et al., described the usefulness of nanoparticles (NPs) in the diagnostic and/or therapeutic sector [4]. The review presents a comprehensive summary of the accessible technologies targeting cell–NP interaction and/or detection in cancer and regenerative medicine. The key nanocarrier-impacting elements (e.g., the typology and functionalization of NPs, the tuning capacity of cells' in vitro and in vivo interaction mechanisms, and labeling with NPs) were analyzed.

The rapid spread of bacteria and antibiotic resistance requires new infection control strategies. In this sense, nanomaterials seem to be promising tools for maximizing drug activity due to their unique size and properties. Quach and co-authors [5] developed an efficient method for improving the antibacterial properties of graphene oxide (GO) by growing nanosilica (NS) on the surface of GO. The silver nanoparticles (AgNPs) that were immobilized on nanosilica to create a composite GO/NS/AgNPs exhibited remarkable antibacterial activity against *Escherichia coli* and *Bacillus subtilis*, suggesting that this system has great potential as an efficient antibacterial coating for medical equipment and other surfaces. In the paper submitted by Lima et al. [6], a new nanoparticle-based approach to modulating the harmful inflammatory consequences of fungal infection for the host using β1,3-glucan-functionalized polystyrene nanoparticles (β-Glc-PS) was presented. Moreover, these nanoparticles were able to down-modulate a *Candida-albicans*-induced proinflammatory response of host immune cells in a size-dependent manner. In the study by Salmerón-Valdés et al. [7], the antibacterial activity against *Streptococcus mutans* and the mechanical properties of conventional and hybrid type I glass ionomers modified with and without halloysite nanotubes loaded with chlorhexidine were investigated. Based on the obtained results, the authors concluded that the addition of nanotubes preloaded with chlorhexidine at concentrations of 5% and 10% effectively inhibited the presence of *S. mutans*, especially regarding the dose–response relationship, while maintaining and improving mechanical properties. This suggests that the addition of halloysite nanotubes to conventional and resin-modified glass ionomer cements could be a new method for counteracting orthodontic ligament injury, offering the advantage of maintaining and improving mechanical properties.

The studies conducted by Xue et al. [8] have shown an opposite color response of a giant polyoxometalate, namely, a brown Keplerate cluster abbreviated as $[Mo_{132}]$ and containing 72 Mo(VI) and 60 Mo(V), to the existing states of the human papillomavirus (HPV) major capsid protein, L1-pentamer (L1-p), and virus-like particles (VLPs). The color responses result from the different binding modes between $[Mo_{132}]$ and the capsid protein. This straightforward colorimetry approach is of importance to estimating the existing states of the HPV capsid protein and could be used in the future to analyze the quality of the HPV vaccine and the existing states of other viruses.

Many studies have shown that CSC chemokine receptor 4 (CXCR4) is a promising target for cancer therapies, and intracellular siRNA delivery to suppress CXCR4 expression in cancer cells is an effective therapeutic strategy. Thus, Cao et al. [9] synthesized carriers, by preparing heptafluorobutyryl-polyethylene glycol-polyethyleneimine (FPP) used to coat magnetic nanoparticles (MNPs) to obtain magnetic nanocarriers, FPP@MNPs, for siRNA delivery and CXCR4 knockdown. The results show that the cellular uptake efficiency of the FPP@MNPs was significantly improved and that they exhibited low cytotoxicity. Furthermore, the siRNA transfection efficiency was validated in various cell lines, with the result showing that the developed nanocarriers could effectively reduce CXCR4 expression on the cell membrane.

Min and co-authors [10] presented a new type of injectable composite hydrogel using glycol chitosan (GCH), a water-soluble derivative of chitosan, and amino-functionalized bioactive glass nanoparticles (ABG NPs) to construct a single crosslinker using genipin (GN) or dual crosslinkers as a combination of GN and poly(ethylene glycol)diglycidyl ether (PEGDE). Using ABG NPs with GCH while employing GN as a single crosslinker, the fabricated ABG/GCH gels exhibited strength and elasticity that was moderately dependent on the spacer length of the ABG NPs. On the other hand, the combination of GN and PEGDE as dual crosslinkers significantly improved the strength and elasticity of the gels, while the gelation time was adjustable. Moreover, some optimally dual-crosslinked ABG/GCH gels were able to support the growth of seeded osteoblast-like cells and enhance matrix deposition. The obtained results indicate that a novel dual-crosslinked hydrogel has potential applications in bone repair.

The work of Tyubaeva et al. [11] explores the effect of hemin on the structure and properties of electrospun nanocomposite materials based on poly-3-hydroxybutyrate (PHB) and on their biocompatibility and antimicrobial activity. The results indicate that the presence of hemin significantly improves the structural properties of the material. The antimicrobial activity of hemin ensured that both Gram-negative and Gram-positive cultures died after contact with the PHB–Hmi fiber materials. These materials could be used for regenerative medicine, as dressing layers, as hygienic agents, as filter materials, and for other clinical products that require an advanced surface combined with antimicrobial properties and biocompatibility.

Ren and coauthors [12] summarized the recent advances in developing lipid-, metal-, carbon-, and polymer-based nanomaterials for antibacterial applications, covering the latest nanotechnologies for the design and development of nano- and nanocomposite materials designed to combat multidrug-resistant bacteria. Moreover, further development of antimicrobial nanomaterials was discussed.

Ding et al. [13] designed protein corona cationic liposomes (CLs) with AT-1002 (TJ regulatory peptide) possessing a core–shell structure based on the characteristics of BSA. Liraglutide was effectively encapsulated in the CLs, with the drug EE% of the liposomes equaling $85 \pm 5\%$ and the average particle size equal to 203 ± 13 nm. The drug itself, liraglutide, showed good structural stability. The protein corona liposomes had a good intestinal internalization effect, longer intestinal absorption time, and good biological safety. The adaptive protein corona liposomes have potential applications in the oral administration of proteins and peptides.

In their review article addressing SARS-CoV-2, a hot topic lately, Kianpout et al. [14] gave an overview of the structure of the virus and the cause of its pathogenicity. Moreover, the authors examine the biotechnological methods for vaccine production and the available nano-based concepts for overcoming this viral pandemic. The optimal conditions for the production of nano-mediated vaccines are discussed, and biotechnological solutions for forthcoming viral strikes are examined.

A new nano-emulsion adjuvant based on squalane (SNA: Span85, Tween60, squalane, polyethene glycol-400) containing CpG was prepared, and its in vivo properties were examined [15]. The SNA particles (diameter ca. 95 nm) showed good stability and biocompatibility. SNA was used for the preparation of a foot-and-mouth disease virus-like particle vaccine. Within 4 weeks in BALB/c mice, the SNA-VLPs vaccine significantly increased specific antibody levels, including IgG1 and IgG2a and, in the immune serum, IFN-γ and IL-1β. In guinea pigs, upon treatment with SNA, a noticeable enhancement of specific and neutralizing antibodies was observed within 4 weeks, thus enabling the proliferation of splenic lymphocytes. Outstandingly, one dose of SNA-VLPs immunized the guinea pigs with a protection rate of up to 83%, which was comparable to the group treated with the ISA-206, indicating that this novel formulation is an effective adjuvant for the FMD-VLP vaccine.

Harada et al., reported a useful technique for protein delivery to macrophages [16]. By applying low-binding mutant $SubAB_{S35A}$ (subtilase cytotoxin; S35A –in B subunit 35th serine mutated to alanine) with poly(D,L-lactide-co-glycolic) acid (PLGA) nanoparticles, the selective delivery of cytotoxin to macrophages, in comparison to epithelial cells, was suggested. This drug delivery system presents anti-inflammatory effects.

In the study by Wen et al. [17], osteoimmunomodulatory nanoparticles for bone regeneration, which is a complex process that involves osteoblasts and osteoclasts (skeletal cells) as well as immune cells, an overview of the most important literature is presented. Namely, the authors highlighted the importance of osteoimmunology in bone regeneration. Moreover, the progress and application of nanoparticle-based methods for bone regeneration and macrophage-targeting drugs, respectively, for advanced osteoimmunomodulation are summarized.

Author Contributions: G.N.K. and N.Đ.P. contributed to the editorial. All authors have read and agreed to the published version of the manuscript.

Funding: This article received no external funding.

Acknowledgments: We are thankful to all the authors who contributed to this Special Issue. We also express our acknowledgments to all referees for reviewing the manuscripts.

Conflicts of Interest: The authors declare no conflict of interest.

References

1. Giusto, E.; Žárská, L.; Beirne, D.F.; Rossi, A.; Bassi, G.; Ruffini, A.; Montesi, M.; Montagner, D.; Ranc, V.; Panseri, S. Graphene oxide nanoplatforms to enhance cisplatin-based drug delivery in anticancer therapy. *Nanomaterials* **2022**, *12*, 2372. [CrossRef] [PubMed]
2. Predarska, I.; Saoud, M.; Drača, D.; Morgan, I.; Komazec, T.; Eichhorn, T.; Mihajlović, E.; Dunđerović, D.; Mijatović, S.; Maksimović-Ivanić, D.; et al. Mesoporous silica nanoparticles enhance the anticancer efficacy of platinum(IV)-phenolate conjugates in breast cancer cell lines. *Nanomaterials* **2022**, *12*, 3767. [CrossRef] [PubMed]
3. Spoială, A.; Ilie, C.-I.; Motelica, L.; Ficai, D.; Semenescu, A.; Oprea, O.-C.; Ficai, A. Smart magnetic drug delivery systems for the treatment of cancer. *Nanomaterials* **2023**, *13*, 876. [CrossRef] [PubMed]
4. Peserico, A.; Di Berardino, C.; Russo, V.; Capacchietti, G.; Di Giacinto, O.; Canciello, A.; Camerano Spelta Rapini, C.; Barboni, B. Nanotechnology-Assisted Cell Tracking. *Nanomaterials* **2022**, *12*, 1414. [CrossRef] [PubMed]
5. Quach, Q.; Abdel-Fattah, T.M. Silver nanoparticles functionalized nanosilica grown over graphene oxide for enhancing antibacterial effect. *Nanomaterials* **2022**, *12*, 3341. [CrossRef] [PubMed]
6. Lima, T.; Gunnarsson, S.B.; Coelho, E.; Evtuguin, D.V.; Correia, A.; Coimbra, M.A.; Cedervall, T.; Vilanova, M. β-Glucan-functionalized nanoparticles down-modulate the proinflammatory response of mononuclear phagocytes challenged with *Candida albicans*. *Nanomaterials* **2022**, *12*, 2475. [CrossRef] [PubMed]
7. Salmerón-Valdés, E.N.; Cruz-Mondragón, A.C.; Toral-Rizo, V.H.; Jiménez-Rojas, L.V.; Correa-Prado, R.; Lara-Carrillo, E.; Morales-Valenzuela, A.A.; Scougall-Vilchis, R.J.; López-Flores, A.I.; Hoz-Rodriguez, L.; et al. Mechanical properties and antibacterial effect on mono-strain of *Streptococcus mutans* of orthodontic cements reinforced with chlorhexidine-modified nanotubes. *Nanomaterials* **2022**, *12*, 2891. [CrossRef] [PubMed]
8. Xue, Y.; Wei, M.; Fu, D.; Wu, Y.; Sun, B.; Yu, X.; Wu, L. A visual discrimination of existing states of virus capsid protein by a giant molybdate cluster. *Nanomaterials* **2022**, *12*, 736. [CrossRef] [PubMed]
9. Cao, Y.; Zhang, S.; Ma, M.; Zhang, Y. Fluorinated PEG-PEI coated magnetic nanoparticles for siRNA delivery and CXCR4 knockdown. *Nanomaterials* **2022**, *12*, 1692. [CrossRef] [PubMed]
10. Min, Q.; Wang, C.; Zhang, Y.; Tian, D.; Wan, Y.; Wu, J. Strong and elastic hydrogels from dual-crosslinked composites composed of glycol chitosan and amino-functionalized bioactive glass nanoparticles. *Nanomaterials* **2022**, *12*, 1874. [CrossRef] [PubMed]
11. Tyubaeva, P.M.; Varyan, I.A.; Nikolskaya, E.D.; Mollaeva, M.R.; Yabbarov, N.G.; Sokol, M.B.; Chirkina, M.V.; Popov, A.A. Biocompatibility and antimicrobial activity of electrospun fibrous materials based on PHB and modified with hemin. *Nanomaterials* **2023**, *13*, 236. [CrossRef] [PubMed]
12. Ren, R.; Lim, C.; Li, S.; Wang, Y.; Song, J.; Lin, T.-W.; Muir, B.W.; Hsu, H.-Y.; Shen, H.-H. Recent advances in the development of lipid-, metal-, carbon-, and polymer-based manomaterials for antibacterial applications. *Nanomaterials* **2022**, *12*, 3855. [CrossRef] [PubMed]
13. Ding, R.; Zhao, Z.; He, J.; Tao, Y.; Zhang, H.; Yuan, R.; Sun, K.; Shi, Y. Preparation, drug distribution, and in vivo evaluation of the safety of protein corona liposomes for liraglutide delivery. *Nanomaterials* **2023**, *13*, 540. [CrossRef] [PubMed]
14. Kianpour, M.; Akbarian, M.; Uversky, V.N. Nanoparticles for coronavirus control. *Nanomaterials* **2022**, *12*, 1602. [CrossRef] [PubMed]
15. Shi, X.; Yang, K.; Song, H.; Teng, Z.; Zhang, Y.; Ding, W.; Wang, A.; Tan, S.; Dong, H.; Sun, S.; et al. Development and efficacy evaluation of a novel nano-emulsion adjuvant for a foot-and-mouth disease virus-like particles vaccine based on squalane. *Nanomaterials* **2022**, *12*, 3934. [CrossRef] [PubMed]

16. Harada, A.; Tsutsuki, H.; Zhang, T.; Yahiro, K.; Sawa, T.; Niidome, T. Controlled delivery of an anti-inflammatory toxin to macrophages by mutagenesis and nanoparticle modification. *Nanomaterials* **2022**, *12*, 2161. [CrossRef] [PubMed]
17. Wen, J.; Cai, D.; Gao, W.; He, R.; Li, Y.; Zhou, Y.; Klein, T.; Xiao, L.; Xiao, Y. Osteoimmunomodulatory nanoparticles for bone regeneration. *Nanomaterials* **2023**, *13*, 692. [CrossRef] [PubMed]

nanomaterials

MDPI

Review

Smart Magnetic Drug Delivery Systems for the Treatment of Cancer

Angela Spoială [1,2], Cornelia-Ioana Ilie [1,2], Ludmila Motelica [1,2], Denisa Ficai [2,3,*], Augustin Semenescu [4,5], Ovidiu-Cristian Oprea [2,3,5] and Anton Ficai [1,2,5]

1 Department of Science and Engineering of Oxide Materials and Nanomaterials, Faculty of Chemical Engineering and Biotechnologies, University Politehnica of Bucharest, 1-7 Gh Polizu Street, 011061 Bucharest, Romania

2 National Centre for Micro and Nanomaterials, and National Centre for Food Safety, Faculty of Chemical Engineering and Biotechnologies, University Politehnica of Bucharest, 313 Spl. Independentei, 060042 Bucharest, Romania

3 Department of Inorganic Chemistry, Physical Chemistry and Electrochemistry, Faculty of Chemical Engineering and Biotechnologies, University Politehnica of Bucharest, 1-7 Gh Polizu Street, 050054 Bucharest, Romania

4 Departament of Engineering and Management for Transports, Faculty of Transports, University Politehnica of Bucharest, 313 Spl. Independentei, 060042 Bucharest, Romania

5 Academy of Romanian Scientists, 3 Street Ilfov, 050045 Bucharest, Romania

* Correspondence: denisa.ficai@upb.ro

Citation: Spoială, A.; Ilie, C.-I.; Motelica, L.; Ficai, D.; Semenescu, A.; Oprea, O.-C.; Ficai, A. Smart Magnetic Drug Delivery Systems for the Treatment of Cancer. *Nanomaterials* **2023**, *13*, 876. https://doi.org/10.3390/nano13050876

Academic Editors: Goran Kaluđerović and Chao Deng

Received: 22 December 2022
Revised: 22 February 2023
Accepted: 23 February 2023
Published: 26 February 2023

Abstract: Cancer remains the most devastating disease, being one of the main factors of death and morbidity worldwide since ancient times. Although early diagnosis and treatment represent the correct approach in the fight against cancer, traditional therapies, such as chemotherapy, radiotherapy, targeted therapy, and immunotherapy, have some limitations (lack of specificity, cytotoxicity, and multidrug resistance). These limitations represent a continuous challenge for determining optimal therapies for the diagnosis and treatment of cancer. Cancer diagnosis and treatment have seen significant achievements with the advent of nanotechnology and a wide range of nanoparticles. Due to their special advantages, such as low toxicity, high stability, good permeability, biocompatibility, improved retention effect, and precise targeting, nanoparticles with sizes ranging from 1 nm to 100 nm have been successfully used in cancer diagnosis and treatment by solving the limitations of conventional cancer treatment, but also overcoming multidrug resistance. Additionally, choosing the best cancer diagnosis, treatment, and management is extremely important. The use of nanotechnology and magnetic nanoparticles (MNPs) represents an effective alternative in the simultaneous diagnosis and treatment of cancer using nano-theranostic particles that facilitate early-stage detection and selective destruction of cancer cells. The specific properties, such as the control of the dimensions and the specific surface through the judicious choice of synthesis methods, and the possibility of targeting the target organ by applying an internal magnetic field, make these nanoparticles effective alternatives for the diagnosis and treatment of cancer. This review discusses the use of MNPs in cancer diagnosis and treatment and provides future perspectives in the field.

Keywords: magnetic nanoparticles; targeting nanoparticles; linkers; passive targeting; active targeting

1. Introduction

Cancer the second most widespread disease as 14.6% of all human deaths are a consequence of cancer. According to the American Cancer Society, worldwide cancer represents one of the major public health problems, currently surpassed only by cardiovascular diseases. The International Agency for Research on Cancer (IARC) has estimated that without increased global investment in cancer research and the application of existing knowledge on cancer control, and without significant efforts to improve global cancer control, cancer deaths could increase to 12.9 million/year by 2030 [1].

Cancer is a disease characterized by uncontrolled, random, invasive cell division. Over the years, special efforts have focused on detecting cancer risk factors. For some types of cancer, the aetiology has been associated primarily with the specific environment, such as radiation and pollution, but also with an unhealthy lifestyle, such as a poorly balanced diet, smoking, stress, and lack of physical activity. All these factors strongly influence the development of different forms of cancer [2,3]. Inherited genetics is another determining factor in cancer occurrence with 5–10% of cases being due to this factor [4]. Advancing age is another crucial risk factor for cancer, and many individual cancers have an increased mutation risk and aggregation of factors connected with age [5].

Cancer treatment includes surgery, radiation therapy, immunotherapy, and chemotherapy. Chemotherapy is used in over 50% of cancer cases as standard treatment, including metastatic cancers [6]. A major drawback of chemotherapy involves the reduced effectiveness of targeted drug delivery to tumour cells, causing unintended penetration of drugs into healthy cells and tissues. They could lead to side effects at a systemic level, including fatigue, hair loss, nausea, vomiting, and increased infections due to low blood cell counts (by affecting the blood-forming cells of the bone marrow). Higher doses of anticancer drugs are used to achieve the necessary drug concentration in the tumour cells, causing even more side effects due to the toxicity of the chemotherapeutic agents on healthy cells and tissues [7]. Another significant disadvantage of chemotherapy is a tumour's intrinsic or acquired resistance to the drug, which often leads to disease reoccurrence and further decreases therapeutic outcomes. During the last decades, treatment resistance has been a major research topic leading to discoveries, such as cancer stem cells, sequence mutations, and bidirectional inter-conversion of cancer stem and non-stem cell populations [8].

Nevertheless, implementing successful cancer treatment protocols that lead to very good outcomes will require surpassing these elements of difficulty by a considerable refinement of our knowledge concerning the treatment, and therefore improving the survival chances of the patients. The current anticancer drugs are very potent killing factors. The next logical step is to manage, and design targeted delivery systems, which can release the drug inside the tumour, blocking the drug's capability to attack the healthy tissue [9]. Such results can be obtained by magnetic carriers, which can be designed to simultaneously assure targeting and triggering, and thus to develop smart systems able to generate personalized therapy.

2. Nanostructured Carriers

In an attempt to replace current cancer treatments, various nanomaterials, such as liposomes [10–12], immunoliposomes [13], MNPs [14–16], polymers, nanogels, etc., are being used in clinical trials to ensure targeted delivery of the biologically active agents into the desired tissue/organ, according to a desired release profile. Table 1 illustrates the most important nanocarriers in the targeted delivery of different chemotherapeutic agents.

Such carriers increase the circulation time in chemotherapeutic agents' bloodstream, improving their accumulation and retention in tumour cells or tissues. It sometimes increases chemotherapeutic agents' release across physiological barriers at the disease site [17,18].

Table 1. Type of nanoparticles used as carriers.

Type of Carriers		Advantage	Disadvantage	References
Inorganic carriers	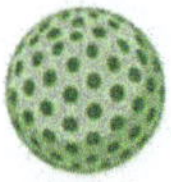 Silica nanoparticle  Quantum dot nanocrystal  Iron oxide nanoparticle  Silver nanoparticle  Gold nanoparticle  Gold nanorod	Small size; Special magnetic, electric, and optic properties; Tuneable size, structure, and functionalization; Suitable for theranostic application.	Low solubility; Toxicity.	[19,20]
Lipid-based carriers	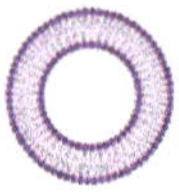 Liposome 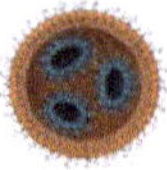 Lipid nanoparticle(3D)  Emulsion	Ease formulation for specific applications; High bioavailability; Hydrophilic and lipophilic carriers; Chemical modification; Improve blood circulation.	Carrier flexibility; Low encapsulation efficiency.	[19,20]
Polymeric carriers	 Polymeric micelle  Nanosphere	Easily controlled; Surface modification; Biodegradable; Hydrophilic and hydrophobic carriers.	Self-aggregation; Toxicity.	[19,20]

As a result of their nanometric dimensions (10 to 100 nm) and the enhanced permeability and retention rate (EPR) of the tumoral cells, these nanocarriers accumulate in tumour cells or tissues much faster than in healthy cells or tissues. This phenomenon can be explained by the fact that the tumour cells are more active, and internalization is also faster because they need more nutrients and oxygen to grow (which is why angiogenesis is faster), leading to increased therapeutic efficacy and reduced side effects [21]. The carrier's size is essential because small particles, such as the quantum dots, facilitate cell internalisation, and the surface charge and chemistry are crucial [22].

Among the various MNPs used as systems for releasing chemotherapeutic agents, special attention has been shown to Fe_3O_4 nanoparticles. They offer opportunities for biomedical applications due to their superparamagnetism [23]. However, there are a series of major disadvantages, such as high susceptibility to acid and oxidative degradation. Additionally, the high degree of agglomeration is due to strong van der Waals and magnetic attractions between particles, which cause the accumulation of MNPs and limit the practical applications of these nanoparticles. To overcome these disadvantages, coating the nanoparticles with an outer protective layer is an effective strategy to maintain the magnetic components' stability and avoid excessive agglomeration [24,25]. An efficient and often used procedure to achieve this is by encapsulating Fe_3O_4 nanoparticles in an inorganic (C, SiO_2, ZnO, etc.) or organic coating (PEG is the most used polymer) to obtain magnetic systems with a core–shell structure [26,27], which can expand their technical application as a result of the unique characteristics of the coating (high stability in biological conditions) and their ability to provide a platform for chelating groups. However, the use of Fe_3O_4@inorganic for cancer treatment is limited. Most inorganic shells are hydrophobic and chemically inert, which disadvantages their applications in an aqueous environment. Indeed, additional surface modification can be done to achieve the proper stability, hydrophilic/hydrophobic ratio, bioaccumulation tendency, and release rate, but also an appropriate reactivity of the surface [28].

Therefore, an organic coating of MNPs can accomplish more than one goal: provide a suitable surface for hydrophilic interactions, loading capacity similar to a sponge, antibody functionalization for targeted delivery, biocompatibility to evade the human body immune system, and capacity to deceive the tumour cell defence mechanisms [29–32]. A promising synthetic material extensively reported in the literature for surface modifications of MNPs is polyethylene glycol (PEG) [33], a hydrophilic, highly water soluble, biocompatible, non-antigenic, and protein-resistant polymer [24]. According to Tai et al. [34], compared to unmodified MNPs, PEG-coated MNPs showed high colloidal stability for up to 21 days. This long-term stability is not always necessary, as cytostatic release occurs within only a few days. Depending on the application, release profile, loading capacity, and targeted delivery are much more important. Therefore, finding suitable capping agents for a particular application is still a limitation/challenge of current approaches.

In oncology, MNPs are specially designed as Trojan horses. They are expected to internalize into the tumoral cells, followed by additional therapies, such as the release of antitumoral agents or radiotherapy [35–37]. This approach is beneficial because, in this way, the systemic toxicity is decreased (the targeting being assured by magnetic fields or by decorating these carriers with specific receptors: folic acid, specific peptides, and antibodies being some of the most studied). These carriers primarily accumulate in the desired cells and tissues [38–40].

3. Antitumoral Agents and Nanomedicine

The National Institute of Health has defined "nanomedicine" as the applications of nanotechnology for the treatment, diagnosis, monitoring, and controlling of biological systems. Scientists have focused on researching the adequate modality to deliver and target pharmaceutical, therapeutic, and diagnostic agents, and they turned towards nanomedicine. Using nanomedicine in cancer means identifying a precise target with specific clinical conditions and choosing suitable nanosystems-drug conjugates to achieve the desired response while minimizing the side effects of anticancer drugs. Therefore, today's nanotechnology and nanomedicine approaches have designed and expanded the basis of drug formulations for humankind's benefit [41].

Nanomedicine has great potential to improve anticancer therapy. Thus, a relatively small number of nanomedicine products are approved for clinical trials that could ensure specific therapeutic benefits for patients. Still, the perspectives in the field are immense. Improving nanomedicine's clinical impact in cancer therapy requires novel perspectives on establishing smart strategies. By integrating clinical trials, pharmaceutical companies, and the authorities in developing innovative anticancer drugs, some crucial achievements are expected, including improved efficiency and lower toxicity [42,43]. Nanoscience has shown that designing various drug formulations with enhanced diagnostic and therapeutic effects could shift into synthesizing a mono-nano-drug [44,45]. Several features must be followed to configure nano-drug formulations. For example, their core is based on organic or inorganic molecules, or a combination must hinge for the intended use. Properties, such as surface charge, tuneable size, and hydrophobicity, could be improved for the desired function [46,47]. In this viewpoint, nanomedicine shifted attention to a significant class of nanoparticles, such as MNPs, with great applicability in developing platforms to combat cancer [48].

MNPs have unique characteristics, being able to be used successfully in diagnosing and treating cancer due to their special physicochemical properties. Due to their easy synthesis, high affinity to surface functionalization, low toxicity, and good biodegradability, MNPs act as outstanding imaging tools and drug delivery carriers in cancer therapies [49–51]. MNPs possess great biomedical potentials, such as biosensing [52], magnetic hyperthermia [53], MRI [54], and controlled drug release [55]. Due to their high magnetization, MNPs have drawn attention towards a new imaging technique.

Moreover, to understand the applications of MNPs in cancer therapy, one must consider their synthesis, characterization, size, shape, and coating (Figure 1). MNPs usually have a magnetic core–shell and a polymeric coating, showing that essential properties are enhanced through functionalization, decoration, and surface coating. The literature includes some traditional syntheses for MNPs, such as co-precipitation, sonochemistry, hydrothermal, or solvothermal methods, reverse microemulsion, pyrolysis, and thermal decomposition [56].

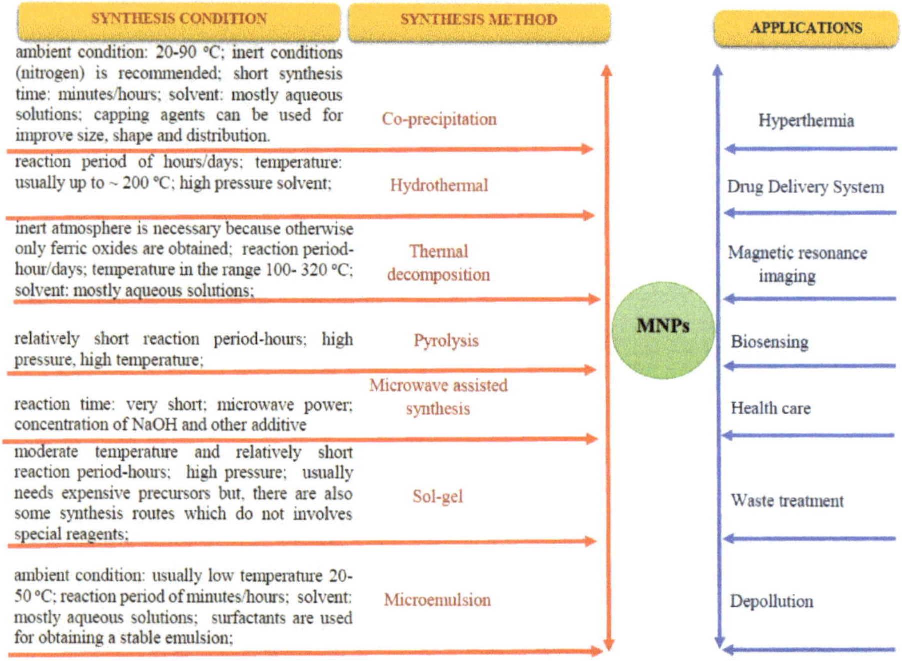

Figure 1. Correlation between synthesis method and application.

Lately, researchers have designed novel synthesis strategies, such as biogenic and microfluidic synthesis. Microfluidic synthesis uses many materials, such as glass, silicon, ceramics, and polymers, which provide significant advantages over the final size, shape, and homogeneity of the formed nanoparticles. For example, Cabrera et al. [57] designed a latex-based microfluidic platform synthesizing gold and iron oxide nanoparticles. It is worth mentioning that the obtained nanoparticles could be mixed and form 10 nm-sized iron oxide NPs decorated with 4 nm AuNPs with monodisperse core sizes.

This new application of MNPs implies the ability of the nanoparticles to detect cancerous cells within diverse necrotic tissues. This capacity could be explained by the fact that the shape, and physical and chemical properties of MNPs are interdependent. The main condition-based advantages of the MNPs are morphology, chemical composition, shape, size, magnetic and functionalization features, which are very important aspects for the desired biomedical usage [58,59]. The magnetic properties of MNPs could be customized with a biocompatible coating to increase their specificity for targeted cancerous tissue. The biocompatible coating can form many surface modifications that could transform into various multi-functionalities approaches of MNPs [60].

Nanoparticles synthesized in the presence of dispersing agents are more reactive than materials synthesized in the absence of dispersing agents due to their high surface/volume ratio [56]. Additionally, they show a high degree of protection against oxidation due to MNPs coating. By covering the MNPs, the degree of toxicity in the body is prevented and reduced in the case of their use in in vivo applications. The choice of the coating materials has to take into account the nature of the coating, especially from the point of view of stability, and the possibility of further physical or chemical functionalization, according to the desired final application. The following examples show a range of coating materials used to develop medical application carriers [61].

The functionalization of the surfaces of magnetic materials using organic linkers is widely used because organic linkers confer specific surface properties to the various biomolecules to be linked. In the specialized literature, a series of organic, such as amines, carboxylic acids, aldehydes, and thiols linkers, are used in the functionalization process of MNPs, the most organic linkers used are those that create electrostatic interactions. This is because the binding strength is relatively easy to manipulate, either by adding ions or changing the pH of the medium. For example, for the use of MNPs as carriers for gene delivery, the surface of the MNPs must be strongly positively charged, as they must ideally bind through electrostatic interactions, a large amount of negatively charged DNA molecules. These electrostatic interactions allow the release of genes after the internalization of MNPs in the cell [62]. For the delivery of drugs, such as ibuprofen and aspirin, which have negatively charged groups (e.g., carboxylate, sulphonate, etc.) in their molecular structure, the surface of MNPs must be functionalized with organic linkers that decorate the surface of these nanoparticles with positive groups, such as ammonium salts. Once these MNP-drug systems reach the target organs, drug delivery will be triggered due to the anion exchange (chlorides and phosphates) [61].

Another material used to modify the surface of MNPs is amorphous silica and/or mesoporous silica. The modification of MNPs with silica is usually carried out as the hydrolysis of tetraethyl orthosilicate (TEOS) at a neutral-slightly alkaline pH by salinization with functionalized silanes such as 3-aminopropyl trimethoxysilan or the neutralization of silicic acid.

Using natural polymers in coating MNPs is a common alternative, due to their high biocompatibility, to use them in medical applications. For example, MNPs covered with dextran are used in the treatment of cancer, the modification of the surfaces of the MNPs being possible by making hydrogen bonds between the -OH groups in the structure and those behind the surface of the MNPs. Dextran was used with other organic (alginate, chitosan, poly-L-lactic acid) and inorganic (silica) polymers to improve the surface properties and generate new ones [61].

The following section of the review is devoted to using superparamagnetic iron oxide materials (SPIONs) in cancer therapy. SPIONs are considered promising nanostructured platforms suitable for drug delivery due to their functionalization ability, good biocompatibility, and enhanced contrast effects for MRI [63].

Zuvin et al. [64] showed the anticancer properties of SPIONs stabilized with polyacrylic acid on breast cancer tumours. The modified NPs successfully provided more effective treatment, showing low toxicity, high stability, and ultra-small particle size.

A study by Kandasamy et al. [65] reported designing a new multifunctional magnetic-polymeric NPs built from ferrofluids by encapsulating hydrophobic SPIONs stabilized with oleylamine into the PLGA-based NPs, with two drugs, curcumin or verapamil. The authors investigated the biocompatibility, magnetic properties, and heating capacity of the formed MF-MPNs. The results showed that PLGA encapsulation significantly improved the stability of SPIONs with minimal toxicity and maximum treatment efficiency.

In this case, polycaprolactone (PCL)-coated SPIONs were used to design a new therapeutic drug with improved thermosensitivity and cytocompatibility. The obtained nanomedicine showed significant advantages, high stability, good dispersion, cytocompatibility, and the possibility of high control when heating [66].

4. Magnetic Hyperthermia

Magnetic hyperthermia is another treatment option, along with chemotherapy and radiation therapy, used in cancer therapy [67]. For therapeutic purposes, hyperthermia therapy (HT) involves exposure to high temperatures of the whole body or specific areas of the body to achieve a therapeutic effect. In recent years, this therapy has begun to be increasingly used as a complementary form of cancer treatment [68].

Techniques currently used to achieve a localized hyperthermic effect are radiofrequency, ultrasound, microwave, laser, and MNPs. The use of MNPs to generate therapeutic hyperthermia is known as magnetic hyperthermia (MHT), and was used as a cancer therapy in 1957 [68]. MHT was first used when Gilchrist et al. [69] selectively heated tumours using magnetic particles by applying an alternating magnetic field (AMF). Later, MHT was introduced clinically as an alternative approach for the local treatment of tumours without affecting the surrounding healthy tissues [70–73].

Hyperthermia treatment generally uses heat from various sources, such as electromagnetic waves or ultrasound, to destroy cancer cells by denaturing the proteins that combine the cell membrane and cytoplasm. Techniques currently used to achieve a localized hyperthermic effect are radiofrequency, ultrasound, microwave, laser, and MNPs. The idea of hyperthermia used as an artificial temperature inducer above the threshold of 46 °C in the human body has been around for decades. Hyperthermia is usually used with other therapies, such as chemotherapy or radiotherapy [5,35–37].

Available MHT techniques do not efficiently direct the heat to the tumour, thus exhibiting low efficiency. Consequently, this lack of efficiency has led to the development of nanomaterials with properties capable of dissipating energy at the tumour site, increasing efficiency and specificity for deeper tumours [6,21]. In addition, the development of diverse strategies for the cellular internalization process and/or paramagnetic nanoparticles [22], such as iron oxide nanoparticles, contributes to increasing the efficiency of this therapy.

Heat destroys tumour cells due to the low dissipation of thermal energy associated with the heterogeneity of oxygen intake and nutrient demand caused by the excessive tortuous branching of blood vessels and the absence of lymphatic vessels. The increase in temperature leads to the modification of the functions of structural and enzymatic proteins in tumour cells, followed by the modification of the cell proliferation index that finally shows their apoptosis/necrosis [74,75].

The heating capacity is closely related to the NPM material's properties and the parameters of the applied field. However, for nanostructured magnetic materials, the heating effectiveness relies on the correlation between the intrinsic time-dependent relaxation processes of the NPM magnetic moments and the time scale of the AMF field vector [76].

The MHT technique is based on two MNP relaxation processes, Néel relaxation and Brownian relaxation, both associated with intracellular and extracellular MHT processes [77]. In the case of nanoparticles internalized in tumour cells, by applying an alternating magnetic field (AMF), only the Neel relaxation contributes to the thermal energy (intracellular MHT). Brownian relaxation does not contribute to the thermal energy due to the high viscosity of the medium, which does not allow the nanoparticles to rotate freely.

For nanoparticles not internalized in tumour cells, the contribution of Néel relaxation and the contribution of Brownian relaxation are relevant to the extracellular MHT process [78].

MHT consists of using the heat generated by MNPs when applied to an alternating magnetic field (AMF) to destroy cancer cells by denaturing the proteins that make up the cell membrane and cytoplasm, resulting in altered physiology of the cancer cell, which ultimately leads to their apoptosis/necrosis (Figure 2).

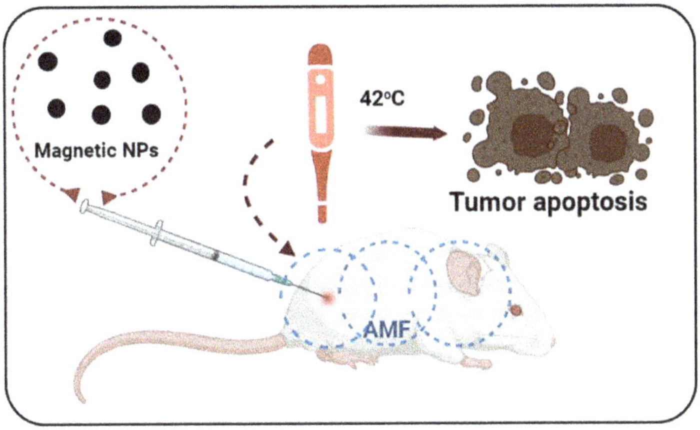

Figure 2. Schematic illustration for the MNPs-mediated MH working mechanisms under AMF.

In exposing iron oxide nanoparticles to alternating magnetic fields, heat is generated by Neel relaxation and Brownian rotation. Ideally, using active or passive targeting processes, MNPs can be targeted to the tumour, where they accumulate. Thus, by applying an external magnetic field, the temperature of the cells around the MNPs is considerably increased compared to the temperature of the cells further away. Thus, cells in the vicinity of MNPs can be selectively destroyed, at the same time reducing the body's exposure period to external stimuli (Figure 2).

MNPs-mediated MH therapeutic treatment (MNPs-MH) helps achieve intracellular hyperthermia [72] by applying an alternating magnetic field. This directly leads to the therapeutic heating of cancer cells, the local and homogeneous heat obtained leads to higher selectivity and efficacy of therapy. The most significant advantage of MNPs-mediated MH (MNPs-MH) is deep tissue penetration and selective killing of cancer cells without affecting healthy cells and tissues. As a result of these advantages, MNPs-MH-based tumour treatment was introduced into clinical trials not long ago, and has been successfully used for the treatment of glioblastoma, pancreatic cancer, and prostate cancer [79]. This technique has proven effective, producing good results in various preclinical studies performed on animals and phase III clinical studies on humans [80,81].

5. Passive Targeting

Passive targeting of a target organ is an essential process because the free accumulation of MNPs in the tumour area leads. In some cases, effective MRI detection facilitates cancer diagnosis and, implicitly, its subsequent treatment. Nanoparticles preferentially accumulate in tumour tissues due to EPR for suitable-sized NPs [82–84]. It has been demonstrated that an NP with d = 10–100 nm may be accumulated predominantly in tumours, compared to normal tissues. Passive targeting is strongly influenced by the nanoparticles' size, surface charge, and hydrophobicity [85]. For example, NPs smaller than 20 nm predominantly accumulate in the kidneys. In comparison, the NPs with d = 30–150 nm predominantly accumulate in the bone marrow, heart, kidneys, and stomach. NPs with d > 150 tend to bio-accumulate in the liver and spleen. Another important property that influences passive targeting is represented by retention times. Due to opsonisation, Gobbo et al. [86] showed that hydrophobic and positively charged structures have short circulation times. In contrast, hydrophilic and negatively charged nanostructures have long circulation times [86,87].

As is known from the literature, tumours develop a series of abnormal vessels through which the blood supply takes place, and abnormal vessels that present cracks through which NPs can accumulate in tumours [86,88]. As observed in Figure 3, there is an essential difference in the blood vessel construction between healthy and tumour tissue. As the tumoral tissue grows, the blood vessels are hastily created to keep pace with the increasing need for nutrients. This generates poorly constructed blood vessels permeable to the magnetic nanoparticles or other nanometre drug delivery systems. In addition, these nanoparticles are preferentially accumulated in the tumoral cells and tissues because more blood vessels irrigate these tissues, and their uptake is upregulated. By this passive targeting, therapeutics only target the tumour cells, leaving the healthy ones unharmed.

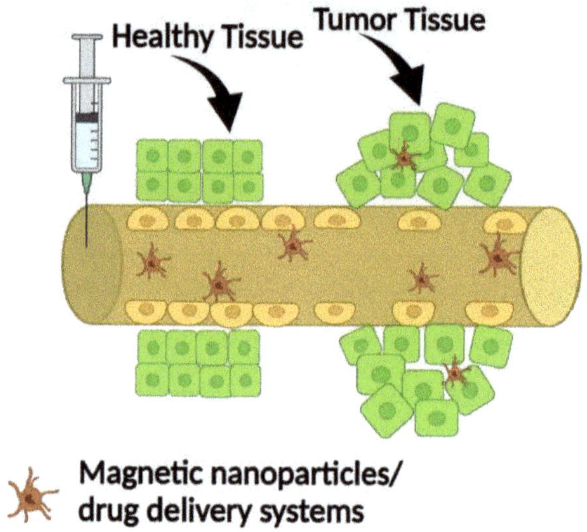

Figure 3. Passive targeting of tumour tissue occurs in blood vessel construction.

6. Active Targeting

In treatments with magnetic support, biologically active molecules are conjugated with the surface of MNPs or could be encapsulated in the MNPs, such as liposomes, micelles, or dendrimers [85,89]. Thus, applying an external magnetic field to the charged MNPs with active molecules should be brought to and maintained in the area of interest. Figure 4 illustrates the action mechanism of passive and active targeting.

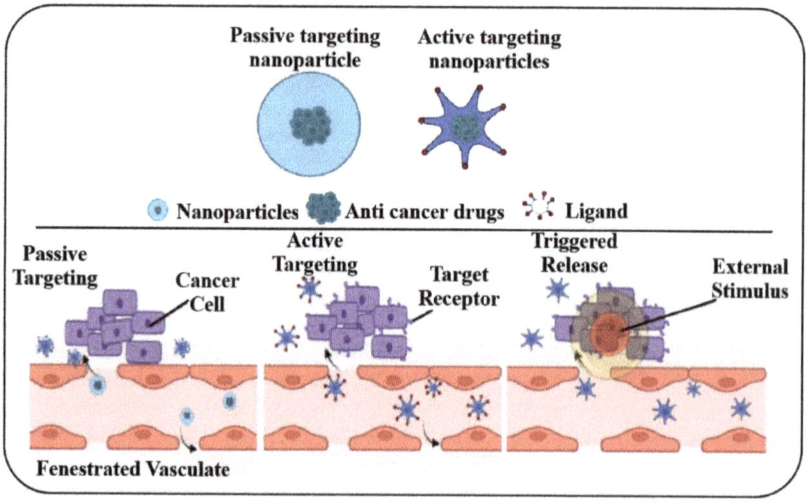

Figure 4. Action mechanism of passive and active targeting.

To understand the mechanism of action of passive and/or active targeting, one must comprehend the specifically targeted organ, and which method is appropriate for the best outcome. As already presented in the previous section, passive targeting is essential and correlated to the accumulation of MNPs and their circulation time. In contrast, active targeting occurs precisely, directly delivering therapeutics to the targeted cancerous cells by ligands to receptors [89,90].

7. Smart Magnetic Drug Delivery Systems

A simple, smart drug delivery system is represented by COLL/HA-Fe_3O_4@cisplatin [91], which was proposed to treat bone cancer. In this case, the multifunctional magnetic composite material loaded with cisplatin is implanted in the bone defect, as presented by Andronescu et al. [92,93]. In this case, such multifunctional materials can be used, and the loco-regional delivery is suitable to avoid systemic toxicity (Figure 5). These systems can be considered smart because they can be externally controlled. Due to the induced hyperthermia, the cisplatin delivery rate can be enhanced; thus, the antitumoral activity is enhanced. Such systems are suitable candidates for assuring personalized therapy, and depending on the evolution of the healing, the active agents' release can be adapted.

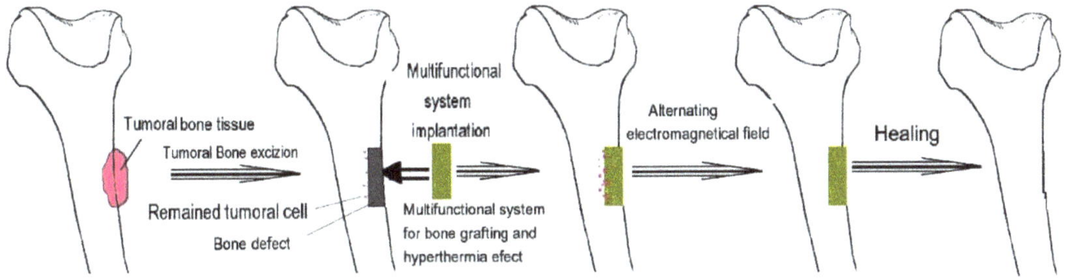

Figure 5. The procedure of treating osteosarcoma using multifunctional materials (with the kind permission of Springer [92]).

Due to its unique properties, such as high stability and the possibility of very easy functionalization via thiol (–SH) linkers [94], gold (and silver) is one of the most frequently used materials as a decorating material [95]. It is well known that various sulphur compounds, such as thiols and more, have a high affinity for gold [61]. Gold-coated MNPs were described in the literature in 2001, when researchers [68], using the reverse micellar mechanism, synthesized multifunctional systems of the "Fe_3O_4@Au" type with a core–shell structure and diameters 18–80 nm. These systems were later functionalized for binding biomolecules using -SH as linkers with an amine functional group [61]. When magnetic core and Au/Au nanoparticles are assembled via the thiolic linkers, the obtained systems can be used in cancer therapy. Magnetite nanoparticles can ensure magnetic targeting and produce hyperthermia. At the same time, these nanoparticles can exhibit photothermic effects and antimicrobial activity while providing the specific binding of receptors. If the silver content is high enough, core@shell@shell structures (Fe_3O_4@Cys@NM, where NM = noble metals, Au/Ag) can be obtained. The as-obtained magnetic systems can be loaded with adequate antitumoral drugs, and their release will be triggered by NIR or magnetic fields (Figure 6) [96].

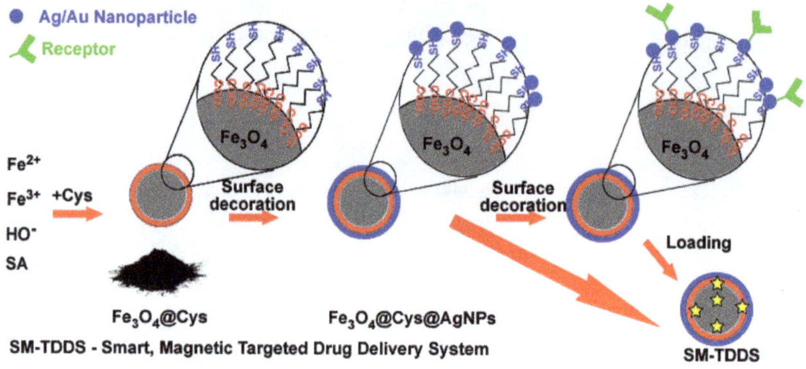

Figure 6. Smart MNP with triggered delivery and targeted capacity [96].

Recently, several studies have developed silica based MNPs for cancer therapy which will be presented next. Hsiao et al. [97] developed an innovative theranostic platform consisting of L-cysteine-grafted mesoporous folic acid-europium-gadolinium-silica (FA-EuGd-MSNs-SS-Cys). This innovative platform has proven to be an effective tool for transporting, imaging, and delivering therapeutic drugs. The advantage of this approach is to function simultaneously as a therapeutic and guided ageing agent in cancer treatment. Another example showed that Fe_3O_4@SiO_2@ tannic acid NPs are used as pH-sensitive drug delivery systems to simultaneously release the anticancer drugs methotrexate (MTX) and doxorubicin (DOX). Findings reported that Fe_3O_4@SiO_2@Tann could be considered a double-shell nano-drug platform for treating cancer [98,99].

Dai et al. [100] developed a DOX-loaded Fe_3O_4@SiO_2 platform by the solvothermal method. The results showed that 82.8% of lung cancer cells were killed by Fe_3O_4@SiO_2@DOX treatment with a drug concentration of only 10 μg/mL of DOX. Furthermore, 81.3% of lung cancer cells were killed during incubation with Fe_3O_4@SiO_2@DOX with a concentration of only 0.5 μg/mL of DOX and 15 min of NIR irradiation, thus proposing a unique synergism between chemotherapy and the photothermal effect [100].

Recently, combining chemotherapy and photothermal therapy with strong theranostic NPs, and integrating diagnostic and therapeutic agents has been a tremendously beneficial alternative in the fight against cancer. Elbialy et al. [101] developed multifunctional magnetic gold NPs (MGNPs) conjugated with PEG and DOX type MGNP_DOX. It was found that by integrating chemo/photothermal treatment, MGNP_DOX had higher efficiency both in vivo and in vitro. Additionally, examination of MGNP-DOX by immunohistochemical and histopathological studies confirmed using it as a theranostic material. Additionally, using MGNP-DOX as an MRI contrast agent for synergistic chemo/photothermal targeted therapies has led to promising results.

As acknowledged worldwide, triple-negative breast cancer (TNBC) is a very aggressive cancer cured via standard chemotherapy. Therefore, it is foremostly required to develop an innovative approach to treating TNBC [102]. In the presented study, Li et al. [103] fabricated a multifunctional magnetic gold hetero nanostructure with photosensitizer Ce_6 (chlorine e6) loading (MF_MGN@Ce_6) for synergic effects of photodynamic and photothermal (PDT/PTT) ability of TNBC. The obtained nano-drug MF_MGN@Ce_6@RT was functionalized with mitochondria-targeting molecular and cell membrane-targeting peptide of cRGD of TPP. Results confirmed the guaranteed efficiency of the MF_MGN@Ce_6@RT to TNBC tumours.

Lately, using natural polysaccharides in nanomedicine applications has gained significant interest. Chitosan (CS) has been used in cancer therapies as a drug delivery nanocarrier with promising achievements in targeted drug delivery [104,105]. Shanavas et al. [106] synthesized a hybrid based on MNPs with a core of PLGA functionalized with a folate-CS shell as an MRI contrast having anticancer features. The folate/CS shell was obtained from carbodiimide covered by SPIONs with a thin layer of PLGA loaded with docetaxel. In conclusion, it has been shown that the biocompatible hybrid core–shell has immense potential for simultaneous MR imaging and cancer treatment [106].

Adimoolam et al. [90] designed MNPs conjugated with DOX via a pH-sensitive imine bond with glutaraldehyde as a cross-linker. Cell viability tests confirmed that the MNPs conjugated DOX presented an improved therapeutic impact. An important outcome of MNPs was the low toxicity to the normal cell, which was assigned to the precise targeting capacity.

Recently, magnetic carbon-based nanomaterials gained significant attention due to their ability to develop platforms in which ligands or drugs are conjugated for cancer therapy applications. An example was Ag@Fe_3O_4@C_PEG_FA (FA-folate) NPs loaded with DOX, which have suitable biocompatibility and stability and exhibit synergistic potential and targeting capacity without any toxic side effects. Therefore, it has been proven that Ag@Fe_3O_4@C_PEG_FA NPs are suitable nano-platforms for chemo/photothermal therapy and imaging contrast [107].

Another example of a nano-drug platform for cancer therapy is discussed next. Kievit et al. [108] synthesized a new formulation based on chitosan-PEG-PEI coated on SPIONs functionalized with chlorotoxin (CTX) and green fluorescent protein (GFP) encoded DNA. The use of chlorotoxin is justified based on the targeting capacity of this ligand to brain tumours, such as glioma. After administering the complex DNA_CTX in the C_6 xenograft tumour in mice, Kievit discovered an increased uptake of the complex DNA_CTX in the targeted tumour compared to the control DNA. Therefore, Kievit et al. [109] demonstrated the importance of copolymer coating of Fe_3O_4 NPs in targeting and gene therapy applications [110].

Even if there have been made significant breakthroughs in clinical trials with various nano-drug based on MNPs platforms towards cancer theranostic applications, further research still needs to be done to overcome many issues related to long-term toxicity and nano-bio human toxicity interactions. However, despite their importance in many clinical studies, especially in cancer therapy, no MNPs formulations are approved for therapeutic use [110]. Additionally, a significant problem was directing the target towards the organ, with each method presenting advantages and disadvantages.

Tumour delivery of macromolecular drugs based on MNPs has attracted great interest in cancer therapy. However, macromolecular drugs could be proteins, peptides, DNA-based constructs, and even lipids. It has been demonstrated that many proteins and peptides have enhanced biological effects, making them perfect therapeutics for developing anticancer agents. It is well known that tumour tissues differ anatomically from normal tissues, which makes them well-distinguished. Using macromolecules for cancer therapy has the advantage of penetrating and accumulating only in tumour tissues, which leads to extensive pharmacological accumulation. To achieve significant development and therapeutic potential of the anticancer drugs, they must attach macromolecules to based-targeting nanoparticles. One of the most commonly used nanoparticles as nanocarriers for targeted therapy are iron oxide nanoparticles, which are used as contrast agents for specific targeted drug delivery [54,111,112].

Yang et al. [113] developed an enzyme-responsive hybrid DOX-SMNPs (silica MNPs) for selective drug delivery and intracellular tumour imaging. The hybrid of enzyme-responsive nanoparticles demonstrated effective DOX release upon specific enzyme interaction in vitro. This multi-responsive hybrid exhibits better cellular tracking of DOX molecules through the help of MRI and fluorescence imaging techniques.

Another study presented an anticancer drug delivery system based on MNPs grafted with carboxymethyl chitosan (CS) and β-cyclodextrin (β-CD). The obtained nano-drug was designed to enhance the delivery of prodigiosin (PG) to cancer cells. The anticancer drug, prodigiosin, loaded into the nanoparticles, was used as a model antitumor drug, targeting aggressive tumour cells. The results show that CS-MNPs presented efficiency and better targeting ability for prodigiosin toxicity effect on cancerous cells than β-CD-MNPs [114].

Wang et al. [115] synthesized polyethylenimine (PEI)-MNPs as nanocarriers for tumour treatment of glioblastoma multiforme, a very aggressive type of malignant brain tumour. Scientists are working to develop effective treatments for curing this bold, incurable brain tumour. The literature provides numerous intriguing insights regarding using surviving as a potential new target for cancerous tumour treatment. Currently, controlling the expression of surviving RNA might be the best approach for cancer research. Additionally, it is a valuable device for specific proteins encoded by mRNA, which become potential cancer therapeutic [116]. Therefore, choosing the appropriate carrier system for small interfering RNA (siRNA) delivery was challenging. MNPs are viable carriers for siRNA delivery due to their unique properties, such as size, nontoxicity, biocompatibility, great stability, and easy functionalization. Therefore, developing PEI-MNPs for siRNA gene delivery has practical applications in brain tumours. Results confirmed that fabricating PEI-MNPs with a cationic polymeric shell could effectively absorb adequate siRNA molecules and protect them from enzymatic media in vitro. This study indicated that the nanocarriers presented successful results in imaging and siRNA delivery for in vitro therapy of glioblastoma multiforme [115].

Another study developed a new magnetic nanovector targeting transgene therapy for oral squamous cell carcinoma (OSCC). The magnetic nanovector fabricated from MNPs modified by PEI polymer was tested as gene transfer vectors. The results indicated that PEI-modified Fe_3O_4 nanoparticles could target Tca83 cell killing and provide a potentially novel method for the future treatment of the OSCC [117].

Tanaka et al. [118] developed an intratumoral injection of immature dendritic cells (DCs) combined with magnetite cationic liposomes (MCL) with the induced effect of hyperthermia in vitro studies. When the DCs were pulsed with mouse B16 melanoma-heated cells, major histocompatibility complex (MHC) I/II and costimulatory molecules (CD80/CD86 and CCR7) were upregulated, concluding in DCs maturation. Additionally, it has been reported that DCs regulate the immune response in tumour cancer and have proven that they were triggered by heat shock proteins (HSPs). Very important to mention is that HSP70 induces antitumor immunity after hyperthermia. Therefore, suggesting that injecting DCs into tumour tissue will release HSP70 after hyperthermia implies that using magnetite nanoparticles is possible for malignant melanoma.

Another example of nanocarriers for cancer therapy was developed by Sun et al. [119], demonstrating that coupling DOX with BMs (bacterial magnetosomes) displayed efficient antitumour properties. In this study, bacterial magnetosomes (BMs) were used as carriers in cancer therapy and coupling DOX with BMs (DBMs) showed compatible results similar to DOX. BMs' potential as drug carriers for enzymes, nucleic acids, and antibodies has been experimentally used for developing magnetic-targeted drug carriers. Results indicated that BMs showed good biocompatibility, and DBMs may offer an innovative target in cancer therapy; thus, clinical studies must be done to overcome any research obstacles.

Colon cancer is considered one of the most aggressive types of cancer. Therefore, studies have shown that the anticancer activity of human cathelicidin LL-37 peptide attached to the surface of MNPs would considerably be improved. Niemirowicz et al. [120] used two colon cancer culture cells (DLD-1 and HT-29 cells), LL-37 antimicrobial peptide and its synthetic analogue ceragenin CSA-13 (mimic peptides). They developed two novel nanosystems based on MNPs, MNP@LL-37 and MNP@CSA-13, to evaluate the effect of MNPs as a drug delivery system. Results proved that combining antimicrobial peptides with MNPs drug systems will decrease the viability of the colorectal cancer cell line. Additionally, studies showed that the ceragenin CSA-13 had enhanced apoptotic properties on colon cancer cells than LL-37. Hence, these results show that both nanosystems are excellent instruments in developing targeted therapy, mainly due to their capability to be incorporated into tumorous cells [121,122].

In another study, Gao et al. [123] developed new nanoplatforms for drug delivery based on nanomagnetic liposomes (Lips)- encapsulated parthenolide (PTL) and glucose oxidase (GOD) with efficient synergic antitumour therapy. This multifunctional drug was synthesized from modified MNPs, PTL and GOD-encapsulated to form the delivery system GOD-PTL-Lips@MNPs, targeting the tumour's acidic environment. The formed nanoplatforms drug delivery system showed a significant antitumour effect, minimising the tissues' toxicity in vivo. Through the synergistic effect of the constituent compounds of the drug delivery system, it has been provided with a possible approach, which could improve the efficiency in targeting and curing cancerous tumours.

8. Conclusions

As cancer is a disease provoked by the uncontrolled division of the cells, any antitumor drug is a killing agent or is a substance that stops the natural multiplication of the cells. Therefore, being an anti-growing substance, it will affect both tumoral and healthy cells, hence the unwanted side effects of the chemotherapy. The new therapies are counting on target delivery of the drugs with the help of appropriate nanocarrier. As the drug is encapsulated, the new paradigm is to manage the delivery only to the tumour tissue. Magnetic nanocarriers present the distinct advantage of following an external magnetic field and concentrating in the tumour zone. In addition, these nanocarriers can generate hyperthermia and stress, which induces apoptosis and/or necrosis of the tumoral cells. However, maybe more importantly, the increased temperature can assure a triggered delivery. Thus, the active agent can be released at the desired moment and according to a desired rate.

Moreover, by decorating these nanoparticles with adequate molecules or applying an appropriate magnetic field, targeted accumulation and cell internalization can be achieved in the desired tumoral tissue/organ. Hyperthermia is controlled externally by using the correct electromagnetic field. The triggering and targeting can ensure an efficient therapy. At the same time, the release can be tuned according to the field characteristics and applied time to create the premises of personalized therapy/treatment.

Author Contributions: Conceptualization, A.S. (Angela Spoială), C.-I.I. and D.F.; funding acquisition, D.F.; investigation, A.S. (Angela Spoială), C.-I.I., L.M., D.F., A.S. (Augustin Semenescu) and A.F.; methodology, D.F. and O.-C.O.; project administration, D.F.; resources, D.F.; software A.S. (Angela Spoială) L.M., A.S. (Augustin Semenescu) and O.-C.O.; supervision, D.F. and A.F.; validation, O.-C.O. and A.F.; visualization, A.S. (Angela Spoială); writing—original draft, A.S. (Angela Spoială), C.-I.I., L.M., D.F. and A.S. (Augustin Semenescu); writing—review and editing, D.F., O.-C.O. and A.F. All authors have read and agreed to the published version of the manuscript.

Funding: The APC was funded by UPB.

Institutional Review Board Statement: Not applicable.

Informed Consent Statement: Not applicable.

Data Availability Statement: These data can be available upon request.

Acknowledgments: The authors acknowledge the financial support of the "Sisteme inteligente cu eliberare controlata pentru tratarea cancerului utilizand o abordare personalizata (SmartACT)" project PN-III-P1-1.1-PED-2021-1342, contract TE 96 din 17/05/2022.

Conflicts of Interest: The authors declare no conflict of interest.

References

1. Daher, M. Cultural beliefs and values in cancer patients. *Ann. Oncol.* **2012**, *23*, 66–69. [CrossRef]
2. Wu, S.; Zhu, W.; Thompson, P.; Hannun, Y.A. Evaluating intrinsic and non-intrinsic cancer risk factors. *Nat. Commun.* **2018**, *9*, 3490. [CrossRef]
3. Quazi, S. Telomerase gene therapy: A remission toward cancer. *Med. Oncol.* **2022**, *39*, 105. [CrossRef] [PubMed]
4. Anand, P.; Kunnumakara, A.B.; Sundaram, C.; Harikumar, K.B.; Tharakan, S.T.; Lai, O.S.; Sung, B.; Aggarwal, B.B. Cancer is a Preventable Disease that Requires Major Lifestyle Changes. *Pharm. Res.* **2008**, *25*, 2097–2116. [CrossRef]
5. Gavas, S.; Quazi, S.; Karpiński, T.M. Nanoparticles for Cancer Therapy: Current Progress and Challenges. *Nanoscale Res. Lett.* **2021**, *16*, 173. [CrossRef] [PubMed]

6. Wu, G.; Wilson, G.; George, J.; Liddle, C.; Hebbard, L.; Qiao, L. Overcoming treatment resistance in cancer: Current understanding and tactics. *Cancer Lett.* **2016**, *387*, 69–76. [CrossRef] [PubMed]
7. Cheung-Ong, K.; Giaever, G.; Nislow, C. DNA-Damaging Agents in Cancer Chemotherapy: Serendipity and Chemical Biology. *Chem. Biol.* **2013**, *20*, 648–659. [CrossRef]
8. Carrère, C. Optimization of an in vitro chemotherapy to avoid resistant tumours. *J. Theor. Biol.* **2017**, *413*, 24–33. [CrossRef]
9. Ardelean, I.L.; Ficai, D.; Sonmez, M.; Oprea, O.; Nechifor, G.; Andronescu, E.; Ficai, A.; Titu, M.A. Hybrid Magnetic Nanostructures For Cancer Diagnosis And Therapy. *Anti-Cancer Agents Med. Chem.* **2019**, *19*, 6–16. [CrossRef] [PubMed]
10. Lacatusu, I.; Badea, N.; Badea, G.; Brasoveanu, L.; Stan, R.; Ott, C.; Opreaa, O.; Meghea, A. Ivy leaves extract based—Lipid nanocarriers and their bio-efficacy on antioxidant and antitumor activities. *RSC Adv.* **2016**, *6*, 77243–77255. [CrossRef]
11. Lacatusu, I.; Badea, N.; Badea, G.; Oprea, O.; Mihaila, M.; Kaya, D.; Stan, R.; Meghea, A. Lipid nanocarriers based on natural oils with high activity against oxygen free radicals and tumor cell proliferation. *Mater. Sci. Eng. C* **2015**, *56*, 88–94. [CrossRef]
12. Istrati, D.; Lacatusu, I.; Bordei, N.; Badea, G.; Oprea, O.; Stefan, L.; Stan, R.; Badea, N.; Meghea, A. Phyto-mediated nanostructured carriers based on dual vegetable actives involved in the prevention of cellular damage. *Mater. Sci. Eng. C* **2016**, *64*, 249–259. [CrossRef] [PubMed]
13. Schwendener, R.A. Liposomes and Immuno-Liposomes as Carriers for Cytostatic Drugs, Magnetic-Resonance Contrast Agents, and Fluorescent Chelates. *Chimia* **1992**, *46*, 69–77. [CrossRef]
14. Ficai, D.; Ficai, A.; Dinu, E.; Oprea, O.; Sonmez, M.; Keler, M.; Sahin, Y.; Ekren, N.; Inan, A.; Daglilar, S.; et al. Magnetic Core Shell Structures: From 0D to 1D assembling. *Curr. Pharm. Des.* **2015**, *21*, 5301–5311. [CrossRef] [PubMed]
15. Ficai, D.; Oprea, O.; Ficai, A.; Holban, A. Metal Oxide Nanoparticles: Potential Uses in Biomedical Applications. *Curr. Proteom.* **2014**, *11*, 139–149. [CrossRef]
16. Ficai, D.; Ficai, A.; Vasile, B.S.; Ficai, M.; Oprea, O.; Guran, C.; Andronescu, E. Synthesis of Rod-Like Magnetite by Using Low Magnetic Field. *Dig. J. Nanomater. Biostructures* **2011**, *6*, 943–951.
17. Niculae, G.; Badea, N.; Meghea, A.; Oprea, O.; Lacatusu, I. Coencapsulation of Butyl-Methoxydibenzoylmethane and Octocrylene into Lipid Nanocarriers: UV Performance, Photostability and *in vitro* Release. *Photochem. Photobiol.* **2013**, *89*, 1085–1094. [CrossRef]
18. Lacatusu, I.; Badea, N.; Murariu, A.; Oprea, O.; Bojin, D.; Meghea, A. Antioxidant Activity of Solid Lipid Nanoparticles Loaded with Umbelliferone. *Soft Mater.* **2013**, *11*, 75–84. [CrossRef]
19. Mohammed, S.A.; Shaaban, E.I.A. Efficacious nanomedicine track toward combating COVID-19. *Nanotechnol. Rev.* **2022**, *11*, 680–698. [CrossRef]
20. Kasina, V.; Mownn, R.J.; Bahal, R.; Sartor, G.C. Nanoparticle delivery systems for substance use disorder. *Neuropsychopharmacology* **2022**, *47*, 1431–1439. [CrossRef]
21. Tanaka, T.; Shiramoto, S.; Miyashita, M.; Fujishima, Y.; Kaneo, Y. Tumor targeting based on the effect of enhanced permeability and retention (EPR) and the mechanism of receptor-mediated endocytosis (RME). *Int. J. Pharmaceut.* **2004**, *277*, 39–61. [CrossRef] [PubMed]
22. Xu, W.; Liu, X.H.; Liu, X.H. Effects of nanoparticle size and shape in clathrin-mediated endocytosis. *J. Appl. Phys.* **2022**, *131*, 134701. [CrossRef]
23. Unsoy, G.; Gunduz, U.; Oprea, O.; Ficai, D.; Sonmez, M.; Radulescu, M.; Alexie, M.; Ficai, A. Magnetite: From Synthesis to Applications. *Curr. Top. Med. Chem.* **2015**, *15*, 1622–1640. [CrossRef] [PubMed]
24. Anbarasu, M.; Anandan, M.; Chinnasamy, E.; Gopinath, V.; Balamurugan, K. Synthesis and characterization of polyethylene glycol (PEG) coated Fe3O4 nanoparticles by chemical co-precipitation method for biomedical applications. *Spectrochim. Acta Part A Mol. Biomol. Spectrosc.* **2015**, *135*, 536–539. [CrossRef] [PubMed]
25. Sonmez, M.; Georgescu, M.; Alexandrescu, L.; Gurau, D.; Ficai, A.; Ficai, D.; Andronescu, E. Synthesis and Applications of Fe3O4/SiO2 Core-Shell Materials. *Curr. Pharm. Design* **2015**, *21*, 5324–5335. [CrossRef]
26. Chircov, C.; Matei, M.-F.; Neacșu, I.A.; Vasile, B.S.; Oprea, O.-C.; Croitoru, A.-M.; Trușcă, R.-D.; Andronescu, E.; Sorescu, I.; Bărbuceanu, F. Iron Oxide–Silica Core–Shell Nanoparticles Functionalized with Essential Oils for Antimicrobial Therapies. *Antibiotics* **2021**, *10*, 1138. [CrossRef] [PubMed]
27. Istrati, D.; Moroșan, A.; Stan, R.; Vasile, B.; Vasilievici, G.; Oprea, O.; Dolete, G.; Purcăreanu, B.; Mihaiescu, D.E. Microwave-Assisted Sol–Gel Preparation of the Nanostructured Magnetic System for Solid-Phase Synthesis. *Nanomaterials* **2021**, *11*, 3176. [CrossRef]
28. Zhang, W.; Lai, E.P.C. Chemical Functionalities of 3-aminopropyltriethoxy-silane for Surface Modification of Metal Oxide Nanoparticles. *Silicon* **2021**, *14*, 6535–6545. [CrossRef]
29. Chircov, C.; Ștefan, R.-E.; Dolete, G.; Andrei, A.; Holban, A.M.; Oprea, O.-C.; Vasile, B.S.; Neacșu, I.A.; Tihăuan, B. Dextran-Coated Iron Oxide Nanoparticles Loaded with Curcumin for Antimicrobial Therapies. *Pharmaceutics* **2022**, *14*, 1057. [CrossRef]
30. Caciandone, M.; Niculescu, A.-G.; Grumezescu, V.; Bîrcă, A.C.; Ghica, I.C.; Vasile, B.; Oprea, O.; Nica, I.C.; Stan, M.S.; Holban, A.M.; et al. Magnetite Nanoparticles Functionalized with Therapeutic Agents for Enhanced ENT Antimicrobial Properties. *Antibiotics* **2022**, *11*, 623. [CrossRef]
31. Spirescu, V.A.; Niculescu, A.-G.; Slave, Ș.; Bîrcă, A.C.; Dorcioman, G.; Grumezescu, V.; Holban, A.M.; Oprea, O.-C.; Vasile, B.; Grumezescu, A.M.; et al. Anti-Biofilm Coatings Based on Chitosan and Lysozyme Functionalized Magnetite Nanoparticles. *Antibiotics* **2021**, *10*, 1269. [CrossRef]

32. Puiu, R.A.; Balaure, P.C.; Constantinescu, E.; Grumezescu, A.M.; Andronescu, E.; Oprea, O.C.; Vasile, B.S.; Grumezescu, V.; Negut, I.; Nica, C.; et al. Anti-Cancer Nanopowders and MAPLE-Fabricated Thin Films Based on SPIONs Surface Modified with Paclitaxel Loaded beta-Cyclodextrin. *Pharmaceutics* **2021**, *13*, 1356. [CrossRef]
33. Caciandone, M.; Niculescu, A.G.; Roșu, A.R.; Grumezescu, V.; Negut, I.; Holban, A.M.; Oprea, O.-C.; Vasile, B.Ș.; Bîrcă, A.C.; Grumezescu, A.M.; et al. PEG-Functionalized Magnetite Nano-particles for Modulation of Microbial Biofilms on Voice Prosthesis. *Antibiotics* **2021**, *11*, 39. [CrossRef]
34. Tai, M.F.; Lai, C.W.; Hamid, S.B.A. Facile Synthesis Polyethylene Glycol Coated Magnetite Nanoparticles for High Colloidal Stability. *J. Nanomater.* **2016**, *2016*, 8612505. [CrossRef]
35. Maier-HauffFrank, K.; Ulrich, F.; Nestler, D.; Niehoff, H.; Wust, P.; Thiesen, B.; Orawa, H.; Budach, V.; Jordan, A. Efficacy and safety of intratumoral thermotherapy using magnetic iron-oxide nanoparticles combined with external beam radiotherapy on patients with recurrent glioblastoma multiforme. *J. Neuro-Oncol.* **2010**, *103*, 317–324. [CrossRef] [PubMed]
36. Maier-Hauff, K.; Rothe, R.; Scholz, R.; Gneveckow, U.; Wust, P.; Thiesen, B.; Feussner, A.; Von Deimling, A.; Waldoefner, N.; Felix, R.; et al. Intracranial thermotherapy using magnetic na-noparticles combined with external beam radiotherapy: Results of a feasibility study on patients with glioblastoma multiforme. *J. Neuro-Oncol.* **2007**, *81*, 53–60. [CrossRef] [PubMed]
37. Jordan, A.; Scholz, R.; Maier-Hauff, K.; van Landeghem, F.; Waldoefner, N.; Teichgraeber, U.; Pinkernelle, J.; Bruhn, H.; Neumann, F.; Thiesen, B.; et al. The effect of thermotherapy using magnetic nanoparticles on rat malignant glioma. *J. Neuro-Oncol.* **2005**, *78*, 7–14. [CrossRef]
38. Jiang, W.; Lai, K.; Wu, Y.; Gu, Z. Protein corona on magnetite nanoparticles and internalization of nanoparticle–protein complexes into healthy and cancer cells. *Arch. Pharmacal Res.* **2013**, *37*, 129–141. [CrossRef]
39. Xing, Z.-C.; Park, M.-J.; Han, S.-J.; Kang, I.-K.; Choi, M.-J.; Lee, B.-H.; Chang, Y. Intracellular uptake of magnetite nanoparticles conjugated with RGDS-peptide. *Macromol. Res.* **2011**, *19*, 897–903. [CrossRef]
40. Movileanu, C.; Anghelache, M.; Turtoi, M.; Voicu, G.; Neacsu, I.A.; Ficai, D.; Trusca, R.; Oprea, O.; Ficai, A.; Andronescu, E.; et al. Folic acid-decorated PEGylated magnetite na-noparticles as efficient drug carriers to tumor cells overexpressing folic acid receptor. *Int. J. Pharm.* **2022**, *625*, 122064. [CrossRef] [PubMed]
41. Moghimi, S.M.; Hunter, A.; Murray, J.C. Nanomedicine: Current status and future prospects. *FASEB J.* **2004**, *19*, 311–330. [CrossRef]
42. van der Meel, R.; Sulheim, E.; Shi, Y.; Kiessling, F.; Mulder, W.J.M.; Lammers, T. Smart cancer nanomedicine. *Nat. Nanotechnol.* **2019**, *14*, 1007–1017. [CrossRef] [PubMed]
43. Wicki, A.; Witzigmann, D.; Balasubramanian, V.; Huwyler, J. Nanomedicine in cancer therapy: Challenges, opportunities, and clinical applications. *J. Control. Release* **2015**, *200*, 138–157. [CrossRef] [PubMed]
44. Xin, Y.; Yin, M.; Zhao, L.; Meng, F.; Luo, L. Recent progress on nanoparticle-based drug delivery systems for cancer therapy. *Cancer Biol. Med.* **2017**, *14*, 228–241. [CrossRef]
45. Mottaghitalab, F.; Farokhi, M.; Fatahi, Y.; Atyabi, F.; Dinarvand, R. New insights into designing hybrid nanoparticles for lung cancer: Diagnosis and treatment. *J. Control. Release* **2019**, *295*, 250–267. [CrossRef] [PubMed]
46. Revia, R.A.; Zhang, M. Magnetite nanoparticles for cancer diagnosis, treatment, and treatment monitoring: Recent advances. *Mater. Today* **2016**, *19*, 157–168. [CrossRef]
47. Kievit, F.M.; Zhang, M. Cancer Therapy: Cancer Nanotheranostics: Improving Imaging and Therapy by Targeted Delivery Across Biological Barriers. *Adv. Mater.* **2011**, *23*, H209. [CrossRef]
48. Wang, F.; Li, C.; Cheng, J.; Yuan, Z. Recent Advances on Inorganic Nanoparticle-Based Cancer Therapeutic Agents. *Int. J. Environ. Res. Public Health* **2016**, *13*, 1182. [CrossRef]
49. Sun, C.; Lee, J.S.H.; Zhang, M. Magnetic nanoparticles in MR imaging and drug delivery. *Adv. Drug Deliv. Rev.* **2008**, *60*, 1252–1265. [CrossRef]
50. Veiseh, O.; Gunn, J.W.; Zhang, M. Design and fabrication of magnetic nanoparticles for targeted drug delivery and imaging. *Adv. Drug Deliv. Rev.* **2010**, *62*, 284–304. [CrossRef]
51. Laurent, S.; Forge, D.; Port, M.; Roch, A.; Robic, C.; Vander Elst, L.; Muller, R.N. Magnetic Iron Oxide Nanoparticles: Synthesis, Stabilization, Vectorization, Physicochemical Characterizations, and Biological Applications. *Chem. Rev.* **2008**, *108*, 2064–2110. [CrossRef] [PubMed]
52. Cheon, H.J.; Adhikari, M.D.; Chung, M.; Tran, T.D.; Kim, J.; Kim, M.I. Magnetic Nanoparticles-Embedded Enzyme-Inorganic Hybrid Nanoflowers with Enhanced Peroxidase-Like Activity and Substrate Channeling for Glucose Biosensing. *Adv. Health Mater.* **2019**, *8*, e1801507. [CrossRef]
53. Garanina, A.S.; Naumenko, V.A.; Nikitin, A.A.; Myrovali, E.; Petukhova, A.Y.; Klimyuk, S.V.; Nalench, Y.A.; Ilyasov, A.R.; Vodopyanov, S.S.; Erofeev, A.S.; et al. Temperature-controlled magnetic nanoparticles hyperthermia inhibits primary tumor growth and metastases dissemination. *Nanomed. Nanotechnol. Biol. Med.* **2020**, *25*, 102171. [CrossRef] [PubMed]
54. Avasthi, A.; Caro, C.; Pozo-Torres, E.; Leal, M.P.; García-Martín, M.L. Magnetic Nanoparticles as MRI Contrast Agents. *Top. Curr. Chem.* **2020**, *378*, 40. [CrossRef] [PubMed]
55. Rahimi, M.; Safa, K.D.; Salehi, R. Co-delivery of doxorubicin and methotrexate by dendritic chitosan-g-mPEG as a magnetic nanocarrier for multidrug delivery in combination chemotherapy. *Polym. Chem.-UK* **2017**, *8*, 7333–7350. [CrossRef]
56. Ali, A.; Zafar, H.; Zia, M.; ul Haq, I.; Phull, A.R.; Ali, J.S.; Hussain, A. Synthesis, characterization, applications, and challenges of iron oxide nanoparticles. *Nanotechnol. Sci. Appl.* **2016**, *9*, 49–67. [CrossRef]

57. Cabrera, F.C.; Melo, A.F.A.A.; de Souza, J.C.P.; Job, A.E.; Crespilho, F.N. A flexible lab-on-a-chip for the synthesis and magnetic sep-aration of magnetite decorated with gold nanoparticles. *Lab Chip.* **2015**, *15*, 1835–1841. [CrossRef] [PubMed]
58. Shasha, C.; Teeman, E.; Krishnan, K.M.; Szwargulski, P.; Knopp, T.; Möddel, M. Discriminating nanoparticle core size using mul-ti-contrast MPI. *Phys. Med. Biol.* **2019**, *64*, 074001. [CrossRef] [PubMed]
59. Liu, S.; Yu, B.; Wang, S.; Shen, Y.; Cong, H. Preparation, surface functionalization and application of Fe3O4 magnetic nanopar-ticles. *Adv. Colloid Interface Sci.* **2020**, *281*, 102165. [CrossRef]
60. Zhu, N.; Ji, H.; Yu, P.; Niu, J.; Farooq, M.U.; Akram, M.W.; Udego, I.O.; Li, H.; Niu, X. Surface Modification of Magnetic Iron Oxide Nanoparticles. *Nanomaterials* **2018**, *8*, 810. [CrossRef]
61. McBain, S.C.; Yiu, H.H.P.; Dobson, J. Magnetic nanoparticles for gene and drug delivery. *Int. J. Nanomed.* **2008**, *3*, 169–180.
62. McBain, S.C.; Yiu, H.H.P.; El Haj, A.; Dobson, J. Polyethyleneimine functionalized iron oxide nanoparticles as agents for DNA delivery and transfection. *J. Mater. Chem.* **2007**, *17*, 2561–2565. [CrossRef]
63. Dulinska-Litewka, J.; Lazarczyk, A.; Halubiec, P.; Szafranski, O.; Karnas, K.; Karewicz, A. Superparamagnetic Iron Oxide Nano-particles-Current and Prospective Medical Applications. *Materials* **2019**, *12*, 617. [CrossRef]
64. Zuvin, M.; Koçak, M.; Ünal, Ö.; Akkoç, Y.; Kutlu, Ö.; Acar, H.Y.; Gözüaçik, D.; Koşar, A. Nanoparticle based induction heating at low magnitudes of magnetic field strengths for breast cancer therapy. *J. Magn. Magn. Mater.* **2019**, *483*, 169–177. [CrossRef]
65. Kandasamy, G.; Sudame, A.; Maity, D.; Soni, S.; Sushmita, K.; Veerapu, N.S.; Bose, S.; Tomy, C. Multifunctional magnetic-polymeric nanoparticles based ferrofluids for multi-modal in vitro cancer treatment using thermotherapy and chemotherapy. *J. Mol. Liq.* **2019**, *293*, 111549. [CrossRef]
66. Hedayatnasab, Z.; Dabbagh, A.; Abnisa, F.; Daud, W.M.A.W. Polycaprolactone-coated superparamagnetic iron oxide nano-particles for in vitro magnetic hyperthermia therapy of cancer. *Eur. Polym. J.* **2020**, *133*, 109789. [CrossRef]
67. Rubia-Rodríguez, I.; Santana-Otero, A.; Spassov, S.; Tombácz, E.; Johansson, C.; De La Presa, P.; Teran, F.J.; Morales, M.D.P.; Veintemillas-Verdaguer, S.; Thanh, N.T.; et al. Whither Magnetic Hyper-thermia? *A Tentat. Roadmap Mater.* **2021**, *14*, 706.
68. Lin, J.; Zhoub, W.; Kumbhar, A.; Wiemann, J.; Fangb, J.; Carpenter, E.; O'Connor, C. Gold-Coated Iron (Fe@Au) Nanoparticles: Synthesis, Characterization, and Magnetic Field-Induced Self-Assembly. *J. Solid State Chem.* **2001**, *159*, 26–31. [CrossRef]
69. Gilchrist, R.K.; Medal, R.; Shorey, W.D.; Hanselman, R.C.; Parrot, J.C.; Taylor, C.B. Selective Inductive Heating of Lymph Nodes. *Ann. Surg.* **1957**, *146*, 596–606. [CrossRef] [PubMed]
70. Li, X.; Xia, S.; Zhou, W.; Ji, R.; Zhan, W. Targeted Fe-doped silica nanoparticles as a novel ultrasound–magnetic resonance du-al-mode imaging contrast agent for HER2-positive breast cancer. *Int. J. Nanomed.* **2019**, *14*, 2397–2413. [CrossRef]
71. Javanbakht, S.; Shadi, M.; Mohammadian, R.; Shaabani, A.; Ghorbani, M.; Rabiee, G.; Amini, M.M. Preparation of Fe3O4@SiO2@Tannic acid double core-shell magnetic nanoparticles via the Ugi multicomponent reaction strategy as a pH-responsive co-delivery of doxorubicin and methotrexate. *Mater. Chem. Phys.* **2020**, *247*, 122857. [CrossRef]
72. Spoială, A.; Ilie, C.-I.; Crăciun, L.N.; Ficai, D.; Ficai, A.; Andronescu, E. Magnetite-Silica Core/Shell Nanostructures: From Surface Functionalization towards Biomedical Applications—A Review. *Appl. Sci.* **2021**, *11*, 11075. [CrossRef]
73. Dai, Z.; Wen, W.; Guo, Z.; Song, X.-Z.; Zheng, K.; Xu, X.; Qi, X.; Tan, Z. SiO2-coated magnetic nano-Fe3O4 photosensitizer for synergistic tumour-targeted chemo-photothermal therapy. *Colloids Surf. B Biointerfaces* **2020**, *195*, 111274. [CrossRef] [PubMed]
74. Roti, J.L.R. Cellular responses to hyperthermia (40–46 degrees C): Cell killing and molecular events. *Int. J. Hyperth.* **2008**, *24*, 3–15. [CrossRef]
75. Sato, I.; Umemura, M.; Mitsudo, K.; Kioi, M.; Nakashima, H.; Iwai, T.; Feng, X.; Oda, K.; Miyajima, A.; Makino, A.; et al. Hyperthermia generated with ferucarbotran (Resovist(A (R))) in an alternating magnetic field enhances cisplatin-induced apoptosis of cultured human oral cancer cells. *J. Physiol. Sci.* **2014**, *64*, 177–183. [CrossRef] [PubMed]
76. Mahmoudi, K.; Bouras, A.; Bozec, D.; Ivkov, R.; Hadjipanayis, C. Magnetic hyperthermia therapy for the treatment of glio-blastoma: A review of the therapy's history, efficacy and application in humans. *Int. J. Hyperth.* **2018**, *34*, 1316–1328. [CrossRef]
77. Jeun, M.; Kim, Y.J.; Park, K.H.; Paek, S.H.; Bae, S. Physical contribution of Néel and Brown relaxation to interpreting intracellular hyperthermia characteristics using superparamagnetic nanofluids. *J. Nanosci. Nanotechnol.* **2013**, *13*, 5719–5725. [CrossRef] [PubMed]
78. Torres, T.E.; Lima, E.; Calatayud, M.P.; Sanz, B.; Ibarra, A.; Fernández-Pacheco, R.; Mayoral, A.; Marquina, C.; Ibarra, M.R.; Goya, G.F. The relevance of Brownian relaxation as power absorption mechanism in Magnetic Hyperthermia. *Sci. Rep.* **2019**, *9*, 3992. [CrossRef]
79. Espinosa, A.; Kolosnjaj-Tabi, J.; Abou-Hassan, A.; Plan Sangnier, A.; Curcio, A.; Silva, A.K.; Di Corato, R.; Neveu, S.; Pellegrino, T.; Liz-Marzán, L.M.; et al. Magnetic (Hyper)Thermia or Pho-tothermia? Progressive Comparison of Iron Oxide and Gold Nanoparticles Heating in Water, in Cells, and In Vivo. *Adv. Funct. Mater.* **2018**, *28*, 1803660. [CrossRef]
80. Mamani, J.; Souza, T.; Nucci, M.; Oliveira, F.; Nucci, L.; Alves, A.; Rego, G.; Marti, L.; Gamarra, L. In Vitro Evaluation of Hyperthermia Magnetic Technique Indicating the Best Strategy for Internalization of Magnetic Nanoparticles Applied in Glioblastoma Tumor Cells. *Pharmaceutics* **2021**, *13*, 1219. [CrossRef]
81. Angrove, D.M. *Magnetite: Structure, Properties, and Applications*; Nova Science Publishers: Hauppauge, NY, USA, 2011.
82. Maeda, H.; Matsumura, Y. Tumoritropic and Lymphotropic Principles of Macromolecular Drugs. *Crit. Rev. Thera-Peutic Drug Carr. Syst.* **1989**, *6*, 193–210.
83. Matsumura, Y.; Maeda, H. A New Concept for Macromolecular Therapeutics in Cancer-Chemotherapy—Mechanism of Tu-moritropic Accumulation of Proteins and the Antitumor Agent Smancs. *Cancer Res.* **1986**, *46*, 6387–6392. [PubMed]

84. Bazak, R.; Houri, M.; EL Achy, S.; Hussein, W.; Refaat, T. Passive targeting of nanoparticles to cancer: A comprehensive review of the literature. *Mol. Clin. Oncol.* **2014**, *2*, 904–908. [CrossRef] [PubMed]
85. Chomoucka, J.; Drbohlavova, J.; Huska, D.; Adam, V.; Kizek, R.; Hubalek, J. Magnetic nanoparticles and targeted drug delivering. *Pharmacol. Res.* **2010**, *62*, 144–149. [CrossRef] [PubMed]
86. Gobbo, O.L.; Sjaastad, K.; Radomski, M.W.; Volkov, Y.; Prina-Mello, A. Magnetic Nanoparticles in Cancer Theranostics. *Theranostics* **2015**, *5*, 1249–1263. [CrossRef] [PubMed]
87. He, C.; Hu, Y.; Yin, L.; Tang, C.; Yin, C. Effects of particle size and surface charge on cellular uptake and biodistribution of polymeric nanoparticles. *Biomaterials* **2010**, *31*, 3657–3666. [CrossRef] [PubMed]
88. Ranganathan, R.; Madanmohan, S.; Kesavan, A.; Baskar, G.; Krishnamoorthy, Y.R.; Santosham, R.; Ponraju, D.; Rayala, S.K.; Venkatraman, G. Nanomedicine: Towards development of patient-friendly drug-delivery systems for oncological applications. *Int. J. Nanomed.* **2012**, *7*, 1043–1060.
89. Krown, S.E.; Northfelt, D.W.; Osoba, D.; Stewart, J.S. Use of liposomal anthracyclines in Kaposi's sarcoma. *Semin. Oncol.* **2004**, *31*, 36–52. [CrossRef] [PubMed]
90. Adimoolam, M.G.; Amreddy, N.; Nalam, M.R.; Sunkara, M.V. A simple approach to design chitosan functionalized Fe3O4 nano-particles for pH responsive delivery of doxorubicin for cancer therapy. *J. Magn. Magn. Mater.* **2018**, *448*, 199–207. [CrossRef]
91. Ficai, A.; Sonmez, M.; Albu, M.G.; Mihaiescu, D.E.; Ficai, D.; Bleotu, C. Antitumoral materials with regenerative function obtained using a layer-by-layer technique. *Drug Des. Dev. Ther.* **2015**, *9*, 1269–1279. [CrossRef]
92. Andronescu, E.; Ficai, M.; Voicu, G.; Ficai, D.; Maganu, M.; Ficai, A. Synthesis and characterization of collagen/hydroxyapatite: Magnetite composite material for bone cancer treatment. *J. Mater. Sci. Mater. Med.* **2010**, *21*, 2237–2242. [CrossRef]
93. Ficai, A.; Marques, C.; Ferreira, J.M.; Andronescu, E.; Ficai, D.; Sonmez, M. Multifunctional materials for bone cancer treatment. *Int. J. Nanomed.* **2014**, *9*, 2713–2725. [CrossRef]
94. Bertilsson, L.; Liedberg, B. Infrared Study of Thiol Monolayer Assemblies on Gold—Preparation, Characterization, and Functionalization of Mixed Monolayers. *Langmuir* **1993**, *9*, 141–149. [CrossRef]
95. Hu, M.; Chen, J.; Li, Z.-Y.; Au, L.; Hartland, G.V.; Li, X.; Marquez, M.; Xia, Y. Gold nanostructures: Engineering their plasmonic properties for biomedical applications. *Chem. Soc. Rev.* **2006**, *35*, 1084–1094. [CrossRef]
96. Ficai, D.; Andronescu, E.; Sonmez, M.; Ficai, A.; Oprea, O.; Vasile, B.S. Process for producing multifunction systems of the magnetite/thio acids/Ag or Au type, to be employed in cancer diagnosis and guided treatment. Bachelor's Thesis, Romanian State Office for Inventions and Trademarks, Bucharest, Romania, 2018; A/00238/2013 (RO129824 (A2); RO129824B1.
97. Hsiao, S.-M.; Peng, B.-Y.; Tseng, Y.S.; Liu, H.-T.; Chen, C.-H.; Lin, H.-M. Preparation and characterization of multifunctional meso-porous silica nanoparticles for dual magnetic resonance and fluorescence imaging in targeted cancer therapy. *Microporous Mesoporous Mater.* **2017**, *250*, 210–220. [CrossRef]
98. Siminzar, P.; Omidi, Y.; Golchin, A.; Aghanejad, A.; Barar, J. Targeted delivery of doxorubicin by magnetic mesoporous silica nanoparticles armed with mucin-1 aptamer. *J. Drug Target.* **2019**, *28*, 92–101. [CrossRef]
99. Demin, A.M.; Vakhrushev, A.V.; Valova, M.S.; Korolyova, M.A.; Uimin, M.A.; Minin, A.S.; Pozdina, V.A.; Byzov, I.V.; Tumashov, A.A.; Chistyakov, K.A.; et al. Effect of the Silica-Magnetite Nano-composite Coating Functionalization on the Doxorubicin Sorption/Desorption. *Pharmaceuticals* **2022**, *14*, 2271. [CrossRef] [PubMed]
100. Maximenko, A.; Depciuch, J.; Łopuszyńska, N.; Stec, M.; Światkowska-Warkocka, Ż.; Bayev, V.; Zieliński, P.M.; Baran, J.; Fedotova, J.; Węglarz, W.P.; et al. Fe3O4@SiO2@Au nanoparticles for MRI-guided chemo/NIR photothermal therapy of cancer cells. *RSC Adv.* **2020**, *10*, 26508. [CrossRef] [PubMed]
101. Elbialy, N.S.; Fathy, M.M.; Reem, A.W.; Darwesh, R.; Abdel-Dayem, U.A.; Aldhahri, M.; Noorwali, A.; Al-Ghamdi, A.A. Multifunctional magnetic-gold nano-particles for efficient combined targeted drug delivery and interstitial photothermal therapy. *Int. J. Pharm.* **2019**, *554*, 256–263. [CrossRef]
102. Padayachee, J.; Singh, M. Therapeutic applications of CRISPR/Cas9 in breast cancer and delivery potential of gold nano-materials. *Nanobiomedicine* **2020**, *7*, 1849543520983196. [CrossRef]
103. Li, B.; Zhou, Q.; Wang, H.; Zha, Y.; Zheng, P.; Yang, T.; Ma, D.; Qiu, L.; Xu, X.; Hu, Y.; et al. Mitochondria-targeted magnetic gold nanoheterostructure for mul-ti-modal imaging guided photothermal and photodynamic therapy of triple-negative breast cancer. *Chem. Eng. J.* **2021**, *403*, 126364. [CrossRef]
104. Baghban, R.; Afarid, M.; Soleymani, J.; Rahimi, M. Were magnetic materials useful in cancer therapy? *Biomed. Pharmacother.* **2021**, *144*, 112321. [CrossRef]
105. Radulescu, M.; Ficai, D.; Oprea, O.; Ficai, A.; Andronescu, E.; Holban, A.M. Antimicrobial Chitosan based formulations with impact on different biomedical applications. *Curr. Pharm. Biotechnol.* **2015**, *16*, 128–136. [CrossRef] [PubMed]
106. Shanavas, A.; Sasidharan, S.; Bahadur, D.; Srivastava, R. Magnetic core-shell hybrid nanoparticles for receptor targeted an-ti-cancer therapy and magnetic resonance imaging. *J. Colloid Interface Sci.* **2017**, *486*, 112–120. [CrossRef]
107. Wang, M.; Liang, Y.; Zhang, Z.; Ren, G.; Liu, Y.; Wu, S.; Shen, J. Ag@Fe3O4@C nanoparticles for multi-modal imaging-guided chemo-photothermal synergistic targeting for cancer therapy. *Anal. Chim. Acta* **2019**, *1086*, 122–132. [CrossRef]
108. Kievit, F.M.; Stephen, Z.R.; Wang, K.; Dayringer, C.J.; Sham, J.G.; Ellenbogen, R.G.; Silber, J.R.; Zhang, M. Nanoparticle mediated silencing of DNA repair sensitizes pediatric brain tumor cells to gamma-irradiation. *Mol. Oncol.* **2015**, *9*, 1071–1080. [CrossRef] [PubMed]

109. Kievit, F.M.; Veiseh, O.; Fang, C.; Bhattarai, N.; Lee, D.; Ellenbogen, R.G.; Zhang, M. Chlorotoxin Labeled Magnetic Nanovectors for Tar-geted Gene Delivery to Glioma. *Acs Nano* **2010**, *4*, 4587–4594. [CrossRef] [PubMed]
110. Mukherjee, S.; Liang, L.; Veiseh, O. Recent Advancements of Magnetic Nanomaterials in Cancer Therapy. *Pharmaceutics* **2020**, *12*, 147. [CrossRef] [PubMed]
111. Torchilin, V. Tumor delivery of macromolecular drugs based on the EPR effect. *Adv. Drug Deliv. Rev.* **2011**, *63*, 131–135. [CrossRef] [PubMed]
112. Luo, Y.; Prestwich, G.D. Cancer-Targeted Polymeric Drugs. *Curr. Cancer Drug Targets* **2002**, *2*, 209–226. [CrossRef]
113. Yang, Y.; Aw, J.; Chen, K.; Liu, F.; Padmanabhan, P.; Hou, Y.; Cheng, Z.; Xing, B. Enzyme-responsive multifunctional magnetic nanoparticles for tumor intracellular drug delivery and imaging. *Chem. An Asian J.* **2011**, *6*, 1381–1389. [CrossRef]
114. Rastegari, B.; Karbalaei-Heidari, H.R.; Zeinali, S.; Sheardown, H. The enzyme-sensitive release of prodigiosin grafted beta-cyclodextrin and chitosan magnetic nanoparticles as an anticancer drug delivery system: Synthesis, characterization and cytotoxicity studies. *Colloids Surf. B Biointerfaces* **2017**, *158*, 589–601. [CrossRef] [PubMed]
115. Wang, X.; Zhu, L.; Hou, X.; Wang, L.; Yin, S. Polyethylenimine mediated magnetic nanoparticles for combined intracellular im-aging, siRNA delivery and anti-tumor therapy. *RSC Adv.* **2015**, *5*, 101569–101581. [CrossRef]
116. Li, F.; Aljahdali, I.; Ling, X. Cancer therapeutics using survivin BIRC5 as a target: What can we do after over two decades of study? *J. Exp. Clin. Cancer Res.* **2019**, *38*, 368. [CrossRef]
117. Miao, L.; Liu, C.; Ge, J.; Yang, W.; Liu, J.; Sun, W.; Yang, B.; Zheng, C.; Sun, H.; Hu, Q. Antitumor effect of TRAIL on oral squamous cell carcinoma using magnetic nanoparticle-mediated gene expression. *Cell Biochem. Biophys.* **2014**, *69*, 663–672. [CrossRef]
118. Tanaka, K.; Ito, A.; Kobayashi, T.; Kawamura, T.; Shimada, S.; Matsumoto, K.; Saida, T.; Honda, H. Intratumoral injection of immature dendritic cells enhances antitumor effect of hyperthermia using magnetic nanoparticles. *Int. J. Cancer* **2005**, *116*, 624–633. [CrossRef]
119. Sun, J.-B.; Duan, J.-H.; Dai, S.-L.; Ren, J.; Guo, L.; Jiang, W.; Li, Y. Preparation and anti-tumor efficiency evaluation of doxorubicin-loaded bacterial magnetosomes: Magnetic nanoparticles as drug carriers isolated from *Magnetospirillum gryphiswaldense*. *Biotechnol. Bioeng.* **2008**, *101*, 1313–1320. [CrossRef] [PubMed]
120. Niemirowicz, K.; Prokop, I.; Wilczewska, A.; Wnorowska, U.; Piktel, E.; Wątek, M.; Savage, P.; Bucki, R. Magnetic nanoparticles enhance the anticancer activity of cathelicidin LL-37 peptide against colon cancer cells. *Int. J. Nanomed.* **2015**, *10*, 3843–3853. [CrossRef] [PubMed]
121. Wilczewska, A.Z.; Niemirowicz, K.; Markiewicz, K.H.; Car, H. Nanoparticles as drug delivery systems. *Pharmacol. Rep.* **2012**, *64*, 1020–1037. [CrossRef]
122. Huang, C.; Tang, Z.; Zhou, Y.; Zhou, X.; Jin, Y.; Li, D.; Yang, Y.; Zhou, S. Magnetic micelles as a potential platform for dual targeted drug delivery in cancer therapy. *Int. J. Pharm.* **2012**, *429*, 113–122. [CrossRef] [PubMed]
123. Gao, W.; Wei, S.; Li, Z.; Li, L.; Zhang, X.; Li, C.; Gao, D. Nano magnetic liposomes-encapsulated parthenolide and glucose oxidase for ultra-efficient synergistic antitumor therapy. *Nanotechnology* **2020**, *31*, 355104. [CrossRef] [PubMed]

nanomaterials

Review

Osteoimmunomodulatory Nanoparticles for Bone Regeneration

Jingyi Wen [1,†], Donglin Cai [2,†], Wendong Gao [1], Ruiying He [3], Yulin Li [4], Yinghong Zhou [5,6], Travis Klein [1,6], Lan Xiao [1,6,*] and Yin Xiao [1,2,6,*]

1 School of Mechanical, Medical and Process Engineering, Centre for Biomedical Technologies, Queensland University of Technology, Brisbane, QLD 4059, Australia
2 School of Medicine and Dentistry, Menzies Health Institute Queensland, Griffith University, Southport, QLD 4222, Australia
3 College of Chemistry and Chemical Engineering, Hubei University, Wuhan 430061, China
4 The Key Laboratory for Ultrafine Materials of Ministry of Education, State Key Laboratory of Bioreactor Engineering, Engineering Research Center for Biomedical Materials of Ministry of Education, School of Materials Science and Engineering, East China University of Science and Technology, Shanghai 200231, China
5 School of Dentistry, The University of Queensland, Herston, QLD 4006, Australia
6 Australia-China Centre for Tissue Engineering and Regenerative Medicine, Queensland University of Technology, Brisbane, QLD 4000, Australia
* Correspondence: l5.xiao@qut.edu.au (L.X.); yin.xiao@qut.edu.au and yin.xiao@griffith.edu.au (Y.X.)
† These authors contributed equally to this work.

Abstract: Treatment of large bone fractures remains a challenge for orthopedists. Bone regeneration is a complex process that includes skeletal cells such as osteoblasts, osteoclasts, and immune cells to regulate bone formation and resorption. Osteoimmunology, studying this complicated process, has recently been used to develop biomaterials for advanced bone regeneration. Ideally, a biomaterial shall enable a timely switch from early stage inflammatory (to recruit osteogenic progenitor cells) to later-stage anti-inflammatory (to promote differentiation and terminal osteogenic mineralization and model the microstructure of bone tissue) in immune cells, especially the M1-to-M2 phenotype switch in macrophage populations, for bone regeneration. Nanoparticle (NP)-based advanced drug delivery systems can enable the controlled release of therapeutic reagents and the delivery of therapeutics into specific cell types, thereby benefiting bone regeneration through osteoimmunomodulation. In this review, we briefly describe the significance of osteoimmunology in bone regeneration, the advancement of NP-based approaches for bone regeneration, and the application of NPs in macrophage-targeting drug delivery for advanced osteoimmunomodulation.

Keywords: nanoparticles; bone regeneration; osteoimmunomodulation; targeted drug delivery; nanomedicine

Citation: Wen, J.; Cai, D.; Gao, W.; He, R.; Li, Y.; Zhou, Y.; Klein, T.; Xiao, L.; Xiao, Y. Osteoimmunomodulatory Nanoparticles for Bone Regeneration. *Nanomaterials* **2023**, *13*, 692. https://doi.org/10.3390/nano13040692

Academic Editors: Goran Kaluđerović and Nebojša Pantelić

Received: 6 January 2023
Revised: 3 February 2023
Accepted: 7 February 2023
Published: 10 February 2023

1. Introduction

Treatments for large bone defects caused by cancer, trauma, infection, and progressive congenital conditions remain challenging for orthopedic surgeons [1,2]. Trauma or disease can cause segmental bone defects, a common and severe clinical condition that can delay the union or non-union of bone [3]. Bone grafting is among the most often utilized surgical approaches to treat bone defects; with almost two million annual surgeries, it is the second most frequent medical procedure worldwide following blood transfusion [4]. Despite the availability of grafts, autologous bone is still the preferred option and gold standard because autologous bone grafts have natural osseointegration, osteoinductivity, and excellent biocompatibility. However, appropriate bone tissue for autologous grafting is generally in short supply, and its harvesting is frequently linked with recipient morbidity [5,6]. Alternatively, bone allografts are the second most popular choice for orthopedic treatment, which have provided feasible alternatives for some complicated bone defects without some of the weaknesses of autografts [7,8]. Bone allografts are mainly osteoconductive, with only demineralized bone matrix (DBM) preparations retaining lower osteoinductivity. Despite this, inferior recovery was detected compared with autologous grafts, and the risk of disease transmission and other infectious agents was also documented [9]. More critically, the typical amounts of naturally available bone graft substitutes are still insufficient to meet therapeutic demands, especially in light of the approaching aging and obesity situations worldwide [10]. Such cases call for an urgent need for artificial bone substitutes.

Biomaterials, especially nanoscale materials with high biocompatibility and plasticity, have been widely utilized in preclinical studies for managing bone-associated diseases. Nanomaterials have shown their potential in guided bone regeneration (GBR) and achieved satisfying biocompatibility, mechanical properties, essential barrier function, and enhanced osteogenesis and angiogenesis [11,12]. Recent studies suggest that the immune microenvironment is critical for biomaterial-regulated bone regeneration. The implanted cells or scaffolds often fail to integrate successfully with the host tissues due to the unfavorable immune response. On the contrary, a biomaterial capable of generating an ideal immune environment for osteogenesis benefits bone regeneration, an effect termed "osteoimmunomodulation". Meanwhile, nanomaterials, especially nanoparticles (NPs), are well-developed in the drug delivery field for multiple disease treatments, which can load and release functional chemicals and proteins to regulate the local immune microenvironments [13]. Multifunctional NPs encapsulated in cell membranes with a wide range of functions are considered as a future-proof platform for targeted drug delivery [14]. Therefore, novel osteoimmunomodulatory nanomaterials are expected to enhance osteoinduction by generating a favorable bone regeneration environment. In this review, we focus on the importance of osteoimmunology in bone regeneration, summarize the effects of using different materials and different modified NPs to further enhance and promote bone regeneration, and discuss the potential application of NPs as osteoimmunomodulatory tools to improve bone regeneration. Primarily, it innovatively focuses on the recent advances in the development of macrophage-targeted nanotherapeutic agents, a novel and popular research field in Material Science and Nanotechnology, pointing out the potential application of this technology in bone healing, and therefore shedding light on future nanomaterial development for advanced osteoimmunomodulation.

2. Bone Regeneration Process

Bone regeneration is a complex, well-coordinated physiological process (Figure 1) [15]. Immediately after fracture, the blood vessels which supply blood to the bone are ruptured, resulting in the formation of a hematoma around the fracture site [1]. This hematoma serves as a temporary framework for healing [1]. Inflammatory cytokines such as interleukins (e.g., IL-1), bone-morphogenetic proteins (BMPs), and tumor necrosis factor-alpha (TNF-α) are released into the injury site. These cytokines attract monocytes, lymphocytes, and macrophages, which work together to eliminate dented, necrotic tissue and produce growth factors such as vascular endothelial growth factors (VEGF) to promote angiogenesis for bone healing. Inside the hematoma, granulation tissue begins to develop. More osteoprogenitor cells/mesenchymal stem cells (MSCs) are attracted to the region, where they start to differentiate into chondroblasts and fibroblasts. As a result, chondrogenesis occurs, a collagen-rich fibrocartilaginous network spans the fracture sites, and hyaline cartilage encloses it. Alongside the periosteal layer, osteoprogenitor cells simultaneously construct a surface of woven bone [16]. Osteocytes, osteoclasts, and chondroblasts are typically stimulated to differentiate during endochondral ossification of the cartilaginous callus. The callus of cartilage is trapped and begins to calcify [16]. Subperiosteally, woven bone is deposited. At the same time, newly formed blood vessels grow, allowing MSCs to migrate. At the end of this process, an abrasive callus of immature bone forms. In a process known as "coupled remodeling", the osteoclasts repeatedly remodel the hard callus [16]. This process involves both osteoblast bone formation and osteoclast resorption [16]. The spongy bone of the soft callus is supplanted by lamellar bone, and the callus center is substituted mainly by compact bone [16]. The vasculature has undergone significant remodeling in addition to these modifications [17].

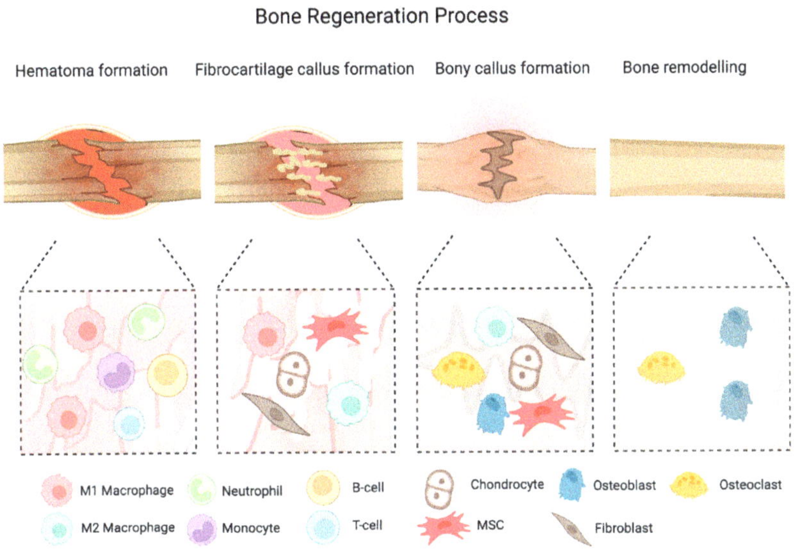

Figure 1. The figure of bone (fracture) healing process. The several phases of bone regeneration are presented, from the initial hematoma formation phase to callus formation and subsequent remodeling. At each phase, the major cell populations are indicated. Created with BioRender.com.

Numerous essential molecules that control the intricate physiological process of bone regeneration have been identified. BMPs are potent and effective osteoinductive factors that have received the most attention. They promote the differentiation of osteoprogenitors into osteoblasts by encouraging their mitogenesis. BMPs, which act as strong osteoinductive constituents in diverse tissue-engineering products, show much promise for clinical cartilage and bone regeneration [18]. Cervical fusion, the repair of lengthy bone deformities, and craniofacial and periodontitis applications are just a few of the current clinical applications. The US FDA recently approved BMP-7 and BMP-2 for specific clinical conditions, which can be administered in absorbable collagen, food, and drugs [19]. Except for BMPs, biological substances such as growth factors derived from platelets (PDGF) and plasma rich in platelets (PRP), have been found to aid in the healing of bone defects [20].

3. Osteoimmunology in Bone Regeneration

Osteoimmunology is defined as the study of the communication between the immune system and skeletal system [1,21]. The skeletal and immune systems appear separate but are integral and closely related [1,22]. The basic framework for immune system regulation is established by the enrichment and different environment provided by bone marrow for the growth of hematopoietic stem cells (HSCs), which are the common progenitors of all immune cell types [22]. The communication between immune and skeletal cells, on the other hand, is critical for the pathogenesis and progression of skeletal damage diseases, postponed bone regeneration, and some other infectious diseases. Osteoclasts, osteoblasts, and immune cells, such as macrophages and T cells, play a crucial role in bone regeneration and healing. They interact with each other and the surrounding microenvironment to regulate bone remodeling balance and determine bone regeneration (Figure 2). As a result, cells from both the immunologic and skeletal systems interfere widely in the same bone microenvironment [22]. The receptor activator of nuclear factor-B (RANK) and RANK ligand (RANKL) osteoprotegerin (OPG) regulates bone homeostasis and the progression of autoimmune bone diseases by recognizing key signals which regulate intercellular communication among bone and immune cells [21]. To initiate differentiation and stimulation programs, RANK present on the surface of osteoclast progenitors should bind to RANKL present on the surface of many other cells (including osteoblasts) inside the bone microenvironment. On the other hand, the activating threshold of the RANK–RANKL axis is influenced by the relative expression of OPG, which intervenes with the RANK–RANKL axis by acting as a coreceptor for RANK. This axis also exists in immune–skeleton interplay, where immune cells can produce RANKL to activate osteoclastogenesis [23]. Importantly, this invention has resulted in the effective treatment of bone loss related to metastasis and osteoporosis, in which RANKL is targeted with a therapeutic neutralizing antibody [24].

In bone injury, immune cells are the first responders at the defect site, restoring vasculature and initiating signal cascades to attract cells to undertake the healing process. T lymphocytes and B lymphocytes are observed at the injury site after three days of injury, and their quantities are diminished when chondrogenesis starts. It has been discovered that T-cell depletion reduces bone health and fracture healing [25]. B lymphocytes are reported to be increased in the injury site and peripheral blood during fracture healing, and reduced production of IL-10 by B cells has been linked to delayed fracture healing. One of the earliest cell types infiltrated in bone healing hematoma is the macrophage, which remains active through the healing process. Derived from the mononuclear phagocyte system (MPS) in the bone marrow, macrophages appear to serve as regulators for the differentiation and function of osteoblasts and osteoclasts, participating in intermodulation as well as interaction to reach equilibrium in bone remodeling, which makes them crucial for bone formation and remodeling [26]. Macrophages have been broadly characterized into unpolarized M0, pro-inflammatory M1 phenotypes (M1a and M1b), and anti-inflammatory M2 phenotypes (M2a, M2b, and M2c) based on local stimulators, surface markers, and different functions (Figure 3) [27]. The M1 macrophages, which can be stimulated by lipopolysaccharide (LPS), interferon-gamma (IFN-γ), or cytokines, including tumor necrosis factor-alpha (TNF-α), primarily infiltrate the site of the bone defects during the early inflammatory stage. In contrast, the M2 macrophages are stimulated by cytokines such as IL-4 and IL-13, which appear during the subacute phase [1]. The function of M1 macrophages includes clearance of intracellular pathogens and secreting pro-inflammatory cytokines, whereas the activation of the M2 phenotype mainly results in anti-inflammatory responses and subsequent tissue healing. Therefore, the M1 phenotype is traditionally considered to induce/enhance inflammation. In contrast, the M2 phenotype can reduce inflammation and promote tissue repair [28,29]. However, some recent researchers have discovered that the presence of M1 macrophages enhances osteogenesis [30], and an excessive exchange to the M2 phenotype leads to fibrous tissue healing [31,32]. Therefore, it is hypothesized that both M1 and M2 are crucial during the bone healing process [1]. During the first stage of healing, the recruited macrophages polarize to pro-inflammatory M1 phenotypes and generally remain at the site of the defect for three–four days, recruiting immune cells and MSCs. Then, they gradually polarize to anti-inflammatory M2 phenotypes along with the healing process, releasing anti-inflammatory cytokines, eliminating inflammation, and promoting tissue restoration [1,33]. Therefore, early and short-term activation of M1 macrophages is essential, as the M1 macrophage depletion or over-inhibition during the initial stages would inhibit tissue healing [34]. Meanwhile, early activation of the M2 macrophages impairs tissue healing and induces fibrous encapsulation. Therefore, it is indispensable to effectively control M1 to M2 polarization at an appropriate time, conduct an osteogenesis-favoring cytokine release pattern, and benefit the subsequent bone formation 2/8/2023 1:04:00 PM.

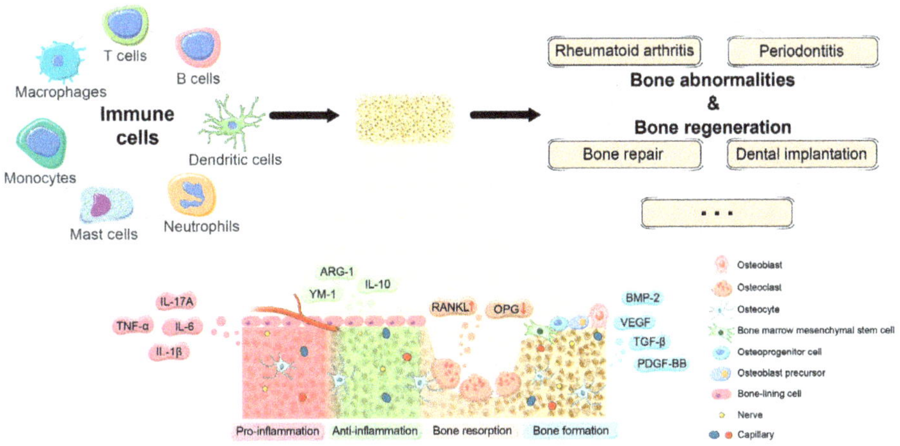

Figure 2. The mechanisms of osteoimmunomodulation in bone regeneration. Reprinted/adapted with permission from Ref. [35]. Copyright 2021 Zhou, Wu, Yu, Tang, Liu, Jia, Yang and Xiang.

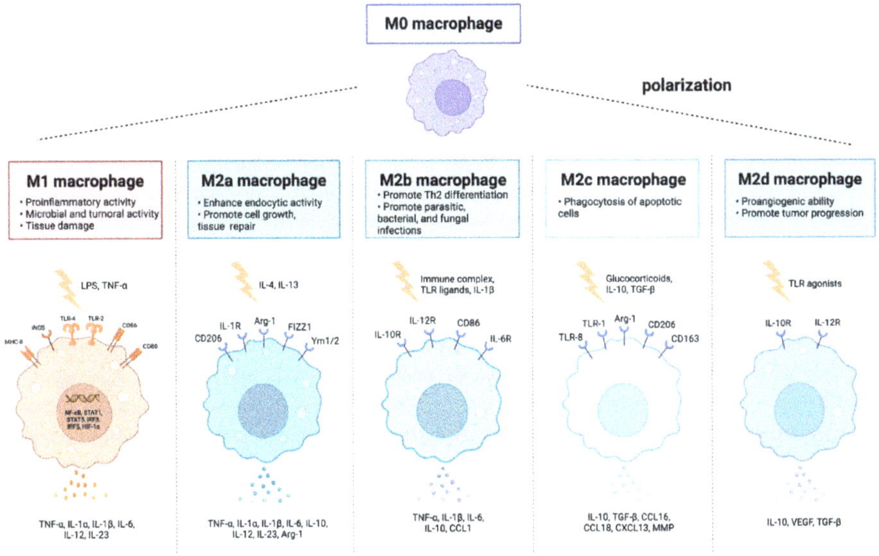

Figure 3. Macrophages have been characterized as unpolarized M0, pro-inflammatory M1 phenotypes, and anti-inflammatory M2 phenotypes (M2a, M2b, and M2c) with different functions. Created with BioRender.com.

4. Bioapplication of Nanoparticles

Biomaterials, including polymers, ceramics, and metals, are usually utilized in bone regeneration treatments, which act as bone substitutes or tissue engineering scaffolds [36]. Biomaterials for bone-associated applications have undergone significant improvement in recent years, intending to generate functionalized materials capable of delivering bioactive chemicals that may directly regulate cell activity [37]. The anatomical intricacy of bone makes bone one-of-a-kind and nearly impossible to replicate in artificial materials, along with the severe mechanical stress to which it is subjected. Nonetheless, certain tactics have been implemented with success [38] via nanotechnology. Nanotechnology has enabled the creation of nanostructures to mimic the structures and sizes found in natural bone. Nanomaterials exhibit unique physical and chemical properties, making them attractive for various applications in various fields, including medicine, electronics, energy, and the environment. The physical and chemical properties of nanomaterials are determined by their size, shape, composition, and surface characteristics. One of the most significant physical properties is their size, which results in a large surface area and enhanced reactivity, making nanomaterials more reactive than their bulk counterparts. The shape of the nanomaterials ranges from spherical, rod-like, or triangular to more complex shapes, which can affect their performance, such as the dispersibility in the liquid base [39]. Chemical properties of nanomaterials, such as composition, surface chemistry, surface charge, solubility, and hydrophobicity/hydrophilicity, can affect their stability, solubility, and reactivity, as well as their interaction with other materials and biological systems. The surface charge of nanomaterials can affect their interaction with other materials and biological systems and can be used to control the release of therapeutic agents [40]. Structural properties determine the size and shape of the nanomaterials and the arrangement of the atoms in the material. For example, the electrical, optical, and magnetic properties of nanomaterials are significantly affected by the performance of atoms in the NP structure [41].

Nanoparticulate systems, bioactive glass, hybrid materials, metal and metal oxide nanomaterials, and carbon-based nanomaterials are categories of osteoimmunomodulatory nanomaterials that have gained significant attention in recent years regarding their potential applications in bone tissue engineering. Nanoparticulate systems, including NPs, liposomes, and dendrimers, have effectively delivered therapeutics to bone tissue [42]. Bioactive glass has osteoinductive and osteoconductive properties, making it valuable for promoting bone growth and repair [43,44]. Hybrid materials combine inorganic and organic materials to enhance biological responses, making them ideal for bone tissue engineering applications [45]. Metal and metal oxide nanomaterials exhibit antibacterial and anti-inflammatory properties, making them useful for preventing infections in bone tissue [46]. Carbon-based nanomaterials, such as graphene and carbon nanotubes, have high mechanical strength and excellent biocompatibility, providing a supportive scaffold for bone cells to grow and proliferate [47]. Different types of NPs and nano-hybrid particles, such as ceramic and metal NPs, are used as material coatings and provide great potential for material modification [48]. Hence, NPs can change the scaffold qualities, resulting in improved attributes such as better mechanical properties, induced osteoinduction, and improved osteoconduction. NPs are prospective biomaterials with sizes smaller than 100 nanometers, which have an essential influence on modern medicine [49] by delivering therapeutics in a controlled and reliant manner [50]. There are two main types of NPs: organic (e.g., liposomes, polymeric NPs) and inorganic NPs such as silica, carbon, magnetic, and metallic NPs (Figure 4).

Liposomes have been used in drug delivery. To achieve drug delivery, the cargo should be included in the liposome structure [51]. Depending on the characteristics of the products to be transported, this process can be carried out in two ways. If the cargo is hydrophobic, it is combined with an organic solvent and incorporated into the hydrophobic portion. However, when the cargo is hydrophilic, it should be supplied as an aqueous medium so that it can be retained in the inner section of the liposome. Liposome size is another critical factor that directly impacts the circulatory period. Liposomes throughout the nanoscale range, in particular, can be used to administrate therapeutics [52]. The major disadvantage of liposome biomedical application is that the reticuloendothelial system can recognize liposomes quickly, which facilitates the removal of liposomes from circulation [53] and impairs their drug delivery efficiency.

Polymers are employed to synthesize polymeric NPs. The self-assembly of adaptive block copolymers could also produce structures with a high degree of complexity. Another benefit of polymeric NPs is their high drug-loading capacity [54]. The loaded molecules can be directly dissolved, distributed, or bonded to polymeric elements through covalent connections. As a result, polymeric NPs are now being used to deliver molecules in various biomedical fields, including vaccination, cancer treatments, inflammation, neurologic diseases, and tissue regeneration [55].

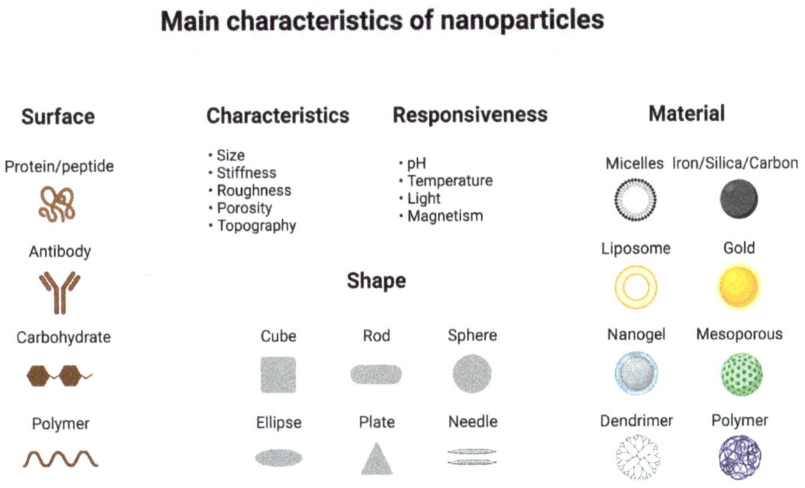

Figure 4. Physical and chemical properties of NPs, including the surface, shapes, characteristics, responsiveness, and nano-based materials. Created with BioRender.com.

Silica is well-known for its biocompatibility, chemical stability, and well-defined surface features. Silica-based NPs, especially mesoporous silica NPs (MSNPs), have been widely applied due to their adjustable particle and pore size, easy surface modification, specific porous structure, high surface area, big pore volume, etc. Consequently, MSNPs can load immense quantities of biomolecules [56]. For bioapplication, MSNPs with pore sizes ranging from 2 nm to 50 nm are ideal choices [57]. Additionally, MSNPs are resistant to degradation by heat, pH, mechanical forces, and dissolution and are thus ideal drug vehicles. Furthermore, their good biocompatibility, ease of production, and excellent binding to multiple antibiotics suggest the good bioapplication potential of MSNPs [58].

In addition to drug delivery, NPs have attracted significant attention in medical imaging. For example, iron oxide NP-based fluorescent probes have been well-accepted [59]. Meanwhile, the versatility of gold NPs makes them appealing for bioimaging procedures. The optical properties of the AuNPs can be adjusted and optimized by engineering the shape and size ratio of the AuNPs [60]. Tailored to absorption nearly in the infrared range, gold NPs allow for better visualization of the deep tissue [61]. Biological applications, including biosensing and diagnostics, can benefit from this technology [62].

5. Application of NPs in Bone Regeneration

As a nanostructured material, bone comprises organic and inorganic components with hierarchical structures ranging from the nano- to the macroscopic level. In addition to traditional treatments, nanomaterials offer a novel strategy for bone repair. Nanostructured scaffolds control cellular proliferation and differentiation, which contributes to the regeneration of healthy tissues, and give cells a more supportive structure comparable to native bone structure [63]. The specific properties of NPs, including their physical properties, chemical properties, and different modifications, as well as their quantum physical mechanisms, make them advantageous over conventional materials [64]. There are plenty of approaches using NPs to regulate bone regeneration. For example, in the initial implantation period, NPs can be an effective enhancer on the surface of biomaterials to acquire good mechanical properties and stability, providing structural function in the injury site for bone healing [65].

NPs can also be incorporated into biomaterials to offer them adjustable mechanical strength (stiffness), stimulating stem cells to take on an extended shape to differentiate preferentially into osteoblasts [66,67]. Meanwhile, a CaP ceramic–magnetic NP (CaP-MNP) composite can use magnetic fields to promote bone healing [68]. Moreover, some NPs themselves can directly improve osteogenesis. For instance, titanium oxide nanotubes of 70 nm diameter induced osteogenic differentiation by regulating H3K4 trimethylation [69]. In the deficiency of any osteoinductive factor, one kind of synthetic silicate nanoplatelet can promote the stem cells' osteogenic differentiation [70]. Another common application of nanotechnology in bone regeneration is to use NPs to load biomolecules/drugs facilitating osteogenesis, including osteoinductive factors (e.g., osteopontin, BMPs, VEGF) [71–73]; drugs reducing bone resorption; and inducing osteogenesis (e.g., alendronate, simvastatin, dexamethasone) [74–76], microRNAs (e.g., miR-590-5p, miR-2861, miR-210) [75,77,78] and others [55,79,80].

Despite delivering one bioactive factor, combining two growth factors can better mimic the natural process of bone healing. For example, stromal cell-derived factor 1 (SDF-1), a significant chemokine for stem cell migration, plays a crucial role in the recruitment of MSCs. Meanwhile, BMP-2 is an inducer of osteogenesis in MSCs. Wang et al. introduced a chitosan oligosaccharide/heparin NPs for delivery. They sustained the release of BMP-2 and SDF-1, which sequentially induced migration of MSCs and promoted their osteogenic differentiation for bone repair, an efficient strategy to avoid the rapid degradation of SDF-1 and BMP-2 [81]. Another research study by Poth et al. also loaded BMP-2 on bio-degradable chitosan-tripolyphosphate NPs to induce bone formation [73].

VEGF is a kind of growth factor that plays a vital role in the process of angiogenesis [82]. VEGF is primarily expressed during the early stages to promote blood vessel formation and re-establish vascularization throughout normal bone repair and healing. Meanwhile, BMPs are uninterruptedly expressed to stimulate bone remodeling and regeneration [83,84]. Many researchers have reported that the synergistic effects of BMP-2 and VEGF would better benefit bone regeneration than one growth factor. VEGF expression in bone defects can upregulate the production of BMP-2, which is indispensable in bone healing [85,86]. As a result, more and more studies focused on the co-delivery of VEGF and BMP-2 using NPs. Geuze et al. created poly(lactic-co-glycolic acid) (PLGA) microparticles for sustained release of BMP-2 and VEGF, which achieved improved osteogenesis [84]. Young Park et al. developed 3D polycaprolactone (PCL) structures with hydrogel decorated with both VEGF and BMP-2 and showed more capillary and bone regeneration compared with the delivery of BMP-2 alone [87]. To achieve sequential release of VEGF and BMP-2, some researchers used microspheres (e.g., PLGA microspheres, O-Carboxymethyl chitosan microspheres) loaded with BMP-2 integrated into scaffolds (e.g., poly(propylene) scaffold, hydroxyapatite collagen scaffold) loaded with VEGF. The scaffolds exhibited a substantial initial strong release of VEGF and a sustained release of BMP-2 over the rest of the implantation period. These studies indicated that it is beneficial for bone formation and remodeling to have a sequential angiogenic and osteogenic growth factor secretion [88,89].

Nanoemulsification is one of the most common and well-known methods for producing NPs. It is characterized by synthesizing nanosized particle dispersions by combining the polar phase with the non-polar phase when a surfactant is available and enables the production of 100 nm, injectable, 3D-printable with a high specific surface area and limited mass transport restrictions NPs. Hydroxyapatite NPs synthesized via nanoemulsion technology are thoroughly explored as inorganic components of composite bone implant materials. The combination of nano-hydroxyapatite with an elastic biodegradable polymer, which mimics the organic materials of bone extracellular matrix, has been demonstrated to enhance viability, adhesion, and proliferation significantly. Osteogenic differentiation of cells seeded onto implants such as human mesenchymal stem cells (hMSCs), which is attributed to osteoinductive properties of hydroxyapatite nanomaterials [90]. Additionally, the NPs synthesized from hydroxyapatite and metal materials have significant bactericidal properties [91]. Therefore, nano-hydroxyapatite has been used to create osteoinductive coating materials for bone implants, a strategy to facilitate their osseointegration with the host tissue [92]. Bone implants modified with silver NPs synthesized by bioreduction techniques have enhanced antibacterial and antioxidant properties [93].

Recently, many endeavors have been devoted to developing NPs that bind specifically to the bone. Such NPs can accumulate at the targeted sites, increasing therapeutic efficiency, limiting the adverse side effects of the drug delivery to other tissues/organs [94] and can be widely used in diagnosis, bone tissue engineering, and treatment of bone disease [95]. Bone-targeting NPs are typically created by modifying them with compounds with high affinity for bone tissue, such as Ca^{2+} ions. Examples of these compounds include bisphosphonates (BP), which comprise two Ca^{2+}-binding phosphonate groups in their molecules [96], and alendronate, an anti-osteoporotic drug that can bind to hydroxyapatite via multiple Ca^{2+} ions [97]. When NPs are functionalized with alendronate, they can selectively target bone, restraining bone resorption and acting as "anchors" to strengthen the interaction of the implant with the host tissue [98,99]. For this reason, alendronate has been widely utilized for the functionalization of NPs for bone regeneration applications such as inorganic (e.g., Fe_3O_4, hydroxyapatite, clay) [80,100,101] and polymer (e.g., poly(g-benzyl-L-glutamate), PLGA) NPs [55,79,99].

NPs have unique properties, such as a high surface area-to-volume ratio, which can make them more efficient delivery vehicles for drugs and other therapeutic agents. However, their unique properties also raise several safety concerns, primarily related to their biocompatibility, immunogenic properties, and toxicity.

NPs are generally considered biocompatible as long as they do not cause obvious inflammation or irritation. Otherwise, the application of NPs can be limited due to their bio-incompatibility. One study showed that 50 nm-sized particles of Fe_2O_3-NP caused severe oxidative stress in HepG2 cells and extreme damage in rat liver [102]. NPs may be immunogenic if they contain foreign proteins or other molecules the body recognizes as threats. Immunogenic NPs can trigger an immune response, leading to inflammation, cell death, and other adverse reactions [103]. The toxicity of NPs depends on their composition and size. Smaller NPs have a larger specific surface area and therefore are more likely to interact with cellular components and are more likely to enter cells and be taken up by organs, which can result in toxicity. For example, in one study, the effects of silver nanoparticles of different sizes (20, 80, 113 nm) on cytotoxicity, inflammation, genotoxicity, and developmental toxicity were compared in in vitro experiments, and 20 nm silver nanoparticles were more toxic than larger nanoparticles [104]. The released Ag^+ endangers cellular functions, causing damage to deoxyribonucleic acid and cell death [105].

NPs have been frequently used in bone regeneration in recent years. Integrating nanotechnology into tissue engineering applications has created a plethora of new potential for researchers and new clinical applications.

6. Applications of NPs in Osteoimmunomodulation

Osteoimmunomodulation refers to the modulation of the immune system to make the local immune environment beneficial for bone regeneration. It aims to use functional materials to regulate the immune cell responses to sequentially modulate the bone remodeling processes, facilitating bone healing [106]. It involves regulating immune cells or cytokines to influence bone remodeling and maintain bone health [107].

Immune suppression benefits certain conditions, such as allergies, autoimmune disorders, and organ transplants. Immunomodulatory or anti-inflammatory characteristics are required for these applications. Several experimental and characterization methods are used to assess the properties of nanomaterials, such as polymers, ceramics, composites, and metals in osteoimmunomodulation (Table 1).

Engineered NPs serve as vehicles for delivering anti-inflammatory drugs to phagocytes, lowering therapeutic doses and immune-related adverse effects [108]. Immune system activation is inevitable when NPs invade. The innate immune cells interact with newly initiated NPs immediately and produce complex immune reactions as a first defense against impending threats to the host. Depending on their physicochemical characteristics, NPs can engage the interactions between proteins and cells to stimulate or inhibit the innate immune response and complement system activation or avoidance. NP size, structure, hydrophobicity, and surface chemistry are the major factors that affect the interactions between the innate immune system and NPs [109].

Table 1. Experimental Approach for Osteoimmunomodulation Characterization.

In Vitro and In Vivo Assays	Physical and Chemical Characterization	Biocompatibility Evaluation and Biomechanical Analysis
Cell culture-based assays (osteoblast, osteoclast, macrophage) Enzyme-linked immunosorbent assays (ELISA) Alkaline phosphatase activity assays Mineralization assays	X-ray diffraction (XRD) Transmission electron microscopy (TEM) Scanning electron microscopy (SEM) Dynamic light scattering (DLS)	Cytotoxicity assays (MTT, LDH) Hemocompatibility assays (hemolysis, platelet activation) Inflammation assays (IL-1β, TNF-α, IL-6)
Implantation studies in animal models (rats, mice, rabbits) Histological analysis (bone formation and resorption) Micro-computed tomography (μCT) Bone density measurements	Fourier transform infrared spectroscopy (FTIR) Energy-dispersive X-ray spectroscopy (EDS) X-ray fluorescence (XRF) Nuclear magnetic resonance spectroscopy (NMR)	Contact angle measurements Zeta potential measurements Surface roughness analysis Compression tests Tensile tests Indentation tests

For bone regeneration, immunomodulation is required to generate an ideal environment for the subsequent osteogenesis, which can be achieved by NPs. As explained in Section 3, macrophage populations are critical regulators of bone regeneration. The pro-inflammatory M1 phenotype of macrophages causes a rise in pro-inflammatory cytokines such as IL-1β, IL-6, and TNF-α, resulting in the inhibition of osteogenesis [110,111] and promoting osteoclastogenesis [112]. Alternatively, the anti-inflammatory M2 phenotype can reverse inflammation and secrete osteogenic cytokines, including BMP2 and VEGF, to encourage bone regeneration [113–115]. Hence, targeting macrophages to induce their M2 polarization has been regarded as an efficient way to enhance bone regeneration, and nanomaterials are shown as effective agents for macrophage polarization (Table 2). Some NPs (Figure 5) can efficiently promote M2 polarization, such as gold, TiO_2, and cerium oxide (CeO_2) NPs [116–118]. Moreover, the nanopore structure and pore size were discovered to affect the inflammatory response and release of pro-osteogenic factors of macrophages by influencing their spreading, cell shape, and adhesion [119,120]. For instance, Chen et al. ascertained that macrophages grown on larger pore size NPs (100 and 200 nm) were highly anti-inflammatory, demonstrating a decrease in pro-inflammatory cytokine and expression of M1 phenotype surface-marker [119]. One study found that silver NPs with different sizes and shapes showed different effects on bone metabolism and immunity, indicating that controlling the size and shape of nanomaterials can affect their osteoimmunomodulatory effects [121]. NPs with rough surfaces also alter macrophage activation and cytokine release. Research indicated that titanium (Ti) with a smooth surface could induce M1 activation and inflammatory cytokines expression, including IL-1β, IL-6, and TNF-α. Meanwhile, Ti with a rough and hydrophilic surface enhances anti-inflammatory macrophage polarization and the secretion of cytokines such as IL-4 and IL-10 [122]. Another way to promote M2 polarization is to modify the composition of NPs surfaces by doping anti-inflammatory elements or decorating bioactive molecules [123–125]. For example, hexapeptides Cys-Leu-Pro-Phe-Phe-Asp [112], peptide arginine-glycine-aspartic acid (RGD) [126], and IL-4 [127] have been successfully conjugated on gold NP surfaces to achieve successful anti-inflammation. Besides, CeO_2 NPs have been coated with hydroxyapatite to promote M2 polarization [128]. A previous study indicated that surface modification of hydroxyapatite nanorods with chitosan reduced macrophage activation and enhanced osteoblast proliferation [129]. Moreover, strontium (Sr) or copper (Cu)-decorated bioactive glass particles have been found to enhance M2 polarization and promote osteogenesis [124,125]. Zhang et al. synthesized strontium-substituted sub-micron bioactive glasses (Sr-SBG), which have been found to advance the proliferation and osteogenic differentiation of mMSCs [130].

As potential drug delivery systems, NPs have been widely used for bioactive molecule delivery, such as cytokines, growth factors, gene-modulators, and signaling pathway regulators, to stimulate the M1-to-M2 polarization. For instance, IL-4, a widely used anti-inflammatory cytokine, has been frequently adopted as cargo delivered by various nanocarriers to induce M2 polarization [131–133]. One research study introduced an IL-4-incorporated nanofibrous heparin-modified gelatin microsphere, which can alleviate chronic inflammation due to diabetes and improve osteogenesis [132]. Sphingosine-1-phosphate (S1P), as a sphingolipid growth factor, can also stimulate macrophages to polarize to the M2 phenotype [134]. Das et al. synthesized nanofibers composed of polycaprolactone (PCL) and poly (D, L-lactide-co-glycolide) (PLGA) for an S1P synthetic analog delivery, which was found to induce macrophage differentiation to M2 phenotypes, facilitating osseous repair in an animal model of the mandibular bone defect [135]. CD163 is an M2 phenotype marker affiliated with the scavenger receptor cysteine-rich (SRCR) family [136]. One study encapsulated CD163 gene plasmid into polyethyleneimine NPs assembled with a mannose ligand for selectively targeting macrophages and inducing CD163 expression, and further transferring macrophages into their anti-inflammatory phenotype [137]. Upregulation of miR-223 can drive the macrophage polarization toward the anti-inflammatory (M2) phenotype, whereas local-targeted delivery of miRNAs is still challenging due to the low stability of miRNA. To solve this problem, Saleh et al. developed an adhesive hydrogel with NPs loaded with miR-223 5p mimic to regulate macrophage polarization to M2 to promote tissue remodeling [138]. Yin et al. loaded an anti-inflammatory drug, resolvin D1, into the gold nanocages (AuNC) coated with cell membranes from LPS-stimulated M1-like macrophages to facilitate M2 polarization. The overexpressed inflammatory cytokine receptors on the cell membrane can competitively bind to the pro-inflammatory cytokines with cell surface receptors, thereby impeding inflammatory responses [139]. The results indicate that this nanosystem could efficiently inhibit inflammatory responses, stimulate an M2-like phenotype polarization, and promote bone regeneration in the femoral defect.

Despite the crucial role of M2 macrophages in promoting bone tissue regeneration, more and more studies have focused on the importance of M1 macrophages in osteoimmunomodulation. As mentioned, M1 macrophages dominate in the early stage of inflammation, enhancing the early commitment and recruitment of angiogenic and osteogenic precursors. In contrast, M2 macrophages function in the later stage of bone regeneration by facilitating osteocyte maturation and determining the microstructure of the newly formed bone tissue [140]. Therefore, a highly orchestrated immune response comprising sequential activation of M1 and M2 macrophages is essential for subsequent bone healing [141]. Thus, a sequential release of therapeutics from NPs to instruct the timely phenotypic switching of macrophages is deemed necessary. For example, as IFN-γ and IL-4 can induce M1 and M2 polarization, Spillar et al. designed a scaffold with a quick release of IFN-γ to increase the M1 phenotype, subsequently with a release of IL-4 to enhance the M2 phenotype. The sequential release feature was achieved by physically adsorbing IFN-γ onto the scaffolds, while loading IL-4 on the material via biotin-streptavidin binding [142]. In another example, miRNA-155 is highly expressed in M1 and less in M2, while the delivery of miRNA-21 can promote macrophage polarization toward M2 phenotypes [143–145]. Li et al. synthesized NPs through free radical polymerization carrying both miRNA-155 and miRNA-21 to induce macrophages first toward M1 sequentially and then M2 polarization, a new strategy for bone regeneration [146]. Zinc (Zn) is an essential trace element in various immune responses. Zn's scarcity and low concentration caused inflammation, while a proper concentration of Zn exhibited an anti-inflammatory effect [147,148]. Therefore, one study fabricated microcrystalline bioactive glass scaffolds with different doses of ZnO to orchestrate the sequential M1-to-M2 macrophage polarization, taking advantage of varying amounts of Zn^{2+} released from the material [149]. Yang et al. incorporated IFN-γ and Sr-substituted nanohydroxyapatite (nano-SrHA) coatings to the surface of native small intestinal submucosa (SIS) membrane, which is widely applied in GBR to direct a sequential M1-M2 macrophage polarization. The nano-SrHA coatings were loaded on the SIS membrane using the sol-gel method, while the IFN-γ was physically deposited. As a result, the physically absorbed IFN-γ released in a burst manner to induce temporary M1 macrophage polarization, then a more sequential release of Sr irons to promote M2 polarization, which intensely improved the vascularization and bone regeneration [150]. Bone marrow macrophages have various receptors on their surface that enable them to recognize molecules such as cytokines, chemokines, lipids, and glycans. NPs to ensure a drug delivery to target bone marrow macrophages can be achieved using strategies such as surface modification of NPs with components interacting with bone marrow macrophage receptors. However, NPs in circulation are removed by the mononuclear phagocyte system (MPS), including the spleen, liver, and Kupffer cells, affecting the NP-based targeted delivery on bone marrow macrophages. Therefore, combining the NPs with bone implants (via approaches such as surface coating, 3D printing, etc.) is suggested instead of systemic administration, which can facilitate the NPs to modulate the local bone healing immune environment and avoid particle clearance due to blood circulation and MPS.

Table 2. Applications of NPs in osteoimunomodulation via modulating macrophage response.

Strategies for Regulating Macrophage Polarization	Applications of NPs in Osteoimunomodulation	References
Intrinsic properties	Gold, TiO_2, and cerium oxide (CeO_2) NPs can enhance M2 polarization.	[116–118]
Nanopore structure and pore size	NPs with pores of larger size (100 and 200 nm) were highly anti-inflammatory and inhibited M1 polarization.	[119]
	The nanoneedle structure induced M2 polarization. The micropattern sizes of 12 μm and 36 μm in the micro/nano hierarchy enhanced M2 polarization.	[120]
Surface roughness	Ti with smooth surface stimulated M1 activation. Ti with rough surface enhanced M2 polarization.	[122]
Composition	Gold NPs fused hexapeptides Cys-Leu-Pro-Phe-Phe-Asp, peptide arginine-glycine-aspartic acid (RGD), and IL-4 could stimulate M2 polarization.	[112,126,127]
	CeO_2 NPs with hydroxyapatite could enhance M2 polarization.	[128]
	Strontium (Sr)- or copper (Cu)-doped bioactive glass particles promoted M2 polarization and enhanced osteogenesis.	[124,125]
Drug delivery	Various nanocarriers have delivered IL-4 (anti-inflammatory cytokine) to induce M2 polarization.	[131–133]
	NPs can deliver S1P synthetic analog to direct macrophage polarization toward M2.	[134]
	CD163 gene has been encapsulated into polyethyleneimine NPs decorated with a mannose ligand to induce CD163 expression and macrophage polarization toward M2.	[137]
	miR-223 5p mimic was delivered to induce macrophage polarization to M2.	[138]
	Resolvin D1-loaded gold nanocages (AuNC) were coated with M1-like macrophage membranes to enhance M2 polarization.	[139]
A sequential release of therapeutics induces the M1-to-M2 phenotype switch during tissue regeneration.	Spillar et al. designed a scaffold that achieved a sequential release of first IFN-γ and then IL-4 to modulate macrophage polarization from early stage M1 to later-stage M2.	[142]
	NPs carry both miRNA-155 and miRNA-21 to sequentially stimulate macrophage polarization first toward M1 and then the M2 phenotype.	[146]
	Microcrystalline bioactive glass scaffolds with different doses of ZnO orchestrate the sequential M1-to-M2 macrophage polarization.	[149]
	Sr-substituted nanohydroxyapatite (nano-SrHA) coatings and IFN-γ to the surface of native SIS membrane control a sequential M1-M2 macrophage transition.	[150]

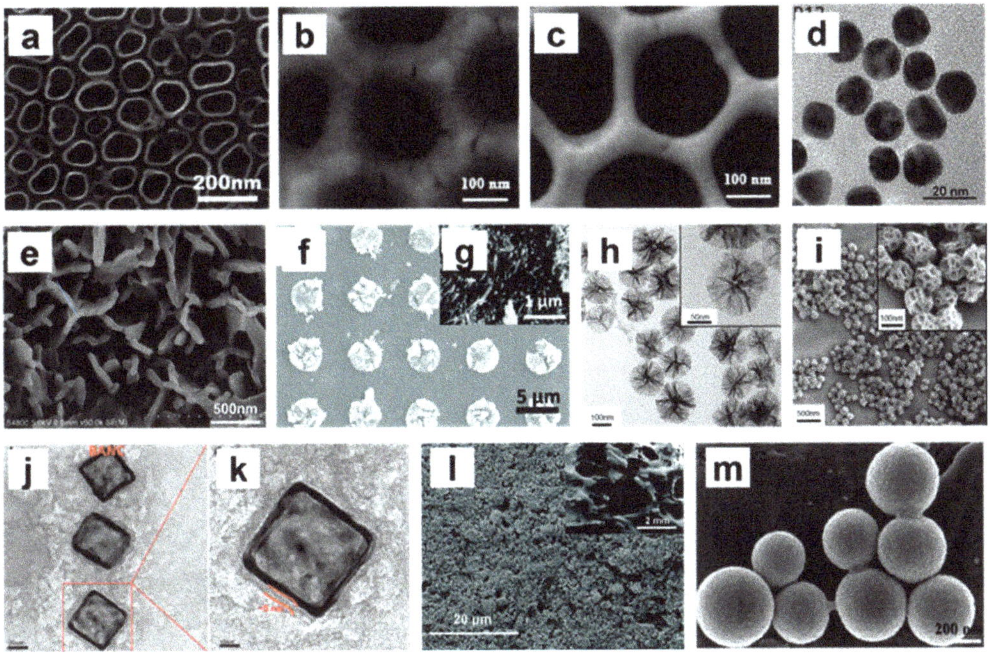

Figure 5. Composite images (TEM and SEM images) of different nanoparticle structures. (**a**) SEM image of 110 nm titania nanotubes (TNTs) [116]; (**b**,**c**) SEM images of anodic alumina structures with different sized pores (100 nm and 200 nm) [119]; (**d**) TEM image of peptide-coated gold NPs (P12) [112]; (**e**) SEM image of surface morphology of SIS/SrHA [150]; (**f**,**g**) SEM images of HA bioceramics with nanoneedle structures. (g: high magnification image) [120]; (**h**) TEM images of 150 nm extra-large pore mesoporous silica NPs (XL-MSNs) (inset: high magnification image) [133]; (**i**) SEM images of 150 nm XL-MSNs (inset: high magnification image) [133]; (**j**,**k**) TEM images of biomimetic anti-inflammatory nano-capsules (BANC) [139]; (**l**) SEM image of 5 wt% ZnO incorporated microcrystalline bioactive glass (5Zn-MCBG) [149]; (**m**) SEM image of strontium-substituted submicrometer bioactive glass (Sr–SBG) [124]. Reprinted/adapted with permission from Ref. [112]. Copyright 2020 Wang, Zhang, Sun, Gao, Xiong, Ma, Liu, Shen, Li and Yang. Ref. [116]. Copyright 2019 Shen, Yu, Ma, Luo, Hu, Li, He, Zhang, Peng and Song. Ref. [119]. Copyright 2017 Chen, Ni, Han, Crawford, Lu, Wei, Chang, Wu and Xiao. Ref. [124]. Copyright 2016 Zhang, Zhao, Huang, Fu, Li and Chen. Ref. [133]. Copyright 2017 Kwon, Cha, Cho, Min, Park, Kang and Kim. Ref. [139]. Copyright 2020 Yin, Zhao, Li, Zhao, Wang, Deng, Zhang, Shen, Li and Zhang. Ref. [149]. Copyright 2021 Bai, Liu, Xu, Ye, Zhou, Berg, Yuan, Li and Xia. Ref. [150]. Copyright 2021 Yang, Zhou, Yu, Yang, Sun, Ji, Xiong and Guo.

Taken together (Figure 6), osteoimmunology is a fascinating field focusing on the interconnected molecular pathways between the immune and skeletal systems. Among all the immune cells, macrophages play the most crucial role, secreting cytokines that determine the immune response and modulate the subsequent bone regeneration. Nanomaterials can assist in regulating immune responses by targeting macrophages and managing their polarization, bringing a new strategy for managing bone-related diseases [151].

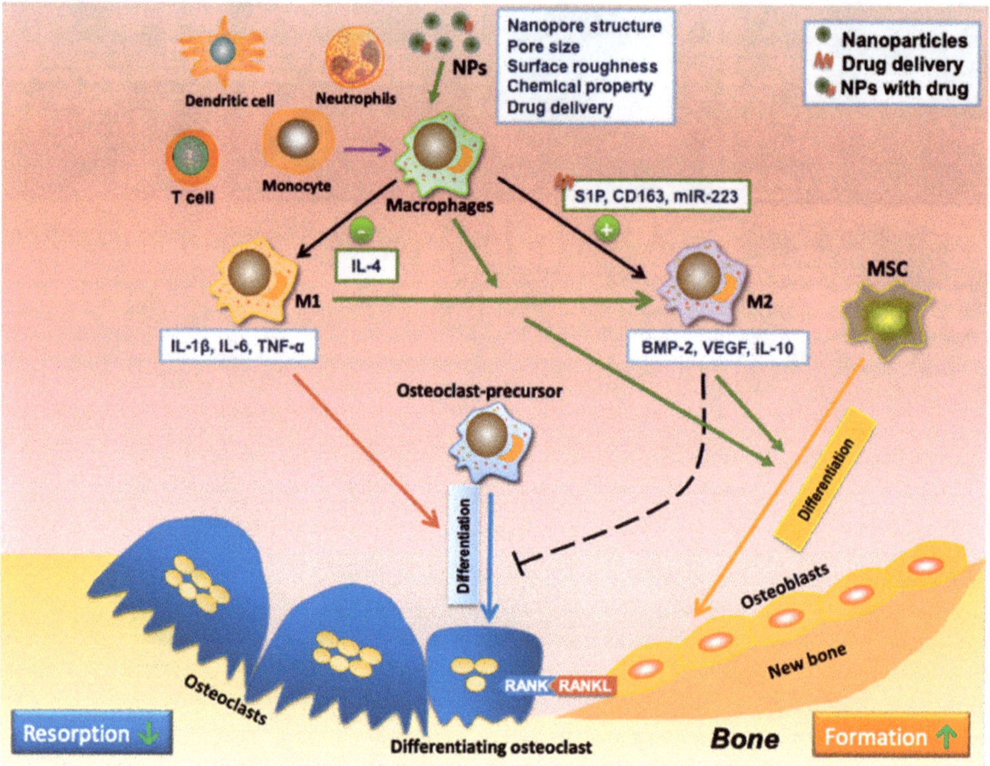

Figure 6. NPs as drug delivery systems to introduce functional osteoimmunomodulation to promote bone regeneration. Ideally, NPs should modulate the immune system to enable the formation of an ideal immune microenvironment for subsequent osteogenesis and bone regeneration. Macrophage polarization is essential in osteoimmunomodulation. The pro-inflammatory M1 phenotype of macrophages could secrete pro-inflammatory cytokines such as IL-1β, IL-6, and TNF-α to promote osteoclastogenesis. The anti-inflammatory M2 phenotype of macrophages could secrete osteogenic cytokines, including BMP2 and VEGF, to enhance bone regeneration. The timely M1-to-M2 phenotype switch is critical in bone regeneration, which can be induced by NP-based drug delivery. NPs can regulate macrophage polarization through different strategies, such as nanopore structure and size, surface roughness, chemical properties, and delivered drugs. NPs can inhibit M1 polarization, promote macrophage polarization to M2, or enhance M1 to M2 polarization, further promoting bone healing.

7. Conclusions and Future Remarks

NPs have been widely applied in bone regeneration and showed great potential in osteoimmunomodulation. However, certain disadvantages, such as biocompatibility, immunogenic properties, and toxicity, limit the clinical application of NPs. Additionally, how to ensure the NPs target the bone marrow macrophages instead of macrophages in other organs (e.g., spleen, liver, etc.) remains a challenge for future research. Meanwhile, the complex multi-stage regenerative process of bone healing, the discrepancy or mismatch between the degradation rate of NPs and the growth rate of bone tissues, the problem of regulating the release rate of therapeutic cargo (drugs, factors, or genes), and other limitations still pose obstacles to the application of NPs, which still need further improvement. The fabrication process and approach of nanotopography should be enhanced and optimized to modify the immune response accurately. As previously stated, ordinary materials have imprecise chemical properties that are typically overlooked. The administration must consider the chemical characteristics of the outermost surface. Plasma polymerization is an excellent technique for creating a persistent and non-pinhole biocompatible coating on diverse nanostructures, allowing for specific chemical adjustment of the outermost material, thus achieving precision-tuned bio-physicochemical and biomechanical surface properties. With the development of nanomaterials and material modification approaches, macrophage-targeting nanotherapeutics can ensure the drugs are delivered more precisely to the therapeutic site, therefore allowing for advanced osteoimmunomodulation to improve bone regeneration. Furthermore, the improvement of NP-based drug delivery systems enables the delivery of multiple drugs to target the different stages of bone regeneration. For example, immunomodulatory therapeutics can be released in the early stage of bone healing to ensure the local environment suits bone regeneration. The osteogenic factors can be sequentially released later to boost bone regeneration. Other approaches, such as environmental-responsive releases of immunomodulators and osteogenic factors, can facilitate personalized osteoimmunomodulatory regulation and bone healing.

In summary, this review introduced the importance of osteoimmunology in bone regeneration, the types and current biomedical applications of NPs, the multiple roles of NPs in osteogenesis, and specifically, the significance of NP application on macrophage-targeting osteoimmunomodulation for advanced bone regeneration. Therefore, it is expected that advanced nanotechnology will shed light on bone tissue engineering and facilitate functional bone repair in the future.

Author Contributions: Conceptualization, L.X. and Y.X.; software, W.G.; validation, Y.L., Y.Z., T.K., L.X. and Y.X.; writing—original draft preparation, J.W., D.C. and R.H.; writing—review and editing, W.G., Y.Z., T.K., L.X. and Y.X.; visualization, W.G.; supervision, Y.Z., T.K., L.X. and Y.X.; project administration, L.X. and Y.X.; funding acquisition, Y.X. All authors have read and agreed to the published version of the manuscript.

Funding: This research was supported by the Joint Research Centre Fund from the Department of Environment and Science (2019–2023), Queensland; Young Researcher Grant (19-066) from the Osteology Foundation, Switzerland, QUT Centre for Biomedical Technologies ECR/MCR grant scheme 2021.

Institutional Review Board Statement: Not applicable.

Informed Consent Statement: Not applicable.

Data Availability Statement: Not applicable.

Conflicts of Interest: The authors declare no conflict of interest.

References

1. Xiao, L.; Ma, Y.; Crawford, R.; Mendhi, J.; Zhang, Y.; Lu, H.; Zhao, Q.; Cao, J.; Wu, C.; Wang, X.; et al. The Interplay between Hemostasis and Immune Response in Biomaterial Development for Osteogenesis. *Mater. Today* **2022**, *54*, 202–224. [CrossRef]
2. Roddy, E.; DeBaun, M.R.; Daoud-Gray, A.; Yang, Y.P.; Gardner, M.J. Treatment of Critical-Sized Bone Defects: Clinical and Tissue Engineering Perspectives. *Eur. J. Orthop. Surg. Traumatol.* **2018**, *28*, 351–362. [CrossRef] [PubMed]
3. Tal, H. (Ed.) *Bone Regeneration*; InTech: Rijeka, Croatia, 2012; ISBN 978-953-51-0487-2.
4. Bhatt, R.A.; Rozental, T.D. Bone Graft Substitutes. *Hand. Clin.* **2012**, *28*, 457–468. [CrossRef] [PubMed]
5. Ball, A.N.; Donahue, S.W.; Wojda, S.J.; McIlwraith, C.W.; Kawcak, C.E.; Ehrhart, N.; Goodrich, L.R. The Challenges of Promoting Osteogenesis in Segmental Bone Defects and Osteoporosis: CHALLENGES OF PROMOTING OSTEOGENESIS. *J. Orthop. Res.* **2018**, *36*, 1559–1572. [CrossRef] [PubMed]
6. Lobb, D.C.; DeGeorge, B.R.; Chhabra, A.B. Bone Graft Substitutes: Current Concepts and Future Expectations. *J. Hand Surg.* **2019**, *44*, 497–505.e2. [CrossRef]
7. Gillman, C.E.; Jayasuriya, A.C. FDA-Approved Bone Grafts and Bone Graft Substitute Devices in Bone Regeneration. *Mater. Sci. Eng. C* **2021**, *130*, 112466. [CrossRef]
8. Baldwin, P.; Li, D.J.; Auston, D.A.; Mir, H.S.; Yoon, R.S.; Koval, K.J. Autograft, Allograft, and Bone Graft Substitutes: Clinical Evidence and Indications for Use in the Setting of Orthopaedic Trauma Surgery. *J. Orthop. Trauma* **2019**, *33*, 203–213. [CrossRef]

9. Schwartz, N.G.; Hernandez-Romieu, A.C.; Annambhotla, P.; Filardo, T.D.; Althomsons, S.P.; Free, R.J.; Li, R.; Wyatt Wilson, W.; Deutsch-Feldman, M.; Drees, M.; et al. Nationwide Tuberculosis Outbreak in the USA Linked to a Bone Graft Product: An Outbreak Report. *Lancet Infect. Dis.* **2022**, *22*, 1617–1625. [CrossRef]
10. Imerb, N.; Thonusin, C.; Chattipakorn, N.; Chattipakorn, S.C. Aging, Obese-Insulin Resistance, and Bone Remodeling. *Mech. Ageing Dev.* **2020**, *191*, 111335. [CrossRef]
11. Zhu, L.; Luo, D.; Liu, Y. Effect of the Nano/Microscale Structure of Biomaterial Scaffolds on Bone Regeneration. *Int. J. Oral. Sci.* **2020**, *12*, 6. [CrossRef]
12. Lyons, J.G.; Plantz, M.A.; Hsu, W.K.; Hsu, E.L.; Minardi, S. Nanostructured Biomaterials for Bone Regeneration. *Front. Bioeng. Biotechnol.* **2020**, *8*, 922. [CrossRef] [PubMed]
13. Chindamo, G.; Sapino, S.; Peira, E.; Chirio, D.; Gonzalez, M.C.; Gallarate, M. Bone Diseases: Current Approach and Future Perspectives in Drug Delivery Systems for Bone Targeted Therapeutics. *Nanomaterials* **2020**, *10*, 875. [CrossRef] [PubMed]
14. Lee, N.-H.; You, S.; Taghizadeh, A.; Taghizadeh, M.; Kim, H.S. Cell Membrane-Cloaked Nanotherapeutics for Targeted Drug Delivery. *Int. J. Mol. Sci.* **2022**, *23*, 2223. [CrossRef] [PubMed]
15. Chen, Z.; Bachhuka, A.; Wei, F.; Wang, X.; Liu, G.; Vasilev, K.; Xiao, Y. Nanotopography-Based Strategy for the Precise Manipulation of Osteoimmunomodulation in Bone Regeneration. *Nanoscale* **2017**, *9*, 18129–18152. [CrossRef] [PubMed]
16. Claes, L.; Recknagel, S.; Ignatius, A. Fracture Healing under Healthy and Inflammatory Conditions. *Nat. Rev. Rheumatol.* **2012**, *8*, 133–143. [CrossRef] [PubMed]
17. Sheen, J.R.; Garla, V.V. Fracture Healing Overview. In *StatPearls*; StatPearls Publishing: Treasure Island, FL, USA, 2022.
18. Dimitriou, R.; Tsiridis, E.; Giannoudis, P.V. Current Concepts of Molecular Aspects of Bone Healing. *Injury* **2005**, *36*, 1392–1404. [CrossRef]
19. Bessa, P.C.; Casal, M.; Reis, R.L. Bone Morphogenetic Proteins in Tissue Engineering: The Road from Laboratory to Clinic, Part II (BMP Delivery). *J. Tissue Eng. Regen. Med.* **2008**, *2*, 81–96. [CrossRef]
20. Carreira, A.C.; Lojudice, F.H.; Halcsik, E.; Navarro, R.D.; Sogayar, M.C.; Granjeiro, J.M. Bone Morphogenetic Proteins: Facts, Challenges, and Future Perspectives. *J. Dent. Res.* **2014**, *93*, 335–345. [CrossRef]
21. *Osteoimmunology: Interactions of the Bone and Immune System | Endocrine Reviews |*; Oxford Academic: Oxford, UK, 2008. Available online: https://academic.oup.com/edrv/article/29/4/403/2354973?login=false (accessed on 25 January 2023).
22. Walsh, M.C.; Takegahara, N.; Kim, H.; Choi, Y. Updating Osteoimmunology: Regulation of Bone Cells by Innate and Adaptive Immunity. *Nat. Rev. Rheumatol.* **2018**, *14*, 146–156. [CrossRef]
23. Crockett, J.C.; Mellis, D.J.; Scott, D.I.; Helfrich, M.H. New Knowledge on Critical Osteoclast Formation and Activation Pathways from Study of Rare Genetic Diseases of Osteoclasts: Focus on the RANK/RANKL Axis. *Osteoporos. Int.* **2011**, *22*, 1–20. [CrossRef]
24. Zaheer, S.; LeBoff, M.; Lewiecki, E.M. Denosumab for the Treatment of Osteoporosis. *Expert. Opin. Drug Metab. Toxicol.* **2015**, *11*, 461–470. [CrossRef] [PubMed]
25. Könnecke, I.; Serra, A.; El Khassawna, T.; Schlundt, C.; Schell, H.; Hauser, A.; Ellinghaus, A.; Volk, H.-D.; Radbruch, A.; Duda, G.N.; et al. T and B Cells Participate in Bone Repair by Infiltrating the Fracture Callus in a Two-Wave Fashion. *Bone* **2014**, *64*, 155–165. [CrossRef] [PubMed]
26. Kitagawa, Y.; Ohkura, N.; Sakaguchi, S. Molecular Determinants of Regulatory T Cell Development: The Essential Roles of Epigenetic Changes. *Front. Immunol.* **2013**, *4*, 106. [CrossRef]
27. Monocyte and Macrophage Biology: An Overview-ClinicalKey. Available online: https://www.clinicalkey.com.au/#!/content/playContent/1-s2.0-S0270929510000525 (accessed on 25 January 2023).
28. Wilson, H.M.; Barker, R.N.; Erwig, L.-P. Macrophages: Promising Targets for the Treatment of Atherosclerosis. *Curr. Vasc. Pharmacol.* **2009**, *7*, 234–243. [CrossRef] [PubMed]
29. *Macrophage Polarization and Bone Formation: A Review*; SpringerLink: Berlin, Germany, 2016. Available online: https://link.springer.com/article/10.1007/s12016-015-8519-2 (accessed on 25 January 2023).
30. Guihard, P.; Danger, Y.; Brounais, B.; David, E.; Brion, R.; Delecrin, J.; Richards, C.D.; Chevalier, S.; Rédini, F.; Heymann, D. Induction of Osteogenesis in Mesenchymal Stem Cells by Activated Monocytes/Macrophages Depends on Oncostatin M Signaling. *Stem. Cells* **2012**, *30*, 762–772. [CrossRef]
31. Brown, B.N.; Badylak, S.F. Expanded Applications, Shifting Paradigms and an Improved Understanding of Host–Biomaterial Interactions. *Acta Biomater.* **2013**, *9*, 4948–4955. [CrossRef]
32. Mokarram, N.; Bellamkonda, R.V. A Perspective on Immunomodulation and Tissue Repair. *Ann. Biomed. Eng.* **2014**, *42*, 338–351. [CrossRef]
33. Murray, P.J.; Allen, J.E.; Biswas, S.K.; Fisher, E.A.; Gilroy, D.W.; Goerdt, S.; Gordon, S.; Hamilton, J.A.; Ivashkiv, L.B.; Lawrence, T.; et al. Macrophage Activation and Polarization: Nomenclature and Experimental Guidelines. *Immunity* **2014**, *41*, 14–20. [CrossRef]
34. Awojoodu, A.O.; Ogle, M.E.; Sefcik, L.S.; Bowers, D.T.; Martin, K.; Brayman, K.L.; Lynch, K.R.; Peirce-Cottler, S.M.; Botchwey, E. Sphingosine 1-Phosphate Receptor 3 Regulates Recruitment of Anti-Inflammatory Monocytes to Microvessels during Implant Arteriogenesis. *Proc. Natl. Acad. Sci. USA* **2013**, *110*, 13785–13790. [CrossRef]
35. Zhou, A.; Wu, B.; Yu, H.; Tang, Y.; Liu, J.; Jia, Y.; Yang, X.; Xiang, L. Current Understanding of Osteoimmunology in Certain Osteoimmune Diseases. *Front. Cell Dev. Biol.* **2021**, *9*, 698068. [CrossRef]
36. Bains, P.S.; Sidhu, S.S.; Bahraminasab, M.; Prakash, C. (Eds.) *Biomaterials in Orthopaedics and Bone Regeneration: Design and Synthesis*; Materials Horizons: From Nature to Nanomaterials; Springer: Singapore, 2019; ISBN 9789811399770.

37. Mieszawska, A.J.; Kaplan, D.L. Smart Biomaterials-Regulating Cell Behavior through Signaling Molecules. *BMC Biol.* **2010**, *8*, 59. [CrossRef] [PubMed]
38. Subramani, K.; Ahmed, W. *Emerging Nanotechnologies in Dentistry: Materials, Processes, and Applications*, 2nd ed.; Elsevier: Amsterdam, The Netherlands, 2018; ISBN 978-0-12-812292-1.
39. Čitaković, N.M. Physical Properties of Nanomaterials. *Encycl. Nanosci. Nanotechnol.* **2019**, *67*, 159–171. [CrossRef]
40. Patil, S.P.; Burungale, V.V. 2-Physical and Chemical Properties of Nanomaterials. In *Nanomedicines for Breast Cancer Theranostics*; Thorat, N.D., Bauer, J., Eds.; Micro and Nano Technologies; Elsevier: Amsterdam, The Netherlands, 2020; pp. 17–31. ISBN 978-0-12-820016-2.
41. Joudeh, N.; Linke, D. Nanoparticle Classification, Physicochemical Properties, Characterization, and Applications: A Comprehensive Review for Biologists. *J. Nanobiotechnol.* **2022**, *20*, 262. [CrossRef] [PubMed]
42. Nanoparticles for Bone Tissue Engineering-Vieira-2017-Biotechnology Progress-Wiley Online Library. Available online: https://aiche.onlinelibrary.wiley.com/doi/full/10.1002/btpr.2469?casa_token=gdFilaTgL7gAAAAA%3AIf2hUMWXZC4FIyGCY-RrsVVjhoeEzfx4r22oy8Z1S-TkhTPN7S7PSRFBTUTKf0I3p13NUcOZsyijjLO2 (accessed on 1 February 2023).
43. Crush, J.; Hussain, A.; Seah, K.T.M.; Khan, W.S. Bioactive Glass: Methods for Assessing Angiogenesis and Osteogenesis. *Front. Cell Dev. Biol.* **2021**, *9*, 643781. [CrossRef]
44. Materials | Free Full-Text | Bioactive Glass and Glass-Ceramic Scaffolds for Bone Tissue Engineering. Available online: https://www.mdpi.com/1996-1944/3/7/3867 (accessed on 1 February 2023).
45. Nanostructured Hybrid Materials for Bone Tissue Regeneration: Ingenta Connect. Available online: https://www.ingentaconnect.com/content/ben/cnano/2006/00000002/00000003/art00003 (accessed on 1 February 2023).
46. Eivazzadeh-Keihan, R.; Bahojb Noruzi, E.; Khanmohammadi Chenab, K.; Jafari, A.; Radinekiyan, F.; Hashemi, S.M.; Ahmadpour, F.; Behboudi, A.; Mosafer, J.; Mokhtarzadeh, A.; et al. Metal-based Nanoparticles for Bone Tissue Engineering. *J. Tissue Eng. Regen. Med.* **2020**, *14*, 1687–1714. [CrossRef] [PubMed]
47. Zheng, S.; Tian, Y.; Ouyang, J.; Shen, Y.; Wang, X.; Luan, J. Carbon Nanomaterials for Drug Delivery and Tissue Engineering. *Front. Chem.* **2022**, *10*, 1106. [CrossRef]
48. Baldelli, A.; Ou, J.; Li, W.; Amirfazli, A. Spray-On Nanocomposite Coatings: Wettability and Conductivity. *Langmuir* **2020**, *36*, 11393–11410. [CrossRef]
49. Murthy, S.K. Nanoparticles in Modern Medicine: State of the Art and Future Challenges. *Int. J. Nanomed.* **2007**, *2*, 129–141.
50. Laschke, M.W.; Strohe, A.; Menger, M.D.; Alini, M.; Eglin, D. In Vitro and in Vivo Evaluation of a Novel Nanosize Hydroxyapatite Particles/Poly(Ester-Urethane) Composite Scaffold for Bone Tissue Engineering. *Acta Biomater.* **2010**, *6*, 2020–2027. [CrossRef]
51. Filipczak, N.; Pan, J.; Yalamarty, S.S.K.; Torchilin, V.P. Recent Advancements in Liposome Technology. *Adv. Drug Deliv. Rev.* **2020**, *156*, 4–22. [CrossRef]
52. Liu, Y.; Castro Bravo, K.M.; Liu, J. Targeted Liposomal Drug Delivery: A Nanoscience and Biophysical Perspective. *Nanoscale Horiz.* **2021**, *6*, 78–94. [CrossRef] [PubMed]
53. Al-Jamal, W.T.; Kostarelos, K. Liposomes: From a Clinically Established Drug Delivery System to a Nanoparticle Platform for Theranostic Nanomedicine. *Acc. Chem. Res.* **2011**, *44*, 1094–1104. [CrossRef] [PubMed]
54. Rao, J.P.; Geckeler, K.E. Polymer Nanoparticles: Preparation Techniques and Size-Control Parameters. *Prog. Polym. Sci.* **2011**, *36*, 887–913. [CrossRef]
55. Thamake, S.I.; Raut, S.L.; Gryczynski, Z.; Ranjan, A.P.; Vishwanatha, J.K. Alendronate Coated Poly-Lactic-Co-Glycolic Acid (PLGA) Nanoparticles for Active Targeting of Metastatic Breast Cancer. *Biomaterials* **2012**, *33*, 7164–7173. [CrossRef] [PubMed]
56. Tang, Y.; Zhao, Y.; Wang, X.; Lin, T. Layer-by-Layer Assembly of Silica Nanoparticles on 3D Fibrous Scaffolds: Enhancement of Osteoblast Cell Adhesion, Proliferation, and Differentiation. *J. Biomed. Mater. Res. A* **2014**, *102*, 3803–3812. [CrossRef]
57. Wang, Y.; Zhao, Q.; Han, N.; Bai, L.; Li, J.; Liu, J.; Che, E.; Hu, L.; Zhang, Q.; Jiang, T.; et al. Mesoporous Silica Nanoparticles in Drug Delivery and Biomedical Applications. *Nanomed. Nanotechnol. Biol. Med.* **2015**, *11*, 313–327. [CrossRef] [PubMed]
58. Gounani, Z.; Asadollahi, M.A.; Pedersen, J.N.; Lyngsø, J.; Skov Pedersen, J.; Arpanaei, A.; Meyer, R.L. Mesoporous Silica Nanoparticles Carrying Multiple Antibiotics Provide Enhanced Synergistic Effect and Improved Biocompatibility. *Colloids Surf. B Biointerfaces* **2019**, *175*, 498–508. [CrossRef]
59. El-Fiqi, A.; Kim, J.-H.; Kim, H.-W. Osteoinductive Fibrous Scaffolds of Biopolymer/Mesoporous Bioactive Glass Nanocarriers with Excellent Bioactivity and Long-Term Delivery of Osteogenic Drug. *ACS Appl. Mater. Interfaces* **2015**, *7*, 1140–1152. [CrossRef]
60. Ali, M.R.K.; Wu, Y.; Chapman, S.; Ding, Y. Synthesis, Structure Evolution, and Optical Properties of Gold Nanobones. *Res. Chem. Intermed* **2019**, *45*, 3973–3983. [CrossRef]
61. Li, J.; Liu, J.; Chen, C. Remote Control and Modulation of Cellular Events by Plasmonic Gold Nanoparticles: Implications and Opportunities for Biomedical Applications. *ACS Nano* **2017**, *11*, 2403–2409. [CrossRef]
62. Kairdolf, B.A.; Qian, X.; Nie, S. Bioconjugated Nanoparticles for Biosensing, in Vivo Imaging, and Medical Diagnostics. *Anal. Chem.* **2017**, *89*, 1015–1031. [CrossRef] [PubMed]
63. Gong, T.; Xie, J.; Liao, J.; Zhang, T.; Lin, S.; Lin, Y. Nanomaterials and Bone Regeneration. *Bone Res.* **2015**, *3*, 15029. [CrossRef] [PubMed]
64. Dyondi, D.; Webster, T.J.; Banerjee, R. A Nanoparticulate Injectable Hydrogel as a Tissue Engineering Scaffold for Multiple Growth Factor Delivery for Bone Regeneration. *Int. J. Nanomed.* **2013**, *8*, 47–59. [CrossRef] [PubMed]

65. Xu, H.H.K.; Weir, M.D.; Simon, C.G. Injectable and Strong Nano-Apatite Scaffolds for Cell/Growth Factor Delivery and Bone Regeneration. *Dent. Mater.* **2008**, *24*, 1212–1222. [CrossRef]
66. Cai, L.; Guinn, A.S.; Wang, S. Exposed Hydroxyapatite Particles on the Surface of Photo-Crosslinked Nanocomposites for Promoting MC3T3 Cell Proliferation and Differentiation. *Acta Biomater.* **2011**, *7*, 2185–2199. [CrossRef]
67. Zhang, T.; Gao, Y.; Cui, W.; Li, Y.; Xiao, D.; Zhou, R. Nanomaterials-Based Cell Osteogenic Differentiation and Bone Regeneration. *Curr. Stem Cell Res. Ther.* **2021**, *16*, 36–47. [CrossRef]
68. Wu, Y.; Jiang, W.; Wen, X.; He, B.; Zeng, X.; Wang, G.; Gu, Z. A Novel Calcium Phosphate Ceramic–Magnetic Nanoparticle Composite as a Potential Bone Substitute. *Biomed. Mater.* **2010**, *5*, 015001. [CrossRef]
69. Lv, L.; Liu, Y.; Zhang, P.; Zhang, X.; Liu, J.; Chen, T.; Su, P.; Li, H.; Zhou, Y. The Nanoscale Geometry of TiO2 Nanotubes Influences the Osteogenic Differentiation of Human Adipose-Derived Stem Cells by Modulating H3K4 Trimethylation. *Biomaterials* **2015**, *39*, 193–205. [CrossRef]
70. Gaharwar, A.K.; Mihaila, S.M.; Swami, A.; Patel, A.; Sant, S.; Reis, R.L.; Marques, A.P.; Gomes, M.E.; Khademhosseini, A. Bioactive Silicate Nanoplatelets for Osteogenic Differentiation of Human Mesenchymal Stem Cells. *Adv. Mater.* **2013**, *25*, 3329–3336. [CrossRef]
71. Kim, B.-S.; Yang, S.-S.; Kim, C.S. Incorporation of BMP-2 Nanoparticles on the Surface of a 3D-Printed Hydroxyapatite Scaffold Using an ε-Polycaprolactone Polymer Emulsion Coating Method for Bone Tissue Engineering. *Colloids Surf. B Biointerfaces* **2018**, *170*, 421–429. [CrossRef]
72. Wang, L.; Xu, W.; Chen, Y.; Wang, J. Alveolar Bone Repair of Rhesus Monkeys by Using BMP-2 Gene and Mesenchymal Stem Cells Loaded Three-Dimensional Printed Bioglass Scaffold. *Sci. Rep.* **2019**, *9*, 18175. [CrossRef] [PubMed]
73. Poth, N.; Seiffart, V.; Gross, G.; Menzel, H.; Dempwolf, W. Biodegradable Chitosan Nanoparticle Coatings on Titanium for the Delivery of BMP-2. *Biomolecules* **2015**, *5*, 3–19. [CrossRef] [PubMed]
74. Moradikhah, F.; Doosti-Telgerd, M.; Shabani, I.; Soheili, S.; Dolatyar, B.; Seyedjafari, E. Microfluidic Fabrication of Alendronate-Loaded Chitosan Nanoparticles for Enhanced Osteogenic Differentiation of Stem Cells. *Life Sci.* **2020**, *254*, 117768. [CrossRef] [PubMed]
75. Liu, J.; Cui, Y.; Kuang, Y.; Xu, S.; Lu, Q.; Diao, J.; Zhao, N. Hierarchically Porous Calcium–Silicon Nanosphere-Enabled Co-Delivery of MicroRNA-210 and Simvastatin for Bone Regeneration. *J. Mater. Chem. B* **2021**, *9*, 3573–3583. [CrossRef] [PubMed]
76. Seddighian, A.; Ganji, F.; Baghaban-Eslaminejad, M.; Bagheri, F. Electrospun PCL Scaffold Modified with Chitosan Nanoparticles for Enhanced Bone Regeneration. *Prog. Biomater.* **2021**, *10*, 65–76. [CrossRef]
77. Balagangadharan, K.; Chandran, S.V.; Arumugam, B.; Saravanan, S.; Venkatasubbu, G.D.; Selvamurugan, N. Chitosan/Nano-Hydroxyapatite/Nano-Zirconium Dioxide Scaffolds with MiR-590-5p for Bone Regeneration. *Int. J. Biol. Macromol.* **2018**, *111*, 953–958. [CrossRef]
78. Bu, W.; Xu, X.; Wang, Z.; Jin, N.; Liu, L.; Liu, J.; Zhu, S.; Zhang, K.; Jelinek, R.; Zhou, D. Ascorbic Acid-PEI Carbon Dots with Osteogenic Effects as MiR-2861 Carriers to Effectively Enhance Bone Regeneration. *ACS Appl. Mater. Interfaces* **2020**, *12*, 50287–50302. [CrossRef]
79. Cenni, E.; Granchi, D.; Avnet, S.; Fotia, C.; Salerno, M.; Micieli, D.; Sarpietro, M.G.; Pignatello, R.; Castelli, F.; Baldini, N. Biocompatibility of Poly(d,l-Lactide-Co-Glycolide) Nanoparticles Conjugated with Alendronate. *Biomaterials* **2008**, *29*, 1400–1411. [CrossRef]
80. Hwang, S.-J.; Lee, J.-S.; Ryu, T.-K.; Kang, R.-H.; Jeong, K.-Y.; Jun, D.-R.; Koh, J.-M.; Kim, S.-E.; Choi, S.-W. Alendronate-Modified Hydroxyapatite Nanoparticles for Bone-Specific Dual Delivery of Drug and Bone Mineral. *Macromol. Res.* **2016**, *24*, 623–628. [CrossRef]
81. Wang, B.; Guo, Y.; Chen, X.; Zeng, C.; Hu, Q.; Yin, W.; Li, W.; Xie, H.; Zhang, B.; Huang, X.; et al. Nanoparticle-Modified Chitosan-Agarose-Gelatin Scaffold for Sustained Release of SDF-1 and BMP-2. *Int. J. Nanomed.* **2018**, *13*, 7395–7408. [CrossRef]
82. Heparin-Regulated Release of Growth Factors in Vitro and Angiogenic Response in Vivo to Implanted Hyaluronan Hydrogels Containing VEGF and BFGF-ScienceDirect. Available online: https://www.sciencedirect.com/science/article/abs/pii/S0142961 20600456X (accessed on 24 September 2022).
83. Kang, W.; Yun, Y.-R.; Lee, D.-S.; Kim, T.-H.; Kim, J.-H.; Kim, H.-W.; Jang, J.-H. Fluorescence-Based Retention Assays Reveals Sustained Release of Vascular Endothelial Growth Factor from Bone Grafts. *J. Biomed. Mater. Res. Part A* **2016**, *104*, 283–290. [CrossRef] [PubMed]
84. Geuze, R.E.; Theyse, L.F.; Kempen, D.H.; Hazewinkel, H.A.; Kraak, H.Y.; Öner, F.C.; Dhert, W.J.; Alblas, J. A Differential Effect of Bone Morphogenetic Protein-2 and Vascular Endothelial Growth Factor Release Timing on Osteogenesis at Ectopic and Orthotopic Sites in a Large-Animal Model. *Tissue Eng. Part A* **2012**, *18*, 2052–2062. [CrossRef]
85. Cao, L.; Kong, X.; Lin, S.; Zhang, S.; Wang, J.; Liu, C.; Jiang, X. Synergistic Effects of Dual Growth Factor Delivery from Composite Hydrogels Incorporating 2-N, 6-O-Sulphated Chitosan on Bone Regeneration. *Artif. Cells Nanomed. Biotechnol.* **2018**, *46*, S1–S17. [CrossRef]
86. Talavera-Adame, D.; Wu, G.; He, Y.; Ng, T.T.; Gupta, A.; Kurtovic, S.; Hwang, J.Y.; Farkas, D.L.; Dafoe, D.C. Endothelial Cells in Co-Culture Enhance Embryonic Stem Cell Differentiation to Pancreatic Progenitors and Insulin-Producing Cells through BMP Signaling. *Stem Cell Rev. Rep.* **2011**, *7*, 532–543. [CrossRef] [PubMed]

87. Young Park, J.; Shim, J.-H.; Choi, S.-A.; Jang, J.; Kim, M.; Hwa Lee, S.; Cho, D.-W. 3D Printing Technology to Control BMP-2 and VEGF Delivery Spatially and Temporally to Promote Large-Volume Bone Regeneration. *J. Mater. Chem. B* **2015**, *3*, 5415–5425. [CrossRef]
88. Kempen, D.H.; Lu, L.; Heijink, A.; Hefferan, T.E.; Creemers, L.B.; Maran, A.; Yaszemski, M.J.; Dhert, W.J. Effect of Local Sequential VEGF and BMP-2 Delivery on Ectopic and Orthotopic Bone Regeneration. *Biomaterials* **2009**, *30*, 2816–2825. [CrossRef] [PubMed]
89. Dou, D.D.; Zhou, G.; Liu, H.W.; Zhang, J.; Liu, M.L.; Xiao, X.F.; Fei, J.J.; Guan, X.L.; Fan, Y.B. Sequential Releasing of VEGF and BMP-2 in Hydroxyapatite Collagen Scaffolds for Bone Tissue Engineering: Design and Characterization. *Int. J. Biol. Macromol.* **2019**, *123*, 622–628. [CrossRef] [PubMed]
90. Kupikowska-Stobba, B.; Kasprzak, M. Fabrication of Nanoparticles for Bone Regeneration: New Insight into Applications of Nanoemulsion Technology. *J. Mater. Chem. B* **2021**, *9*, 5221–5244. [CrossRef]
91. Nirmala, R.; Sheikh, F.A.; Kanjwal, M.A.; Lee, J.H.; Park, S.-J.; Navamathavan, R.; Kim, H.Y. Synthesis and Characterization of Bovine Femur Bone Hydroxyapatite Containing Silver Nanoparticles for the Biomedical Applications. *J. Nanopart. Res.* **2011**, *13*, 1917–1927. [CrossRef]
92. Baba Ismail, Y.M.; Ferreira, A.M.; Bretcanu, O.; Dalgarno, K.; El Haj, A.J. Polyelectrolyte Multi-Layers Assembly of SiCHA Nanopowders and Collagen Type I on Aminolysed PLA Films to Enhance Cell-Material Interactions. *Colloids Surf. B Biointerfaces* **2017**, *159*, 445–453. [CrossRef]
93. Narciso, A.M.; da Rosa, C.G.; Nunes, M.R.; Sganzerla, W.G.; Hansen, C.M.; de Melo, A.P.Z.; Paes, J.V.; Bertoldi, F.C.; Barreto, P.L.M.; Masiero, A.V. Antimicrobial Green Silver Nanoparticles in Bone Grafts Functionalization for Biomedical Applications. *Biocatal. Agric. Biotechnol.* **2021**, *35*, 102074. [CrossRef]
94. Cheng, H.; Chawla, A.; Yang, Y.; Li, Y.; Zhang, J.; Jang, H.L.; Khademhosseini, A. Development of Nanomaterials for Bone-Targeted Drug Delivery. *Drug Discov. Today* **2017**, *22*, 1336–1350. [CrossRef] [PubMed]
95. Zhou, X.; Cornel, E.J.; He, S.; Du, J. Recent Advances in Bone-Targeting Nanoparticles for Biomedical Applications. *Mater. Chem. Front.* **2021**, *5*, 6735–6759. [CrossRef]
96. Ossipov, D.A. Bisphosphonate-Modified Biomaterials for Drug Delivery and Bone Tissue Engineering. *Expert Opin. Drug Deliv.* **2015**, *12*, 1443–1458. [CrossRef]
97. Leu, C.-T.; Luegmayr, E.; Freedman, L.P.; Rodan, G.A.; Reszka, A.A. Relative Binding Affinities of Bisphosphonates for Human Bone and Relationship to Antiresorptive Efficacy. *Bone* **2006**, *38*, 628–636. [CrossRef]
98. Zhang, J.; Liu, X.; Deng, T.; Yao, P.; Song, H.; Zhou, S.; Yan, W. Development of Drug Loaded Nanoparticles Binding to Hydroxyapatite Based on a Bisphosphonate Modified Nonionic Surfactant. *J. Nanomater.* **2015**, *16*, 145. Available online: https://dl.acm.org/doi/abs/10.1155/2015/393968 (accessed on 24 September 2022). [CrossRef]
99. de Miguel, L.; Noiray, M.; Surpateanu, G.; Iorga, B.I.; Ponchel, G. Poly(γ-Benzyl-l-Glutamate)-PEG-Alendronate Multivalent Nanoparticles for Bone Targeting. *Int. J. Pharm.* **2014**, *460*, 73–82. [CrossRef]
100. Lee, M.-S.; Su, C.-M.; Yeh, J.-C.; Wu, P.-R.; Tsai, T.-Y.; Lou, S.-L. Synthesis of Composite Magnetic Nanoparticles Fe3O4 with Alendronate for Osteoporosis Treatment. *Int. J. Nanomed.* **2016**, *11*, 4583–4594. [CrossRef]
101. Piao, H.; Kim, M.H.; Cui, M.; Choi, G.; Choy, J.-H. Alendronate-Anionic Clay Nanohybrid for Enhanced Osteogenic Proliferation and Differentiation. *J. Korean Med. Sci.* **2019**, *34*, e37. [CrossRef]
102. Sadeghi, L.; Tanwir, F.; Yousefi Babadi, V. Antioxidant Effects of Alfalfa Can Improve Iron Oxide Nanoparticle Damage: Invivo and Invitro Studies. *Regul. Toxicol. Pharmacol.* **2016**, *81*, 39–46. [CrossRef]
103. Malachowski, T.; Hassel, A. Engineering Nanoparticles to Overcome Immunological Barriers for Enhanced Drug Delivery. *Eng. Regen.* **2020**, *1*, 35–50. [CrossRef]
104. Park, M.V.D.Z.; Neigh, A.M.; Vermeulen, J.P.; de la Fonteyne, L.J.J.; Verharen, H.W.; Briedé, J.J.; van Loveren, H.; de Jong, W.H. The Effect of Particle Size on the Cytotoxicity, Inflammation, Developmental Toxicity and Genotoxicity of Silver Nanoparticles. *Biomaterials* **2011**, *32*, 9810–9817. [CrossRef] [PubMed]
105. Cao, H. *Silver Nanoparticles for Antibacterial Devices: Biocompatibility and Toxicity*; Taylor & Francis Group: London, UK, 2017; ISBN 978-1-315-35347-0.
106. Chen, Z.; Klein, T.; Murray, R.Z.; Crawford, R.; Chang, J.; Wu, C.; Xiao, Y. Osteoimmunomodulation for the Development of Advanced Bone Biomaterials. *Mater. Today* **2016**, *19*, 304–321. [CrossRef]
107. Xie, Y.; Hu, C.; Feng, Y.; Li, D.; Ai, T.; Huang, Y.; Chen, X.; Huang, L.; Tan, J. Osteoimmunomodulatory Effects of Biomaterial Modification Strategies on Macrophage Polarization and Bone Regeneration. *Regen. Biomater.* **2020**, *7*, 233–245. [CrossRef]
108. Song, G.; SPetschauer, J.; JMadden, A.; CZamboni, W. Nanoparticles and the Mononuclear Phagocyte System: Pharmacokinetics and Applications for Inflammatory Diseases. *Curr. Rheumatol. Rev.* **2014**, *10*, 22–34. [CrossRef]
109. Liu, Y.; Hardie, J.; Zhang, X.; Rotello, V.M. Effects of Engineered Nanoparticles on the Innate Immune System. *Semin. Immunol.* **2017**, *34*, 25–32. [CrossRef]
110. Zhao, Y.-J.; Gao, Z.-C.; He, X.-J.; Li, J. The Let-7f-5p–Nme4 Pathway Mediates Tumor Necrosis Factor α-Induced Impairment in Osteogenesis of Bone Marrow-Derived Mesenchymal Stem Cells. *Biochem. Cell Biol.* **2021**, *99*, 488–498. [CrossRef]
111. Ono, T.; Takayanagi, H. Osteoimmunology in Bone Fracture Healing. *Curr. Osteoporos. Rep.* **2017**, *15*, 367–375. [CrossRef]
112. Wang, L.; Zhang, H.; Sun, L.; Gao, W.; Xiong, Y.; Ma, A.; Liu, X.; Shen, L.; Li, Q.; Yang, H. Manipulation of Macrophage Polarization by Peptide-Coated Gold Nanoparticles and Its Protective Effects on Acute Lung Injury. *J. Nanobiotechnol.* **2020**, *18*, 38. [CrossRef]

113. Rifas, L. T-Cell Cytokine Induction of BMP-2 Regulates Human Mesenchymal Stromal Cell Differentiation and Mineralization. *J. Cell. Biochem.* **2006**, *98*, 706–714. [CrossRef]
114. Freytes, D.O.; Wan, L.Q.; Vunjak-Novakovic, G. Geometry and Force Control of Cell Function. *J. Cell. Biochem.* **2009**, *108*, 1047–1058. [CrossRef] [PubMed]
115. Schett, G. Osteoimmunology in Rheumatic Diseases. *Arthritis Res. Ther.* **2009**, *11*, 210. [CrossRef] [PubMed]
116. Shen, X.; Yu, Y.; Ma, P.; Luo, Z.; Hu, Y.; Li, M.; He, Y.; Zhang, Y.; Peng, Z.; Song, G. Titania Nanotubes Promote Osteogenesis via Mediating Crosstalk between Macrophages and MSCs under Oxidative Stress. *Colloids Surf. B Biointerfaces* **2019**, *180*, 39–48. [CrossRef] [PubMed]
117. Lee, C.-H.; Kim, Y.-J.; Jang, J.-H.; Park, J.-W. Modulating Macrophage Polarization with Divalent Cations in Nanostructured Titanium Implant Surfaces. *Nanotechnology* **2016**, *27*, 085101. [CrossRef] [PubMed]
118. Chigurupati, S.; Mughal, M.R.; Okun, E.; Das, S.; Kumar, A.; McCaffery, M.; Seal, S.; Mattson, M.P. Effects of Cerium Oxide Nanoparticles on the Growth of Keratinocytes, Fibroblasts and Vascular Endothelial Cells in Cutaneous Wound Healing. *Biomaterials* **2013**, *34*, 2194–2201. [CrossRef]
119. Chen, Z.; Ni, S.; Han, S.; Crawford, R.; Lu, S.; Wei, F.; Chang, J.; Wu, C.; Xiao, Y. Nanoporous Microstructures Mediate Osteogenesis by Modulating the Osteo-Immune Response of Macrophages. *Nanoscale* **2017**, *9*, 706–718. [CrossRef]
120. Yang, C.; Zhao, C.; Wang, X.; Shi, M.; Zhu, Y.; Jing, L.; Wu, C.; Chang, J. Stimulation of Osteogenesis and Angiogenesis by Micro/Nano Hierarchical Hydroxyapatite via Macrophage Immunomodulation. *Nanoscale* **2019**, *11*, 17699–17708. [CrossRef]
121. Almatroudi, A. Silver Nanoparticles: Synthesis, Characterisation and Biomedical Applications. *Open Life Sci.* **2020**, *15*, 819–839. [CrossRef]
122. Hotchkiss, K.M.; Reddy, G.B.; Hyzy, S.L.; Schwartz, Z.; Boyan, B.D.; Olivares-Navarrete, R. Titanium Surface Characteristics, Including Topography and Wettability, Alter Macrophage Activation. *Acta Biomater.* **2016**, *31*, 425–434. [CrossRef]
123. Cai, D.; Gao, W.; Li, Z.; Zhang, Y.; Xiao, L.; Xiao, Y. Current Development of Nano-Drug Delivery to Target Macrophages. *Biomedicines* **2022**, *10*, 1203. [CrossRef]
124. Zhang, W.; Zhao, F.; Huang, D.; Fu, X.; Li, X.; Chen, X. Strontium-Substituted Submicrometer Bioactive Glasses Modulate Macrophage Responses for Improved Bone Regeneration. *ACS Appl. Mater. Interfaces* **2016**, *8*, 30747–30758. [CrossRef]
125. Lin, R.; Deng, C.; Li, X.; Liu, Y.; Zhang, M.; Qin, C.; Yao, Q.; Wang, L.; Wu, C. Copper-Incorporated Bioactive Glass-Ceramics Inducing Anti-Inflammatory Phenotype and Regeneration of Cartilage/Bone Interface. *Theranostics* **2019**, *9*, 6300–6313. [CrossRef] [PubMed]
126. Bartneck, M.; Ritz, T.; Keul, H.A.; Wambach, M.; Bornemann, J.; Gbureck, U.; Ehling, J.; Lammers, T.; Heymann, F.; Gassler, N. Peptide-Functionalized Gold Nanorods Increase Liver Injury in Hepatitis. *Acs Nano* **2012**, *6*, 8767–8777. [CrossRef]
127. Raimondo, T.M.; Mooney, D.J. Functional Muscle Recovery with Nanoparticle-Directed M2 Macrophage Polarization in Mice. *Proc. Natl. Acad. Sci. USA* **2018**, *115*, 10648–10653. [CrossRef]
128. Li, K.; Shen, Q.; Xie, Y.; You, M.; Huang, L.; Zheng, X. Incorporation of Cerium Oxide into Hydroxyapatite Coating Regulates Osteogenic Activity of Mesenchymal Stem Cell and Macrophage Polarization. *J. Biomater. Appl.* **2017**, *31*, 1062–1076. [CrossRef] [PubMed]
129. Nanostructured Surface Modification to Bone Implants for Bone Reg...: Ingenta Connect. Available online: https://www.ingentaconnect.com/contentone/asp/jbn/2018/00000014/00000004/art00002 (accessed on 1 February 2023).
130. Zhang, W.; Huang, D.; Zhao, F.; Gao, W.; Sun, L.; Li, X.; Chen, X. Synergistic Effect of Strontium and Silicon in Strontium-Substituted Sub-Micron Bioactive Glass for Enhanced Osteogenesis. *Mater. Sci. Eng. C* **2018**, *89*, 245–255. [CrossRef]
131. *IL-4 Administration Exerts Preventive Effects via Suppression of Underlying Inflammation and TNF-α-Induced Apoptosis in Steroid-Induced Osteonecrosis*; SpringerLink: Berlin, Germany, 2016. Available online: https://link.springer.com/article/10.1007/s00198-015-3474-6 (accessed on 24 September 2022).
132. He, X.-T.; Li, X.; Xia, Y.; Yin, Y.; Wu, R.-X.; Sun, H.-H.; Chen, F.-M. Building Capacity for Macrophage Modulation and Stem Cell Recruitment in High-Stiffness Hydrogels for Complex Periodontal Regeneration: Experimental Studies in Vitro and in Rats. *Acta Biomater.* **2019**, *88*, 162–180. [CrossRef]
133. Kwon, D.; Cha, B.G.; Cho, Y.; Min, J.; Park, E.-B.; Kang, S.-J.; Kim, J. Extra-Large Pore Mesoporous Silica Nanoparticles for Directing in Vivo M2 Macrophage Polarization by Delivering IL-4. *Nano Lett.* **2017**, *17*, 2747–2756. [CrossRef]
134. Hughes, J.E.; Srinivasan, S.R.; Hedrick, C.C. Sphingosine-1-Phosphate Induces an Anti-Inflammatory Phenotype in Macrophages During Inflammation. *FASEB J.* **2017**, *21*, A772. Available online: https://faseb.onlinelibrary.wiley.com/doi/abs/10.1096/fasebj.21.6.A772-b (accessed on 24 September 2022).
135. Das, A.; Segar, C.E.; Hughley, B.B.; Bowers, D.T.; Botchwey, E.A. The Promotion of Mandibular Defect Healing by the Targeting of S1P Receptors and the Recruitment of Alternatively Activated Macrophages. *Biomaterials* **2013**, *34*, 9853–9862. [CrossRef]
136. Etzerodt, A.; Moestrup, S.K. CD163 and Inflammation: Biological, Diagnostic, and Therapeutic Aspects. *Antioxid. Redox Signal.* **2013**, *18*, 2352–2363. [CrossRef] [PubMed]
137. Alvarado-Vazquez, P.A.; Bernal, L.; Paige, C.A.; Grosick, R.L.; Vilrriales, C.M.; Ferreira, D.W.; Ulecia-Morón, C.; Romero-Sandoval, E.A. Macrophage-Specific Nanotechnology-Driven CD163 Overexpression in Human Macrophages Results in an M2 Phenotype under Inflammatory Conditions. *Immunobiology* **2017**, *222*, 900–912. [CrossRef]

138. Saleh, B.; Dhaliwal, H.K.; Portillo-Lara, R.; Shirzaei Sani, E.; Abdi, R.; Amiji, M.M.; Annabi, N. Local Immunomodulation Using an Adhesive Hydrogel Loaded with MiRNA-Laden Nanoparticles Promotes Wound Healing. *Small* **2019**, *15*, 1902232. [CrossRef] [PubMed]
139. Yin, C.; Zhao, Q.; Li, W.; Zhao, Z.; Wang, J.; Deng, T.; Zhang, P.; Shen, K.; Li, Z.; Zhang, Y. Biomimetic Anti-Inflammatory Nano-Capsule Serves as a Cytokine Blocker and M2 Polarization Inducer for Bone Tissue Repair. *Acta Biomater.* **2020**, *102*, 416–426. [CrossRef] [PubMed]
140. Wang, S.; Xiao, L.; Prasadam, I.; Crawford, R.; Zhou, Y.; Xiao, Y. Inflammatory Macrophages Interrupt Osteocyte Maturation and Mineralization via Regulating the Notch Signaling Pathway. *Mol. Med.* **2022**, *28*, 102. [CrossRef] [PubMed]
141. Qiao, W.; Xie, H.; Fang, J.; Shen, J.; Li, W.; Shen, D.; Wu, J.; Wu, S.; Liu, X.; Zheng, Y. Sequential Activation of Heterogeneous Macrophage Phenotypes Is Essential for Biomaterials-Induced Bone Regeneration. *Biomaterials* **2021**, *276*, 121038. [CrossRef]
142. Spiller, K.L.; Nassiri, S.; Witherel, C.E.; Anfang, R.R.; Ng, J.; Nakazawa, K.R.; Yu, T.; Vunjak-Novakovic, G. Sequential Delivery of Immunomodulatory Cytokines to Facilitate the M1-to-M2 Transition of Macrophages and Enhance Vascularization of Bone Scaffolds. *Biomaterials* **2015**, *37*, 194–207. [CrossRef]
143. Cai, X.; Yin, Y.; Li, N.; Zhu, D.; Zhang, J.; Zhang, C.-Y.; Zen, K. Re-Polarization of Tumor-Associated Macrophages to pro-Inflammatory M1 Macrophages by MicroRNA-155. *J. Mol. Cell Biol.* **2012**, *4*, 341–343. [CrossRef]
144. Jablonski, K.A.; Gaudet, A.D.; Amici, S.A.; Popovich, P.G.; Guerau-de-Arellano, M. Control of the Inflammatory Macrophage Transcriptional Signature by MiR-155. *PloS ONE* **2016**, *11*, e0159724. [CrossRef]
145. *MicroRNA-155 Inhibits Polarization of Macrophages to M2-Type and Suppresses Choroidal Neovascularization*; SpringerLink: Berlin, Germany, 2018. Available online: https://link.springer.com/article/10.1007/s10753-017-0672-8 (accessed on 25 September 2022).
146. Li, X.; Xue, S.; Zhan, Q.; Sun, X.; Chen, N.; Li, S.; Zhao, J.; Hou, X.; Yuan, X. Sequential Delivery of Different MicroRNA Nanocarriers Facilitates the M1-to-M2 Transition of Macrophages. *ACS Omega* **2022**, *7*, 8174–8183. [CrossRef]
147. Gao, H.; Dai, W.; Zhao, L.; Min, J.; Wang, F. The Role of Zinc and Zinc Homeostasis in Macrophage Function. *J. Immunol. Res.* **2018**, *2018*, e6872621. [CrossRef] [PubMed]
148. Liu, M.-J.; Bao, S.; Gálvez-Peralta, M.; Pyle, C.J.; Rudawsky, A.C.; Pavlovicz, R.E.; Killilea, D.W.; Li, C.; Nebert, D.W.; Wewers, M.D.; et al. ZIP8 Regulates Host Defense through Zinc-Mediated Inhibition of NF-KB. *Cell Rep.* **2013**, *3*, 386–400. [CrossRef] [PubMed]
149. Bai, X.; Liu, W.; Xu, L.; Ye, Q.; Zhou, H.; Berg, C.; Yuan, H.; Li, J.; Xia, W. Sequential Macrophage Transition Facilitates Endogenous Bone Regeneration Induced by Zn-Doped Porous Microcrystalline Bioactive Glass. *J. Mater. Chem. B* **2021**, *9*, 2885–2898. [CrossRef]
150. Yang, L.; Zhou, J.; Yu, K.; Yang, S.; Sun, T.; Ji, Y.; Xiong, Z.; Guo, X. Surface Modified Small Intestinal Submucosa Membrane Manipulates Sequential Immunomodulation Coupled with Enhanced Angio-and Osteogenesis towards Ameliorative Guided Bone Regeneration. *Mater. Sci. Eng. C* **2021**, *119*, 111641. [CrossRef] [PubMed]
151. Díez-Pascual, A.M. Surface Engineering of Nanomaterials with Polymers, Biomolecules, and Small Ligands for Nanomedicine. *Materials* **2022**, *15*, 3251. [CrossRef]

nanomaterials

MDPI

Article

Preparation, Drug Distribution, and In Vivo Evaluation of the Safety of Protein Corona Liposomes for Liraglutide Delivery

Ruihuan Ding [1], Zhenyu Zhao [2], Jibiao He [1], Yuping Tao [1], Houqian Zhang [2], Ranran Yuan [2], Kaoxiang Sun [1,*] and Yanan Shi [2,*]

[1] School of Pharmacy, Key Laboratory of Molecular Pharmacology and Drug Evaluation, Ministry of Education, Collaborative Innovation Center of Advanced Drug Delivery System and Biotech Drugs in Universities of Shandong, Yantai University, Yantai 261400, China

[2] School of Life Science, Yantai University, Yantai 261400, China

* Correspondence: sunkx@ytu.edu.cn (K.S.); shiyanan001@163.com (Y.S.)

Abstract: The development of oral drug delivery systems is challenging, and issues related to the mucus layer and low intestinal epithelial permeability have not yet been surmounted. The purpose of this study was to develop a promising formulation that is more adapted to in vivo absorption and to facilitate the administration of oral liraglutide. Cationic liposomes (CLs) linked to AT-1002 were prepared using a double-emulsion method, and BSA was adsorbed on the surface of the AT-CLs, resulting in protein corona cationic liposomes with AT-1002 (Pc-AT-CLs). The preparation method was determined by investigating various process parameters. The particle size, potential, and encapsulation efficiency (EE%) of the Pc-AT-CLs were 202.9 ± 12.4 nm, 1.76 ± 4.87 mV, and 84.63 ± 5.05%, respectively. The transmission electron microscopy (TEM) imaging revealed a nearly spherical structure of the Pc-AT-CLs, with a recognizable coating. The circular dichroism experiments confirmed that the complex preparation process did not affect the secondary structure of liraglutide. With the addition of BSA and AT-1002, the mucosal accumulation of the Pc-AT-CLs was nearly two times lower than that of the AT-CLs, and the degree of enteric metaplasia was 1.35 times higher than that of the PcCLs. The duration of the intestinal absorption of the Pc-AT-CLs was longer, offering remarkable biological safety.

Keywords: liposome; oral drug delivery systems; osmoregulatory peptide; mucus layers; intestinal epithelial permeability

Citation: Ding, R.; Zhao, Z.; He, J.; Tao, Y.; Zhang, H.; Yuan, R.; Sun, K.; Shi, Y. Preparation, Drug Distribution, and In Vivo Evaluation of the Safety of Protein Corona Liposomes for Liraglutide Delivery. *Nanomaterials* **2023**, *13*, 540. https://doi.org/10.3390/nano13030540

Academic Editors: Goran Kaluđerović and Nebojša Pantelić

Received: 23 December 2022
Revised: 17 January 2023
Accepted: 17 January 2023
Published: 29 January 2023

1. Introduction

Protein and peptide drugs possess both chemical and protein properties and offer the advantages of high activity, safety, specificity, selectivity, and good drug-formation capability. Due to patient compliance and ease, oral administration is the most preferred route. However, due to the key physical and chemical properties of proteins and peptides, including their vulnerability to enzyme degradation and poor permeability through the intestinal mucosa, their absorption is incomplete [1]. Due to their poor oral bioavailability, many drugs must be injected intravenously rather than administered orally, rendering such treatments expensive [2]. Currently, polypeptide drugs are widely used in the treatment of many diseases, including diabetes, which is a metabolic disease. Diabetes is mainly caused by insufficient insulin secretion and is mainly treated by insulin or glucagon-like peptide-1 (GLP-1) analogs and other drugs. Among these drugs, liraglutide has 97% structural homology to endogenous human GLP-1 and can stably resist metabolic degradation by dipeptidyl peptidase-4 (DPP-4). Its half-life after subcutaneous administration is 13 h, implying that it needs to be administered once daily [3,4]. Its most common adverse reactions include gastrointestinal problems, which range from mild to moderate. Treatment costs are high. Without insurance, the average out-of-pocket expenses per month exceed USD 1000. Moreover, liraglutide can only be administered by subcutaneous injection, which may pose difficulties for some patients [5]. Thus, it is necessary to develop an oral preparation of liraglutide. The improvement in oral bioavailability is a major challenge faced by formulators in the development of successful products [6]. This endeavor requires a deep understanding of the behavior of an oral drug delivery system in the gastrointestinal tract during the development of the preparation process, and it is recommended that this preparatory process must be conducted according to the drug's nature; indeed, it is important to predict the drug's in vivo behavior [7]. The pressing questions concerning the development of oral drug delivery systems are as follows: (1) How can drug design be guided by the information obtained from oral delivery barrier characteristics? (2) What are the stability-related issues of peptide drugs before they arrive at the site of action? (3) Which structural features of nano-formulations affect their intestinal absorption efficiency?

Concerning oral drug delivery systems, the mucus layer is the first barrier to the passage of exogenous substances. Mucus is produced by specialized cells and secreted apically as a protective and lubricating fluid [8]. The second barrier is the phospholipid bilayer of the epithelial cell itself and the intracellular environment [9]. Nanoparticles initially penetrate through the mucus layer to gain entry to the surface of epithelial cells, and most nanocarrier transport relies on endocytosis (the involvement of endosomal capture can lead to inefficient delivery) [10,11]. Additionally, nanoparticles can also enter cells through the paracellular route [12]. The paracellular pathway is composed of tight junctions (TJs) maintained by a complex network of protein interactions. TJs are selectively permeable barriers that often represent a rate-limiting step in paracellular trafficking [13]. However, a TJ chain is a dynamic structure, and nanocarriers can temporarily and reversibly open intercellular TJs by using auxiliary agents or smart formulations to penetrate cells [14]. Existing transcellular penetration enhancers (PEs) acting on the paracellular pathway have been shown to enhance the permeability of macromolecules [15]. For example, researchers have constructed citric acid cross-linked shells made of carboxymethylcellulose wrapped around core nanoparticles for efficient transcellular transport by reversibly opening tightly connected paracellular transport pathways [16]. Anthocyanins from strawberries can enhance the intestinal permeability of drugs, as evidenced by the screening of polyphenolic compounds isolated from plants [17].

AT-1002 is a hexamer synthetic peptide (FCIGRL) and is the active domain of the second toxin of Vibrio cholerae, namely, the zonal closure toxin (Zot). Zot can bind to putative surface receptors, activate intracellular signaling, and, finally, decompose TJs. Due to the expression and purity of prokaryotic toxins, some side effects can occur; thus, synthetic peptide AT-1002 was developed to enhance intestinal permeability and avoid the abovementioned problems [18]. AT-1002 can cause the TJ opening of the Caco-2 cell monolayer [19,20], and this effect is reversible. The mechanism of action of AT-1002 can be reversed by octapeptide azapeptide [21], which is a reversible TJ regulator that functions via the inhibition of zonlin [22]. In our previous study, chitosan (CS) nanoparticles linked to AT-1002 as a core and poly-N-(2-hydroxypropyl) methacrylamide (pHPMA) as a smart escape in a core–shell structure overcame the pH and mucus barriers, demonstrating that this delivery system significantly enhanced the oral hypoglycemic effect of liraglutide [23].

A liposome is a self-assembled spherical vesicle system containing water-borne nuclei, showing the characteristics of biodegradability and high biocompatibility. The physical and chemical properties of liposomes can be changed, including particle size, surface charge, lipid composition, and bilayer fluidity, and can be used to design effective drug carriers for diseases [24]. In recent years, cationic liposomes (CLs) have been used as oral drug carriers, owing to their increased permeability through intestinal cells via their electrostatic interactions with the intestinal epithelial cell membrane. Therefore, CLs were selected as the oral drug delivery agents in this study. Cationic amphiphilic compounds are the main components of CLs, which are usually used for their positively charged lipids and polymers. The use of polymer carriers to modify the surface of liposomes to design more effective drug delivery systems is an effective preparation method [25].

Bovine serum albumin (BSA) is a plasma protein widely distributed in bovine blood, with a relative molecular weight and an isoelectric point of 66.5 kDa and 4.7, respectively. As endogenous macromolecules from a wide range of sources, these plasma proteins are widely used in the preparation of albumin-derived prodrugs and albumin-loaded nanocarriers [26]. BSA, as a part of the mucus-penetrating carrier, is degraded by trypsin in the process of penetrating the mucus, thereby exposing the core structure for the oral delivery of insulin [27]. CLs can be prepared by adding cationic materials to the nano-prescription of liposomes. Because albumin is negatively charged at a neutral pH, it can be adsorbed on the surface of the cationic nanocarrier through electrostatic attraction, changing the positive charge surface of the cationic nanocarrier and resulting in electrically neutral and hydrophilic nanoparticles.

There are few related studies for improving the oral delivery of peptides that involve both protein-crowned liposomes combined with cellular bypass-regulating peptides. Therefore, in this study, we used BSA as a protein coating with CLs attached to an osmoregulatory peptide as the core to improve mucus permeability and enhance absorption efficiency in the intestine. First, the physicochemical properties, encapsulation efficiency, drug release, and stability of the optimal formulation of the liraglutide protein corona CLs were evaluated. Through the mucus and intestinal uptake experiments, the protein coating was found to significantly improve intestinal absorption efficiency. The IVIS spectral imaging system and hematoxylin–eosin (HE) staining were used to evaluate the distribution and biological safety of the liraglutide protein corona cationic liposomes in vivo.

2. Materials and Methods

2.1. Materials

Distearoyl phosphatidylcholine (DSPC) was purchased from Penske Biologicals (Shanghai, China). Chol-PEG-AT-1002 and FITC-liraglutide were purchased from the Shanghai Qiang Yao Biochemical Co., Ltd. Cholesterol (Chol) was purchased from the RVT Pharmaceutical Technology Co., Ltd. (Shanghai, China). Fluorescein isothiocyanate (FITC) and mucin were purchased from the Yuanye Biochemical Co., Ltd. (Shanghai, China). N-acetyl-l-cysteine (NAC) was procured from the Shanghai Macklin Biochemical Co., Ltd. (Shanghai, China). Mice were purchased from the Gem Pharmatech Co., Ltd. (Jiangsu, China).

2.2. Prescription Screening

DSPC is a synthetic phospholipid that is widely used in pharmaceutical preparations. In clinical liposome products, Onivyde (irinotecan liposome injection) [28] and DaunoXome (daunorubicin liposome) [29] with 6.81 mg/mL of DSPC and 65% DSPC in the membrane component, respectively, were used. Chol is an amphiphilic substance, which can be used as a medicinal excipient, including liposome membrane materials, especially in liposome preparations. (2,3-Dioleyl-propyl)-trimethylamine (DOTAP) is a commonly used cationic lipid.

As shown in Table 1, we investigated the formulation process of the preparation by using dynamic light scanning (DLS) to evaluate the particle size, potential, polydispersity index (PDI), and particle size stability within three days. In the follow-up investigation, the optimal value obtained in the previous step was used to formulate the optimal prescription.

Table 1. Single factor list for prescription screening.

Investigation Factors	Single Factor Ratio		
DSPC: Chol (mg/mg)	2:1	3:1	4:1
DOTAP: Chol (mg/mg)	1:1	2:1	3:1
Drug: Lipid (mg/mg)	1:5	1:10	1:15
Concentration of BSA (mg/mL)	5	10	15
BSA: CLs (*v*/*v*)	1:1	2:1	3:1
Internal water phase to oil phase (*v*/*v*)	100:1000	200:1000	300:1000

DSPC: 1,2-Dioctadecanoyl-sn-glycero-3-phophocholine; Chol: Cholesterol; DOTAP: (2,3-Dioleoyloxy-propyl)-trimethylammonium-chloride; BSA: Bovine serum albumin.

2.3. *Preparations of Protein Corona Liposomes*

CLs were prepared by a double-emulsion method as follows: the membrane materials, including 4 mg of DSPC, 1 mg of Chol, and 1 mg of DOTAP, were dissolved in 1 mL of dichloromethane (DCM); a total of 200 μL of phosphate-buffered saline (PBS) containing 0.6 mg of liraglutide was added and sonicated at 80 W for 1 min (ultrasonic 60 times, 1 s bursts, 1 s intervals). Colostrum was added to a 4 mL polyvinyl alcohol (PVA) solution (1% *w*/*v*). Double milk was formed at 250 W ultrasound for 1 min. The organic solvent was removed by stirring overnight to obtain the CL suspension.

The CL suspension was mixed with 10 mg/mL of BSA (dissolved in 0.9% NaCl solution) in a volume ratio of 1:1. The incubation was carried out at 37 °C for 2 h. This was followed by centrifugation at 15,000 rpm for 20 min. The precipitate was resuspended using PBS, washed by centrifugation, and repeated three times. The PcCL suspension was obtained by resuspending in 0.9% NaCl solution.

Following the above method, the Pc-AT-CL suspension was obtained by replacing Chol with Chol-PEG-AT-1002. The preparation process is shown in Figure 1.

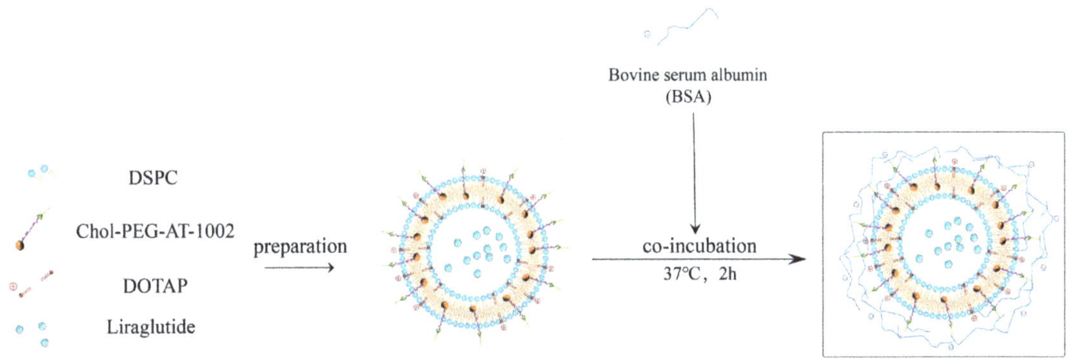

Figure 1. Schematic diagram of Pc-AT-CL preparation.

2.4. *Characterization of Protein Corona Liposomes*

2.4.1. Characterization of Liposome Nanoparticles

The particle sizes and zeta potentials of the nanoparticles were characterized by DLS. Free liraglutide was separated by ultrafiltration, and the concentrated liposome solution in the upper chamber was demulsified. The absorbance of the liposome after demulsification was measured by using an ultraviolet–visible (UV-VIS) spectroscopy (UV-2450). For PcCLs/Pc-AT-CLs, trypsin was used to degrade and shed the protein canopy, and methanol was used for demulsification. The amount of protein coating was measured using the BCA kit. The encapsulation efficiency (EE%) and the drug-loading capacity (LC%) of liraglutide were calculated using the following equations:

$$\text{EE\%} = (\text{Demulsification liposome liraglutide content})/(\text{Total amount of liraglutide}) \times 100\% \quad (1)$$

$$\text{LC\%} = (\text{Demulsification liposome liraglutide content})/(\text{Total amount of nanocarriers}) \times 100\% \quad (2)$$

2.4.2. Study on Co-Localization

To investigate whether the protein coating had coated the surface of CLs/AT-CLs well, we first performed fluorescence microscopy to observe the relative positional relationship between protein coating and CLs/AT-CLs. 1,1′-dioctadecyl-3,3,3′,3′-tetramethylindodicarbocyanine,4-chlorobenzenesulfonate salt (Did) was used to label CLs and AT-CLs, and FITC-BSA fluorescence-labeled protein corona was added to observe the co-localization relationship between the Did-AT-CLs and FITC-BSA by using fluorescence microscopy analysis.

2.4.3. Preparation Characterization by Scanning Electron Microscopy (SEM) and Transmission Electron Microscopy (TEM)

The liposomes were freeze-dried into powder form, and an appropriate amount of the liposome samples were fixed on a plate using conductive glue. Gold was sprayed under a vacuum, and the morphology of the preparation was observed by using SEM.

The morphology of the liposomes was studied by using TEM following a standard procedure. Briefly, a drop of the liposome dispersion was placed on a carbon support film and allowed to adsorb. The surplus was removed using blotting paper. A drop of 1% phosphotungstic acid was added, and the liposomes were stained for 60 s. The stained liposomes were allowed to dry under ambient conditions, and TEM analysis was performed (JEM-1230; JEOL, Tokyo, Japan).

2.5. *In Vitro Release and Evaluation of Peptide Stability*

Drug release performances were studied in a simulated gastric fluid (SGF) (pH = 1.2) and a simulated intestinal fluid (SIF) (pH = 6.8). The same concentration of different liposome nano-solutions was placed in dialysis bags (100 kD, Spectrum) at 37 °C. It was placed in the SGF (10 mL) for the first 2 h and then transferred to the SIF (10 mL). At specified time points (0, 1, 2, 3, 4, 6, 8, and 10 h), 1.0 mL of the solution was removed from the release medium and replaced with an equal volume of the fresh release medium. The samples were analyzed using high-performance liquid chromatography (HPLC), and the cumulative release profile was determined.

The liraglutide solutions or liposomal nanoparticles were put into the dialysis medium and incubated overnight at 37 °C in a shaker at 100 r/min. The dialysate was taken out, concentrated to the same concentration, and subjected to circular dichroism (CD). A 400 µL sample was placed in a micro quartz dish, and the spectrum was recorded on a CD spectrophotometer with a scanning speed of 1 nm/s (190–260 nm). The sample temperature was set at 20 °C. The sample background was recorded and subtracted from the values for different samples. After the baseline subtraction, a 6-point curve smoothing was performed on the spectra to plot the CD curve.

2.6. *Mucus Co-interaction Measurement*

The binding rate of mucin to different NPs was determined as follows: first, disodium hydrogen phosphate and potassium dihydrogen phosphate were used to configure the buffer solution at pH = 6.5, and an appropriate amount of mucin was weighed to prepare the mucin solutions with concentrations of 0.1%, 0.3%, and 0.5% *w*/*v*. The Did-labeled CLs, Pc-CLs, AT-CLs, and Pc-AT-CLs were added to different concentrations of mucin solution, and three groups were parallelly established. The mixture was transferred to 37 °C and incubated for 30 min; it was then centrifuged at 1500 rpm for 10 min, and the supernatant was collected. Next, the aggregation of nanoparticles and mucins was studied by measuring the fluorescence intensity of the supernatant.

2.7. *Transmucosal Transport*

In situ absorption studies in rats were performed to investigate the influences of the mucus barrier on the internalization of nanoparticles in the small intestines. The SD rats were forced to fast overnight before the experiments and given free access to water. The rats were anesthetized by an intraperitoneal injection of chloral hydrate (4 mg/kg), and a midline laparotomy was performed to expose the jejunum. A 4–5 cm intestinal loop was created and ligated at both ends. The mucus layer was removed at pre-treatment or post-treatment with N-acetyl-L-cysteine (NAC). Subsequently, the pre-treated, post-treated, and un-treated intestinal tissues were mixed with 0.5 mL of the Did-labeled CLs, AT-CLs, PcCLs, and Pc-AT-CLs (3 µg/mL) for 2 h. The intestinal tissues were homogenized on a high-speed shear machine, and each sample group was quantified.

2.8. Biosafety Evaluation

Healthy rats were administered the drug once daily (540 μg/kg) for one week to investigate the long-term safety of drug administration. The control group did not receive any treatment. All animals were fed with normal feed and given free access to drinking water. After one week, the rats in each group were sacrificed, and their hearts, livers, spleens, lungs, and kidneys were excised after perfusion, fixed in 4% paraformaldehyde, paraffin-embedded, and stained with hematoxylin and eosin (HE). The histological status of the different tissues was observed using a light microscopy.

2.9. In Vivo Distribution

Twelve mice, weighing 25–30 g (whole body hair removal), were randomized into two groups of six mice each, and 1,1-dioctadecyl-3,3,3,3-tetramethylindotricarbocyanine iodide (Dir) was used label to PcCLs and Pc-AT-CLs were injected via the intra-jejunum route. The mice were dissected at 2, 4, 8, 12, 14, and 24 h after injection, and the gastrointestinal tract, heart, liver, spleen, lung, and kidney were imaged on a real-time imager (IVIS Spectrum Imaging System S091, PerkinElmer, Waltham, MA, USA). Two mice were randomly selected and anesthetized intraperitoneally. The distribution of the nanoparticles in vivo was observed at 2, 4, 8, 12, and 24 h.

2.10. Statistical Analysis

All data are expressed as mean ± standard deviation (SD). The statistical analysis was performed using the IBM SPSS Statistics software (version 26). One-way analysis of variance (ANOVA) was used to compare and analyze the differences among three or more groups. A $p < 0.05$ was considered statistically significant.

3. Results and Discussion

3.1. Characterization of Protein Corona Liposomes

3.1.1. Process and Prescription Optimization

The application of liposomes containing a protein canopy in the field of proteins and peptides has gained traction for the use of these carriers to deliver liraglutide orally. These liposomes have good mucus penetration and intestinal cell uptake [27]. The size and surface charge of nanoparticles are known to severely affect drug delivery efficiency in vitro and in vivo [30]. To penetrate the mucus barrier and intestinal epithelial barrier and to exert a good hypoglycemic effect, there must be appropriate particle size, zeta potential, and better encapsulation efficiency.

When investigating the prescription process of CLs, small particle size, obvious positive charge, and good stability need to be considered. We first examined the mass ratio of DSPC to Chol. As shown in Figure 2, DSPC: Chol at a mass ratio of 4:1 has a particle size (Figure 2a) of 108.1 nm and a potential (Figure 2b) of 29.05 mV, showing the best particle size stability (Figure 2c) over three days and the lowest volatility. When examining the mass ratio of DOTAP to Chol, the prescription was screened based on a 4:1 mass ratio of DSPC:Chol. The three evaluation criteria of particle size (Figure 2d), potential (Figure 2e), and stability (Figure 2f) were combined, and the observations are as follows: the particle size is too large for a 3:1 ratio and the stability is poor for a 2:1 ratio. Thus, a 1:1 mass ratio was selected. Based on the drug-to-lipid ratio examination, the particle size (Figure 2g) is 128.9 nm, the potential (Figure 2h) is 28.56 mV, and the stability (Figure 2i) is the best at a ratio of 1:10. In the ultrasonic process of colostrum, due to its small volume and excessive power, it is easy to cause liquid splash. Therefore, in the colostrum process, smaller power was used to minimize the loss of excipients and drugs during the preparation. Therefore, we mainly studied the ultrasonic power of double milk. At a constant power for the colostrum, the particle size decreases with an increase in the ultrasonic power of the compound emulsion. Simultaneously, as shown in Table 2, EE% is the highest when the power is 250 W. Through the prescription and process investigation, the particle size was reduced to nearly 200 nm, the potential was close to neutral, and the best EE% was selected. In summary, the prescription process of CLs is DSPC:Chol:DOTAP = 4:1:1, drug:lipid = 1:10, a colostrum ultrasound power of 80 W, and a compound milk ultrasound power of 250 W.

Table 2. Investigation using ultrasonic power.

Ultrasonic Power of Colostrum (W)	Ultrasonic Power of Double Milk (W)	Zeta Potential (mV)	Size (nm)	Polydisperse Index (PDI)	Encapsulation Efficiency (EE%)
150	150	24.78	144.9	0.26	73%
80	150	21.54	140	0.24	54%
80	200	21.44	122.7	0.24	78%
80	250	27.51	120	0.29	87%

By using the CLs prepared above, the prescribed dosage of the protein canopy was examined. Nanoparticles with small particle sizes, near electric neutrality, and better stability are required for penetrating the mucus layer. As shown in Figure 3, at a BSA concentration of 10 mg/mL and a volume ratio of 1:1 with CLs, the particle size is approximately 200 nm, the potential is close to electroneutrality, and the formulation's stability is the best.

The physical and chemical properties of the AT-CLs were verified by replacing all Chol with Chol-PEG-AT 1002; the results in Table 3 show no difference in particle size potential, encapsulation rate, and drug loading with CLs.

Table 3. Size (nm), zeta potential (mV), and encapsulation efficiency (%) of the nanoparticles.

Formations	Size (nm)	Zeta (mV)	Encapsulation Efficiency (EE%)	Drug Loading (DL%)
CLs	127 ± 10.0	36.09 ± 2.7	85.85 ± 1.79	8.19 ± 0.17
AT-CLs	119.6 ± 5.6	36.14 ± 0.16	84.63 ± 5.05	8.08 ± 0.48
PcCLs	209.0 ± 9.6	0.67 ± 0.36	85.85 ± 1.79	2.92 ± 1.67
Pc-AT-CLs	202.9 ± 12.4	1.76 ± 4.87	84.63 ± 5.05	2.08 ± 0.30

3.1.2. Characterization of Liposome Nanoparticles

In this study, PcCLs/Pc-AT-CLs were successfully prepared and used for the oral treatment of diabetes. Due to the electrostatic interaction, BSA and CLs/AT-CLs formed a core–shell structure of the nanoparticles. Liraglutide was encapsulated during liposome formation, and BSA was introduced to form a protein coating on the surface of CLs/AT-CLs. This protected the peptides that penetrated the mucosal and intestinal epithelial cells.

As shown in Table 3, the particle size of the Pc-AT-CLs is 202.9 ± 12.4 nm, and the potential is closer to the neutral charge (1.76 ± 4.87 mV), which may be due to the incompletely uniformity of the BSA-coating layer, and the zeta potential measurement is sensitive. However, the slight change in the particle size between the Pc-AT-CLs (202.9 ± 12.4) and PcCLs (209.0 ± 9.6), as well as the AT-CLs (119.6 ± 5.6) and CLs (127 ± 10.0), may be due to the enhanced interaction force of the cell-penetrating peptide AT-1002 with CLs, resulting in a slight decrease in particle size. The second goal of this preparation was to achieve high loading; the EE% in CLs is 84.63 ± 5.05%, and the drug-loading efficiency is 8.08 ± 0.48%. However, during protein coating, the drug-loading efficiency of the protein corona liposomes is 2.08 ± 0.30% due to the loss of the CL solution and the addition of protein coating.

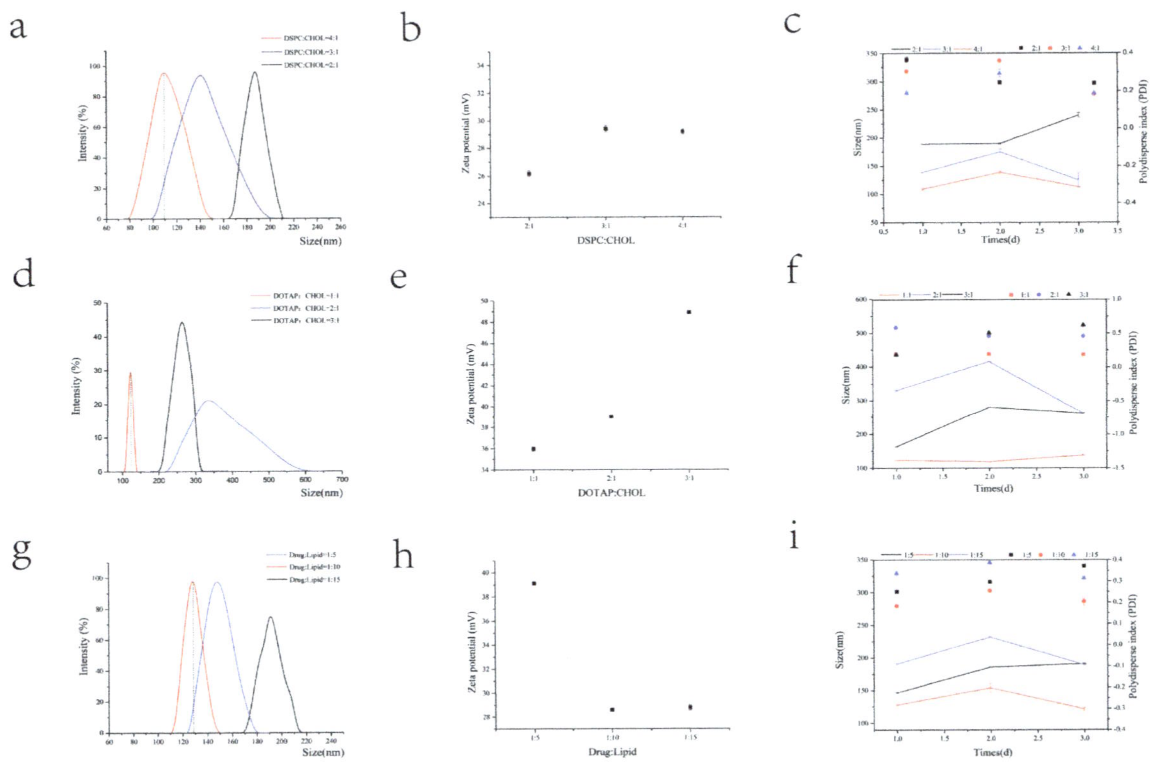

Figure 2. Prescription screening of cationic liposomes. The ratio of DSPC to Chol was changed while keeping other factors fixed. The particle size (**a**), potential (**b**), and particle size and PDI stability (**c**) were investigated. The optimal ratio of DSPC to Chol was adopted, and other factors were unchanged, while the ratio of DOTAP to Chol was changed. The particle size (**d**), potential (**e**), and particle size and PDI stability (**f**) were investigated. The optimal ratios of DSPC to Chol and of DOTAP to Chol were adopted, and other factors were unchanged, while the ratio of drug to lipid was changed. The particle size (**g**), potential (**h**), and particle size and PDI stability (**i**) were investigated.

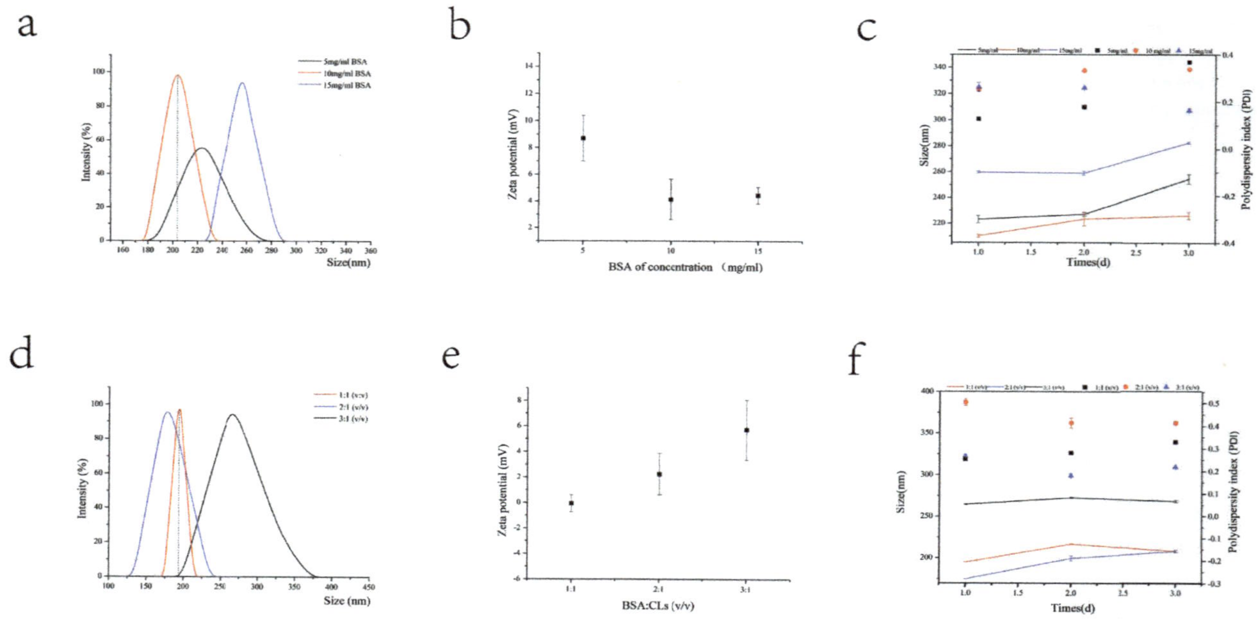

Figure 3. Prescription screening for protein canopy. The concentration of BSA was changed while keeping other factors fixed. The particle size (**a**), potential (**b**), and particle size and PDI stability (**c**) were investigated. The optimal concentration of BSA was adopted, and while other factors were unchanged, the volume ratio of protein solution to cationic liposome solution was changed. The particle size (**d**), potential (**e**), and particle size and PDI stability (**f**) were investigated.

3.1.3. Co-localization Relationship

The relative positional relationship between the BSA and the inner core liposomes was investigated by using fluorescence microscopy analysis. The inner core liposomes were tracked by wrapping the red fluorescent dye, Did, inside the liposomes by drug-loading, and the protein coating was tracked by labeling BSA with the green fluorescent dye, FITC. As shown in Figure 4a, the FITC-BSA and Did-AT-CLs/Did-CLs show green and red fluorescence, respectively. Using Image J analysis, the yellow fluorescence shows their co-localization. Thus, the BSA and inner core liposomes are in the same position.

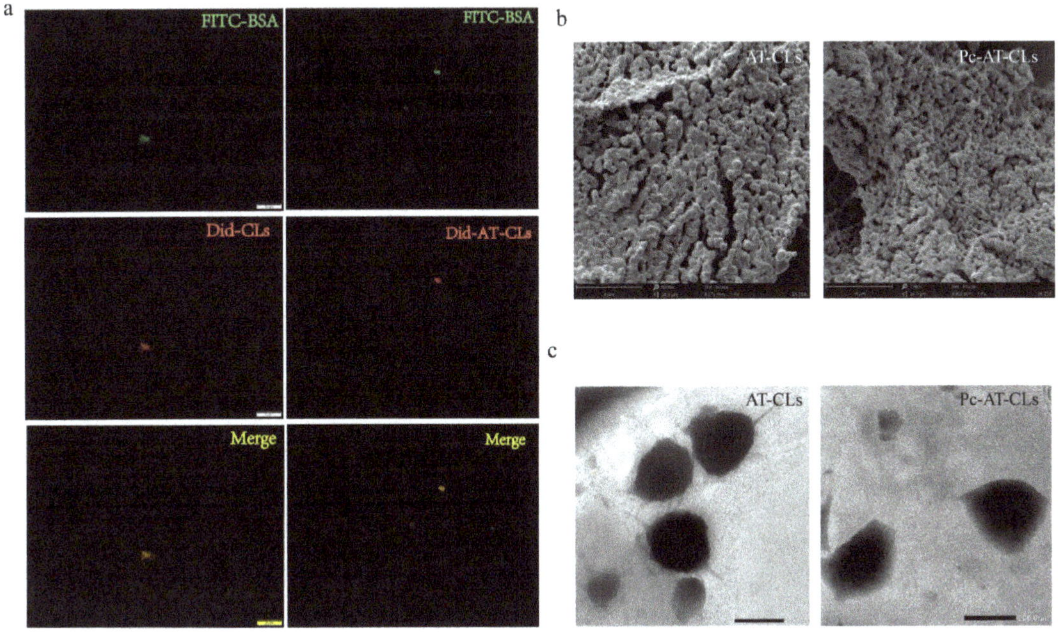

Figure 4. Preparation characterization of PcCL and Pc-AT-CL nanoparticles. (**a**) Visualization of the double-labeled Pc-AT-CLs by fluorescence microscopy. (**b**) SEM images of the AT-CLs and Pc-AT-CLs. (**c**) TEM images of the AT-CLs and Pc-AT-CLs.

3.1.4. Preparation Characterization by SEM and TEM

Since the protein crown liposome is a nano-preparation, SEM and TEM were used to determine its morphology. As shown in Figure 4b, the lyophilized AT-CLs and Pc-AT-CLs are observed by the SEM, and both are spherical. The TEM results show a distinct protein coating on the surface of the Pc-AT-CLs compared to the AT-CLs (Figure 4c), confirming successful protein crown coating.

3.2. In Vitro Release and Evaluation of Peptide Stability

The liraglutide release profiles from different formulations were tested by an in vitro assay. The pH was set to 1.2 and 6.8 to simulate the acidic environment of gastric and intestinal fluids in vivo, respectively [31]. As shown in Figure 5a, the amount of liraglutide released from CLs/AT-CLs is approximately 80% at 8 h, unlike the liraglutide released from the PcCLs/Pc-AT-CLs, which is significantly slower at approximately 55%. The same method was used to verify the release of pure liraglutide. The release rate of pure liraglutide was linear and rapid in the first 3 h, gradually reaching an equilibrium.

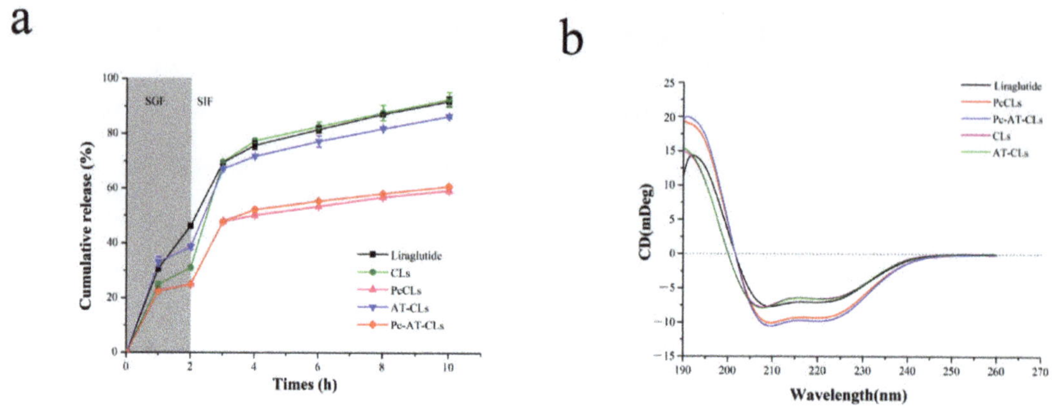

Figure 5. In vitro liraglutide release profiles and stability studies. (**a**) Liraglutide release profiles of the CLs, PcCLs, AT-CLs, and Pc-AT-CLs in SGF (2 h) and SIF (8 h). (**b**) Circular dichroism spectra of the liraglutide released from the CLs, PcCLs, AT-CLs, and Pc-AT-CLs (data are presented as mean ± SD, n = 3).

Liraglutide may be sensitive to interfacial pressure when subjected to ultrasound, so it is important to ensure the stability of liraglutide in the final product. To characterize the secondary structure of liraglutide, it was determined in the far ultraviolet region (190~260 nm) using the CD experiments (Figure 5b). The original liraglutide has two negative absorption peaks at 209 nm and 220 nm. The AT-CLs/CLs have two negative absorption peaks at 208 nm and 220 nm with a slight change; the characteristic peaks of the PcCLs/Pc-AT-CLs are unchanged. This indicates that the secondary structure of liraglutide is preserved, and the ultrasonic pressure did not cause its instability.

3.3. *Mucus Co-Interaction*

Nanoparticles with a near-neutral or negative charge and a smaller absolute value can effectively reduce the electrostatic interactions with mucus [32,33]. The lumen pH of the proximal small intestine (duodenum and jejunum) is about 6.0–7.0 [34–36]. Herein, we chose pH = 6.5 to model the pH in the intestinal mucus environment. To detect the effect of the protein corona modification on mucus adsorption, various concentrations of mucin solutions were prepared to evaluate the aggregation of different nanoparticles in the mucin solutions. As shown in Figure 6a, the aggregation of the PcCLs and Pc-AT-CLs decreases significantly, indicating that adsorption and aggregation occur between mucins and the CLs/AT-CLs, while the capture ability of the PcCLs/Pc-AT-CLs weakens relatively. The coating of the protein cap reverses the surface charge of the original CLs, and the nanoparticles close to electrical neutrality are more beneficial in penetrating the mucus.

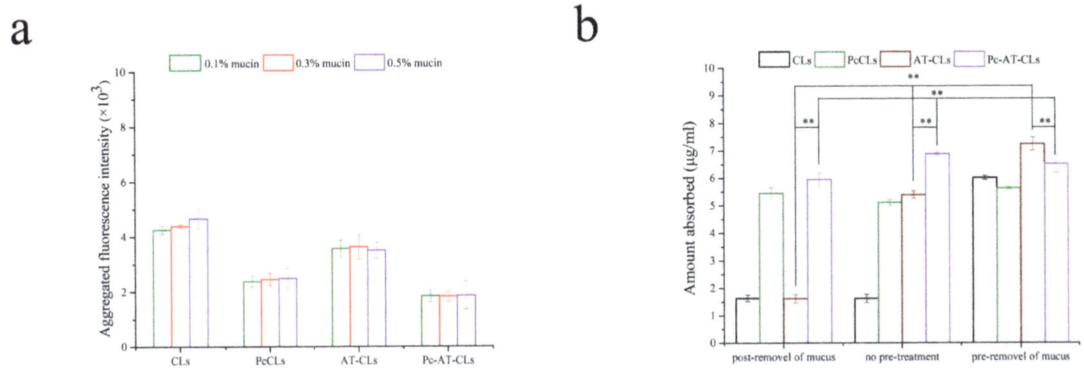

Figure 6. Mucus aggregation and transmucosal transport. (**a**) The binding rate of CLs, PcCLs, AT-CLs, and Pc-AT-CLs with different concentrations of mucus (data are presented as mean ± SD, n = 3). (**b**) Transmucosal transport of CLs, PcCLs, AT-CLs, and Pc-AT-CLs with or without a pretreatment or post-treatment process to remove mucus (data are presented as mean ± SD, n = 3, ** $p < 0.05$).

3.4. Transmucosal Transport

To further investigate the obstructive influence of mucus on the absorption of the PcCLs and Pc-AT-CLs, in situ absorption studies were undertaken. Specifically, the mucus layer adherent to the surface of the intestine epithelium was removed at pre-treatment or post-treatment with 0.2% NAC, because NAC is a widely used mucolytic agent in clinical setting to break hydrogen and disulfide bonds of mucus [37,38]. The internalized amounts of Did are shown in Figure 6b. In the presence of mucus, the intestinal absorption of the Pc-AT-CLs is always higher compared to the AT-CLs, indicating that hydrophilic and electrically neutral preparations are more beneficial properties for mucus penetration. In the case of NAC pre-treatment, compared to the Pc-AT-CLs, the amount of intestinal internalization of the AT-CLs increases significantly, confirming that positive and hydrophobic preparations are more conducive to the internalization and absorption by intestinal epithelial cells. Simultaneously, compared to the CLs, the AT-CLs are more easily absorbed in the intestine, proving that AT-1002 plays an important role in the absorption of nanoparticles. In the case of NAC post-treatment, the uptake of the CLs and AT-CLs reduces significantly compared to the untreated samples, indicating that these are easily trapped by the mucus layer. However, the mucus layer has little effect on their absorption of PcCLs and Pc-AT-CLs, proving that the protein coating facilitates easy penetration to the mucus layer, ultimately increasing the efficiency of intestinal internalization.

3.5. Biosafety Evaluation

Biosafety assessment is a prerequisite for evaluating utility in future biological applications [39]. After continuous administration for a week, the viscera were dissected and stained with H&E, and the histopathological changes were observed. The microscopic examination results of the H&E staining are shown in Figure 7. There are no obvious pathological changes in the heart, liver, spleen, lung, or kidney. In general, compared to the control group, no signs of inflammatory reactions were observed in the experimental group. Thus, the oral administration of the nanoparticles prepared in this study had no obvious toxicity in the experimental animals. The PcCLs and Pc-AT-CLs are safe and effective with good application prospects.

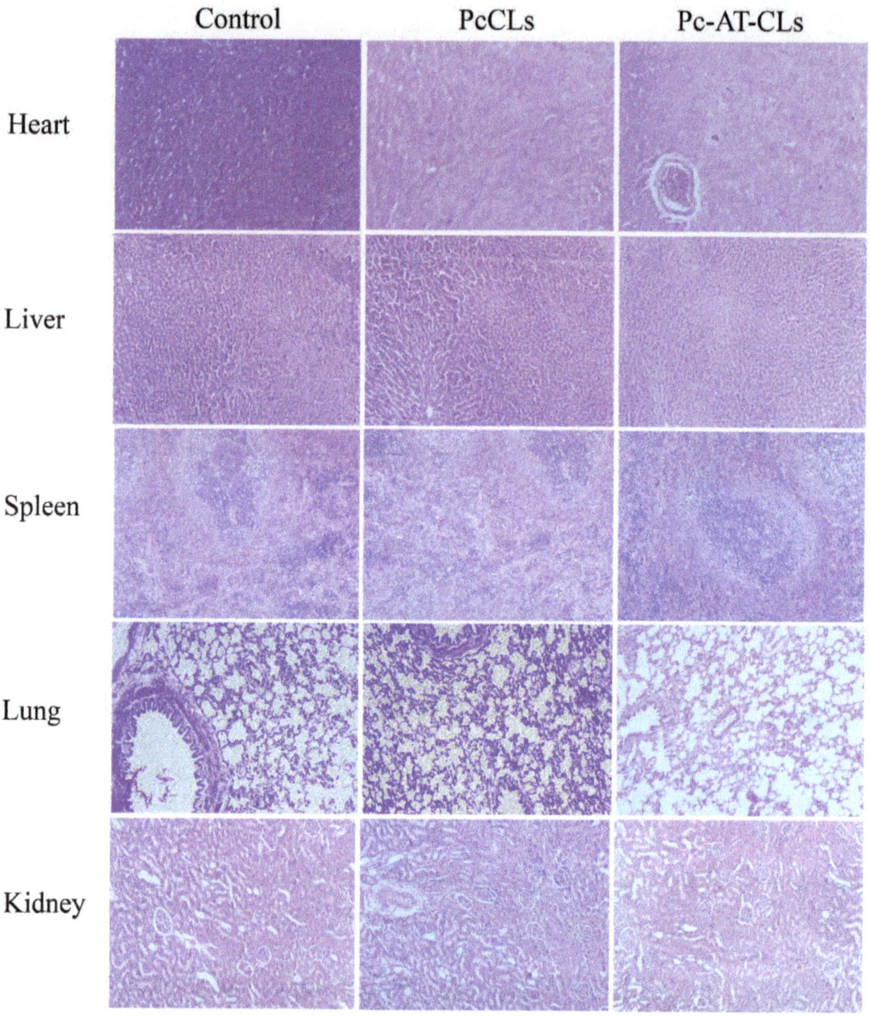

Figure 7. Histopathological images of the heart, liver, spleen, lung, and kidney on the 7th day, obtained using an upright fluorescence microscope (scale bar = 50 μm).

3.6. In Vivo Distribution

The small animal in vivo imaging system was used to visualize the distribution of the nanoparticles in the animal models, and the in vivo distribution of the PcCLs and Pc-AT-CLs was mainly evaluated based on the luminescence intensity. As shown in Figure 8a, after administration, the fluorescence intensity of the PcCLs and Pc-AT-CLs decreases with the extension of time, and at 24 h after administration, the fluorescence intensity of both groups is weak, indicating that the drug has been digested and degraded or excreted in vitro. As shown in Figure 8b, the fluorescence intensities from other major organs (heart, liver, spleen, lung, and kidney) increase. The fluorescence intensity attenuation in the PcCL group is greater than that of the Pc-AT-CL group 2–8 h after administration, indicating that the latter has a longer absorption time. This was also verified by comparing the renal fluorescence intensity; the renal fluorescence intensity of the PcCL group is higher than that of the Pc-AT-CL group.

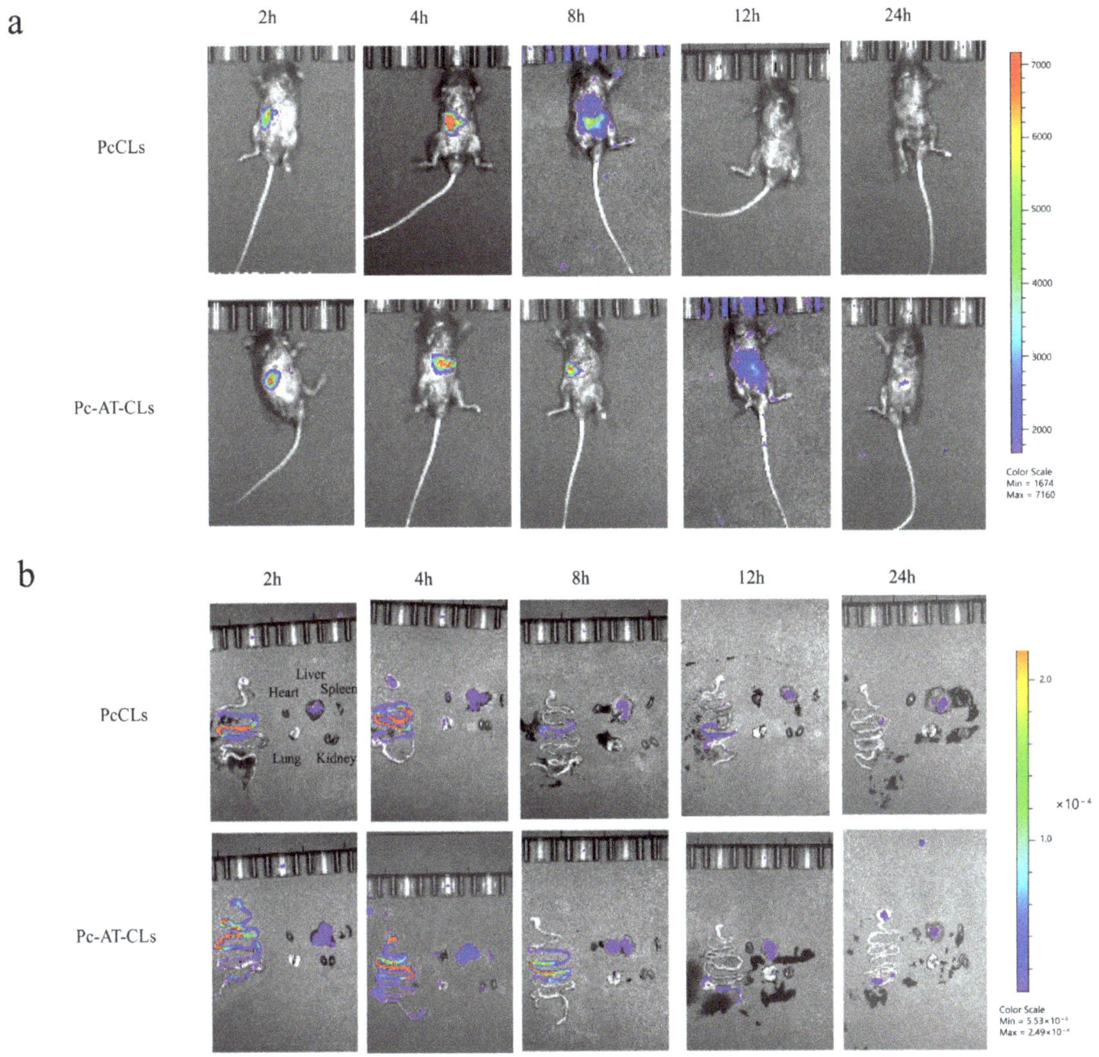

Figure 8. Fluorescence images of mouse organs. (**a**) After the administration of PcCLs and Pc-AT-CLs labeled with Dir, the in vivo distribution was observed at different time points. (**b**) In vivo fluorescence images of the stomach, intestine, heart, liver, spleen, lungs, and kidneys after the administration of Dir-labeled PcCLs and Pc-AT-CLs at different time points.

4. Conclusions

According to the properties of the mucus and intestinal epithelial barriers during oral delivery, a protein corona liposome with neutral hydrophilic properties and a core–shell structure was designed based on the characteristics of BSA and the TJ regulatory peptide, AT-1002. After penetrating the mucus layer, the outer shell gradually peeled off and exposed the core CLs that crossed the intestinal epithelial cell layer efficiently. First, liraglutide was effectively encapsulated in the CLs by the emulsion solvent evaporation method. By optimizing the preparation process, the drug EE% was improved to the best possible extent. Under the optimum conditions, the EE% of the liposomes was 84.63 ± 5.05%, and the average particle size was 202.9 ± 12.4 nm. The in vitro release of liposomes and the stability of liraglutide after release were investigated in vitro. Liraglutide showed good structural stability. The intestinal absorption potential, drug distribution, and biological safety of the adaptive protein crown liposomes were studied in vivo. The results showed that the protein corona liposomes had a good intestinal internalization effect, longer intestinal absorption time, and good biological safety. The adaptive protein corona liposomes have potential application for the oral administration of proteins and peptides.

Author Contributions: Conceptualization, R.D., K.S. and Y.S.; Data curation, R.D.; Funding acquisition, K.S. and Y.S.; Methodology, R.D., Z.Z., Y.T., J.H., H.Z., R.Y. and Y.S.; Project administration, K.S. and Y.S.; Resources, Y.S.; Supervision, K.S. and Y.S.; Visualization, K.S.; Writing–original draft, R.D.; Writing–review and editing, R.D., K.S. and Y.S. All authors have read and agreed to the published version of the manuscript.

Funding: This work was supported by grants from the Doctoral Program of Yantai University [SM20B35] and the Natural Science Foundation of Shandong Province [ZR2021MH395].

Institutional Review Board Statement: All animal experiments were approved by the Experimental Animal Ethics Committee of the College of Pharmacy, Yantai University (Approval number, YTDX 20180124).

Data Availability Statement: The data presented in this study are provided in the article.

Acknowledgments: We thank the laboratory of Yantai University School of Pharmacy for their support.

Conflicts of Interest: The authors declare no conflict of interest.

References

1. Sonia, T.A.; Sharma, C.P. An Overview of Natural Polymers for Oral Insulin Delivery. *Drug Discov. Today* **2012**, *17*, 784–792. [CrossRef]
2. Zhong, J.; Xia, B.; Shan, S.; Zheng, A.; Zhang, S.; Chen, J.; Liang, X.-J. High-quality Milk Exosomes as Oral Drug Delivery System. *Biomaterials* **2021**, *277*, 121126. [CrossRef]
3. Agerso, H.; Jensen, L.B.; Elbrond, B.; Rolan, P.; Zdravkovic, M. The Pharmacokinetics; Pharmacodynamics, Safety and Tolerability of NN2211, a New Long-acting GLP-1 Derivative, in Healthy Men. *Diabetologia* **2002**, *45*, 195–202. [CrossRef] [PubMed]
4. Degn, K.B.; Juhl, C.B.; Sturis, J.; Jakobsen, G.; Brock, B.; Chandramouli, V.; Rungby, J.; Landau, B.R.; Schmitz, O. One Week's Treatment with the Long-acting Glucagon-like Peptide 1 Derivative Liraglutide (NN2211) Markedly Improves 24-h Glycemia and Alpha- and Beta-cell Function and Reduces Endogenous Glucose Release in Patients with Type 2 Diabetes. *Diabetes* **2004**, *53*, 1187–1194. [CrossRef] [PubMed]
5. Nuffer, W.A.; Trujillo, J.M. Liraglutide: A New Option for the Treatment of Obesity. *Pharmacotherapy* **2015**, *35*, 926–934. [CrossRef] [PubMed]
6. Maji, I.; Mahajan, S.; Sriram, A.; Medtiya, P.; Vasave, R.; Khatri, D.K.; Kumar, R.; Singh, S.B.; Madan, J.; Singh, P.K. Solid Self Emulsifying Drug Delivery System: Superior Mode for Oral Delivery of Hydrophobic Cargos. *J. Control. Release* **2021**, *337*, 646–660. [CrossRef]
7. Pouton, C.W. Lipid Formulations for Oral Administration of Drugs: Non-emulsifying, Self-Emulsifying and 'Self-microemulsifying' Drug Delivery Systems. *Eur. J. Pharm. Sci.* **2000**, *11*, S93–S98. [CrossRef]
8. Camilleri, M.; Madsen, K.; Spiller, R.; Meerveld-Van, B.; Verne, G.N. Intestinal Barrier Function in Health and Gastrointestinal Disease. *Neurogastroenterol. Motil.* **2012**, *24*, 503–512. [CrossRef]
9. Brunner, J.; Ragupathy, S.; Borchard, G. Target Specific Tight Junction Modulators. *Adv. Drug Delivery Rev.* **2021**, *171*, 266–288. [CrossRef]
10. Fan, W.; Xia, D.; Zhu, Q.; Hu, L.; Gan, Y. Intracellular Transport of Nanocarriers Across the Intestinal Epithelium. *Drug Discov. Today* **2016**, *21*, 856–863. [CrossRef]
11. Chen, N.; He, Y.; Zang, M.; Zhang, Y.; Lu, H.; Zhao, Q.; Wang, S.; Gao, Y. Approaches and materials for endocytosis-independent intracellular delivery of proteins. *Biomaterials* **2022**, *286*, 121567. [CrossRef] [PubMed]
12. Han, X.; Zhang, E.; Shi, Y.; Song, B.; Du, H.; Cao, Z. Biomaterial-tight Junction Interaction and Potential Impacts. *J. Mater. Chem. B* **2019**, *7*, 6310–6320. [CrossRef] [PubMed]
13. Buckley, A.; Turner, J.R. Cell Biology of Tight Junction Barrier Regulation and Mucosal Disease. *Cold Spring Harbor Perspect. Biol.* **2018**, *10*, a029314. [CrossRef] [PubMed]
14. Hashimoto, Y.; Tachibana, K.; Krug, S.M.; Kunisawa, J.; Fromm, M.; Kondoh, M. Potential for Tight Junction Protein-directed Drug Development Using Claudin Binders and Angubindin-1. *Int. J. Mol. Sci.* **2019**, *20*, 4016. [CrossRef] [PubMed]
15. Danielsen, E.M.; Hansen, G.H. Probing Paracellular -versus Transcellular Tissue Barrier Permeability Using a Gut Mucosal Explant Culture System. *Tissue Barriers* **2019**, *7*, 1601955. [CrossRef] [PubMed]

16. Li, C.; Yuan, L.; Zhang, X.; Zhang, A.; Pan, Y.; Wang, Y.; Qu, W.; Hao, H.; Algharib, S.A.; Chen, D.; et al. Core-shell nanosystems designed for effective oral delivery of polypeptide drugs. *J. Control. Release* **2022**, *352*, 540–555. [CrossRef] [PubMed]
17. Lamson, N.G.; Fein, K.C.; Gleeson, J.P.; Newby, A.N.; Xian, S.; Cochran, K.; Chaudhary, N.; Melamed, J.R.; Ball, R.L.; Suri, K.; et al. The strawberry-derived permeation enhancer pelargonidin enables oral protein delivery. *Proc. Natl. Acad. Sci. USA* **2022**, *119*, e2207829119. [CrossRef]
18. Gopalakrishnan, S.; Pandey, N.; Tamiz, A.P.; Vere, J.; Carrasco, R.; Somerville, R.; Tripathi, A.; Ginski, M.; Paterson, B.M.; Alkan, S.S. Mechanism of Action of ZOT-derived Peptide AT-1002, a Tight Junction Regulator and Absorption Enhancer. *Int. J. Pharmacol.* **2009**, *365*, 121–130. [CrossRef]
19. Motlekar, N.A.; Fasano, A.; Wachtel, M.S.; Youan, B.B. Zonula Occludens Toxin Synthetic Peptide Derivative AT1002 Enhances in Vitro and in Vivo Intestinal Absorption of Low Molecular Weight Heparin. *J. Drug Target.* **2006**, *14*, 321–329. [CrossRef]
20. Song, K.-H.; Fasano, A.; Eddington, N.D. Enhanced Nasal Absorption of Hydrophilic Markers after Dosing with AT1002, a Tight Junction Modulator. *Eur. J. Pharm. Biopharm.* **2008**, *69*, 231–237. [CrossRef]
21. Gopalakrishnan, S.; Durai, M.; Kitchens, K.; Tamiz, A.P.; Somerville, R.; Ginski, M.; Paterson, B.M.; Murray, J.A.; Verdu, E.F.; Alkan, S.S.; et al. Larazotide Acetate Regulates Epithelial Tight Junctions In Vitro and In Vivo. *Peptides* **2012**, *35*, 86–94. [CrossRef] [PubMed]
22. Yoosuf, S.; Makharia, G.K. Evolving Therapy for Celiac Disease. *Front. Pediatr.* **2019**, *7*, 193. [CrossRef] [PubMed]
23. Shi, Y.; Liu, L.; Yin, M.; Zhao, Z.; Liang, Y.; Sun, K.; Li, Y. Mucus- and pH-mediated controlled release of core-shell chitosan nanoparticles in the gastrointestinal tract for diabetes treatment. *J. Drug Target.* **2023**, *31*, 65–73. [CrossRef]
24. Rahman, M.A.; Ahuja, A.; Baboota, S.; Bhavna; Bali, V.; Saigal, N.; Ali, J. Recent Advances in Pelletization Technique for Oral Drug Delivery: A Review. *Curr. Drug Deliv.* **2009**, *6*, 122–129. [CrossRef] [PubMed]
25. Nag, O.K.; Awasthi, V. Surface Engineering of Liposomes for Stealth Behavior. *Pharmaceutics* **2013**, *5*, 542–569. [CrossRef]
26. Yin, X.Q.; Huo, M.R.; Zhou, J.P.; Zou, A.F.; Li, J.; Peng, X.L. Synthesis, Characterization and Drug Loading Capacity of Dodecyl Serum Albumin for Insoluble Antitumor Drugs. *J. China Pharm. Univ.* **2011**, *42*, 319–323.
27. Wang, A.; Yang, T.; Fan, W.; Yang, Y.; Zhu, Q.; Guo, S.; Zhu, C.; Yuan, Y.; Zhang, T.; Gan, Y. Protein Corona Liposomes Achieve Efficient Oral Insulin Delivery by Overcoming Mucus and Epithelial Barriers. *Adv. Healthcare Mater.* **2019**, *8*, e1801123. [CrossRef]
28. Passero, F.C., Jr.; Grapsa, D.; Syrigos, K.N.; Saif, M.W. The Safety and Efficacy of Onivyde (Irinotecan Liposome Injection) for the Treatment of Metastatic Pancreatic Cancer Following Gemcitabine-based Therapy. *Expert Rev. Anticancer Ther.* **2016**, *16*, 697–703. [CrossRef]
29. Christin, E.P.; Dittmer, D.P. Liposomal Daunorubicin as Treatment for Kaposi's Sarcoma. *Int. J. Nanomed.* **2007**, *2*, 277–288.
30. Zhao, Z.; Ukidve, A.; Krishnan, V.; Mitragotri, S. Effect of physicochemical and surface properties on in vivo fate of drug nanocarriers. *Adv. Drug Deliv. Rev.* **2019**, *143*, 3–21. [CrossRef]
31. Zhang, H.; Gu, Z.; Li, W.; Guo, L.; Wang, L.; Guo, L.; Ma, S.; Han, B.; Chang, J. pH-sensitive O-carboxymethyl chitosan/sodium alginate nanohydrogel for enhanced oral delivery of insulin. *Int. J. Biol. Macromol.* **2022**, *223*, 433–445. [CrossRef] [PubMed]
32. Xu, Q.G.; Ensign, L.M.; Boylan, N.J.; Schön, A.; Gong, X.Q.; Yang, J.C.; Lamb, N.W.; Cai, S.T.; Yu, T.; Freire, E.; et al. Impact of Surface Polyethylene Glycol (peg) Density on Biodegradable Nanoparticle Transport in Mucus ex Vivo and Distribution in Vivo. *ACS Nano* **2015**, *9*, 9217–9227. [CrossRef]
33. Akkus, Z.B.; Nazir, I.; Jalil, A.; Tribus, M.; Bernkop-Schnurch, A. Zeta Potential Changing Polyphosphate Nanoparticles: A Promising Approach to Overcome the Mucus and Epithelial Barrier. *Mol. Pharmacol.* **2019**, *16*, 2817–2825. [CrossRef] [PubMed]
34. Sonaje, K.; Lin, Y.H.; Juang, J.H.; Wey, S.P.; Chen, C.T.; Sung, H.W. In Vivo Evaluation of Safety and Efficacy of Self-Assembled Nanoparticles for Oral Insulin Delivery. *Biomaterials* **2009**, *30*, 2329–2339. [CrossRef] [PubMed]
35. Fallingborg, J. Intraluminal pH of the Human Gastrointestinal Tract. *Dan. Med. Bull.* **1999**, *46*, 183–196. [PubMed]
36. Sung, H.W.; Sonaje, K.; Liao, Z.-X.; Hsu, L.W.; Chuang, E.Y. pH-Responsive Nanoparticles Shelled with Chitosan for Oral Delivery of Insulin: From Mechanism to Therapeutic Applications. *Acc. Chem. Res.* **2011**, *45*, 619–629. [CrossRef]
37. Takatsuka, S.; Kitazawa, T.; Morita, T.; Horikiri, Y.; Yoshino, H. Enhancement of Intestinal Absorption of Poorly Absorbed Hydrophilic Compounds by Simultaneous Use of Mucolytic Agent and Non-Ionic Surfactant. *Eur. J. Pharm. Biopharm.* **2006**, *62*, 52–58. [CrossRef]
38. Guo, S.; Liang, Y.; Liu, L.; Yin, M.; Wang, A.; Sun, K.; Li, Y.; Shi, Y. Research on the Fate of Polymeric Nanoparticles in the Process of the Intestinal Absorption Based on Model Nanoparticles with Various Characteristics: Size, Surface Charge and Pro-hydrophobics. *J. Nanobiotechnol.* **2021**, *19*, 32. [CrossRef]
39. Hu, S.; Yang, Z.; Wang, S.; Wang, L.; He, Q.; Tang, H.; Chen, T. Zwitterionic polydopamine modified nanoparticles as an efficient nanoplatform to overcome both the mucus and epithelial barriers. *Chem. Eng. J.* **2022**, *428*, 132107. [CrossRef]

nanomaterials

Article

Biocompatibility and Antimicrobial Activity of Electrospun Fibrous Materials Based on PHB and Modified with Hemin

Polina M. Tyubaeva [1,2,*], Ivetta A. Varyan [1,2], Elena D. Nikolskaya [1], Mariia R. Mollaeva [1], Nikita G. Yabbarov [1], Maria B. Sokol [1], Margarita V. Chirkina [1] and Anatoly A. Popov [1,2]

[1] Emanuel Institute of Biochemical Physics, Russian Academy of Sciences, 4 Kosygina Street, 119334 Moscow, Russia
[2] Academic Department of Innovational Materials and Technologies Chemistry, Plekhanov Russian University of Economics, 36 Stremyanny Per., 117997 Moscow, Russia
* Correspondence: polina-tyubaeva@yandex.ru; Tel.: +7-9268805508

Abstract: The effect of the hemin (Hmi) on the structure and properties of nanocomposite electrospun materials based on poly-3-hydroxybutyrate (PHB) is discussed in the article. The additive significantly affected the morphology of fibers allowed to produce more elastic material and provided high antimicrobial activity. The article considers also the impact of the hemin on the biocompatibility of the nonwoven material based on PHB and the prospects for wound healing.

Keywords: poly-3-hydroxybutyrate; hemin; electrospun fibrous materials; biocompatibility; antimicrobial activity

Citation: Tyubaeva, P.M.; Varyan, I.A.; Nikolskaya, E.D.; Mollaeva, M.R.; Yabbarov, N.G.; Sokol, M.B.; Chirkina, M.V.; Popov, A.A. Biocompatibility and Antimicrobial Activity of Electrospun Fibrous Materials Based on PHB and Modified with Hemin. *Nanomaterials* **2023**, *13*, 236. https://doi.org/10.3390/nano13020236

Academic Editors: Goran Kaluđerović and Nebojša Pantelić

Received: 21 December 2022
Revised: 30 December 2022
Accepted: 1 January 2023
Published: 5 January 2023

1. Introduction

The interest in materials from renewable resources for formulation of innovative biomedical materials is growing rapidly [1]. High attention is paid to biopolymers—for instance, to polyhydroxyalkanoates (PHA)—which is a class of sustainable aliphatic polyesters produced by various microorganisms [2].

Poly(3-hydroxybutyrate) (PHB) has become the most widespread biopolymer among all PHA due to the large number of advantages. PHB is biodegradable, biocompatible, and thermoplastic polymer [3]. Figure 1a shows the monomeric link of the PHB. PHB particles could be extracted from microorganisms, which synthesize, store, and able to degrade this polymer as a natural source of energy [4]. PHB decay products are nontoxic [4]. Moreover, PHB is able to be decomposed in a short period—Singh and coauthors reported full PHB decay during 30 days in 25% humidified compost [5]. PHB has found wide application in biomedicine [6]: scaffolds [7] and implants [8] design in tissue engineering; nanoparticles for controlled drug release [9] and delivery [10].

The PHB-based composites become popular in the biomedical application due to high biocompatibility [11,12]. Biomedical PHB-based materials with a large surface area including highly porous films [13] and fibrous materials [14] possess high similarity with the structures and surfaces of live organisms and promotes cell adhesion, viability, migration and growth [15]. PHB has shown also high efficiency in design of the new materials for wound healing [16,17].

However, industrial and commercial application of PHB-based materials is limited due to poor mechanical properties—low tensile strength and elongation [18]. Nevertheless, many collectives developed effective methods for PHB modification [19]. Especial efficacy demonstrated various nanocomposites based on PHB and biopolymers or additives of natural origin: poly(ethylene glycol) [20], polylactide [21], polycaprolactone [22], chitosan [23], nanoparticles [24], catalysts and enzymes [25], bioactive molecules [26].

A number of efforts were made by the different collectives in order to improve or endow PHB with specific properties: control of hydrophobicity [26,27] and permeability [28], increase mechanical [29] and antimicrobial [30,31] properties.

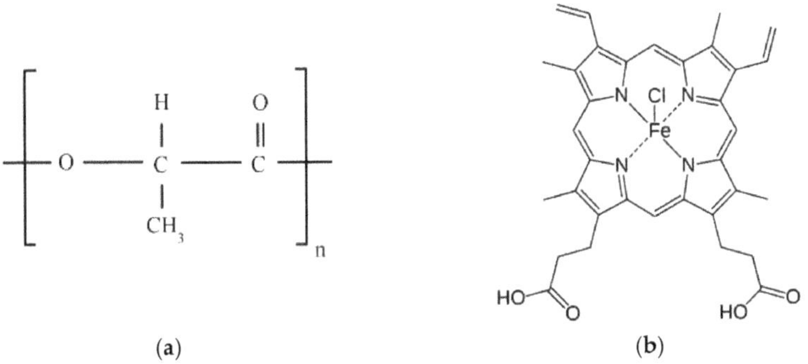

Figure 1. Structural formulas of PHB (a) and hemin (b).

Among the number of additives, porphyrins represent a particular interest. Researchers apply widely synthetic and natural porphyrins in biomedicine [32,33], photo- and chemotherapy [34–36]. Most of the porphyrins are biocompatible, chemically and thermally stable [37]. Moreover, the porphyrins demonstrated a high antimicrobial and antiviral activity [38]. Earlier, the several groups developed the porphyrin-polymer systems through hydrogen bonding, weak interactions (hydrophobic or electrostatic) or coordination bounding [39,40].

The researchers display especial interest to natural porphyrins, among which is hemin (Hmi). Figure 1b shows the structural formula of Hmi. Hmi is applied in various biomedical materials: as a moiety promoting protein-polymer binding [41]; as a container for bioactive molecules [42]; as a biocatalyst [43]. The chemical structure explains a wide variety of applications and unique properties of Hmi, among which, probably the most important—biocompatibility [44] and high antimicrobial activity [45].

In this research, we formulated PHB–Hmi composites by electrospinning method (ES) [46]. ES allowed to obtain fibrous materials with a large surface area and constant distribution of Hmi in polymer matrix, which is very valuable in the production of biocompatible materials [47]. A number of reports are described application of electrospun for PHB-porphyrin composite materials fromulation: polystyrene/polyhydroxybutyrate/graphene/tetraphenylporphyrin [48], polyhydroxybutyrate/Hemin [49], polyhydroxybutyrate/tetraphenylporphyrin with Fe [50], and polyhydroxybutyrate/5,10,15,20-tetrakis(4-hydroxyphenyl)-21H,23H-porphine [51].

In our previous study, we described the nature of Hmi effect on the supramolecular structure of PHB formation [52]. Comparing the formulated composite with the HPB fibers supplemented with a synthetic Fe^{3+} porphyrin complex, we revealed the high potential of these composite materials based on PHB–Hmi fibers [53].

The main goal of this study was assessment of the changes in the structure and properties of PHB under the influence of hemin, and influence evaluation of the Hmi molecular complexes on the biocompatibility and antimicrobial activity.

2. Materials and Methods

2.1. Materials

Polyester of natural origin—poly-3-hydroxybutyrate (PHB) was used in the work [49,53]. PHB was used in the form of a finely dispersed powder (16F series, BIOMER, Schwalbach am Taunus, Germany), characterized by 59% of crystalline phase, 206 kDa of molecular weight, 1.248 g/cm^3 of density (Figure 1a). A tetrapyrrole complex of natural origin—hemin (Hmi) was used in the work (Figure 1b) [54]. Hmi was obtained by the extraction method from the bovine blood (production by Aldrich Sigma, Saint Louis, MO, USA). Phosphate buffered saline (PBS) (Biolot, St. Petersburg, Russia); 3-(4,5-dimethyl-2-thiazolyl)-2,5-diphenyl-2H-tetrasolium bromide (MTT) and Mowiol (Sigma-Aldrich, St. Louis, MO, USA); 96% ethanol (Chimmed, Moscow, Russia); Dulbecco's modified Eagle's medium (DMEM)

(Gibco, Waltham, MA, USA); fetal bovine serum (FBS) (Gibco, Waltham, MA, USA); 0.9% saline (PanEco, Moscow, Russia); dimethyl sulfoxide (DMSO) (Amreso, Solon, OH, USA); 0.02% EDTA; 0.05% trypsin solutions (Gibco, Waltham, MA, USA); and gentamycin (PanEco, Moscow, Russia) were used for the experiments with cell cultures.

2.2. *Methods*

2.2.1. Preparation of the Electrospun Materials

Electrospinning (ES) method was used for obtaining the fibrous materials based on PHB–Hmi [49,52]. The laboratory unit EFV-1 (Moscow, Russia) was single-capillary. Conditions of the ES process are given in the Table 1.

Table 1. Conditions of the ES method.

Diameter of Capillary, mm	Voltage, kV	Distance between the Electrodes, mm	Gas Pressure on the Solution, kg(f)/cm^2
0.1	17–20	190–200	10–14

Since Hmi is soluble in N,N-dimethylformamide and PHB is soluble in chloroform, the method of double-solution electrospinning was used for obtaining PHB–Hmi fibers [55,56]. For preparation of forming solutions PHB powder was dissolved in chloroform at a temperature of 60 °C and Hmi powder was dissolved in N,N-dimethylformamide at a temperature of 25 °C. Both solutions were homogenized and were used 12 h after manufacture. The properties of the forming solutions based on PHB–Hmi are given in Table 2.

Table 2. The properties of the forming solution.

Content PHB, wt. %	Content of Hmi, wt. % of PHB Mass	Electrical Conductivity, µS/cm	Viscosity, Pa s
7	0	10	1.0
7	1	11	1.4
7	3	13	1.7
7	5	14	1.9

2.2.2. Scanning Electron Microscopy

Images of electrospun materials based on PHB–Hmi were obtained by scanning electron microscopy using the Tescan VEGA3 (Brno, Czech Republic) on the samples with a platinum layer.

2.2.3. Mechanical Analysis

Tensile strength and elongation at break were obtained by the mechanical test on the tensile compression testing machine Devotrans DVT GP UG 5 (Istanbul, Turkey) on the samples 10 × 40 mm at the stretching speed was 25 mm/min without preload pressure. All data were averaged on the ten samples. Tensile strength was registered by the Devotrans software with the average statistical error in measuring thermal effects was ±0.02 MPa. Elongation at break, ε, was calculated as:

$$\varepsilon = \frac{\Delta l}{l_0} \times 100\% \quad (1)$$

where Δl—the difference between the final and initial length of the sample; l_0—the initial length of the sample. The average statistical error in measuring thermal effects was ±0.2%.

2.2.4. X-ray Diffraction Analysis

Degree of crystallinity of PHB and the average sizes of crystallites were obtained by X-ray diffraction analysis on the HZG4 diffractometer (Freiberger Präzisionsmechanik,

Germany) HZG4 diffractometer (Freiberger Präzisionsmechanik, Germany). To calculate the degree of crystallinity, the method was used [57].

Average sizes of PHB crystallites, L_{020}, were calculated from diffractograms obtained with the Bragg–Brentano method using the Selyakov–Scherrer formula, the method was used [58].

2.2.5. Differential Scanning Calorimetry

Thermal properties of the PHB–Hmi samples were obtained by differential scanning calorimeter (DSC) using Netzsch 214 Polyma (Selb, Germany), in an argon atmosphere, with a heating rate of 10° K/min and with a cooling rate of 10° K/min with samples' weight 6–7 mg. The DSC temperature program included 2 heating from 20 °C to 220 °C and 2 cooling to 20 °C with average statistical error 2.5%.

Enthalpy of melting, ΔH, was calculated by NETZSCH Proteus software according to the standard technique [59].

Crystallinity degree, χ, was defined from the melting peak as:

$$\chi = \frac{\Delta H}{H_{PHB}} \times 100\% \quad (2)$$

where ΔH—melting enthalpy; H_{PHB}—melting enthalpy of the ideal crystal of the PHB; 146 J/g [60]; C—the content of the PHB in the composition.

2.2.6. Wetting Contact Angle

Wetting contact angle is a measure of wettability of the surface of the PHB–Hmi samples. Water drops (2 µL) were applied to three different areas of the nonwoven material's surface by an automatic dispenser. Measurements were prepared using an optical microscope M9 No. 63649, lens FMA050 (Moscow, Russia) by Altami studio 3.4 Software. The relative measurement error was ±0.5%

2.2.7. Permeability to Air

Permeability to air characterizes barrier properties of porous nonwoven material. Air permeability of the PHB–Hmi porous samples was measured according to the standard protocol according to Gurley method [61,62]. The pressure was 1.22 MPa, volume of the air was 100 mL, and the test sample's area was 6.5 cm^2. The relative measurement error was ±5%

2.2.8. Antimicrobial Tests

The antimicrobial activity of PHB–Hmi samples was studied by biomedical tests on cellular material of *Staphylococcus aureus* p 209, *Salmonella typhimurium* and *Escherichia coli* 1257. Meat-peptone agar was used for cultures of microorganisms, incubation time was 24 h at 37 °C. Concentration of microbial cells in the saline solution was 5×10^5 CFU per mL. The crops were incubated for 48 h at 37 °C after preparation PHB–Hmi samples in Petri dishes with meat–peptone agar. In parallel, the test culture suspensions used in the experiment were seeded to control the concentration of viable microorganisms. The colonies of viable microorganisms grown on the surface of the agar were counted.

2.2.9. Hemin Release Studies

Hemin release study allowed to evaluate time-dependently the amount of additive released from the material. Electrospun materials containing 5% of Hmi (10×10 mm^2) were poured in 1.5 mL of 0.1 M PBS (phosphate-buffered saline, pH 7.4). Samples were incubated under constant shaking of 180 rpm at 37 °C during 48 h. The supernatant samples were picked at 0 h (just after the films soaking), 5 h, 24 h, and 48 h. The released hemin absorbance was determined by UV spectrophotometry (SHIMADZU UV-1800 (Shimadzu, Kyoto, Japan)) at 292 nm. The release data are presented as the average value of five specimens with the standard deviation.

2.2.10. Cell Culture

The immortalized human fibroblasts BJ-5ta cell line was maintained in 25 cm^2 polystyrene flasks in the DMEM medium supplemented with 10% FBS and gentamycin (50 μg/mL) at 37 °C in a humidified atmosphere containing 5% CO_2. The cells were replated using trypsin-EDTA solution twice per week.

2.2.11. Cytotoxic Activity Analysis

To assess the cytotoxic activity and biocompatibility, the cells were seeded into 24-well plates (20,000 cells per well) directly before experiment on film samples and incubated under standard conditions for 72 h. Cells photo were taken at 24, 48, 72 h of incubation by Nikon Diaphot phase contrast microscope at 40x magnification and a Levenhuk M1400Plus camera. We applied standard MTT assay to evaluate cells survival [63]. Each well was supplemented with 250 μL of MTT solution (1 mg/mL) in the serum-free DMEM and incubated during 4 h. Next, the medium was aspirated, precipitated formazan crystals in each well were dissolved in 400 μL of DMSO, and the light absorption was measured at 540 nm. Survival curves plotting, IC_{50} values calculation, and statistical analysis were performed in Excel (Microsoft Corporation, Redmond, WA, USA) and OriginPro (version 2020b, OriginLab Corp., Northampton, MA, USA).

3. Results and Discussion

3.1. Characterization of PHB–Hmi Fibers

Addition of hemin to the poly(3-hydroxybutyrate) fibers is a good approach to modify its surface and properties. The introduction of Hmi into the forming solution increased the electrical conductivity by 10–40% and the viscosity by 40–90%, that contributed to a significant improvement in the fibers' quality. All key parameters of ES process such as the flow rate of the polymer solution, the curing rate of the fibers, and the trajectory of the thread were more stable due to addition of Hmi to the forming solution. As shown in Figure 2, all the PHB–Hmi fibers displayed uniform and randomly orientated structure. It fully corresponded to the type of the structure produced by the ES method during formation of fibrous layer [64]. One of the important parameters of ES is the distance between the capillary and the collection zone. First of all, this distance affects the size of the ES area, as well as the diameter of the formed fibers, which makes a significant contribution to the formation of a uniform layer of nonwoven material [65]. The optimal distance was selected experimentally taking into account the optimal molding conditions for the PHB solution to obtain a uniform Taylor cone during the molding process [66]. Another significant aspect in the formation of composites by the ES method is the contribution of the solvent. There are a large number of approaches to the implementation of double-solution electrospinning [55,56]. The main contribution of the two-solution ES process of PHB–Hmi composites is due to the fact that PHB is not soluble in N,N-dimethylformamide, and Hmi is not soluble in chloroform. At the same time, the solutions mix well, forming a sufficiently homogeneous system for forming fibrous materials with a uniform distribution of Hmi in the structure [52,53]. Moreover, the introduction of Hmi contributed to changes in the structure of the fibers. Characteristics of the nonwoven materials are presented in Table 3.

The surface density of the material was reduced by 30–40% due to an increase in porosity. The average fibers' diameter was reduced by 40–50%. With an increase in the concentration of Hmi the number of defects on the surface of the fibers noticeably decreased. Thickenings, gluings, and spherical formations were almost completely absent at 5% wt. of Hmi.

The reduction in the number of defects and the formation of more uniform fibers contributed to the growth of mechanical properties of PHB-based materials. The tensile strength increased by 3.2 times, and the elongation at break increased by 1.7 times. Typical tensile stress—strain curves of electrospun PHB–Hmi materials are shown on the Figure 3. The addition of higher concentrations of Hmi caused weakening of mechanical properties of the material.

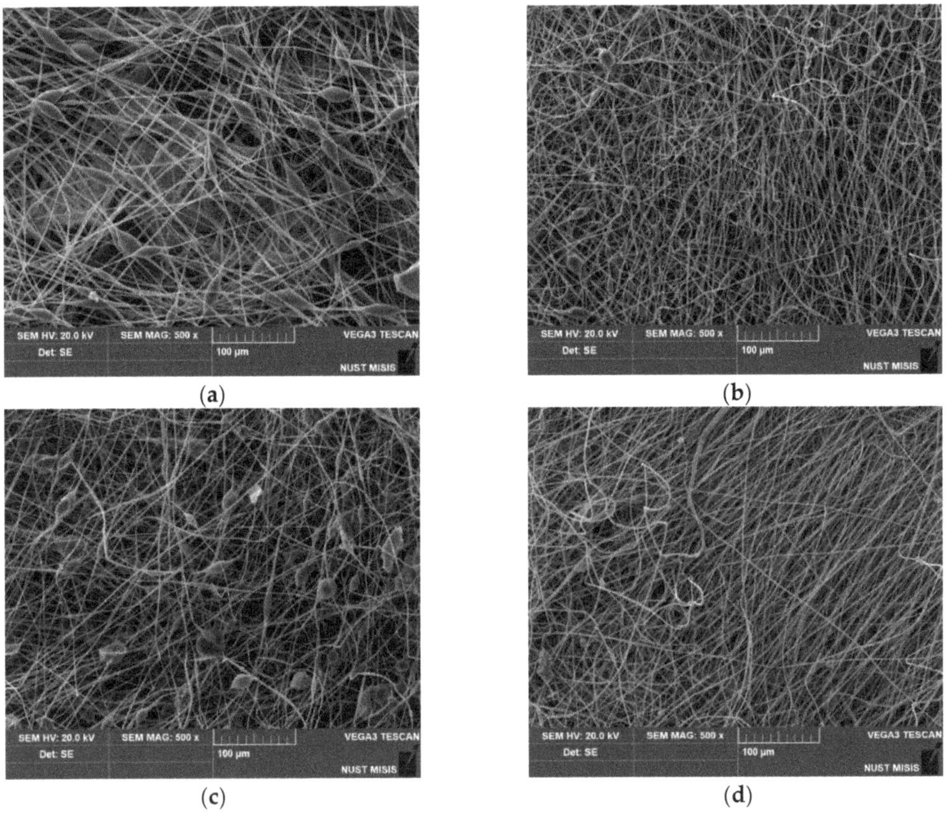

Figure 2. SEM images of electrospun materials based on PHB with different content of Hmi: 0% wt. (**a**), 1% wt. (**b**), 3% wt., (**c**) and 5% wt. (**d**).

Table 3. Material properties of PHB–Hmi fibers [a].

Sample	Density, g/cm^3 (Mean ± SD, n = 10)	Average Diameter, μm (Mean ± SD, n = 100)	Tensile Strength, MPa (Mean ± SD = 0.05, n = 10)	Elongation at Break, % (Mean ± SD = 0.2, n = 10)
PHB 0% wt.	0.30 ± 0.01	3.50 ± 0.08	1.7	3.6
PHB with 1% wt. of Hmi	0.20 ± 0.02	2.06 ± 0.07	0.7	4.7
PHB with 3% wt. of Hmi	0.20 ± 0.01	1.77 ± 0.04	1.9	4.7
PHB with 5% wt. Hmi	0.17 ± 0.01	1.77 ± 0.04	5.5	6.1

[a] Density and average diameter were calculated per area 400 × 600 μm^2.

The supramolecular structure of the polymer plays a significant role in the key properties of the material including biocompatibility, degradation, stability under different environmental conditions [67]. PHB is a semi-crystalline polymer with the orthorhombic crystal lattice (a = 0.576 nm, b = 1.320 nm, c = 0.596 nm, and space group symmetry of $2_1 2_1 2_1$) [68]. Hmi did not affect these parameters of native crystalline phase of PHB. However, Hmi significantly affected the degree of crystallinity and the size of the crystallites of PHB (Figure 4).

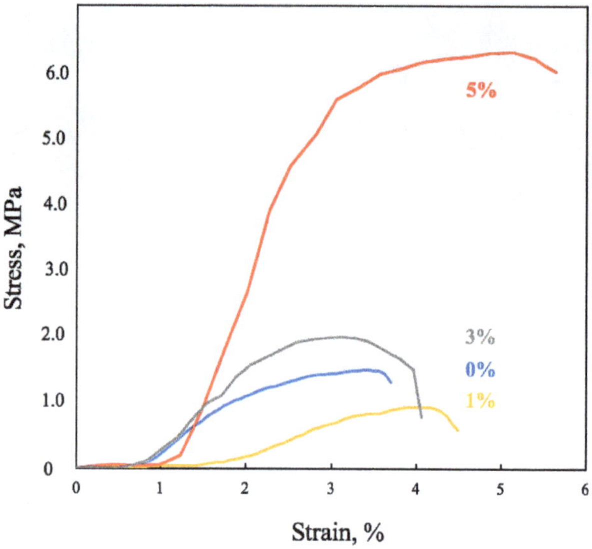

Figure 3. Typical tensile stress-strain curves of electrospun materials based on PHB with different content of Hmi: 0% wt. (**blue**), 1% wt. (**yellow**), 3% wt., (**grey**) and 5% wt. (**red**).

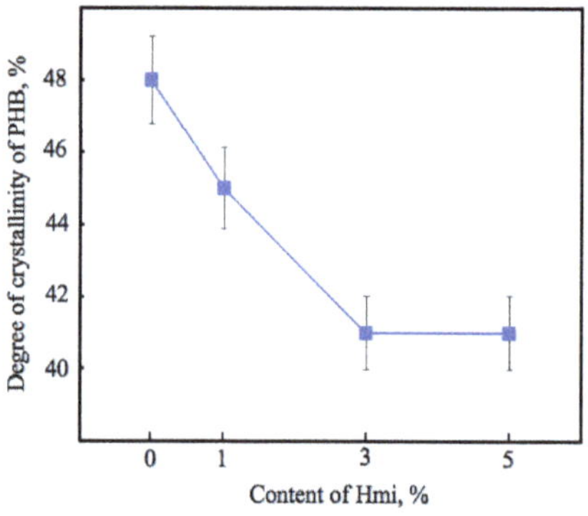

(**a**)

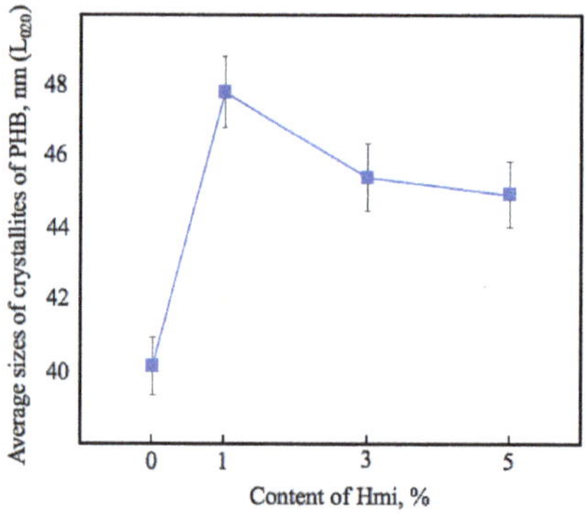

(**b**)

Figure 4. Degree of crystallinity (**a**) and average sizes of PHB crystallites L_{020} (**b**) of PHB–Hmi composites.

The introduction of Hmi led to decrease in the proportion of the crystalline phase by 6–15%, however, the size of the crystallites increased by 26–15%. Probably, hemin could act as a crystallization center during the curing of the forming solution. Thus, PHB was able to form more regular and larger crystallites, which also contributed to the mechanical properties of the material.

These results were consistent with the changes in thermal properties of PHB–Hmi (Table 4). While Hmi very slightly effected on the melting temperature of the crystalline phase, the melting enthalpy varied according to the changes in the degree of crystallinity. During the first melting, it decreased by 12–19%, and during the second one by 13–20%. so slight differences between the first and second heating showed that the polymer in the molding solution had time to crystallize sufficiently, and the fibrous structure had little effect on the phase distribution [69].

Table 4. Thermal properties of PHB–Hmi, where χ—crystallinity degree Δ ± 2.5%, ΔH—melting enthalpy Δ ± 2.5%, and T_m—melting temperature Δ ± 2%.

Sample	Concentration of Additive, %	First Heating Run		Second Heating Run	
		T_m, °C	ΔH, J/g	T_m, °C	ΔH, J/g
PHB	0	175	93.1	170	90.8
PHB–Hmi	1	172	81.8	168	78.7
PHB–Hmi	3	173	77.8	170	75.4
PHB–Hmi	5	174	75.3	170	72.7

These characteristics showed the significant positive contribution of Hmi to the formation of nanomaterials. Exerting a significant influence on the crystallization of PHB, this modifying additive allowed to obtain the material devoid of the disadvantages of pure PHB (PHB–Hmi was more durable, with fewer defects, more uniform fibers). In addition, the significant influence of the Hmi on the molding properties of the solution makes it possible to obtain a material with a more predictable structure, which could not be obtained using other metal-containing modifying additives [53].

3.2. The Barrier Properties of PHB–Hmi Fibers

PHB is a hydrophobic material, which can make it difficult for cells to consolidate in a living organism and slow down the wound healing process. The control of hydrophobicity is an important task. Figure 5 shows the impact of Hmi on the hydrophobicity of the nonwoven material.

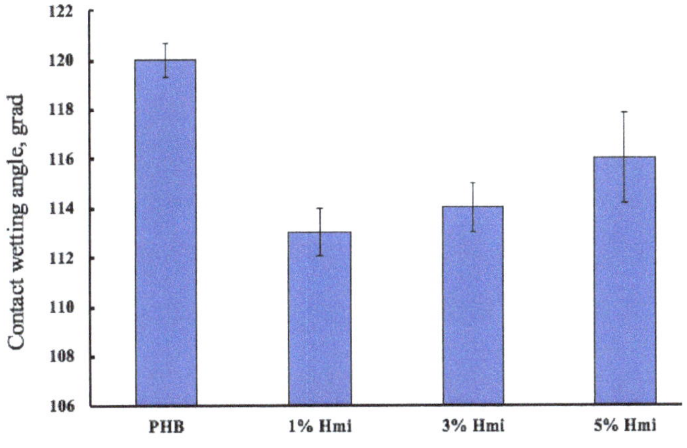

Figure 5. Contact wetting angles of the fibrous materials based on PHB–Hmi.

The introduction of Hmi has a hydrophilic effect due to the polar groups—COOH (Figure 1b) located in the structure of the tetrapyrrole ring. Moreover, it is known that tetrapyroll complexes tend to mutual aggregation [70], and with increasing the Hmi concentration, this effect could be observed. The wetting angle decreases slightly due to the smaller number of hydrophilic sites that are freely available on the surface of the fibers with the growth of the Hmi concentration.

Another important aspect of wound healing is the permeability to air [71]. The introduction of Hmi made a significant contribution to the control of the breathability of nonwoven fabric (Figure 6).

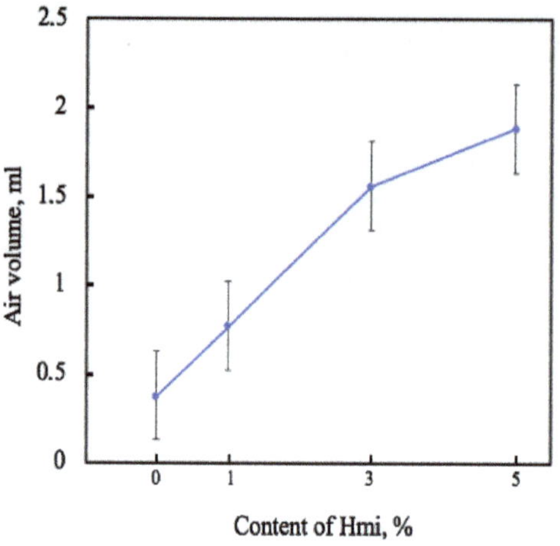

Figure 6. Air volume that passed through the fibrous materials based on PHB–Hmi according to the Gurley method.

The key parameter affecting the permeability of the nanofibrous material is the morphology of the fibrous layer. With an increase in porosity and with a decrease in the number of glues and engagements, the material becomes more accessible for air transfer. Air permeability control is extremely important for the formation of a reliable environment in the wound healing zone. Thus, we observed five-fold air permeability increment with the addition of 5% of Hmi.

3.3. The Antimicrobial Tests of PHB–Hmi Fibers

The antimicrobial activity of hemin against *S. aureus* is well known [72]. The results of the antimicrobial efficacy of PHB–Hmi electrospun materials against Gram-positive and Gram-negative cultures are shown in Table 5.

Table 5. Antimicrobial activity of electrospun fibrous materials based on PHB–Hmi composites.

Test Culture	Initial Test Culture, CFU/mL	Sample, CFU/mL	Control, CFU/mL
		PHB with 1 % wt. Hmi	
S. aureus 209	2.0×10^4	4.5×10^3	8.6×10^3
E. coli 1257	2.0×10^4	8.5×10^2	9.8×10^3
S. typhimurium	2.0×10^4	7.2×10^3	8.1×10^3
		PHB with 3 % wt. Hmi	
S. aureus 209	2.1×10^4	1.8×10^3	8.6×10^3
E. coli 1257	2.0×10^4	$<1 \times 10^2$	9.8×10^3
S. typhimurium	2.0×10^4	2.1×10^3	8.1×10^3
		PHB with 5 % wt. Hmi	
S. aureus 209	2.0×10^4	0.9×10^3	8.6×10^3
E. coli 1257	2.0×10^4	$<1 \times 10^2$	9.8×10^3
S. typhimurium	2.0×10^4	2.0×10^3	8.1×10^3

It is known that pure PHB-based materials have no antibacterial activity [73]. On the contrary, PHB is able to be a good substrate in view of its microbiological origin.

Table 5 shows that the increasing of the Hmi concentration leaded to the growth of the antimicrobial activity of the fibrous material. 1% of Hmi leaded to the *S. aureus* CFU decrease by 47%, and of *E. coli* by 90%. In relation to *S. typhimurium*, a small concentration of Hmi was less effective reducing the number of colony-forming units within 15%, and 3 and 5% of Hmi provided high activity against *S. aureus* and *S. typhimurium* inhibiting CFU by 79–89% and 74–75% correspondingly. Moreover, 3 and 5% of Hmi displayed almost 98% CFU inhibition against *E. coli*.

Probably, the antimicrobial effect is explained by the gradual Hmi release from the PHB matrix. The Hmi release profile from the electrospun sample containing 5% of Hmi was recorded from the immersion in PBS solution (pH 7.4, 37 °C) for 48 h. Figure 7 shows the gradual hemin release from PHB containing 5% Hmi. We observed 0.84% Hmi release after 5 h of incubation: 1.72% after 24 h and 4.03% after 48 h. Thus, the released Hmi could explain the average antimicrobial properties of nonwoven materials, which may be beneficial during future applications of these nonwoven materials.

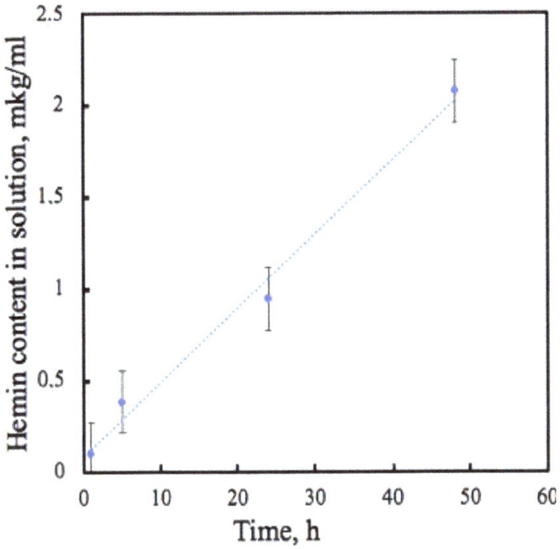

Figure 7. The changes in the Hmi concentration in PBS solution (pH 7.4, 37 °C).

3.4. *Cytotoxic Activity Analysis and Biocompatibility of PHB–Hmi Fibers*

During microscopic examination, we observed the fibroblasts distribution along the PHB-based fiber samples after 24, 48 and 72 h of the cultivation. The microscopic images of control PHB with different hemin content in cells-free DMEM are represented in Figure 8.

The cells exhibited a flattened morphology and demonstrated a good adherence to the polymeric plate surface in presence of PHB membranes. The normal morphology and proliferation rates were comparable with control cells (Figure 9a) evidencing the lack of noticeable cytotoxic effect of the PHB-based fibers and good potential biocompatibility.

According to Figure 9, the cells revealed high viability after 24–72 h of incubation in presence of PHB with different hemin content, indicating low PHB toxicity.

The MTT test results (Figure 10) also evidenced the absence of pronounced cytotoxic effects of PHB-based fibers.

Summarizing, there were no significant differences in morphology, cells shape, adherence, or survival rate between groups after 24, 48, and 72 h of incubation of BJ-5ta cells with PHB with different hemin content (Figures 9 and 10). The lack of ruptures, deformations, or other phenomena confirmed the good biocompatibility of the examined PHBs. Good biocompatibility is one of the basics of the materials applied in medicine. Thus, these preliminary results evidence that all examined PHB-based fibers lack toxic effects on cell viability and morphology.

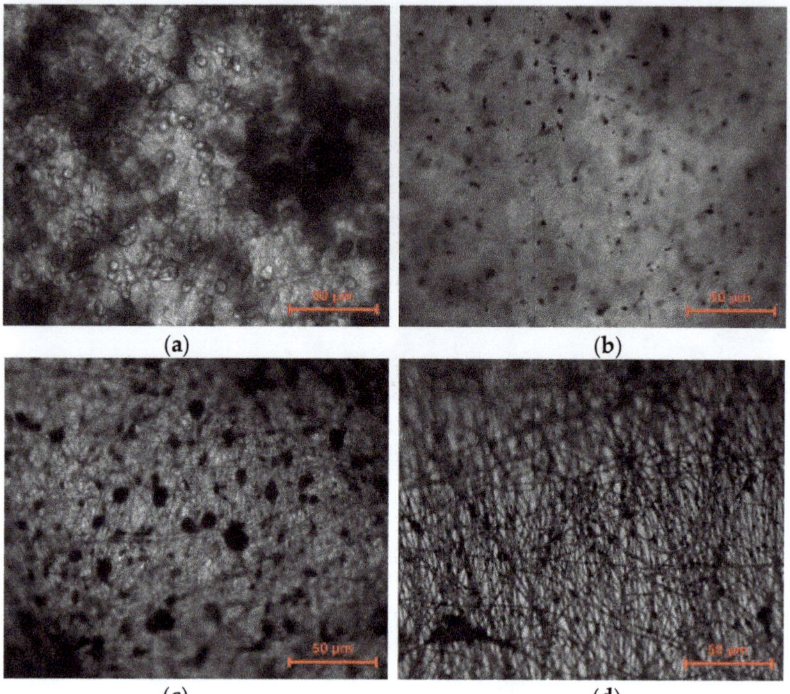

(a) (b) (c) (d)

Figure 8. The optical microscopic images of PHB with different hemin content: 0% wt. (**a**), 1% wt. (**b**), 3% wt. (**c**), and 5% wt. (**d**) before fibroblasts seeding BJ-5ta cells. Bar—50 μm.

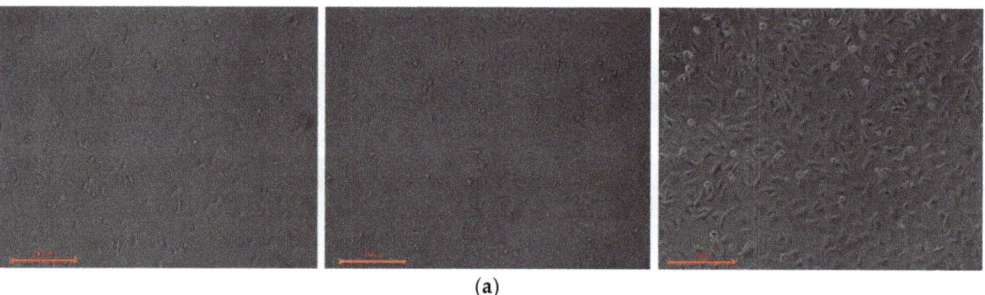

(**a**)

Figure 9. *Cont.*

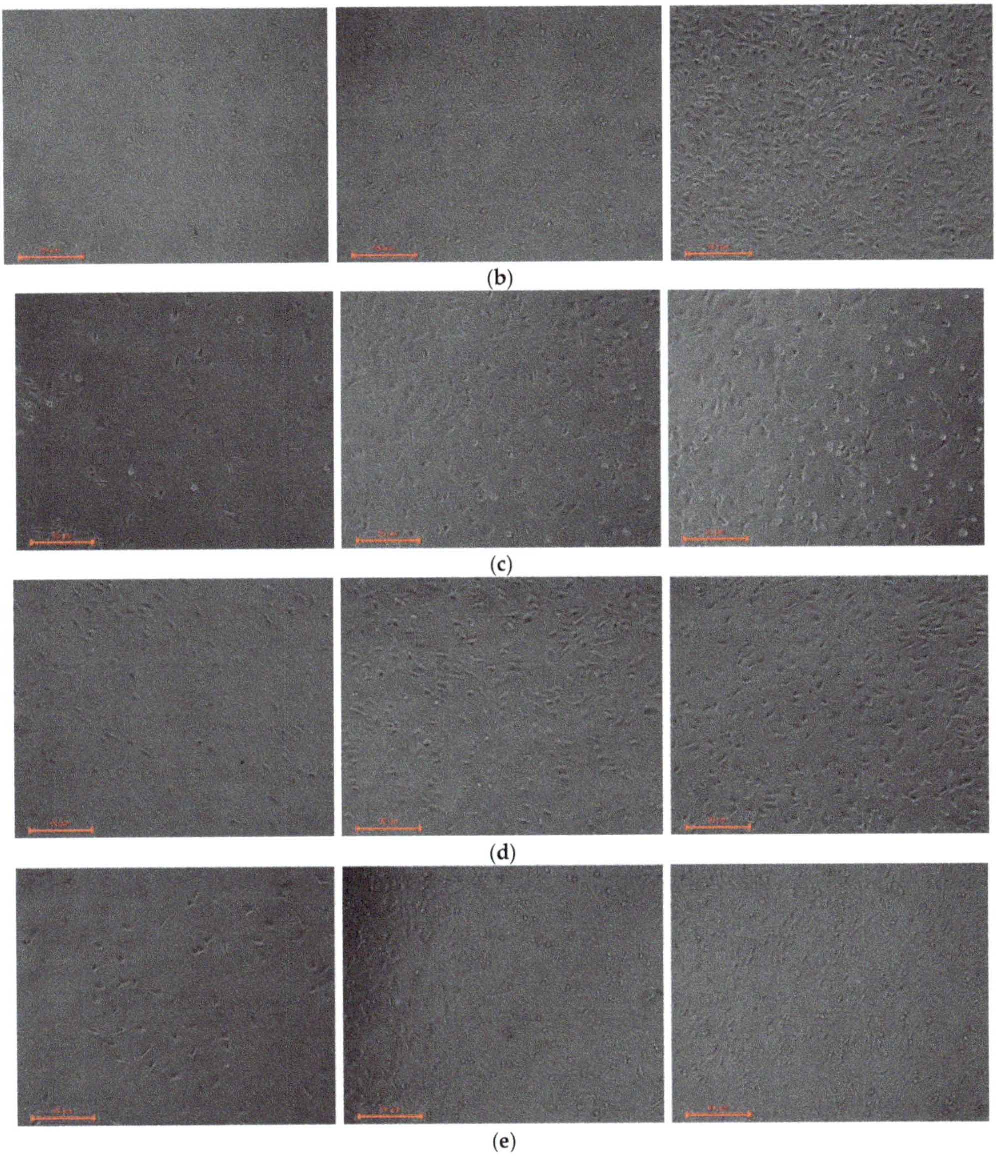

Figure 9. The optical microscopic images of the fibroblasts BJ-5ta after 24, 48 and 72 h (the images aligned left to right respectively) of cultivation in presence of PHB with different content of the hemin: 0% wt. (**b**), 1% wt. (**c**), 3% wt. (**d**), and 5% wt. (**e**); (**a**) control. Bar—50 μm.

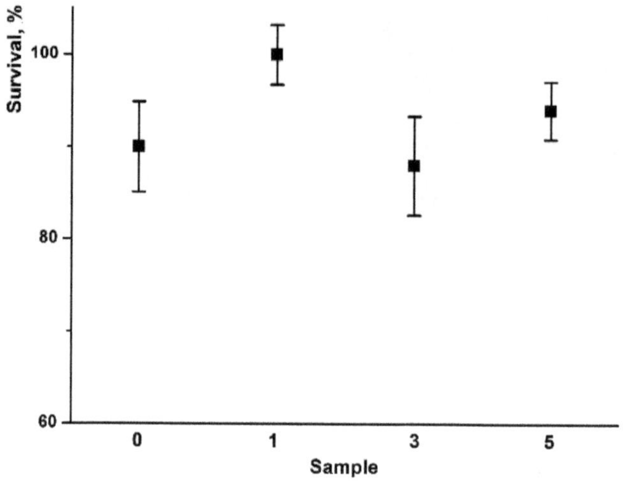

Figure 10. Survival of BJ-5ta cells after 72 h incubation with the PHB with different content of the hemin (0—0% wt., 1—1% wt., 3—3% wt., and 5—5% wt). (mean ± SD, n = 3).

4. Conclusions

We evaluated the effect of 1–5% wt. of the hemin molecular complexes on the structure and properties of the composite materials based on poly-3-hydroxybutyrate in this research. The Hmi made it possible to obtain fibers with improved morphology. The presence of an iron atom in the Hmi structure significantly improved the properties of the solution during the ES process, which had a positive effect on the structural characteristics of the material. As a result, the tensile strength of the fibrous layer increased by 3.2 times, and the elongation at break increased by 1.7 times with the introduction of 5% wt. of Hmi. Although it should be mentioned that the high modulus of elasticity of the samples, which may impose restrictions on the use of film samples in clinical products, however, electrospun nonwoven samples are characterized by high softness and elasticity, due to the high degree of freedom of the fibers relative to each other. Moreover, Hmi acted as a crystallization center, which allowed the formation of a more favorable crystal structure of the polymer. The size of PHB crystallites increased, and their total fraction decreased. The Hmi antimicrobial activity ensured the death of both Gram-negative and Gram-positive cultures after contact with the PHB–Hmi fibrous material. Cytotoxic activity results demonstrated that formulated PHB-based fibers characterized with high potential safety and could be promising vehicles for regenerative medicine applications. Despite the low statistical differences in the MTT results, we could assume influence of the localization and behavior of Hmi in the material on the cells survival. As mentioned earlier, tetrapyrroles are prone to aggregation processes which are directly related to their concentration [70,74]. The minimum of the aggregation was detected at 1% of Hmi and corresponded to a high survival rate. The maximum of aggregation was detected at 3% of Hmi and corresponded to a low survival rate, while at 5% of Hmi, there were both aggregated and free components. This assumption is consistent with the previously described trends in the accumulation of iron atoms obtained by the EDX atomic analysis of an iron atom [49]. Thus, we can assume presence of dependence between Hmi aggregation degree and cells survival, what, of course, have to be tested further. Thus, summarizing the data obtained, we can recommend PHB–Hmi materials for regenerative medicine as wound dressing layers, and among alternative applications, we can offer hygienic agents, filter materials, and other clinical products that require a highly developed surface in combination with antimicrobial properties and biocompatibility.

Author Contributions: Conceptualization, P.M.T. and I.A.V.; methodology, P.M.T.; software, E.D.N. and M.R.M.; validation, I.A.V., M.V.C. and M.B.S.; formal analysis, P.M.T. and I.A.V.; investigation, P.M.T., I.A.V., N.G.Y., M.R.M., M.V.C. and E.D.N.; resources, I.A.V.; data curation, P.M.T.; writing—original draft preparation, P.M.T., I.A.V., N.G.Y. and A.A.P.; writing—review and editing, P.M.T., A.A.P. and I.A.V.; visualization, P.M.T.; supervision, P.M.T.; and project administration, A.A.P. All authors have read and agreed to the published version of the manuscript.

Funding: This research was funded from the Russian science foundation (Agreement No. 22-73-00038).

Data Availability Statement: Data are contained within the article.

Acknowledgments: The study was carried out using scientific equipment of the Center of Shared Usage «New Materials and Technologies» of Emanuel Institute of Biochemical Physics and the Common Use Centre of Plekhanov Russian University of Economics.

Conflicts of Interest: The authors declare no conflict of interest.

References

1. Arif, U.; Haider, S.; Haider, A.; Khan, N.; Alghyamah, A.A.; Jamila, N.; Khan, M.I.; Almasry, W.A.; Kang, I.K. Biocompatible Polymers and their Potential Biomedical Applications: A Review. *Curr. Pharm. Des.* **2019**, *25*, 3608–3619. [CrossRef]
2. Li, M.; Wilkins, M.R. Recent advances in polyhydroxyalkanoate production: Feedstocks, strains and process developments. *Int. J. Biol. Macromol.* **2020**, *156*, 691–703. [CrossRef]
3. Kushwah, B.S.; Kushwah, A.V.S.; Singh, V. Towards understanding polyhydroxyalkanoates and their use. *J. Polym. Res.* **2016**, *23*, 153. [CrossRef]
4. Zhao, J.; Weng, G.; Li, J.; Zhu, J.; Zhao, J. Polyester-based nanoparticles for nucleic acid delivery. *Mater. Sci. Eng. C* **2018**, *92*, 983–994. [CrossRef]
5. Singh, G.; Kumari, A.; Mittal, A.; Yadav, A.; Aggarwal, N.K. Polyβ-Hydroxybutyrate Production by Bacillus subtilisNG220 Using Sugar Industry Waste Water. *BioMed Res. Int.* **2013**, *2013*, 952641. [CrossRef] [PubMed]
6. Kai, D.; Loh, X.J. Polyhydroxyalkanoates: Chemical Modifications Toward Biomedical Applications. *ACS Sustain. Chem. Eng.* **2014**, *2*, 106–119. [CrossRef]
7. Bakhtiari, S.S.E.; Karbasi, S.; Toloue, E.B. Modified Poly(3-hydroxybutyrate)-based scaffolds in tissue engineering applications: A Review. *Int. J. Biol. Macromol.* **2021**, *166*, 986–988. [CrossRef]
8. Chen, G.; Wang, Y. Medical applications of biopolyesters polyhydroxyalkanoates. *Chin. J. Polym. Sci.* **2013**, *31*, 719–736. [CrossRef]
9. Babos, G.; Rydz, J.; Kawalec, M.; Klim, M.; Fodor-Kardos, A.; Trif, L.; Feczkó, T. Poly(3-Hydroxybutyrate)-Based Nanoparticles for Sorafenib and Doxorubicin Anticancer Drug Delivery. *Int. J. Mol. Sci.* **2020**, *21*, 7312. [CrossRef] [PubMed]
10. Fan, F.; Tan, D.; Shang, S.; Wu, X.; Zhao, J.; Ran, G.; Lu, X. Poly (3-hydroxybutyrate-co-3-hydroxyhexanoate) Biopolyester Based Nanoparticles as NVP-BEZ235 Delivery Vehicle for Tumor Targeting Therapy. *Biomacromolecules* **2019**, *20*, 3313–3323. [CrossRef]
11. Shishatskaya, E.I.; Khlusov, I.A.; Volova, T.G. A hybrid PHB–hydroxyapatite composite for biomedical application: Production, in vitro and in vivo investigation. *J. Biomater. Sci. Polym. Ed.* **2006**, *17*, 481–498. [CrossRef] [PubMed]
12. Kai, D.; Zhang, K.; Liow, S.S.; Loh, X.J. New Dual Functional PHB-Grafted Lignin Copolymer: Synthesis, Mechanical Properties, and Biocompatibility Studies. *ACS Appl. Bio Mater.* **2018**, *2*, 127–134. [CrossRef]
13. Zhijiang, C. Biocompatibility and Biodegradation of novel PHB porous substrates with controlled multi-pore size by emulsion templates method. *J. Mater. Sci. Mater. Med.* **2006**, *17*, 1297–1303. [CrossRef] [PubMed]
14. Zarei, M.; Karbasi, S.; Sari Aslani, F.; Zare, S.; Koohi-Hosseinabad, O.; Tanideh, N. In Vitro and In Vivo Evaluation of Poly (3-hydroxybutyrate)/Carbon Nanotubes Electrospun Scaffolds for Periodontal Ligament Tissue Engineering. *J. Dent. (Shiraz)* **2020**, *21*, 18–30. [CrossRef]
15. Zhijiang, C.; Qin, Z.; Xianyou, S.; Yuanpei, L. Zein/Poly(3-hydroxybutyrate- co -4-hydroxybutyrate) electrospun blend fiber scaffolds: Preparation, characterization and cytocompatibility. *Mater. Sci. Eng. C* **2017**, *71*, 797–806. [CrossRef]
16. Sanhueza, C.; Hermosilla, J.; Bugallo-Casal, A.; Da Silva-Candal, A.; Taboada, C.; Millán, R.; Acevedo, F. One-step electrospun scaffold of dual-sized gelatin/poly-3-hydroxybutyrate nano/microfibers for skin regeneration in diabetic wound. *Mater. Sci. Eng. C* **2020**, *2020*, 111602. [CrossRef] [PubMed]
17. Knowles, J.; Hastings, G.; hta, H.; Niwa, S.; Boeree, N. Development of a degradable composite for orthopaedic use: In vivo biomechanical and histological evaluation of two bioactive degradable composites based on the polyhydroxybutyrate polymer. *Biomaterials* **1992**, *13*, 491–496. [CrossRef]
18. Rajan, K.P.; Thomas, S.P.; Gopanna, A.; Chavali, M. Polyhydroxybutyrate (PHB): A Standout Biopolymer for Environmental Sustainability. *Handb. Ecomater.* **2017**, 1–23. [CrossRef]
19. Sreedevi, S.; Unni, K.N.; Sajith, S.; Priji, P.; Josh, M.S.; Benjamin, S. Bioplastics: Advances in Polyhydroxybutyrate Research. *Adv. Polym. Sci.* **2014**. [CrossRef]

20. Arrieta, M.P.; López, J.; Hernández, A.; Rayón, E. Ternary PLA–PHB–Limonene blends intended for biodegradable food packaging applications. *Eur. Polym. J.* **2014**, *50*, 255–270. [CrossRef]
21. Kumara, B.P.; Maruthi, Y.P.; Pratap, S.; Sudha, K. Development and characterization of polycaprolactone (PCL)/poly ((R)-3-hydroxybutyric acid) (PHB) blend microspheres for tamoxifen drug relese studies. *Int. J. Pharm. Pharmac. Sci.* **2015**, *7*, 95–100.
22. Karimi, A.; Karbasi, S.; Razavi, S.; Zargar, E.N. Poly(hydroxybutyrate)/chitosan Aligned Electrospun Scaffold as a Novel Substrate for Nerve Tissue Engineering. *Adv. Biomed. Res.* **2018**, *7*, 44. [CrossRef]
23. Kim, G.-M.; Wutzler, A.; Radusch, H.-J.; Michler, G.H.; Simon, P.; Sperling, R.A.; Parak, W.J. One-Dimensional Arrangement of Gold Nanoparticles by Electrospinning. *Chem. Mater.* **2005**, *17*, 4949–4957. [CrossRef]
24. Zeng, J.; Aigner, A.; Czubayko, F.; Kissel, T.; Wendorff, J.H.; Greiner, A. Poly(vinyl alcohol) Nanofibers by Electrospinning as a Protein Delivery System and the Retardation of Enzyme Release by Additional Polymer Coatings. *Biomacromolecules* **2005**, *6*, 1484–1488. [CrossRef] [PubMed]
25. Joung, Y.K.; Bae, J.W.; Park, K.D. Controlled release of heparin-binding growth factors using heparin-containing particulate systems for tissue regeneration. *Expert Opin. Drug Deliv.* **2008**, *5*, 1173–1184. [CrossRef] [PubMed]
26. Bhattacharjee, A.; Kumar, K.; Arora, A.; Katti, D.S. Fabrication and characterization of Pluronic modified poly(hydroxybutyrate) fibers for potential wound dressing applications. *Mater. Sci. Eng. C* **2016**, *63*, 266–273. [CrossRef]
27. Seoane, I.; Manfredi, L.; Cyras, V.; Torre, L.; Fortunati, E.; Puglia, D. Effect of Cellulose Nanocrystals and Bacterial Cellulose on Disintegrability in Composting Conditions of Plasticized PHB Nanocomposites. *Polymers* **2017**, *9*, 561. [CrossRef] [PubMed]
28. Patel, M.; Hansson, F.; Pitkänen, O.; Geng, S.; Oksman, K. Biopolymer Blends of Poly(lactic acid) and Poly(hydroxybutyrate) and Their Functionalization with Glycerol Triacetate and Chitin Nanocrystals for Food Packaging Applications. *ACS Appl Polym Mater.* **2022**, *9*, 6592–6601. [CrossRef]
29. Díez-Pascual, A.M. Effect of Graphene Oxide on the Properties of Poly(3-Hydroxybutyrate-co-3-Hydroxyhexanoate). *Polymers* **2021**, *13*, 2233. [CrossRef]
30. Xavier, J.R.; Babusha, S.T.; George, J.; Ramana, K.V. Material Properties and Antimicrobial Activity of Polyhydroxybutyrate (PHB) Films Incorporated with Vanillin. *Biotechnol. Appl. Biochem.* **2015**, *176*, 1498–1510. [CrossRef]
31. Salama, H.E.; Saad, G.R.; Sabaa, M.W. Synthesis, characterization and antimicrobial activity of biguanidinylated chitosan- g -poly[(R)-3-hydroxybutyrate]. *Int. J. Biol. Macromol.* **2017**, *101*, 438–447. [CrossRef] [PubMed]
32. Wu, J.; Li, S.; Wei, H. Integrated nanozymes: Facile preparation and biomedical applications. *ChemComm* **2018**, *54*, 6520–6530. [CrossRef] [PubMed]
33. Imran, M.; Ramzan, M.; Qureshi, A.; Khan, M.; Tariq, M. Emerging Applications of Porphyrins and Metalloporphyrins in Biomedicine and Diagnostic Magnetic Resonance Imaging. *Biosensors* **2018**, *8*, 95. [CrossRef]
34. Waghorn, P.A. Radiolabelled porphyrins in nuclear medicine. *J. Label. Comp. Radiopharm.* **2014**, *57*, 304–309. [CrossRef]
35. Chen, Z.; Mai, B.; Tan, H.; Chen, X. Nucleic Acid Based Nanocomposites and Their Applications in Biomedicine. *Compos. Commun.* **2018**, *10*, 194–204. [CrossRef]
36. Yu, W.; Zhen, W.; Zhang, Q.; Li, Y.; Luo, H.; He, J.; Liu, Y.-M. Porphyrin-Based Metal-Organic Frameworks compounds as a promising nanomedicine in photodynamic therapy. *ChemMedChem* **2020**, *15*, 1766–1775. [CrossRef]
37. Zhu, Y.; Chen, J.; Kaskel, S. Porphyrin-Based Metal-Organic Frameworks for Biomedical Applications. *Angew. Chem. Int. Ed.* **2020**, *60*, 5010–5035. [CrossRef]
38. Stojiljkovic, I.; Evavold, B.D.; Kumar, V. Antimicrobial properties of porphyrins. *Expert Opin. Investig. Drugs* **2001**, *10*, 309–320. [CrossRef]
39. Ruthard, C.; Schmidt, M.; Gröhn, F. Porphyrin-Polymer Networks, Worms, and Nanorods: pH-triggerable Hierarchical Self-assembly. *Macromol. Rapid Commun.* **2011**, *32*, 706–711. [CrossRef]
40. Zhao, L.; Qu, R.; Li, A.; Ma, R.; Shi, L. Cooperative self-assembly of porphyrins with polymers possessing bioactive functions. *ChemComm* **2016**, *52*, 13543–13555. [CrossRef] [PubMed]
41. Lu, Y.; Berry, S.M.; Pfister, T.D. Engineering novel metalloproteins: Design of metal-binding sites into native protein scaffolds. *Chem. Rev.* **2001**, *101*, 3047–3080. [CrossRef]
42. Zhang, Y.; Xu, C.; Li, B. Self-assembly of hemin on carbon nanotube as highly active peroxidase mimetic and its application for biosensing. *RSC Adv.* **2013**, *3*, 6044. [CrossRef]
43. Qu, R.; Shen, L.; Chai, Z.; Jing, C.; Zhang, Y.; An, Y.; Shi, L. Hemin-block copolymer micelle as an artificial peroxidase and its applications in chromogenic detection and biocatalysis. *ACS Appl. Mater. Interfaces* **2014**, *6*, 19207–19216. [CrossRef]
44. Alsharabasy, A.M.; Pandit, A.; Farràs, P. Recent advances in the design and sensing applications of hemin/coordination polymer-based nanocomposites. *Adv. Mater.* **2020**, *33*, 2003883. [CrossRef]
45. Nitzan, Y.; Ladan, H.; Gozansky, S.; Malik, Z. Characterization of hemin antibacterial action on Staphylococcus aureus. *FEMS Microbiol. Lett.* **1987**, *48*, 401–406. [CrossRef]
46. Walker, B.W.; Lara, R.P.; Mogadamd, E.; Yub, C..; Kimball, W.; Annabi, N. Rational design of microfabricated electroconductive hydrogels for biomedical applications. *Prog. Polym. Sci.* **2019**, *92*, 135–157. [CrossRef] [PubMed]

47. Munj, H.R.; Nelson, M.T.; Karandikar, P.S.; Lannutti, J.J.; Tomasko, D.L. Biocompatible electrospun polymer blends for biomedical applications. *J. Biomed. Mater. Res. Part B Appl. Biomater.* **2014**, *102*, 1517–1527. [CrossRef] [PubMed]
48. Avossa, J.; Paolesse, R.; Di Natale, C.; Zampetti, E.; Bertoni, G.; De Cesare, F.; Macagnano, A. Electrospinning of Polystyrene/Polyhydroxybutyrate Nanofibers Doped with Porphyrin and Graphene for Chemiresistor Gas Sensors. *Nanomaterials* **2019**, *9*, 280. [CrossRef]
49. Tyubaeva, P.; Varyan, I.; Lobanov, A.; Olkhov, A.; Popov, A. Effect of the Hemin Molecular Complexes on the Structure and Properties of the Composite Electrospun Materials Based on Poly(3-hydroxybutyrate). *Polymers* **2021**, *13*, 4024. [CrossRef]
50. Ol'khov, A.A.; Tyubaeva, P.M.; Zernova, Y.N.; Kurnosov, A.S.; Karpova, S.G.; Iordanskii, A.L. Structure and Properties of Biopolymeric Fibrous Materials Based on Polyhydroxybutyrate–Metalloporphyrin Complexes. *Russ. J. Gen. Chem.* **2021**, *91*, 546–553. [CrossRef]
51. Pramual, S.; Assavanig, A.; Bergkvist, M.; Batt, C.A.; Sunintaboon, P.; Lirdprapamongkol, K.; Niamsiri, N. Development and characterization of bio-derived polyhydroxyalkanoate nanoparticles as a delivery system for hydrophobic photodynamic therapy agents. *J. Mater. Sci. Mater. Med.* **2015**, *27*, 40. [CrossRef] [PubMed]
52. Tyubaeva, P.M.; Varyan, I.A.; Zykova, A.K.; Yarysheva, A.Y.; Ivchenko, P.V.; Olkhov, A.A.; Arzhakova, O.V. Bioinspired Electropun Fibrous Materials Based on Poly-3-Hydroxybutyrate and Hemin: Preparation, Physicochemical Properties, and Weathering. *Polymers* **2022**, *14*, 4878. [CrossRef] [PubMed]
53. Tyubaeva, P.; Varyan, I.; Krivandin, A.; Shatalova, O.; Karpova, S.; Lobanov, A.; Olkhov, A.; Popov, A. The Comparison of Advanced Electrospun Materials Based on Poly(-3-hydroxybutyrate) with Natural and Synthetic Additives. *J. Funct. Biomater.* **2022**, *13*, 23. [CrossRef] [PubMed]
54. Adler, A.D.; Longo, F.R.; Kampas, F.; Kim, J. On the preparation of metalloporphyrins. *J. Radioanal. Nucl. Chem.* **1970**, *32*, 2443–2445. [CrossRef]
55. Lubasova, D.; Martinova, L. Controlled Morphology of Porous Polyvinyl Butyral Nanofibers. *J. Nanomater.* **2011**, *2011*, 292516. [CrossRef]
56. You, Y.; Youk, J.H.; Lee, S.W.; Min, B.M.; Lee, S.J.; Park, W.H. Preparation of porous ultrafine PGA fibers via selective dissolution of electrospun PGA/PLA blend fibers. *Mater. Lett.* **2006**, *60*, 757–760. [CrossRef]
57. Krivandin, A.V.; Solov'eva, A.B.; Glagolev, N.N.; Shatalova, O.V.; Kotova, S.L. Structure alterations of perfluorinated sulfocationic membranes under the action of ethylene glycol (SAXS and WAXS studies). *Polymer* **2003**, *44*, 5789–5796. [CrossRef]
58. Shibryaeva, L.S.; Shatalova, O.V.; Krivandin, A.V.; Tertyshnaya, Y.V.; Solovova, Y.V. Specific structural features of crystalline regions in biodegradable composites of poly-3-hydroxybutyrate with chitosan. *Russ. J. Appl. Chem.* **2017**, *90*, 1443–1453. [CrossRef]
59. Vyazovkin, S.; Koga, N.; Schick, C.V. *Handbook of Thermal Analysis and Calorimetry, Applications to Polymers and Plastics*; Elsevier: Amsterdam, The Netherlands; Boston, MA, USA; London, UK, 2002.
60. Scandola, M.; Focarete, M.L.; Adamus, G.; Sikorska, W.; Baranowska, I.; Świerczek, S.; Jedliński, Z. Polymer blends of natural poly(3-hydroxybutyrate-co-3-hydroxyvalerate) and a synthetic atactic poly(3-hydroxybutyrate). characterization and biodegradation studies. *Macromolecules* **1997**, *30*, 2568–2574. [CrossRef]
61. Hantel, M.M.; Armstrong, M.J.; Rosa, F.; l' Abee, R. Characterization of Tortuosity in Polyetherimide Membranes Based on Gurley and Electrochemical Impedance Spectroscopy. *J. Electrochem. Soc.* **2016**, *164*, A334–A339. [CrossRef]
62. Arzhakova, O.V.; Nazarov, A.I.; Solovei, A.R.; Dolgova, A.A.; Kopnov, A.Y.; Chaplygin, D.K.; Tyubaeva, P.M.; Yarysheva, A.Y. Mesoporous Membrane Materials Based on Ultra-High-Molecular-Weight Polyethylene: From Synthesis to Applied Aspects. *Membranes* **2021**, *11*, 834. [CrossRef]
63. Mossman, T. Rapid colorimetric assay for cellular growth and survival: Application to proliferation and cytotoxicity assays. *J. Immunol. Meth.* **1983**, *65*, 55–63. [CrossRef] [PubMed]
64. Domaschke, S.; Zündel, M.; Mazza, E.; Ehret, A.E. A 3D computational model of electrospun networks and its application to inform a reduced modelling approach. *Int. J. Solids Struct.* **2019**, *178*, 76–89. [CrossRef]
65. Hekmati, A.H.; Rashidi, A.; Ghazisaeidi, R.; Drean, J.-Y. Effect of needle length, electrospinning distance, and solution concentration on morphological properties of polyamide-6 electrospun nanowebs. *Text. Res. J.* **2013**, *83*, 1452–1466. [CrossRef]
66. Acevedo, F.; Villegas, P.; Urtuvia, V.; Hermosilla, J.; Navia, R.; Seeger, M. Bacterial polyhydroxybutyrate for electrospun fiber production. *Int. J. Biol. Macromol.* **2018**, *106*, 692–697. [CrossRef]
67. Szewczyk, P.K.; Stachewicz, U. The impact of relative humidity on electrospun polymer fibers: From structural changes to fiber morphology. *Adv. Colloid Interface Sci.* **2020**, *2020*, 102315. [CrossRef]
68. Cobntbekt, J.; Mabchessault, R.H. Physical properties of poly-β-hydroxybutyrate. *J. Mol. Biol.* **1972**, *71*, 735–756. [CrossRef]
69. Yeo, J.C.C.; Muiruri, J.K.; Thitsartarn, W.; Li, Z.; He, C. Recent advances in the development of biodegradable PHB-based toughening materials: Approaches, advantages and applications. *Mater. Sci. Eng. C* **2018**, *92*, 1092–1116. [CrossRef]
70. Ricchelli, F.; Gobbo, S.; Moreno, G.; Salet, C.; Brancaleon, L.; Mazzini, A. Photophysical properties of porphyrin planar aggregates in liposomes. *Eur. J. Biochem.* **1992**, *253*, 760–765. [CrossRef]
71. Lin, X.; Li, S.; Jung, J.; Ma, W.; Li, L.; Ren, X.; Huang, T.-S. PHB/PCL fibrous membranes modified with SiO_2@TiO_2-based core@shell composite nanoparticles for hydrophobic and antibacterial applications. *RSC Adv.* **2019**, *9*, 23071–23080. [CrossRef]
72. Zozulia, O.; Korendovych, I.V. Semi-rationally designed short peptides self-assemble and bind hemin to promote cyclopropanation. *Angew. Chem. Int. Ed.* **2020**, *59*, 8108–8112. [CrossRef] [PubMed]

73. Dias, Y.J.; Robles, J.R.; Sinha-Ray, S.; Abiade, J.; Pourdeyhimi, B.; Niemczyk-Soczynska, B.; Yarin, A.L. Solution-Blown Poly(hydroxybutyrate) and ε-Poly-l-lysine Submicro- and Microfiber-Based Sustainable Nonwovens with Antimicrobial Activity for Single-Use Applications. *ACS Biomat. Sci. Eng.* **2021**, *7*, 3980–3992. [CrossRef] [PubMed]
74. Pasternack, R.F.; Giannetto, A.; Pagano, P.; Gibbs, E.J. Self-assembly of porphyrins on nucleic acids and polypeptides. *J. Am. Chem. Soc.* **1991**, *113*, 7799–7800. [CrossRef]

nanomaterials

Article

Development and Efficacy Evaluation of a Novel Nano-Emulsion Adjuvant for a Foot-and-Mouth Disease Virus-like Particles Vaccine Based on Squalane

Xiaoni Shi [1,2,3], Kun Yang [2], Hetao Song [2], Zhidong Teng [2], Yun Zhang [2], Weihao Ding [2], Aofei Wang [2], Shuzhen Tan [2], Hu Dong [2], Shiqi Sun [2], Yonghao Hu [1,*] and Huichen Guo [2,*]

1 College of Veterinary Medicine, Gansu Agricultural University, Lanzhou 730070, China
2 State Key Laboratory of Veterinary Etiological Biology, College of Veterinary Medicine, Lanzhou University, Lanzhou Veterinary Research Institute, Chinese Academy of Agricultural Sciences, Lanzhou 730046, China
3 School of Chemical Engineering, Lanzhou City University, Lanzhou 730070, China
* Correspondence: yhh0817@126.com (Y.H.); guohuichen@caas.cn (H.G.)

Abstract: The successful development of foot-and-mouth disease virus-like particles (FMD-VLPs) has opened a new direction for researching a novel subunit vaccine for foot-and-mouth disease (FMD). Therefore, it is urgent to develop an adjuvant that is highly effective and safe to facilitate a better immune response to be pair with the FMD-VLP vaccine. In this research, we prepared a new nano-emulsion adjuvant based on squalane (SNA) containing CpG using the pseudo-ternary phase diagram method and the phase transformation method. The SNA consisted of Span85, Tween60, squalane, polyethene glycol-400 (PEG400) and CpG aqueous solution. The average particle diameter of the SNA was about 95 nm, and it exhibited good resistance to centrifugation, thermal stability, and biocompatibility. Then, SNA was emulsified as an adjuvant to prepare foot-and-mouth disease virus-like particles vaccine, *BALB/c* mice and guinea pigs were immunized, and we evaluated the immunization effect. The immunization results in mice showed that the SNA-VLPs vaccine significantly increased specific antibody levels in mice within 4 weeks, including higher levels of IgG1 and IgG2a. In addition, it increased the levels of IFN-γ and IL-1β in the immune serum of mice. Meanwhile, guinea pig-specific and neutralizing antibodies were considerably increased within 4 weeks when SNA was used as an adjuvant, thereby facilitating the proliferation of splenic lymphocytes. More importantly, in guinea pigs immunized with one dose of SNA-VLPs, challenged with FMDV 28 days after immunization, the protection rate can reach 83.3%, which is as high as in the ISA-206 control group. In conclusion, the novel squalane nano-emulsion adjuvant is an effective adjuvant for the FMD-VLPs vaccine, indicating a promising adjuvant for the future development of a novel FMD-VLPs vaccine.

Keywords: nano-emulsion; adjuvants; FMD virus-like particles; squalane; immune responses

Citation: Shi, X.; Yang, K.; Song, H.; Teng, Z.; Zhang, Y.; Ding, W.; Wang, A.; Tan, S.; Dong, H.; Sun, S.; et al. Development and Efficacy Evaluation of a Novel Nano-Emulsion Adjuvant for a Foot-and-Mouth Disease Virus-like Particles Vaccine Based on Squalane. *Nanomaterials* **2022**, *12*, 3934. https://doi.org/10.3390/nano12223934

Academic Editors: Goran Kaluđerović and Nebojša Pantelić

Received: 16 September 2022
Accepted: 5 November 2022
Published: 8 November 2022

1. Introduction

Foot-and-mouth disease (FMD) is a disease that spreads rapidly in even-toed animals [1]. In most countries, animals are immunized with the whole virus inactivated vaccines to control the virus; however, safety is a concern [2]. Adjuvants provide an important way to improve the efficacy of FMD vaccines. Hence, the search for specific and targeted adjuvants combined with protective antigens is a new direction for developing novel FMD vaccines [3].

The development of virus-like particles (VLPs) technology has strongly impacted modern vaccinology. Though morphologically similar to native viral particles, VLPs show higher and safer efficiency in stimulating the immune system because they lack replicable viral genetic material [4]. VLPs can be generated using recombinant DNA technology in various exogenous gene expression systems, including yeast, bacteria, mammalian cells, baculovirus systems, plant cell cultures, or plant organisms. VLPs-based vaccines are not only particularly effective and safe, but also have a low cost and can be produced at scale [5]. Therefore, VLPs are expected to be ideal candidates for vaccine development [6]. Compared to inactivated vaccines, VLPs alone do not induce a sufficient specific immune response; they also require the appropriate adjuvants to enhance their immune response. Compared to inactivated and attenuated vaccines, the FMD virus-like particle vaccine is a new safe and effective vaccine [7]. Therefore, it is necessary to identify a new adjuvant with high efficiency, low cost, and low toxicity for FMD-VLPs.

Emulsion adjuvants are usually formulated from oils, such as mineral oil (e.g., Montanide) and squalene (e.g., MF59) and surfactants. They are available in water-in-oil, oil-in-water and water-in-oil-in-water dispersion forms [8]. The oil emulsion adjuvant has a high antigen adsorption capacity and can bind to different types of antigens [9]. By adding immune boosters, nano-emulsion adjuvants can stimulate both humoral and cellular immunity in the body; therefore, they have become one of the more widely used adjuvants in animal vaccines. Montanide ISA-206, a mineral oil-based adjuvant, is produced by Seppic (Shanghai) Chemical Specialities Co., Ltd. (Shanghai, China)., and is presently used for formulating FMD vaccines in many South American and Asian countries.

Squalane is a saturated aliphatic hydrocarbon with low toxicity, derived from the hydrogenation of squalene, which is found in cod liver oil, rice, olives, and soybeans [10]. Based on its strong stability and biocompatibility, squalane is currently used in many vaccines and drug delivery emulsions [11]. For instance, emulsions MF59 (Novartis, Basel, Switzerland), AS03 (GlaxoSmithKline, Brentford, UK), and AF03 (Sanofi, Paris, France) are squalene-based and have been used as adjuvants in anti-influenza virus vaccines [12]. Whether squalane can act as an effective adjuvant to enhance the immune response in an FMD-VLPs vaccine remains unclear.

Compared with conventional emulsions, nano-emulsion has a low viscosity, small particle size, good stability and fewer toxic side effects [13]. As a new type of drug carrier, nano-emulsion has many advantages that are incomparable to other drug carriers. Due to these characteristics, they show attractive prospects for development in the field of biologics.

In this study, a basic formulation of a squalane nano-emulsion adjuvant (SNA) was developed by a pseudo-ternary phase diagram, and CpG was added as an immune booster to the water phase. The physicochemical properties of the adjuvant were tested. Furthermore, SNA was emulsified with FMD VLPs, and animal experiments were performed to evaluate the immune response of these VLPs.

2. Materials and Methods

2.1. Selection of the Surfactant and Cosurfactant

Based on preliminary experiments, we chose squalane (Acmec, Shanghai, China) as the oil phase, Span85 (Aladdin, Shanghai, China) and Tween60 (Sigma-Aldrich, Saint Louis, MO, USA) as the surfactant, polyethene glycol-400 (PEG-400, Acmec, Shanghai, China) as the cosurfactant, and deionized water as the aqueous phase. We used 0.8 g of PEG-400 and 1.2 g of the Span85 and Tween60. In the reaction system, the surfactant and co-surfactant were blended with the oil phase at ratios of 1:9, 2:8, 3:7, 4:6, 5:5, 6:4, 7:3, 8:2, and 9:1. In the surfactant, Span85 and Tween 60 were blended in the ratios of 1:1, 1:2, and 2:1; the appropriate ratio of Span85 to Tween60 was chosen by recording the maximum area of the pseudo-ternary phase diagram [14]. Origin 6.0 software (Origin Lab, Northampton, MA, USA) was used to plot a pseudo-ternary phase diagram, compare the size of the nano-emulsion area, and select the optimal ratio.

After determining the mass ratio of the two surfactants, the mass ratio of the surfactant to the co-surfactant (Km) was further determined. The surfactant (Span85 and Tween60) was coupled with PEG-400 in three groups under different fixed mass ratios of Km (1:1, 2:1, and 1:2), and the total quantity of the mixture (the surfactant and co-surfactant) was maintained at 2.0 g. For each group, squalane was added and mixed well with the surfactant and cosurfactant. The mass ratio of the surfactant and cosurfactant to oil ranged from 1:9, 2:8, 3:7, 4:6, 5:5, 6:4, 7:3, 8:2, to 9:1. The phase diagrams under different Km were developed.

Selection of Span85 and Tween60 ratios, and selection of surfactant and co-surfactant ratios information is provided in the Supplementary Materials.

2.2. Preparation of the SNA

From the pseudoternary-diagram, the precise composition was selected for the nano-emulsion. Briefly, we made a mixture by weighing a certain mass of Span85, Tween60, PEG-400, and squalane in proportion to each other. The aqueous phase (with the addition of CpG) was added to the mixture while stirring at 1000 rpm at room temperature until a clear and transparent emulsion formed; then, stirring was continued to ensure that the nano-emulsion was stable.

2.3. Characterization of the SNA

The ultrastructure and morphology of the SNA were observed by transmission electron microscope (TEM; HT7700, Hitachi, Tokyo, Japan). The SNA for analysis was diluted 100 times by water and magnetically stirred well; then, 10 μL of drops were placed on a carbon copper grid (300 mesh; Pelco, CA, USA). Next, the samples were allowed to stand at room temperature for 5 min; then, we added 10 μL of 1% phosphotungstic acid (pH 7.4, solarbio, Beijing, China) solution, let it dry naturally and observed it. The average size and zeta potential were measured by dynamic light scattering (DLS) with a Zetasizer-Nano (Malvern Zetasizer Nano ZS90; Worcestershire, UK).

2.4. Stability Assessment of the SNA

2.4.1. High-Speed Centrifuge Stability and Thermodynamic Stability

Thermodynamic stability was tested according to the methods described previously [15,16]. Briefly, the SNA prepared freshly were centrifuged at 13,000× *g* for 30 min at 25 °C and 4 °C, maintain for 48 h. Six cycles of centrifugation were performed. After centrifugation, we observed whether phase separation and precipitation occurred.

2.4.2. Long-Term Stability Test

The SNA was stored at 25 °C for 12 months in the dark. Samples were taken at 0, 3, 6, 9 and 12 months to observe the stratification, precipitation and turbidity. We measured the average particle size, zeta potential, PDI and pH, and the microscopic morphology was observed by transmission electron microscopy.

2.5. Biocompatibility Evaluation of the SNA

2.5.1. Tissue Toxicity

Balb/c female mice were purchased from the Experimental Animal Center of Lanzhou Veterinary Research Institute (Lanzhou, China) (age = 8 weeks; weight = 15–20 g). Two experimental groups were designed in this study: the PBS control group (each mouse was injected with 100 μg sterile PBS) and the SNA experimental group (each mouse was injected with 100 μg SNA) (n = 8 in each group). We observed whether the injection site was red and swollen, and recorded the weekly weight gain of the mice. Approximately 28 days later, the mice were anaesthetized with ether. Then, the heart, liver, spleen, lung and kidney tissues were removed, preserved in 4% paraformaldehyde, embedded in paraffin, and stained with H&E.

2.5.2. Cytotoxicity

PK-15 (Porcine kidney) cells were inoculated into 96-well cell culture plates. Approximately 24 h later, different concentrations of SNA (0 μg/mL, 25 μg/mL, 50 μg/mL, 100 μg/mL, 200 μg/mL, 400 μg/mL, and 800 μg/mL) were added to the cells at 10 μL per well, which were incubated for 24 h. Then, 10 μL MTS (Promega, Madison, WI, USA) reagent was added to each well and incubated for 4 h. The absorbance was measured by an enzyme marker. The cell viability was calculated with the following equation:

$$\text{Cell survival } (\%) = \frac{\text{absorbance value of treatment group}}{\text{absorbance value of control group}} \times 100\%$$

2.6. *Preparation of the FMD-VLP Vaccine with SNA*

The expression, purification and assembly of FMD-VLPs in the *Escherichia coli* system have been described in our previous studies [7,17]. Expression, purification, and assembly of FMD-VLPs can be found in the Supplementary Materials. Initially, the mass ratios of adjuvant to antigen phase (FMD-VLP protein solution) was selected as 1:1, 2:3, 1:2, and 3:2 and emulsified by a shear machine at room temperature. After one week, the non-stratified one was selected as having the best emulsification ratio. After several experiments, we finally determined the mass ratio of the adjuvant-to-antigen phase to be 1:1.

2.7. *Immunization Studies in BALB/c Female Mice*

BALB/c female mice (6–8 weeks old, 15–20 g) were maintained in a specific pathogen-free (SPF) laboratory and separated into four groups, each including five animals: Group1 was immunized with sterile phosphate-buffered saline (PBS); Group2 was immunized with the 50 μg FMD VLPs; Group3 was immunized with the vaccine containing 50 μg of FMD VLPs and emulsified with the same volume of SNA; Group4 was immunized with a vaccine emulsified with ISA206 adjuvant (containing 50 μg of FMD VLPs). All mice were immunized by intramuscular injection, and serum was collected 7, 14, 21 and 28 days after immunization. The specific antibody levels were detected by using a liquid phase blocking ELISA kit (Lanzhou Veterinary Research Biotechnology Co, Lanzhou, China).

Indirect ELISA evaluated the levels of the specific antibodies, IgG1 and IgG2a in serum. Briefly, microtiter plates (Coster, Corning, NY, USA) were coated with FMDV VLPs (2 μg/mL) in coating buffer (0.05 M CBS, pH 9.6) at 4 °C overnight and then blocked with BSA (1%, m). Then, the plates were washed with PBST (10 mM PBS containing 0.05% Tween 20, pH 7.4) and dried for the subsequent procedure. For determination of IgG1 and IgG2a sera were diluted 1/200 and incubated at 37 °C for 1 h; goat anti-mouse IgG1 andIgG2a (Sigma-Aldrich, Saint Louis, MO, USA) were diluted 1/1000 and incubated at 37 °C for 60 min; HRP-conjugated rabbit anti goat IgG (Sigma-Aldrich, Saint Louis, MO, USA) was diluted 1/5000 and incubated at RT for 30 min; the enzyme substrate 3,3′,5,5′-tetramethylbenzidine (TMB, Surmodics IVD Inc., MN, USA) was added as described by the manufacturer. Then, the absorbance was measured at 450 nm with a microplate reader (Bio-Tek, Winooski, VT, USA) after the reaction was stopped with the stop buffer of sulfuric acid.

The cytokine levels of the IL-1β and IFN-γ in the serum were detected by Quantikine® ELISA kits (R&D Systems, Inc., Minneapolis, MN, USA).

2.8. *Guinea Pig Immunization with FMD-VLP Vaccine with SNA*

2.8.1. Animal Vaccination

Twenty-four guinea pigs, 250–300 g each, were purchased from the laboratory animal centre of Lanzhou Veterinary Laboratory Experimental, China, which were randomly divided into three groups, each containing eight guinea pigs: Group A, SNA-VLPs (FMD-VLPs emulsified with SNA); Group B, ISA206-VLPs (FMD-VLPs emulsified with ISA206); and Group C, phosphate buffered saline (PBS, pH 7.4). Each group received 50 μg of FMD-VLPs (except the PBS group).

2.8.2. Detection of Specific Antibodies and Cytokine Levels

Specific antibody titres of immunized guinea pigs were determined by indirect ELISA as previously described [7]. Briefly, O-type inactivated FMDV was diluted with a coating solution (0.05 M bicarbonate buffer, pH 9.6), and 100 μL was added to each well of a 96-well plate overnight at 4 °C. Then, 100 μL of PBS containing 5% BSA was added to each well for 1 h at 37 °C to be blocked, washed and drained. One-hundred fold dilutions of tested sera were added to the 96-well plate and were incubated at 37 °C for 1 h. Afterwards, the sample sera were removed and washed. Horseradish peroxidase (HRP)-conjugated anti-guinea pig antibody (1:2000) (Sigma, St. Louis, MO, USA) was added and incubated at 37 °C for 1 h. Then, 50 μL of the enzyme substrate o-phenylenediamine (OPD, Sigma, St. Louis, MO, USA) in sodium citrate was added to each well and incubated for 15 min at room temperature. Finally, the reaction was stopped with 50 μL of 2 M H_2SO_4 and the OD value was read at 492 nm. Antibody reactivity was reported as OD values.

The cytokine secretion levels of the IL-1β and IFN-γ were measured with an ELISA kit (Shanghai MLBIO Biotechnology Co. Ltd, Shanghai, China), according to the manufacturer's instructions.

2.8.3. Detection of the Neutralizing Antibodies

The guinea pig serum to be tested was inactivated at 56 °C for 30 min, with eight gradients of fold dilution, starting at 1:4 in 96-well cell culture plates, with two replicates for each dilution. We added 100 TCID50 of the FMDV type O strain O/China99 to each well of diluted serum and incubate for 1 h at 37 °C. A positive control, negative control, cell control, and virus regression control were also designed. Then, 100 μL of a Baby Hamster Syrian Kidney (BHK-21) cells mixture was added to each well and incubated in a CO_2 incubator for 72 h. Finally, cell lesions were observed under an inverted microscope. The Reed–Munch method calculated the highest dilution of the serum that protected 50% of the cells from cytopathic lesions. The dilution reflected the potency of the serum to neutralize the antibody.

2.8.4. T-Cell Proliferation Assay

The lymphocyte proliferation assay was performed four weeks after immunization, as previously described [18].

2.8.5. Challenge Protocols

The challenge protocols in guinea pigs were performed 28 days after immunization; all guinea pigs were subcutaneously and intradermally challenged with 0.2 mL 100 ID50 of live virus (FMDV/O/China99) on the left back sole. After the attack, the guinea pigs were kept in isolation and observed for more than 7 days. If the guinea pig had no lesions on either hind foot, the set-up was indicated as "protected". If lesions were present on both hind feet, it was indicated as "no protected" [19].

$$\text{Protection rate } (\%) = \frac{\text{Full protection guinea pigs}}{\text{Total number of guinea pigs}} \times 100\%$$

2.9. *Statistical Analysis*

We analysed all data using SPSS 22.0. The *t*-test was used to determine the difference between the two sample groups. Differences at $0.01 < p \leq 0.05$ were considered statistically significant, whereas those at $p \leq 0.01$ were considered highly significant.

3. Results

3.1. *Preparation of the SNA*

In the pseudo-ternary phase diagram, the first axis represents the oil phase, the second axis represents the aqueous phase, and the last axis represents the mixture of surfactant and co-surfactant. The size of the enclosed space formed by the three components is related to the ability to form nano-emulsions, with larger areas indicating greater ability to form nano-emulsions. According to Figure 1, the size of the area for the Smix (Smix represents the ratio of Span85 and Tween60) = 2:1 (Figure 1c) was larger than those for the Smix = 1:1

(Figure 1a) and Smix = 1:2 (Figure 1b). Based on the largest black area in the phase diagram (Figure 1d), Smix = 2:1 was accepted and applied for the preparation of nano-emulsions.

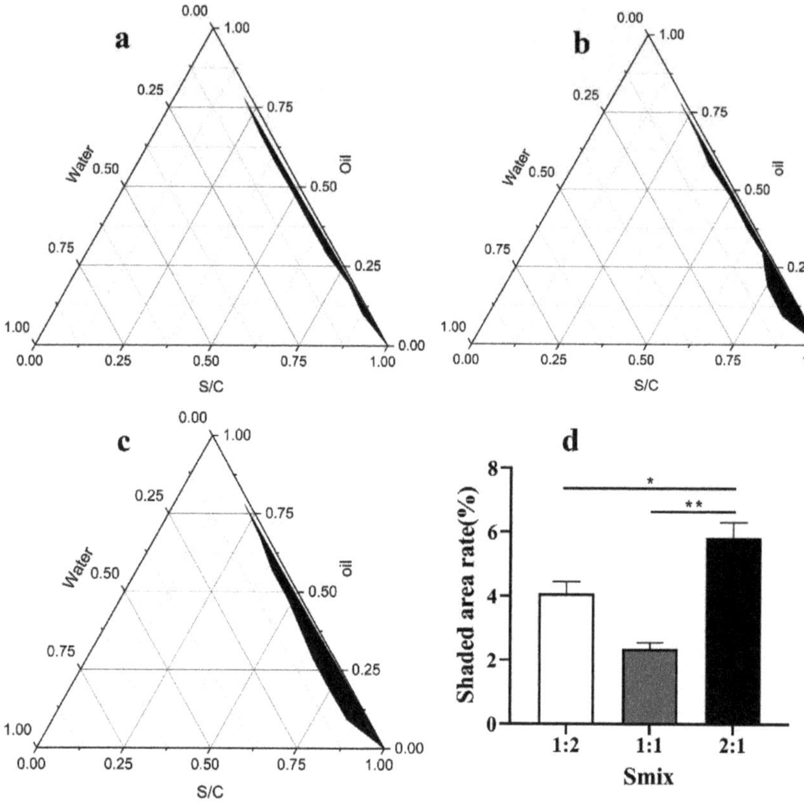

Figure 1. Pseudo-ternary phase diagram obtained with different ratios of Span85 and Tween60. (**a**) Span85:Tween60 = 1:1. (**b**) Span85:Tween60 = 1:2. (**c**) Span85:Tween60 = 2:1. (**d**) The ratio of the shaded area to the area of the ternary phase diagram for the different Smix values (based on the (**a**–**c**)). * $p < 0.05$, ** $p < 0.01$. Notes: S/C represents the total mass of the surfactant and co-surfactant. Smix represents the relative ratio of Span85 and Tween60. Span85 and Tween60 form a surfactant complex.

The pseudo-ternary phase diagrams for different Km (Km represents the ratio of surfactants and co-surfactants) are shown in Figure 2. The area for Km = 3:2 (Figure 2c) was larger than those for Km = 1:1 (Figure 2a) and Km = 2:1 (Figure 2b). Based on the largest black area in the phase diagram (Figure 2d), Km = 3:2 was accepted and applied for the formation of the nano-emulsions.

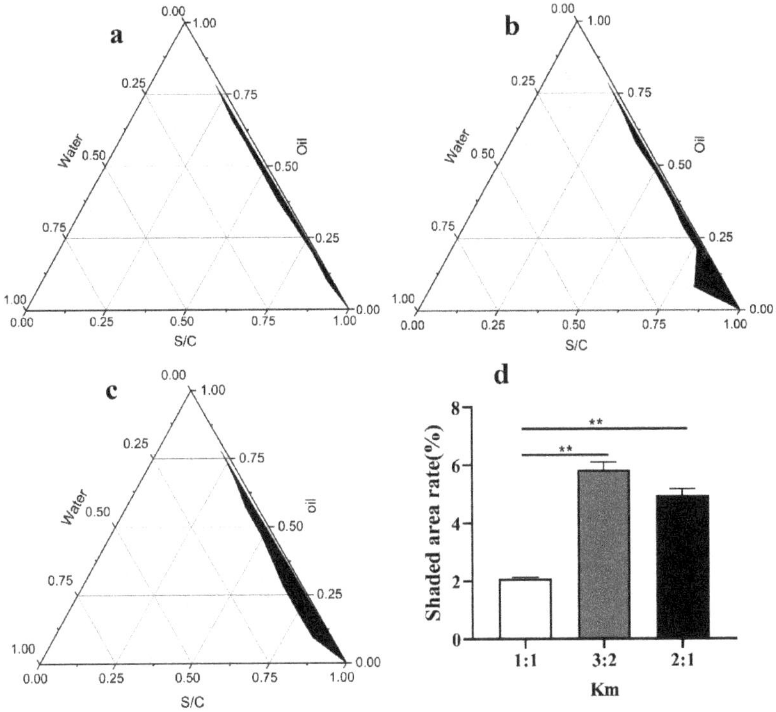

Figure 2. Pseudo-ternary phase diagram for different Km values (**a**) Km = 1:1. (**b**) Km = 2:1. (**c**) Km = 3:2. (**d**) The ratio of the shaded area to the area of the ternary phase diagram for the different Km values. (based on the (**a**–**c**)). ** $p < 0.01$. Notes: Km indicates the relative ratio of the surfactant (Span85 and Tween60) and the co-surfactant (PEG-400).

3.2. *Characterization of the SNA*

The nano-emulsions observed by transmission electron microscopy were uniformly sized circles (Figure 3a), with a particle size of about 100 nm. The morphology of FMD-VLPs is shown in Figure 3c. The SNA prepared by the phase conversion method is shown in Figure 3d, and the FMDV-VLP vaccine emulsified with SNA is shown in Figure 3e.

The average particle size of SNA was 95 nm; about 75% of the nano-emulsion particles were concentrated between 68 nm and 105 nm, with a very narrow particle size distribution (Figure 3b), which were consistent with the results obtained from TEM. In addition, we determined the polydispersity index (PDI) of nano-emulsion SNA with a value of 0.4; this PDI value indicates that dispersion is within acceptable limits, and the average potential of the nano-emulsion was −25.76 mV.

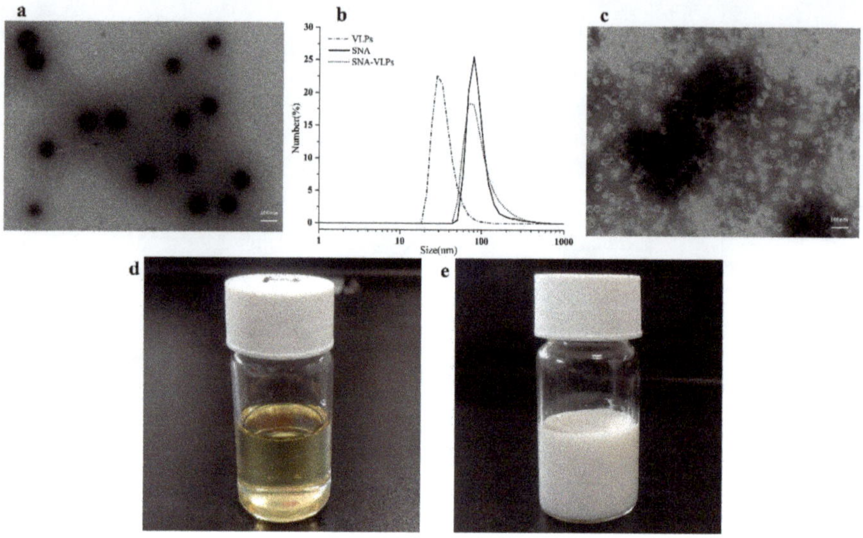

Figure 3. Basic characteristics of the SNA adjuvant. (**a**) The TEM images of the freshly formulated SNA, (**b**) The particle size of the SNA, SNA-VLPs, and FMD-VLPs were measured by DLS. (**c**) The TEM images of the FMD-VLPs. (**d**) A new preparation of SNA. (**e**) A new preparation of the FMD-VLP-SNA vaccine.

3.3. Stability of the SNA

Long-term stability is an essential basis for product preservation conditions and validity. Figure 4a–f show the transmission electron micrographs of the SNA with preservation periods of 0, 90, 180, 270, and 360 days, respectively; Figure 4f–h and Figure 4i show the particle size zeta potential, PDI, and pH for different preservation days, respectively. These results showed that the SNA prepared in this experiment did not change significantly in appearance properties when left at 25 °C for 360 d in the dark. This emulsion was stable at room temperature.

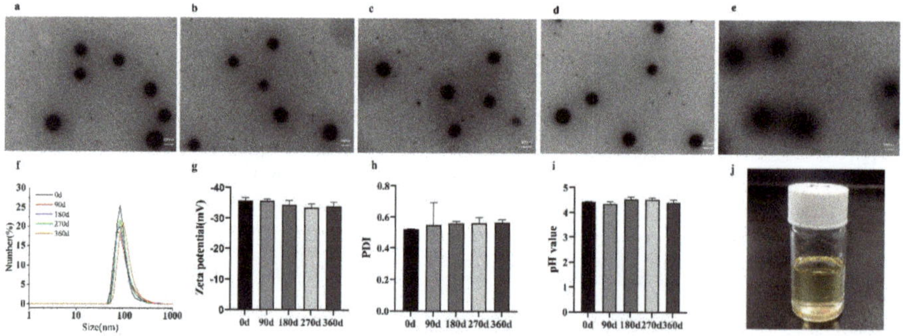

Figure 4. Stability testing of the SNA adjuvants. (**a**) SNA transmission electron micrograph of 0 days. (**b**) SNA transmission electron micrograph of 90 days. (**c**) SNA transmission electron micrograph of 180 days. (**d**) SNA transmission electron micrograph of 270 days. (**e**) SNA transmission electron micrograph of 360 days. (**f**) The particle size for different preservation periods. (**g**) The zeta potential for different preservation periods. (**h**) The PDI for different preservation periods. (**i**) The pH for different preservation periods. (**j**) The SNA after the high−speed centrifugal and thermodynamic stability tests.

The high-speed centrifugation and thermodynamic stability test confirmed that the nano-emulsion was as transparent and homogeneous as the fresh nano-emulsion, with no turbidity and no precipitation appearing (Figure 4j).

3.4. Safety Evaluation of the SNA

Within 48 h after injection, the mice took food and water as normal, and there was no redness or swelling at the injection site. Furthermore, the weight gain trends of the mice in different injection groups were similar, indicating that the injection of SNA had no significant effect on the weight gain of the mice (Figure 5a).

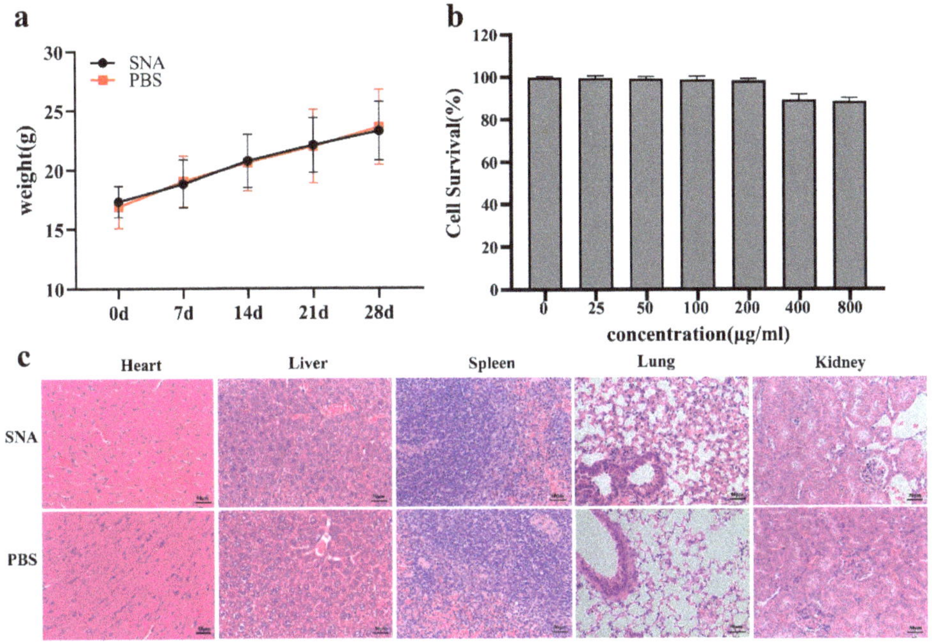

Figure 5. Safety evaluation of the SNA adjuvants. (**a**) The body weight of the immunized mice at 0, 7, 14, 21, and 28 days after immunization. (**b**) The cytotoxicity of different concentration of SNA on the PK-15 cells. (**c**) Histopathological changes in the heart, liver, spleen, lung and kidney of mice after intramuscular injection of SNA and PBS.

Twenty-eight days after injection, the pathological changes of the tissue organs were analysed. Figure 5c shows that no tissue structure injuries or neutrophil infiltration occurred in the SNA group compared to the PBS-immunized group. This indicates that the SNA injection into mice did not cause significant pathological histological toxicity. The PK-15 cells were incubated with different dilution concentrations of SNA, and the results are shown in Figure 5b. The survival rate of the PK-15 cells all remained high when the SNA dilution concentration was as high as 400 μg/mL, which indicates that the SNA had little cytotoxicity in the PK-15 cells (Figure 5b). All the results confirmed that the prepared SNA had good biocompatibility.

3.5. Effects of the FMDV-VLP Vaccine with SNA on the Antibody Production in BALB/c Mice

After several experiments, we finally determined the mass ratio of the adjuvant to antigen phase to be 1:1; based on this ratio, we prepared the SNA for the FMD-VLP vaccine and immunized the animals.

The specific antibody test results showed that the SNA-adjuvant vaccine produced higher antibodies than the VLPs and PBS groups at weeks 2–4, comparable to those produced by the ISA206-adjuvant group. This result indicates that the combination of SNA and FMDV-VLPs can significantly enhance the immune response in *BALB/c* mice as compared to VLPs alone (Figure 6a).

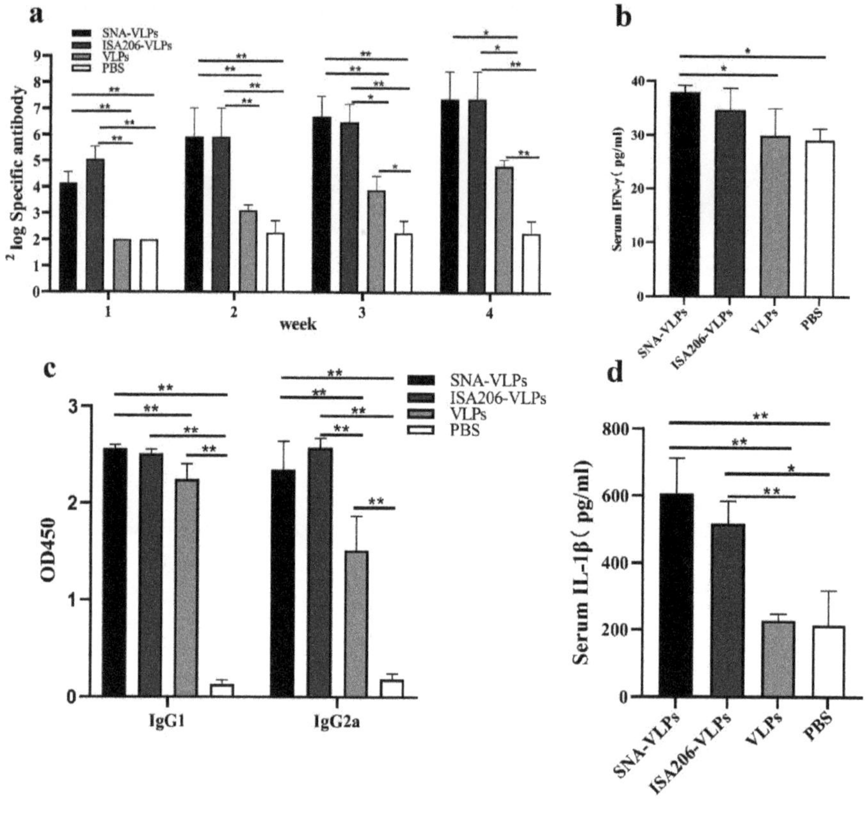

Figure 6. Results of the immunization experiments in mice. (**a**) Specific antibody levels in the mice immunized with various vaccines. (**b**) IFN-γ levels in serum. (**c**) Levels of IgG1 and IgG2a. (**d**) IL-1β levels in serum. * $p < 0.05$, ** $p < 0.01$.

The results showed that, at a serum dilution of 1:100, the SNA vaccine group produced a relatively high IgG1, comparable to the ISA206 group, and produced a higher IgG2a than the VLPs and PBS groups and slightly lower than the ISA206 group (Figure 6c). The IgG2a antibody levels reflect the Th1 immune responses, and the IgG1 antibody levels reflect the Th2 immunity. This result also indicates that the SNA-adjuvant vaccination mainly induced humoral immunity in the body but also induced different levels of cellular immune responses.

In the present mouse immunization experiment, the SNA vaccine group produced higher levels of IFN-γ than the ISA206 vaccine group and the VLPs antigen group (Figure 6b). This result again demonstrates that the SNA-adjuvant vaccine induced a Th1-type immune response. Meanwhile, the serum IL-1β in the SNA adjuvant immunization group was higher than in the ISA206-adjuvant group and the VLPs antigen group (Figure 6d).

3.6. Evaluation of Immunization Effect after Immunization of Guinea Pigs

As shown in Figure 7 and Table 1, the specific and neutralizing antibody titres were significantly higher in the SNA-VLPs immunized guinea pigs compared to the PBS group, and T lymphocyte proliferation was also promoted. However, compared to the ISA206-VLPs group, there were no significant differences in the specific antibody titres, T lymphocyte proliferation levels and protection rates in the SNA immunization group.

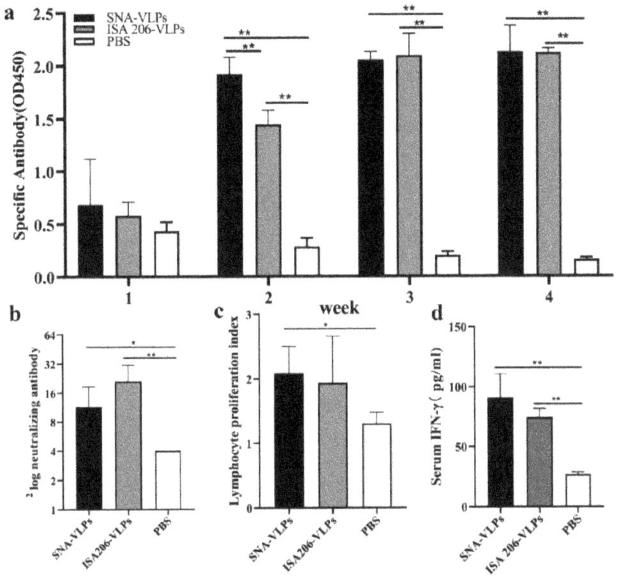

Figure 7. The results of the immunization experiments in guinea pigs. (**a**) The specific antibody levels in the guinea pigs immunized with various vaccines. (**b**) The levels of the neutralizing antibody. (**c**) The lymphocyte proliferation index. (**d**) The IFN-γ levels in serum. * $p < 0.05$, ** $p < 0.01$.

Table 1. Protection of guinea pigs after challenge with FMDV.

Group	Guinea Pigs Number	Protection	Rate of Protection (%)
FMD VLPs-SNA	6	5	83.3% (5/6)
FMD VLPs-ISA206	6	5	83.3% (5/6)
PBS	6	0	0 (0/6)

Note: When neither the inoculated limb nor the uninoculated limb and vesicles were swollen, this was judged as protection; When both the inoculated limb and the uninoculated parts were infected, this was judged as unprotected.

To confirm whether the SNA induced cellular immunity, we measured the expression levels of the related cytokines IFN-γ. The result indicated that the levels of IFN-γ (Figure 7d) were upregulated compared with the PBS group; the IFN-γ levels in the SNA-VLP group were comparable to those in the ISA206-VLP group, and there was no significant difference.

4. Discussion

Vaccination is the most effective and economical method with which to control and prevent FMD. FMD whole-virus-inactivated vaccines have been widely used in various countries, but there are inherent risks. The successful development of virus-like particles (VLPs) has opened a new direction for modern vaccines. Recombinant VLPs are a good alternative to traditional vaccines for FMD because they are non-infectious, require low production conditions, and can also be modified to improve their stability [19].

Nanomaterials with targeted delivery and high bioavailability hold promise as a new delivery system for biocontrol, disease prevention, and chemical antimicrobials [20,21]. Compared with other conventional emulsions, nano-emulsions have a particle size of 20–200 nm, good stability, easy storage, easy penetration of tissue barriers, easier uptake by dendritic cells, and fewer toxic side effects [22,23]; so, it is important to apply nano-emulsions as adjuvants in the research and development of vaccines. Joyappa et al. immunized mice and guinea pigs with a foot-and-mouth disease P1-3CD DNA vaccine loaded onto calcium phosphate nanoparticles and induced significant cellular and humoral immunity while effectively protecting immunized mice and guinea pigs from attack by foot-and-mouth disease virus [24]. Huang et al. developed a novel nano-emulsion PELC and immunized mice by combining PELC with inactivated influenza virus vaccine and found that the PELC significantly enhanced the proliferative activity of T lymphocytes, increased the secretion levels of IFN-γ and IL-4, and induced higher levels of specific antibodies [25].

Squalene oil-based adjuvants, such as MF59 and AS03, are components in commercially available influenza vaccines in Europe, with average droplet sizes in the subnanometre range of 155–160 nm [26]. In this study, we successfully prepared the squalane nano-emulsion adjuvant, SNA, with an average particle size of 95 nm, which was smaller than that of the MF59 and AS03 adjuvants. We also used it for the first time as an adjuvant for the FMD-VLP vaccine and conducted immunization experiments in mice and guinea pigs.

The key to the preparation of nano-emulsions is the selection of a suitable oil, surfactant, and co-surfactant, as well as the appropriate ratio between the oil and the mixed surfactant (surfactant and co-surfactant). It has been noted that water/oil nano-emulsions formulated with non-ionic surfactants have no phase separation over a more extended period time, are more stable, and have greater biological efficacy [27]. Therefore, in this study, we chose Span85 and Tween60 as surfactants and determined the exact ratios between the two surfactants and between the surfactant and the co-surfactant using pseudo-ternary phase diagrams. We used a simple low-energy emulsification method (phase change method) to prepare the nano-emulsions by adding the aqueous phase (antigen phase) dropwise to a mixture of oil, surfactant, and co-surfactant and gently stirring until a clear clarified emulsion appears. Therefore, this method is easy and has great advantages for large-scale production.

The stability of the oil–emulsion adjuvant is an essential indicator of the excellent adjuvant properties; we chose the high-speed centrifugation method [28], the thermodynamic stability method, and the long-term storage method to assess the stability of the SNA comprehensively. The results of all three methods proved that the SNA adjuvant had good stability with uniform particles, good dispersion, and no significant changes in zeta potential and particle size after 12 months of storage.

The safety evaluation of vaccine adjuvants is a very important link before clinical application. At present, there are not many normative documents for the safety evaluation of adjuvants in the world, and no unified evaluation standard has been formed. Currently, the safety evaluation of vaccine adjuvants mainly refers to the evaluation method of clinical drugs. In this study, the safety of the SNA was evaluated from two aspects: a cytotoxicity test in vitro and a tissue toxicity test in vivo. The cytotoxicity test showed that the SNA did not affect the growth of the PK-15 cells at medium and low concentrations. In the mouse experiment, the growth condition of mice was not affected after the nano-emulsion injection, and no pathological changes were found in the tissue sections of essential organs. Thus, these data indicate that the SNA is safe as an adjuvant for injection.

CpG has a good safety profile, enhances the antigen-presenting function of DCs, monocytes and macrophages [29], induces B-cell proliferation, indirectly stimulates the immune activity of NK cells, and significantly favours the immune response to the Th1 type [30,31]. Hence, we added CpG to this new nano-emulsion to obtain a better immune effect. It has been documented that the immunization of guinea pigs with CpG as a vaccine adjuvant encapsulated in chitosan-coated poly (lactic acid)-glycolic acid nanoparticles provided ideal protection [32]. It has also been reported that the addition of CpG to the influenza virus VLP vaccine offers complete protection against the 1918 influenza virus [33].

In mice immune experiments, we determined that the SNA adjuvant with the added CpG component increased the serum levels of IgG2a and IFN-γ, with a significant tendency to induce Th1 cell immune responses and was comparable to ISA206 in promoting IL-1β secretion, lymphocyte proliferation and attack protection, showing the same immune-enhancing effect as the positive control ISA206. The immunization and challenge experiments in guinea pigs showed that SNA as an adjuvant for the FMD-VLPs vaccine induced a stronger specific immune response against FMDV. In particular, there was no significant difference in antibody levels and protection between the SNA-VLPs-immunized and ISA206-VLPs-immunized groups. Pervaiz et al. evaluated the adjuvant role of three oil adjuvants (GAHOL, Montanide ISA-206 and ISA-201) in FMD inactivated vaccines. Among them, GAHOL is a home-made oil adjuvant. Thirty days after vaccination, 100% (6/6) of cattle immunized with the Montanide-201 adjuvanted vaccine were protected in a homologous FMD challenge, which was superior to cattle vaccinated with the ISA-206 (66.6%, 4/6) or GAHOL adjuvanted vaccine (50%, 3/6) [34]. In conclusion, compared with the results of previous studies, the SNA adjuvant prepared in the present study with FMD VLPs produced a protection rate in guinea pigs within the ideal range.

5. Conclusions

The present study indicated that the SNA, a novel nano-emulsion adjuvant of squalene, is simple to prepare and easy to produce on a large scale, has good biocompatibility and a relatively comprehensive immune-enhancing effect, and can be an effective adjuvant for FMD VLPs vaccines. Therefore, the current study indicates that SNA may become a new conventional adjuvant with good application prospects for FMD therapy. However, whether SNA is suitable for other subunit vaccines and its underlying immune mechanism need further investigation, which is our future work.

Supplementary Materials: The following supporting information can be downloaded at: https://www.mdpi.com/article/10.3390/nano12223934/s1.

Author Contributions: X.S., Y.H. and H.G. was involved in the experimental design, data analysis, and original manuscript writing. K.Y., H.S. and Z.T. assisted in writing, reviewing and editing, organizing experimental data. S.T. and Y.Z. assisted in the collection of experimental materials and editing of the manuscript. W.D., A.W. and S.S. assisted in completing animal experiments and analysing data. H.D. was involved in the collection and collation of all data and analysis. All authors have read and agreed to the published version of the manuscript.

Funding: This research was supported by grants from the National Key Research and Development Program (2021YFD1800300), and Science and Technology Major Project of Gansu Provine (21ZD3NA001), and the National Natural Science Foundation of China (32072847, 31873023, 32072859, 32002272), and Science and Technology Talents and Platform Program (202205AF150007).

Institutional Review Board Statement: All animal experiments were approved by the Animal Ethical and Experimental Committee of Lanzhou Veterinary Research Institute, Chinese Academy of Agricultural Sciences (approval number: LVRIAEC2018-008).

Data Availability Statement: Not applicable.

Acknowledgments: We are grateful for the technical support from the technical team of foot-and-mouth disease prevention and control at the Lanzhou Institute of Veterinary Medicine, Chinese Academy of Agricultural Sciences.

Conflicts of Interest: The authors declare no conflict of interest.

References

1. Perry, B.D.; Rich, K.M. Poverty impacts of foot-and-mouth disease and the poverty reduction implications of its control. *Vet. Record.* **2007**, *160*, 238–241. [CrossRef] [PubMed]

2. Pan, L.; Zhang, Z.W.; Lv, J.L.; Zhou, P.; Hu, W.F.; Fang, Y.Z.; Chen, H.T.; Liu, X.S.; Shao, J.J.; Zhao, F.R.; et al. Induction of mucosal immune responses and protection of cattle against direct-contact challenge by intranasal delivery with foot-and-mouth disease virus antigen mediated by nanoparticles. *Int. J. Nanomed.* **2014**, *9*, 5603–5618. [CrossRef] [PubMed]
3. Cao, Y.M. Adjuvants for foot-and-mouth disease virus vaccines: Recent progress. *Expert Rev. Vaccines* **2014**, *13*, 1377–1385. [CrossRef] [PubMed]
4. Tan, M.; Jiang, X. Subviral particle as vaccine and vaccine platform. *Curr. Opin. Virol.* **2014**, *6*, 24–33. [CrossRef] [PubMed]
5. Noad, R.; Roy, P. Virus-like particles as immunogens. *Trends Microbiol.* **2003**, *11*, 438–444. [CrossRef]
6. Qian, C.Y.; Liu, X.L.; Xu, Q.; Wang, Z.P.; Chen, J.; Li, T.T.; Zheng, Q.B.; Yu, H.; Gu, Y.; Li, S.W.; et al. Recent Progress on the Versatility of Virus-Like Particles. *Vaccines* **2020**, *8*, 139. [CrossRef]
7. Guo, H.C.; Sun, S.Q.; Jin, Y.; Yang, S.L.; Wei, Y.Q.; Sun, D.H.; Yin, S.H.; Ma, J.W.; Liu, Z.X.; Guo, J.H.; et al. Foot-and-mouth disease virus-like particles produced by a SUMO fusion protein system in Escherichia coli induce potent protective immune responses in guinea pigs, swine and cattle. *Vet. Res.* **2013**, *44*, 48. [CrossRef]
8. Shah, R.R.; O'Hagan, D.T.; Amiji, M.M.; Brito, L.A. The impact of size on particulate vaccine adjuvants. *Nanomedicine* **2014**, *9*, 2671–2681. [CrossRef]
9. Xing, Q.; Song, J.; You, X.H.; Dong, L.; Wang, K.X.; Song, J.Q.; Guo, Q.; Li, P.Y.; Wu, C.B.; Hu, H.Y. Microemulsions containing long-chain oil ethyl oleate improve the oral bioavailability of piroxicam by increasing drug solubility and lymphatic transportation simultaneously. *Int. J. Pharm.* **2016**, *11*, 709–718. [CrossRef]
10. Popa, O.; Băbeanu, N.E.; Popa, I.; Sultana, N.; Dinu-Pârvu, C.E. Methods for obtaining and determination of squalene from natural sources. *Biomed Res. Int.* **2015**, *2015*, 367202. [CrossRef]
11. Schmidt, R.; Holznagel, E.; Neumann, B.; Alex, N.; Sawatsky, B.; Enkirch, T.; Pfeffermann, K.; Kruip, C.; Messling, V.V.; Wagner, R. Squalene-containing licensed adjuvants enhance strain-specific antibody responses against the influenza hemagglutinin and induce subtype-specific antibodies against the neuraminidase. *Vaccine* **2016**, *34*, 5329–5335. [CrossRef] [PubMed]
12. Nguyen-Contant, P.; Sangster, M.Y.; Topham, D.J. Squalene-Based Influenza Vaccine Adjuvants and Their Impact on the Hemagglutinin-Specific B Cell Response. *Pathogens* **2021**, *10*, 355. [CrossRef] [PubMed]
13. Gregor, C.; Ulrich, V. Nanotechnology and the transdermal route: A state of the art review and critical appraisal. *J. Control. Release* **2010**, *141*, 277–299. [CrossRef]
14. Sun, H.W.; Liu, K.Y.; Liu, W.; Wang, W.X.; Guo, C.L.; Tang, B.; Gu, J.; Zhang, J.Y.; Li, H.B.; Mao, X.H.; et al. Development and characterization of a novel nanoemulsion drug-delivery system for potential application in oral delivery of protein drugs. *Int. J. Nanomed.* **2012**, *7*, 5529–5530. [CrossRef] [PubMed]
15. Sun, H.W.; Wei, C.; Liu, B.S.; Jing, H.M.; Feng, Q.; Tong, Y.N.; Yang, Y.; Yang, L.Y.; Zuo, Q.F.; Zhang, Y.; et al. Induction of systemic and mucosal immunity against methicillin-resistant Staphylococcus aureus infection by a novel nanoemulsion adjuvant vaccine. *Int. J. Nanomed.* **2015**, *10*, 7275–7290. [CrossRef]
16. Myc, A.; Kukowska-Latallo, J.F.; Bielinska, A.U.; Cao, P.; Myc, P.P.; Janczak, K.; Sturm, T.R.; Grabinski, M.S.; Landers, J.J.; Young, S.K.; et al. Development of immune response that protects mice from viral pneumonitis after a single intranasal immunization with influenza A virus and nanoemulsion. *Vaccine* **2003**, *21*, 3801–3814. [CrossRef]
17. Teng, Z.D.; Sun, S.Q.; Luo, X.; Zhang, Z.H.; Seo, H.; Xu, X.Y.; Huang, J.; Dong, H.; Mu, S.Y.; Du, P.; et al. Bi-functional gold nanocages enhance specific immunological responses of foot-and-mouth disease virus-like particles vaccine as a carrier and adjuvant. *Nanomedicine* **2021**, *33*, 102358. [CrossRef]
18. Teng, Z.D.; Sun, S.Q.; Chen, H.; Huang, J.; Du, P.; Dong, H.; Xu, X.Y.; Mu, S.Y.; Zhang, Z.J.; Guo, C.H. Golden-star nanoparticles as adjuvant effectively promotes immune response to foot-and-mouth disease virus-like particles vaccine. *Vaccine* **2018**, *36*, 6752–6760. [CrossRef]
19. Bidart, J.; Mignaqui, A.; Kornuta, C.; Lupi, G.; Gammella, M.; Soria, I.; Galarza, R.; Ferella, A.; Cardillo, S.; Langellotti, C.; et al. FMD empty capsids combined with the Immunostant Particle Adjuvant -ISPA or ISA206 induce protective immunity against foot and mouth disease virus. *Virus. Res.* **2021**, *297*, 198339. [CrossRef]
20. Suresh Kumar, R.S.; Shiny, P.J.; Anjali, C.H.; Jerobin, J.; Goshen, K.M.; Mukherjee, A.; Mukherjee, A.; Chandrasekaran, N. Distinctive effects of nano-sized permethrin in the environment. *Environ. Sci. Pollut. Res. Int.* **2013**, *20*, 2593–2602. [CrossRef]
21. Liu, W.J.; Zhang, G.L.; Wu, J.R.; Zhang, Y.L.; Liu, J.; Luo, H.Y.; Shao, L.Q. Insights into the angiogenic effects of nanomaterials: Mechanisms involved and potential applications. *J. Nanobiotechnol.* **2020**, *18*, 9. [CrossRef] [PubMed]
22. Bachmann, M.F.; Jennings, G.T. Vaccine delivery: A matter of size, geometry, kinetics and molecular patterns. *Nat. Rev. Immunol.* **2010**, *10*, 787–796. [CrossRef] [PubMed]
23. Smith, D.M.; Simon, J.K.; Baker, J.R. Applications of nanotechnology for immunology. *Nat. Rev. Immunol.* **2013**, *13*, 592–605. [CrossRef] [PubMed]
24. Joyappa, D.H.; Kumar, C.A.; Banumathi, N.; Reddy, G.R.; Suryanarayana, V.V.S. Calcium phosphate nanoparticle prepared with foot and mouth disease virus P1-3CD gene construct protects mice and guinea pigs against the challenge virus. *Vet. Microbiol.* **2009**, *139*, 58–66. [CrossRef] [PubMed]
25. Huang, M.H.; Huang, C.Y.; Lin, S.C.; Chen, J.H.; Ku, C.C.; Chou, A.H.; Liu, S.J.; Chen, H.W.; Chong, P.; Leng, C.H. Enhancement of potent antibody and T-cell responses by a single-dose, novel nanoemulsion-formulated pandemic influenza vaccine. *Microbes Infect.* **2009**, *11*, 654–660. [CrossRef] [PubMed]

26. Shah, R.R.; Taccone, M.; Monaci, E.; Brito, L.A.; Bonci, A.; O'Hagan, D.T.; Amiji, M.M.; Seubert, A. The droplet size of emulsion adjuvants has significant impact on their potency, due to differences in immune cell-recruitment and activation. *Sci. Rep.* **2019**, *9*, 11520. [CrossRef]
27. Anjali, C.H.; Sharma, Y.; Mukherjee, A.; Chandrasekaran, N. Neem oil (*Azadirachta indica*) nanoemulsion-a potent larvicidal agent against Culex quinquefasciatus. *Pest Manag. Sci.* **2012**, *68*, 158–163. [CrossRef]
28. Kumar, S.; Chaturvedi, V.K.; Kumar, B.; Kumar, P.; Sudha Rani Somarajan, S.R.; Mishra, A.K.; Sharma, B. Effect of alum co-adjuvantation of oil adjuvant vaccine on emulsion stability and immune responses against haemorhagic septicaemia in mice. *Iran J. Microbiol.* **2015**, *7*, 79–87.
29. Bode, C.; Zhao, G.; Steinhagen, F.; Kinjo, T.; Klinman, D.M. CpG DNA as a vaccine adjuvant. *Expert Rev. Vaccines* **2011**, *10*, 499–511. [CrossRef]
30. Iho, S.; Maeyama, J.; Suzuki, F. CpG oligodeoxynucleotides as mucosal adjuvants. *Hum. Vaccine Immunother.* **2015**, *11*, 755–760. [CrossRef]
31. Kubo, S.; Kobayashi, M.; Masunaga, Y.; Ishii, H.; Hirano, Y.; Takahashi, K.; Shimizu, Y. Cytokine and chemokine expression in cigarette smoke-induced lung injury in Guinea pigs. *Eur. Respir. J.* **2005**, *26*, 993–1001. [CrossRef] [PubMed]
32. Zheng, H.B.; Pan, L.; Lv, J.L.; Zhang, Z.W.; Wang, Y.Y.; Hu, W.F.; Liu, X.S.; Zhou, P.; Wang, Y.L.; Zhang, Y.G. Comparison of immune responses in guinea pigs by intranasal delivery with different nanoparticles-loaded FMDV DNA vaccine. *Microb. Pathog.* **2020**, *142*, 104061. [CrossRef] [PubMed]
33. Galarza, J.M.; Latham, T.; Cupo, A. Virus-like particle vaccine conferred complete protection against a lethal influenza virus challenge. *Viral. Immunol.* **2015**, *18*, 365–372. [CrossRef] [PubMed]
34. Pervaiz, D.; Ramya, K.; Nuru, S.; Bedaso, M.; Subodh, K.; Ganesh, K.; Kondabattula, G. Montanide ISATM 201 adjuvanted FMD vaccine induces improved immune responses and protection in cattle. *Vaccine* **2013**, *31*, 3327–3332. [CrossRef]

nanomaterials

MDPI

Review

Recent Advances in the Development of Lipid-, Metal-, Carbon-, and Polymer-Based Nanomaterials for Antibacterial Applications

Ruohua Ren [1,†], Chiaxin Lim [1,2,†], Shiqi Li [1], Yajun Wang [3], Jiangning Song [2], Tsung-Wu Lin [4], Benjamin W. Muir [5], Hsien-Yi Hsu [6,*] and Hsin-Hui Shen [1,2,*]

1 Department of Materials Science and Engineering, Faculty of Engineering, Monash University, Clayton, VIC 3800, Australia
2 Biomedicine Discovery Institute, Department of Biochemistry and Molecular Biology, Monash University, Clayton, VIC 3800, Australia
3 College of Chemistry & Materials Engineering, Wenzhou University, Wenzhou 325035, China
4 Department of Chemistry, Tunghai University, No.1727, Sec.4, Taiwan Boulevard, Xitun District, Taichung 40704, Taiwan
5 CSIRO, Manufacturing, Clayton, VIC 3169, Australia
6 School of Energy and Environment, Department of Materials Science and Engineering, City University of Hong Kong, Kowloon Tong, Hong Kong 518057, China
* Correspondence: sam.hyhsu@cityu.edu.hk (H.-Y.H.); hsin-hui.shen@monash.edu (H.-H.S.); Tel.: +61-399-029-518 (H.-H.S.)
† These authors contributed equally to this work.

Abstract: Infections caused by multidrug-resistant (MDR) bacteria are becoming a serious threat to public health worldwide. With an ever-reducing pipeline of last-resort drugs further complicating the current dire situation arising due to antibiotic resistance, there has never been a greater urgency to attempt to discover potential new antibiotics. The use of nanotechnology, encompassing a broad range of organic and inorganic nanomaterials, offers promising solutions. Organic nanomaterials, including lipid-, polymer-, and carbon-based nanomaterials, have inherent antibacterial activity or can act as nanocarriers in delivering antibacterial agents. Nanocarriers, owing to the protection and enhanced bioavailability of the encapsulated drugs, have the ability to enable an increased concentration of a drug to be delivered to an infected site and reduce the associated toxicity elsewhere. On the other hand, inorganic metal-based nanomaterials exhibit multivalent antibacterial mechanisms that combat MDR bacteria effectively and reduce the occurrence of bacterial resistance. These nanomaterials have great potential for the prevention and treatment of MDR bacterial infection. Recent advances in the field of nanotechnology are enabling researchers to utilize nanomaterial building blocks in intriguing ways to create multi-functional nanocomposite materials. These nanocomposite materials, formed by lipid-, polymer-, carbon-, and metal-based nanomaterial building blocks, have opened a new avenue for researchers due to the unprecedented physiochemical properties and enhanced antibacterial activities being observed when compared to their mono-constituent parts. This review covers the latest advances of nanotechnologies used in the design and development of nano- and nanocomposite materials to fight MDR bacteria with different purposes. Our aim is to discuss and summarize these recently established nanomaterials and the respective nanocomposites, their current application, and challenges for use in applications treating MDR bacteria. In addition, we discuss the prospects for antimicrobial nanomaterials and look forward to further develop these materials, emphasizing their potential for clinical translation.

Keywords: nanomaterials; multidrug-resistant bacteria; antimicrobial; drug delivery systems; nanoparticles

Citation: Ren, R.; Lim, C.; Li, S.; Wang, Y.; Song, J.; Lin, T.-W.; Muir, B.W.; Hsu, H.-Y.; Shen, H.-H. Recent Advances in the Development of Lipid-, Metal-, Carbon-, and Polymer-Based Nanomaterials for Antibacterial Applications. *Nanomaterials* **2022**, *12*, 3855. https://doi.org/10.3390/nano12213855

Academic Editors: Goran Kaluđerović and Nebojša Pantelić

Received: 10 October 2022
Accepted: 28 October 2022
Published: 1 November 2022

1. Introduction

Antibiotics have been the primary treatment choice for use on bacterial infections due to their cost efficiency and powerful and fast-acting outcomes. However, bacteria possess the intrinsic ability to evolve rapidly through mutations in developing resistance to these treatments. In addition, bacteria can transfer drug-resistant genes among their community through horizontal gene transfer, resulting in the emergence of multidrug-resistant (MDR) bacteria, which are widely known as superbugs as defined by the medical and research communities [1]. Since bacterial resistance emerges and spreads via the acquisition of genetic material from resistant bacterial cells, the evolution of antibiotic resistance is unstoppable [2]. Recent projections indicate that a post-antibiotic era is approaching, and this will result in approximately 10 million annual deaths by 2050 from MDR bacterial infections [3]. Studies have shown that infections caused by multidrug-resistant bacteria cause more harm and higher patient mortality than infections caused by susceptible strains of the same species [4]. A continual increase in the numbers of infections resulting from such resistant strains poses a serious threat globally [5].

The antibiotic resistance crisis is further complicated by a lack of new antibacterial agents to act as last-line defenders for the treatment of MDR bacterial infections. For instance, the World Health Organization (WHO) has identified 80 antibacterial agents that are under clinical development to treat top-priority MDR bacteria up to November 2021, but most of these are modifications of current antibiotics and will act merely as short-term solutions [6]. Only seven of these antibacterial agents are novel chemical entities that will contribute to expanding the current antibiotic pipeline [6]. Due to economic and regulatory hurdles, the biopharmaceutical industry has largely withdrawn from developing new antibiotics, further exacerbating the situation [7]. This has triggered initiatives worldwide to discover and exploit novel antibacterial agents in order to prevent these infections from happening and to overcome the current challenges faced from MDR infections [8]. Promising solutions for the prevention and treatment of MDR bacterial infections are under investigation, such as nanotechnology and biomaterials [9].

Nanotechnology serves as an alternative promising solution for the prevention and treatment of MDR bacterial infection. Nanotechnology plays an important role in this area by covering a broad range of nanostructured materials that possess inherent antibacterial activity. Nanomaterials also show significant potential for delivering drugs to specific targeted sites in vivo [10]. Nanomaterials have at least one dimension in the nano range (1–100 nm) that convey particular and variable physiochemical properties from their bulk constituents [11]. The nanosized scale of these nanomaterials can result in multivalent interactions with bacteria, including electrostatic attractions, hydrophobic and receptor–ligand interactions, and van der Waals (hydrophobic) forces [12]. This offers particular advantages compared to small molecule antibiotics that typically result in a single mode of interaction. The ease of functionalization and engineering of nanomaterials confers them with additional advantages for mechanistically overcoming bacterial resistance [13].

Nanomaterials can be broadly classified into organic nanomaterials and inorganic nanomaterials [14]. Recent advances of nanotechnology have brought novel understandings in using nanosized building blocks to design and create new nanocomposites or nanohybrid materials with unprecedented physical properties and enhanced antibacterial activity [15]. A variety of nanomaterials can be combined to develop new nanocomposite materials, with the most-established examples being depicted in the section below. In this review, we illustrate that each category of these antibacterial nanomaterials has its own distinctive characteristics and properties which are being applied to various antibacterial applications. We present recent advances in developing the use of these nanomaterials in combating MDR bacterial infections. However, the use of nanocomposites is still at an early stage and more research and investment is needed towards these efforts before we start seeing outcomes from their clinical translation. Based on these, this review summarizes previous research progress on nanotechnology in antibacterial aspects which focuses on the last 5 years, including a detailed summary and comparison of the most promising and interesting nanomaterials (Table 1). The aim is to inspire future research ideas in this field by identifying gaps or inconsistencies in the body of knowledge.

Table 1. Summary table of the nanomaterials.

Nanomaterials	Classes	Advantages	Disadvantages	References
Lipid	Organic	• Dual functional role as antibacterial agent and nanocarrier • Ease of industrial manufacturing for commercialization • Good biocompatibilities	• Poor colloidal stability for long-term storage • Relatively weaker antibacterial activity	[16,17]
Polymer	Organic	• Dual functional role as antibacterial agent and nanocarrier • Strong bactericidal activity • Good colloidal integrity and stability	• Poor biocompatibilities	[18,19]
Carbon	Organic	• Dual functional role as antibacterial agent and nanocarrier • Highest drug loading capacity • Strong bactericidal activity with physical and chemical antibacterial mechanism	• Higher tendency of agglomeration • Low water solubility	[20–22]
Metal	Inorganic	• Strong bactericidal activity • Multiple antibacterial applications for dry (coating) and wet environment (disinfectant) • Ease of industrial manufacturing for commercialization	• Higher tendency of agglomeration • Poor biocompatibilities • Lack of delivery ability	[23–25]
Metal oxide	Inorganic	• Good biocompatibilities • Photosensitizing agents with multiple antibacterial mechanisms • Ease of industrial manufacturing for commercialization	• Higher tendency of agglomeration • Lack of delivery ability • Environmental hazards especially to aquatic environment	[26,27]

2. Organic Nanomaterials

Organic nanomaterials usually comprise carbon and hydrogen atoms that form, most simply, hydrocarbon-based molecules. Organic nanomaterials can be designed to those that may self-assemble into nanostructures with different dimensionalities or desired characteristics by utilizing the weak intermolecular interactions of organic molecular structures [28]. Organic nanomaterials can be classified into lipid-based, polymer-based, and carbon-based nanomaterials, and these nanomaterials can be designed to act as nanocarriers or antibacterial agents in antibacterial applications.

2.1. Lipid-Based Nanomaterials

A variety of lipid candidates, including free fatty acids, phospholipids, glycolipids, sphingolipids, fatty alcohols, glycerol esters, and waxes, can be utilized to nanoformulate into different classes of lipid-based nanoparticles including liposomes, emulsions, solid-lipid nanoparticles, and nanostructured lipid carriers [29]. A detailed review of these aforementioned nanoparticles has been described elsewhere [30–32] and will not be discussed here. Lipid-based nanoparticles are the most-established nanocarriers investigated for the delivery of a variety of pharmaceutical agents with different solubilities and pharmacokinetic behaviors. In addition to the role of nanocarrier, lipid-based nanoparticles have an emerging role as antibacterial agents against MDR bacteria. In short, these can be classified into lipidic nanocarriers and lipidic nanoparticles. Lipidic nanocarriers contain and deliver antibacterial agents including antibiotics and antimicrobial peptides, whilst lipidic nanoparticles themselves display inherent antibacterial properties.

2.1.1. Lipidic Nanocarriers as Delivery Vehicles for Antimicrobial Agents

Lipidic nanocarriers are the most-established nanocarriers utilized for delivery of a variety of pharmaceutical agents with different solubilities. Lipidic nanocarriers are composed of colloidal dispersions of physiological or physiological-related lipids (natural or synthetic lipids that have the similar chemical structure to physiological lipids) in aqueous solution. Generally, these dispersions are stabilized by an emulsifier or surfactant which intercalates on the lipid nanoparticle's surfaces. This provides the nanoparticle stability by conferring steric stabilization in between the nanoparticles and reducing the interfacial energy between the lipidic nanoparticles and the aqueous phase [33]. In brief, lipidic nanocarriers such as liposomes [16,34–37], micelles [38–40], nanocapsules [41–44], emulsions [45,46], and solid lipid nanoparticles [29,47] have several advantages for delivering antimicrobial agents [48]. Lipidic nanocarriers can exhibit good biocompatibility and non-immunogenic properties due to the analogous behavior of the physiological or physiological-related lipids to biological membranes as seen in the new SARS-CoV-2 lipid-based mRNA vaccines. The encapsulation of drugs enhances their bioavailability, increases the feasibility for various routes of administration, reduces associated drug toxicity, and protects the drugs from metabolic degradation. Furthermore, drug encapsulation into lipid-based nanoparticles also improves the pharmacokinetic and pharmacodynamic profiles, which lowers the required dosages and improves the therapeutic index. Lastly, surface modifications of lipid-based nanoparticles can be achieved for various purposes, such as targeted therapies, improved cellular uptake, and increased circulation times and half-lives.

The use of these lipidic nanocarriers for delivery can have some limitations, including occasional poor colloidal or thermodynamic/kinetic stability for long-term storage, high membrane permeability that accounts for drug leakage, and low entrapment efficiency for certain hydrophobic drugs [17]. This often leads to costly and restricted preparation conditions that allows reconstitution of lipidic nanocarriers in solution prior to administration [49]. One of the potential solutions is to combine lipid-based nanoparticles with polymeric nanomaterials, forming lipid–polymer hybrid nanoparticles for delivery purposes, which will be described in Section 4.5. Despite being the most widely explored nanoparticulate delivery system for various pharmaceutical products, the role of lipidic nanocarriers in antibacterial application is limited. Currently, only one liposomal formulation (amikacin liposome inhalation suspension, Arikayce) is approved by the Food and Drug Administration (FDA) (ClinicalTrials.gov Identifier: NCT01316276) for the treatment of mycobacterial lung infection. In addition, there are a few liposomal nanoformulations delivering antibacterial agents that are undergoing clinical trials, with the details shown in Table 2.

Table 2. Liposomal nanoformulation in clinical development for antibacterial therapy.

Product Name	Encapsulating Materials	ClinicalTrials.gov Identifier	Description
-	AP10-602/ GLA-SE	NCT02508376	Trial on the safety, tolerability, and immunogenicity of the vaccine candidates for the protection against tuberculosis
-	CAL02	NCT02583373	Trial on broad-spectrum antitoxin agent CAL02 that neutralizes bacterial toxins to protect against infection severity and deadly complications
Pulmaquin	Ciprofloxacin	NCT02104245	Trial on Pulmaquin® in the management of chronic lung infections in patients with non-cystic fibrosis bronchiectasis
MAT2501	Amikacin	-	Orally administered amikacin liposomal formulation for various MDR infections that completed Phase 1 study
CAF01	Tuberculosis Subunit Vaccine Ag85B-ESAT-6	NCT00922363	Trial on the safety of new liposomal vaccine adjuvant for protection against tuberculosis

As compared with liposomes, other lipidic nanocarriers are still in the early stages of development [50–53]. Recently, non-lamellar lyotropic liquid crystalline nanoparticles including cubosomes and hexosomes have emerged to be the next generation of smart lipidic nanoparticles [54–57] for antimicrobial therapeutics. The antibiotic potential of cubosomes with a series of magnetite (Fe_3O_4), copper oxide (Cu_2O), and silver (Ag) nanocrystals were developed by Meikle et al. [58]. The results showed that Ag nanocrystal-embedded cubosomes displayed exhibitory activity against both Gram-positive and Gram-negative bacteria, with observed minimum inhibitory concentration values ranging from 15.6–250 μg/mL. Recent studies have shown that polymyxin-loaded cubosomes can enhance antibacterial potency against Gram-negative bacteria, including polymyxin-resistant strains, and enable an alternative strategy for treating pathogens by combining cubosomes with polymyxins as a combination therapy [57]. To overcome the difficulty of using antimicrobial peptides in antibiotic therapies due to their lack of specificity and their susceptibility to in vivo proteolysis, Boge et al. used cubosomes to topically deliver the antimicrobial peptides, LL-37, to inhibit *S. aureus*. They found that the pre-loading preparation where incorporation of LL-37 into liquid crystal gels followed by dispersion into nanoparticles was most effective in killing *S. aureus* [55]. Additional studies have been reported investigating the use of cubosomes as drug delivery vehicles for LL-37. It was observed that the cubosomes successfully protected LL-37 from proteolytic degradation with significantly enhanced bactericidal effects against Gram-negative strains [59]. Meikle et al. explored the potential of cubosomes as delivery vehicles for six different antimicrobial peptides, including gramicidin A, alamethicin, melittin, indolicidin, pexiganan, and cecropin A [60], wherein it was observed that by adding physiological concentrations of anionic lipids or NaCl to screen the electrostatic charge of peptides, the antimicrobial peptides loading efficiency of the cubosomes was significantly improved, and encapsulation in the cubosome carriers was shown to enhance the antimicrobial activity of certain formulations [60]. Notably, there are fundamental differences in the mechanism of cubosomes uptake between Gram-positive and Gram-negative bacteria. For Gram-positive bacteria, the cubosomes adhere to the exopeptidoglycan layer and slowly internalize into the bacteria, while for Gram-negative bacteria, the interaction occurs in two stages: the cubosomes fuse with the outer lipid membrane and then pass through the inner wall via diffusion [61].

2.1.2. Lipidic Nanoparticles with Inherent Antibacterial Activities

Antimicrobial lipids composed of a carboxylic acid group and a saturated or unsaturated carbon chain (Figure 1) can act as surfactants via a membrane lytic mechanism [62]. Antimicrobial lipids possess broad-spectrum antibacterial activities and serve as new and attractive candidates to fight the antibiotic resistance crisis. However, some technical challenges impede the in vivo activity of antimicrobial lipids in bulk form. These include poor aqueous solubility and weaker in vivo bactericidal activity due to in vivo oxidation, esterification, and lipid–protein complexation [10,63,64]. This can be overcome by developing lipid nanoparticle technologies to encapsulate antimicrobial lipids and convert them into different nanoformulations with inherent antimicrobial activities. The resulting antimicrobial lipidic nanoparticles using nanocarriers have excellent water solubility, can provide high concentrations of antibacterial lipids, and protect antibacterial lipids from degradation, which highlights the great potential for improving the therapeutic ability of antibacterial lipids [10,62]. Several reviews have been published elsewhere to understand the composition, mechanism, and characterization of this class of lipidic nanoparticles [65–69].

Liposomal formulations are the most-studied candidate so far for the emerging role as antimicrobial agents which are spherical closed lipid bilayers that can self-assemble in aqueous solutions and have a water core [16,70]. Antimicrobial lipids such as lauric acid and oleic acid can be incorporated to form antimicrobial liposomal formulations against *Propionibacterium acnes* and methicillin-resistant *S. aureus* (MRSA), respectively [34,71]. Among these different fatty acids, liposomal linolenic acids have received considerable attention by exhibiting particularly high levels of inhibitory activity [70]. Liposomal linolenic acid (LLA, Figure 2) that comprised liposomal nanoparticles made from linolenic acid, phospholipids, and cholesterols eradicated *Helicobacter pylori* clinical isolates including metronidazole-resistant *H. pylori* [72]. Furthermore, the bacteria did not appear to develop resistance to LLA at the sub-bactericidal concentrations used when compared with metronidazole and free linolenic acid. The fusion between the LLA and bacterial membrane, which directly inserts the linolenic acid into the bacterial membranes for subsequent membrane lysis, is suggested to be the bactericidal mechanism [72]. The in vivo efficacy of LLA in treating *H. pylori* infection was further investigated [73]. LLA penetrated into the mucus layer of a murine stomach, which led to reduced bacterial load and proinflammatory cytokines. In addition, a significant portion of LLA remained in the stomach at 24 h post-treatment, showing the long-last effects of LLA. Lastly, the in vivo toxicity showed no significant increase in gastric epithelial apoptosis and no changes of the murine gastric tissue under histological analysis, indicating the excellent biocompatibility of LLA in the stomach of control mice [73].

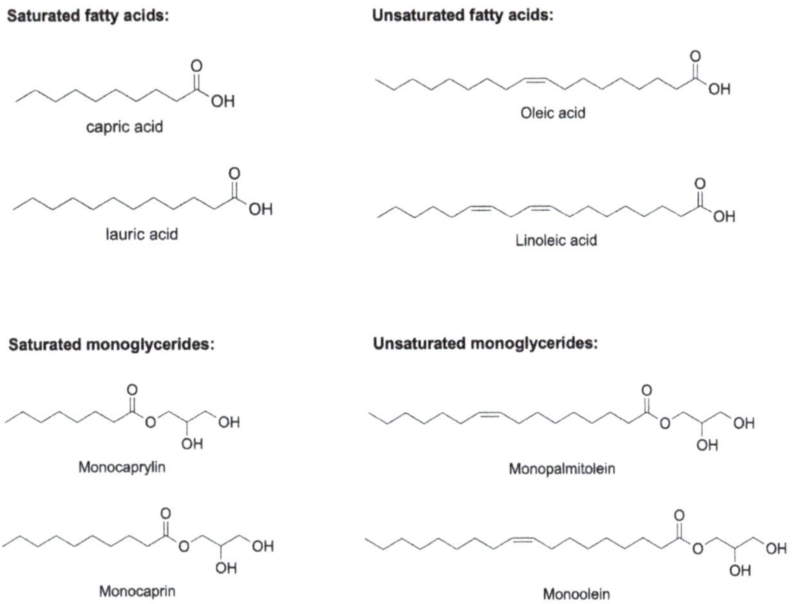

Figure 1. Chemical structures of some potentially antimicrobial fatty acids and monoglycerides. Saturated fatty acids include capric acid and lauric acid, unsaturated fatty acids include oleic acid and elaidic acid, saturated monoglycerides include monocaprylin and monocaprin, and unsaturated monoglycerides include monopalmitolein and monoolein.

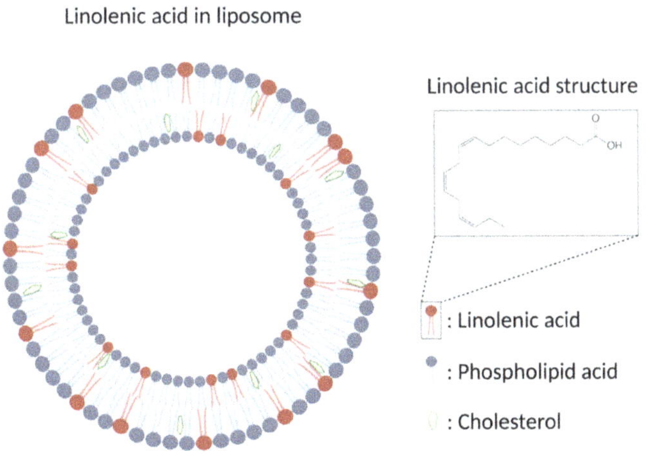

Figure 2. Schematic drawing showing the structure of LLA.

In addition to antimicrobial liposomal formulations, other antimicrobial lipid-based nanoparticle systems, including emulsions and solid lipid nanoparticles, have shown promising antibacterial effects against MRSA and *Pseudomonas aeruginosa*, respectively [47,74]. Sadiq et al. encapsulated nisin in monolaurin nano-emulsions and demonstrated their ability of effectively inhibiting *S. aureus* in vitro [46]. Studies have found that solid lipid nanoparticles loaded with retinoic acid and lauric acid inhibited the growth of *Staphylococcus epidermidis, P. acnes,* and *S. aureus* [75]. A second generation of lipid nanoparticles that can improve the loading capacity and inhibit the excretion of bioactive compounds, called nanostructured lipid carriers, was recently developed from a mixture of solid lipids and liquid lipids [76–80]. Compared to the crystalline lipid core of solid lipid nanoparticles, the structural imperfections of nanostructured lipid carriers with less ordered crystalline arrangement can further improve the loading capacity and prevent the drug leakage for better antibacterial activity. Previous research comparing the antibacterial activity of docosahexaenoic acid (DHA) coated by nanostructured lipid carriers and DHA itself has found the incorporation of DHA into the nanostructured lipid carriers greatly enhanced bactericidal effect against *H. pylori* [81]. However, studies of emulsions, solid lipid nanoparticles, nanostructured lipid carriers, etc., as antimicrobial lipid-based nanoparticle systems are still in the early phases compared with the simplest form of liposomes.

2.2. Biodegradable Polymeric Nanomaterials

Biodegradable polymeric nanosystems can be classified into polymeric nanoparticles for the purposes of a delivery nanocarrier and antimicrobial polymers. The tailored design of polymeric chains confers versatile functions to the biodegradable polymeric nanomaterials including antibacterial activity, enhancing stability, biocompatibility, long circulation, and specific bacterial recognition of the polymeric nanomaterials [82]. Antimicrobial cationic polymers are the most-studied organic nanomaterials that have already entered clinical trials and hold great promise in replacing some antibiotics [83]. Biodegradable polymeric nanoparticles also offer an attractive delivery system which can improve the safety and efficacy of other ingredients by modulating the rate, timing, and location of release compared to lipid-based nanoparticles [84]. Furthermore, the functional groups on the polymer chain serve as a promising matrix to interact with other nanomaterials, forming polymer-based nanocomposites [85]. This paves the way for researchers to synthesize different polymer-based nanocomposites with improved or novel properties, which will be discussed further in Section 4 below.

2.2.1. Polymeric Nanoparticles as Delivery Nanocarriers

For the use of biodegradable polymeric nanoparticles with encapsulated antibacterial agents, they can be classified into four distinct classes including polymeric micelles, vesicles, nanocapsules, and nanospheres, depending on the polymer composition and the final structure of the polymeric nanosystems. A detailed review of the aforementioned nanoparticles has been described elsewhere [85–88]. Currently, over 80 clinical trials are underway or have been completed using polymeric nanoparticles in cancer therapy, highlighting the potential utility of polymeric nanoparticles in drug delivery [89].

Polymeric nanoparticles and lipid-based nanoparticles share similar advantages as drug delivery vehicles, but polymeric nanoparticles have some perceived advantages over lipid-based nanoparticles. This includes higher structural integrity and stability under biological and storage conditions, and controlled release capabilities conferred via the polymer cytoskeleton [18,90]. Among them, the use of stimuli-responsive biodegradable polymer nanoparticles to prepare drug delivery systems has great potential for controlled drug delivery [91,92]. It has been demonstrated that polymer degradation can be controlled by changing the external stimuli (e.g., pH, ultrasound, temperature, IR radiation, magnetic field, etc.), allowing stacked polymer nanoparticles to degrade in a controlled manner and release a drug on demand [90,93,94]. Qiu et al. successfully developed phosphatidylcholine–chitosan hybrid nanoparticles loaded with a gentamicin antibiotic and demonstrated that this synthetic system was able to inhibit the growth and membrane formation of Gram-positive and Gram-negative bacteria [95]. Studies have shown that by encapsulating vancomycin antibiotics in nanovesicles composed of long fatty acids grafted with hydrophilic polymers, these nanocarriers have the ability to self-assemble into spherical drug carriers and are effective against MRSA [96]. However, the polymer degradation products and clearance might cause potential toxicity as lipid-based nanoparticles typically have higher biocompatibilities than polymeric nanoparticles, which makes the application of polymeric nanoparticles in delivering antimicrobial agents a challenge. Hence, the field is still at an early development stage [19].

2.2.2. Antimicrobial Cationic Polymeric Nanoparticles

Over the last decade, synthetic biodegradable antimicrobial cationic polymers have been a promising solution to combat bacteria. The cationic charges of these synthetic polymers selectively act and are attracted to negative-charged bacterial membranes on zwitterionic mammalian cell membranes, in a mechanism similar to natural antimicrobial peptides [97,98]. Antimicrobial cationic polymers have attracted tremendous attention owing to their facile synthesis in bulk quantities at much lower costs, broad spectrum efficacy of their antibacterial activity with membrane disruptive mechanism, as well as a low propensity for inducing bacterial resistance [99,100]. Of note are the natural antimicrobial peptide-mimicking antimicrobial cationic polymers brilacidin (ClinicalTrials.gov Identifier: NCT02324335) and LTX-109 (ClinicalTrials.gov Identifier: NCT01803035), which have completed phase 2 clinical trials.

The antibacterial mechanism of cationic polymers requires contact with a bacterial membrane's outer surfaces, which induces a globally amphiphilic conformational change to sequester cationic and lipophilic side chains [101]. This property is known as facial amphiphilicity and is shown in Figure 3. The cationic subunits are responsible for interacting with the bacterial membrane, whereas the lipophilic side chains insert into bacterial membranes for subsequent membrane disruption. [102]. This leads to cytoplasmic leakage, membrane depolarization, lysis, and ultimately cell death, showing the promising antibacterial activity of these polymers [103]. It remains challenging to achieve proper facial amphiphilicity of cationic polymers. The majority of antimicrobial cationic polymers that are generated from uncontrolled polymeric self-assembly do not comprise truly facial amphiphilicity, which greatly affects antibacterial activity and can lead to nonspecific toxicity in mammalian cells [104]. Manipulation of the sequence of hydrophobic and hydrophilic subunits of antimicrobial polymers is an important factor in achieving facial amphiphilicity for antibacterial activity. A recent study combining vancomycin with the cationic polymer Eudragit E100 ® (Eu) against *P. aeruginosa* showed that *P. aeruginosa* was eradicated within 3–6 h of exposure with this combination treatment [105]. Although bacterial envelope permeabilization and morphological changes after exposure to Eu were not sufficient to cause bacterial death, they allowed vancomycin to enter the target site, thereby enhancing the activity of an otherwise inactive vancomycin against *P. aeruginosa*.

The formulation of antimicrobial polymeric nanoparticles has overcome the aforementioned problems associated with antimicrobial polymers. The first antimicrobial polymer that self-assembled into cationic micellar nanoparticles by dissolution in water was reported by Nederberg [106]. A strong bactericidal activity of the cationic micellar nanoparticles was observed against MRSA and *Enterococcus faecalis* [106]. The polymeric nano-architecture was critical for effective bactericidal activity of the antimicrobial polymer molecules [106]. Unlike conventional antimicrobial polymers, the self-assembled antimicrobial polymeric nanoparticle does not require contact with the bacterial membrane for the formation of the secondary structure. It is hypothesized that the nanoparticle architecture increases the local concentration of cationic charge and polymer mass, leading to strong interactions between the polymer and cell membrane, which translate into effective antibacterial activities. Self-assembled antimicrobial polymeric nanoparticles have demonstrated minimal toxicity along with promising antibacterial activity, highlighting their potential in antibacterial applications and clinically relevant therapies [107–110]. Chin and colleagues reported a class of degradable guanidine-functionalized polycarbonates with a unique mechanism that does not induce drug resistance, which has great potential in the prevention and treatment of multidrug-resistant systemic infections [111]. The team optimized the structure of the polymer for treating multidrug-resistant *Klebsiella pneumoniae* pulmonary infections. In vivo experiments showed that the polymer backbone (pEt_20) self-assembles into micelles at high concentrations, which can alleviate lung infection with *K. pneumoniae* without causing damage to the major organs in mammals [112].

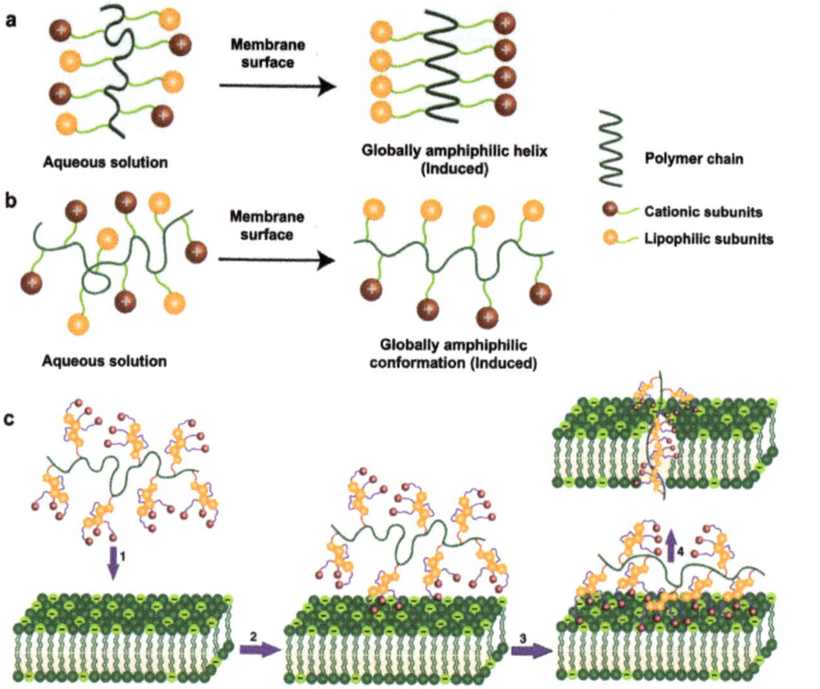

Figure 3. Modes of action upon contact with bacterial membrane surfaces. (**a**) Global amphiphilic helical conformation adopted by host-defense peptides; (**b**) global amphiphilic random conformation adopted by synthetic antimicrobial polymers. (**c**) Proposed antibacterial mechanism of synthetic antimicrobial polymers: (1) diffusion, (2) surface binding via cationic subunits, (3) membrane insertion via lipophilic subunits and (4) membrane disruption. (Adapted from Rahman et al., [104] 2018 with modifications.)

Another breakthrough study that benefits from this nanotechnology is star-shaped peptide polymer nanoparticles [113]. This is the first example of a synthetic antimicrobial polymer that efficiently kills colistin-resistant and multidrug-resistant Gram-negative pathogens, including *Acinetobacter baumannii, K. pneumoniae,* and *P. aeruginosa* [113]. The star-shaped peptide polymer nanoparticles eradicate these Gram-negative bacteria via destabilization and fragmentation of the bacterial outer membrane, disruption of cytoplasmic membrane, and induction of bacterial apoptosis [113]. Singh and her colleagues investigated the antimicrobial activities against clinical and drug-resistant strains (MDR-PA and MRSA) through indole-3-butyryl-polyethyleneimine nanostructured self-assembly in aqueous systems [114]. The amphiphilic indole-3-butyryl-polyethyleneimine polymer nanostructures have positively charged hydrophilic polyethyleneimine on the surface, while the hydrophobic indole-3-butyryl moiety is located inside the core, which showed enhanced antibacterial effects against all drug-resistant strains [114].

2.3. *Carbon-Based Antimicrobial Nanomaterials*

As a novel class of nanomaterials, carbon-based nanomaterials (CNMs) have received significant interest due to their remarkable properties including inherent antibacterial effects, extraordinary mechanical properties, excellent electrical conductivity and thermal conductivity, incredibly high surface area to volume ratios, photoluminescent and photocatalytic activities, and good stabilities [115]. These unique properties make carbon nanoarchitectures promising for a wide range of antibacterial applications including drug delivery, bone and tissue engineering, biosensors, photothermal therapy, and potential new antibacterial agents, which have been discussed elsewhere [116–121]. Due to its valency, carbon is able to form several allotropes that leads to a broad range of nanostructures of different dimensions, shapes, and properties, as depicted in Figure 4.

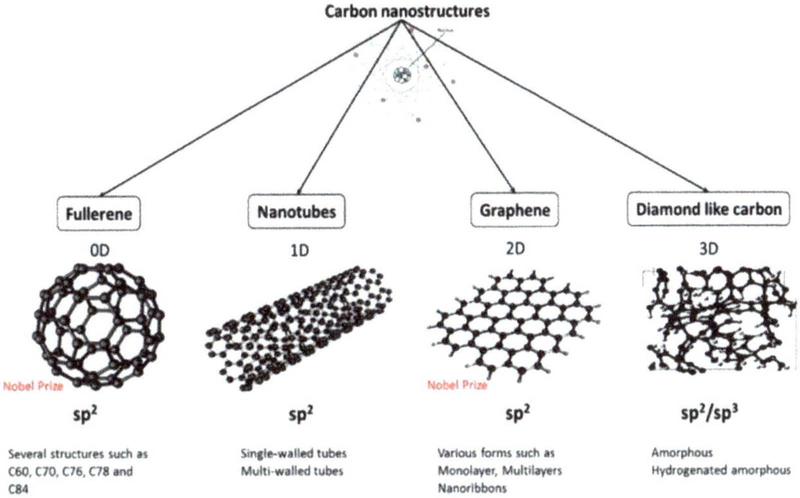

Figure 4. Carbon-based nanomaterials categorized via different dimensionalities (D). (Adapted with permission from Ref. [116]. 2017, Al-Jumaili et al. More details on "Copyright and Licensing" are available via the following link: https://www.mdpi.com/ethics#10 (accessed on 10 October 2022).)

Among carbon-based nanomaterials, the preferential use of graphene-based materials, especially graphene oxide (GO), for antibacterial applications is due to the following reasons: The highly oxygenated surface of GO, bearing hydroxyl, epoxide, diol, and carbonyl functional groups, provides a versatile platform for drug delivery applications or further functionalization [122]. The good aqueous solubility of GO makes it suitable for in vivo antibacterial applications compared with poorly water-soluble fullerenes and nanotubes [20,123]. Another advantage of graphene oxide is its ability to act as a barrier or overlay, delaying and controlling the release of biomolecules over time [124–126]. Lastly, GO synthesis can be devoid of any metallic impurities and these materials can exhibit tolerable toxicity [127,128].

2.3.1. Graphene Oxide as an Antimicrobial Delivery Nanocarrier

The use of graphene oxide as a nanocarrier for drug delivery has received significant attention for several reasons, such as its good biocompatibility with tolerable toxicities [129,130]. In addition, the ease of functionalization provides possibilities in synthesizing novel functional nanohybrids or nanocomposites for specific purposes including targeted drug delivery, as shown in Figure 5 [131]. The extremely large surface area coupled with a two-dimensional planar structure provides a huge drug-loading capacity. In fact, a significant rate of drug loading has been reported previously with a GO-based delivery system [21]. The high mechanical and chemical stability of GO makes it particularly suitable for different delivery environments [132]. Currently, GO nanoparticles have been experimentally used in various biomedical applications including gene delivery [133], drug delivery [134,135], photodynamic therapy [136], anticancer therapies [137,138], and antibacterial therapies [139]. Nevertheless, studies of GO-based delivery systems with bacterial infection are still at a preliminary stage, involving investigations in antibiotic absorption efficacy and the respective in vitro antibacterial activity [140,141]. In a recent study, polyethylene-glycol-functionalized GO nanoparticles loaded with *Nigella sativa* seed extract were tested as a drug delivery system to disrupt bacteria by penetrating bacterial nucleic acid and cytoplasmic membranes, successfully demonstrating potential antibacterial activity against *S. aureus* and *Escherichia coli* [142]. Pan et al. used GO as a carrier to load N-halamine compounds, which not only displayed an antibacterial effect against *S. aureus* and *E. coli*, but also had slow-release properties and good storage stability [143]. In general, with the aforementioned advantages conferred by GO-based delivery systems, the potential of GO as a delivery platform for antimicrobial agents should not be neglected.

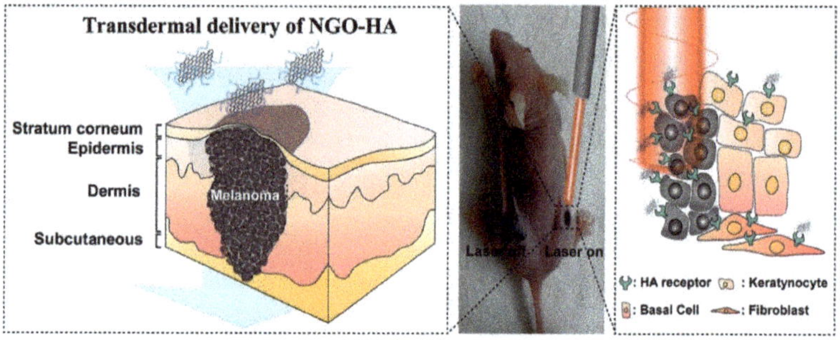

Figure 5. Schematic illustration for the transdermal delivery of nanographene oxide–hyaluronic acid (NGO–HA) conjugates into melanoma skin cancer cells and the following photothermal ablation therapy using a near-infrared laser. (Adapted with permission from Ref. [131]. 2014, Jung et al. More details on "Copyright and Licensing" are available via the following link: https://www.mdpi.com/ethics#10 (accessed on 10 October 2022).)

2.3.2. Graphene Oxide with Inherent Antibacterial Properties

GO has received tremendous attention as a novel antibacterial agent compared with its role as a delivery nanocarrier in antibacterial applications due to its broad-spectrum antibacterial activity and low cytotoxicity at low concentrations [144]. GO has been reported to exhibit strong antibacterial activity against a variety of Gram-positive and Gram-negative bacteria, such as *E. coli*, *S. aureus*, *E. faecalis*, *P. aeruginosa*, and *Candida albicans* [22,139,145–149]. In addition, Di Giulio et al. also reported significant antibiofilm efficacy against biofilms produced by *S. aureus*, *P. aeruginosa*, and *C. albicans* [150]. Recently, attention has also been given to GO-based combination antibacterial therapies. The ternary nanocomposites obtained by combining GO with hydroxyapatite and copper oxide have inhibitory effects on Gram-negative *E. coli* and Gram-positive *S. aureus* [151]. Innovative bionanomaterials composed of GO, agarose, and hydroxyapatite have also shown the ability to significantly reduce *S. aureus* [152].

The strong antibacterial activity of GO is associated with both physical and chemical damages. The physical interactions of GO with bacteria that are reported to date include interactions via direct contact of its sharp edges, lipid extraction, bacteria isolation from their nutrient environment by wrapping, and photothermal/photocatalytic effects owing to the semiconductor properties of graphene [139,153–156]. Interestingly, the mechanism of action of GO on Gram-positive and Gram-negative bacteria appears to be distinct. Pulingam et al. reported that cell entrapment via mechanical wrapping was mainly observed for the Gram-positive bacteria *S. aureus* and *E. faecalis*, whereas with Gram-negative bacteria *E. coli* and *P. aeruginosa* it was observed that membrane rupture due to physical contact [22] was the predominant mechanism. Chemical damages are additionally caused via oxidative stress and generation of reactive oxygen species (ROS) and charge transfer, thereby inhibiting bacterial metabolism, disrupting cellular functions, causing inactivation of intracellular and subcellular proteins, and inducing lipid peroxidation, leading to cellular inactivation [147,157–159]. Zhang et al. recently highlighted electrical conductivity as a key property of GO that may be underestimated in terms of its antibacterial activity role [160]. Research by Chong et al. proposed that sunlight irradiation could increase the antibacterial activity of GO due to enhancing the electron transportation of antioxidants [161]. GO's diverse physicochemical properties including sheet size, shape, number of layers, surface charge, defect density, and the presence of surface functional groups and oxygen content have a strong impact on its antibacterial activity and biological performance [116,162]. However, the physicochemical properties related to antibacterial activity are not fully elucidated yet. Deepening the understanding of the physicochemical properties related to antibacterial activity is a crucial step in designing GO-based nanomaterials for optimized antibacterial activity. Together, these studies provide important insights into the way forwards.

3. Antibacterial Inorganic Nanomaterials

Inorganic nanomaterials do not contain either carbon or hydrogen atoms that are associated with biological matter. As an alternative, inorganic nanomaterials comprise metallic and non-metallic elemental compounds that have weak intermolecular interactions which form nanostructures with higher dimensionality. Among two classes of the inorganic nanomaterials, metallic inorganic nanomaterials (Ag, Au, Zn, Cu, Bi) have attracted significant attention over non-metallic nanomaterials (S, Si, B, Te and Se). This is particularly due to the inherent water insolubility of the non-metallic inorganic nanomaterials that restricted their use in antibacterial applications. Therefore, this review primarily focuses on the development of inorganic metallic nanomaterials in antibacterial applications. Other inorganic nanomaterials such as fluoride in oral treatment has also been researched for a long time. For instance, in the prevention of dental caries, the addition of fluoride has a significant antibacterial effect on *Streptococcus mutans, Lactobacillus acidophilus, E. faecalis, Actinomyces naeslundii,* and *Parvimonas micra* [163–166].

Inorganic metallic nanomaterials do not readily self-assemble into 1D nanowires, nanotubes, nanoribbons, and 2D nanowalls and nanofilms [167]. Therefore, complicated synthesis methodologies are required to promote the "growth" of inorganic metallic nanomaterials into nanostructures with higher dimensionality [168,169]. Zero-dimensional metal and metal oxide nanoparticles are the most popular candidates that can readily be synthesized for antibacterial applications, which are discussed in Sections 3.1 and 3.2 below.

3.1. Metal Nanoparticles

Metal nanoparticles are the most promising candidate in this class of materials with inherently strong antibacterial activities amongst the nanomaterials. A summary of the possible bactericidal effects of metal nanoparticles on different bacteria is shown in Table 3. Researchers reviewed a variety of metal nanoparticles including silver, gold, copper, zinc, and super-paramagnetic iron which demonstrated promising antibacterial effects, with silver nanoparticles (AgNPs) being the most effective against bacteria [169–172]. AgNPs exhibit bactericidal activity at concentrations well below their cytotoxicity and exhibit synergistic antibacterial efficacy with conventional antibiotics when used against MDR bacteria [173–176].

The antibacterial mechanisms of AgNPs are still poorly understood despite extensive studies [171,176]. Currently accepted antibacterial mechanisms include cell wall penetration and membrane damage, toxicity associated with metal ion release, and induction of oxidative stress [177–180]. AgNPs have already been used in various biomedical and antibacterial applications and products, including surface coatings on medical devices, topical treatments, wound dressings, dental fillings, personal care products with sanitizing effects, disinfectants, and detergents [180,181]. A recent study demonstrated the potential of AgNP-containing disinfectants as active ingredients for disinfecting surgical masks, effectively improving mask protection by inhibiting the growth of *E. coli, K. pneumoniae,* and *S. aureus* [23]. Over the past few decades, the AgNP market has been growing steadily, with an estimated annual production of more than 500 tons of nanoparticles, which also reflects the widespread interest of AgNPs [182].

Despite the promising antibacterial effects and the wide use of metal nanoparticles in different applications, metal nanoparticles suffer from several drawbacks. The potential toxicity of metal nanoparticles affects the basic functioning of mammalian cells as metal nanoparticles or released metal ions via direct uptake from mammalian cells [24]. Colloidal metal nanoparticles tend to aggregate over time [25]. The aggregation along with increased particle size reduces their peculiar properties at the nanoscale, including their antibacterial activities. Bacteria have the ability to develop resistance to metal nanoparticles by using adhesive flagellin [183]. Phenotypic changes of adhesive flagellin production triggers the aggregation of metal nanoparticles, thereby reducing their antibacterial activity. Metal nanoparticles are potential environmental hazards and difficult to recover or deactivate in solid-waste incineration plants or wastewater treatment systems [184–186]. To overcome the above limitations, metal nanoparticles could be incorporated into other nanomaterials to form nanocomposites, which show a greater dispersion of metal nanoparticles, improved antibacterial activity, and reduced toxicity [187–189]. This will be further discussed in Section 4 below.

Table 3. Antibacterial activity of metal nanoparticles against different bacteria.

NPs	Target Bacteria	References
Ag	*Acinetobacter baumannii, Salmonella typhi, Vibrio cholerae, Bacillus subtilis, S. aureus,* MDR *E. coli, Streptococcus pyogenes, P. aeruginosa,* coagulase-negative *S. epidermis, E. faecalis, K. pneumoniae, Listeria monocytogenes, Proteus mirabilis, Micrococcus luteus*	[23,190–193]
Au	*E. coli, S. aureus, B. subtilis, K. pneumoniae, S. epidermidis, P. aeruginosa, L. monocytogenes, Salmonella typhimurium*	[194–199]
Cu	*Enterobacter aerogenes, E. coli, Klebsiella oxytoca, S. aureus, S. pyogenes, B. subtilis*	[200–204]
Bi	*Streptococcus mutans, C. albicans, E. faecalis*	[205–208]
Cu/Zn bimetal NPs	*E. coli, S. aureus,* MRSA, *Alcaligenes faecalis, Citrobacter freundii, K. pneumoniae, Clostridium perfringens*	[209–211]
Ag/Cu bimetal NPs	*E. coli, S. aureus, A. faecalis, C. freundii, K. pneumoniae, C. perfringens, P. aeruginosa, B. subtilis*	[211–213]
Superparamagnetic iron oxide NPs coated with Ag or Au	*E. coli, S. aureus, P. aeruginosa, E. faecalis, S. epidermidis*	[214]

Abbreviations: NPs, nanoparticles; MDR, multidrug resistant; MRSA, methicillin-resistant *Staphylococcus aureus*.

3.2. Metal Oxide Nanoparticles

Metal oxide nanoparticles offer another alternative promising solution against MDR bacteria. A variety of metal oxide nanoparticles including titanium dioxide, zinc oxide, magnesium oxide, copper oxide, and aluminium oxide have been demonstrated to exhibit antibacterial effects [215], which will not be discussed in detail here. Zinc oxide (ZnO) nanoparticles are the most-established candidate of these metal oxide nanoparticles due to the following reasons: ZnO is one of the most important metal oxide nanoparticles with widespread applications [26] and worldwide production is up to 1 million tons per year [216]. ZnO can be easily biodegraded and absorbed in the body and has been listed as a Generally Recognized as Safe (GRAS) material by the FDA. ZnO nanoparticles have a higher biocompatibility and lower toxicity than other metal oxide nanoparticles [217,218]. The semiconductor properties of ZnO nanoparticles with a wide band-gap energy readily absorb ultraviolet (UV) light. This allows them to act as potential photosensitizing agents for various antibacterial applications [27]. ZnO nanoparticles exhibit multiple antibacterial mechanisms, as described by Figure 6. Current postulated mechanisms include photo-triggered production of ROS [219] and Zn^{2+} ions mediating a poisoning effect [220]. In a study by Azam et al., the potential of ZnO as an antibacterial agent was demonstrated against Gram-positive bacteria (*B. subtilis, S. aureus*) and Gram-negative bacteria (*P. aeruginosa, Campylobacter jejuni, E. coli*) [213]. In another study, ZnO nanoparticles were also shown to be significantly inhibitory against Gram-positive (*S. aureus*) and Gram-negative (*E. coli* and *P. aeruginosa*) strains [221].

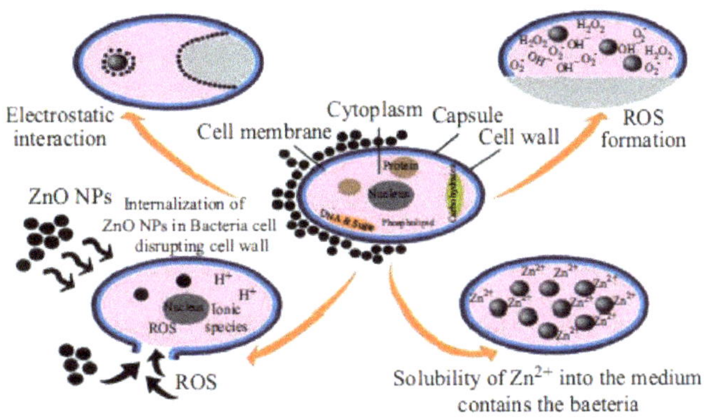

Figure 6. Multiple antibacterial mechanisms that are associated with ZnO nanoparticles. (Adapted with permission from Ref. [222]. 2015, Sirelkhatim et al. More details on "Copyright and Licensing" are available via the following link: https://www.mdpi.com/ethics#10 (accessed on 10 October 2022)).

The aggregation of metal oxide nanoparticles over time remains a problem, which limits their use in vivo applications [223]. One approach to this problem is to disperse metal oxide nanoparticles into a polymer matrix, forming polymer–metal oxide nanocomposites. Polymer–metal oxide nanocomposites have been widely investigated and applied in the textile and polymer industries for various antibacterial applications, which will be discussed in the following section. [224].

4. Nanocomposite/Nanohybrid Antibacterial Materials

Nanocomposites or nanohybrids are a novel class of multiphase materials that exhibit a hierarchical structure, where one phase of the material has at least one dimension in the nanometer range [225]. They have attracted significant attention due to their unprecedented properties compared with their mono-constituent parts, largely attributed to strong reinforcing effects of additional materials. Currently, synthesizing a variety of nanocomposites with unprecedented physical properties and enhanced antibacterial activities is the main focus in the field over the last 10 years. Despite the large volume of studies on nanocomposites, the understanding of the structure–property–activity changes remains in its infancy [226]. It is important to understand the mechanisms behind the property's changes within nanocomposites in order to design materials with enhanced improvements in the desired properties for a specific purpose. Sections 4.1–4.5 discusses the development and potential applications of the nanocomposite antibacterial materials. Of note, graphene oxide is especially emphasized in the section below due to its versatility to form nanocomposites with organic and inorganic metallic nanomaterials, respectively.

4.1. Polymer–Metal Nanocomposite Nanoparticles

Polymer–metal composite nanoparticles are another promising solution to achieve a greater dispersion of metal nanoparticles and prevent metal nanoparticle aggregation. Polymer–metal composite nanoparticles comprise a metal nanoparticle core surrounded by a polymer shell with the alkyl tail arranged toward the surrounding environment. Polymer–metal composites are essentially insoluble in water and the colloidal stability of the polymer–metal composite nanoparticles in aqueous environments has a huge potential for in vivo antibacterial therapies [227]. The polymer macromolecular matrix acts as a reaction chamber for metal nanoparticle synthesis, a capping agent to prevent nanoparticle aggregation and as a scaffold for nanoparticle immobilization [228]. Moreover, synergies between the polymer and the metal nanoparticles confer the nanocomposite with unprecedented performance and improved antibacterial properties [229]. Tamayo et al. summarized the synthesis, properties, and recent applications of polymer composites with metal nanoparticles [230].

The incorporated metal nanoparticles are focused on the use of gold (AuNPs) and silver (AgNPs) due to their antimicrobial properties, catalytic activity, and conductivity properties enabling a wide range of applications.

4.1.1. Development of Synthesis Approaches for Polymer–Metal Nanocomposites

Both in situ and ex situ approaches can be employed to synthesize polymer–metal composite nanoparticles and polymer-matrix metal nanocomposites. In the last decade, several studies have been conducted to develop and improve synthetic methods at higher efficiencies resulting in improved antimicrobial outcomes.

By using in situ methods, the precursor of the nanoparticle is required to be dispersed in a monomeric solution before polymerization. The metal ions can be reduced into the polymer matrix or simultaneous metal ion reduction and polymerization can occur [230]. In general, these in situ reduction methods need a relatively long time to produce nanocomposite films. Kazuhiko et al. developed a rapid and scalable synthetic method exploiting use of a mid-infrared laser, CO_2 laser, at 10.6 μm without the use of reducing agents [231]. The polymer film (polyvinyl alcohol (PVA) or polyethylene glycol (PEG)) containing Ag ions were coated on a glass substrate and then the CO_2 laser was used to heat the substrate. Subsequently, the thermal energy was absorbed by the polymer film, causing Ag ion reduction. Eventually, Ag-PVA or Ag-PEG nanocomposite films were formed in several seconds. This process is industrially scalable by increasing the power of the CO_2 laser.

Ultrasound is a promising tool to be applied in for the in situ production of polymer–metal nanocomposites. Ultrasound radiation was employed as a homogenizing tool to fabricate composites with homogeneously dispersed metal nanoparticles [232]. The ultrasound was used to disperse organic liquids of polymerizing monomer (pyrrole) in the aqueous solution of the oxidizer (Ag^+ or $AuCl^-$). The aqueous solution was placed in an ultrasonic chamber and droplets of the organic solution were added continuously until achieving a volume ratio of 4:1. After polymerization, the nanocomposites could be obtained at the liquid–liquid interface [232]. Wan et al. used ultrasound as both an initiating and reducing agent in the nanocomposite preparation process, shown in Figure 7 [233]. Tertiary amine-containing polymeric nanoparticles were produced by ultrasound-initiated polymerization-induced self-assembly (sono-PISA), following which, the metal ions (Au and Pd) were reduced in situ by radicals generated via the sonolysis of water, forming polymer–metal composites.

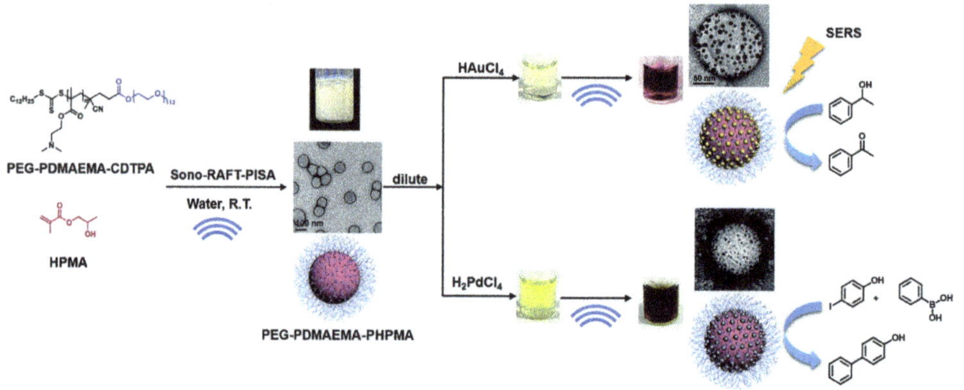

Figure 7. Synthesis of tertiary amine-containing polymeric nanoparticles, and in situ formation of the Au and Pd nanocomposite by ultrasound. (Adapted with permission from Ref. [233]. 2021, Wan et al. More details on "Copyright and Licensing" are available via the following link: https://www.mdpi.com/ethics#10 (accessed on 10 October 2022).)

In contrast, metal nanoparticles are synthesized before they are incorporated into the polymer via an ex situ method. The subsequent deposition of the nanoparticles into the polymer can exploit processes such as melt compounding or solution blending [230]. However, a significant drawback exists when using ex situ methods, which is the fact that the nanoparticles are not optimally distributed in the polymer.

4.1.2. Synergistic or Combined Antibacterial Effects When Using More Than Just a Metal Nanoparticle Agent

A combination of metal nanoparticles and cationic polymers also facilitates the enhanced antibacterial activity of a composite nanoparticle, possibly due to the synergism between the antibacterial mechanisms from two different nanomaterials [234–237]. Nanoparticle formation was also expected to increase the local density of cationic polymer, leading to stronger binding on the negatively charged bacterial membranes [234]. This polyvalent interaction between cationic polymers and bacterial membranes is followed by the synergistic antibacterial mechanisms, including bacterial membrane disruption, internalization of composite nanoparticles, inhibition of intracellular enzymatic activity, and eventual cell death [234]. Imidazole-capped chitosan–gold nanocomposites exhibited enhanced antimicrobial activity to eradicate staphylococcal biofilms in a rabbit wound infection model [238]. The antibacterial mechanism of the composite nanoparticles also involved the binding of cationic polymer to the bacterial surface, and the subsequent synergistic effects from the gold nanoparticles, imidazole, and the chitosan polymer to strongly eradicate the biofilm [238].

Polymer–metal composite nanoparticles could potentially solve the environmental hazards associated with many metal nanoparticles. Richter et al. postulated that a metallic core is not necessary for the antimicrobial action [187]. Instead of synthesizing the entire nanoparticle of metal, the composite nanoparticle can be produced by infusion with a minimum amount of silver ions to the biodegradable lignin core, followed by surface functionalization with a layer of cationic polyelectrolyte [187]. The resulting composite nanoparticles exhibited broad-spectrum bactericidal activity against *E. coli*, *P. aeruginosa*, and a quaternary-amine-resistant *Ralstonia*. This is attributed to the enhanced binding to bacterial membranes by the polyelectrolyte shell and the synergistic antibacterial activity between the silver ions and the polyelectrolyte. The required silver ions were ten times lesser than when using conventional silver nanoparticles. The gradual diffusion of silver ions from the silver-infused lignin core composite nanoparticles into water will rapidly lose their post-utilization activity and be biodegradable in the environment after disposal [187].

4.1.3. On the Potential Clinical Use of Antibacterial Polymer-Matrix Metal Nanocomposites

Biocompatible and safe antibacterial materials are constantly sought to avoid inflammatory syndromes in patients. The formulation of polymer–metal composite nanoparticles generally improves their biocompatibility and reduces the toxicity associated with metal nanoparticles owing to protection by the polymeric shell. For instance, the viability of NIH3T3 cells was not affected at a dosage exceeding 20 times that of the minimum inhibitory concentration (MIC) of the polymer–silver composite nanoparticle, showing the low toxicity of these materials to mammalian cells [234]. Lu et al. further demonstrated that imidazole-capped chitosan–gold nanocomposites did not display hemolytic activity and significant toxicity towards L929 cell line [238]. Pryjmaková et al. modified the surface of polyethylene naphthalate (PEN) by a 248 nm KrF excimer laser and subsequently, Ag and Au nanowires were incorporated onto the modified PEN surface by vacuum evaporation [239]. The resulted nanocomposites displayed antibacterial effects against Gram-negative bacteria (*E. coli*) and Gram-positive bacteria (*S. epidermidis*) via a 24-hr incubation drop plate test and were suggested as a non-toxic material by a WST-1 cytotoxicity test. In addition to the complete eradication of the biofilm, accelerated wound healing by a composite nanoparticle was demonstrated in a rabbit model [238]. These studies demonstrate the great potential of composite nanoparticles as novel antibacterial agents against bacterial infections.

4.2. *Polymer-Matrix Metal Oxide Nanocomposites*

Fine dispersions of metal oxide nanoparticles can be achieved by forming polymer-matrix metal oxide nanocomposites in a manufacturing process similar to polymer-matrix metal nanocomposites. There is an increasing interest in using metal oxide nanoparticles to replace metal nanoparticles for synthesizing polymer-matrix metal oxide nanocomposites [240–243]. Metal oxide nanoparticles, especially zinc oxide nanoparticles, are more desirable than metal nanoparticles as nanofillers in forming polymer-matrix nanocomposites [244–247]. Zinc oxide nanoparticles feature new UV-absorption and photosensitizer characteristics, with higher biocompatibilities and lower toxicities than metal nanoparticles [218,248]. Furthermore, the low cost of production and high stability of zinc oxide nanoparticles present advantages over conventional metal nanoparticles, even in extreme synthesis conditions [249]. The advantages of metal oxide nanoparticles as nanofillers in forming polymer-matrix metal oxide nanocomposites has open a new avenue for research into novel bio-nanocomposites for use as antimicrobial surfaces in various antibacterial applications, which will be depicted in following sections.

Polymer-matrix metal nanocomposites exhibit several advantages. Polymer flexibility allows the final product to be fabricated into complex structures or forms for various antibacterial applications, of which the details are shown in Table 4. The polymeric matrix immobilizes metal nanoparticles and prevents their aggregation, thus extending the antibacterial activity of the metal nanoparticles [228]. Localized release of metal nanoparticles to the desired application site can also be achieved, reducing in vivo toxicity and the environmental hazards caused by undesirable release of metal nanoparticles [250]. Synergistic antibacterial activity between metal nanoparticles and polymers is obtained with inherent antibacterial activities [229]. The strong interfacial binding and intermolecular interactions between the well dispersed metal nanoparticles and the polymer matrix further enhance the mechanical properties of the polymer, including the tensile strength, Young's modulus, yield stress, and ductility [251].

Table 4. Versatility of polymer-matrix metal nanocomposites for various antibacterial applications.

Structures/Forms	Potential Antibacterial Applications	References
Film	Surface coating	[252]
	Food packaging	[253]
	Wound dressing	[254]
Scaffold	Bone tissue engineering	[255]
	Wound dressing	[256]
Membrane	Wastewater treatment/water filtration	[257]
Sponge	Wound dressing	[258]
Gel	Antifouling/surface coating	[259]
	Tissue engineering	[260]
	Wound healing	[261]

4.2.1. Development of Synthesis Approaches for the Industrial Production of Polymer-Matrix Metal Nanocomposites

A prerequisite to the aforementioned advantages exhibited by polymer-matrix metal nanocomposites is the formation of homogenous dispersions of metal nanoparticles in the polymer matrix without metal nanoparticle aggregation [262]. The delicate synthesis conditions to meet this prerequisite is often time-consuming, laborious, and difficult to be industrialized [263]. Therefore, recent studies have focused on developing facile and convenient synthesis approaches, aiming to produce polymer-matrix metal nanocomposites on industrial scales [228,250,264]. Other recent review papers have summarized developments, so we will not discuss these in detail [240–242].

For instance, Tran et al., developed a simple one-pot synthesis method in producing polymer matrix silver nanocomposites [228]. The ionic liquid medium, butylmethylimmidazolium chloride, was utilized as the only reaction medium for dissolving the biopolymer keratin and cellulose, and reduction of a silver ion precursor in the polymeric matrix. The synthesized polymer-matrix silver nanocomposite was found to retain the enhanced mechanical strength by cellulose and controlled release of silver nanoparticles by keratin, with a homogenous dispersion of silver nanoparticles. At 0.48 mmol of silver content, the nanocomposite demonstrated good biocompatibility and excellent antibacterial activity against *E. coli*, *P. aeruginosa*, MRSA, and vancomycin-resistant *E. faecalis* (VRE). An in vitro release assay demonstrated that less than 0.02% of the silver nanoparticles were released from the nanocomposite even after 7 days of soaking in solution, indicating good immobilization of silver nanoparticles using this simple one-pot synthesis method [228].

A scalable approach was recently developed to produce a silver-nanoparticles-doped nanoclay–polylactic acid composite nanocomposite, which involved doped nanoclay with minimal alteration to the fabrication processes and industry standard equipment [264]. Loading the nanoclay can significantly reduce the affinity of the nanocomposites for bacterial adhesion. With this synthesis method, only a 0.1 wt % of silver loading content was required to have satisfactory antibacterial activity. *S. aureus* and *E. coli* numbers were reduced by 91.3% and 90.7% after 48 h of incubation. The material costs associated with the silver-loading content is dramatically reduced compared with other studies which utilize at least 1 wt % of silver nanoparticles in the polymer-matrix metal nanocomposite to achieve a 90% reduction in bacterial numbers [265,266]. This is a great advantage for industrial production, in which the high costs associated with higher loading amounts of metal nanoparticles is a considerable problem. In addition, 3D printing is a promising method to produce metal oxide nanocomposites, with the advantages of keeping the integrity and functionality of the materials and reduce waste from traditional manufacturing methods [267].

4.2.2. The Application of Polymer-Matrix Metal Oxide Nanocomposites as Self-Sterilizing Antimicrobial Surfaces in Healthcare Environments

There are some recent literature reviews on the application of polymer-matrix metal oxide nanocomposites as self-sterilizing antimicrobial surfaces in healthcare environments which will not be repeated here [268–270]. The antibacterial and photosensitizing activity of ZnO nanoparticles has been well exploited in the absence or presence of light irradiation [271]. With the exploitation of their light absorption characteristic, ROS are produced from ZnO nanoparticles to act on bacteria, leading to a self-cleaning or self-sterilizing effect of the polymer-matrix zinc oxide nanocomposites [272]. For instance, Sehmi et al. and Ozkan et al. have developed self-sterilizing surfaces that was coupled with light-activated photodynamic therapy in killing bacteria [272,273]. Both studies showed the polymer matrix zinc oxide nanocomposites demonstrated lethal photosensitization of *E. coli* and *S. aureus* under white light irradiation that has a similar light intensity to that in a clinical setting [272,273]. This could potentially lower the rates of healthcare-associated infections by eliminating bacterial transfer in healthcare environments.

4.2.3. Wound Healing Applications of Polymer-Matrix Metal Oxide Nanocomposites

Polymer-matrix metal oxide nanocomposites are an attracting candidate in wound healing applications. Gobi et al. summarized the recent applications of nanocomposites in wound dressings [274]. A novel polymer-matrix metal oxide nanocomposite comprising a castor oil polymeric matrix reinforced with a chitosan-modified ZnO nanocomposite was recently developed [275]. This novel bio-nanocomposite showed enhanced mechanical properties, porosity, water absorption, hydrophilicity, water vapor transmission rate, and oxygen permeability [275]. These enhanced properties are important for wound healing, which provides porosity to absorb wound exudates and water, enables a moist wound healing environment, a cooling effect for pain alleviation, and gases to exchange for ventilation. ZnO nanofillers also enhanced the antibacterial activity and keratinocyte migration of a polymer-matrix zinc oxide nanocomposite, leading to stronger antibacterial activity that prevented the reoccurrence of a bacterial infection and promoted healing [276]. Recent research has reported a prepared nanocomposite consisting of a Lawsone-loaded o-carboxymethyl chitosan and ZnO which was evaluated against bacterial strains such as *Salmonella*, *S. aureus*, *P. aeruginosa*, and *E. coli* [277]. This prepared nanocomposite tended to prevent the evolution of these harmful bacteria compared to an o-carboxymethyl chitosan or nano-zinc oxide alone, further supporting this advantageous strategy of using a polymer-matrix ZnO nanocomposite in wound dressings. Finally, a polymer-matrix ZnO nanocomposite demonstrated promising in vivo efficacy, biodegradability, cytocompatibility, and promoted cell attachment on the material [278,279]. Taken as a whole, polymer-matrix metal oxide nanocomposites could possibly satisfy all the required standards as wound materials, highlighting their huge potential in wound healing applications.

4.2.4. Food Packaging Applications of Polymer-Matrix Metal Oxide Nanocomposites

For food packaging applications, the addition of ZnO nanoparticles as nanofillers to biodegradable polymeric materials greatly enhances the physiochemical properties and antibacterial activities of the resulting bio-nanocomposites to protect the environment [270,280,281]. ZnO nanoparticles create a barrier effect to hinder the diffusion of the decomposition products from the polymer matrix to the gas phase, which further improve polymer thermal stability and avoid thermal degradation under the wide polymer melt processing window [282]. In addition, ZnO nanoparticles act as a nucleating agent in raising the crystallinity level of the polymer matrix [283]. The combined effect of the increased crystallinity of the polymer along with the barrier effect of ZnO nanoparticles creates a highly tortuous path for the gases, water vapors, and organic compounds [284].

The barrier properties to gases, water vapors, and organic compounds subsequently improve the product quality and shelf life by blocking the diffusion of moisture and oxygen. Mechanical performance such as stiffness, glass transition temperature, tensile strength, and toughness is also enhanced because of the strong polymer matrix–ZnO nanofiller interactions [282,285]. The antibacterial and UV-absorption properties of ZnO nanoparticles inhibit the growth of food-borne pathogens and prevent the photo-oxidative degradation of food, respectively [286]. Taken together, polymer matrix–zinc oxide nanocomposites are promising materials to be used as cutlery, overwrap films, and containers in preventing growth of food-borne pathogens and achieving good quality packaged food with extended shelf-life.

4.3. Graphene Oxide–Metal Nanocomposites

As mentioned in Section 3, the shortcomings of metal nanoparticles limit their potential for medical applications. To overcome these issues, many nanocomposites composed of metal nanoparticles and graphene have been prepared experimentally and studied against various bacterial strains [287–292]. Some of the available literature reviewed the development of graphene–metal matrix nanocomposites which will not be repeated here [293–296]. Among them, a combination of graphene oxide and silver nanoparticles to form nanocomposites has attracted a lot of attention as antibacterial agents in antibacterial therapies, since Ag and its compounds have been used since the time of the ancient Egyptians. The antibacterial and antiviral properties of Ag, Ag ions, and Ag-based compounds have been thoroughly researched [202,297,298]. With the incorporation of GO as the supporting matrix, silver nanoparticles could be dispersed in aqueous solution while minimizing the aggregation problem that would otherwise greatly affect the antibacterial activity of the silver nanoparticles [299]. The large surface area and abundant functional groups on the basal plane of GO allows GO to interact with silver ions or silver nanoparticles through electrostatic interactions, charge–transfer interactions and physical absorption [300]. This allows GO–Ag nanocomposites to be synthesized through loading of pre-synthesized silver nanoparticles into GO (ex situ approach) or via reduction of silver ions in a graphene matrix to form silver nanoparticles in situ [301–305].

4.3.1. Development of Synthetic Approaches for Improving the In Vivo Performance of Graphene Oxide–Metal Nanocomposites

The synthesis process of GO/reduced GO (rGO)–Ag nanocomposites involves harsh conditions, as well as highly toxic reducing agents and organic solvents, which minimizes their use in biomedical applications, which is an unmet area of need that requires more exploration [306,307]. Despite the good dispersity of GO–Ag in aqueous solution, it has been discovered that GO aggregates irreversibly in physiological solutions over time [308]. This greatly affect its bioavailability, significantly weakening its antibacterial efficiency and long-term effectiveness. In an effort to solve the problem, the (polyethylene glycol) PEGylation of GO was carried out for long term antibacterial activity and stability of GO–Ag in physiological solution [309]. The PEGylated GO–Ag nanocomposite remained stable in a series of complex media over one month and resisted centrifugation (Figure 8). In contrast, non-PEGylated GO–Ag aggregated to varying degrees in the media after 1 h, and complete precipitation was observed after 1 week of equilibration. GO−PEG−Ag nanocomposites displayed remarkable long-term antibacterial activity after 1 week of storage in physiological saline, preserving >99% antibacterial activity against *S. aureus* and >95% antibacterial activity against *E. coli*. GO−PEG−Ag inhibited bacterial growth in nutrient rich Luria–Bertani (LB) broth for at least one week, and the repeated usage of GO−PEG−Ag up to three times did not reduce the antibacterial efficacy. In contrast, unmodified GO−Ag exhibited a >60% decline in antibacterial activity after 1 week of storage in physiological saline. This study provides a direct solution for the synthesis of homogenously dispersed and stable GO–Ag nanocomposites under physiological conditions. This result was also confirmed by a subsequent study, in which ternary hybrids of PEG-functionalized GO with Ag nanoparticles exhibited excellent bactericidal effects against *E. coli*, and it was found that those with smaller Ag nanoparticles (8 nm) showed better antibacterial activity than those with larger nanoparticles (50 nm) [310]. Furthermore, modification of GO with polyethyleneimine polymers dramatically enhanced the long-term antibacterial activity and stability of the GO–Ag nanocomposite [311]. In addition to this, Parandhaman and his colleagues recently designed GO–Ag nanocomposites functionalized with the natural antimicrobial peptide poly-L-lysine with remarkably improved stability and adhesion to *S. aureus* biofilms [312]. Notably, poly-L-lysine functionalization prevented the leaching of anions, thereby reducing the cytotoxicity of the graphene–silver nanocomposites. In order to obtain nanomaterials with long-term and stable antibacterial activity, a facile and green method has also been proposed to prepare AgNPs/polymer/GO composites with catalytic and antibacterial activities via the incorporation of furan-functional poly(styrene-*alt*-maleic anhydride) [313].

4.3.2. Potential of Graphene Oxide–Metal Nanocomposites for In Vivo Therapies

In terms of biological activity, GO–Ag nanocomposites have been shown to demonstrate synergistic antibacterial activities against planktonic bacteria and biofilms, with low cytotoxicity and good biocompatibilities [314–317]. The enhanced antibacterial activity of GO–Ag nanocomposites is often ascribed to the synergistic activity of GO and Ag; however, the full antibacterial mechanism remains to be elucidated [299]. Malik et al. have also demonstrated that GO–Ag nanocomposites exhibit significantly enhanced growth inhibition of *E. coli*, *S. aureus*, and *P. aeruginosa* relative to silver nanoparticles alone [291]. Recent studies have accomplished the surface functionalization of GO and Ag nanoparticles by using lantana plant extract, and the results also affirmed the potential of GO–Ag nanocomposites as antibacterial agents against biological pollutants [290]. The negatively charged oxygen-containing groups of graphene oxide can absorb Ag ions through electron absorption, which can improve the confinement of Ag nanoparticle agglomeration and burst release, and synergistically enhance their antibacterial properties [292]. Intriguingly,

GO–Ag nanocomposites exhibit species-specific bactericidal mechanisms, with cell wall disruption being observed against *E. coli* and inhibition of cell division against *S. aureus* [318].

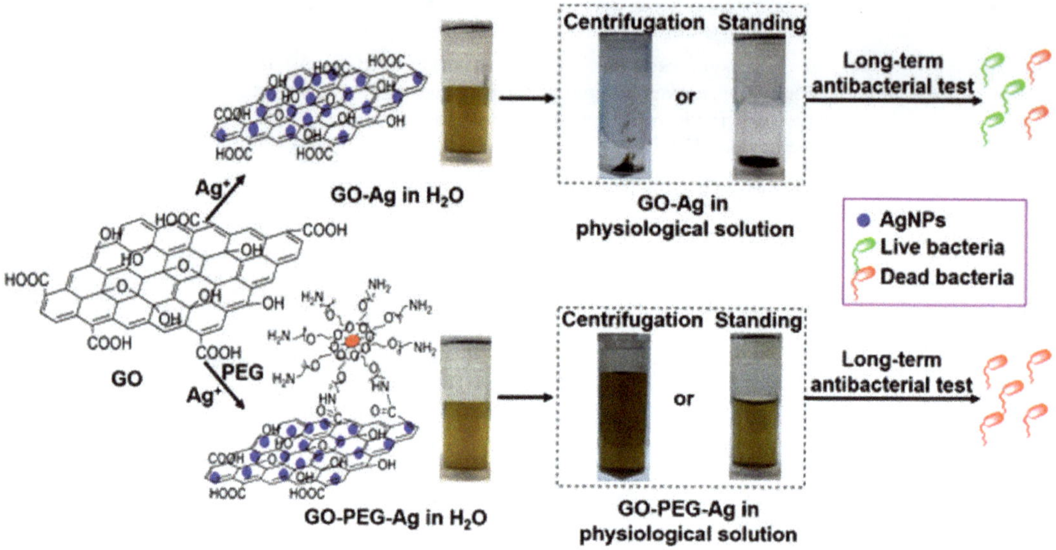

Figure 8. Long-term stability and antibacterial effectiveness of a PEGylated GO–Ag nanocomposite. (Adapted with permission from Ref. [309]. 2017, Zhao et al. More details on "Copyright and Licensing" are available via the following link: https://www.mdpi.com/ethics#10 (accessed on 10 October 2022).)

The promising physical properties of GO, along with its synergistic activity with Ag nanoparticles, hold great potential for a targeted nanocomposite system [319]. A photothermal nanocomposite was produced which was composed of hyaluronic-acid-coated Ag nanoparticles that were integrated with GO [319]. Hyaluronic-acid-coated Ag nanoparticles confer additional protection by preventing the release of metal ions to surrounding mammalian cells. Upon encountering bacteria that secrete hyaluronidase, such as *S. aureus*, hyaluronic acid is degraded, followed by interaction of the GO–Ag nanocomposite and bacteria to further enhance the antibacterial action. Together with the photocatalytic characteristic of GO, local photothermal therapy under light irradiation could be achieved to further enhance the antibacterial activity of the GO–Ag nanocomposite [291].

4.3.3. Potential of Graphene Oxide–Metal Nanocomposites to Reduce Membrane Biofouling Issues for Water Decontamination and Filtration

The intrinsic characteristics of GO, including its availability as single-atomic-thick sheets, high hydrophilicity, extraordinary electrical, thermal, mechanical, structural properties, and low systemic toxicity could potentially reduce membrane biofouling issues for water decontamination or filtration [320,321]. GO–Ag nanocomposites are also a promising membrane surface modifier that contributed to enhanced membrane hydrophilicity, wettability and permeability, and good water influx [322–325]. Surface modification of membranes using GO/Ag nanocomposites exhibited stronger antimicrobial activities than AgNP-modified membranes and GO-modified membranes, without significantly altering the membrane transport properties [326]. The feasibility of using GO–Ag nanocomposites in membrane regeneration for a long-term anti-biofouling effect was demonstrated by conducting, in situ, the Ag-formation procedure to regenerate AgNPs on GO–Ag-modified membranes [327]. This potentially solves the problem of weakening biofouling properties of the functionalized GO–Ag nanocomposite membranes over time due to constant leaching of silver ions throughout the process. More complex membranes containing GO, Ag, and metal-organic frameworks (MOFs) in PES were prepared for water treatments, exploiting the synergistic effects of graphene oxide and silver to enhance the anti-biofouling properties of the membranes [328]. In conclusion, the metal oxide/graphene nanocomposites exhibit enhanced antibacterial properties under visible light irradiation and have great potential as photocatalysts in the field of water purification [321]. Nevertheless, more studies are required to examine the long-term usage, membrane reusability, and regeneration potential of functionalized GO–Ag nanocomposite membranes.

4.4. Graphene Oxide–Polymer Nanocomposites

Graphene oxide–polymer nanocomposites exhibit enhanced antibacterial activity, biocompatibility, hemocompatibility, hydrophilicity, and stability, compared with the polymeric based nanomaterials [329–331]. In addition, GO can greatly reinforce the mechanical properties of GO–polymer nanocomposites, including their breaking strength, Young's modulus, compressive strength, flexural strength, and tensile strength [332–334]. The reinforcing effect is usually explained via strong interactions and bonding between the homogenously dispersed GO and polymeric components [332,333].

The alignment of GO sheets on the polymer film has been suggested to greatly affect the antibacterial activity of the resulting graphene oxide-based polymer nanocomposite [335]. Lu et al. synthesized graphene oxide–polymer nanocomposites by aligning GO in planar, vertical, and random orientations with the aid of a magnetic field (Figure 9A). GO was then immobilized by cross-linking with the surrounding polymer matrix, followed by oxidative etching to expose GO on the surface (Figure 9B). The vertically aligned GO nanosheets on the polymer film exhibited enhanced antibacterial activity compared with the random and horizontal orientations. Mechanistic examinations revealed that direct, edge-mediated contact with bacteria was the major mechanism in causing a greater physical disruption of the bacteria membranes (Figure 9C) [335]. Subsequently, greater levels of intracellular electron donors, for instance, glutathione, would release into the external environment upon membrane disruption, favoring GO to induce antibacterial activity via an oxidative stress mechanism [335]. This study highlights the importance of GO alignment and provides direct implications for the designing of GO–polymer nanocomposite films with enhanced antibacterial activities.

The abundant functional groups present on GO–polymer nanocomposites provide various interactions with nanoparticles, creating GO–polymer-based metal nanocomposites with superior characteristics. For instance, GO–chitosan nanocomposites have been demonstrated to act as both nucleation sites for calcium phosphates mineralization and absorption sites for nanoparticles [336]. The resulting GO/chitosan nanocomposites comprise micro- and nanohierarchical porous structures that allow cell attachment and proliferation after biomineralization [336]. In addition, the immobilization of the AgNPs and growth-factor-encapsulated nanoparticles on the GO–chitosan nanocomposites greatly enhances the antibacterial activity and osteo-inductivity, respectively [336]. Taken together, GO–polymer nanocomposites have great potential for use as multifunctional nanocomposite materials in various antibacterial applications. This is due to the substantial property enhancements and their ability to interact with various metal or metal oxide nanoparticles. Díez-Pascual and Luceño-Sánchez described several antibacterial applications of GO–polymer nanocomposites [325].

4.4.1. Development of Synthetic Approaches for the Production and Use of Graphene Oxide–Polymer Nanocomposites

Various synthesis routes can be applied to prepare graphene oxide–polymer nanocomposite with covalent or non-covalent interactions [337]. Shahryari et al. summarized a range of synthesis routes in a thorough review [338]. Generally, three common methods are utilized: solution blending, melt blending, and in situ polymerization [338,339]. The solution blending approach is the most commonly used method to synthesize graphene oxide–polymer nanocomposites in a dispersion form, due to its convenience and ease of implementation. With this method, facile synthesis is achieved by simply blending the polymer with GO in solution, followed by sonication, or magnetic stirring, or shear mixing to obtain homogenous dispersions of the nanocomposite products as depicted in Figure 10 [340,341]. In Rusakova et al.'s study, an ultrasonic bath was used to disperse GO in a solution of styrene or a polyester resin in styrene, adding toluene and benzoyl peroxide and then, the mixture was placed into a special mold for polymerizing the nanocomposite [342].

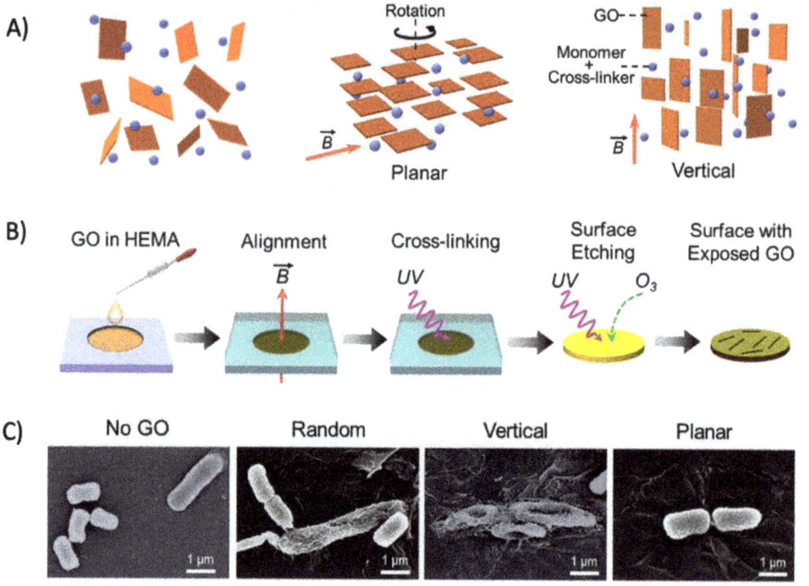

Figure 9. (**A**) Schematic illustration of the GO film with different orientations. From left to right: random, planar, and vertical orientations. (**B**) Schematic illustration of the film fabrication procedure. A magnetic field is applied to control the orientation of dispersed GO nanosheets, with the orientation preserved by photo-cross-linking the dispersing agents. (**C**) SEM micrographs pictured the intact *E. coli* morphology treated by no-GO film, retained *E. coli* morphological integrity treated by randomly aligned- or planar aligned-GO film and flattened and wrinkled *E. coli* morphologies after being treated by a vertically aligned-GO film. (Adapted with permission from Ref. [335]. 2017, Lu et al. More details on "Copyright and Licensing" are available via the following link: https://www.mdpi.com/ethics#10 (accessed on 10 October 2022).)

Melt blending is applied commonly in industry due to its low cost and scalability. Unlike solution blending, melt blending eliminates the use of any toxic solvents for the dispersion. Following melting the polymer at high temperature, the GO nanofillers are dispersed into a polymer matrix by mechanical shear forces. However, it is likely to induce particle aggregation by thermal heating and local mechanical stresses during melt blending. A study employed wet phase inversion to prepare a sponge-like structure of a polymer/GO/solvent mixture as an exfoliating agent. It was subsequently ground into powder form and mixed with a polymer using melt blending [343]. The hybrid method was exemplified with two polymers, polyamide 6 (PA6) and poly (ethylene-co-vinyl acetate) (EVA). The produced nanocomposites exhibited enhanced dispersions with improved mechanical and dynamic–mechanical properties, compared with the nanocomposites prepared via melt blending [343].

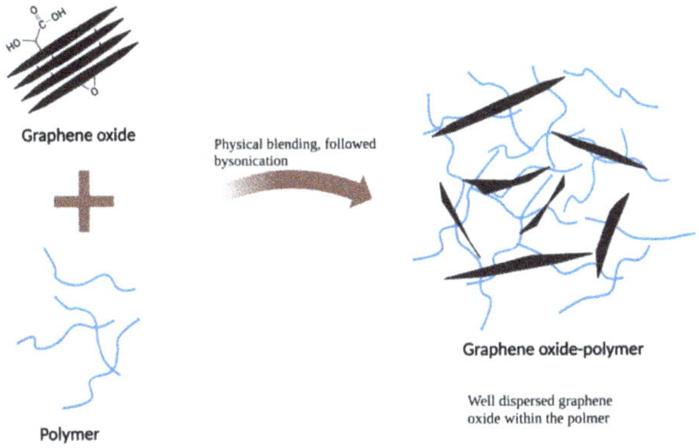

Figure 10. Formation of a well-dispersed graphene oxide–polymer nanocomposite with graphene being embedded in polymeric matrix.

In situ polymerization can lead to a more homogeneous dispersion of the graphene derivatives within the polymer matrix than the two methods mentioned above. In this technique, GO was initially mixed with the monomers, followed by polymerization. The polymerization can be initiated by heat or radiation. Microwave heating has been demonstrated to be a particularly efficient method to produce well-dispersed rGO polymer nanocomposites [344,345]. Hou et al. exploited microwave heating to simultaneously reduce GO and conducted a nitroxide-mediated polymerization of styrene, forming rGO-polystyrene nanocomposites [346].

The resulting nanocomposite could be further modified into various forms and shapes for different antibacterial applications [336,347,348].

4.4.2. Application of Graphene Oxide–Polymer-Based Metal Nanocomposites in Wound Healing

Graphene oxide–polymer-based metal nanocomposites have a potential emerging role in wound healing application as in situ-forming hydrogels. Yan et al. have synthesized a novel GO–polymer-based metal nanocomposite (PEP-Ag@GO) for a wound healing application [349]. PEP-Ag@GO comprises a poly(Nisopropylacrylamide$_{166}$-co-n-butyl acrylate$_9$)-poly (ethyleneglycol)-poly(N-isopropylacrylamide$_{166}$-co-n-butyl acrylate$_9$) copolymer, denoted as PEP and AgNPs decorated on reduced GO nanosheets, denoted as Ag@GO. An aqueous mixture of the PEP-Ag@GO could transit to a hydrogel immediately in situ upon contact with the skin area that has a higher temperature than the transition temperature of PEP-Ag@GO (30 °C). The in situ formation of the hydrogel allows the treatment of wound areas that are difficult to access and minimizes tissue damage that is associated with changing of the wound dressing material [350]. More importantly, the strong interactions and bonding that arises from the PEP-GO give rise to good stability of the composite network. Therefore, the PEP-Ag@GO hydrogel resisted a transition back to liquid form at lower temperatures, even at 5 °C. This is particularly advantageous as commonly synthesized thermo-responsive hydrogels would phase transition back to a liquid at lower temperatures [351]. In vitro and in vivo experiments have also demonstrated good biocompatibility and enhanced antibacterial activity of a PEP-Ag@GO against MRSA (Mu50), leading to a much faster wound healing rate of an MRSA-infected skin defect [349].

4.4.3. Application of Graphene Oxide–Polymer-Based Nanocomposites in Water Treatment

Membrane technologies have been widely applied in water treatments, but the growth of biofilms causes deterioration of water filtration membranes. Therefore, GO–polymer-based nanocomposites are promising materials for use in water treatments due to their antibacterial and antifouling properties. A recent study discussed newly developed GO-based nanocomposites in water treatments and identified limitations for future improvements [352]. Zeng et al. modified poly(vinylidene fluoride) (PVDF) membranes by covalently immobilizing graphene oxide quantum dots (GOQDs) to exert the antibiofouling and antibacterial properties and maintain excellent permeation properties of PVDF membranes [353]. It was found that the GOQDs–PVDF membrane inhibited the growth of *E. coli* and *S. aureus* more effectively than two-dimensional GO sheets. Cheng et al. evaluated the performances of GO-coated and GO-blended polysulfone ultrafiltration membranes and GO-coated membranes presented lower declines in water flux and higher flux recoveries than GO-blended membranes [354]. They showed strong antibacterial activity and biofouling resistance against *E. coli*. To enhance the antibacterial activity, metal nanoparticles can also be incorporated into GO–polymer-based nanocomposites. Mahmoudi et al. incorporated Ag-decorated GO nanoplates into polysulfone membranes, which demonstrated superior antibacterial properties and inhibited the formation of biofouling [355].

4.4.4. Application of Graphene Oxide–Polymer-Based Nanocomposites in Food Packaging

Graphene oxide–polymer-based membranes are attractive candidates to be applied in food packaging due to the fact that their incorporation increases the permeability, selectivity, barrier, and antibacterial activities of packaging and prolongs the durability of the material. For example, GO was cross-linked with chitosan at 120 °C to form nanocomposite films which improved the tensile strength and thermal stability and exhibited antimicrobial properties against *E. coli* and *B. subtilis* [356]. As a result of these enhancements, the films are suitable for use in food packaging. Multiple reviews have been conducted on the development and applications of graphene oxide nanocomposites in food packaging [357–360]. However, the current state of GO–polymer composite membranes in food packaging applications are yet to be commercialized to the best of our knowledge.

4.5. Lipid Polymer Hybrid Nanoparticles

Lipid polymer hybrid nanoparticles (LPHNPs) have emerged as a potentially superior nanocomposite delivery system by combining the advantages and mitigating the limitations associated with liposomes and polymeric nanoparticles alone [361–366]. LPHNPs comprise a biodegradable polymeric core for drug encapsulation, an inner lipid layer surrounding the polymeric core, and an outer polymeric stealth layer (Figure 11). The polymeric core provides high structural integrity, mechanical stability, a narrow size distribution, and higher lipophilic drug loading capacities of the LPHNPs [367]. The inner lipid layer confers the biocompatibility and delays the polymeric degradation of LPHNPs by limiting inward water diffusion, contributing to the sustained release of the composite system [368]. Finally, the outer polymeric stealth layer provides steric stabilization of the nanoparticles, acting as a stealth coating that enhances the in vivo circulation time and protect the composite system from immune recognition [369].

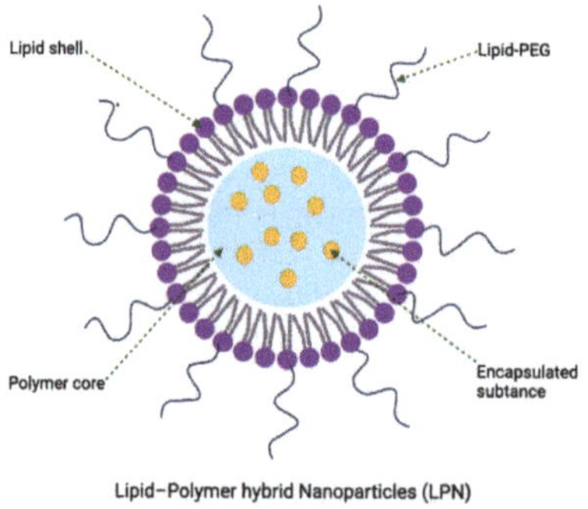

Figure 11. Structure of a lipid–polymer hybrid nanoparticle that comprises a polymeric core, an inner lipid layer, and an outer lipid–polyethylene glycol (PEG) shell (polymeric stealth layer).

4.5.1. Development of Lipid Polymer Hybrid Nanoparticles Using a Quality-by-Design Approach

LPHNPs are still in their early developmental stage for antibacterial applications despite the advances witnessed in cancer therapeutics [370]. The synthesis conditions of antibiotic loaded LPHNPs with desired particle characteristics is highly challenging due to the strong interplay between process variables. Therefore, a quality-by-design approach is often utilized to customize the LPHNPs in meeting the requirements or desired particle characteristics [371–375]. A quality-by-design approach is a statistical method that applies multiple factorial concepts and modelling to determine the interactions between two or more process variables and the desired and observed response conditions [374,376–378]. For instance, Dave et al. developed and optimized norfloxacin loaded LPHNPs for a topical drug delivery application [372]. They utilized a Box–Behnken design to determine the effect of process conditions including the concentration of soya lecithin (lipid) and the concentration of a polylactic acid (polymer) on the response conditions. This included parameters such as the entrapment efficiency, particle size, and cumulative drug release. It was found that the optimal norfloxacin-loaded LPHNPs have a high drug encapsulation efficiency, desired particle size with a narrow distribution range, and an improved drug release profile and stability [372]. In a study aimed at optimizing LPHNPs, Thanki et al. customized LPHNPs composed of lipidoid 5 (cationic lipid-like molecule) and poly(DL-lactic-co-glycolic acid) (PLGA) for loading antisense oligonucleotide-mediated luciferase gene (Luc-ASO) transcripts and they achieved efficient cellular delivery by using a quality-by-design approach [374]. In their study, they determined the effect of process conditions including the lipidoid 5 content and the lipidoid 5: Luc-ASO ratio against the response conditions (intensity-weighted average hydrodynamic diameter, polydispersity index, zeta potential, Luc-ASO encapsulation efficiency, Luc-ASO loading, in vitro splice-correction efficiency, and in vitro cell viability) to achieve efficient cell delivery [374]. A recent study by Ma et al. involved efficient delivery of hydroxycamptothecin (HCPT) via PEGylated LPHNPs [375]. A quality-by-design strategy was used to optimize HCPT-loaded LPHNPs with desired properties, among them, the factors representing key process conditions were the lipid/polymer mass ratio, polymer concentration, medium chain triglyceride volume, water-solvent ratio, and poly(D,L-lactide-co-glycolide) molecular weight, and the response conditions are particle size, particle size distribution, and drug-loading capacity [375]. The experimental results showed that the optimal LPHNPs had greater controlled release behavior and good stability in plasma, and effectively increased the loading of HCPT [375].

4.5.2. Potential of Lipid Polymer Hybrid Nanoparticles as Antibacterial Delivery Vehicles

Recently, LPHNPs have received great attention in antibacterial applications as efficient drug delivery systems due to their combined advantages of liposomes and polymer nanoparticles [361,363–366,379]. Cai et al. have investigated the application of LPHNPs to act as promising antibacterial delivery vehicles for biofilm eradication [379]. In this study, the lipid layer was designed to contain mixed lipids of phospholipids and rhamnolipids, which acted as anti-adhesive and disrupting agents against the biofilms. The inner polymer core comprises multidrug regimens including antibiotic amoxicillin to exert antibacterial activity and the amoxicillin potentiator pectin sulfate that prevent the re-adherence of *H. pylori*. As a result, a complete eradication of *H. pylori* biofilm with impaired antibacterial resistance was observed under in vitro conditions. The performance of vancomycin-loaded LPHNPs was enhanced via the synthesis using multiple lipid excipients, including glyceryl triplamitate, oleic acid, polymer excipients Eudragit RS100, chitosan, and sodium alginate [361]. Compared to LPHNPs using mono-constituents of the lipid and polymer, the LPHNPs with multiple co-excipients demonstrated higher drug loading capacities and enhanced antibacterial activity against both sensitive strains of *S. aureus* and MRSA [361]. In one recent experiment, Jaglal et al. designed a pH-responsive LPHNP system for co-delivery of vancomycin and 18β-glycyrrhetinic acid, showing its ability to eliminate 75% of MRSA in less than 12 h with the advantage of sustained and rapid release of vancomycin in acidic conditions [363]. LPHNPs consisting of a poly(lactic-co-glycolic acid) core and a dioleoyl-3-trimethylpropane lipid shell were developed for loading vancomycin and were shown to have prominent antibacterial effect against planktonic *S. aureus* cells [364]. In this study, enhanced interactions with bacterial cells and penetration into biofilms was due to the presence of lipid shells. These studies indicate that LPHNPs could be a useful strategy to deliver and enhance the antibacterial activity of the loaded drugs against planktonic cells and biofilms of diverse species. However, more studies are needed to accelerate the clinical translation of the LPHNPs in antibacterial application.

5. Conclusion, Bottlenecks, and Future Perspective of Nanotechnologies Being Developed for Antibacterial Applications

This review mainly focuses on the progress and development of the prospects of nanomaterials in antibacterial applications. As described, the use of nanomaterials to combat bacterial infections has great potential for human health and medical development. Over the past few decades, significant progress has been made in understanding the antibacterial activity and potential of different classes of nanomaterials as drug carriers, leading to the discovery of a number of particularly promising candidate nanoparticle systems. Moreover, Chieruzzi et al. discussed that nanomodification such as incorporating fluoroapatite nanobioceramics into traditional clinical treatment materials, such as dental restorative glass ionomer cement, can lead to significant changes in the mechanical properties of materials, which is a very noteworthy direction for the development of new nano antibacterial materials [380]. Given their vast therapeutic potential and wide range of antibacterial applications of these nanomaterials and nanocomposites, we anticipate that more studies with emphasis on aiding the clinical translation and subsequent clinical trials of these nanotechnology-associated products will increase rapidly in the next decade. Several challenges remain to be addressed. Despite the importance of these nanomaterials as therapeutics and use in the field of biomedicine, their current limitations on human health cannot be ignored. The high-efficiency properties of nanomaterials as antibacterial agents or drug delivery vehicles allow their diffusion in different bodily organs, and sometimes the accumulation of these nanomaterials in various cells and tissues may cause negative health effects. It should be noted that the establishment of the consensus of nanomaterials physiochemical properties leading to maximum antibacterial activity and minimum toxicity is of utmost importance. With the understanding of the structure–property–activity relationship, researchers are able to reduce off-target effects of nanomaterials and effectively deliver nanotherapeutics to a desired infected tissue. For example, the use of single nanoparticles such as metal particles have many drawbacks due to their inherent toxicity. Therefore, combining a variety of nanomaterials to develop a new type of nanocomposite or nanohybrid material to obtain enhanced antibacterial activity has become a major research trend. In addition, exploring the mechanisms behind the antibacterial and physical properties of nanomaterials and nanocomposites should not be neglected. The more we learn, the better we as a community can devise new strategies to combat antimicrobial resistance. More studies in assessing the dose calibration and identification of the appropriate routes of administration for a wide range of nanomaterials are also needed. This will greatly speed up the progress towards clinical trial progression and commercialization of nanotechnology-associated end products.

Author Contributions: Conceptualization, R.R., C.L. and S.L; writing—original draft preparation, R.R., C.L. and S.L.; writing—review and editing, Y.W., J.S., T.-W.L., B.W.M., H.-Y.H. and H.-H.S.; supervision, H.-H.S.; funding acquisition, H.-H.S. All authors have read and agreed to the published version of the manuscript.

Funding: This work was supported by the NHMRC Career Development Fellowship GNT1106798.

Institutional Review Board Statement: Not applicable.

Informed Consent Statement: Not applicable.

Conflicts of Interest: The authors declare no conflict of interest.

Abbreviations

MDR	Multidrug-resistant
WHO	World Health Organization
SARS-CoV-2	Severe acute respiratory coronavirus 2
mRNA	Messenger RNA
FDA	Food and Drug Administration
LLA	Liposomal linolenic acid
S. aureus	*Staphylococcus aureus*
MRSA	Methicillin-resistant *Staphylococcus aureus*
H. pylori	*Helicobacter pylori*
P. acnes	*Propionibacterium acnes*
DHA	Docosahexaenoic acid
IR	Infrared radiation
P. aeruginosa	*Pseudomonas aeruginosa*
K. pneumoniae	*Klebsiella pneumoniae*
MDR-PA	Multidrug-resistant *Pseudomonas aeruginosa*
CNM	Carbon-based nanomaterial
GO	Graphene oxide
E. coli	*Escherichia coli*
NGO–HA	Nanographene oxide–hyaluronic acid
E. faecalis	*Enterococcus faecalis*
C. albicans	*Candida albicans*
ROS	Reactive oxygen species
NPs	Nanoparticles
B. subtilis	*Bacillus subtilis*
S. epidermidis	*Staphylococcus epidermidis*
L. monocytogenes	*Listeria monocytogenes*
GRAS	Generally Recognized as Safe
UV	Ultraviolet
PVA	Polyvinyl alcohol
PEG	Polyethylene glycol
MIC	Minimum inhibitory concentration
PEN	Polyethylene naphthalate
KrF	Krypton fluoride
VRE	Vancomycin-resistant *E. faecalis*
rGO	Reduced graphene oxide
PEG	Polyethylene glycol
LB	Luria–Bertani
MOFs	Metal-organic frameworks
PES	Polyethersulfone
SEM	Scanning electron microscope
PA6	Polyamide 6
EVA	Ethylene-co-vinyl acetate
PEP	Poly(Nisopropylacrylamide$_{166}$-co-n-butyl acrylate$_9$)-poly(ethyleneglycol)-poly (N-isopropylacrylamide$_{166}$-co-n-butyl acrylate$_9$)
PVDF	Poly(vinylidene fluoride)
GOQDs	Graphene oxide quantum dots
LPHNPs	Lipid polymer hybrid nanoparticles
PLGA	Poly(DL-lactic-co-glycolic acid)
Luc	Luciferase gene
ASO	Antisense oligonucleotide
HCPT	Hydroxycamptothecin

References

1. Davis, M.; Whittaker, A.; Lindgren, M.; Djerf-Pierre, M.; Manderson, L.; Flowers, P. Understanding media publics and the antimicrobial resistance crisis. *Glob. Public Health* **2018**, *13*, 1158–1168. [CrossRef] [PubMed]
2. European Centre for Disease Prevention and Control. *Antimicrobial Resistance in the EU/EEA (EARS-Net)—Annual Epidemiological Report 2019*; ECDC: Solna kommun, Stockholm, 2020.
3. Kwon, J.H.; Powderly, W.G. The post-antibiotic era is here. *Science* **2021**, *373*, 471. [CrossRef] [PubMed]
4. Kang, C.-I.; The Korean Network for Study of Infectious Diseases (KONSID); Chung, D.R.; Ko, K.S.; Peck, K.R.; Song, J.-H. Risk factors for infection and treatment outcome of extended-spectrum β-lactamase-producing Escherichia coli and Klebsiella pneumoniae bacteremia in patients with hematologic malignancy. *Ann. Hematol.* **2012**, *91*, 115–121. [CrossRef] [PubMed]
5. World Health Organization. *Antimicrobial Resistance Global Report on Surveillance: 2014 Summary*; World Health Organization: Geneva, Switzerland, 2017.
6. World Health Organization. *2021 Antibacterial Agents in Clinical and Preclinical Development: An Overview and Analysis*; World Health Organization: Geneva, Switzerland, 2022.
7. Gao, W.; Zhang, L. Nanomaterials arising amid antibiotic resistance. *Nat. Rev. Genet.* **2021**, *19*, 5–6. [CrossRef]
8. Blackman, L.D.; Sutherland, T.D.; De Barro, P.J.; Thissen, H.; Locock, K.E.S. Addressing a future pandemic: How can non-biological complex drugs prepare us for antimicrobial resistance threats? *Mater. Horiz.* **2022**, *9*, 2076–2096. [CrossRef]
9. Lardani, L.; Derchi, G.; Marchio, V.; Carli, E. One-Year Clinical Performance of Activa™ Bioactive-Restorative Composite in Primary Molars. *Children* **2022**, *9*, 433. [CrossRef]
10. Zhang, Q.; Wu, W.; Zhang, J.; Xia, X. Antimicrobial lipids in nano-carriers for antibacterial delivery. *J. Drug Target.* **2020**, *28*, 271–281. [CrossRef]
11. Baig, N.; Kammakakam, I.; Falath, W. Nanomaterials: A review of synthesis methods, properties, recent progress, and challenges. *Mater. Adv.* **2021**, *2*, 1821–1871. [CrossRef]
12. Gupta, A.; Mumtaz, S.; Li, C.-H.; Hussain, I.; Rotello, V.M. Combatting antibiotic-resistant bacteria using nanomaterials. *Chem. Soc. Rev.* **2019**, *48*, 415–427. [CrossRef]
13. Hochvaldová, L.; Večeřová, R.; Kolář, M.; Prucek, R.; Kvítek, L.; Lapčík, L.; Panáček, A. Antibacterial nanomaterials: Upcoming hope to overcome antibiotic resistance crisis. *Nanotechnol. Rev.* **2022**, *11*, 1115–1142. [CrossRef]
14. Kankala, R.K. Organic- or Inorganic-based Nanomaterials: Opportunities and Challenges in the Selection for Biomedicine. *Curr. Pharm. Des.* **2022**, *28*, 208–215. [CrossRef] [PubMed]
15. Matharu, R.K.; Ciric, L.; Edirisinghe, M. Nanocomposites: Suitable alternatives as antimicrobial agents. *Nanotechnology* **2018**, *29*, 282001. [CrossRef] [PubMed]
16. Shah, S.; Dhawan, V.; Holm, R.; Nagarsenker, M.S.; Perrie, Y. Liposomes: Advancements and innovation in the manufacturing process. *Adv. Drug Deliv. Rev.* **2020**, *154–155*, 102–122. [CrossRef] [PubMed]
17. Grimaldi, N.; Andrade, F.; Segovia, N.; Ferrer-Tasies, L.; Sala, S.; Veciana, J.; Ventosa, N. Lipid-based nanovesicles for nanomedicine. *Chem. Soc. Rev.* **2016**, *45*, 6520–6545. [CrossRef]
18. Hadinoto, K.; Sundaresan, A.; Cheow, W.S. Lipid–polymer hybrid nanoparticles as a new generation therapeutic delivery platform: A review. *Eur. J. Pharm. Biopharm.* **2013**, *85*, 427–443. [CrossRef]
19. Singh, R.; RamaRao, P. Accumulated Polymer Degradation Products as Effector Molecules in Cytotoxicity of Polymeric Nanoparticles. *Toxicol. Sci.* **2013**, *136*, 131–143. [CrossRef]
20. Liu, Z.; Sun, X.; Nakayama-Ratchford, N.; Dai, H. Supramolecular Chemistry on Water-Soluble Carbon Nanotubes for Drug Loading and Delivery. *ACS Nano* **2007**, *1*, 50–56. [CrossRef]
21. Liu, Z.; Robinson, J.T.; Sun, X.; Dai, H. PEGylated Nanographene Oxide for Delivery of Water-Insoluble Cancer Drugs. *J. Am. Chem. Soc.* **2008**, *130*, 10876–10877. [CrossRef]
22. Pulingam, T.; Thong, K.L.; Ali, E.; Appaturi, J.N.; Dinshaw, I.J.; Ong, Z.Y.; Leo, B.F. Graphene oxide exhibits differential mechanistic action towards Gram-positive and Gram-negative bacteria. *Colloids Surfaces B Biointerfaces* **2019**, *181*, 6–15. [CrossRef]
23. Valdez-Salas, B.; Beltran-Partida, E.; Cheng, N.; Salvador-Carlos, J.; Valdez-Salas, E.A.; Curiel-Alvarez, M.; Ibarra-Wiley, R. Promotion of Surgical Masks Antimicrobial Activity by Disinfection and Impregnation with Disinfectant Silver Nanoparticles. *Int. J. Nanomed.* **2021**, *16*, 2689–2702. [CrossRef]
24. Milić, M.; Leitinger, G.; Pavicic, I.; Avdičević, M.Z.; Dobrović, S.; Goessler, W.; Vrček, I.V. Cellular uptake and toxicity effects of silver nanoparticles in mammalian kidney cells. *J. Appl. Toxicol.* **2015**, *35*, 581–592. [CrossRef]
25. Bélteky, P.; Rónavári, A.; Igaz, N.; Szerencsés, B.; Tóth, I.Y.; Pfeiffer, I.; Kiricsi, M.; Kónya, Z. Silver nanoparticles: Aggregation behavior in biorelevant conditions and its impact on biological activity. *Int. J. Nanomed.* **2019**, *14*, 667–687. [CrossRef] [PubMed]
26. Osmond, M.J.; Mccall, M.J. Zinc oxide nanoparticles in modern sunscreens: An analysis of potential exposure and hazard. *Nanotoxicology* **2009**, *4*, 15–41. [CrossRef]
27. Li, Y.; Zhang, W.; Niu, J.; Chen, Y. Mechanism of Photogenerated Reactive Oxygen Species and Correlation with the Antibacterial Properties of Engineered Metal-Oxide Nanoparticles. *ACS Nano* **2012**, *6*, 5164–5173. [CrossRef] [PubMed]
28. Yao, W.; Yan, Y.; Xue, L.; Zhang, C.; Li, G.; Zheng, Q.; Zhao, Y.S.; Jiang, H.; Yao, J. Controlling the Structures and Photonic Properties of Organic Nanomaterials by Molecular Design. *Angew. Chem. Int. Ed.* **2013**, *52*, 8713–8717. [CrossRef]
29. Gordillo-Galeano, A.; Mora-Huertas, C.E. Solid lipid nanoparticles and nanostructured lipid carriers: A review emphasizing on particle structure and drug release. *Eur. J. Pharm. Biopharm.* **2018**, *133*, 285–308. [CrossRef] [PubMed]

30. Akbarzadeh, A.; Rezaei-Sadabady, R.; Davaran, S.; Joo, S.W.; Zarghami, N.; Hanifehpour, Y.; Samiei, M.; Kouhi, M.; Nejati-Koshki, K. Liposome: Classification, preparation, and applications. *Nanoscale Res. Lett.* **2013**, *8*, 102. [CrossRef]
31. Ghasemiyeh, P.; Mohammadi-Samani, S. Solid lipid nanoparticles and nanostructured lipid carriers as novel drug delivery systems: Applications, advantages and disadvantages. *Res. Pharm. Sci.* **2018**, *13*, 288–303. [CrossRef]
32. Mishra, D.K.; Shandilya, R.; Mishra, P.K. Lipid based nanocarriers: A translational perspective. *Nanomed. Nanotechnol. Biol. Med.* **2018**, *14*, 2023–2050. [CrossRef]
33. Leonardi, A.; Bucolo, C.; Romano, G.L.; Platania, C.B.M.; Drago, F.; Puglisi, G.; Pignatello, R. Influence of different surfactants on the technological properties and in vivo ocular tolerability of lipid nanoparticles. *Int. J. Pharm.* **2014**, *470*, 133–140. [CrossRef]
34. Yang, D.; Pornpattananangkul, D.; Nakatsuji, T.; Chan, M.; Carson, D.; Huang, C.-M.; Zhang, L. The antimicrobial activity of liposomal lauric acids against Propionibacterium acnes. *Biomaterials* **2009**, *30*, 6035–6040. [CrossRef] [PubMed]
35. Pornpattananangkul, D.; Fu, V.; Thamphiwatana, S.; Zhang, L.; Chen, M.; Vecchio, J.; Gao, W.; Huang, C.-M.; Zhang, L. In Vivo Treatment of *Propionibacterium acnes* Infection with Liposomal Lauric Acids. *Adv. Health Mater.* **2013**, *2*, 1322–1328. [CrossRef] [PubMed]
36. Madan, S.; Nehate, C.; Barman, T.K.; Rathore, A.S.; Koul, V. Design, preparation, and evaluation of liposomal gel formulations for treatment of acne: In vitro and in vivo studies. *Drug Dev. Ind. Pharm.* **2019**, *45*, 395–404. [CrossRef] [PubMed]
37. Li, X.-X.; Shi, S.; Rong, L.; Feng, M.-Q.; Zhong, L. The impact of liposomal linolenic acid on gastrointestinal microbiota in mice. *Int. J. Nanomed.* **2018**, *13*, 1399–1409. [CrossRef] [PubMed]
38. Hwang, D.; Ramsey, J.D.; Kabanov, A.V. Polymeric micelles for the delivery of poorly soluble drugs: From nanoformulation to clinical approval. *Adv. Drug Deliv. Rev.* **2020**, *156*, 80–118. [CrossRef]
39. Owen, S.C.; Chan, D.P.; Shoichet, M.S. Polymeric micelle stability. *Nano Today* **2012**, *7*, 53–65. [CrossRef]
40. Tran, T.-Q.; Hsieh, M.-F.; Chang, K.-L.; Pho, Q.-H.; Nguyen, V.-C.; Cheng, C.-Y.; Huang, C.-M. Bactericidal Effect of Lauric Acid-Loaded PCL-PEG-PCL Nano-Sized Micelles on Skin Commensal Propionibacterium acnes. *Polymers* **2016**, *8*, 321. [CrossRef]
41. Huynh, N.; Passirani, C.; Saulnier, P.; Benoit, J. Lipid nanocapsules: A new platform for nanomedicine. *Int. J. Pharm.* **2009**, *379*, 201–209. [CrossRef]
42. Umerska, A.; Cassisa, V.; Matougui, N.; Joly-Guillou, M.-L.; Eveillard, M.; Saulnier, P. Antibacterial action of lipid nanocapsules containing fatty acids or monoglycerides as co-surfactants. *Eur. J. Pharm. Biopharm.* **2016**, *108*, 100–110. [CrossRef]
43. Umerska, A.; Cassisa, V.; Bastiat, G.; Matougui, N.; Nehme, H.; Manero, F.; Eveillard, M.; Saulnier, P. Synergistic interactions between antimicrobial peptides derived from plectasin and lipid nanocapsules containing monolaurin as a cosurfactant against Staphylococcus aureus. *Int. J. Nanomed.* **2017**, *12*, 5687–5699. [CrossRef]
44. Rozenbaum, R.T.; Su, L.; Umerska, A.; Eveillard, M.; Håkansson, J.; Mahlapuu, M.; Huang, F.; Liu, J.; Zhang, Z.; Shi, L.; et al. Antimicrobial synergy of monolaurin lipid nanocapsules with adsorbed antimicrobial peptides against Staphylococcus aureus biofilms in vitro is absent in vivo. *J. Control. Release* **2019**, *293*, 73–83. [CrossRef]
45. Anton, N.; Benoit, J.-P.; Saulnier, P. Design and production of nanoparticles formulated from nano-emulsion templates—A review. *J. Control. Release* **2008**, *128*, 185–199. [CrossRef] [PubMed]
46. Sadiq, S.; Imran, M.; Habib, H.; Shabbir, S.; Ihsan, A.; Zafar, Y.; Hafeez, F.Y. Potential of monolaurin based food-grade nano-micelles loaded with nisin Z for synergistic antimicrobial action against Staphylococcus aureus. *LWT Food Sci. Technol.* **2016**, *71*, 227–233. [CrossRef]
47. Taylor, E.N.; Kummer, K.M.; Dyondi, D.; Webster, T.J.; Banerjee, R. Multi-scale strategy to eradicate Pseudomonas aeruginosa on surfaces using solid lipid nanoparticles loaded with free fatty acids. *Nanoscale* **2014**, *6*, 825–832. [CrossRef]
48. Hallaj-Nezhadi, S.; Hassan, M. Nanoliposome-based antibacterial drug delivery. *Drug Deliv.* **2015**, *22*, 581–589. [CrossRef] [PubMed]
49. El-Nesr, O.H.; Yahiya, S.A.; El-Gazayerly, O.N. Effect of formulation design and freeze-drying on properties of fluconazole multilamellar liposomes. *Saudi Pharm. J.* **2010**, *18*, 217–224. [CrossRef] [PubMed]
50. Orizondo, R.A.; Fabiilli, M.L.; Morales, M.A.; Cook, K.E. Effects of Emulsion Composition on Pulmonary Tobramycin Delivery During Antibacterial Perfluorocarbon Ventilation. *J. Aerosol Med. Pulm. Drug Deliv.* **2016**, *29*, 251–259. [CrossRef]
51. Severino, P.; Silveira, E.F.; Loureiro, K.; Chaud, M.V.; Antonini, D.; Lancellotti, M.; Sarmento, V.H.; da Silva, C.F.; Santana, M.H.A.; Souto, E.B. Antimicrobial activity of polymyxin-loaded solid lipid nanoparticles (PLX-SLN): Characterization of physicochemical properties and in vitro efficacy. *Eur. J. Pharm. Sci.* **2017**, *106*, 177–184. [CrossRef] [PubMed]
52. Vieira, A.C.; Magalhães, J.; Rocha, S.; Cardoso, M.S.; Santos, S.G.; Borges, M.; Pinheiro, M.; Reis, S. Targeted macrophages delivery of rifampicin-loaded lipid nanoparticles to improve tuberculosis treatment. *Nanomedicine* **2017**, *12*, 2721–2736. [CrossRef]
53. Walduck, A.; Sangwan, P.; Vo, Q.A.; Ratcliffe, J.; White, J.; Muir, B.W.; Tran, N. Treatment of *Staphylococcus aureus* skin infection *in vivo* using rifampicin loaded lipid nanoparticles. *RSC Adv.* **2020**, *10*, 33608–33619. [CrossRef]
54. Tran, N.; Mulet, X.; Hawley, A.M.; Fong, C.; Zhai, J.; Le, T.C.; Ratcliffe, J.; Drummond, C.J. Manipulating the Ordered Nanostructure of Self-Assembled Monoolein and Phytantriol Nanoparticles with Unsaturated Fatty Acids. *Langmuir* **2018**, *34*, 2764–2773. [CrossRef]
55. Boge, L.; Hallstensson, K.; Ringstad, L.; Johansson, J.; Andersson, T.; Davoudi, M.; Larsson, P.T.; Mahlapuu, M.; Håkansson, J.; Andersson, M. Cubosomes for topical delivery of the antimicrobial peptide LL-37. *Eur. J. Pharm. Biopharm.* **2019**, *134*, 60–67. [CrossRef] [PubMed]

56. Lai, X.; Ding, Y.; Wu, C.-M.; Chen, X.; Jiang, J.-H.; Hsu, H.-Y.; Wang, Y.; Le Brun, A.P.; Song, J.; Han, M.-L.; et al. Phytantriol-Based Cubosome Formulation as an Antimicrobial against Lipopolysaccharide-Deficient Gram-Negative Bacteria. *ACS Appl. Mater. Interfaces* **2020**, *12*, 44485–44498. [CrossRef]
57. Lai, X.; Han, M.-L.; Ding, Y.; Chow, S.H.; Le Brun, A.P.; Wu, C.-M.; Bergen, P.J.; Jiang, J.-H.; Hsu, H.-Y.; Muir, B.W.; et al. A polytherapy based approach to combat antimicrobial resistance using cubosomes. *Nat. Commun.* **2022**, *13*, 343. [CrossRef] [PubMed]
58. Meikle, T.G.; Dyett, B.P.; Strachan, J.B.; White, J.; Drummond, C.J.; Conn, C.E. Preparation, Characterization, and Antimicrobial Activity of Cubosome Encapsulated Metal Nanocrystals. *ACS Appl. Mater. Interfaces* **2020**, *12*, 6944–6954. [CrossRef]
59. Boge, L.; Umerska, A.; Matougui, N.; Bysell, H.; Ringstad, L.; Davoudi, M.; Eriksson, J.; Edwards, K.; Andersson, M. Cubosomes post-loaded with antimicrobial peptides: Characterization, bactericidal effect and proteolytic stability. *Int. J. Pharm.* **2017**, *526*, 400–412. [CrossRef] [PubMed]
60. Meikle, T.G.; Dharmadana, D.; Hoffmann, S.V.; Jones, N.C.; Drummond, C.J.; Conn, C.E. Analysis of the structure, loading and activity of six antimicrobial peptides encapsulated in cubic phase lipid nanoparticles. *J. Colloid Interface Sci.* **2021**, *587*, 90–100. [CrossRef]
61. Dyett, B.P.; Yu, H.; Sarkar, S.; Strachan, J.B.; Drummond, C.J.; Conn, C.E. Uptake Dynamics of Cubosome Nanocarriers at Bacterial Surfaces and the Routes for Cargo Internalization. *ACS Appl. Mater. Interfaces* **2021**, *13*, 53530–53540. [CrossRef]
62. Jackman, J.A.; Yoon, B.K.; Li, D.; Cho, N.-J. Nanotechnology Formulations for Antibacterial Free Fatty Acids and Monoglycerides. *Molecules* **2016**, *21*, 305. [CrossRef]
63. Desbois, A.P.; Smith, V.J. Antibacterial free fatty acids: Activities, mechanisms of action and biotechnological potential. *Appl. Microbiol. Biotechnol.* **2010**, *85*, 1629–1642. [CrossRef]
64. Thorn, C.R.; Thomas, N.; Boyd, B.J.; Prestidge, C.A. Nano-fats for bugs: The benefits of lipid nanoparticles for antimicrobial therapy. *Drug Deliv. Transl. Res.* **2021**, *11*, 1598–1624. [CrossRef] [PubMed]
65. Boushehri, M.A.S.; Dietrich, D.; Lamprecht, A. Nanotechnology as a Platform for the Development of Injectable Parenteral Formulations: A Comprehensive Review of the Know-Hows and State of the Art. *Pharmaceutics* **2020**, *12*, 510. [CrossRef] [PubMed]
66. Babadi, D.; Dadashzadeh, S.; Osouli, M.; Daryabari, M.S.; Haeri, A. Nanoformulation strategies for improving intestinal permeability of drugs: A more precise look at permeability assessment methods and pharmacokinetic properties changes. *J. Control. Release* **2020**, *321*, 669–709. [CrossRef] [PubMed]
67. Mitchell, M.J.; Billingsley, M.M.; Haley, R.M.; Wechsler, M.E.; Peppas, N.A.; Langer, R. Engineering precision nanoparticles for drug delivery. *Nat. Rev. Drug Discov.* **2021**, *20*, 101–124. [CrossRef]
68. Hou, X.; Zaks, T.; Langer, R.; Dong, Y. Lipid nanoparticles for mRNA delivery. *Nat. Rev. Mater.* **2021**, *6*, 1078–1094. [CrossRef] [PubMed]
69. Ahmed, K.S.; Hussein, S.A.; Ali, A.; Korma, S.A.; Lipeng, Q.; Jinghua, C. Liposome: Composition, characterisation, preparation, and recent innovation in clinical applications. *J. Drug Target.* **2019**, *27*, 742–761. [CrossRef] [PubMed]
70. Tan, J.Y.B.; Yoon, B.K.; Cho, N.-J.; Lovrić, J.; Jug, M.; Jackman, J.A. Lipid Nanoparticle Technology for Delivering Biologically Active Fatty Acids and Monoglycerides. *Int. J. Mol. Sci.* **2021**, *22*, 9664. [CrossRef] [PubMed]
71. Huang, C.-M.; Chen, C.-H.; Pornpattananangkul, D.; Zhang, L.; Chan, M.; Hsieh, M.-F.; Zhang, L. Eradication of drug resistant Staphylococcus aureus by liposomal oleic acids. *Biomaterials* **2011**, *32*, 214–221. [CrossRef]
72. Obonyo, M.; Zhang, L.; Thamphiwatana, S.; Pornpattananangkul, D.; Fu, V.; Zhang, L. Antibacterial Activities of Liposomal Linolenic Acids against Antibiotic-Resistant *Helicobacter pylori*. *Mol. Pharm.* **2012**, *9*, 2677–2685. [CrossRef]
73. Thamphiwatana, S.; Gao, W.; Obonyo, M.; Zhang, L. In vivo treatment of *Helicobacter pylori* infection with liposomal linolenic acid reduces colonization and ameliorates inflammation. *Proc. Natl. Acad. Sci. USA* **2014**, *111*, 17600–17605. [CrossRef]
74. Sonawane, S.J.; Kalhapure, R.S.; Jadhav, M.; Rambharose, S.; Mocktar, C.; Govender, T. Transforming linoleic acid into a nanoemulsion for enhanced activity against methicillin susceptible and resistant Staphylococcus aureus. *RSC Adv.* **2015**, *5*, 90482–90492. [CrossRef]
75. Silva, E.L.; Carneiro, G.; De Araújo, L.A.; Trindade, M.D.J.V.; Yoshida, M.I.; Oréfice, R.L.; Farias, L.D.M.; De Carvalho, M.A.R.; Dos Santos, S.G.; Goulart, G.A.C.; et al. Solid Lipid Nanoparticles Loaded with Retinoic Acid and Lauric Acid as an Alternative for Topical Treatment of Acne Vulgaris. *J. Nanosci. Nanotechnol.* **2015**, *15*, 792–799. [CrossRef] [PubMed]
76. Teixeira, M.; Carbone, C.; Souto, E. Beyond liposomes: Recent advances on lipid based nanostructures for poorly soluble/poorly permeable drug delivery. *Prog. Lipid Res.* **2017**, *68*, 1–11. [CrossRef] [PubMed]
77. Katouzian, I.; Esfanjani, A.F.; Jafari, S.M.; Akhavan, S. Formulation and application of a new generation of lipid nano-carriers for the food bioactive ingredients. *Trends Food Sci. Technol.* **2017**, *68*, 14–25. [CrossRef]
78. Da Silva Santos, V.; Ribeiro, A.P.B.; Santana, M.H.A. Solid lipid nanoparticles as carriers for lipophilic compounds for applications in foods. *Food Res. Int.* **2019**, *122*, 610–626. [CrossRef]
79. Brandelli, A.; Pola, C.C.; Gomes, C.L. Antimicrobial delivery systems. In *Antimicrobials in Food*; CRC Press: Boca Raton, FL, USA, 2020; pp. 665–694.
80. Pinilla, C.; Lopes, N.; Brandelli, A. Lipid-Based Nanostructures for the Delivery of Natural Antimicrobials. *Molecules* **2021**, *26*, 3587. [CrossRef]

81. Seabra, C.L.; Nunes, C.; Gomez-Lazaro, M.; Correia, M.; Machado, J.C.; Gonçalves, I.C.; Reis, C.A.; Reis, S.; Martins, M.C.L. Docosahexaenoic acid loaded lipid nanoparticles with bactericidal activity against Helicobacter pylori. *Int. J. Pharm.* **2017**, *519*, 128–137. [CrossRef]
82. Wang, Y.; Sun, H. Polymeric Nanomaterials for Efficient Delivery of Antimicrobial Agents. *Pharmaceutics* **2021**, *13*, 2108. [CrossRef]
83. Kamaruzzaman, N.F.; Tan, L.P.; Hamdan, R.H.; Choong, S.S.; Wong, W.K.; Gibson, A.J.; Chivu, A.; Pina, M.F. Antimicrobial Polymers: The Potential Replacement of Existing Antibiotics? *Int. J. Mol. Sci.* **2019**, *20*, 2747. [CrossRef]
84. Sidhu, A.K.; Verma, N.; Kaushal, P. Role of Biogenic Capping Agents in the Synthesis of Metallic Nanoparticles and Evaluation of Their Therapeutic Potential. *Front. Nanotechnol.* **2022**, *3*, 801620. [CrossRef]
85. León-Buitimea, A.; Garza-Cárdenas, C.R.; Román-García, M.F.; Ramírez-Díaz, C.A.; Ulloa-Ramírez, M.; Morones-Ramírez, J.R. Nanomaterials-Based Combinatorial Therapy as a Strategy to Combat Antibiotic Resistance. *Antibiotics* **2022**, *11*, 794. [CrossRef] [PubMed]
86. Letchford, K.; Burt, H. A review of the formation and classification of amphiphilic block copolymer nanoparticulate structures: Micelles, nanospheres, nanocapsules and polymersomes. *Eur. J. Pharm. Biopharm.* **2007**, *65*, 259–269. [CrossRef]
87. Ding, X.; Wang, A.; Tong, W.; Xu, F.J. Biodegradable Antibacterial Polymeric Nanosystems: A New Hope to Cope with Multidrug-Resistant Bacteria. *Small* **2019**, *15*, e1900999. [CrossRef]
88. Iyisan, B.; Landfester, K. Modular Approach for the Design of Smart Polymeric Nanocapsules. *Macromol. Rapid Commun.* **2019**, *40*, e1800577. [CrossRef] [PubMed]
89. Anselmo, A.C.; Mitragotri, S. Nanoparticles in the clinic. *Bioeng. Transl. Med.* **2016**, *1*, 10–29. [CrossRef]
90. Adhikari, C. Polymer nanoparticles-preparations, applications and future insights: A concise review. *Polym. Technol. Mater.* **2021**, *60*, 1996–2024. [CrossRef]
91. Carmona-Ribeiro, A.M. Biomimetic Lipid Polymer Nanoparticles for Drug Delivery. In *Nanoparticles in Biology and Medicine: Methods and Protocols*; Ferrari, E., Soloviev, M., Eds.; Springer US: New York, NY, USA, 2020; pp. 45–60.
92. Deirram, N.; Zhang, C.; Kermaniyan, S.S.; Johnston, A.P.R.; Such, G.K. pH-Responsive Polymer Nanoparticles for Drug Delivery. *Macromol. Rapid Commun.* **2019**, *40*, 1800917. [CrossRef]
93. Saini, R.K.; Bagri, L.P.; Bajpai, A.K.; Mishra, A. Responsive polymer nanoparticles for drug delivery applications. In *Stimuli Responsive Polymeric Nanocarriers for Drug Delivery Applications*; Elsevier: Amsterdam, The Netherlands, 2018; Volume 1, pp. 289–320.
94. Dararatana, N.; Seidi, F.; Hamela, J.; Crespy, D. Controlling release kinetics of pH-responsive polymer nanoparticles. *Polym. Chem.* **2020**, *11*, 1752–1762. [CrossRef]
95. Qiu, Y.; Xu, D.; Sui, G.; Wang, D.; Wu, M.; Han, L.; Mu, H.; Duan, J. Gentamicin decorated phosphatidylcholine-chitosan nanoparticles against biofilms and intracellular bacteria. *Int. J. Biol. Macromol.* **2020**, *156*, 640–647. [CrossRef]
96. Walvekar, P.; Gannimani, R.; Salih, M.; Makhathini, S.; Mocktar, C.; Govender, T. Self-assembled oleylamine grafted hyaluronic acid polymersomes for delivery of vancomycin against methicillin resistant Staphylococcus aureus (MRSA). *Colloids Surfaces B Biointerfaces* **2019**, *182*, 110388. [CrossRef]
97. Ergene, C.; Yasuhara, K.; Palermo, E.F. Biomimetic antimicrobial polymers: Recent advances in molecular design. *Polym. Chem.* **2018**, *9*, 2407–2427. [CrossRef]
98. Lin, M.; Sun, J. Antimicrobial peptide-inspired antibacterial polymeric materials for biosafety. *Biosaf. Health* **2022**, *4*, 269–279. [CrossRef]
99. Samal, S.K.; Dash, M.; Van Vlierberghe, S.; Kaplan, D.L.; Chiellini, E.; van Blitterswijk, C.; Moroni, L.; Dubruel, P. Cationic polymers and their therapeutic potential. *Chem. Soc. Rev.* **2012**, *41*, 7147–7194. [CrossRef] [PubMed]
100. Si, Z.; Zheng, W.; Prananty, D.; Li, J.; Koh, C.H.; Kang, E.-T.; Pethe, K.; Chan-Park, M.B. Polymers as advanced antibacterial and antibiofilm agents for direct and combination therapies. *Chem. Sci.* **2022**, *13*, 345–364. [CrossRef]
101. Mowery, B.P.; Lindner, A.H.; Weisblum, B.; Stahl, S.S.; Gellman, S.H. Structure−activity Relationships among Random Nylon-3 Copolymers That Mimic Antibacterial Host-Defense Peptides. *J. Am. Chem. Soc.* **2009**, *131*, 9735–9745. [CrossRef]
102. Ganewatta, M.S.; Tang, C. Controlling macromolecular structures towards effective antimicrobial polymers. *Polymer* **2015**, *63*, A1–A29. [CrossRef]
103. Takahashi, H.; Caputo, G.A.; Vemparala, S.; Kuroda, K. Synthetic Random Copolymers as a Molecular Platform To Mimic Host-Defense Antimicrobial Peptides. *Bioconjug. Chem.* **2017**, *28*, 1340–1350. [CrossRef]
104. Rahman, M.A.; Bam, M.; Luat, E.; Jui, M.S.; Ganewatta, M.S.; Shokfai, T.; Nagarkatti, M.; Decho, A.W.; Tang, C. Macromolecular-clustered facial amphiphilic antimicrobials. *Nat. Commun.* **2018**, *9*, 5231. [CrossRef]
105. Corti, M.B.; Campagno, L.P.; Romero, V.L.; Gutierrez, S.; Alovero, F.L. Cationic polymer contributes to broaden the spectrum of vancomycin activity achieving eradication of Pseudomonas aeruginosa. *Arch. Microbiol.* **2022**, *204*, 507. [CrossRef]
106. Nederberg, F.; Zhang, Y.; Tan, J.P.K.; Xu, K.; Wang, H.; Yang, C.; Gao, S.; Guo, X.D.; Fukushima, K.; Li, L.; et al. Biodegradable nanostructures with selective lysis of microbial membranes. *Nat. Chem.* **2011**, *3*, 409–414. [CrossRef]
107. Chen, J.; Wang, F.; Liu, Q.; Du, J. Antibacterial polymeric nanostructures for biomedical applications. *Chem. Commun.* **2014**, *50*, 14482–14493. [CrossRef] [PubMed]
108. Hisey, B.; Ragogna, P.J.; Gillies, E.R. Phosphonium-Functionalized Polymer Micelles with Intrinsic Antibacterial Activity. *Biomacromolecules* **2017**, *18*, 914–923. [CrossRef] [PubMed]

109. Zhou, C.; Yuan, Y.; Zhou, P.; Wang, F.; Hong, Y.; Wang, N.; Xu, S.; Du, J. Highly Effective Antibacterial Vesicles Based on Peptide-Mimetic Alternating Copolymers for Bone Repair. *Biomacromolecules* **2017**, *18*, 4154–4162. [CrossRef] [PubMed]
110. Sun, H.; Hong, Y.; Xi, Y.; Zou, Y.; Gao, J.; Du, J. Synthesis, Self-Assembly, and Biomedical Applications of Antimicrobial Peptide-Polymer Conjugates. *Biomacromolecules* **2018**, *19*, 1701–1720. [CrossRef]
111. Chin, W.; Zhong, G.; Pu, Q.; Yang, C.; Lou, W.; De Sessions, P.F.; Periaswamy, B.; Lee, A.; Liang, Z.C.; Ding, X.; et al. A macromolecular approach to eradicate multidrug resistant bacterial infections while mitigating drug resistance onset. *Nat. Commun.* **2018**, *9*, 917. [CrossRef]
112. Yang, C.; Lou, W.; Zhong, G.; Lee, A.; Leong, E.; Chin, W.; Ding, B.; Bao, C.; Tan, J.P.; Pu, Q.; et al. Degradable antimicrobial polycarbonates with unexpected activity and selectivity for treating multidrug-resistant Klebsiella pneumoniae lung infection in mice. *Acta Biomater.* **2019**, *94*, 268–280. [CrossRef]
113. Lam, S.J.; O'Brien-Simpson, N.M.; Pantarat, N.; Sulistio, A.; Wong, E.H.H.; Chen, Y.-Y.; Lenzo, J.C.; Holden, J.A.; Blencowe, A.; Reynolds, E.C.; et al. Combating multidrug-resistant Gram-negative bacteria with structurally nanoengineered antimicrobial peptide polymers. *Nat. Microbiol.* **2016**, *1*, 16162. [CrossRef]
114. Singh, R.; Jha, D.; Dhawan, U.; Gautam, H.K.; Kumar, P. Therapeutic Applications of Self-assembled Indole-3-butanoyl-polyethylenimine Nanostructures. *Indian J. Microbiol.* **2022**, *62*, 411–418. [CrossRef]
115. Nasir, S.; Hussein, M.Z.; Zainal, Z.; Yusof, N.A. Carbon-Based Nanomaterials/Allotropes: A Glimpse of Their Synthesis, Properties and Some Applications. *Materials* **2018**, *11*, 295. [CrossRef]
116. Al-Jumaili, A.; Alancherry, S.; Bazaka, K.; Jacob, M.V. Review on the Antimicrobial Properties of Carbon Nanostructures. *Materials* **2017**, *10*, 1066. [CrossRef]
117. Rao, N.; Singh, R.; Bashambu, L. Carbon-based nanomaterials: Synthesis and prospective applications. *Mater. Today Proc.* **2021**, *44*, 608–614. [CrossRef]
118. Xin, Q.; Shah, H.; Nawaz, A.; Xie, W.; Akram, M.Z.; Batool, A.; Tian, L.; Jan, S.U.; Boddula, R.; Guo, B.; et al. Antibacterial Carbon-Based Nanomaterials. *Adv. Mater.* **2019**, *31*, e1804838. [CrossRef] [PubMed]
119. Díez-Pascual, A.M. State of the Art in the Antibacterial and Antiviral Applications of Carbon-Based Polymeric Nanocomposites. *Int. J. Mol. Sci.* **2021**, *22*, 10511. [CrossRef] [PubMed]
120. Cui, F.; Li, T.; Wang, D.; Yi, S.; Li, J.; Li, X. Recent advances in carbon-based nanomaterials for combating bacterial biofilm-associated infections. *J. Hazard. Mater.* **2022**, *431*, 128597. [CrossRef] [PubMed]
121. Zhou, H.; Zou, F.; Koh, K.; Lee, J. Antibacterial Activity of Graphene-Based Nanomaterials. In *Multifaceted Biomedical Applications of Graphene*; Han, D.-W., Hong, S.W., Eds.; Springer Singapore: Singapore, 2022; pp. 233–250.
122. Yousefi, M.; Dadashpour, M.; Hejazi, M.; Hasanzadeh, M.; Behnam, B.; de la Guardia, M.; Shadjou, N.; Mokhtarzadeh, A. Anti-bacterial activity of graphene oxide as a new weapon nanomaterial to combat multidrug-resistance bacteria. *Mater. Sci. Eng. C* **2017**, *74*, 568–581. [CrossRef]
123. Kojima, C.; Toi, Y.; Harada, A.; Kono, K. Aqueous Solubilization of Fullerenes Using Poly(amidoamine) Dendrimers Bearing Cyclodextrin and Poly(ethylene Glycol). *Bioconjugate Chem.* **2008**, *19*, 2280–2284. [CrossRef]
124. Li, W.; Zhang, G.; Wei, X. Lidocaine-loaded reduced graphene oxide hydrogel for prolongation of effects of local anesthesia: In vitro and in vivo analyses. *J. Biomater. Appl.* **2021**, *35*, 1034–1042. [CrossRef]
125. Luo, S.; Jin, S.; Yang, T.; Wu, B.; Xu, C.; Luo, L.; Chen, Y. Sustained release of tulobuterol from graphene oxide laden hydrogel to manage asthma. *J. Biomater. Sci. Polym. Ed.* **2021**, *32*, 524–535. [CrossRef]
126. Choi, M.; Chung, J.-H.; Cho, Y.; Hong, B.Y.; Hong, J. Nano-film modification of collagen hydrogels for controlled growth factor release. *Chem. Eng. Sci.* **2015**, *137*, 626–630. [CrossRef]
127. Szunerits, S.; Boukherroub, R. Antibacterial activity of graphene-based materials. *J. Mater. Chem. B* **2016**, *4*, 6892–6912. [CrossRef]
128. Rhazouani, A.; Gamrani, H.; Achaby, M.E.; Aziz, K.; Gebrati, L.; Uddin, M.S.; Aziz, F. Synthesis and Toxicity of Graphene Oxide Nanoparticles: A Literature Review of In Vitro and In Vivo Studies. *BioMed Res. Int.* **2021**, *2021*, 5518999. [CrossRef] [PubMed]
129. Zhang, X.; Yin, J.; Peng, C.; Hu, W.; Zhu, Z.; Li, W.; Fan, C.; Huang, Q. Distribution and biocompatibility studies of graphene oxide in mice after intravenous administration. *Carbon* **2011**, *49*, 986–995. [CrossRef]
130. Sun, X.; Liu, Z.; Welsher, K.; Robinson, J.T.; Goodwin, A.; Zaric, S.; Dai, H. Nano-graphene oxide for cellular imaging and drug delivery. *Nano Res.* **2008**, *1*, 203–212. [CrossRef] [PubMed]
131. Jung, H.S.; Lee, M.-Y.; Kong, W.H.; Do, I.H.; Hahn, S.K. Nano graphene oxide–hyaluronic acid conjugate for target specific cancer drug delivery. *RSC Adv.* **2014**, *4*, 14197–14200. [CrossRef]
132. Zhao, X.; Li, Y.; Wang, J.; Ouyang, Z.; Li, J.; Wei, G.; Su, Z. Interactive Oxidation–Reduction Reaction for the in Situ Synthesis of Graphene–Phenol Formaldehyde Composites with Enhanced Properties. *ACS Appl. Mater. Interfaces* **2014**, *6*, 4254–4263. [CrossRef]
133. Singh, V.; Sagar, P.; Kaul, S.; Sandhir, R.; Singhal, N.K. Liver Phosphoenolpyruvate Carboxykinase-1 Downregulation via siRNA-Functionalized Graphene Oxide Nanosheets Restores Glucose Homeostasis in a Type 2 Diabetes Mellitus In Vivo Model. *Bioconjugate Chem.* **2021**, *32*, 259–278. [CrossRef]
134. Yang, Y.; Liu, Y.; Shen, Y. Plasmonic-Enhanced Graphene Oxide-Based Aquatic Robot for Target Cargo Delivery. *ACS Appl. Mater. Interfaces* **2021**, *13*, 1503–1510. [CrossRef] [PubMed]
135. Sharma, H.; Mondal, S. Functionalized Graphene Oxide for Chemotherapeutic Drug Delivery and Cancer Treatment: A Promising Material in Nanomedicine. *Int. J. Mol. Sci.* **2020**, *21*, 6280. [CrossRef]

136. Deng, X.; Liang, H.; Yang, W.; Shao, Z. Polarization and function of tumor-associated macrophages mediate graphene oxide-induced photothermal cancer therapy. *J. Photochem. Photobiol. B Biol.* **2020**, *208*, 111913. [CrossRef] [PubMed]
137. Ashrafizadeh, M.; Saebfar, H.; Gholami, M.H.; Hushmandi, K.; Zabolian, A.; Bikarannejad, P.; Hashemi, M.; Daneshi, S.; Mirzaei, S.; Sharifi, E.; et al. Doxorubicin-loaded graphene oxide nanocomposites in cancer medicine: Stimuli-responsive carriers, co-delivery and suppressing resistance. *Expert Opin. Drug Deliv.* **2022**, *19*, 355–382. [CrossRef]
138. Barrera, C.C.; Groot, H.; Vargas, W.L.; Narváez, D.M. Efficacy and Molecular Effects of a Reduced Graphene Oxide/Fe(3)O(4) Nanocomposite in Photothermal Therapy against Cancer. *Int. J. Nanomed.* **2020**, *15*, 6421–6432. [CrossRef] [PubMed]
139. Matharu, R.; Tabish, T.A.; Trakoolwilaiwan, T.; Mansfield, J.; Moger, J.; Wu, T.; Lourenco, C.; Chen, B.; Ciric, L.; Parkin, I.P.; et al. Microstructure and antibacterial efficacy of graphene oxide nanocomposite fibres. *J. Colloid Interface Sci.* **2020**, *571*, 239–252. [CrossRef]
140. Zhang, X.; Shen, J.; Zhuo, N.; Tian, Z.; Xu, P.; Yang, Z.; Yang, W. Interactions between Antibiotics and Graphene-Based Materials in Water: A Comparative Experimental and Theoretical Investigation. *ACS Appl. Mater. Interfaces* **2016**, *8*, 24273–24280. [CrossRef] [PubMed]
141. Gao, Y.; Wu, J.; Ren, X.; Tan, X.; Hayat, T.; Alsaedi, A.; Cheng, C.; Chen, C. Impact of graphene oxide on the antibacterial activity of antibiotics against bacteria. *Environ. Sci. Nano* **2017**, *4*, 1016–1024. [CrossRef]
142. Jihad, M.; Noori, F.; Jabir, M.; Albukhaty, S.; AlMalki, F.; Alyamani, A. Polyethylene Glycol Functionalized Graphene Oxide Nanoparticles Loaded with *Nigella sativa* Extract: A Smart Antibacterial Therapeutic Drug Delivery System. *Molecules* **2021**, *26*, 3067. [CrossRef] [PubMed]
143. Pan, N.; Wang, Y.; Ren, X.; Huang, T.-S.; Kim, I.S. Graphene oxide as a polymeric N-halamine carrier and release platform: Highly-efficient, sustained-release antibacterial property and great storage stability. *Mater. Sci. Eng. C* **2019**, *103*, 109877. [CrossRef]
144. Chen, H.; Leng, S. Rapid synthesis of hollow nano-structured hydroxyapatite microspheres via microwave transformation method using hollow CaCO3 precursor microspheres. *Ceram. Int.* **2015**, *41*, 2209–2213. [CrossRef]
145. Liu, S.; Zeng, T.H.; Hofmann, M.; Burcombe, E.; Wei, J.; Jiang, R.; Kong, J.; Chen, Y. Antibacterial Activity of Graphite, Graphite Oxide, Graphene Oxide, and Reduced Graphene Oxide: Membrane and Oxidative Stress. *ACS Nano* **2011**, *5*, 6971–6980. [CrossRef]
146. Li, J.; Wang, G.; Zhu, H.; Zhang, M.; Zheng, X.; Di, Z.; Liu, X.; Wang, X. Antibacterial activity of large-area monolayer graphene film manipulated by charge transfer. *Sci. Rep.* **2014**, *4*, 4359. [CrossRef]
147. Gurunathan, S.; Han, J.W.; Dayem, A.A.; Eppakayala, V.; Kim, J.-H. Oxidative stress-mediated antibacterial activity of graphene oxide and reduced graphene oxide in Pseudomonas aeruginosa. *Int. J. Nanomed.* **2012**, *7*, 5901–5914. [CrossRef]
148. Al-Thani, R.F.; Patan, N.K.; Al-Maadeed, S. Graphene oxide as antimicrobial against two gram-positive and two gram-negative bacteria in addition to one fungus. *Online J. Biol. Sci.* **2014**, *14*, 230–239. [CrossRef]
149. Hu, M.; Cui, Z.; Li, J.; Zhang, L.; Mo, Y.; Dlamini, D.S.; Wang, H.; He, B.; Li, J.; Matsuyama, H. Ultra-low graphene oxide loading for water permeability, antifouling and antibacterial improvement of polyethersulfone/sulfonated polysulfone ultrafiltration membranes. *J. Colloid Interface Sci.* **2019**, *552*, 319–331. [CrossRef] [PubMed]
150. Di Giulio, M.; Zappacosta, R.; Di Lodovico, S.; DI Campli, E.; Siani, G.; Fontana, A.; Cellini, L. Antimicrobial and Antibiofilm Efficacy of Graphene Oxide against Chronic Wound Microorganisms. *Antimicrob. Agents Chemother.* **2018**, *62*, e00547-18. [CrossRef]
151. Afifi, M.; Ahmed, M.; Ibrahium, H.A.; Awwad, N.S.; Abdel-Fattah, E.; Alshahrani, M.Y. Improvement of physicochemical properties of ternary nanocomposites based on hydroxyapatite/CuO/graphene oxide for biomedical usages. *Ceram. Int.* **2022**, *48*, 3993–4004. [CrossRef]
152. Khosalim, I.P.; Zhang, Y.Y.; Yiu, C.K.Y.; Wong, H.M. Synthesis of a graphene oxide/agarose/hydroxyapatite biomaterial with the evaluation of antibacterial activity and initial cell attachment. *Sci. Rep.* **2022**, *12*, 1971. [CrossRef]
153. Akhavan, O.; Ghaderi, E.; Esfandiar, A. Wrapping Bacteria by Graphene Nanosheets for Isolation from Environment, Reactivation by Sonication, and Inactivation by Near-Infrared Irradiation. *J. Phys. Chem. B* **2011**, *115*, 6279–6288. [CrossRef]
154. Tu, Y.; Lv, M.; Xiu, P.; Huynh, T.; Zhang, M.; Castelli, M.; Liu, Z.; Huang, Q.; Fan, C.; Fang, H.; et al. Destructive extraction of phospholipids from Escherichia coli membranes by graphene nanosheets. *Nat. Nanotechnol.* **2013**, *8*, 594–601. [CrossRef]
155. Akhavan, O.; Ghaderi, E. Toxicity of graphene and graphene oxide nanowalls against bacteria. *ACS Nano* **2010**, *4*, 5731–5736. [CrossRef]
156. Wu, M.-C.; Deokar, A.R.; Liao, J.-H.; Shih, P.-Y.; Ling, Y.-C. Graphene-Based Photothermal Agent for Rapid and Effective Killing of Bacteria. *ACS Nano* **2013**, *7*, 1281–1290. [CrossRef]
157. Xie, C.; Zhang, P.; Guo, Z.; Li, X.; Pang, Q.; Zheng, K.; He, X.; Ma, Y.; Zhang, Z.; Lynch, I. Elucidating the origin of the surface functionalization—dependent bacterial toxicity of graphene nanomaterials: Oxidative damage, physical disruption, and cell autolysis. *Sci. Total Environ.* **2020**, *747*, 141546. [CrossRef]
158. Wang, Y.; Basdogan, Y.; Zhang, T.; Lankone, R.S.; Wallace, A.N.; Fairbrother, D.H.; Keith, J.A.; Gilbertson, L.M. Unveiling the Synergistic Role of Oxygen Functional Groups in the Graphene-Mediated Oxidation of Glutathione. *ACS Appl. Mater. Interfaces* **2020**, *12*, 45753–45762. [CrossRef] [PubMed]
159. Wang, L.; Gao, F.; Wang, A.; Chen, X.; Li, H.; Zhang, X.; Zheng, H.; Ji, R.; Li, B.; Yu, X.; et al. Defect-Rich Adhesive Molybdenum Disulfide/rGO Vertical Heterostructures with Enhanced Nanozyme Activity for Smart Bacterial Killing Application. *Adv. Mater.* **2020**, *32*, 2005423. [CrossRef] [PubMed]

160. Zhang, P.; Guo, Z.; Chen, C.; Lynch, I. Uncertainties in the antibacterial mechanisms of graphene family materials. *Nano Today* **2022**, *43*, 101436. [CrossRef]
161. Chong, Y.; Ge, C.; Fang, G.; Wu, R.; Zhang, H.; Chai, Z.; Chen, C.; Yin, J.-J. Light-Enhanced Antibacterial Activity of Graphene Oxide, Mainly via Accelerated Electron Transfer. *Environ. Sci. Technol.* **2017**, *51*, 10154–10161. [CrossRef]
162. Perreault, F.; de Faria, A.F.; Nejati, S.; Elimelech, M. Antimicrobial Properties of Graphene Oxide Nanosheets: Why Size Matters. *ACS Nano* **2015**, *9*, 7226–7236. [CrossRef]
163. Wassel, M.O.; Khattab, M.A. Antibacterial activity against Streptococcus mutans and inhibition of bacterial induced enamel demineralization of propolis, miswak, and chitosan nanoparticles based dental varnishes. *J. Adv. Res.* **2017**, *8*, 387–392. [CrossRef]
164. Soleimani, B.; Goli, H.; Naranjian, M.; Mousavi, S.J.; Nahvi, A. Comparison of Antimicrobial Activity of Fluoride Varnishes against Streptococcusmutans and Lactobacillus acidophilus: An In Vitro Study. *Iran. J. Pediatr.* **2021**, *in press*.
165. Briseño-Marroquín, B.; Ismael, Y.; Callaway, A.; Tennert, C.; Wolf, T.G. Antibacterial effect of silver diamine fluoride and potassium iodide against E. faecalis, A. naeslundii and P. micra. *BMC Oral Health* **2021**, *21*, 175. [CrossRef]
166. Nathanael, A.J.; Oyane, A.; Nakamura, M.; Mahanti, M.; Koga, K.; Shitomi, K.; Miyaji, H. Rapid and area-specific coating of fluoride-incorporated apatite layers by a laser-assisted biomimetic process for tooth surface functionalization. *Acta Biomater.* **2018**, *79*, 148–157. [CrossRef]
167. Rao, B.C.N.R.; Govindaraj, A.; Vivekchand, S.R.C. Inorganic nanomaterials: Current status and future prospects. *Annu. Rep. Sect. A Inorganic Chem.* **2006**, *102*, 20–45. [CrossRef]
168. Gasparotto, A.; Barreca, D.; Maccato, C.; Tondello, E. Manufacturing of inorganic nanomaterials: Concepts and perspectives. *Nanoscale* **2012**, *4*, 2813–2825. [CrossRef]
169. Kannan, P.K.; Late, D.J.; Morgan, H.; Rout, C.S. Recent developments in 2D layered inorganic nanomaterials for sensing. *Nanoscale* **2015**, *7*, 13293–13312. [CrossRef] [PubMed]
170. Hemeg, H.A. Nanomaterials for alternative antibacterial therapy. *Int. J. Nanomed.* **2017**, *12*, 8211–8225. [CrossRef]
171. Slavin, Y.N.; Asnis, J.; Häfeli, U.O.; Bach, H. Metal nanoparticles: Understanding the mechanisms behind antibacterial activity. *J. Nanobiotechnol.* **2017**, *15*, 65. [CrossRef]
172. Alavi, M.; Rai, M. Recent advances in antibacterial applications of metal nanoparticles (MNPs) and metal nanocomposites (MNCs) against multidrug-resistant (MDR) bacteria. *Expert Rev. Anti-infective Ther.* **2019**, *17*, 419–428. [CrossRef] [PubMed]
173. Panáček, A.; Smékalová, M.; Kilianová, M.; Prucek, R.; Bogdanová, K.; Večeřová, R.; Kolář, M.; Havrdová, M.; Płaza, G.A.; Chojniak, J.; et al. Strong and Nonspecific Synergistic Antibacterial Efficiency of Antibiotics Combined with Silver Nanoparticles at Very Low Concentrations Showing No Cytotoxic Effect. *Molecules* **2015**, *21*, 26. [CrossRef] [PubMed]
174. Riaz, M.; Mutreja, V.; Sareen, S.; Ahmad, B.; Faheem, M.; Zahid, N.; Jabbour, G.; Park, J. Exceptional antibacterial and cytotoxic potency of monodisperse greener AgNPs prepared under optimized pH and temperature. *Sci. Rep.* **2021**, *11*, 2866. [CrossRef]
175. Urnukhsaikhan, E.; Bold, B.-E.; Gunbileg, A.; Sukhbaatar, N.; Mishig-Ochir, T. Antibacterial activity and characteristics of silver nanoparticles biosynthesized from Carduus crispus. *Sci. Rep.* **2021**, *11*, 21047. [CrossRef]
176. Baptista, P.V.; McCusker, M.P.; Carvalho, A.; Ferreira, D.A.; Mohan, N.M.; Martins, M.; Fernandes, A.R. Nano-strategies to fight multidrug resistant bacteria—"A Battle of the Titans". *Front. Microbiol.* **2018**, *9*, 1441. [CrossRef]
177. Kittler, S.; Greulich, C.; Diendorf, J.; Köller, M.; Epple, M. Toxicity of Silver Nanoparticles Increases during Storage Because of Slow Dissolution under Release of Silver Ions. *Chem. Mater.* **2010**, *22*, 4548–4554. [CrossRef]
178. Mukha, I.P.; Eremenko, A.M.; Smirnova, N.P.; Mikhienkova, A.I.; Korchak, G.I.; Gorchev, V.F.; Chunikhin, A.Y. Antimicrobial activity of stable silver nanoparticles of a certain size. *Appl. Biochem. Microbiol.* **2013**, *49*, 199–206. [CrossRef]
179. Choi, O.; Hu, Z. Size Dependent and Reactive Oxygen Species Related Nanosilver Toxicity to Nitrifying Bacteria. *Environ. Sci. Technol.* **2008**, *42*, 4583–4588. [CrossRef] [PubMed]
180. Bruna, T.; Maldonado-Bravo, F.; Jara, P.; Caro, N. Silver Nanoparticles and Their Antibacterial Applications. *Int. J. Mol. Sci.* **2021**, *22*, 7202. [CrossRef] [PubMed]
181. Sim, W.; Barnard, R.T.; Blaskovich, M.A.T.; Ziora, Z.M. Antimicrobial silver in medicinal and consumer applications: A patent review of the past decade (2007–2017). *Antibiotics* **2018**, *7*, 93. [CrossRef] [PubMed]
182. Yaqoob, A.A.; Umar, K.; Ibrahim, M.N.M. Silver nanoparticles: Various methods of synthesis, size affecting factors and their potential applications–a review. *Appl. Nanosci.* **2020**, *10*, 1369–1378. [CrossRef]
183. Panáček, A.; Kvítek, L.; Smékalová, M.; Večeřová, R.; Kolář, M.; Röderová, M.; Dyčka, F.; Šebela, M.; Prucek, R.; Tomanec, O.; et al. Bacterial resistance to silver nanoparticles and how to overcome it. *Nat. Nanotechnol.* **2018**, *13*, 65–71. [CrossRef]
184. León-Silva, S.; Fernández-Luqueño, F.; López-Valdez, F. Silver Nanoparticles (AgNP) in the Environment: A Review of Potential Risks on Human and Environmental Health. *Water Air Soil Pollut.* **2016**, *227*, 306. [CrossRef]
185. Qamar, S.U.R.; Ahmad, J.N. Nanoparticles: Mechanism of biosynthesis using plant extracts, bacteria, fungi, and their applications. *J. Mol. Liq.* **2021**, *334*, 116040. [CrossRef]
186. Ramzan, U.; Majeed, W.; Hussain, A.A.; Qurashi, F.; Qamar, S.U.R.; Naeem, M.; Uddin, J.; Khan, A.; Al-Harrasi, A.; Razak, S.I.A.; et al. New Insights for Exploring the Risks of Bioaccumulation, Molecular Mechanisms, and Cellular Toxicities of AgNPs in Aquatic Ecosystem. *Water* **2022**, *14*, 2192. [CrossRef]
187. Shumbula, N.P.; Nkabinde, S.S.; Ndala, Z.B.; Mpelane, S.; Shumbula, M.P.; Mdluli, P.S.; Njengele-Tetyana, Z.; Tetyana, P.; Hlatshwayo, T.; Mlambo, M.; et al. Evaluating the antimicrobial activity and cytotoxicity of polydopamine capped silver and silver/polydopamine core-shell nanocomposites. *Arab. J. Chem.* **2022**, *15*, 103798. [CrossRef]

188. Peña-Juarez, M.G.; Sanchez-Vargas, L.O.; Flores-Gonzalez, L.A.; Almendarez-Camarillo, A.; Gutierrez-Castañeda, E.J.; Navarrete-Damian, J.; Pérez, E.; Gonzalez-Calderon, J.A. Mechanical, antibacterial, and non-cytotoxic performance of polypropylene nanocomposites reinforced with $sTiO_2$ deposited with AgNPs mediated by quercetin biomolecule. *Polym. Bull.* **2022**, 1–27. [CrossRef]
189. Richter, A.P.; Brown, J.S.; Bharti, B.; Wang, A.; Gangwal, S.; Houck, K.; Hubal, E.A.C.; Paunov, V.; Stoyanov, S.; Velev, O. An environmentally benign antimicrobial nanoparticle based on a silver-infused lignin core. *Nat. Nanotechnol.* **2015**, *10*, 817–823. [CrossRef] [PubMed]
190. Dakal, T.C.; Kumar, A.; Majumdar, R.S.; Yadav, V. Mechanistic basis of antimicrobial actions of silver nanoparticles. *Front. Microbiol.* **2016**, *7*, 1831. [CrossRef]
191. Rai, M.K.; Deshmukh, S.D.; Ingle, A.P.; Gade, A.K. Silver nanoparticles: The powerful nanoweapon against multidrug-resistant bacteria. *J. Appl. Microbiol.* **2012**, *112*, 841–852. [CrossRef]
192. Rizzello, L.; Pompa, P.P. Nanosilver-based antibacterial drugs and devices: Mechanisms, methodological drawbacks, and guidelines. *Chem. Soc. Rev.* **2014**, *43*, 1501–1518. [CrossRef] [PubMed]
193. El-Azizi, M.M.; El Din, S.N.; El-Tayeb, T.A.; Aisha, K.A. In vitro and in vivo antimicrobial activity of combined therapy of silver nanoparticles and visible blue light against Pseudomonas aeruginosa. *Int. J. Nanomed.* **2016**, *11*, 1749–1758. [CrossRef] [PubMed]
194. Shamaila, S.; Zafar, N.; Riaz, S.; Sharif, R.; Nazir, J.; Naseem, S. Gold Nanoparticles: An Efficient Antimicrobial Agent against Enteric Bacterial Human Pathogen. *Nanomaterials* **2016**, *6*, 71. [CrossRef]
195. Abdel-Raouf, N.; Al-Enazi, N.M.; Ibraheem, I.B. Green biosynthesis of gold nanoparticles using Galaxaura elongata and characterization of their antibacterial activity. *Arab. J. Chem.* **2017**, *10*, S3029–S3039. [CrossRef]
196. Lanh, L.T.; Hoa, T.T.; Cuong, N.D.; Khieu, D.Q.; Quang, D.T.; Van Duy, N.; Hoa, N.D.; Van Hieu, N. Shape and size controlled synthesis of Au nanorods: H 2 S gas-sensing characterizations and antibacterial application. *J. Alloy. Compd.* **2015**, *635*, 265–271. [CrossRef]
197. Lee, K.; Nagajyothi, P.; Sreekanth, T.; Park, S. Eco-friendly synthesis of gold nanoparticles (AuNPs) using Inonotus obliquus and their antibacterial, antioxidant and cytotoxic activities. *J. Ind. Eng. Chem.* **2015**, *26*, 67–72. [CrossRef]
198. Vanaraj, S.; Jabastin, J.; Sathiskumar, S.; Preethi, K. Production and Characterization of Bio-AuNPs to Induce Synergistic Effect against Multidrug Resistant Bacterial Biofilm. *J. Clust. Sci.* **2017**, *28*, 227–244. [CrossRef]
199. Boomi, P.; Poorani, G.P.; Selvam, S.; Palanisamy, S.; Jegatheeswaran, S.; Anand, K.; Balakumar, C.; Premkumar, K.; Prabu, H.G. Green biosynthesis of gold nanoparticles using Croton sparsiflorus leaves extract and evaluation of UV protection, antibacterial and anticancer applications. *Appl. Organomet. Chem.* **2020**, *34*, e5574. [CrossRef]
200. Chatterjee, A.K.; Chakraborty, R.; Basu, T. Mechanism of antibacterial activity of copper nanoparticles. *Nanotechnology* **2014**, *25*, 135101. [CrossRef] [PubMed]
201. Esparza-González, S.; Sánchez-Valdés, S.; Ramírez-Barrón, S.; Loera-Arias, M.; Bernal, J.; Meléndez-Ortiz, H.I.; Betancourt-Galindo, R. Effects of different surface modifying agents on the cytotoxic and antimicrobial properties of ZnO nanoparticles. *Toxicol. Vitr.* **2016**, *37*, 134–141. [CrossRef] [PubMed]
202. Yoon, K.-Y.; Byeon, J.H.; Park, J.-H.; Hwang, J. Susceptibility constants of Escherichia coli and Bacillus subtilis to silver and copper nanoparticles. *Sci. Total Environ.* **2007**, *373*, 572–575. [CrossRef]
203. Raffi, M.; Mehrwan, S.; Bhatti, T.M.; Akhter, J.I.; Hameed, A.; Yawar, W.; ul Hasan, M.M. Investigations into the antibacterial behavior of copper nanoparticles against Escherichia coli. *Ann. Microbiol.* **2010**, *60*, 75–80. [CrossRef]
204. Chatterjee, A.K.; Sarkar, R.K.; Chattopadhyay, A.P.; Aich, P.; Chakraborty, R.; Basu, T. A simple robust method for synthesis of metallic copper nanoparticles of high antibacterial potency against *E. coli*. *Nanotechnology* **2012**, *23*, 85103. [CrossRef]
205. Cabral-Romero, C.; Hernandez-Delgadillo, R.; Velasco-Arias, D.; Diaz, D.; Arevalo-Niño, K.; A De la Garza-Ramos, M. Zerovalent bismuth nanoparticles inhibit Streptococcus mutans growth and formation of biofilm. *Int. J. Nanomed.* **2012**, *7*, 2109–2113. [CrossRef]
206. Vazquez-Munoz, R.; Arellano-Jimenez, M.J.; Lopez-Ribot, J.L. Bismuth nanoparticles obtained by a facile synthesis method exhibit antimicrobial activity against Staphylococcus aureus and Candida albicans. *BMC Biomed. Eng.* **2020**, *2*, 11. [CrossRef]
207. Azad, A.; Rostamifar, S.; Modaresi, F.; Bazrafkan, A.; Rezaie, Z. Assessment of the Antibacterial Effects of Bismuth Nanoparticles against Enterococcus faecalis. *BioMed Res. Int.* **2020**, *2020*, 5465439. [CrossRef]
208. Jawad, K.H.; Marzoog, T.R.; Hasoon, B.A.; Sulaiman, G.M.; Jabir, M.S.; Ahmed, E.M.; Khalil, K.A.A. Antibacterial Activity of Bismuth Oxide Nanoparticles Compared to Amikacin against Acinetobacter baumannii and Staphylococcus aureus. *J. Nanomater.* **2022**, *2022*, 8511601. [CrossRef]
209. Ashfaq, M.; Verma, N.; Khan, S. Copper/zinc bimetal nanoparticles-dispersed carbon nanofibers: A novel potential antibiotic material. *Mater. Sci. Eng. C* **2016**, *59*, 938–947. [CrossRef] [PubMed]
210. Cruces, E.; Arancibia-Miranda, N.; Manquián-Cerda, K.; Perreault, F.; Bolan, N.; Azócar, M.I.; Cubillos, V.; Montory, J.; Rubio, M.A.; Sarkar, B. Copper/Silver Bimetallic Nanoparticles Supported on Aluminosilicate Geomaterials as Antibacterial Agents. *ACS Appl. Nano Mater.* **2022**, *5*, 1472–1483. [CrossRef]
211. Merugu, R.; Gothalwal, R.; Deshpande, P.K.; De Mandal, S.; Padala, G.; Chitturi, K.L. Synthesis of Ag/Cu and Cu/Zn bimetallic nanoparticles using toddy palm: Investigations of their antitumor, antioxidant and antibacterial activities. *Mater. Today Proc.* **2021**, *44*, 99–105. [CrossRef]

212. Khatak, S.; Wadhwa, N.; Jain, P. Monometallic Zinc and Bimetallic Cu-Zn Nanoparticles Synthesis Using Stem Extracts of Cissusquadrangularis (Haddjod) and Proneness as Alternative Antimicrobial Agents. *Biosci. Biotechnol. Res. Asia* **2021**, *17*, 763–774. [CrossRef]
213. Perdikaki, A.; Galeou, A.; Pilatos, G.; Karatasios, I.; Kanellopoulos, N.K.; Prombona, A.; Karanikolos, G.N. Ag and Cu Monometallic and Ag/Cu Bimetallic Nanoparticle–Graphene Composites with Enhanced Antibacterial Performance. *ACS Appl. Mater. Interfaces* **2016**, *8*, 27498–27510. [CrossRef]
214. Liakos, I.; Grumezescu, A.M.; Holban, A.M. Magnetite Nanostructures as Novel Strategies for Anti-Infectious Therapy. *Molecules* **2014**, *19*, 12710–12726. [CrossRef]
215. Azam, A.; Ahmed, A.S.; Oves, M.; Khan, M.S.; Habib, S.S.; Memic, A. Antimicrobial activity of metal oxide nanoparticles against Gram-positive and Gram-negative bacteria: A comparative study. *Int. J. Nanomed.* **2012**, *7*, 6003–6009. [CrossRef]
216. Jacobsen, N.R.; Stoeger, T.; van den Brule, S.; Saber, A.T.; Beyerle, A.; Vietti, G.; Mortensen, A.; Szarek, J.; Budtz, H.C.; Kermanizadeh, A.; et al. Acute and subacute pulmonary toxicity and mortality in mice after intratracheal instillation of ZnO nanoparticles in three laboratories. *Food Chem. Toxicol.* **2015**, *85*, 84–95. [CrossRef]
217. Padmavathy, N.; Vijayaraghavan, R. Enhanced bioactivity of ZnO nanoparticles—An antimicrobial study. *Sci. Technol. Adv. Mater.* **2008**, *9*, 35004. [CrossRef]
218. Asture, A.; Rawat, V.; Srivastava, C.; Vaya, D. Investigation of properties and applications of ZnO polymer nanocomposites. *Polym. Bull.* **2022**, 1–39. [CrossRef]
219. Dutta, R.; Nenavathu, B.P.; Gangishetty, M.K.; Reddy, A. Studies on antibacterial activity of ZnO nanoparticles by ROS induced lipid peroxidation. *Colloids Surfaces B Biointerfaces* **2012**, *94*, 143–150. [CrossRef] [PubMed]
220. Pasquet, J.; Chevalier, Y.; Pelletier, J.; Couval, E.; Bouvier, D.; Bolzinger, M.-A. The contribution of zinc ions to the antimicrobial activity of zinc oxide. *Colloids Surfaces A Physicochem. Eng. Asp.* **2014**, *457*, 263–274. [CrossRef]
221. Dadi, R.; Azouani, R.; Traore, M.; Mielcarek, C.; Kanaev, A. Antibacterial activity of ZnO and CuO nanoparticles against gram positive and gram negative strains. *Mater. Sci. Eng. C* **2019**, *104*, 109968. [CrossRef]
222. Sirelkhatim, A.; Mahmud, S.; Seeni, A.; Kaus, N.H.M.; Ann, L.C.; Bakhori, S.K.M.; Hasan, H.; Mohamad, D. Review on Zinc Oxide Nanoparticles: Antibacterial Activity and Toxicity Mechanism. *Nano-Micro Lett.* **2015**, *7*, 219–242. [CrossRef]
223. Tso, C.-P.; Zhung, C.-M.; Shih, Y.-H.; Tseng, Y.-M.; Wu, S.-C.; Doong, R.-A. Stability of metal oxide nanoparticles in aqueous solutions. *Water Sci. Technol.* **2010**, *61*, 127–133. [CrossRef]
224. Kumar, R.; Reddy, P.; Shankar, K.; Rambabu, D.; Venkateswarulu, M.; Kumbam, L.R.; Sagara, P.; Nakka, N.; Yogesh, M. 9—Surface coating and functionalization of metal and metal oxide nanoparticles for biomedical applications. In *Metal Oxides for Biomedical and Biosensor Applications*; Mondal, K., Ed.; Elsevier: Amsterdam, The Netherlands, 2022; pp. 205–231.
225. Armentano, I.; Puglia, D.; Luzi, F.; Arciola, C.R.; Morena, F.; Martino, S.; Torre, L. Nanocomposites Based on Biodegradable Polymers. *Materials* **2018**, *11*, 795. [CrossRef]
226. Jancar, J.; Douglas, J.; Starr, F.; Kumar, S.; Cassagnau, P.; Lesser, A.; Sternstein, S.; Buehler, M. Current issues in research on structure–property relationships in polymer nanocomposites. *Polymer* **2010**, *51*, 3321–3343. [CrossRef]
227. Li, X.; Robinson, S.M.; Gupta, A.; Saha, K.; Jiang, Z.; Moyano, D.F.; Sahar, A.; Riley, M.A.; Rotello, V.M. Functional Gold Nanoparticles as Potent Antimicrobial Agents against Multi-Drug-Resistant Bacteria. *ACS Nano* **2014**, *8*, 10682–10686. [CrossRef]
228. Tran, C.D.; Prosenc, F.; Franko, M.; Benzi, G. One-Pot Synthesis of Biocompatible Silver Nanoparticle Composites from Cellulose and Keratin: Characterization and Antimicrobial Activity. *ACS Appl. Mater. Interfaces* **2016**, *8*, 34791–34801. [CrossRef]
229. Dai, T.; Wang, C.; Wang, Y.; Xu, W.; Hu, J.; Cheng, Y. A Nanocomposite Hydrogel with Potent and Broad-Spectrum Antibacterial Activity. *ACS Appl. Mater. Interfaces* **2018**, *10*, 15163–15173. [CrossRef] [PubMed]
230. Tamayo, L.; Palza, H.; Bejarano, J.; Zapata, P.A. 8—Polymer Composites With Metal Nanoparticles: Synthesis, Properties, and Applications. In *Polymer Composites with Functionalized Nanoparticles*; Pielichowski, K., Majka, T.M., Eds.; Elsevier: Amsterdam, The Netherlands, 2019; pp. 249–286.
231. Kashihara, K.; Uto, Y.; Nakajima, T. Rapid in situ synthesis of polymer-metal nanocomposite films in several seconds using a CO2 laser. *Sci. Rep.* **2018**, *8*, 14719. [CrossRef] [PubMed]
232. Gniadek, M.; Malinowska, S.; Rapecki, T.; Stojek, Z.; Donten, M. Synthesis of polymer–metal nanocomposites at liquid–liquid interface supported by ultrasonic irradiation. *Synth. Met.* **2014**, *187*, 193–200. [CrossRef]
233. Wan, J.; Fan, B.; Thang, S.H. Sonochemical preparation of polymer–metal nanocomposites with catalytic and plasmonic properties. *Nanoscale Adv.* **2021**, *3*, 3306–3315. [CrossRef]
234. Mei, L.; Lu, Z.; Zhang, X.; Li, C.; Jia, Y. Polymer-Ag Nanocomposites with Enhanced Antimicrobial Activity against Bacterial Infection. *ACS Appl. Mater. Interfaces* **2014**, *6*, 15813–15821. [CrossRef]
235. Dai, X.; Chen, X.; Zhao, J.; Zhao, Y.; Guo, Q.; Zhang, T.; Chu, C.; Zhang, X.; Li, C. Structure–Activity Relationship of Membrane-Targeting Cationic Ligands on a Silver Nanoparticle Surface in an Antibiotic-Resistant Antibacterial and Antibiofilm Activity Assay. *ACS Appl. Mater. Interfaces* **2017**, *9*, 13837–13848. [CrossRef]
236. Manoswini, M.; Bhattacharya, D.; Sen, P.; Ganguly, N.; Mohanty, P.S. Antibacterial and cytotoxic activity of polymer-metal hybrid nanoparticle. *Adv. Nat. Sci. Nanosci. Nanotechnol.* **2021**, *12*, 25003. [CrossRef]
237. Tamayo, L.; Azócar, M.; Kogan, M.; Riveros, A.; Páez, M. Copper-polymer nanocomposites: An excellent and cost-effective biocide for use on antibacterial surfaces. *Mater. Sci. Eng. C* **2016**, *69*, 1391–1409. [CrossRef]

238. Lu, B.; Lu, F.; Ran, L.; Yu, K.; Xiao, Y.; Li, Z.; Dai, F.; Wu, D.; Lan, G. Imidazole-molecule-capped chitosan–gold nanocomposites with enhanced antimicrobial activity for treating biofilm-related infections. *J. Colloid Interface Sci.* **2018**, *531*, 269–281. [CrossRef]
239. Pryjmaková, J.; Kaimlová, M.; Vokatá, B.; Hubáček, T.; Slepička, P.; Švorčík, V.; Siegel, J. Bimetallic Nanowires on Laser-Patterned PEN as Promising Biomaterials. *Nanomaterials* **2021**, *11*, 2285. [CrossRef]
240. Prasanna, S.R.V.S.; Balaji, K.; Pandey, S.; Rana, S. Chapter 4—Metal Oxide Based Nanomaterials and Their Polymer Nanocomposites. In *Nanomaterials and Polymer Nanocomposites*; Karak, N., Ed.; Elsevier: Amsterdam, The Netherlands, 2019; pp. 123–144.
241. Soytaş, S.H.; Oğuz, O.; Menceloğlu, Y.Z. 9—Polymer Nanocomposites With Decorated Metal Oxides. In *Polymer Composites with Functionalized Nanoparticles*; Pielichowski, K., Majka, T.M., Eds.; Elsevier: Amsterdam, The Netherlands, 2019; pp. 287–323.
242. Shameem, M.M.; Sasikanth, S.; Annamalai, R.; Raman, R.G. A brief review on polymer nanocomposites and its applications. *Mater. Today Proc.* **2021**, *45*, 2536–2539. [CrossRef]
243. Aktitiz, I.; Aydın, K.; Darıcık, F.; Topcu, A. Production of different metal oxide nanoparticle embedded polymer matrix composite structures by the additive manufacturing technology and investigation of their properties. *Polym. Compos.* **2022**. [CrossRef]
244. Li, L.; Yu, T. Curing comparison and performance investigation of polyurethane concrete with retarders. *Constr. Build. Mater.* **2022**, *326*, 126883. [CrossRef]
245. Pholnak, C.; Sirisathitkul, C.; Soontaranon, S.; Rugmai, S. UV–Vis Absorption and Small Angle X-ray Scattering Spectra of Commercial Polyurethane Coating Filled with Zinc Oxide. *Natl. Acad. Sci. Lett.* **2016**, *39*, 125–128. [CrossRef]
246. Tavakoli, A.; Sohrabi, M.; Kargari, A. A review of methods for synthesis of nanostructured metals with emphasis on iron compounds. *Chem. Pap.* **2007**, *61*, 151–170. [CrossRef]
247. Bahadur, A.; Iqbal, S.; Alsaab, H.O.; Awwad, N.S.; Ibrahium, H.A. Thermal degradation study of polymethylmethacrylate with AlI_3 nanoadditive. *Microsc. Res. Tech.* **2022**, *85*, 1494–1501. [CrossRef]
248. Zhou, J.; Xu, N.S.; Wang, Z.L. Dissolving Behavior and Stability of ZnO Wires in Biofluids: A Study on Biodegradability and Biocompatibility of ZnO Nanostructures. *Adv. Mater.* **2006**, *18*, 2432–2435. [CrossRef]
249. Rai, P.; Kwak, W.-K.; Yu, Y.-T. Solvothermal Synthesis of ZnO Nanostructures and Their Morphology-Dependent Gas-Sensing Properties. *ACS Appl. Mater. Interfaces* **2013**, *5*, 3026–3032. [CrossRef]
250. Poyraz, S.; Cerkez, I.; Huang, T.S.; Liu, Z.; Kang, L.; Luo, J.; Zhang, X. One-Step Synthesis and Characterization of Polyaniline Nanofiber/Silver Nanoparticle Composite Networks as Antibacterial Agents. *ACS Appl. Mater. Interfaces* **2014**, *6*, 20025–20034. [CrossRef]
251. Hashemi, A.; Jouault, N.; Williams, G.A.; Zhao, D.; Cheng, K.J.; Kysar, J.W.; Guan, Z.; Kumar, S.K. Enhanced Glassy State Mechanical Properties of Polymer Nanocomposites via Supramolecular Interactions. *Nano Lett.* **2015**, *15*, 5465–5471. [CrossRef]
252. Nikitin, D.; Madkour, S.; Pleskunov, P.; Tafiichuk, R.; Shelemin, A.; Hanuš, J.; Gordeev, I.; Sysolyatina, E.; Lavrikova, A.; Ermolaeva, S.; et al. Cu nanoparticles constrain segmental dynamics of cross-linked polyethers: A trade-off between non-fouling and antibacterial properties. *Soft Matter* **2019**, *15*, 2884–2896. [CrossRef]
253. Kostic, D.; Sekulic, M.V.; Armentano, I.; Torre, L.; Obradovic, B. Multifunctional ternary composite films based on PLA and Ag/alginate microbeads: Physical characterization and silver release kinetics. *Mater. Sci. Eng. C* **2019**, *98*, 1159–1168. [CrossRef] [PubMed]
254. Hernández-Rangel, A.; Silva-Bermudez, P.; España-Sánchez, B.; Luna-Hernández, E.; Almaguer-Flores, A.; Ibarra, C.; Garcia-Perez, V.; Velasquillo, C.; Luna-Barcenas, G. Fabrication and in vitro behavior of dual-function chitosan/silver nanocomposites for potential wound dressing applications. *Mater. Sci. Eng. C* **2019**, *94*, 750–765. [CrossRef] [PubMed]
255. Hasan, A.; Waibhaw, G.; Saxena, V.; Pandey, L.M. Nano-biocomposite scaffolds of chitosan, carboxymethyl cellulose and silver nanoparticle modified cellulose nanowhiskers for bone tissue engineering applications. *Int. J. Biol. Macromol.* **2018**, *111*, 923–934. [CrossRef] [PubMed]
256. Mehrabani, M.G.; Karimian, R.; Mehramouz, B.; Rahimi, M.; Kafil, H.S. Preparation of biocompatible and biodegradable silk fibroin/chitin/silver nanoparticles 3D scaffolds as a bandage for antimicrobial wound dressing. *Int. J. Biol. Macromol.* **2018**, *114*, 961–971. [CrossRef] [PubMed]
257. Díez, B.; Santiago-Morales, J.; Martínez-Bueno, M.J.; Fernández-Alba, A.R.; Rosal, R. Antimicrobial organic–inorganic composite membranes including sepiolite-supported nanometals. *RSC Adv.* **2017**, *7*, 2323–2332. [CrossRef]
258. Huang, X.; Bao, X.; Wang, Z.; Hu, Q. A novel silver-loaded chitosan composite sponge with sustained silver release as a long-lasting antimicrobial dressing. *RSC Adv.* **2017**, *7*, 34655–34663. [CrossRef]
259. Baek, K.; Liang, J.; Lim, W.T.; Zhao, H.; Kim, D.H.; Kong, H. In Situ Assembly of Antifouling/Bacterial Silver Nanoparticle-Hydrogel Composites with Controlled Particle Release and Matrix Softening. *ACS Appl. Mater. Interfaces* **2015**, *7*, 15359–15367. [CrossRef]
260. García-Astrain, C.; Chen, C.; Burón, M.; Palomares, T.; Eceiza, A.; Fruk, L.; Corcuera, M.Á.; Gabilondo, N. Biocompatible Hydrogel Nanocomposite with Covalently Embedded Silver Nanoparticles. *Biomacromolecules* **2015**, *16*, 1301–1310. [CrossRef]
261. GhavamiNejad, A.; Park, C.H.; Kim, C.S. In Situ Synthesis of Antimicrobial Silver Nanoparticles within Antifouling Zwitterionic Hydrogels by Catecholic Redox Chemistry for Wound Healing Application. *Biomacromolecules* **2016**, *17*, 1213–1223. [CrossRef]
262. Liu, J.; Gao, Y.; Cao, D.; Zhang, L.; Guo, Z. Nanoparticle Dispersion and Aggregation in Polymer Nanocomposites: Insights from Molecular Dynamics Simulation. *Langmuir* **2011**, *27*, 7926–7933. [CrossRef]

263. Zhong, X.; Song, Y.; Yang, P.; Wang, Y.; Jiang, S.; Zhang, X.; Li, C. Titanium Surface Priming with Phase-Transited Lysozyme to Establish a Silver Nanoparticle-Loaded Chitosan/Hyaluronic Acid Antibacterial Multilayer via Layer-by-Layer Self-Assembly. *PLoS ONE* **2016**, *11*, e0146957. [CrossRef] [PubMed]
264. Cai, S.; Pourdeyhimi, B.; Loboa, E.G. High-Throughput Fabrication Method for Producing a Silver-Nanoparticles-Doped Nanoclay Polymer Composite with Novel Synergistic Antibacterial Effects at the Material Interface. *ACS Appl. Mater. Interfaces* **2017**, *9*, 21105–21115. [CrossRef] [PubMed]
265. Mohiti-Asli, M.; Pourdeyhimi, B.; Loboa, E.G. Skin Tissue Engineering for the Infected Wound Site: Biodegradable PLA Nanofibers and a Novel Approach for Silver Ion Release Evaluated in a 3D Coculture System of Keratinocytes and *Staphylococcus aureus*. *Tissue Eng. Part C Methods* **2014**, *20*, 790–797. [CrossRef] [PubMed]
266. Majumdar, A.; Butola, B.S.; Thakur, S. Development and performance optimization of knitted antibacterial materials using polyester–silver nanocomposite fibres. *Mater. Sci. Eng. C* **2015**, *54*, 26–31. [CrossRef]
267. Abudula, T.; Qurban, R.O.; Bolarinwa, S.O.; Mirza, A.A.; Pasovic, M.; Memic, A. 3D Printing of Metal/Metal Oxide Incorporated Thermoplastic Nanocomposites With Antimicrobial Properties. *Front. Bioeng. Biotechnol.* **2020**, *8*, 568186. [CrossRef] [PubMed]
268. Liu, M.; Bauman, L.; Nogueira, C.L.; Aucoin, M.G.; Anderson, W.A.; Zhao, B. Antimicrobial polymeric composites for high-touch surfaces in healthcare applications. *Curr. Opin. Biomed. Eng.* **2022**, *22*, 100395. [CrossRef]
269. Murthy, P.S.; Pandiyan, V.; Das, A. Potential of Metal Oxide Nanoparticles and Nanocomposites as Antibiofilm Agents: Leverages and Limitations. In *Emerging Nanomaterials for Advanced Technologies*; Krishnan, A., Ravindran, B., Balasubramanian, B., Swart, H.C., Panchu, S.J., Prasad, R., Eds.; Springer International Publishing: Cham, Switzerland, 2022; pp. 163–209.
270. Choudhury, M.; Bindra, H.S.; Singh, K.; Singh, A.K.; Nayak, R. Antimicrobial polymeric composites in consumer goods and healthcare sector: A healthier way to prevent infection. *Polym. Adv. Technol.* **2022**, *33*, 1997–2024. [CrossRef]
271. Prasanna, V.L.; Vijayaraghavan, R. Insight into the Mechanism of Antibacterial Activity of ZnO: Surface Defects Mediated Reactive Oxygen Species Even in the Dark. *Langmuir* **2015**, *31*, 9155–9162. [CrossRef]
272. Ozkan, E.; Allan, E.; Parkin, I.P. White-Light-Activated Antibacterial Surfaces Generated by Synergy between Zinc Oxide Nanoparticles and Crystal Violet. *ACS Omega* **2018**, *3*, 3190–3199. [CrossRef]
273. Sehmi, S.K.; Noimark, S.D.; Pike, S.; Bear, J.C.; Peveler, W.J.; Williams, C.K.; Shaffer, M.; Allan, E.; Parkin, I.P.; MacRobert, A.J. Enhancing the Antibacterial Activity of Light-Activated Surfaces Containing Crystal Violet and ZnO Nanoparticles: Investigation of Nanoparticle Size, Capping Ligand, and Dopants. *ACS Omega* **2016**, *1*, 334–343. [CrossRef] [PubMed]
274. Gobi, R.; Ravichandiran, P.; Babu, R.; Yoo, D. Biopolymer and Synthetic Polymer-Based Nanocomposites in Wound Dressing Applications: A Review. *Polymers* **2021**, *13*, 1962. [CrossRef] [PubMed]
275. Díez-Pascual, A.M.; Díez-Vicente, A.L. Wound Healing Bionanocomposites Based on Castor Oil Polymeric Films Reinforced with Chitosan-Modified ZnO Nanoparticles. *Biomacromolecules* **2015**, *16*, 2631–2644. [CrossRef]
276. Mohandas, A.; Kumar P T, S.; Raja, B.; Lakshmanan, V.K.; Jayakumar, R. Exploration of alginate hydrogel/nano zinc oxide composite bandages for infected wounds. *Int. J. Nanomed.* **2015**, *10* (Suppl. S1), 53–66. [CrossRef]
277. Sakthiguru, N.; Sithique, M.A. Preparation and In Vitro Biological Evaluation of Lawsone Loaded O-Carboxymethyl Chitosan/Zinc Oxide Nanocomposite for Wound-Healing Application. *ChemistrySelect* **2020**, *5*, 2710–2718. [CrossRef]
278. Kumar, P.T.S.; Lakshmanan, V.-K.; Anilkumar, T.; Ramya, C.; Reshmi, P.; Unnikrishnan, A.; Nair, S.V.; Jayakumar, R. Flexible and Microporous Chitosan Hydrogel/Nano ZnO Composite Bandages for Wound Dressing: In Vitro and In Vivo Evaluation. *ACS Appl. Mater. Interfaces* **2012**, *4*, 2618–2629. [CrossRef] [PubMed]
279. Raafat, A.I.; El-Sawy, N.M.; Badawy, N.A.; Mousa, E.A.; Mohamed, A.M. Radiation fabrication of Xanthan-based wound dressing hydrogels embedded ZnO nanoparticles: In vitro evaluation. *Int. J. Biol. Macromol.* **2018**, *118*, 1892–1902. [CrossRef]
280. Siracusa, V.; Rocculi, P.; Romani, S.; Rosa, M.D. Biodegradable polymers for food packaging: A review. *Trends Food Sci. Technol.* **2008**, *19*, 634–643. [CrossRef]
281. Jaiswal, L.; Shankar, S.; Rhim, J.-W. Chapter 3—Applications of nanotechnology in food microbiology. In *Methods in Microbiology*; Gurtler, V., Ball, A.S., Soni, S., Eds.; Academic Press: Cambridge, MA, USA, 2019; pp. 43–60.
282. Díez-Pascual, A.M.; Díez-Vicente, A.L. ZnO-Reinforced Poly(3-hydroxybutyrate-co-3-hydroxyvalerate) Bionanocomposites with Antimicrobial Function for Food Packaging. *ACS Appl. Mater. Interfaces* **2014**, *6*, 9822–9834. [CrossRef]
283. Arfat, Y.A.; Ahmed, J.; Al Hazza, A.; Jacob, H.; Joseph, A. Comparative effects of untreated and 3-methacryloxypropyltrimethoxysilane treated ZnO nanoparticle reinforcement on properties of polylactide-based nanocomposite films. *Int. J. Biol. Macromol.* **2017**, *101*, 1041–1050. [CrossRef]
284. Pantani, R.; Giuliana, G.; Vigliottab, G.; Murariuc, M.; Duboisc, P. PLA-ZnO nanocomposite films: Water vapor barrier properties and specific end-use characteristics. *Eur. Polym. J.* **2013**, *49*, 3471–3482. [CrossRef]
285. Zahedi, Y.; Fathi-Achachlouei, B.; Yousefi, A.R. Physical and mechanical properties of hybrid montmorillonite/zinc oxide reinforced carboxymethyl cellulose nanocomposites. *Int. J. Biol. Macromol.* **2018**, *108*, 863–873. [CrossRef] [PubMed]
286. Shankar, S.; Wang, L.-F.; Rhim, J.-W. Incorporation of zinc oxide nanoparticles improved the mechanical, water vapor barrier, UV-light barrier, and antibacterial properties of PLA-based nanocomposite films. *Mater. Sci. Eng. C* **2018**, *93*, 289–298. [CrossRef] [PubMed]
287. Li, C.; Wang, X.; Chen, F.; Zhang, C.; Zhi, X.; Wang, K.; Cui, D. The antifungal activity of graphene oxide–silver nanocomposites. *Biomaterials* **2013**, *34*, 3882–3890. [CrossRef]

288. Das, M.R.; Sarma, R.K.; Saikia, R.; Kale, V.S.; Shelke, M.V.; Sengupta, P. Synthesis of silver nanoparticles in an aqueous suspension of graphene oxide sheets and its antimicrobial activity. *Colloids Surfaces B Biointerfaces* **2011**, *83*, 16–22. [CrossRef]
289. Bao, Q.; Zhang, D.; Qi, P. Synthesis and characterization of silver nanoparticle and graphene oxide nanosheet composites as a bactericidal agent for water disinfection. *J. Colloid Interface Sci.* **2011**, *360*, 463–470. [CrossRef]
290. Salunkhe, A.; Tandon, S.; Dudhwadkar, S. Surface Functionalization of Graphene Oxide with Silver Nanoparticles Using Phyto Extract and its Antimicrobial Properties against Biological Contaminants. *Arab. J. Sci. Eng.* **2022**, 1–15. [CrossRef]
291. Malik, S.B.; Saggu, J.I.; Gul, A.; Abbasi, B.A.; Iqbal, J.; Waris, S.; Bin Jardan, Y.A.; Chalgham, W. Synthesis and Characterization of Silver and Graphene Nanocomposites and Their Antimicrobial and Photocatalytic Potentials. *Molecules* **2022**, *27*, 5184. [CrossRef]
292. Yu, Z.; Xu, Y.; Tian, X. Silver-modified graphene oxide nanosheets for antibacterial performance of bone scaffold. *AIP Adv.* **2022**, *12*, 15024. [CrossRef]
293. Ranjan, R.; Bajpai, V. Graphene-based metal matrix nanocomposites: Recent development and challenges. *J. Compos. Mater.* **2021**, *55*, 2369–2413. [CrossRef]
294. Zhang, X.; Kong, H.; Yang, G.; Zhu, D.; Luan, X.; He, P.; Wei, G. Graphene-Based Functional Hybrid Membranes for Antimicrobial Applications: A Review. *Appl. Sci.* **2022**, *12*, 4834. [CrossRef]
295. Xia, M.-Y.; Xie, Y.; Yu, C.-H.; Chen, G.-Y.; Li, Y.-H.; Zhang, T.; Peng, Q. Graphene-based nanomaterials: The promising active agents for antibiotics-independent antibacterial applications. *J. Control. Release* **2019**, *307*, 16–31. [CrossRef] [PubMed]
296. Kumar, P.; Huo, P.; Zhang, R.; Liu, B. Antibacterial Properties of Graphene-Based Nanomaterials. *Nanomaterials* **2019**, *9*, 737. [CrossRef] [PubMed]
297. Ge, L.; Li, Q.; Wang, M.; Ouyang, J.; Li, X.; Xing, M.M. Nanosilver particles in medical applications: Synthesis, performance, and toxicity. *Int J Nanomedicine* **2014**, *9*, 2399–2407. [CrossRef] [PubMed]
298. Oka, H.; Tomioka, T.; Tomita, K.; Nishino, A.; Ueda, S. Inactivation of Enveloped Viruses by a Silver-Thiosulfate Complex. *Met. Drugs* **1994**, *1*, 511. [CrossRef]
299. Shao, W.; Liu, X.; Min, H.; Dong, G.; Feng, Q.; Zuo, S. Preparation, Characterization, and Antibacterial Activity of Silver Nanoparticle-Decorated Graphene Oxide Nanocomposite. *ACS Appl. Mater. Interfaces* **2015**, *7*, 6966–6973. [CrossRef]
300. Hui, K.; Dinh, D.; Tsang, C.; Cho, Y.; Zhou, W.; Hong, X.; Chun, H.-H. Green synthesis of dimension-controlled silver nanoparticle–graphene oxide with in situ ultrasonication. *Acta Mater.* **2014**, *64*, 326–332. [CrossRef]
301. Liu, L.; Liu, J.; Wang, Y.; Yan, X.; Sun, D.D. Facile synthesis of monodispersed silver nanoparticles on graphene oxide sheets with enhanced antibacterial activity. *New J. Chem.* **2011**, *35*, 1418–1423. [CrossRef]
302. Das, M.R.; Sarma, R.K.; Borah, S.C.; Kumari, R.; Saikia, R.; Deshmukh, A.B.; Shelke, M.V.; Sengupta, P.; Szunerits, S.; Boukherroub, R. The synthesis of citrate-modified silver nanoparticles in an aqueous suspension of graphene oxide nanosheets and their antibacterial activity. *Colloids Surf. B Biointerfaces* **2013**, *105*, 128–136. [CrossRef]
303. Ganjouzadeh, F.; Khorrami, S.; Gharbi, S. Controlled cytotoxicity of Ag-GO nanocomposite biosynthesized using black peel pomegranate extract against MCF-7 cell line. *J. Drug Deliv. Sci. Technol.* **2022**, *71*, 103340. [CrossRef]
304. Dat, N.M.; Quan, T.H.; Nguyet, D.M.; Anh, T.N.M.; Thinh, D.B.; Diep, T.C.; Huy, L.A.; Tai, L.T.; Hai, N.D.; Khang, P.T.; et al. Hybrid graphene oxide-immobilized silver nanocomposite with optimal fabrication route and multifunctional application. *Appl. Surf. Sci.* **2021**, *551*, 149434. [CrossRef]
305. Gautam, S.; Sharma, S.; Sharma, B.; Jain, P. Antibacterial efficacy of poly (vinyl alcohol) nanocomposites reinforced with graphene oxide and silver nanoparticles for packaging applications. *Polym. Compos.* **2021**, *42*, 2829–2837. [CrossRef]
306. Tseng, K.-H.; Ku, H.-C.; Tien, D.-C.; Stobinski, L. Novel Preparation of Reduced Graphene Oxide–Silver Complex using an Electrical Spark Discharge Method. *Nanomaterials* **2019**, *9*, 979. [CrossRef] [PubMed]
307. Fan, B.; Li, Y.; Han, F.; Su, T.; Li, J.; Zhang, R. Synthesis of Ag/rGO composite materials with antibacterial activities using facile and rapid microwave-assisted green route. *J. Mater. Sci. Mater. Med.* **2018**, *29*, 69. [CrossRef] [PubMed]
308. Xu, X.; Mao, X.; Wang, Y.; Li, D.; Du, Z.; Wu, W.; Jiang, L.; Yang, J.; Li, J. Study on the interaction of graphene oxide-silver nanocomposites with bovine serum albumin and the formation of nanoparticle-protein corona. *Int. J. Biol. Macromol.* **2018**, *116*, 492–501. [CrossRef]
309. Zhao, R.; Lv, M.; Li, Y.; Sun, M.; Kong, W.; Wang, L.; Song, S.; Fan, C.; Jia, L.; Qiu, S.; et al. Stable Nanocomposite Based on PEGylated and Silver Nanoparticles Loaded Graphene Oxide for Long-Term Antibacterial Activity. *ACS Appl. Mater. Interfaces* **2017**, *9*, 15328–15341. [CrossRef]
310. Han, F.; Lv, S.; Li, Z.; Jin, L.; Fan, B.; Zhang, J.; Zhang, R.; Zhang, X.; Han, L.; Li, J. Triple-synergistic 2D material-based dual-delivery antibiotic platform. *NPG Asia Mater.* **2020**, *12*, 15. [CrossRef]
311. Zhao, R.; Kong, W.; Sun, M.; Yang, Y.; Liu, M.; Lv, M.; Song, S.; Wang, L.; Song, H.; Hao, R. Highly Stable Graphene-Based Nanocomposite (GO-PEI-Ag) with Broad-Spectrum, Long-Term Antimicrobial Activity and Antibiofilm Effects. *ACS Appl Mater Interfaces* **2018**, *10*, 17617–17629. [CrossRef]
312. Parandhaman, T.; Choudhary, P.; Ramalingam, B.; Schmidt, M.; Janardhanam, S.; Das, S.K. Antibacterial and Antibiofouling Activities of Antimicrobial Peptide-Functionalized Graphene–Silver Nanocomposites for the Inhibition and Disruption of *Staphylococcus aureus* Biofilms. *ACS Biomater. Sci. Eng.* **2021**, *7*, 5899–5917. [CrossRef]
313. Nguyen, N.T.T.; An, T.N.M.; Pham, S.Q.T.; Cao, X.T. Synthesis of silver nanoparticles stabilized polymer/graphene oxide for catalytic and antibacterial application. *Mol. Cryst. Liq. Cryst.* **2022**, *742*, 56–63. [CrossRef]

314. Kulshrestha, S.; Qayyum, S.; Khan, A.U. Antibiofilm efficacy of green synthesized graphene oxide-silver nanocomposite using Lagerstroemia speciosa floral extract: A comparative study on inhibition of gram-positive and gram-negative biofilms. *Microb. Pathog.* **2017**, *103*, 167–177. [CrossRef] [PubMed]
315. Wang, X.; Han, Q.; Yu, N.; Wang, T.; Wang, C.; Yang, R. GO-AgCl/Ag nanocomposites with enhanced visible light-driven catalytic properties for antibacterial and biofilm-disrupting applications. *Colloids Surf B Biointerfaces* **2018**, *162*, 296–305. [CrossRef] [PubMed]
316. Wu, X.; Li, H.; Xiao, N. Advancement of Near-infrared (NIR) laser interceded surface enactment of proline functionalized graphene oxide with silver nanoparticles for proficient antibacterial, antifungal and wound recuperating therapy in nursing care in hospitals. *J. Photochem. Photobiol. B Biol.* **2018**, *187*, 89–95. [CrossRef] [PubMed]
317. Liu, S.; Cao, S.; Guo, J.; Luo, L.; Zhou, Y.; Lin, C.; Shi, J.; Fan, C.; Lv, M.; Wang, L. Graphene oxide–silver nanocomposites modulate biofilm formation and extracellular polymeric substance (EPS) production. *Nanoscale* **2018**, *10*, 19603–19611. [CrossRef]
318. Tang, J.; Chen, Q.; Xu, L.; Zhang, S.; Feng, L.; Cheng, L.; Xu, H.; Liu, Z.; Peng, R. Graphene Oxide–Silver Nanocomposite As a Highly Effective Antibacterial Agent with Species-Specific Mechanisms. *ACS Appl. Mater. Interfaces* **2013**, *5*, 3867–3874. [CrossRef]
319. Ran, X.; Du, Y.; Wang, Z.; Wang, H.; Pu, F.; Ren, J.; Qu, X. Hyaluronic Acid-Templated Ag Nanoparticles/Graphene Oxide Composites for Synergistic Therapy of Bacteria Infection. *ACS Appl. Mater. Interfaces* **2017**, *9*, 19717–19724. [CrossRef]
320. Chae, H.-R.; Lee, J.; Lee, C.-H.; Kim, I.-C.; Park, P.-K. Graphene oxide-embedded thin-film composite reverse osmosis membrane with high flux, anti-biofouling, and chlorine resistance. *J. Membr. Sci.* **2015**, *483*, 128–135. [CrossRef]
321. Singh, P.; Shandilya, P.; Raizada, P.; Sudhaik, A.; Rahmani-Sani, A.; Hosseini-Bandegharaei, A. Review on various strategies for enhancing photocatalytic activity of graphene based nanocomposites for water purification. *Arab. J. Chem.* **2020**, *13*, 3498–3520. [CrossRef]
322. Chen, G.-E.; Wu, Q.; Sun, W.-G.; Xu, Z.-L.; Xu, S.-J.; Zhu, W.-W.; Zheng, X.-P. Synergy of graphene oxide–silver nanocomposite and amphiphilic co-polymer F127 on antibacterial properties and permeability of PVDF membrane. *RSC Adv.* **2016**, *6*, 100334–100343. [CrossRef]
323. Naeem, H.; Ajmal, M.; Qureshi, R.B.; Muntha, S.T.; Farooq, M.; Siddiq, M. Facile synthesis of graphene oxide–silver nanocomposite for decontamination of water from multiple pollutants by adsorption, catalysis and antibacterial activity. *J. Environ. Manag.* **2019**, *230*, 199–211. [CrossRef]
324. Yu, L.; Zhou, W.; Li, Y.; Zhou, Q.; Xu, H.; Gao, B.; Wang, Z. Antibacterial Thin-Film Nanocomposite Membranes Incorporated with Graphene Oxide Quantum Dot-Mediated Silver Nanoparticles for Reverse Osmosis Application. *ACS Sustain. Chem. Eng.* **2019**, *7*, 8724–8734. [CrossRef]
325. Díez-Pascual, A.M.; Luceño-Sánchez, J.A. Antibacterial Activity of Polymer Nanocomposites Incorporating Graphene and Its Derivatives: A State of Art. *Polymers* **2021**, *13*, 2105. [CrossRef] [PubMed]
326. Soroush, A.; Ma, W.; Silvino, Y.; Rahaman, S. Surface modification of thin film composite forward osmosis membrane by silver-decorated graphene-oxide nanosheets. *Environ. Sci. Nano* **2015**, *2*, 395–405. [CrossRef]
327. Soroush, A.; Ma, W.; Cyr, M.; Rahaman, S.; Asadishad, B.; Tufenkji, N. In Situ Silver Decoration on Graphene Oxide-Treated Thin Film Composite Forward Osmosis Membranes: Biocidal Properties and Regeneration Potential. *Environ. Sci. Technol. Lett.* **2016**, *3*, 13–18. [CrossRef]
328. Firouzjaei, M.D.; Shamsabadi, A.A.; Aktij, S.A.; Seyedpour, S.F.; Sharifian Gh., M.; Rahimpour, A.; Esfahani, M.R.; Ulbricht, M.; Soroush, M. Exploiting Synergetic Effects of Graphene Oxide and a Silver-Based Metal–Organic Framework To Enhance Antifouling and Anti-Biofouling Properties of Thin-Film Nanocomposite Membranes. *ACS Appl. Mater. Interfaces* **2018**, *10*, 42967–42978. [CrossRef]
329. He, C.; Shi, Z.-Q.; Cheng, C.; Lu, H.-Q.; Zhou, M.; Sun, S.-D.; Zhao, C.-S. Graphene oxide and sulfonated polyanion co-doped hydrogel films for dual-layered membranes with superior hemocompatibility and antibacterial activity. *Biomater. Sci.* **2016**, *4*, 1431–1440. [CrossRef]
330. Xu, M.; Zhu, J.; Wang, F.; Xiong, Y.; Wu, Y.; Wang, Q.; Weng, J.; Zhang, Z.; Chen, W.; Liu, S. Improved In Vitro and In Vivo Biocompatibility of Graphene Oxide through Surface Modification: Poly(Acrylic Acid)-Functionalization is Superior to PEGylation. *ACS Nano* **2016**, *10*, 3267–3281. [CrossRef]
331. Shams, E.; Yeganeh, H.; Naderi-Manesh, H.; Gharibi, R.; Hassan, Z.M. Polyurethane/siloxane membranes containing graphene oxide nanoplatelets as antimicrobial wound dressings: In vitro and in vivo evaluations. *J. Mater. Sci. Mater. Med.* **2017**, *28*, 75. [CrossRef]
332. Yadav, S.K.; Jung, Y.C.; Kim, J.H.; Ko, Y.-I.; Ryu, H.J.; Yadav, M.K.; Kim, Y.A.; Cho, J.W. Mechanically Robust, Electrically Conductive Biocomposite Films Using Antimicrobial Chitosan-Functionalized Graphenes. *Part. Part. Syst. Charact.* **2013**, *30*, 721–727. [CrossRef]
333. Díez-Pascual, A.M.; Díez-Vicente, A.L. Poly(propylene fumarate)/Polyethylene Glycol-Modified Graphene Oxide Nanocomposites for Tissue Engineering. *ACS Appl. Mater. Interfaces* **2016**, *8*, 17902–17914. [CrossRef]
334. Khan, Y.H.; Islam, A.; Sarwar, A.; Gull, N.; Khan, S.M.; Munawar, M.A.; Zia, S.; Sabir, A.; Shafiq, M.; Jamil, T. Novel green nano composites films fabricated by indigenously synthesized graphene oxide and chitosan. *Carbohydr. Polym.* **2016**, *146*, 131–138. [CrossRef]
335. Lu, X.; Feng, X.; Werber, J.R.; Chu, C.; Zucker, I.; Kim, J.-H.; Osuji, C.O.; Elimelech, M. Enhanced antibacterial activity through the controlled alignment of graphene oxide nanosheets. *Proc. Natl. Acad. Sci. USA* **2017**, *114*, E9793–E9801. [CrossRef] [PubMed]

336. Xie, C.; Lu, X.; Han, L.; Xu, J.; Wang, Z.; Jiang, L.; Wang, K.; Zhang, H.; Ren, F.; Tang, Y. Biomimetic Mineralized Hierarchical Graphene Oxide/Chitosan Scaffolds with Adsorbability for Immobilization of Nanoparticles for Biomedical Applications. *ACS Appl. Mater. Interfaces* **2016**, *8*, 1707–1717. [CrossRef] [PubMed]
337. Tegou, E.; Magana, M.; Katsogridaki, A.E.; Ioannidis, A.; Raptis, V.; Jordan, S.; Chatzipanagiotou, S.; Chatzandroulis, S.; Ornelas, C.; Tegos, G.P. Terms of endearment: Bacteria meet graphene nanosurfaces. *Biomaterials* **2016**, *89*, 38–55. [CrossRef] [PubMed]
338. Shahryari, Z.; Yeganeh, M.; Gheisari, K.; Ramezanzadeh, B. A brief review of the graphene oxide-based polymer nanocomposite coatings: Preparation, characterization, and properties. *J. Coatings Technol. Res.* **2021**, *18*, 945–969. [CrossRef]
339. Tripathi, S.N.; Rao, G.S.S.; Mathur, A.B.; Jasra, R. Polyolefin/graphene nanocomposites: A review. *RSC Adv.* **2017**, *7*, 23615–23632. [CrossRef]
340. Mejias Carpio, I.E.; Santos, C.M.; Wei, X.; Rodrigues, D.F. Toxicity of a polymer-graphene oxide composite against bacterial planktonic cells, biofilms, and mammalian cells. *Nanoscale* **2012**, *4*, 4746–4756. [CrossRef]
341. Li, P.; Gao, Y.; Sun, Z.; Chang, D.; Gao, G.; Dong, A. Synthesis, Characterization, and Bactericidal Evaluation of Chitosan/Guanidine Functionalized Graphene Oxide Composites. *Molecules* **2016**, *22*, 12. [CrossRef]
342. Rusakova, H.V.; Fomenko, L.S.; Lubenets, S.V.; Dolbin, A.V.; Vinnikov, N.A.; Basnukaeva, R.M.; Khlistyuck, M.V.; Blyznyuk, A.V. Synthesis and micromechanical properties of graphene oxide-based polymer nanocomposites. *Low Temp. Phys.* **2020**, *46*, 276–284. [CrossRef]
343. Maio, A.; Fucarino, R.; Khatibi, R.; Rosselli, S.; Bruno, M.; Scaffaro, R. A novel approach to prevent graphene oxide re-aggregation during the melt compounding with polymers. *Compos. Sci. Technol.* **2015**, *119*, 131–137. [CrossRef]
344. Aldosari, M.A.; Othman, A.A.; Alsharaeh, E.H. Synthesis and Characterization of the in Situ Bulk Polymerization of PMMA Containing Graphene Sheets Using Microwave Irradiation. *Molecules* **2013**, *18*, 3152–3167. [CrossRef] [PubMed]
345. Alsharaeh, E.H.; Othman, A.A.; Aldosari, M.A. Microwave Irradiation Effect on the Dispersion and Thermal Stability of RGO Nanosheets within a Polystyrene Matrix. *Materials* **2014**, *7*, 5212–5224. [CrossRef] [PubMed]
346. Hou, D.; Bostwick, J.E.; Shallenberger, J.R.; Zofchak, E.S.; Colby, R.H.; Liu, Q.; Hickey, R.J. Simultaneous Reduction and Polymerization of Graphene Oxide/Styrene Mixtures To Create Polymer Nanocomposites with Tunable Dielectric Constants. *ACS Appl. Nano Mater.* **2020**, *3*, 962–968. [CrossRef]
347. Lee, J.-H.; Jo, J.-K.; Kim, D.-A.; Patel, K.D.; Kim, H.-W.; Lee, H.-H. Nano-graphene oxide incorporated into PMMA resin to prevent microbial adhesion. *Dent. Mater.* **2018**, *34*, e63–e72. [CrossRef]
348. Konwar, A.; Kalita, S.; Kotoky, J.; Chowdhury, D. Chitosan–Iron Oxide Coated Graphene Oxide Nanocomposite Hydrogel: A Robust and Soft Antimicrobial Biofilm. *ACS Appl. Mater. Interfaces* **2016**, *8*, 20625–20634. [CrossRef] [PubMed]
349. Yan, X.; Fang, W.-W.; Xue, J.; Sun, T.-C.; Dong, L.; Zha, Z.; Qian, H.; Song, Y.-H.; Zhang, M.; Gong, X.; et al. Thermoresponsive *in Situ* Forming Hydrogel with Sol–Gel Irreversibility for Effective Methicillin-Resistant *Staphylococcus aureus* Infected Wound Healing. *ACS Nano* **2019**, *13*, 10074–10084. [CrossRef]
350. Konieczynska, M.D.; Villa-Camacho, J.C.; Ghobril, C.; Perez-Viloria, M.; Tevis, K.M.; Blessing, W.A.; Nazarian, A.; Rodriguez, E.K.; Grinstaff, M.W. On-Demand Dissolution of a Dendritic Hydrogel-based Dressing for Second-Degree Burn Wounds through Thiol-Thioester Exchange Reaction. *Angew. Chem. Int. Ed.* **2016**, *55*, 9984–9987. [CrossRef]
351. Yeh, M.-Y.; Zhao, J.-Y.; Hsieh, Y.-R.; Lin, J.-H.; Chen, F.-Y.; Chakravarthy, R.D.; Chung, P.-C.; Lin, H.-C.; Hung, S.-C. Reverse thermo-responsive hydrogels prepared from Pluronic F127 and gelatin composite materials. *RSC Adv.* **2017**, *7*, 21252–21257. [CrossRef]
352. Ng, L.Y.; Chua, H.S.; Ng, C.Y. Incorporation of graphene oxide-based nanocomposite in the polymeric membrane for water and wastewater treatment: A review on recent development. *J. Environ. Chem. Eng.* **2021**, *9*, 105994. [CrossRef]
353. Zeng, Z.; Yu, D.; He, Z.; Liu, J.; Xiao, F.-X.; Zhang, Y.; Wang, R.; Bhattacharyya, D.; Tan, T. Graphene Oxide Quantum Dots Covalently Functionalized PVDF Membrane with Significantly-Enhanced Bactericidal and Antibiofouling Performances. *Sci. Rep.* **2016**, *6*, 20142. [CrossRef]
354. Cheng, W.; Lu, X.; Kaneda, M.; Zhang, W.; Bernstein, R.; Ma, J.; Elimelech, M. Graphene Oxide-Functionalized Membranes: The Importance of Nanosheet Surface Exposure for Biofouling Resistance. *Environ. Sci. Technol.* **2020**, *54*, 517–526. [CrossRef] [PubMed]
355. Mahmoudi, E.; Ng, L.Y.; Ba-Abbad, M.M.; Mohammad, A. Novel nanohybrid polysulfone membrane embedded with silver nanoparticles on graphene oxide nanoplates. *Chem. Eng. J.* **2015**, *277*, 1–10. [CrossRef]
356. Grande, C.D.; Mangadlao, J.; Fan, J.; De Leon, A.; Delgado-Ospina, J.; Rojas, J.G.; Rodrigues, D.F.; Advincula, R. Chitosan Cross-Linked Graphene Oxide Nanocomposite Films with Antimicrobial Activity for Application in Food Industry. *Macromol. Symp.* **2017**, *374*, 1600114. [CrossRef]
357. Rossa, V.; Ferreira, L.E.M.; Vasconcelos, S.d.C.; Shimabukuro, E.T.T.; Madriaga, V.G.d.C.; Carvalho, A.P.; Pergher, S.B.C.; Silva, F.D.C.D.; Ferreira, V.F.; Junior, C.A.C.; et al. Nanocomposites based on the graphene family for food packaging: Historical perspective, preparation methods, and properties. *RSC Adv.* **2022**, *12*, 14084–14111. [CrossRef]
358. Ahmed, J. Use of Graphene/Graphene Oxide in Food Packaging Materials: Thermomechanical, Structural and Barrier Properties. *Reference Module in Food Science.* **2019**, 452–473. [CrossRef]
359. Zeng, L.; Zhu, Z.; Sun, D.-W. Novel graphene oxide/polymer composite membranes for the food industry: Structures, mechanisms and recent applications. *Crit. Rev. Food Sci. Nutr.* **2022**, *62*, 3705–3722. [CrossRef]

360. Lyn, F.H.; Hanani, Z.A.N. Graphene-based polymer nanocomposites in food packaging and factors affecting the behaviour of graphene-based materials: A review. *J. Nanoparticle Res.* **2022**, *24*, 179. [CrossRef]
361. Seedat, N.; Kalhapure, R.S.; Mocktar, C.; Vepuri, S.; Jadhav, M.; Soliman, M.; Govender, T. Co-encapsulation of multi-lipids and polymers enhances the performance of vancomycin in lipid–polymer hybrid nanoparticles: In vitro and in silico studies. *Mater. Sci. Eng. C* **2016**, *61*, 616–630. [CrossRef]
362. Mukherjee, A.; Waters, A.K.; Kalyan, P.; Achrol, A.S.; Kesari, S.; Yenugonda, V.M. Lipid-polymer hybrid nanoparticles as a next-generation drug delivery platform: State of the art, emerging technologies, and perspectives. *Int. J. Nanomed.* **2019**, *14*, 1937–1952. [CrossRef]
363. Jaglal, Y.; Osman, N.; Omolo, C.A.; Mocktar, C.; Devnarain, N.; Govender, T. Formulation of pH-responsive lipid-polymer hybrid nanoparticles for co-delivery and enhancement of the antibacterial activity of vancomycin and 18β-glycyrrhetinic acid. *J. Drug Deliv. Sci. Technol.* **2021**, *64*, 102607. [CrossRef]
364. Lee, H.W.; Kharel, S.; Loo, S.C.J. Lipid-Coated Hybrid Nanoparticles for Enhanced Bacterial Biofilm Penetration and Antibiofilm Efficacy. *ACS Omega* **2022**. *Research Square*. [CrossRef]
365. Mohanty, A.; Uthaman, S.; Park, I.-K. Chapter 12—Lipid–polymer hybrid nanoparticles as a smart drug delivery platform. In *Stimuli-Responsive Nanocarriers*; Gajbhiye, V., Gajbhiye, K.R., Hong, S., Eds.; Academic Press: Cambridge, MA, USA, 2022; pp. 319–349.
366. Xu, L.; Wang, X.; Liu, Y.; Yang, G.; Falconer, R.J.; Zhao, C.X. Lipid Nanoparticles for Drug Delivery. *Adv. NanoBiomed Res.* **2022**, *2*, 2100109. [CrossRef]
367. Grabnar, P.A.; Kristl, J. The manufacturing techniques of drug-loaded polymeric nanoparticles from preformed polymers. *J. Microencapsul.* **2011**, *28*, 323–335. [CrossRef] [PubMed]
368. Bose, R.J.; Ravikumar, R.; Karuppagounder, V.; Bennet, D.; Rangasamy, S.; Thandavarayan, R.A. Lipid–polymer hybrid nanoparticle-mediated therapeutics delivery: Advances and challenges. *Drug Discov. Today* **2017**, *22*, 1258–1265. [CrossRef] [PubMed]
369. Zhang, L.I.; Zhang, L. Lipid–polymer hybrid nanoparticles: Synthesis, characterization and applications. *Nano Life* **2010**, *1*, 163–173. [CrossRef]
370. Garg, N.K.; Tandel, N.; Jadon, R.S.; Tyagi, R.K.; Katare, O.P. Lipid–polymer hybrid nanocarrier-mediated cancer therapeutics: Current status and future directions. *Drug Discov. Today* **2018**, *23*, 1610–1621. [CrossRef]
371. Bachhav, S.S.; Dighe, V.D.; Kotak, D.; Devarajan, P.V. Rifampicin Lipid-Polymer hybrid nanoparticles (LIPOMER) for enhanced Peyer's patch uptake. *Int. J. Pharm.* **2017**, *532*, 612–622. [CrossRef]
372. Dave, V.; Yadav, R.B.; Kushwaha, K.; Yadav, S.; Sharma, S.; Agrawal, U. Lipid-polymer hybrid nanoparticles: Development & statistical optimization of norfloxacin for topical drug delivery system. *Bioact. Mater.* **2017**, *2*, 269–280. [CrossRef]
373. Thakur, K.; Sharma, G.; Singh, B.; Chhibber, S.; Patil, A.; Katare, O.P. Chitosan-tailored lipidic nanoconstructs of Fusidic acid as promising vehicle for wound infections: An explorative study. *Int. J. Biol. Macromol.* **2018**, *115*, 1012–1025. [CrossRef]
374. Thanki, K.; Papai, S.; Lokras, A.; Rose, F.; Falkenberg, E.; Franzyk, H.; Foged, C. Application of a Quality-By-Design Approach to Optimise Lipid-Polymer Hybrid Nanoparticles Loaded with a Splice-Correction Antisense Oligonucleotide: Maximising Loading and Intracellular Delivery. *Pharm. Res.* **2019**, *36*, 37. [CrossRef]
375. Ma, Z.; Liu, J.; Li, X.; Xu, Y.; Liu, D.; He, H.; Wang, Y.; Tang, X. Hydroxycamptothecin (HCPT)-loaded PEGlated lipid–polymer hybrid nanoparticles for effective delivery of HCPT: QbD-based development and evaluation. *Drug Deliv. Transl. Res.* **2022**, *12*, 306–324. [CrossRef]
376. Cunha, S.; Costa, C.P.; Moreira, J.N.; Sousa Lobo, J.M.; Silva, A.C. Using the quality by design (QbD) approach to optimize formulations of lipid nanoparticles and nanoemulsions: A review. *Nanomed. Nanotechnol. Biol. Med.* **2020**, *28*, 102206. [CrossRef] [PubMed]
377. Soni, G.; Kale, K.; Shetty, S.; Gupta, M.; Yadav, K.S. Quality by design (QbD) approach in processing polymeric nanoparticles loading anticancer drugs by high pressure homogenizer. *Heliyon* **2020**, *6*, e03846. [CrossRef] [PubMed]
378. Beg, S.; Rahman, M.; Kohli, K. Quality-by-design approach as a systematic tool for the development of nanopharmaceutical products. *Drug Discov. Today* **2019**, *24*, 717–725. [CrossRef] [PubMed]
379. Cai, J.; Huang, H.; Song, W.; Hu, H.; Chen, J.; Zhang, L.; Li, P.; Wu, R.; Wu, C. Preparation and evaluation of lipid polymer nanoparticles for eradicating H. pylori biofilm and impairing antibacterial resistance in vitro. *Int. J. Pharm.* **2015**, *495*, 728–737. [CrossRef] [PubMed]
380. Chieruzzi, M.; Pagano, S.; Lombardo, G.; Marinucci, L.; Kenny, J.M.; Torre, L.; Cianetti, S. Effect of nanohydroxyapatite, antibiotic, and mucosal defensive agent on the mechanical and thermal properties of glass ionomer cements for special needs patients. *J. Mater. Res.* **2018**, *33*, 638–649. [CrossRef]

nanomaterials

Article

Mesoporous Silica Nanoparticles Enhance the Anticancer Efficacy of Platinum(IV)-Phenolate Conjugates in Breast Cancer Cell Lines

Ivana Predarska [1,2], Mohamad Saoud [3], Dijana Drača [4], Ibrahim Morgan [3], Teodora Komazec [4], Thomas Eichhorn [2], Ekatarina Mihajlović [4], Duško Dunđerović [5], Sanja Mijatović [4], Danijela Maksimović-Ivanić [4], Evamarie Hey-Hawkins [1,*] and Goran N. Kaluđerović [2,3,*]

1 Faculty of Chemistry and Mineralogy, Institute of Inorganic Chemistry, Universität Leipzig, Johannisallee 29, 04103 Leipzig, Germany
2 Department of Engineering and Natural Sciences, University of Applied Sciences Merseburg, Eberhard-Leibnitz-Str. 2, 06217 Merseburg, Germany
3 Department of Bioorganic Chemistry, Leibniz Institute of Plant Biochemistry, Weinberg 3, 06120 Halle (Saale), Germany
4 Institute for Biological Research "Siniša Stanković", National Institute of Republic of Serbia, University of Belgrade, Bulevar despota Stefana 142, 11060 Belgrade, Serbia
5 Institute of Pathology, School of Medicine, University of Belgrade, dr Subotića 1, 11000 Belgrade, Serbia
* Correspondence: hey@uni-leipzig.de (E.H.-H.); goran.kaluderovic@hs-merseburg.de (G.N.K.); Tel.: +49-341-97-36151 (E.H.-H.); +49-3461-46-2012 (G.N.K.)

Citation: Predarska, I.; Saoud, M.; Drača, D.; Morgan, I.; Komazec, T.; Eichhorn, T.; Mihajlović, E.; Dunđerović, D.; Mijatović, S.; Maksimović-Ivanić, D.; et al. Mesoporous Silica Nanoparticles Enhance the Anticancer Efficacy of Platinum(IV)-Phenolate Conjugates in Breast Cancer Cell Lines. *Nanomaterials* **2022**, *12*, 3767. https://doi.org/10.3390/nano12213767

Academic Editor: Cesare Malagù

Received: 29 September 2022
Accepted: 17 October 2022
Published: 26 October 2022

Abstract: The main reasons for the limited clinical efficacy of the platinum(II)-based agent cisplatin include drug resistance and significant side effects. Due to their better stability, as well as the possibility to introduce biologically active ligands in their axial positions constructing multifunctional prodrugs, creating platinum(IV) complexes is a tempting strategy for addressing these limitations. Another strategy for developing chemotherapeutics with lower toxicity relies on the ability of nanoparticles to accumulate in greater quantities in tumor tissues through passive targeting. To combine the two approaches, three platinum(IV) conjugates based on a cisplatin scaffold containing in the axial positions derivatives of caffeic and ferulic acid were prepared and loaded into SBA-15 to produce the corresponding mesoporous silica nanoparticles (MSNs). The free platinum(IV) conjugates demonstrated higher or comparable activity with respect to cisplatin against different human breast cancer cell lines, while upon immobilization, superior antiproliferative activity with markedly increased cytotoxicity (more than 1000-fold lower IC_{50} values) compared to cisplatin was observed. Mechanistic investigations with the most potent conjugate, cisplatin-diacetyl caffeate (**1**), and the corresponding MSNs (SBA-15|**1**) in a 4T1 mouse breast cancer cell line showed that these compounds induce apoptotic cell death causing strong caspase activation. In vivo, in BALB/c mice, **1** and SBA-15|**1** inhibited the tumor growth while decreasing the necrotic area and lowering the mitotic rate.

Keywords: platinum(IV) conjugates; cisplatin; phenolic acid; nanoparticles; drug delivery; breast cancer

1. Introduction

Caffeic acid (CA) and ferulic acid (FA) belong to the most common phenolic acids, naturally occurring in fruits, grains, vegetables, spices, tea and coffee [1]. Apart from the key functions that these secondary plant metabolites exhibit on plant growth, development and defence, they are also found to have highly beneficial effects on humans. They are known for being potent natural antioxidants that are important for a variety of biological and pharmacological processes, such as cancer prevention, inflammation reduction, bacterial and viral infections prevention, blood clotting prevention, liver protection and many others [2–8]. The antioxidant properties of these phenolic acids are accredited to their chemical structure which includes one (ferulic acid) or two (caffeic acid) free phenolic hydroxyl groups and a double bond in the aliphatic chain enabling the formation of a phenoxy radical stabilized through resonance. In addition to the free radical scavenging via hydrogen atom donation [9], these phenolic acids are also found to modulate enzymatic activity, affect signal transduction as well as activate transcription factors and gene expression. As an example, alkyl esters of caffeic and ferulic acid are found to demonstrate significant inhibition on the activity of cyclooxygenase enzymes (COX-1 and COX-2) [10–13] responsible for the formation of key proinflammatory mediators. Additionally, COX-2 is overexpressed in a variety of tumors [14], making it a desirable target for the creation of anticancer medications. Therefore, in the past decade, these phenolic acids have attracted extensive attention as potential chemopreventive and chemotherapeutic agents. In that sense, multiple groups have pointed to the ability of caffeic acid and its derivatives to inhibit the growth of colon cancer [15–18], express anti-hepatocellular carcinoma activity [19], suppress the cancer cell proliferation in the human HT-1080 fibrosarcoma cell line [20] as well as human cervical cancer cells (HeLa and Me-180) [21] and reduce not only the growth of breast cancer, but also the ability of the breast cancer MCF-7 cells to migrate [22,23] which is a crucial factor for cancer metastasis. Ferulic acid as well has been found to have tumor-suppression potential in colon and breast cancers [4,24]. Additionally, both caffeic and ferulic acid have been found to increase the therapeutic efficacy of known anticancer drugs or lower the drug-related side effects. In this regard, the combined therapy involving caffeic acid and paclitaxel has been shown to have synergistic anticancer activity on H1299 lung cancer cells leading to enhanced apoptosis in vitro and in vivo [25]. Furthermore, caffeic acid elevates the antitumor effect of metformin in HTB-34 human metastatic cervical carcinoma cells [26] and increases the susceptibility of gastric cancer cells to cisplatin and doxorubicin [27]. Synergistic effects are also found applying cisplatin and caffeic acid in combination resulting in apoptotic mode of cell death in human cervical cancer cells [28]. A very interesting finding has been made by Sirota et al. who have discovered that in ovarian carcinoma cell lines the timing of caffeic acid treatment determines whether they become cisplatin-sensitive or -resistant [29]. Their findings demonstrate that preincubation with caffeic acid before treatment with cisplatin resulted in acquired resistance to cisplatin and decreased DNA binding, whereas simultaneous treatment with cisplatin and caffeic acid can increase cytotoxicity of cisplatin and platinum binding to nuclear DNA [29].

Based on all these findings and bearing in mind the different pharmacokinetic profiles of phenolic acids and cisplatin, including rising uncertainties whether these agents can reach the tumors within the required time frame in vivo, we have designed conjugates that combine the two agents into a single platinum(IV) prodrug. In this way, together with the molecules of the platinum component, the molecules of the respective phenolic acid can also be delivered into the tumor cells. Upon intracellular activation of these prodrugs via 2-electron reduction and protonation, the two moieties can be released and together carry out their biological functions in a synergistic manner [30]. Another reason which makes this approach attractive is that the coordinatively saturated platinum(IV) complexes are more stable than platinum(II) complexes, such as cisplatin, thus preventing its premature activation occurring by hydrolysis of the labile platinum chlorido ligand bonds, which is associated with many of its drawbacks such as the severe side effects, fast detoxification and innate or acquired resistance [31,32].

In an effort to create chemotherapeutics with lower toxicity, we have also used an SBA-15 mesoporous silica as nanocarrier for the platinum(IV) conjugates, aiming to take advantage of the accumulation of nanoparticles in tumor tissues through the passive enhanced permeability and retention (EPR) effect [33]. An additional advantage of this strategy is the possibility to further protect the drug from premature interaction with biomolecules on the way to cancer cells as well as provide tailorable drug release. The mesoporous silica SBA-15 material was chosen as a drug delivery vehicle because it is nontoxic to cells [34] and it has advantageous properties, including high pore volume with tuneable nanoscale pores as well as large surface area [35–37]. The properties of the inorganic mesoporous silica materials that make them interesting for biomedical applications, especially as nanoplatforms for drug delivery, have been highlighted in several reviews in the recent years [38–44]. The use of SBA-15 as a cisplatin carrier in earlier studies has been associated with increased cytotoxicity in leukemic cells [45], a twenty-fold greater antiproliferative effect in HT-29 colon cancer cells [46] and activation of pathways that change the phenotype of cancer cells (i.e., melanoma B16 tumor cells differentiate into senescent cells) [47]. Formerly reported SBA-15 nanoparticles loaded with platinum(IV) conjugates have also demonstrated very high antiproliferative activity against different breast cancer cell lines, while maintaining the structural integrity and allowing for relatively slow release of the active compounds [48].

Thus, we report here the preparation of three platinum(IV) derivatives of cisplatin bearing phenolates as axial ligands and their immobilization in SBA-15 nanoparticles. In these conjugates, both axial positions are occupied with acetyl-protected caffeate (**1**) or ferulate (**2**), as well as ferulate (**3**) (Figure 1). It should be noted that a cisplatin-ferulate conjugate has previously been reported [49]. However, in our view, the spectra included in the supplementary information of the corresponding paper do not confirm or support the structure of the desired product [49]. The prepared platinum(IV) complexes and their respective nanomaterials demonstrated remarkable antiproliferative activity in four different human breast cancer cell lines with different characteristics. In addition, synthesized conjugates and appropriate nanomaterials showed notable anticancer potential against a mouse breast cancer cell line as well. The activity of conjugate **1** and its corresponding MSNs SBA-15|**1**, which proved to be the most potent compounds in vitro, was further assessed in vivo, in a BALB/c mouse model with breast tumor induced by inoculation of 4T1 cells. **1** and SBA-15|**1** demonstrated potential to inhibit the tumor growth while decreasing the necrotic area and lowering the mitotic rate.

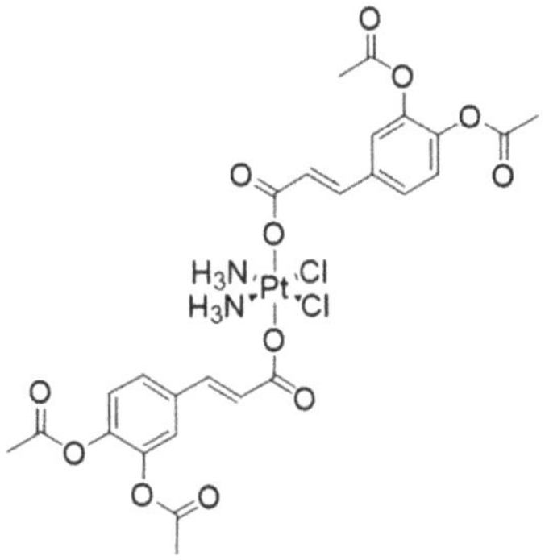

1

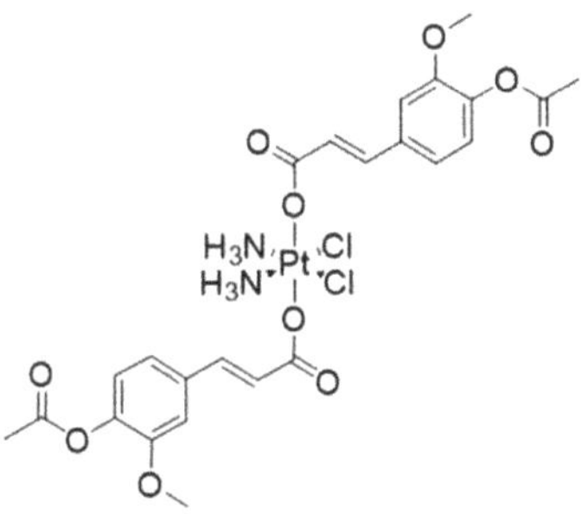

2

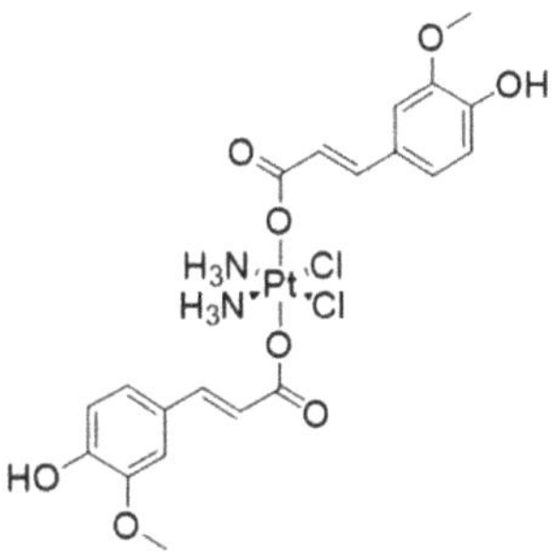

3

Figure 1. Platinum(IV)-phenolate conjugates: cisplatin-acetyl-protected caffeate (**1**), cisplatin-acetyl-protected ferulate (**2**) and cisplatin-ferulate (**3**).

2. Materials and Methods

Standard Schlenk methods and anhydrous, degassed solvents were used to conduct all reactions in a nitrogen atmosphere. Acetone was first refluxed with $KMnO_4$, then distilled from drierite (anhydrous $CaSO_4$), pyridine was distilled from CaH_2; both solvents were stored over activated 3 and 4 Å molecular sieves, respectively; MBraun Solvent Purification System (MB SPS-800) was used to obtain dry toluene, dichloromethane (DCM) and dimethylformamide (DMF), which were afterwards kept over activated 4 Å molecular sieves. Chemicals (cisplatin (Carbolution, St. Ingbert, Germany), caffeic acid (Fluorochem, Dublin, Ireland), ferulic acid, oxalyl chloride, tetraethyl orthosilicate (TEOS), pluronic 123 (P123) and acetic anhydride (Sigma-Aldrich Chemie GmbH, Steinheim, Germany)) were used as supplied. The synthesis of oxoplatin, *cis,trans,cis*-$[PtCl_2(OH)_2(NH_3)_2]$, was performed according to an already reported procedure [50]. In short, cisplatin was oxidized with 30% (*w*/*w*) aqueous solution of hydrogen peroxide. Protection of the free phenolic hydroxy groups by acylation was done with acetic anhydride employing a literature-known procedure [51]. The acyl chloride derivative of acetyl-protected caffeic acid was prepared with oxalyl chloride and catalytic amounts of DMF in DCM [51], while for the preparation of the acyl chloride derivative of acetyl-protected ferulic acid, toluene was used as solvent.

A 400 MHz NMR spectrometer (BRUKER Avance III HD, Karlsruhe, Germany) was used for recording of ^{1}H (400.13 MHz, tetramethylsilane (TMS) as an internal reference), ^{13}C (100.63 MHz, internal reference TMS) and ^{195}Pt (85.85 MHz, external reference $Na_2[PtCl_6]$), as well as two-dimensional $^{1}H-^{1}H$ COSY, $^{1}H-^{13}C$ HSQC, $^{1}H-^{13}C$ HMBC NMR spectra at 25 °C. The values of chemical shifts δ are given in parts per million (ppm). Electrospray ionization mass spectrometry was performed by using ESI-qTOF Impact II in positive mode (Bruker Daltonics GmbH, Bremen, Germany). Elemental analyses (C, H and N) were carried out with a microanalyser (VARIO EL, Heraeus Group, Hanau, Germany). VEGA3 (Tescan, Brno, Czech Republic) was employed for collection of scanning electron microscopy (SEM) images, as well as to conduct energy-dispersive X-ray spectroscopy (EDX) experiments (element detector EDAX Inc, Mahwah, NJ, USA). For nitrogen sorption measurements an Autosorb iQ/ASiQwin (Quantachrome Instruments, Anton Paar, QuantaTec Inc., Boynton Beach, FL, USA) was used. Small angle X-ray scattering (SAXS) measurements were performed on a D8 ADVANCE (Bruker, Karlsruhe, Germany) X-ray diffraction system. Inductively coupled plasma-optical emission spectrometry (ICP-OES) analysis was achieved on an Optima 7000 DV spectrometer (PerkinElmer, Waltham, MA, USA) using WinLab32 software.

2.1. *Synthesis of Platinum(IV) Conjugates* **1** *and* **2**

Pyridine (0.48 mL, 6.0 mmol, 10 eq.) was added to an oxoplatin (0.2 g, 0.6 mmol, 1 eq.) suspension in acetone (12 mL). After 10 min stirring, acyl chloride derivative, dissolved in acetone (12 mL), of either acetyl-protected caffeic acid or acetyl-protected ferulic acid (0.85 g or 0.76 g, respectively, each 3.0 mmol, 5 eq.) was added. The reaction mixture was refluxed at 75 °C (48 h). The precipitate formed was filtrated and thoroughly washed (water: 30 mL × 3; ethanol: 10 mL × 3; diethyl ether: 10 mL × 3) to eliminate the pyridinium salt. After drying in air, the pure products were collected.

1, pale yellow powder. Yield: 0.32 g (64%).

^{1}H NMR (DMSO-d_6, ppm): δ = 7.63 (d, $^{4}J_{HH}$ = 2 Hz, 2H, CH_{aryl}), 7.60 (dd, $^{3}J_{HH}$ = 8 Hz, $^{4}J_{HH}$ = 2 Hz, 2H, CH_{aryl}), 7.38 (d, $^{3}J_{HH}$ = 16 Hz, 2H, CH_{vinyl}), 7.30 (d, $^{3}J_{HH}$ = 8 Hz, 2H, CH_{aryl}), 6.80–6.50 (br, 6H, NH_3), 6.59 (d, $^{3}J_{HH}$ = 16 Hz, 2H, CH_{vinyl}), 2.29 (s, 12H, $OCOCH_3$).

^{13}C{^{1}H} NMR (DMSO-d_6, ppm): δ = 174.0 (qC, COO), 168.7 (qC, COO), 143.4 (qC, C_{aryl}), 142.9 (qC, C_{aryl}), 139.9 (CH, C_{vinyl}), 133.9 (qC, C_{aryl}), 126.7 (CH, C_{aryl}), 124.6 (CH, C_{aryl}), 123.3 (CH, C_{aryl}), 122.8 (CH, C_{vinyl}), 20.9 (CH_3, $OCOCH_3$).

^{195}Pt{^{1}H} NMR (DMSO-d_6, ppm): δ = 1213 (br s).

HR-ESI-MS (positive mode, CH_3OH): m/z $[2M + H]^+$: calc. for $C_{52}H_{57}Cl_4N_4O_{24}Pt_2$: 1653.139, found: 1653.135; m/z $[M + H]^+$: calc. for $C_{26}H_{29}Cl_2N_2O_{12}Pt$: 827.073, found: 827.070.

Elemental analysis: Found: C, 37.95; H, 3.2; N 3.5. Calc. for $C_{26}H_{28}Cl_2N_2O_{12}Pt$: C, 37.8; H, 3.4; N, 3.4%.

2, yellow powder. Yield: 0.27 g (59%).

^{1}H NMR (DMSO-d_6, ppm): δ = 7.45 (d, $^{4}J_{HH}$ = 2 Hz, 2H, CH_{aryl}), 7.39 (d, $^{3}J_{HH}$ = 16 Hz, 2H, CH_{vinyl}), 7.21 (dd, $^{3}J_{HH}$ = 8 Hz, $^{4}J_{HH}$ = 2 Hz, 2H, CH_{aryl}), 7.11 (d, $^{3}J_{HH}$ = 8 Hz, 2H, CH_{aryl}), 6.80–6.50 (br, 6H, NH_3), 6.65 (d, $^{3}J_{HH}$ = 16 Hz, 2H, CH_{vinyl}), 3.83 (s, 6H, OCH_3), 2.27 (s, 6H, $OCOCH_3$).

^{13}C{^{1}H} NMR (DMSO-d_6, ppm): δ = 174.2 (qC, COO), 168.9 (qC, COO), 151.5 (qC, C_{aryl})148.4 (qC, C_{aryl}), 140.9 (qC, C_{vinyl}), 133.9 (qC, C_{aryl}), 123.6 (CH, C_{aryl}), 121.9 (CH, C_{vinyl}), 121.3 (CH, C_{aryl}), 111.7 (CH, C_{aryl}), 59.3 (CH_3, OCH_3), 20.8 (CH_3, $OCOCH_3$).

^{195}Pt{^{1}H} NMR (DMSO-d_6, ppm): δ = 1212 (br s).

HR-ESI-MS (positive mode, CH_3OH): m/z $[2M + H]^+$: calc. for $C_{48}H_{57}Cl_4N_4O_{20}Pt_2$: 1541.159, found: 1541.141; m/z $[M + H]^+$: calc. for $C_{24}H_{29}Cl_2N_2O_{10}Pt$: 771.083, found: 771.086; m/z $[M + Na]^+$: calc. for $C_{24}H_{28}Cl_2N_2NaO_{10}Pt$: 793.065, found: 793.068.

Elemental analysis: Found: C, 37.6; H, 3.5; N, 3.9. Calc. for $C_{24}H_{28}Cl_2N_2O_{10}Pt$: C, 37.4; H, 3.6; N, 3.65%.

2.2. *Deprotection of* **2** *to Obtain Platinum(IV) Conjugate* **3**

In a mixture of DCM, methanol and acetone (20 mL of each solvent) conjugate **2** (0.3 g, 0.39 mmol) was dissolved. Conc. HCl was added dropwise (10 mL) to the colorless solution, and a color change to light yellow was noticeable after a few minutes. To track the progress of the reaction TLC (ethyl acetate:*n*-hexane = 3:1) was used. After 3 h stirring at r.t., the starting material was no longer present. The volume of the solvents was reduced to ca. 15 mL, subsequently water (50 mL) was added. The reaction mixture was filtrated and the isolated product was washed (water: 10 mL × 3; ethanol: 10 mL × 3; diethyl ether: 10 mL × 3) and dried in air.

3, dark yellow powder. Yield: 0.25 g (93%).

^{1}H NMR (DMSO-d_6, ppm): δ = 9.45 (s, 2H, O*H*), 7.30 (d, $^3J_{HH}$ = 16 Hz, 2H, CH_{vinyl}), 7.23 (d, $^4J_{HH}$ = 2 Hz, 2H, CH_{aryl}), 7.02 (dd, $^3J_{HH}$ = 8 Hz, $^4J_{HH}$ = 2 Hz, 2H, CH_{aryl}), 6.80–6.50 (br, 6H, NH_3), 6.78 (d, $^3J_{HH}$ = 8 Hz, 2H, CH_{aryl}), 6.42 (d, $^3J_{HH}$ = 16 Hz, 2H, CH_{vinyl}), 3.82 (s, 6H, OCH_3).

^{13}C{^{1}H} NMR (DMSO-d_6, ppm): δ = 174.9 (qC, COO), 149.1 (qC, C_{aryl}), 148.4 (qC, C_{aryl}), 142.1 (CH, C_{vinyl}), 126.5 (qC, C_{vinyl}), 122.7 (CH, C_{aryl}), 118.4 (CH, C_{vinyl}), 115.9 (CH, C_{aryl}), 111.1 (CH, C_{aryl}), 56.1 (CH_3, OCH_3).

^{195}Pt{^{1}H} NMR (DMSO-d_6, ppm): δ = 1216 (br s).

HR-ESI-MS (positive mode, CH_3OH): m/z $[2M + H]^+$: calc. for $C_{40}H_{49}Cl_4N_4O_{16}Pt_2$: 1373.116, found: 1373.098; m/z $[2M + Na]^+$: calc. for $C_{40}H_{48}Cl_4N_4NaO_{16}Pt_2$: 1395.098, found: 1395.068; m/z $[2M + K]^+$: calc. for $C_{40}H_{48}Cl_4KN_4O_{16}Pt_2$: 1411.072, found: 1411.035; m/z $[M + H]^+$: calc. for $C_{20}H_{25}Cl_2N_2O_8Pt$: 687.062, found: 687.060.

Elemental analysis: Found: C, 35.1; H, 3.5; N, 4.2. Calc. for $C_{20}H_{24}Cl_2N_2O_8Pt$: C, 35.0; H, 3.5; N, 4.1%.

2.3. *Solubility and Stability of Conjugates* **1–3**

For determination of their water solubility, compounds **1–3** (5 mg, each) were suspended in distilled water (5 mL) and stirred for 12 h. After that, the undissolved solid was separated by filtration and the amount of dissolved conjugate was evaluated after evaporation of the solvent from the filtrate.

To deduce the stability of **1–3** in DMSO, time-resolved (0, 3, 6, 12, 24, 48 and 72 h) ^{1}H NMR spectroscopy was applied.

2.4. *Preparation of SBA-15*

The SBA-15 silica mesoporous nanoparticles were prepared according to a reported procedure [52]. Briefly, to a solution of P123 (48.4 g) in a mixture of HCl (2 M, 1400 mL) and H_2O (360 mL) at 35 °C, TEOS (102 g) was added dropwise. The reaction mixture was vigorously stirred at 35 °C (20 h), then at 80 °C (24 h). The solid material was isolated by filtration and dried at 90 °C (14 h). Finally, the material was heated to 500 °C (1 °C min^{-1}) and kept for 24 h at this temperature for calcification.

Yield: 30.2 g; BET surface: 517 m^2 g^{-1}; wall thickness: 3.66 nm; pore diameter: 4.74 nm; pore volume: 0.72 cm^3 g^{-1}; lattice parameter: 8.4 nm; XRD (2θ in °, Miller indices): 1.0802 (100), 1.8032 (111), 2.0645 (200).

2.5. *Preparation of SBA-15 Material Loaded with Platinum(IV) Conjugates, SBA-15|***1***–SBA-15|***3**

The literature procedure was followed for loading the conjugates **1–3** in the MSN mesopores [53,54]. Namely, to the activated SBA-15 nanoparticles (vacuum, 150 °C, 16 h), a suspension of **1–3** in toluene was added and the reaction mixture was agitated at 80 °C

(48 h). Filtration was used to separate the solid material, which was then thoroughly washed using toluene and *n*-pentane. The MSNs loaded with drug (**1–3**) were obtained and dried at r.t.. Physical parameters of prepared MSNs are given in Table 1.

SBA-15 | **1**: 165 mg of **1**, 200 mg of SBA-15; yield: 340 mg.

SBA-15 | **2**: 231 mg of **2**, 300 mg of SBA-15; yield: 498 mg.

SBA-15 | **3**: 103 mg of **3**, 150 mg of SBA-15; yield: 244 mg.

Table 1. Selected physical parameters of the MSNs.

MSNs	SBET $[m^2\ g^{-1}]$	Wall Thickness [nm]	Pore Diameter [nm]	Pore Volume $[cm^3\ g^{-1}]$	Lattice Parameter *a* [nm]	XRD 2θ [°]	*hkl*
SBA-15 [48]	517	3.66	4.74	0.72	8.4	1.0802 1.8032 2.0645	100 111 200
SBA-15 | 1	245	3.34	4.75	0.34	8.1	1.1404 1.8543 2.1378	100 111 200
SBA-15 | 2	260	4.26	4.46	0.37	8.7	1.0311 1.7450 2.0075	100 111 200
SBA-15 | 3	292	5.03	3.68	0.37	8.7	1.0149 1.7603 2.0228	100 111 200

2.6. Drug Release Studies

ICP-OES was used to track the release of the conjugates **1–3** from the MSNs in phosphate-buffered saline (PBS) depending on the amount of platinum in the liquid phase. Experiments were performed by suspending the drug-loaded MSN material (2 mg) in PBS (1 mL) and sampling at defined times (0.08, 0.5, 1, 3, 6, 12, 24 and 72 h). Quantification of platinum in the PBS was done by ICP-OES upon separation of the solid material by centrifugation. External platinum standard solutions, obtained from an ICP platinum standard reference solution (Pt 1000 µg mL^{-1}, Specpure®, Alfa Aesar GmbH & Co KG, Germany), were used for the calibration of the ICP-OES device. The measurement was carried out on the ^{194}Pt isotope on a previously chosen emission line of platinum of 265.9 nm. Using different mathematical models (zero/first order, Korsmeyer-Peppas and Higuchi), the drug release kinetics were analyzed.

In addition, SBA-15 | **1** and SBA-15 | **2** (2 mg each) were each suspended in PBS (1 mL). The PBS was exchanged with fresh PBS (1 mL) every hour for 10 h consecutively. Again, the platinum present in the liquid phase was quantified using ICP-OES.

2.7. In Vitro Studies

ATCC (Manassas, VA, USA) provided all four human breast cancer cell lines. The mouse breast cancer cell line 4T1 (derived from the mammary gland tissue of a mouse BALB/c strain), was a kind gift from Prof. Nebojša Arsenijević from the Faculty of Medical Sciences, University of Kragujevac, Serbia. PBS, fetal calf serum (FCS) and basal cell culture media Roswell Park Memorial Institute (RPMI) were purchased from both Sigma-Aldrich (St. Louis, MO, USA) and Capricorn Scientific GmbH (Ebsdorfergrund, Germany). Trypsin/ethylenediaminetetraacetic acid (EDTA) and L-glutamine were bought from Capricorn Scientific GmbH (Ebsdorfergrund, Germany). Paraformaldehyde (PFA) was obtained from Serva (Heidelberg, Germany). DMSO, carboxyfluorescein diacetate succinimidyl ester (CFSE), propidium iodide (PI) and crystal violet (CV) were purchased from Sigma-Aldrich (St. Louis, MO, USA). The penicillin/streptomycin solution was obtained from Biological Industries (Cromwell, CT, USA). Acridine orange (AO) was obtained from Labo-Moderna

(Paris, France). PAN Biotech GmbH (Aidenbach, Germany) supplied us with Endopan 3 with supplementing kits. ApoStat was purchased from R&D Systems (Minneapolis, MN, USA) while Annexin V-FITC (AnnV) was from Biolegend (San Diego, CA, USA). 4-Amino-5-methylamino-2′,7′-difluorofluorescein diacetate (DAF-FM diacetate) and dihydrorhodamine 123 (DHR) were bought from Thermo Fisher Scientific (Waltham, MA, USA). The multi-well plates, culture flasks and other cell culture plastics were obtained from TPP (Trasadingen, Switzerland) and Greiner Bio-One GmbH (Frickenhausen, Germany).

2.8. Cell Culture

In this work, the following human breast cancer cell lines were used: triple-negative breast adenocarcinoma (MDA-MB-468 and HCC1937), ER- and HER2-positive breast cancer (BT-474 and MCF-7) [55] as well as one mouse-derived breast cancer cell line (4T1). RPMI-1640 medium containing 2 mM L-glutamine, heat-inactivated FCS (10%), and penicillin/streptomycin (1%) was used to sustain all cell lines. Prior to use or further subculturing, the cell lines were cultured in T-75 flasks at 37 °C in a humidified environment with 5% CO_2. Before cell passaging and seeding, the adherent cells were washed with PBS and detached by using trypsin/EDTA (0.05% in PBS) [54].

2.9. Cell Viability Studies

Using the above-mentioned cell growth media, cells were seeded in 96-well plates in a density of 6000 cells per 100 µL per well for the in vitro investigations in human breast cancer cell lines. Following seeding, cells were given a 24-h period to adhere before being treated with platinum(IV) conjugates **1–3** and MSNs loaded with **1–3**, namely SBA-15 | **1**, SBA-15 | **2** and SBA-15 | **3**. Stock solutions of platinum(IV) conjugates **1–3**, phenolic acids (ligand precursors) and cisplatin were prepared in DMSO. Serial dilution was afterwards done in standard growth media to reach the following concentrations: 10, 5, 1, 0.5, 0.1, 0.01 and 0.001 µM for the conjugates, 100, 50, 25, 12.5, 6.25, 3.125 and 1.6 µM for the phenolic acids, and 300, 100, 30, 10, 3, 1, and 0.1 µM for cisplatin. In a repeated experiment, HCC1937, MCF-7 and BT-474 cells were treated with various concentrations of the conjugates **1–3**, namely 50, 25, 12.5, 6.25, 3.125, 1.6, 0.8 and 0.4 µM. Stock suspensions of SBA-15 and drug-loaded SBA-15 were prepared in PBS and serially diluted with standard growth media to achieve final concentrations of 50, 10, 5, 1, 0.5, 0.1, 0.01 and 0.001 µg mL^{-1} for drug-loaded SBA-15, and 100, 50, 25, 12.5, 6.25, 3.125 and 1.6 µg mL^{-1} for SBA-15 alone. Each 96-well plate contained a positive control using digitonin (100 µM). Three independent experiments were performed in quadruplicate.

For the in vitro studies in the mouse-derived breast cancer cell line, cells were seeded in a density of 2×10^3 cells per well (96-well plates) using the previously indicated cell growth media. Following seeding, cells were given a 24 h adhesion period before being treated with conjugate **1** and SBA-15 | **1**. Using a DMSO stock solution of **1**, prepared at a concentration of 10 mM and retained at −20 °C for 3 days, a serial dilution was made in standard growth media to achieve concentrations of 25, 12.5, 6.3, 3.12, 1.56, 0.78 and 0.39 µM. A stock suspension of SBA-15 | **1** was prepared in PBS and serially diluted in standard growth media until final concentrations of 50, 25, 12.5, 6.25, 3.12, 1.56 and 0.78 µg mL^{-1} were reached. A DMSO solution of cisplatin at a concentration of 10 mM was prepared freshly before usage, and serial dilutions were made in culture medium to reach concentrations of 100, 50, 25, 12.5, 6.25, 3.12 and 1.56 µM. To determine the outcome of induced autophagy, cells were treated with an IC_{50} dose of conjugate **1** and SBA-15 | **1** and concomitantly with 3-methyladenine (3-MA, 1 mM).

Cell viability was evaluated in each case after a 72-h incubation period. Colorimetric MTT- (3-(4,5-dimethylthiazol-2-yl)-2,5-diphenyltetrazolium bromide) and CV (crystal violet)-based cell viability assays were used to examine the compounds' potential cytotoxicity. The MTT test was carried out in the following way: cells were rinsed with PBS before being incubated at 37 °C in a humid environment with 5% CO_2 with an MTT standard solution (0.5 mg mL^{-1} MTT in culture medium). After one hour, the MTT solution was discarded and DMSO was used to dissolve the formazan that had formed. The absorbance of this formazan, along with the reference/background signal, could be measured at wavelengths of 540 nm and 670 nm. Measurements were taken utilizing SpectraMax M5 multi-well plate reader (MolecularDevices, San Jose, CA, USA). For the CV assay, cells were fixed with 4% PFA for 20 min at room temperature after being washed once with PBS. The PFA solution was then removed and the cells were left to dry for 10 min. Afterwards, cells were stained with 0.2% crystal violet solution for 20 min followed by removal of the staining solution, washing of the cells with water and letting them dry overnight at room temperature. Finally, upon addition of acetic acid (33% in aqua bidest.) to the stained cells, the absorbance was measured at reference wavelengths of 540 nm and 670 nm as described earlier [56]. A four-parametric logistic function was used to obtain the mean value of the cell viability, which is expressed as a percentage in comparison to untreated cells [57]. SigmaPlot 14.0 and Microsoft Excel 2013 were employed for data analyses and IC_{50} and MC_{50} calculation.

2.10. *Flow Cytometry*

Six-well plates were seeded with 4T1 cells (5×10^4 per well). After overnight adherence, cells were exposed to an IC_{50} dose of **1** and an MC_{50} dose of SBA-15 | **1** and analyzed by flow cytometry. There were several staining procedures used: (1) CFSE for observing the influence on cellular proliferation, (2) AnnV/PI for identifying apoptotic cells, (3) ApoStat for caspase activity detection, (4) AO for autophagy detection, (5) DHR for reactive oxygen/nitrogen species (ROS/RNS) measurement, (6) DAF-FM for intracellular nitric oxide (NO) measurement, and (7) PI for cell cycle analysis. Results were acquired by CyFlow® Space Partec using the PartecFloMax® software v1.0.0 (Partec GmbH, Münster, Germany), with an exception for the results of cell cycle analysis which were obtained using BD FACSAria III and BD FACSDiva software v8.3 (BD Biosciences, Franklin Lakes, NJ, USA). Three separate replicates of the experiment were run. Depending on the particular staining agent, channels FL1 (green emission), FL2 (orange emission), and/or FL3 (dark red emission) were used for fluorescence detection.

For AnnV/PI, ApoStat, and AO staining, after 72 h treatment with the investigated agents, the cells were trypsinized and rinsed with PBS. Afterwards, cells staining according to the manufacturer's protocols was performed, with AnnV/PI in AnnV-binding buffer (15 min, room temperature), ApoStat in PBS 5% FBS (30 min, 37 °C) and AO 10 μM in PBS (15 min, 37 °C). Finally, upon washing, cells were resuspended in PBS (or in AnnV-binding buffer for AnnV/PI), and analyzed. For CFSE staining, pre-staining with PBS solution of CFSE (1 μM; 10 min, 37 °C) followed by washing and seeding of the cells was done, after which they were exposed to experimental agents over a period of 72 h. Before analysis, cells were once again washed, trypsinized and suspended in PBS. Similarly, pre-staining with 1 μM DHR for 20 min at 37 °C was first done also for DHR staining. For DAF-FM staining, the cells were first subjected to the experimental agents for 72 h, washed with PBS, and then stained with 5 μM DAF-FM diacetate in phenol red-free RPMI-1640 for 1 h at 37 °C. Thereafter, the dye was discarded and the cells were washed with PBS. Additional incubation of the cells in fresh RPMI-1640 not containing phenol red and serum, was done for 15 min to finish the reaction of de-esterification. Afterwards, cells were trypsinized, resuspended in PBS and analyzed. For analysis of cell distribution across cell cycle phases, after the treatments, 4T1 cells were fixed with 70% ethanol overnight at 4 °C, and then stained with PI (20 μg mL^{-1}) in the presence of RNase (0.1 mg mL^{-1}) during 45 min at 37 °C.

2.11. *In Vivo Studies*

Inbred female BALB/c mice (6–8 weeks old) from the Institute for Biological Research "Siniša Stanković"—National Institute of Republic of Serbia (IBISS) were used for this study. Animals were kept in standard laboratory conditions, free of non-specific pathogens, with unlimited access to food and water. The handling of animals and the study protocol adhered to national regulations established by the Law on Animal Welfare of the Republic of Serbia (Official Gazette of the Republic of Serbia No. 41/2009) and European Ethical Normative (Directive 2010/63/EU) on the protection of animals used for experimental and other scientific purposes. The national licensing committee at the Department of Animal Welfare, Veterinary Directorate, Ministry of Agriculture, Forestry and Water Management of Republic of Serbia granted approval for the experimental protocols (permission No. 323-07-07906/2022-05).

2.12. *Induction of Tumors and In Vivo Treatment*

BALB/c mice were orthotopically inoculated with 4T1 cells (2×10^4 cells in 50 µL PBS) in the fat pad region of the fourth mammary gland region. On the fifth day following cell implantation when tumors were palpable, random assignment of the animals into different treatment groups was made. The administration regime consisted of i.p. application of the appropriate agent three times a week. Cisplatin was applied in a dose of 2 mg kg^{-1} in 2% DMF/PBS, conjugate **1** at 5 and 10 mg kg^{-1} in 2% DMSO/PBS, while SBA-15|**1** was applied in PBS at a dose of 17.5 and 35 mg kg^{-1} for 25 days. Control mice were receiving 2% DMSO/PBS and 2% DMF/PBS as vehicle. Tumor growth was monitored every second day and on the 29th day after cell inoculation, mice were sacrificed. Tumors were extracted and measured in three dimensions. The tumor volume was calculated based on the following equation:

$$V = 0.52 \times a \times b^2$$

where "a" is the longest and "b" the shortest diameter. On the day of sacrifice, urine was collected from the animals and different biochemical parameters were analyzed with Multistix 10 SG (Bayer, Leverkusen, Germany).

2.13. *Histopathological Examination*

Harvested tissues of sacrificed animals (tumor, liver and kidney) were macroscopically examined, measured and sectioned to 3 mm thick slices through the largest tissue plane. Tissues were processed in automatic tissue processor (Milestone SRL LOGOS ONE, Sorisole, BG–Italy). Embedding in paraffin blocks was done on embedding console (SAKURA Tissue-Tek TEC 5, Sakura Finetek, CA, USA). Sections 4 µm thick were cut from the tissue using microtome (LEICA RM 2245, (Leica Biosystems, Nussloch, Germany), and the slices were mounted on glass slides. The slides were then stained with hematoxilin eosin stain (H/E) in automated slide stainer MYREVA SS-30H (Especialidades Médicas MYR, S.L., Tarragona, Spain). After H/E staining, all glass slides were examined and microscopically analyzed under an Olympus BX43 microscope (OLYMPUS EUROPA HOLDING GMBH, Hamburg, Germany). Additionally, all slides were digitalized with a Leica Aperio AT2 slide scanner (Leica Biosystems, Nussloch GmbH, Germany) for analysis and documentation purposes. Morphometric analysis was done with a Leica Aperio ImageScope (version 12.4.6, Leica Biosystems, Nussloch GmbH, Germany).

2.14. *Statistical Analysis*

Differences between treatments were assessed using analysis of variance (ANOVA) accompanied with a Student-Newman-Keuls test. Results of in vivo experiments were evaluated by a Mann-Whitney test. Differences in histopathological changes between treatments were analyzed by a Oneway ANOVA test followed by a Tukey test. Statistical significance was considered if p-value was 0.05 or less. Statistical analyses were done using the program Statistica 10.

3. Results and Discussion

3.1. *Synthesis and Characterization of Platinum(IV) Conjugates* **1–3**

In this study, three cisplatin derivatives with acetyl-protected caffeate, acetyl-protected ferulate or free ferulate in the axial positions (complexes **1**, **2** and **3**, respectively, Figure 1) were prepared. Complexes **1** and **2** were synthesized by reacting oxoplatin (afforded by cisplatin oxidation with hydrogen peroxide) [50] with an excess of the corresponding acyl chloride derivative of the acetyl-protected phenolic acid in the presence of pyridine as base. Acetyl-protected phenolic acids were used for the synthesis in order to avoid side reactions involving the hydroxyl groups present in their structure. Additionally, as reported by other authors, the reduction potential of the phenolic acids is directly linked to the existence as well as the number of hydroxyl groups [58,59]. Therefore, using caffeic and ferulic acid with free hydroxyl groups could induce reduction of the platinum(IV) in oxoplatin making the formation of the desired products impossible [60,61]. After complexes **1** and **2** were successfully obtained in yields of 64% and 59%, respectively, deprotection of **2** was done with conc. HCl to afford the cisplatin-ferulate conjugate **3**. The three obtained platinum(IV) conjugates were characterized by ^{1}H, ^{13}C and ^{195}Pt NMR spectroscopy and mass spectrometry (ESI, Figures S1–S18) and their purity was further confirmed through elemental analysis. Deprotection of **1** was also attempted, employing different procedures and reagents, from conc. HCl to ones which have been reported to be mild and highly selective for deprotection of aromatic acetates such as guanidine·HCl [62], dibutyltin oxide [63] and ammonium acetate [64]. However, the desired cisplatin-caffeate conjugate could not be obtained; instead, degradation of the complex resulting in very complex ^{1}H NMR spectra was observed in each deprotecting approach. Finally, we also attempted the deprotection using a commercially available enzyme Amano lipase A from *Aspergillus niger* (Sigma-Aldrich Chemie GmbH, Steinheim, Germany) which has been reported to cause full *O*-deacetylation of different acetylated biomolecules under mild conditions [65]. Closely following the reaction by time-resolved ^{1}H NMR spectroscopy (ESI, Figure S19), we noticed that as soon as the deacetylation of **1** begins, chemical shifts corresponding to free caffeate appear. This means that after deprotection, the free hydroxyl groups might reduce the platinum(IV) to platinum(II), leading to separation of the caffeate ligand from the cisplatin scaffold. This confirms the capacity of phenolic acids to induce metal reduction and the higher reduction potential of caffeic acid, which bears two hydroxyl groups in *ortho* position to each other, by comparison with ferulic acid, which has only one free hydroxyl group. As the deprotection of the acetyl protecting groups can occur under acidic conditions, it could be anticipated that the acidic tumor microenvironment would facilitate such a reaction in vivo. The free hydroxyl groups would then ease the reduction of the platinum(IV) to the cytotoxic platinum(II) and release the axial ligands, making both species available to act upon their respective targets.

The platinum(IV) conjugates are particularly poorly soluble in water (**1–3**: 0.09 g, 0.04 g and 0.56 g, respectively, in 1 L H_2O) owing to the presence of lipophilic moieties. However, in polar aprotic solvents like DMF and DMSO, they are highly soluble.

As confirmed by time-resolved ^{1}H NMR spectroscopy, the conjugates **1–3** are stable in DMSO solution over 72 h. During this time, no ligand exchange or complex degradation was observed (Figures S20–S22).

3.2. Synthesis and Characterization of the Mesoporous Silica Materials

For the preparation of SBA-15 nanoparticles, a synthesis described in the literature, with TEOS as a silica source and P123 serving as a structure-directing agent, was employed [52]. Calcination was performed to remove the organic template, resulting in rod-shaped particles with consistent morphology as demonstrated by SEM imaging (Figure 2). Narrow size distribution of the particles ranging from 200–400 × 600–800 nm was observed.

Nitrogen physisorption experiments that resulted in a type IV isotherm with a hysteresis loop showing capillary condensation characteristic for highly organized mesoporous structures (Figure 3A) [66] were used to confirm the material's mesoporous nature. Carrying out these analyses also upon loading of the SBA-15 material with the platinum(IV) complexes **1–3** which afforded the three materials SBA-15|**1**, SBA-15|**2** and SBA-15|**3**, it was shown that the conjugates have not significantly influenced the SBA-15 structure, so the shape, morphology as well as the mesoporous nature of the materials are retained.

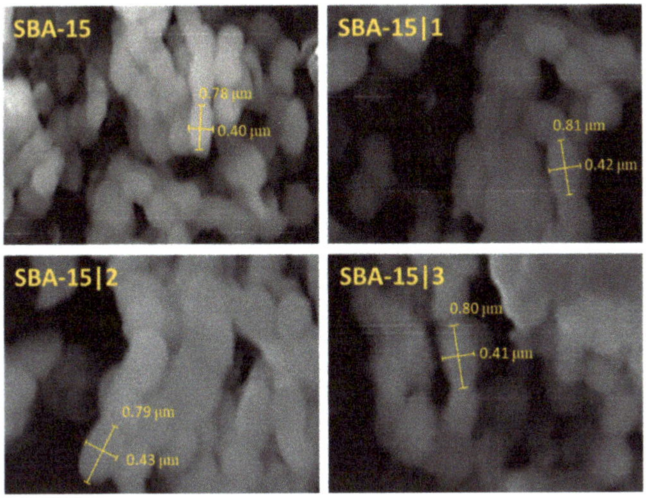

Figure 2. SEM images of MSNs: SBA-15, SBA-15 | **1**, SBA-15 | **2** and SBA-15 | **3**.

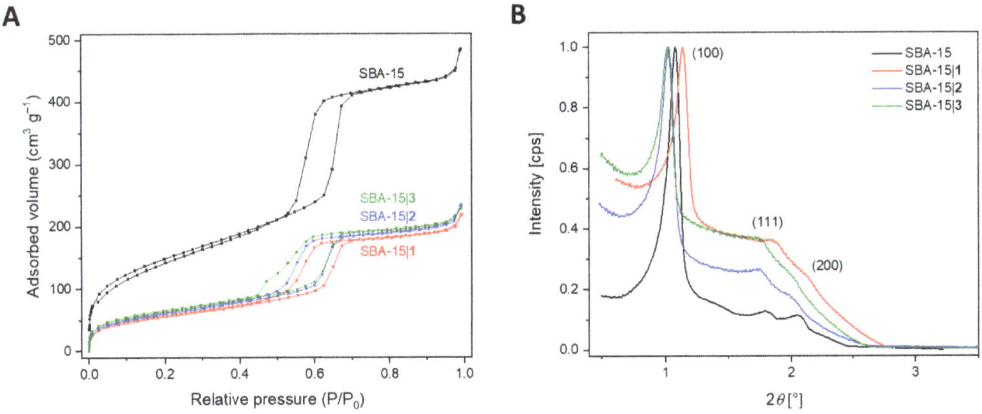

Figure 3. (**A**) N_2 adsorption-desorption isotherms and (**B**) SAXS patterns of MSNs: SBA-15, SBA-15 | **1**–SBA-15 | **3**.

As expected, after loading of the cisplatin derivatives, the high specific surface area, S_{BET} (calculated based on the Brunauer–Emmett–Teller (BET) model), and pore volume, V_p, which were corroborated for the SBA-15 material (517 $m^2\ g^{-1}$ and 0.72 $cm^3\ g^{-1}$, respectively), were reduced to 245−292 $m^2\ g^{-1}$ for the specific surface area and 0.34–0.37 $cm^3\ g^{-1}$ for the pore volume. Additionally, there was a change of the pore diameter D_p from 4.74 nm as determined for the pure SBA-15 to 3.68−4.75 nm for the loaded MSNs. On the contrary, the wall thickness W_p, calculated by deducting the pore diameter from the lattice parameter a, increased upon immobilization of conjugates **2** and **3**, (3.66 nm 4.26 and 5.03 nm (SBA-15, SBA-15 | **2** and SBA-15 | **3**, respectively). The lattice parameter itself was attained by SAXS, and specifies the repetition rate of the hexagonal pores of the MSNs. Thus, the acquired data showing similar values for a for all pure and loaded MSNs are consistent with what is expected. Finally, as shown in Figure 3B, the SAXS patterns of pure SBA-15 and SBA-15 | **1**–SBA-15 | **3** exhibit diffraction peaks (see Table 1) typical for hexagonally ordered MSNs, highlighting the particles' unaltered structure once more. All corresponding values for a and 2θ are presented in Table 1.

The amount of immobilized platinum(IV) complexes **1–3** in the MSNs was calculated based on the amount of platinum detected with EDX analysis. Successful loading of all three compounds **1–3** with load contents of 7.51, 9.23 and 10.33 wt% Pt, respectively, was confirmed. The corresponding encapsulation efficiency was calculated to be 70.4%, 83.8% and 89.2%, for SBA-15 | **1**–SBA-15 | **3**, respectively (see ESI, Figure S23 and Tables S1 and S2).

Despite the successful production of the mesoporous material with the applied method, which resulted in MSNs with optimal drug loading, it is worth mentioning that significant advances have been made in the engineering of MSNs [67], including some greener approaches with less undesirable environmental impact [68], which should be considered in the future.

3.3. Release of Platinum(IV) Conjugates from MSNs

The in vitro release profiles of the platinum(IV) complexes **1–3** encapsulated in silica nanoparticles were determined in PBS (pH = 7.4) by assessing the platinum content in solution (ICP-OES) at predetermined time points (0.08, 0.5, 1, 3, 6, 12, 24, 48 and 72 h). As presented in Figure 4A, SBA-15 | **1** and SBA-15 | **2** exhibited very low release of only 18 and 5%, respectively, over 72 h. On the contrary, SBA-15 | **3** released over 70% of its cargo over the same period of time. As these results are consistent with the solubility profiles of the three conjugates in water, it is probable that in this case drug release is driven by solubility of the drugs in the surrounding media. In the first hours, saturation of the solution is achieved and, therefore, prolonged contact with the fluid does not lead to increase the amount of platinum(IV) conjugate released. By exchanging the medium each hour in the time interval of up to 10 h, saturation of PBS with the drugs from nanomaterials SBA-15 | **1** and SBA-15 | **2** is avoided. As initially observed, the release is low, but continuous (Figure 4B). The obtained results indicate that the drug-loaded nanoparticles' structural integrity is maintained under the simulated physiological setting, preventing immediate leakage of the platinum(IV) complexes. In vivo, this could, in addition to the favorable EPR effect of MSNs, be very beneficial as it could imply minimal drug release during circulation, thus lowering the possibility of off-target interactions and side effects, and higher accumulation in the tumor, where the slow release could ensure prolonged drug exposure.

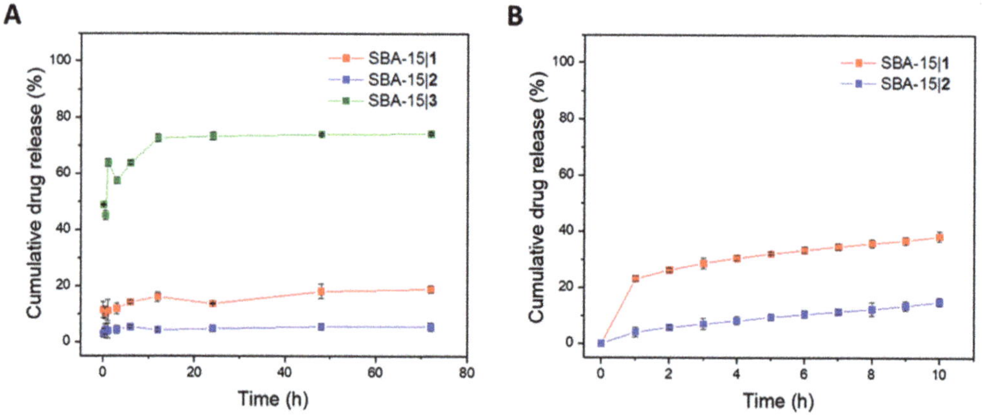

Figure 4. (**A**) Cumulative drug release (%) of the respective platinum(IV) conjugate from SBA-15 | **1**, SBA-15 | **2** and SBA-15 | **3** over 72 h; (**B**) Cumulative drug release (%) of the respective platinum(IV) conjugate from SBA-15 | **1** and SBA-15 | **2** with exchange of the PBS solution every hour for 10 h.

Mathematical models *viz.* zero/first order, Korsmeyer–Peppas and Higuchi (see ESI, Figures S24–S26) were employed for a study of the kinetics of the drug release [69]. Data was quantitatively correlated with these models and the most suitable model was determined based on the degree of correlation R^2 (see ESI, Table S3). The results obtained suggest that although the release of the cisplatin derivatives from the corresponding MSNs does not follow perfectly the principles of any of the investigated release kinetics ($R^2 < 0.9$), diffusion-controlled drug release kinetics (Higuchi for SBA-15 | **1** and Korsmeyer–Peppas for SBA-15 | **2** and SBA-15 | **3**) are the most fitting [70]. The value *n* obtained from the Korsmeyer–Peppas model (in all cases $n < 0.45$) further ascertained Fickian diffusion as prime release mechanism [71].

3.4. Antiproliferative Activity

The antiproliferative activity of the free platinum(IV) complexes **1**, **2** and **3** and the corresponding MSNs (SBA-15, SBA-15 | **1**–SBA-15 | **3**) was first determined against four human breast tumor cell lines, involving two triple-negative cell lines, namely MDA-MB-468 (human breast adenocarcinoma with COX-2 expression) and HCC1937 (human breast carcinoma with BRCA1 mutation), and the two ER- and PR-positive cell lines MCF-7 (with COX-1 expression) and BT-474 (both invasive ductal breast carcinoma). After 72 h of treatment with the examined compounds and MSNs, the cell viability was assessed using the two distinct assays, CV and MTT, and their activity was compared with that of clinically used cisplatin, as well as the phenolic acids used as ligands. IC_{50} and MC_{50} values are presented in Table 2. As expected, the ligands alone were found to be inactive (>100 μM) against all treated tumor cell lines. A similar observation was made for the SBA-15 material (>100 μg mL^{-1}). On the contrary, the platinum(IV) conjugates **1–3** demonstrated antitumor activity in the range similar to (against MDA-MB-468 and HCC1937 cell lines) or higher (against MCF-7 and BT-474 cell lines) than cisplatin. Certain discrepancies were observed between the results acquired by MTT and CV assays against the MCF-7 cell line upon treatment with the platinum(IV) conjugates and against BT-474, HCC1937 and MDA-MB-468 cell lines upon treatment with cisplatin, implying that in these cases the mechanism through which cytotoxicity is achieved might be involving cell metabolism pathways [72–75].

Table 2. IC_{50} [μM] and MC_{50} [μg mL^{-1}] values of cisplatin, **1**–**3**, free ligands, SBA-15, and SBA-15 | **1**–SBA-15 | **3** determined with CV and MTT assays (72 h).

Compound/Material	CV					MTT				
	MDA-MB-468	HCC1937	MCF-7	BT-474	4T1	MDA-MB-468	HCC1937	MCF-7	BT-474	4T1
	IC_{50} [μM]					IC_{50} [μM]				
Acetyl-protected caffeic acid			>100					>100		
Acetyl-protected ferulic acid			>100					>100		
Ferulic acid			>100					>100		
Cisplatin	3.3 ± 0.2	7.6 ± 0.9	33.6 ± 4.8	54.9 ± 6.0	3.2 ± 0.3	0.6 ± 0.1	4.3 ± 0.7	32.0 ± 4.3	70.3 ± 8.5	1.3 ± 0.1
Cisplatin-acetyl protected caffeate conjugate (**1**)	2.6 ± 0.7	8.9 ± 0.3	16.0 ± 2.3	5.4 ± 0.3	4.4 ± 0.7	2.7 ± 0.8	9.9 ± 0.5	> 50	4.5 ± 0.4	3.3 ± 0.4
Cisplatin-acetyl protected ferulate conjugate (**2**)	3.5 ± 0.5	10.4 ± 1.1	8.4 ± 2.2	5.9 ± 0.4		3.8 ± 0.8	5.5 ± 4.3	27.6 ± 2.6	9.7 ± 1.2	
Cisplatin-ferulate conjugate (**3**)	7.1 ± 0.3	11.1 ± 0.8	19.9 ± 1.3	>50		8.3 ± 0.7	15.1 ± 2.1	>50	>50	
SBA-15	**1** [(a)]	0.3 ± 0.03	1.7 ± 0.1	1.6 ± 0.1	0.0001 ± 0.00001	1.0 ± 0.2	0.1 ± 0.001	0.3 ± 0.02	0.3 ± 0.1	0.1 ± 0.001	0.8 ± 0.04
SBA-15	**2** [(a)]	0.1 ± 0.01	1.0 ± 0.1	1.0 ± 0.1	0.0017 ± 0.0001		0.02 ± 0.002	0.2 ± 0.02	/	0.1 ± 0.02	
SBA-15	**3** [(a)]	1.6 ± 0.1	2.6 ± 0.5	4.4 ± 0.5	0.7 ± 0.1		0.4 ± 0.01	1.0 ± 0.2	3.7 ± 0.2	1.3 ± 0.2	
	MC_{50} [μg mL^{-1}]					MC_{50} [μg mL^{-1}]				
SBA-15			>100					>100		
SBA-15	**1**	3.0 ± 0.5	19.9 ± 1.0	18.5 ± 1.0	0.0017 ± 0.0002	11.9 ± 1.9	0.6 ± 0.02	3.3 ± 0.3	7.3 ± 0.8	0.7 ± 0.02	9.7 ± 0.5
SBA-15	**2**	3.6 ± 0.4	32.9 ± 2.1	35.0 ± 3.0	0.6 ± 0.006		0.6 ± 0.1	6.5 ± 0.7	>50	4.3 ± 0.7	
SBA-15	**3**	3.9 ± 0.2	6.4 ± 1.2	10.7 ± 1.3	1.7 ± 0.3		1.1 ± 0.03	2.3 ± 0.4	5.0 ± 0.5	3.3 ± 0.6	

[(a)] recalculated based on platinum content (EDX) and drug release (ICP-OES).

Immobilization of the platinum(IV) complexes **1–3** in MSNs led to a significant improvement of the antiproliferative activity of all three conjugates against all cell lines involved in the study. For comparative evaluation with the activity of cisplatin and the free platinum(IV) complexes, recalculation of the IC_{50} values for the drug-loaded MSNs (SBA-15 | **1**, SBA-15 | **2** and SBA-15 | **3**) was made based on the obtained MC_{50} values considering the platinum content obtained by EDX, as well as the release of the drugs obtained by ICP-OES (detailed explanation in ESI). Submicromolar IC_{50} values which were at least two times and in some cases several thousand times lower than those for cisplatin were obtained. All drug-loaded MSNs showed the highest activity against the BT-474 cell line responsible for the development of invasive ductal breast cancer and known to be resistant to the conventional treatment with tamoxifen [76,77]. Among the three drug-loaded MSNs, SBA-15 | **1** was found to have the highest potency against this cell line with an IC_{50} value of only 0.1 nM.

Considering these very promising results, further investigations were necessary in an appropriate in vivo breast cancer model; thus, an orthotopic model of breast cancer by inoculation of 4T1 cells in BALB/c mice was selected. Therefore, the antiproliferative activity of the most active conjugate **1** and appropriate nanomaterial SBA-15 | **1** was first assessed in vitro in this triple-negative mouse cell line. After 72 h of incubation, the viability of the cells was determined by measurement of total cell culture respiration or number of adherent cells using MTT and CV assays. The results summarized in Table 2 reveal that conjugate **1** strongly affected the cell viability of 4T1 cells similarly as cisplatin alone. Higher activity exceeding the one of cisplatin and **1** was observed upon drug-loading into SBA-15 particles, which is consistent with the results seen in the human breast cancer cell lines. Cell viability under **1–3**, SBA-15 | **1**–SBA-15 | **3** and cisplatin exposure is presented in the ESI, Figures S27–S31.

3.5. Molecular Mechanism of Action

Aiming to get an insight in the plausible mechanism of drug-induced cytotoxicity, flow cytometric evaluation of apoptotic cell death, cell cycle perturbation, rate of cell proliferation, caspase activation, and induction of autophagic response was employed upon treatment of the 4T1 cells with IC_{50} doses of **1** and cisplatin and an MC_{50} dose of SBA-15 | **1**. The obtained results reveal that in all applied treatments, induction of massive apoptosis was in the background of diminished cell viability (Figure 5A). In concordance, significant accumulation of fragmented DNA in subG compartment of the cell cycle (Figure 5B), together with a strong caspase activation (Figure 5C) was found in cultures exposed to **1**, SBA-15 | **1** and cisplatin. In parallel, surviving cell subpopulation lost the dividing potential (CFSE assay, Figure 5D). Since the process of autophagy is deeply involved in the cells' response to toxic stimuli, varying from protective to destructive, it is very important to define the eventual presence of autophagosomes in cultures exposed to the experimental therapeutics [78,79].

An intensified autophagic process (AO assay, Figure 5E) in all treated cultures was noticed. However, additional decrease in cell viability was observed also upon neutralization of autophagy through concomitant exposure of the treated cells to the specific inhibitor 3-MA (Figure 5F). Thus, the obtained data suggest that autophagy has a protective role, defending the cell from destructive stimuli.

In addition, treatment with conjugate **1** and SBA-15 | **1** triggered increased production of nitric oxide (DAF-FM assay, Figure 6A), as well as ROS/RNS (DHR assay, Figure 6B).

Altogether, platinum(IV) conjugate **1**, free and loaded in SBA-15, clearly showed potential to downregulate cell growth through inhibition of proliferation and apoptotic cell induction which is even amplified with autophagy suppression. Behind the observed effect could be the enhanced production of reactive species which affect different cellular processes and subsequently the cell viability [80,81].

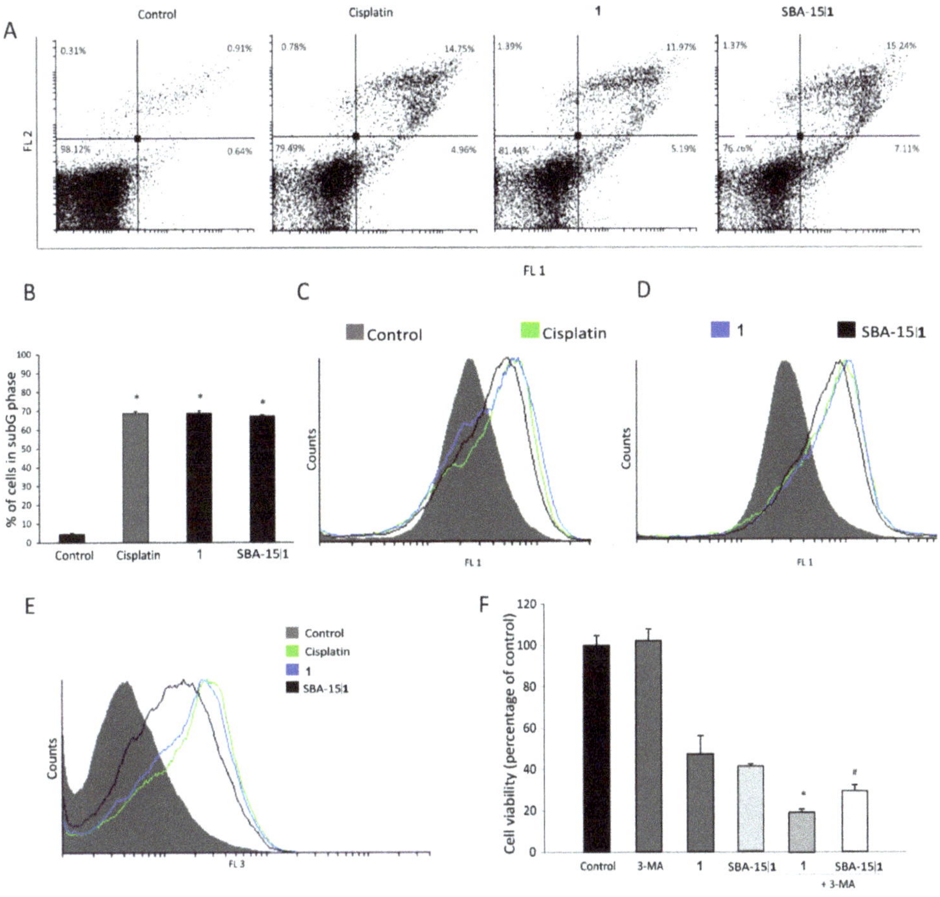

Figure 5. Effect of platinum(IV) conjugate **1** and corresponding MSNs SBA-15|**1** on 4T1 mouse breast cancer cells: (**A**) apoptosis (Annexin V-FITC/propidium iodide, (**B**) cell cycle—subG phase (* $p < 0.05$ refers to untreated cells), (**C**) caspase activation (Apostat), (**D**) cell proliferation potential (CFSE), (**E**) induction of autophagy (AO staining), and (**F**) effect of 3-MA addition on viability of 4T1 cells (MTT assay) (* $p < 0.05$ refers to culture treated with **1**, # $p < 0.05$ refers to culture treated with SBA-15|**1**).

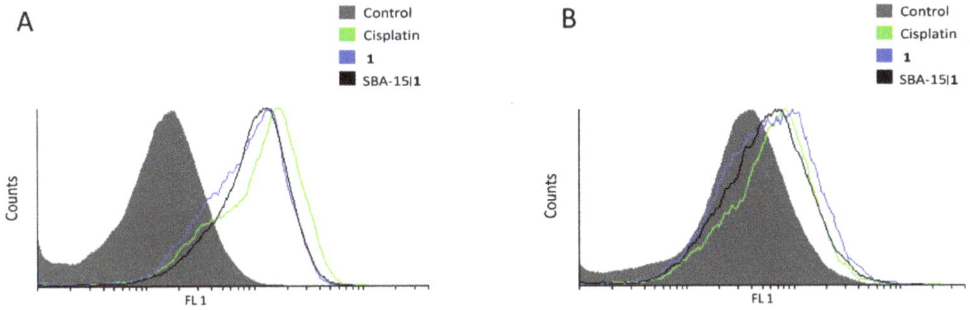

Figure 6. Effect of platinum(IV) conjugate **1** and corresponding MSNs SBA-15|**1** on 4T1 mouse breast cancer cells: (**A**) intracellular NO production (DAF-FM) and (**B**) ROS/RNS production (DHR).

3.6. *In Vivo Decrease of Breast Cancer Growth*

With the aim to check the effectiveness of conjugate **1** and SBA-15 | **1** in vivo, an orthotopic model of breast cancer was used. Tumors were induced by inoculation of 4T1 cells in fat pad of the fourth breast in syngeneic BALB/c mice, and treatment started when tumors became palpable. Cisplatin was applied in a therapeutic dose [82], while **1** or SBA-15 | **1** were applied in an equimolar dose with regards to cisplatin. Results revealed approximately 50% tumor volume reduction upon treatment with cisplatin and SBA-15 | **1**, while conjugate **1** alone decreased the tumor volume by 20% (Figure 7A). As statistical significance was not reached after the application of the mentioned doses, the experiment was repeated with a dose of 10 mg kg^{-1} for the free conjugate **1** and 35 mg kg^{-1} for SBA-15 | **1**. The obtained results revealed remarkable inhibition of tumor growth in all three experimental groups (Figure 7B). The experiment was concluded with the sacrifice of the animals and histopathological examination of their tumors, liver and kidney.

Average percentage of tumor's necrosis on tissue cross section was the largest in the control group (30%), while all treatments dramatically decreased necrotic area surface (conjugate **1** (14%), SBA-15 | **1** (14%) and cisplatin (12%), Figure 7C). In addition, mitotic index, represented as mitotic figure count per one square millimeter, again was the highest in the control group, while rate of mitosis was diminished in tumor tissue sections of all treated groups (Figure 7B,D). Having in mind that tumors with large necrotic areas are recognized as highly aggressive and associated with poor prognosis [83], reduced necrotic zones together with lower mitotic rate upon the treatments indicates decreased neoplastic potential of the tumor tissue and underlines the advantage of the applied treatments.

In liver, findings characteristic for 4T1 metastatic breast cancer model with presence of extramedullary hematopoiesis, mainly represented by cells of myeloid lineage and in lesser amount by megakaryocytes, were observed [84,85]. Hematopoietic cells were present around and within the portal tracts, as well as in the sinusoid of the liver. The highest level of hematopoiesis was found in the control group, while upon treatment, in all treated groups it was almost absent (Figure 8). This fact led to the conclusion that all three treatments were effective in a way that inhibits hematopoiesis which was conditioned with the amount of tumor's cell burden.

No relevant signs of hepatotoxicity were found in any of the groups. There was no inflammation nor necrosis in this experimental setting (Figure 8). Some degree of dilatation of small blood vessels (mainly venules) mostly in the SBA-15 | **1** group, without knowledge of functional derangement, was noted.

In kidney tissue, there were no significant signs of nephrotoxicity in all groups. Normal histomorphology was dominantly registered (Figure 8). Concordantly, all parameters (blood, bilirubin, pH, ketone, protein, specific gravity, leukocyte, urobilinogen and nitrite) tested in urine of animals which were exposed to treatment, showed no difference to the control group of untreated animals (ESI, Table S4). In the group treated with cisplatin, however, protein casts were found in the tubules, and dilatation of small periglomerular blood vessels was observed. This indicates that despite the fact that previous findings suggested the absence of toxicity, cisplatin has a nephrotoxic input [86] which was not observed for **1** and SBA-15 | **1**.

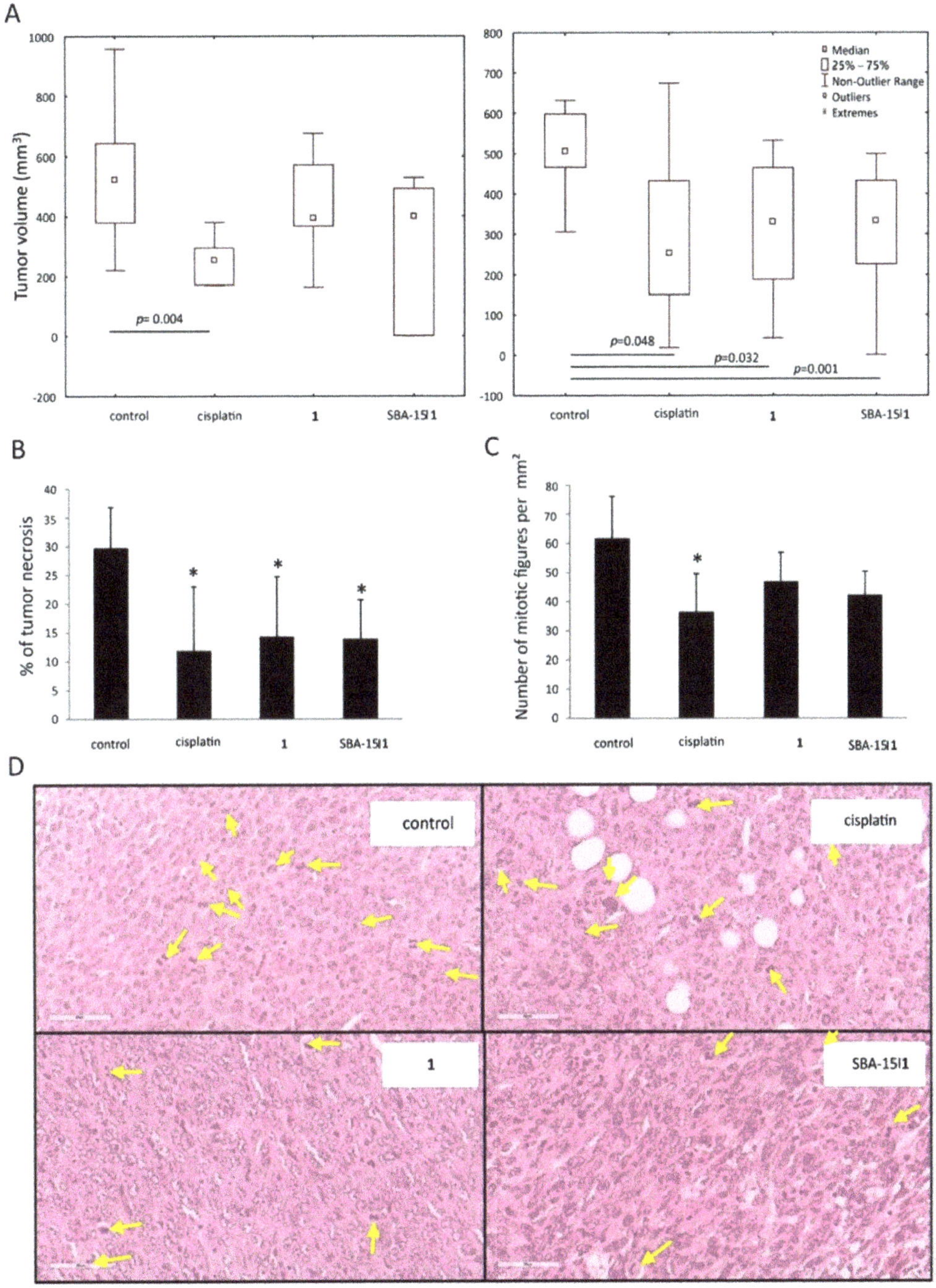

Figure 7. In vivo effect of platinum(IV) conjugate **1** and the corresponding MSNs SBA-15 | **1** on 4T1-inoculated breast tumors: (**A**) tumors' growth suppression, (**B**) tumor necrotic areas, (**C**) mitotic index, and (**D**) mitotic figures (marked with yellow arrows, H/E stain, 400× magnification).

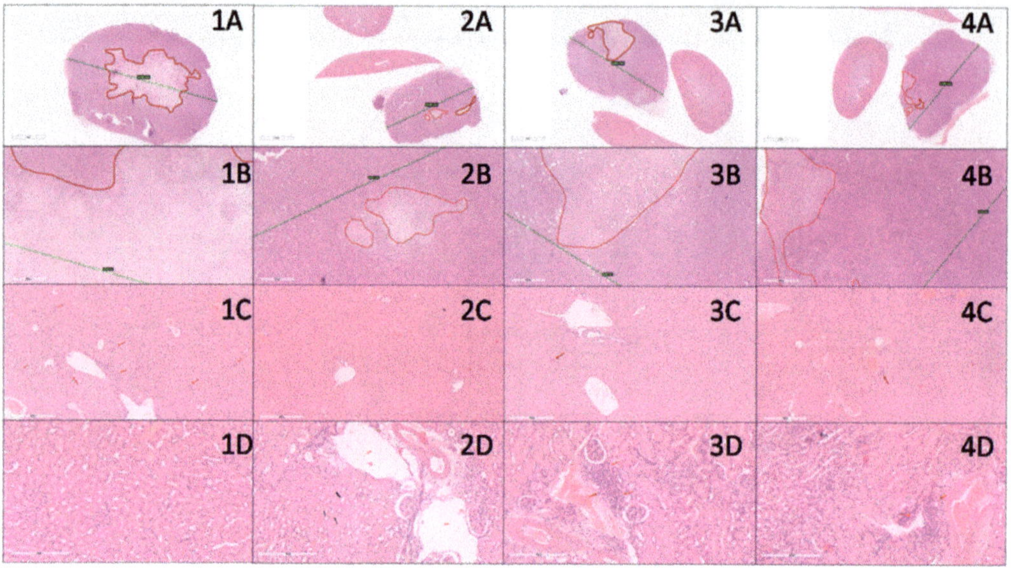

Figure 8. In vivo effect of platinum(IV) conjugate **1** and the corresponding MSNs SBA-15 | **1** on 4T1-inoculated breast tumors—histopathological evaluation: tissues in columns 1, 2, 3, and 4 represent control, cisplatin, **1** and SBA-15 | **1** groups, respectively. Tissues in rows **A** and **B** are tumor tissues (8× and 40× magnification, respectively) with red lines delineating necrotic areas and green lines denoting largest diameter of tumor. Row **C** is liver tissue with red arrows pointing to hematopoietic cells (40× magnification). Row **D** is kidney tissue (200× magnification). In picture **2D**, black arrows point to protein cast in kidney tubules and red arrows point to dilated blood vessels. Red arrows in **3D** and **4D** point to areas of inflammatory infiltrate.

4. Conclusions

In summary, we report on three cisplatin derivatives bearing acetyl-protected caffeate (**1**), acetyl-protected ferulate (**2**) or ferulate (**3**) as axial ligands. Upon loading of these platinum(IV) complexes into mesoporous silica SBA-15 material, the MSNs SBA-15 | **1**, SBA-15 | **2** and SBA-15 | **3** were prepared. Under simulated physiological settings, the MSNs maintained their structural integrity, enabling a moderate (for SBA-15 | **3**) to extremely slow release (for SBA-15 | **1** and SBA-15 | **2**) of the appropriate platinum(IV) complex. This delayed release is particularly advantageous in vivo because it results in low drug release during circulation and longer drug exposure upon accumulation in the tumor.

The free platinum(IV) conjugates **1**–**3** demonstrated higher or similar antiproliferative activity with respect to cisplatin in four human breast cancer cell lines (BT-474, MCF-7, MDA-MB-468 and HCC1937). Great anticancer potential was, however, observed for all three complexes upon their loading into the nanostructured SBA-15. Thus, SBA-15 | **1**, SBA-15 | **2** and SBA-15 | **3** demonstrated significantly higher cytotoxicity as compared to cisplatin with nanomolar IC_{50} values in all four cell lines.

For the most potent complex **1** and the corresponding SBA-15 | **1**, mechanistic investigations made in the mouse breast cancer cell line 4T1 indicated the potential of drug and nanomaterial to downregulate cell growth through inhibition of proliferation and apoptotic cell induction. Drug toxicity can be further amplified via autophagy suppression. All observed effects could be due to the enhanced production of NO and ROS/RNS, which affect different cellular processes and, consequently, the cell viability. Furthermore, **1** and SBA-15 | **1**, caused in vivo inhibition of tumor growth in BALB/c mice, decreasing the necrotic area and lowering the mitotic rate. Importantly, nephrotoxicity present in mice treated with cisplatin, was not observed when **1** or SBA-15 | **1** were used.

Supplementary Materials: The following supporting information can be downloaded at https://www.mdpi.com/article/10.3390/nano12213767/s1. Figure S1. ^{1}H NMR spectrum of **1** in DMSO-d_6; Figure S2. ^{13}C{^{1}H} NMR spectrum of **1** in DMSO-d_6; Figure S3. ^{195}Pt{^{1}H} NMR spectrum of **1** in DMSO-d_6; Figure S4. HR-ESI-MS (positive mode, CH_3OH) of **1**, *m*/*z* [2M+H]$^+$; Figure S5. HR-ESI-MS (positive mode, CH_3OH) of **1**, *m*/*z* [M+H]$^+$; Figure S6. ^{1}H NMR spectrum of **2** in DMSO-d_6; Figure S7. ^{13}C{^{1}H} NMR spectrum of **2** in DMSO-d_6; Figure S8. ^{195}Pt{^{1}H} NMR spectrum of **2** in DMSO-d_6; Figure S9. HR-ESI-MS (positive mode, CH_3OH) of **2**, *m*/*z* [2M+H]$^+$; Figure S10. HR-ESI-MS (positive mode, CH_3OH) of **2**, *m*/*z* [M+H]$^+$; Figure S11. HR-ESI-MS (positive mode, CH_3OH) of **2**, *m*/*z* [M+Na]$^+$; Figure S12. ^{1}H NMR spectrum of **3** in DMSO-d_6; Figure S13. ^{13}C{^{1}H} NMR spectrum of **3** in DMSO-d_6; Figure S14. ^{195}Pt{^{1}H} NMR spectrum of **3** in DMSO-d_6; Figure S15. HR-ESI-MS (positive mode, CH_3OH) of **3**, *m*/*z* [M+H]$^+$; Figure S16. HR-ESI-MS (positive mode, CH_3OH) of **3**, m/z [2M+H]$^+$; Figure S17. HR-ESI-MS (positive mode, CH_3OH) of **3**, *m*/*z* [2M+Na]$^+$; Figure S18. HR-ESI-MS (positive mode, CH_3OH) of **3**, *m*/*z* [2M+K]$^+$; Figure S19. Attempted *O*-deacetylation of **1** with enzyme Amano lipase A from *Aspergillus niger*; time-resolved ^{1}H NMR spectra in DMSO-d_6, aromatic region section; Figure S20. Stability of **1** in DMSO-d_6 over 72 h; time-resolved ^{1}H NMR spectra; Figure S21. Stability of **2** in DMSO-d_6 over 72 h; time-resolved ^{1}H NMR spectra; Figure S22 Stability of **3** in DMSO-d_6 over 72 h; time-resolved ^{1}H NMR spectra; Figure S23. Spatial distribution of silicon and platinum in SBA 15 | **1**, SBA 15 | **2** and SBA 15 | **3** determined through EDX mapping; Figure S24. Zero order, first order, Higuchi and Korsmeyer-Peppas kinetic release of **1** from SBA-15 | **1**; Figure S25. Zero order, first order, Higuchi and Korsmeyer-Peppas kinetic release of **2** from SBA-15 | **2**; Figure S26. Zero order, first order, Higuchi and Korsmeyer-Peppas kinetic release of **3** from SBA-15 | **3**; Figure S27. Cell viability of cisplatin, **1**, **2** and **3** determined by CV and MTT assays in MDA-MB-468 and HCC1937 human breast cancer cell lines; Figure S28. Cell viability of cisplatin, **1**, **2** and **3** determined by CV and MTT assays in MCF-7 and BT-474 human breast cancer cell lines; Figure S29. Cell viability of SBA-15 | **1**, SBA-15 | **2** and SBA-15 | **3** determined by CV and MTT assays in MDA-MB-468 and HCC1937 human breast cancer cell lines; Figure S30. Cell viability of SBA-15 | **1**, SBA-15 | **2** and SBA-15 | **3** determined by CV and MTT assays in MCF-7 and BT-474 human breast cancer cell lines; Figure S31. Cell viability of cisplatin, **1** and SBA-15 | **1** determined by CV and MTT assays in mouse-derived 4T1 breast cancer cell line; Table S1. Relative weight % of silicon and platinum at six points in SBA-15 | **1**, SBA-15 | **2** and SBA-15 | **3** determined by EDX analysis; Table S2. Platinum load (wt%) and encapsulation efficiency (%) in SBA-15 | **1**, SBA-15 | **2** and SBA-15 | **3**; Table S3. Constants and coefficient of determinations (R^2) for each model; Table S4. Urine parameters of untreated animals (control) and animals exposed to treatment with cisplatin, **1** and SBA-15 | **1**.

Author Contributions: I.P., S.M., D.M.-I., E.H.-H. and G.N.K. conceptualized the project. I.P., M.S., I.M., T.E., T.K., E.M., D.D. (Dijana Drača) and D.D. (Duško Dunđerović) did investigation and formal analysis of data. I.P., S.M. and D.M.-I. did data curation. I.P., D.D. (Duško Dunđerović), S.M. and D.M.-I. wrote the original draft. S.M., D.M.-I., E.H.-H. and G.N.K. supervised the work and reviewed and edited the original draft. S.M., E.H.-H. and G.N.K. acquired the funding and administrated the project. All authors have read and agreed to the published version of the manuscript.

Funding: This work was financially supported through the program FEM POWER (funded by the operational program ESF Saxony-Anhalt WISSENSCHAFT Chancengleichheit with funds from the European Union; doctoral grant for I.P.), the Graduate School "Building with Molecules and Nano-objects (BuildMoNa)", the Research Academy Leipzig and the Ministry of Education, Science and Technological Development of the Republic of Serbia, grant number 451-03-68/2022-14/200007.

Institutional Review Board Statement: The animal study protocol was approved by the national licensing committee of Department of Animal Welfare, Veterinary Directorate, Ministry of Agriculture, Forestry and Water Management of Republic of Serbia (permission No. 323 07 07906/2022 05).

Informed Consent Statement: Not applicable.

Data Availability Statement: Not applicable.

Acknowledgments: The authors thank Prof. Dr. Valentin Cepus for the ICP-OES measurements and Prof. Dr. Stefan Ebbinghaus for the SAXS measurements.

Conflicts of Interest: The authors declare no conflict of interest.

References

1. Shahidi, F.; Naczk, M. *Food Phenolics: Sources, Chemistry, Effects, Applications*; Technomic Pub. Co.: Lancaster, PA, USA, 1995; ISBN 1-56676-279-0.
2. Kumar, N.; Goel, N. Phenolic Acids: Natural Versatile Molecules with Promising Therapeutic Applications. *Biotechnol. Rep.* **2019**, *24*, e00370. [CrossRef] [PubMed]
3. Kumar, N.; Gupta, S.; Chand Yadav, T.; Pruthi, V.; Kumar Varadwaj, P.; Goel, N. Extrapolation of Phenolic Compounds as Multi-Target Agents against Cancer and Inflammation. *J. Biomol. Struct. Dyn.* **2019**, *37*, 2355–2369. [CrossRef] [PubMed]
4. Mori, H.; Kawabata, K.; Yoshimi, N.; Tanaka, T.; Murakami, T.; Okada, T.; Murai, H. Chemopreventive Effects of Ferulic Acid on Oral and Rice Germ on Large Bowel Carcinogenesis. *Anticancer Res.* **1999**, *19*, 3775–3778. [PubMed]
5. Middleton, E.J.; Kandaswami, C.; Theoharides, T.C. The Effects of Plant Flavonoids on Mammalian Cells: Implications for Inflammation, Heart Disease, and Cancer. *Pharmacol. Rev.* **2000**, *52*, 673–751. [PubMed]
6. Rocha, L.D.; Monteiro, M.; Teodoro, A.J. Anticancer Properties of Hydroxycinnamic Acids -A Review. *Cancer Clin. Oncol.* **2012**, *1*, 109. [CrossRef]
7. Działo, M.; Mierziak, J.; Korzun, U.; Preisner, M.; Szopa, J.; Kulma, A. The Potential of Plant Phenolics in Prevention and Therapy of Skin Disorders. *Int. J. Mol. Sci.* **2016**, *17*, 160. [CrossRef]
8. Sytar, O.; Hemmerich, I.; Zivcak, M.; Rauh, C.; Brestic, M. Comparative Analysis of Bioactive Phenolic Compounds Composition from 26 Medicinal Plants. *Saudi J. Biol. Sci.* **2018**, *25*, 631–641. [CrossRef]
9. Son, S.; Lewis, B.A. Free Radical Scavenging and Antioxidative Activity of Caffeic Acid Amide and Ester Analogues: Structure−Activity Relationship. *J. Agric. Food Chem.* **2002**, *50*, 468–472. [CrossRef]
10. Kumar, N.; Pruthi, V. Structural Elucidation and Molecular Docking of Ferulic Acid from Parthenium Hysterophorus Possessing COX-2 Inhibition Activity. *3 Biotech* **2015**, *5*, 541–551. [CrossRef]
11. Jayaprakasam, B.; Vanisree, M.; Zhang, Y.; Dewitt, D.L.; Nair, M.G. Impact of Alkyl Esters of Caffeic and Ferulic Acids on Tumor Cell Proliferation, Cyclooxygenase Enzyme, and Lipid Peroxidation. *J. Agric. Food Chem.* **2006**, *54*, 5375–5381. [CrossRef]
12. Michaluart, P.; Masferrer, J.L.; Carothers, A.M.; Subbaramaiah, K.; Zweifel, B.S.; Koboldt, C.; Mestre, J.R.; Grunberger, D.; Sacks, P.G.; Tanabe, T.; et al. Inhibitory Effects of Caffeic Acid Phenethyl Ester on the Activity and Expression of Cyclooxygenase-2 in Human Oral Epithelial Cells and in a Rat Model of Inflammation. *Cancer Res.* **1999**, *59*, 2347–2352. [PubMed]
13. Kang, N.J.; Lee, K.W.; Shin, B.J.; Jung, S.K.; Hwang, M.K.; Bode, A.M.; Heo, Y.-S.; Lee, H.J.; Dong, Z. Caffeic Acid, a Phenolic Phytochemical in Coffee, Directly Inhibits Fyn Kinase Activity and UVB-Induced COX-2 Expression. *Carcinogenesis* **2009**, *30*, 321–330. [CrossRef] [PubMed]
14. Müller-Decker, K.; Fürstenberger, G. The Cyclooxygenase-2-Mediated Prostaglandin Signaling Is Causally Related to Epithelial Carcinogenesis. *Mol. Carcinog.* **2007**, *46*, 705–710. [CrossRef] [PubMed]
15. Chiang, E.-P.I.; Tsai, S.-Y.; Kuo, Y.-H.; Pai, M.-H.; Chiu, H.-L.; Rodriguez, R.L.; Tang, F.-Y. Caffeic Acid Derivatives Inhibit the Growth of Colon Cancer: Involvement of the PI3-K/Akt and AMPK Signaling Pathways. *PLoS ONE* **2014**, *9*, e99631. [CrossRef] [PubMed]
16. Jaganathan, S.K. Growth Inhibition by Caffeic Acid, One of the Phenolic Constituents of Honey, in HCT 15 Colon Cancer Cells. *Sci. World J.* **2012**, *2012*, 372345. [CrossRef]
17. Murad, L.D.; Soares ND, C.P.; Brand, C.; Monteiro, M.C.; Teodoro, A.J. Effects of Caffeic and 5-Caffeoylquinic Acids on Cell Viability and Cellular Uptake in Human Colon Adenocarcinoma Cells. *Nutr. Cancer* **2015**, *67*, 532–542. [CrossRef]
18. Tang, H.; Yao, X.; Yao, C.; Zhao, X.; Zuo, H.; Li, Z. Anti-Colon Cancer Effect of Caffeic Acid p-Nitro-Phenethyl Ester in Vitro and in Vivo and Detection of Its Metabolites. *Sci. Rep.* **2017**, *7*, 7599. [CrossRef]
19. Espíndola, K.M.M.; Ferreira, R.G.; Narvaez, L.E.M.; Silva Rosario, A.C.R.; da Silva, A.H.M.; Silva, A.G.B.; Vieira, A.P.O.; Monteiro, M.C. Chemical and Pharmacological Aspects of Caffeic Acid and Its Activity in Hepatocarcinoma. *Front. Oncol.* **2019**, *9*, 541. [CrossRef]
20. Rajendra Prasad, N.; Karthikeyan, A.; Karthikeyan, S.; Reddy, B.V. Inhibitory Effect of Caffeic Acid on Cancer Cell Proliferation by Oxidative Mechanism in Human HT-1080 Fibrosarcoma Cell Line. *Mol. Cell. Biochem.* **2011**, *349*, 11–19. [CrossRef]
21. Kanimozhi, G.; Prasad, N.R. Chapter 73-Anticancer Effect of Caffeic Acid on Human Cervical Cancer Cells. In *Coffee in Health and Disease Prevention*; Preedy, V.R., Ed.; Academic Press: San Diego, CA, USA, 2015; pp. 655–661. ISBN 978-0-12-409517-5.
22. Kabała-Dzik, A.; Rzepecka-Stojko, A.; Kubina, R.; Wojtyczka, R.D.; Buszman, E.; Stojko, J. Caffeic Acid Versus Caffeic Acid Phenethyl Ester in the Treatment of Breast Cancer MCF-7 Cells: Migration Rate Inhibition. *Integr. Cancer Ther.* **2018**, *17*, 1247–1259. [CrossRef]
23. Rosendahl, A.H.; Perks, C.M.; Zeng, L.; Markkula, A.; Simonsson, M.; Rose, C.; Ingvar, C.; Holly, J.M.P.; Jernström, H. Caffeine and Caffeic Acid Inhibit Growth and Modify Estrogen Receptor and Insulin-like Growth Factor I Receptor Levels in Human Breast Cancer. *Clin. Cancer Res.* **2015**, *21*, 1877–1887. [CrossRef] [PubMed]
24. Hudson, E.A.; Dinh, P.A.; Kokubun, T.; Simmonds, M.S.; Gescher, A. Characterization of Potentially Chemopreventive Phenols in Extracts of Brown Rice That Inhibit the Growth of Human Breast and Colon Cancer Cells. *Cancer Epidemiol. Biomarkers Prev.* **2000**, *9*, 1163–1170. [PubMed]
25. Min, J.; Shen, H.; Xi, W.; Wang, Q.; Yin, L.; Zhang, Y.; Yu, Y.; Yang, Q.; Wang, Z.-N. Synergistic Anticancer Activity of Combined Use of Caffeic Acid with Paclitaxel Enhances Apoptosis of Non-Small-Cell Lung Cancer H1299 Cells in Vivo and in Vitro. *Cell. Physiol. Biochem.* **2018**, *48*, 1433–1442. [CrossRef] [PubMed]

26. Tyszka-Czochara, M.; Konieczny, P.; Majka, M. Caffeic Acid Expands Anti-Tumor Effect of Metformin in Human Metastatic Cervical Carcinoma HTB-34 Cells: Implications of AMPK Activation and Impairment of Fatty Acids De Novo Biosynthesis. *Int. J. Mol. Sci.* **2017**, *18*, 462. [CrossRef]
27. Matsunaga, T.; Tsuchimura, S.; Azuma, N.; Endo, S.; Ichihara, K.; Ikari, A. Caffeic Acid Phenethyl Ester Potentiates Gastric Cancer Cell Sensitivity to Doxorubicin and Cisplatin by Decreasing Proteasome Function. *Anticancer Drugs* **2019**, *30*, 251–259. [CrossRef]
28. Koraneekit, A.; Limpaiboon, T.; Sangka, A.; Boonsiri, P.; Daduang, S.; Daduang, J. Synergistic Effects of Cisplatin-Caffeic Acid Induces Apoptosis in Human Cervical Cancer Cells via the Mitochondrial Pathways. *Oncol. Lett.* **2018**, *15*, 7397–7402. [CrossRef]
29. Sirota, R.; Gibson, D.; Kohen, R. The Timing of Caffeic Acid Treatment with Cisplatin Determines Sensitization or Resistance of Ovarian Carcinoma Cell Lines. *Redox Biol.* **2017**, *11*, 170–175. [CrossRef]
30. Kelland, L. The Resurgence of Platinum-Based Cancer Chemotherapy. *Nat. Rev. Cancer* **2007**, *7*, 573–584. [CrossRef]
31. Neumann, W.; Crews, B.C.; Marnett, L.J.; Hey-Hawkins, E. Conjugates of Cisplatin and Cyclooxygenase Inhibitors as Potent Antitumor Agents Overcoming Cisplatin Resistance. *ChemMedChem* **2014**, *9*, 1150–1153. [CrossRef]
32. Neumann, W.; Crews, B.C.; Sárosi, M.B.; Daniel, C.M.; Ghebreselasie, K.; Scholz, M.S.; Marnett, L.J.; Hey-Hawkins, E. Conjugation of Cisplatin Analogues and Cyclooxygenase Inhibitors to Overcome Cisplatin Resistance. *ChemMedChem* **2015**, *10*, 183–192. [CrossRef]
33. Matsumura, Y.; Maeda, H. A New Concept for Macromolecular Therapeutics in Cancer Chemotherapy: Mechanism of Tumoritropic Accumulation of Proteins and the Antitumor Agent Smancs. *Cancer Res.* **1986**, *46*, 6387–6392. [PubMed]
34. Hagman, E.; Elimam, A.; Kupferschmidt, N.; Ekbom, K.; Rössner, S.; Iqbal, M.N.; Johnston, E.; Lindgren, M.; Bengtsson, T.; Danielsson, P. Oral Intake of Mesoporous Silica Is Safe and Well Tolerated in Male Humans. *PLoS ONE* **2020**, *15*, e0240030. [CrossRef] [PubMed]
35. Sweeney, S.K.; Luo, Y.; O'Donnell, M.A.; Assouline, J. Nanotechnology and Cancer: Improving Real-Time Monitoring and Staging of Bladder Cancer with Multimodal Mesoporous Silica Nanoparticles. *Cancer Nanotechnol.* **2016**, *7*, 3. [CrossRef] [PubMed]
36. Tang, F.; Li, L.; Chen, D. Mesoporous Silica Nanoparticles: Synthesis, Biocompatibility and Drug Delivery. *Adv. Mater.* **2012**, *24*, 1504–1534. [CrossRef] [PubMed]
37. Bharti, C.; Nagaich, U.; Pal, A.K.; Gulati, N. Mesoporous Silica Nanoparticles in Target Drug Delivery System: A Review. *Int. J. Pharm. Investig.* **2015**, *5*, 124. [CrossRef] [PubMed]
38. Manzano, M.; Vallet-Regí, M. Mesoporous Silica Nanoparticles for Drug Delivery. *Adv. Funct. Mater.* **2020**, *30*, 1902634. [CrossRef]
39. Sábio, R.M.; Meneguin, A.B.; Ribeiro, T.C.; Silva, R.R.; Chorilli, M. New Insights towards Mesoporous Silica Nanoparticles as a Technological Platform for Chemotherapeutic Drugs Delivery. *Int. J. Pharm.* **2019**, *564*, 379–409. [CrossRef]
40. Castillo, R.R.; Lozano, D.; Vallet-Regí, M. Mesoporous Silica Nanoparticles as Carriers for Therapeutic Biomolecules. *Pharmaceutics* **2020**, *12*, 432. [CrossRef]
41. Taleghani, A.S.; Nakhjiri, A.T.; Khakzad, M.J.; Rezayat, S.M.; Ebrahimnejad, P.; Heydarinasab, A.; Akbarzadeh, A.; Marjani, A. Mesoporous Silica Nanoparticles as a Versatile Nanocarrier for Cancer Treatment: A Review. *J. Mol. Liq.* **2021**, *328*, 115417. [CrossRef]
42. Yu, A.; Dai, X.; Wang, Z.; Chen, H.; Guo, B.; Huang, L. Recent Advances of Mesoporous Silica as a Platform for Cancer Immunotherapy. *Biosensors* **2022**, *12*, 109. [CrossRef]
43. Yuan, Z.; Pan, Y.; Cheng, R.; Sheng, L.; Wu, W.; Pan, G.; Feng, Q.; Cui, W. Doxorubicin-Loaded Mesoporous Silica Nanoparticle Composite Nanofibers for Long-Term Adjustments of Tumor Apoptosis. *Nanotechnology* **2016**, *27*, 245101. [CrossRef] [PubMed]
44. Edeler, D.; Arlt, S.; Petković, V.; Ludwig, G.; Drača, D.; Maksimović-Ivanić, D.; Mijatović, S.; Kaluđerović, G.N. Delivery of [Ru(η^6-*p*-Cymene)Cl_2{$Ph_2P(CH_2)_3SPh$-κ*P*}] Using Unfunctionalized and Mercapto Functionalized SBA-15 Mesoporous Silica: Preparation, Characterization and in Vitro Study. *J. Inorg. Biochem.* **2018**, *180*, 155–162. [CrossRef] [PubMed]
45. Tao, Z.; Toms, B.; Goodisman, J.; Asefa, T. Mesoporous Silica Microparticles Enhance the Cytotoxicity of Anticancer Platinum Drugs. *ACS Nano* **2010**, *4*, 789–794. [CrossRef]
46. Lin, C.H.; Cheng, S.H.; Liao, W.N.; Wei, P.R.; Sung, P.J.; Weng, C.F.; Lee, C.H. Mesoporous Silica Nanoparticles for the Improved Anticancer Efficacy of Cis-Platin. *Int. J. Pharm.* **2012**, *429*, 138–147. [CrossRef]
47. Edeler, D.; Kaluđerović, M.R.; Dojčinović, B.; Schmidt, H.; Kaluđerović, G.N. SBA-15 Mesoporous Silica Particles Loaded with Cisplatin Induce Senescence in B16F10 Cells. *RSC Adv.* **2016**, *6*, 111031–111040. [CrossRef]
48. Predarska, I.; Saoud, M.; Morgan, I.; Eichhorn, T.; Kaluđerović, G.N.; Hey-Hawkins, E. Cisplatin−cyclooxygenase Inhibitor Conjugates, Free and Immobilised in Mesoporous Silica SBA-15, Prove Highly Potent against Triple-Negative MDA-MB-468 Breast Cancer Cell Line. *Dalton Trans.* **2022**, *51*, 857–869. [CrossRef]
49. Tan, M.-X.; Wang, Z.-F.; Qin, Q.-P.; Zou, B.-Q.; Liang, H. Complexes of Oxoplatin with Rhein and Ferulic Acid Ligands as Platinum(IV) Prodrugs with High Anti-Tumor Activity. *Dalton Trans.* **2020**, *49*, 1613–1619. [CrossRef] [PubMed]
50. Hall, M.D.; Dillon, C.T.; Zhang, M.; Beale, P.; Cai, Z.; Lai, B.; Stampfl, A.P.; Hambley, T.W. The Cellular Distribution and Oxidation State of Platinum(II) and Platinum(IV) Antitumour Complexes in Cancer Cells. *J. Biol. Inorg. Chem.* **2003**, *8*, 726–732. [CrossRef] [PubMed]
51. Touaibia, M.; Guay, M. Natural Product Total Synthesis in the Organic Laboratory: Total Synthesis of Caffeic Acid Phenethyl Ester (CAPE), A Potent 5-Lipoxygenase Inhibitor from Honeybee Hives. *J. Chem. Educ.* **2011**, *88*, 473–475. [CrossRef]

52. García-Peñas, A.; Gómez-Ruiz, S.; Pérez-Quintanilla, D.; Paschke, R.; Sierra, I.; Prashar, S.; del Hierro, I.; Kaluđerović, G.N. Study of the Cytotoxicity and Particle Action in Human Cancer Cells of Titanocene-Functionalized Materials with Potential Application against Tumors. *J. Inorg. Biochem.* **2012**, *106*, 100–110. [CrossRef]
53. Kaluđerović, G.N.; Pérez-Quintanilla, D.; Sierra, I.; Prashar, S.; del Hierro, I.; Žižak, Ž.; Juranić, Z.D.; Fajardo, M.; Gómez-Ruiz, S. Study of the Influence of the Metal Complex on the Cytotoxic Activity of Titanocene-Functionalized Mesoporous Materials. *J. Mater. Chem.* **2010**, *20*, 806–814. [CrossRef]
54. Smolko, L.; Smolková, R.; Samoľová, E.; Morgan, I.; Saoud, M.; Kaluđerović, G.N. Two Isostructural Co(II) Flufenamato and Niflumato Complexes with Bathocuproine: Analogues with a Different Cytotoxic Activity. *J. Inorg. Biochem.* **2020**, *210*, 111160. [CrossRef]
55. Subik, K.; Lee, J.-F.; Baxter, L.; Strzepek, T.; Costello, D.; Crowley, P.; Xing, L.; Hung, M.-C.; Bonfiglio, T.; Hicks, D.G.; et al. The Expression Patterns of ER, PR, HER2, CK5/6, EGFR, Ki-67 and AR by Immunohistochemical Analysis in Breast Cancer Cell Lines. *Breast Cancer* **2010**, *4*, 35–41. [CrossRef]
56. Khan, M.F.; Nasr, F.A.; Noman, O.M.; Alyhya, N.A.; Ali, I.; Saoud, M.; Rennert, R.; Dube, M.; Hussain, W.; Green, I.R.; et al. Cichorins D–F: Three New Compounds from Cichorium Intybus and Their Biological Effects. *Molecules* **2020**, *25*, 4160. [CrossRef]
57. Drača, D.; Edeler, D.; Saoud, M.; Dojčinović, B.; Dunđerović, D.; Đmura, G.; Maksimović-Ivanić, D.; Mijatović, S.; Kaluđerović, G.N. Antitumor Potential of Cisplatin Loaded into SBA-15 Mesoporous Silica Nanoparticles against B16F1 Melanoma Cells: In Vitro and in Vivo Studies. *J. Inorg. Biochem.* **2021**, *217*, 111383. [CrossRef] [PubMed]
58. Rice-Evans, C.A.; Miller, N.J.; Paganga, G. Structure-Antioxidant Activity Relationships of Flavonoids and Phenolic Acids. *Free Radic. Biol. Med.* **1996**, *20*, 933–956. [CrossRef]
59. Spiegel, M.; Kapusta, K.; Kołodziejczyk, W.; Saloni, J.; Żbikowska, B.; Hill, G.A.; Sroka, Z. Antioxidant Activity of Selected Phenolic Acids–Ferric Reducing Antioxidant Power Assay and QSAR Analysis of the Structural Features. *Molecules* **2020**, *25*, 3088. [CrossRef]
60. Scampicchio, M.; Wang, J.; Blasco, A.J.; Sanchez Arribas, A.; Mannino, S.; Escarpa, A. Nanoparticle-Based Assays of Antioxidant Activity. *Anal. Chem.* **2006**, *78*, 2060–2063. [CrossRef]
61. Amini, S.M.; Akbari, A. Metal Nanoparticles Synthesis through Natural Phenolic Acids. *IET Nanobiotechnol.* **2019**, *13*, 771–777. [CrossRef]
62. LeBlanc, L.M.; Paré, A.F.; Jean-François, J.; Hébert, M.J.G.; Surette, M.E.; Touaibia, M. Synthesis and Antiradical/Antioxidant Activities of Caffeic Acid Phenethyl Ester and Its Related Propionic, Acetic, and Benzoic Acid Analoguesc. *Molecules* **2012**, *17*, 14637–14650. [CrossRef]
63. Wang, S.-M.; Zhang, Y.-B.; Liu, H.-M.; Yu, G.-B.; Wang, K.-R. Mild and Selective Deprotection Method of Acetylated Steroids and Diterpenes by Dibutyltin Oxide. *Steroids* **2007**, *72*, 26–30. [CrossRef] [PubMed]
64. Ramesh, C.; Mahender, G.; Ravindranath, N.; Das, B. A Mild, Highly Selective and Remarkably Easy Procedure for Deprotection of Aromatic Acetates Using Ammonium Acetate as a Neutral Catalyst in Aqueous Medium. *Tetrahedron* **2003**, *59*, 1049–1054. [CrossRef]
65. Dunne, A.; Palomo, J.M. Efficient and Green Approach for the Complete Deprotection of O-Acetylated Biomolecules. *RSC Adv.* **2016**, *6*, 88974–88978. [CrossRef]
66. Sing, K.S. Reporting Physisorption Data for Gas/Solid Systems with Special Reference to the Determination of Surface Area and Porosity (Recommendations 1984). *Pure Appl. Chem.* **1985**, *57*, 603–619. [CrossRef]
67. Kankala, R.K.; Han, Y.-H.; Na, J.; Lee, C.-H.; Sun, Z.; Wang, S.-B.; Kimura, T.; Ok, Y.S.; Yamauchi, Y.; Chen, A.-Z.; et al. Nanoarchitectured Structure and Surface Biofunctionality of Mesoporous Silica Nanoparticles. *Adv. Mat.* **2020**, *32*, 1907035. [CrossRef]
68. Dolete, G.; Purcăreanu, B.; Mihaiescu, D.E.; Ficai, D.; Oprea, O.-C.; Bîrcă, A.C.; Chircov, C.; Vasile, B.Ş.; Vasilievici, G.; Ficai, A.; et al. A Comparative Loading and Release Study of Vancomycin from a Green Mesoporous Silica. *Molecules* **2022**, *27*, 5589. [CrossRef]
69. Costa, P.; Sousa Lobo, J.M. Modeling and Comparison of Dissolution Profiles. *Eur. J. Pharm. Sci.* **2001**, *13*, 123–133. [CrossRef]
70. Korsmeyer, R.W.; Gurny, R.; Doelker, E.; Buri, P.; Peppas, N.A. Mechanisms of Solute Release from Porous Hydrophilic Polymers. *Inter. J. Pharm.* **1983**, *15*, 25–35. [CrossRef]
71. Langer, R.S.; Peppas, N.A. Present and Future Applications of Biomaterials in Controlled Drug Delivery Systems. *Biomaterials* **1981**, *2*, 201–214. [CrossRef]
72. Vistica, D.T.; Skehan, P.; Scudiero, D.; Monks, A.; Pittman, A.; Boyd, M.R. Tetrazolium-Based Assays for Cellular Viability: A Critical Examination of Selected Parameters Affecting Formazan Production. *Cancer Res.* **1991**, *51*, 2515–2520.
73. Shoemaker, M.; Cohen, I.; Campbell, M. Reduction of MTT by Aqueous Herbal Extracts in the Absence of Cells. *J. Ethnopharmacol.* **2004**, *93*, 381–384. [CrossRef]
74. Wang, P.; Henning, S.M.; Heber, D. Limitations of MTT and MTS-Based Assays for Measurement of Antiproliferative Activity of Green Tea Polyphenols. *PLoS ONE* **2010**, *5*, e10202. [CrossRef] [PubMed]
75. Bruggisser, R.; von Daeniken, K.; Jundt, G.; Schaffner, W.; Tullberg-Reinert, H. Interference of Plant Extracts, Phytoestrogens and Antioxidants with the MTT Tetrazolium Assay. *Planta Med.* **2002**, *68*, 445–448. [CrossRef] [PubMed]
76. Chang, M. Tamoxifen Resistance in Breast Cancer. *Biomol. Ther.* **2012**, *20*, 256–267. [CrossRef] [PubMed]

77. Ali, S.; Rasool, M.; Chaoudhry, H.; N Pushparaj, P.; Jha, P.; Hafiz, A.; Mahfooz, M.; Abdus Sami, G.; Azhar Kamal, M.; Bashir, S.; et al. Molecular Mechanisms and Mode of Tamoxifen Resistance in Breast Cancer. *Bioinformation* **2016**, *12*, 135–139. [CrossRef] [PubMed]
78. Chavez-Dominguez, R.; Perez-Medina, M.; Lopez-Gonzalez, J.S.; Galicia-Velasco, M.; Aguilar-Cazares, D. The Double-Edge Sword of Autophagy in Cancer: From Tumor Suppression to Pro-Tumor Activity. *Front. Oncol.* **2020**, *10*, 578418. [CrossRef]
79. Jung, S.; Jeong, H.; Yu, S.-W. Autophagy as a Decisive Process for Cell Death. *Exp. Mol. Med.* **2020**, *52*, 921–930. [CrossRef]
80. Perillo, B.; Di Donato, M.; Pezone, A.; Di Zazzo, E.; Giovannelli, P.; Galasso, G.; Castoria, G.; Migliaccio, A. ROS in Cancer Therapy: The Bright Side of the Moon. *Exp. Mol. Med.* **2020**, *52*, 192–203. [CrossRef]
81. Mijatović, S.; Savić-Radojević, A.; Plješa-Ercegovac, M.; Simić, T.; Nicoletti, F.; Maksimović-Ivanić, D. The Double-Faced Role of Nitric Oxide and Reactive Oxygen Species in Solid Tumors. *Antioxidants* **2020**, *9*, 374. [CrossRef]
82. Maksimović-Ivanić, D.; Mijatović, S.; Mirkov, I.; Stošić-Grujičić, S.; Miljković, D.; Sabo, T.J.; Trajković, V.; Kaluđerović, G.N. Melanoma Tumor Inhibition by Tetrachlorido(*O,O′*-Dibutyl-Ethylenediamine-*N,N′*-Di-3-Propionate)Platinum(IV) Complex: In Vitro and in Vivo Investigations. *Metallomics* **2012**, *4*, 1155–1159. [CrossRef]
83. Bredholt, G.; Mannelqvist, M.; Stefansson, I.M.; Birkeland, E.; Hellem Bø, T.; Øyan, A.M.; Trovik, J.; Kalland, K.-H.; Jonassen, I.; Salvesen, H.B.; et al. Tumor Necrosis Is an Important Hallmark of Aggressive Endometrial Cancer and Associates with Hypoxia, Angiogenesis and Inflammation Responses. *Oncotarget* **2015**, *6*, 39676. [CrossRef] [PubMed]
84. duPre', S.A.; Hunter, K.W. Murine Mammary Carcinoma 4T1 Induces a Leukemoid Reaction with Splenomegaly: Association with Tumor-Derived Growth Factors. *Exp. Mol. Pathol.* **2007**, *82*, 12–24. [CrossRef] [PubMed]
85. DuPre´, S.A.; Redelman, D.; Hunter Jr, K.W. The Mouse Mammary Carcinoma 4T1: Characterization of the Cellular Landscape of Primary Tumours and Metastatic Tumour Foci. *Int. J. Exp.Pathol.* **2007**, *88*, 351–360. [CrossRef] [PubMed]
86. McSweeney, K.R.; Gadanec, L.K.; Qaradakhi, T.; Ali, B.A.; Zulli, A.; Apostolopoulos, V. Mechanisms of Cisplatin-Induced Acute Kidney Injury: Pathological Mechanisms, Pharmacological Interventions, and Genetic Mitigations. *Cancers* **2021**, *13*, 1572. [CrossRef] [PubMed]

nanomaterials

MDPI

Article

Silver Nanoparticles Functionalized Nanosilica Grown over Graphene Oxide for Enhancing Antibacterial Effect

Qui Quach and Tarek M. Abdel-Fattah *

Applied Research Center at Thomas Jefferson National Accelerator Facility, Department of Molecular Biology and Chemistry, Christopher Newport University, Newport News, VA 23606, USA
* Correspondence: fattah@cnu.edu

Abstract: The continuous growth of multidrug-resistant bacteria due to the overuse of antibiotics and antibacterial agents poses a threat to human health. Silver nanoparticles, silica-based materials, and graphene-based materials have become potential antibacterial candidates. In this study, we developed an effective method of enhancing the antibacterial property of graphene oxide (GO) by growing nanosilica (NS) of approximately 50 nm on the graphene oxide (GO) surface. The structures and compositions of the materials were characterized through powdered X-ray diffraction (P-XRD), transmission electron microscopy (TEM), scanning electron microscopy coupled with energy dispersive X-ray spectroscopy (SEM-EDS), ultraviolet–visible spectroscopy (UV–VIS), dynamic light scattering (DLS), Raman spectroscopy (RM), Fourier-transform infrared spectroscopy (FTIR), Brunauer–Emmet–Teller (BET) surface area, and pore size determination. The silver nanoparticles (AgNPs) with an average diameter of 26 nm were functionalized on the nanosilica (NS) surface. The composite contained approximately 3% of silver nanoparticles. The silver nanoparticles on nanosilica supported over graphene oxide (GO/NS/AgNPs) exhibited a 7-log reduction of *Escherichia coli* and a 5.2-log reduction of *Bacillus subtilis* within one hour of exposure. Both GO/NS and GO/NS/AgNPs exhibited substantial antimicrobial effects against *E. coli* and *B. subtilis*

Keywords: silver nanoparticles; nanosilica; graphene oxide composites; antibacterial effect; *Escherichia coli*; bacillus subtilis

Citation: Quach, Q.; Abdel-Fattah, T.M. Silver Nanoparticles Functionalized Nanosilica Grown over Graphene Oxide for Enhancing Antibacterial Effect. *Nanomaterials* **2022**, *12*, 3341. https://doi.org/10.3390/nano12193341

Academic Editors: Goran Kaluđerović and Nebojsa Pantelic

Received: 18 August 2022
Accepted: 20 September 2022
Published: 25 September 2022

1. Introduction

Multidrug-resistant bacteria have become a global health threat in the 21st century [1]. Antibiotic resistance has caused common infections to become more difficult to treat and has led to expensive healthcare costs [1]. In order to reduce the dependency on antibiotic treatments which are ineffective against resistant bacterial strains, various studies have been conducted to develop novel materials to treat and prevent the growth of these pathogens [2–5]. Among those materials, silver nanoparticles appear to have the most potential and be the most attractive solution due to their antibacterial properties [6,7]. However, some studies have shown that bacteria can develop resistance to silver nanoparticles after repeated exposure [8,9]. The resistant strains produced adhesive flagellum that caused the nanoparticles to aggregate, resulting in a reduction in the surface area and antibacterial effect [8,9].

In previously reported studies, support templates or capping agents were often used to improve the size, shape, and stability of nanoparticles [10–14]. It is important to control the surface properties, morphology, and functionality for bio-medical applications [15–17]. Among the materials, silica-based and graphene materials are excellent support templates for enhancing antimicrobial activity [18,19]. These materials are well-known in medicine and biotechnology [19–21]. In a study from Liu et al., silver nanoparticles were loaded on to the graphene oxide and a large silicate template, but the synthesis process required heating which caused the graphene oxide to reduce [22]. Previous research reported that graphene oxide had a higher antibacterial ability than that of the reduced form [23]. Additionally, the large mesoporous silica template was less effective than small-sized silica nanoparticles in improving the antimicrobial property of the composite [24]. Based on different review studies, the combination of graphene oxide, nanosilica, and silver nanoparticles was neither synthesized nor tested for antibacterial activity [25,26]. Furthermore, the method which we applied in this study to grow nanosilica on graphene oxide was different from other reported methods [27–29]. In a Yang et al. study, the SiO_2 was functionalized with an amine group and then loaded on to the graphene oxide; the pH of the synthesis process was required to be 7 for the coupling process to be successful [27]. Another study by Kou and Gao grew the silica nanoparticles on the graphene oxide with the assistance of ammonia, but the resulting product had to be stored in alcohol [28]. In our study, our product was formed in a basic condition and was stored without using any solvent.

In our study, we grew the nanosilica (NS) on graphene oxide (GO) with a modified method to avoid the formation of a large silicate sandwich-like layer. The silver nanoparticles (AgNPs) were then loaded on to the nanosilica (NS) to form the composite (GO/NS/AgNPs). The structure and composition of GO/NS and GO/NS/AgNPs composites were characterized via P-XRD, TEM, SEM-EDS, UV–VIS, DLS, Raman spectroscopy, FTIR, and BET surface area measurements. The antibacterial ability of GO/NS and GO/NS/AgNPs against *Escherichia coli* (*E. coli*) and *Bacillus subtilis* (*B. subtilis*) was examined using the spread plate method and zone inhibition test.

2. Experimental

2.1. Synthesis of Nanosilica Grown on Graphene Oxide

Graphene oxide (GO) was synthesized by following the method reported by Bhawal et al. [30]. We modified the Liu et al. method to synthesize nanosilica (NS) on GO [31]. A total of 50 mL of GO solution was prepared at a 2:1 ratio of GO and deionized water (DI, 18 MΩ). An amount of 500 mg of CTAB (Acros Organics, Geel, Belgium, 99%) and 20 mg of sodium hydroxide (Sigma Aldrich, St. Louis, MO, USA, 97%) were mixed together with the GO solution through sonication. Then, 0.5 mL of tetraethyl orthosilicate (TEOS) (Tokyo Chemical Industry, Tokyo, Japan, 96%) was added to the above mixture and heated in a shaking water bath overnight at 40 °C. The solution was refluxed in 50 mL of hydrogen chloride and ethanol (1% v/v) for 3 h to remove the CTAB template. The collected GO/NS was washed with ethanol and deionized water (DI, 18 MΩ).

2.2. Synthesis of Silver Nanoparticles and Silver Nanoparticles Supported over Nanosilica/Graphene Oxide Composites

A total of 50 mL of silver nitrate (Sigma-Aldrich, Burlington, Massachusetts, 99%) (0.001 M) was prepared and heated until boiling. Then, 5 mL of 1% w/w sodium citrate was added slowly to the boiling solution. After the solution turned a yellow color, the solution was cooled to room temperature (293 K) and then centrifuged at 10,000 rpm for 10 min to remove unreacted solutes. Then, 2 mL of colloidal silver nanoparticles (AgNPs) was added to 10 mg of GO/NS. The mixture was sonicated for 15 min and then dried at ambient temperature (293 K) to obtain silver nanoparticles supported over GO/NS (GO/NS/AgNPs).

2.3. Characterization

The silver nanoparticles were examined via ultraviolet–visible spectroscopy (UV–VIS, Shimadzu UV-2600). Approximately 3.5 mL of colloidal AgNPs was added to a cuvette. The sample was then scanned from 600 nm to 300 nm.

The composite's crystal structures were characterized by using powdered X-ray diffraction (P-XRD, Rigaku Miniflex II, Cu Kα X-ray, nickel filters). Each sample was spread flat on a sample holder. The P-XRD of each material was scanned from 5° to 90°.

The functional groups of GO, GO/NS (before and after removing CTAB), and GO/NS/AgNPs were confirmed through Fourier-transform infrared spectroscopy (FTIR, Shimadzu

IR-Tracer 100). A small amount of the sample was placed on the ATR attachment of FTIR, and then each sample was scanned from 4000 cm^{-1} to 500 cm^{-1}. The chemical structures of GO/NS and GO/NS/AgNPs were also verified by using a Raman microscope and spectrometer (Renishaw, ISC3-1233). The samples were spread flat on the sample slides. The Raman microscope was used to locate the samples as well as adjusting the resolution. After that, each sample was measured from 3000 cm^{-1} to 150 cm^{-1}.

In TEM analysis, each sample was dispersed and sonicated in 5 mL of DI water. A few drops of the solution were added to a Cu grid (300 mesh). The grid was dried in the oven at 80 °C overnight. Then, each grid was scanned via transmission electron microscopy (TEM, JEM-2100F).

The size and dispersity of graphene, silicate nanoparticles, and silver nanoparticles were analyzed via dynamic light scattering (DLS, Brookhaven Instrument, Nanobrook 90Plus). Before analyzing via DLS, the sample was prepared by mixing it with DI water (1 mg/mL) and sonicating it for an hour to fully dissolve the sample. Then, the prepared sample was filtered to avoid undissolved particles. The solutions were then added to the quartz curvette and then scanned through DLS.

The surface structure image of GO/NS/AgNPs was obtained via scanning electron microscopy (SEM, JEOL JSM-6060LV). The weight ratios among elements and elemental mapping images were determined through energy-dispersive X-ray spectroscopy (EDS, Thermo Scientific UltraDry). The nitrogen adsorption–desorption isotherm, BET surface area, and pore size of each composite were measured by using an Accelerated Surface Area and Porosimetry System (Micromeritics -ASAP 2020) at 77.4 K. The composites were degassed for approximately 12 h before being analyzed with the ASAP system.

2.4. Antibacterial Activity Assay

The method by Rajapaksha et al. was applied to test the antibacterial ability of GO/NS and GO/NS/AgNPs against *E. coli* and *B. subtilis* [32]. All tools and glassware were sterilized by autoclaving at 120 °C. The sterile LB broth medium was inoculated with either E. coli or B. subtilis overnight at 37 °C in a shaking incubator. A total of 10 mg of the composite was mixed with 9 mL of 1x PBS buffer. The control only contained 9 mL of 1x PBS buffer. An amount of 1 mL of bacterial solution was transferred to both the control tube and the tubes that contained composites. A serial dilution method was conducted by transferring 1 mL of stock solution from the control tubes or composite tubes to tubes containing 9 mL of PBS solution until it reached the fifth diluent. Then, 0.1 mL of the solution in the diluent was transferred and spread on agar plates which were incubated overnight at 37 °C. The colony forming unit (CFU) was determined and calculated via CFU scope v1.6 software. The images of colony forming unit (CFU) plates for the control and each material are shown in the supplementary material Figures S1 and S2. The test was repeated twice. The data were statistically analyzed by ANOVA.

The zone inhibition test against *E. coli* and *B. subtilis* followed the Kirby–Bauer disk diffusion susceptibility test protocol. An average of 5.5 mg of each sample (GO/NS, GO/NS/AgNPs) was loaded on to the sterile disk. The control disk only received sterile DI water. The disks were placed on to the agar plates that were inoculated with bacteria and streaked. The plates were incubated overnight at 37 °C. The zone of inhibition of each composite disk was measured using a caliper and compared to the control disk. The images of the disk diffusion experiment for the control and each material are depicted in the Supplementary Material Figure S3.

3. Result and Discussion

Figure 1 shows the P-XRD spectra of GO/MS and GO/NS/AgNPs. The peaks at 10° and 20° are indicative of the (001) plane of GO [33]. The nanosilica (NS) had peaks at 10° and 21° which corresponded with the (001) and (002) planes (JCPDS 29-0085). The planes were consistent with other literature [33,34]. After coupling with the AgNPs, the characteristic peaks of the AgNPs can be found at 38°, 44°, 64°, and 77°. These peaks corresponded with the (111), (200), (220), and (311) lattice planes of AgNPs (JCPDS 65-2871). The results were consistent with previously reported studies [35,36]. The UV–VIS spectrum in Figure 2 shows absorbance at 437 nm, which confirms the presence of AgNPs [35].

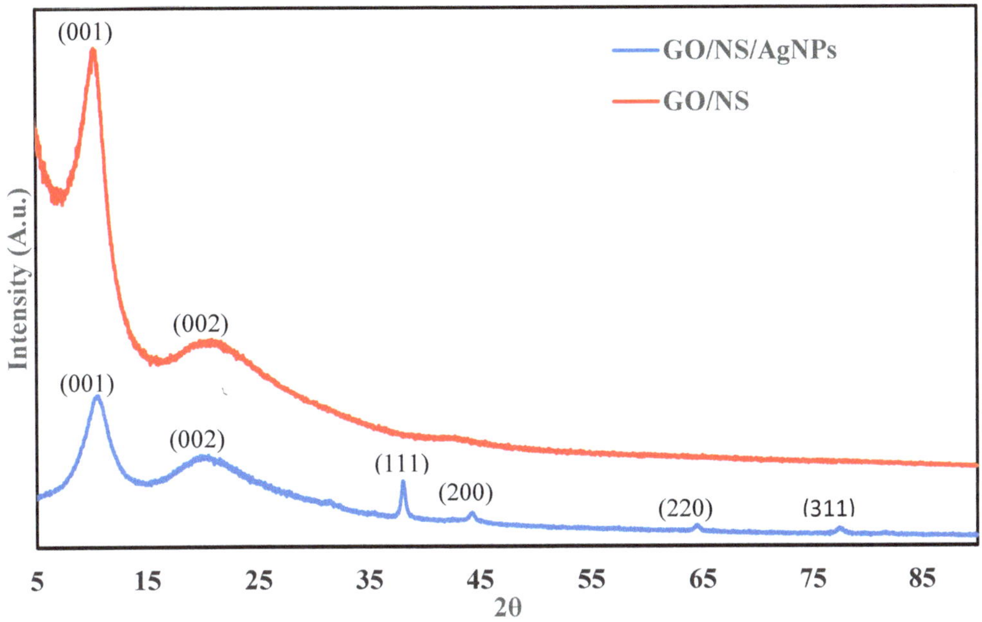

Figure 1. P-XRD of GO/NS/AgNPs and GO/NS.

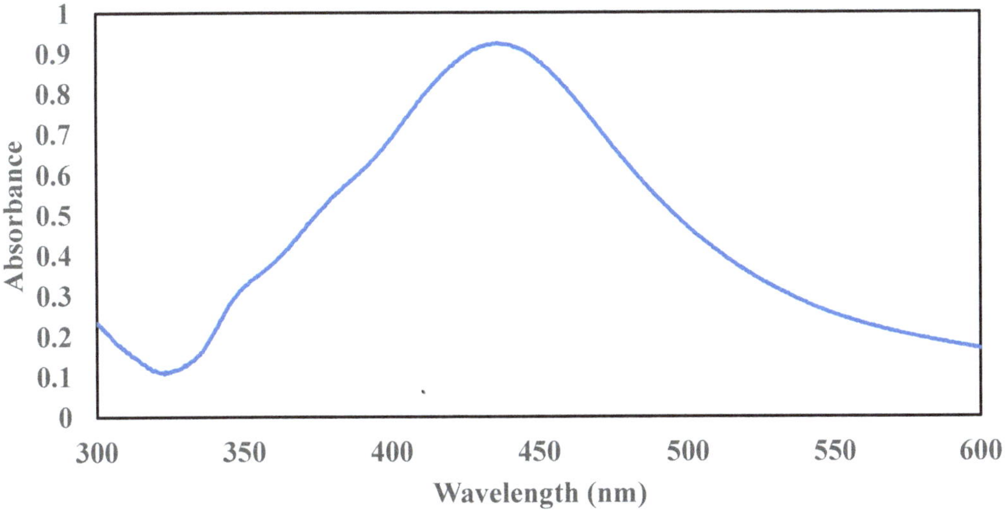

Figure 2. UV–VIS spectrum of AgNPs.

Figure 3 depicts the functional groups of GO, GO/NS, and GO/NS/AgNPs. GO showed broad peaks at 3410 cm^{-1} which corresponds with the O-H stretching group. The peaks at 1627 cm^{-1} and 1049 cm^{-1} indicate the presence of C=C and C-O stretching groups. These results were consistent with the characterization of GO reported in previous literature [37,38]. The peak at 1049 cm^{-1} of GO/NS and GO/NS/AgNPs was sharper due to the presence of Si-O-Si. This phenomenon was reported in a Nodeh et al. study [39]. Before removing the CTAB template, the material showed bands at 2916 cm^{-1} and 2846 cm^{-1} which were contributed to by C-H stretching vibration of the methyl and methylene group of CTAB [40]. The small band at 1473 cm^{-1} was attributed to the C-H bending vibration of CTAB [40]. After removing the CTAB, those three bands (2916 cm^{-1}, 2846 cm^{-1}, and 1473 cm^{-1}) disappeared. The result indicated that the reflux method completely eliminated the CTAB.

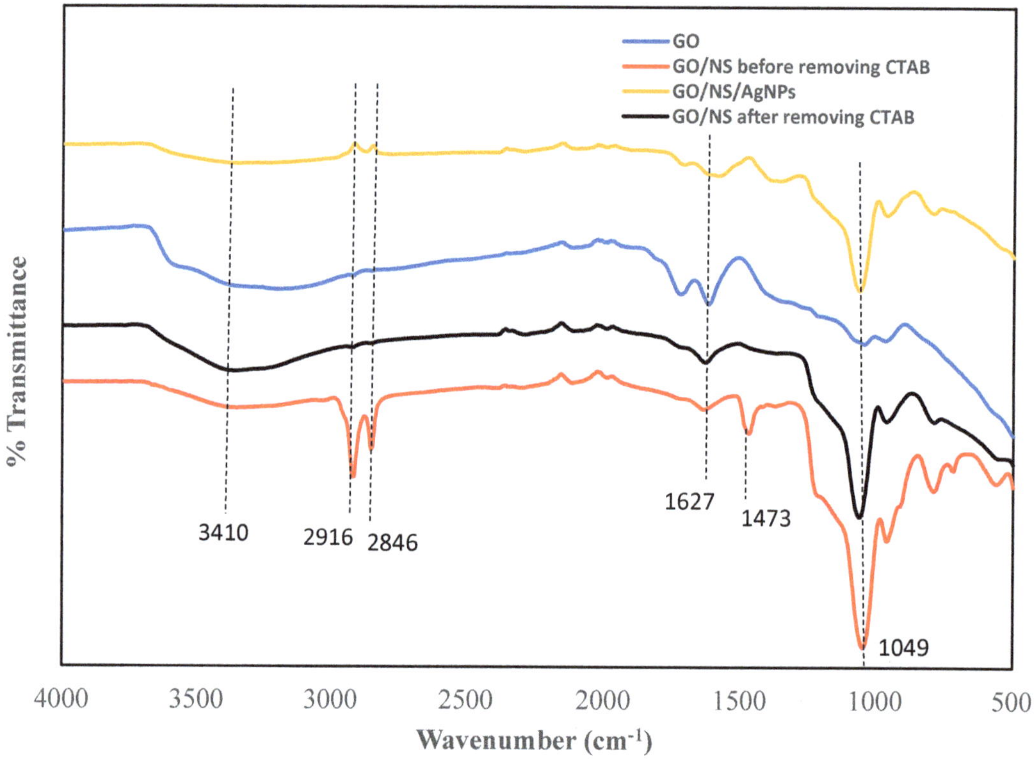

Figure 3. FTIR of GO, GO/NS before and after removing CTAB, and GO/NS/AgNPs.

The chemical structures of GO/NS and GO/NS/AgNPs were further analyzed using Raman spectroscopy. Figure 4 shows the Raman spectra of GO/NS and GO/NS/AgNPs. The G band and the D band of GO were observed at 1607 cm^{-1} and 1360 cm^{-1}. In a study of Perumbilavil et al., the G band of GO ranged from 1607 cm^{-1} to 1595 cm^{-1}, while the D band of GO ranged from 1365 cm^{-1} to 1355 cm^{-1} [41]. The G band was contributed to by the C-C stretching. The D bands formed and sharpened after the graphite was oxidized and caused a reduction in the sp^2 carbon domain [34]. Two bands at 2943 cm^{-1} and 2716 cm^{-1} were attributed to the silicate nanoparticles. In the research of Carboni et al., these mesoporous silicates exhibited bands at 2945 cm^{-1} and 2706 cm^{-1} [42]. The bands of silver nanoparticles were tiny due to their low concentration within the composite. The peaks at 1147 cm^{-1} and 815 cm^{-1} corresponded to the C-H bending of AgNPs' capping agent. The Raman spectrum in a study by Kora et al. showed the C-H groups in their synthesized nanoparticles appeared in the range of 1165 cm^{-1}–803 cm^{-1} [43].

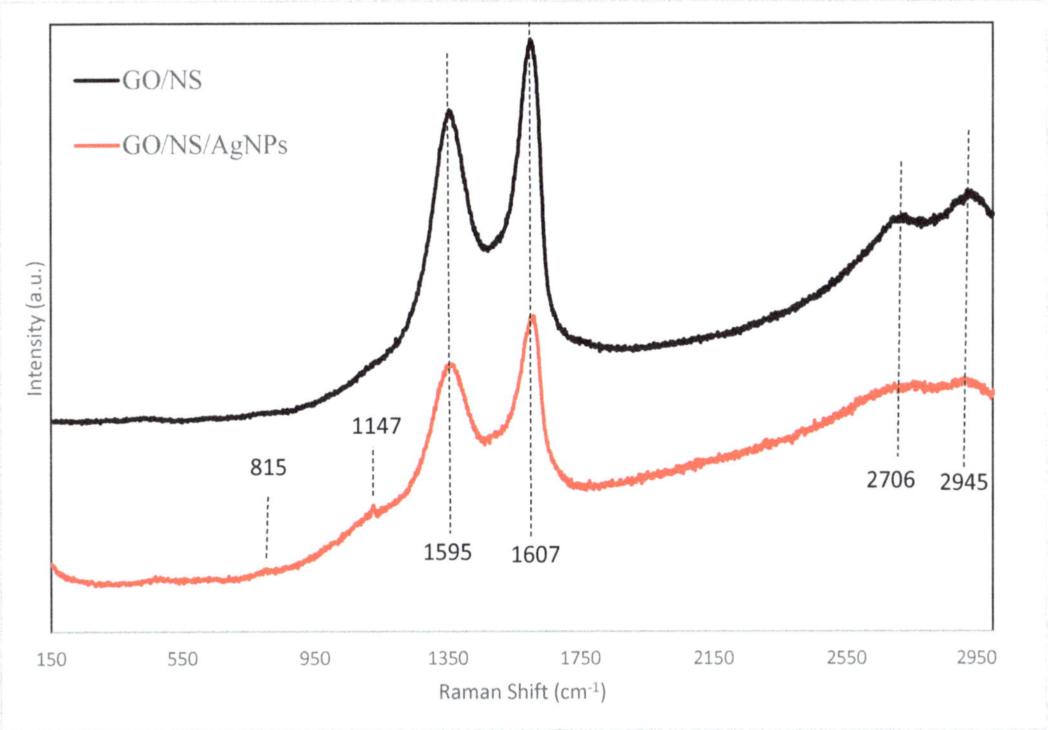

Figure 4. Raman spectra of GO/NS and GO/NS/AgNPs.

The SEM image (Figure 5a) depicts the micrograph of GO/NS/AgNPs. The result of EDS (Figure 5b) shows that the composite contained approximately 3% (wt %) of AgNPs. The ratios of AgNPs to GO and NS were 1:5 and 1:10, respectively. The elemental mapping images across the surface of GO/NS/AgNPs are represented in Figure 6, which shows that both Si (silica nanoparticles) (Figure 6c) and Ag (silver nanoparticles) (Figure 6d) were homogeneously distributed. Figure 6b,c depict similar distribution patterns which were due to the Si-O-Si bonding of the silicate nanoparticles.

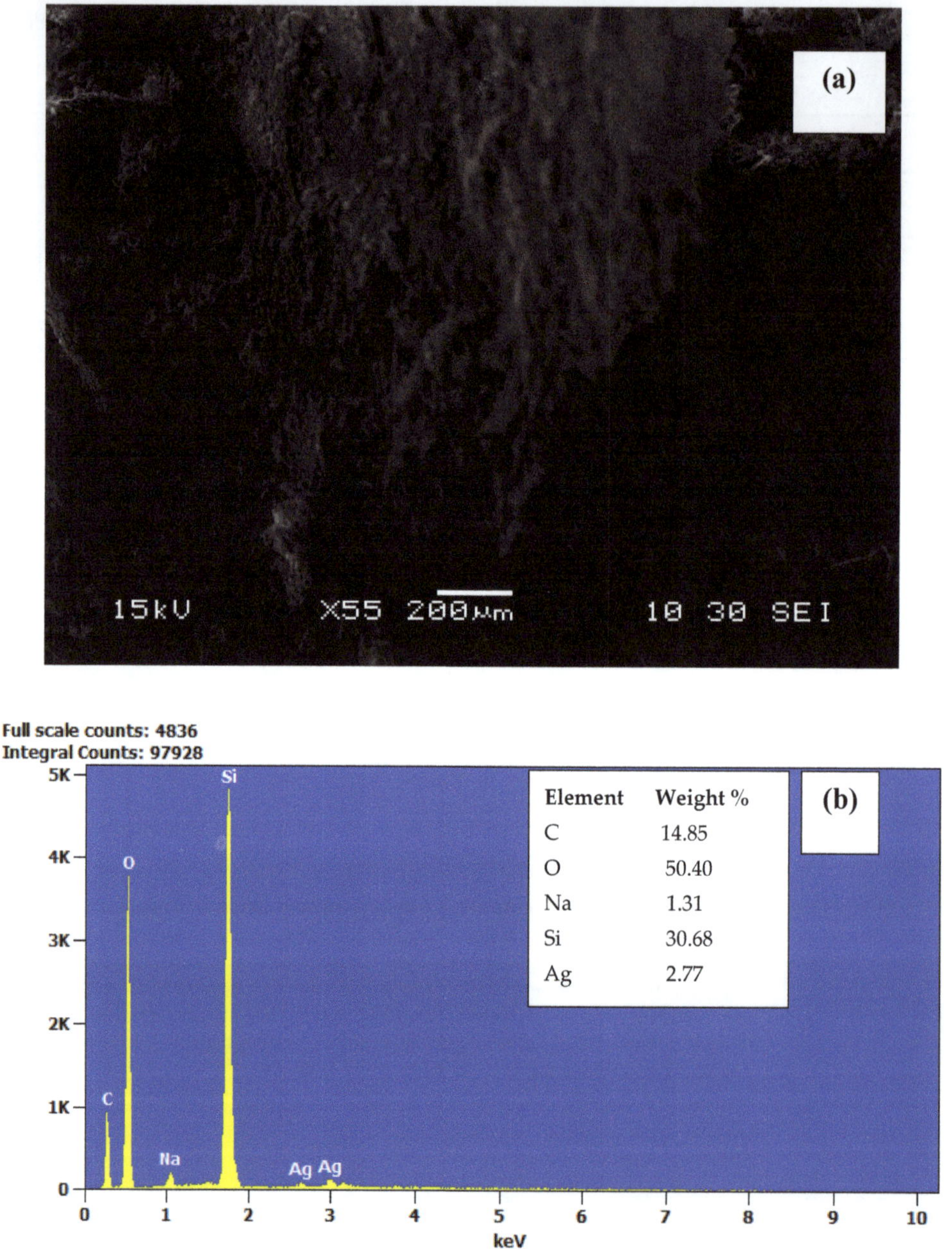

Figure 5. SEM image of (**a**) GO/NS/AgNPs (scale bar of 200 µm) and (**b**) EDS of GO/NS/AgNPs.

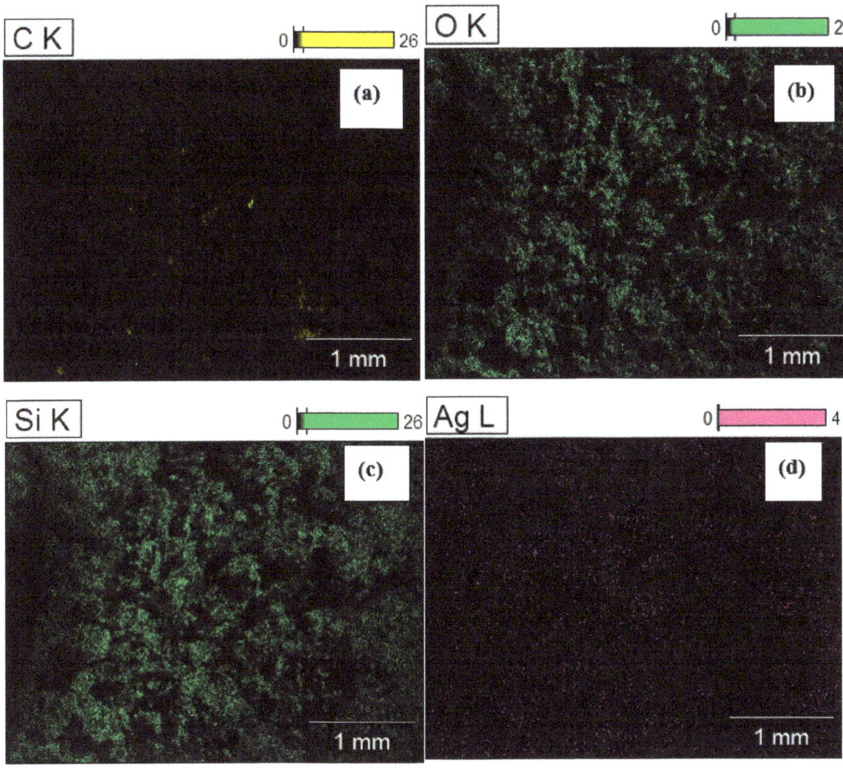

Figure 6. Elemental mapping of carbon (**a**), oxygen (**b**), silica (**c**), and silver (**d**) of GO/NS/AgNPs.

Figure 7a,b depict the TEM of silicate nanoparticles grown on graphene oxide. Figure 7c,d show the TEM images of AgNPs dispersed on the NS at scales of 20 nm and 10 nm, respectively. The images indicate that AgNPs were well-dispersed and exhibited no aggregation. It appeared that the materials highly supported and evenly distributed the AgNPs. The particle size distribution of GO/NS and GO/NS/AgNPs is represented in Figure 8a,b. Figure 8a consisted of two peaks which represented the silicate nanoparticles and graphene oxide. Silicate nanoparticles and graphene oxide had average sizes of 85 nm and 527 nm. In Figure 8b, there were three peaks that were contributed to by silver nanoparticles, silicate nanoparticles, and graphene oxide. The first peak indicated that silver nanoparticles had an average diameter of 26 nm. The second and third peaks represented the silicate nanoparticles and graphene oxide with sizes of 109 nm and 731 nm, respectively. Figure 8a,b indicated that the GO had a size range from 527 nm to 731 nm, while the size of silicate nanoparticles ranged from 85 nm to 109 nm. The polydispersity indexes of GO/NS and GO/NS/AgNPs were reported to be 0.218 and 0.294, respectively. Based on ISO standard classification (ISO 22412:2017), the particles of GO/NS and GO/NS/AgNPs were monodispersed.

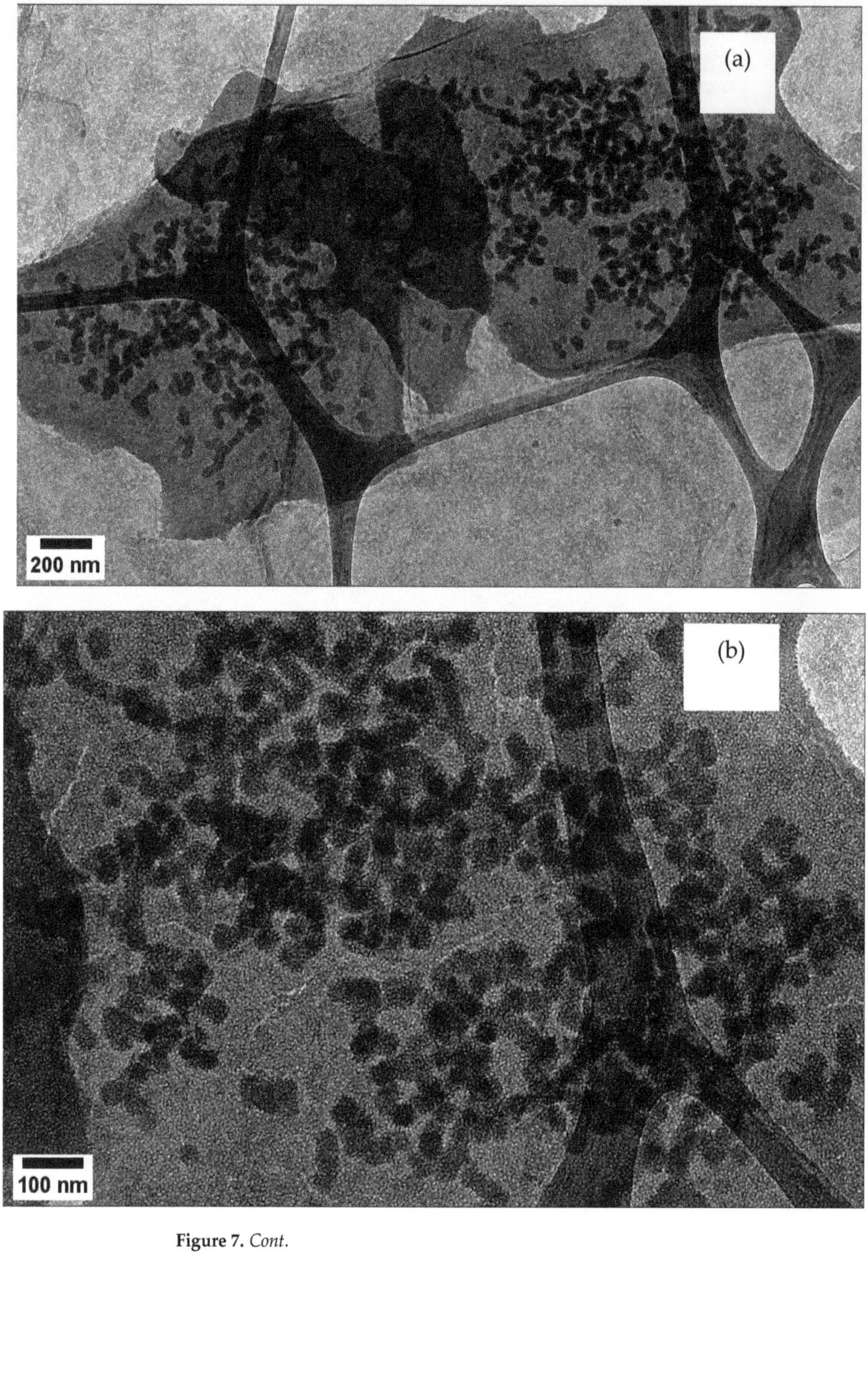

Figure 7. *Cont.*

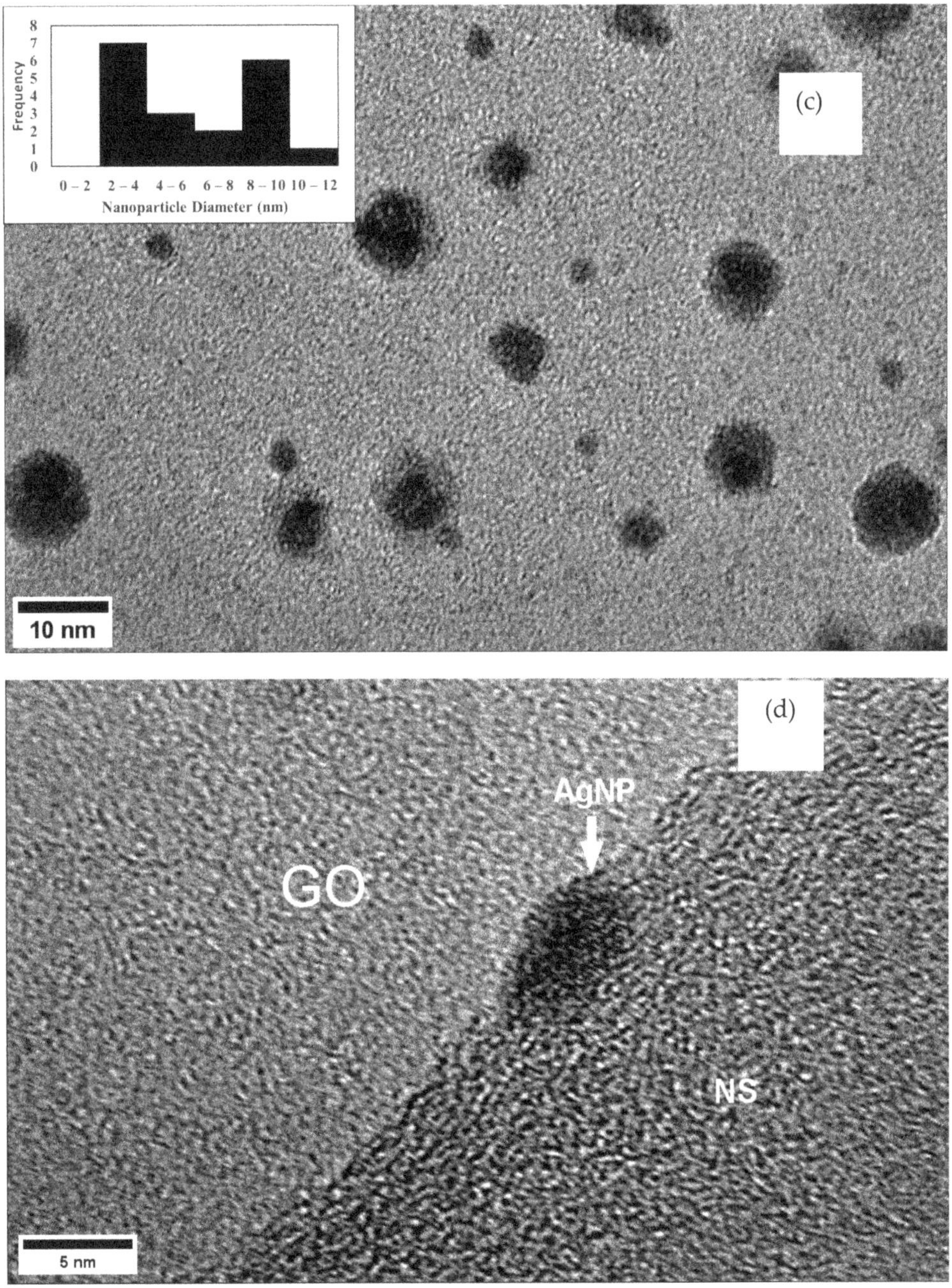

Figure 7. (**a**) TEM depicts NS grown on GO (scale bar of 200 nm); (**b**) TEM of NS grown on GO (scale bar of 100 nm); (**c**) TEM depicts the distribution of AgNPs on the nanosilica (scale bar of 10 nm); (**d**) TEM shows single AgNP on the nanosilica (scale bar of 5 nm).

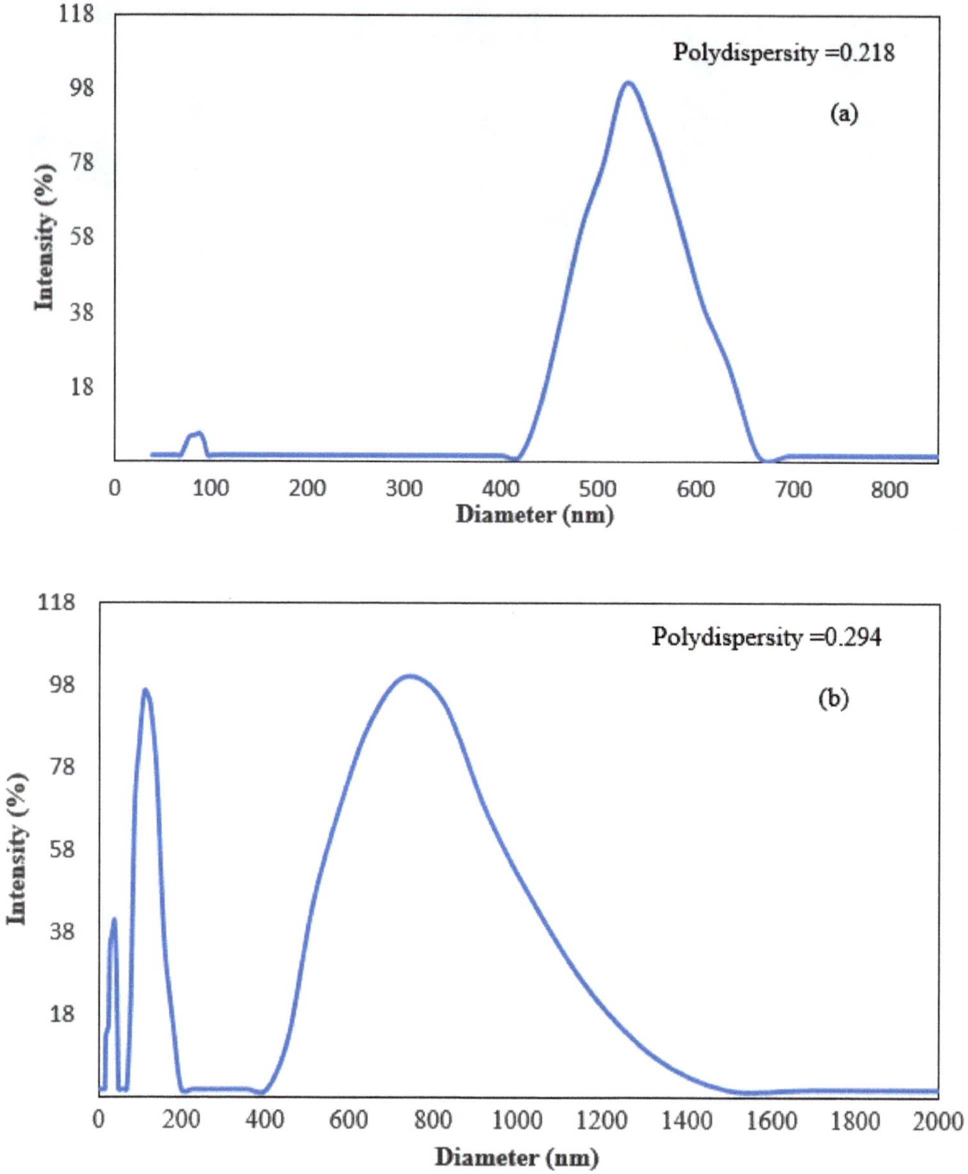

Figure 8. Dynamic light scattering (DLS) spectrums of (**a**) GO/NS and (**b**) GO/NS/AgNPs.

The accelerated surface area and porosimetry system was applied to measure the BET surface area and the porosity of GO/NS and GO/NS/AgNPs. The data are represented in Table 1. The BET surface area of GO/NS was higher than that of GO/NS/AgNPs. It was possible that the AgNPs blocked the pores of GO/NS and caused a reduction in pore volume and pore size. The data also showed that the BJH adsorption pore volume (0.11 cm^3/g) and pore size (7.77 nm) of GO/NS were higher than that of GO/NS/AgNPs. Based on the IUPAC classification, both GO/NS and GO/NS/AgNPs are mesoporous materials.

Table 1. Brunauer–Emmett–Teller (BET) surface areas, Barret–Joyner–Halenda (BJH) pore volumes, and BJH pore sizes of GO/NS and GO/NS/AgNPs.

Composite	BET Surface Area (m^2/g)	BJH Adsorption Pore Volume (cm^3/g)	BJH Adsorption Pore Size (nm)
GO/NS	62.09	0.11	7.77
GO/NS/AgNPs	25.83	0.04	7.08

Figure 9a shows the antibacterial effect over time of GO/NS and GO/NS/AgNPs against *E. coli*. During the first hours, both materials achieved approximately a 7-log reduction of *E. coli*. After two hours, both materials achieved an 8-log reduction. After three hours, GO/NS's antibacterial effect appeared to reduce and dropped back to 6-log, while GO/NS/AgNPs maintained the 8-log reduction. Based on the ANOVA analysis, both GO/NS and GO/NS/AgNPs were significantly different from the control ($F(2,11) = 9.10$, $p = 0.01$). These results indicate that GO/NS and GO/NS/AgNPs significantly reduced the concentration of *E. coli*. The average *E. coli* log reduction of GO/NS/AgNPs was higher than that of GO/NS after 3 h.

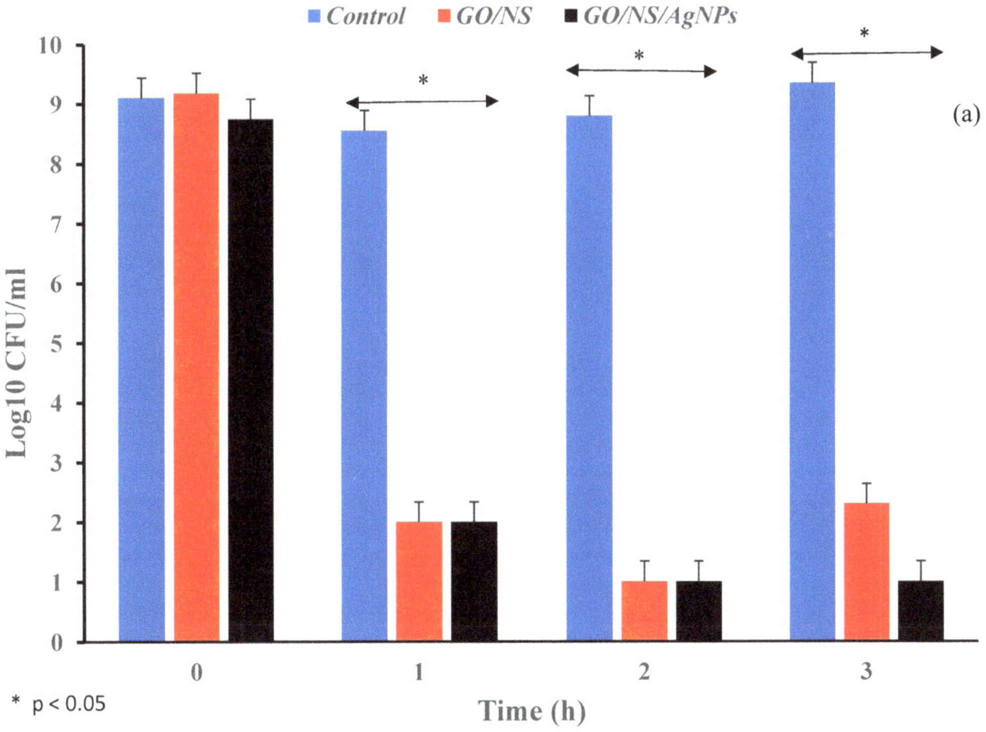

Figure 9. *Cont.*

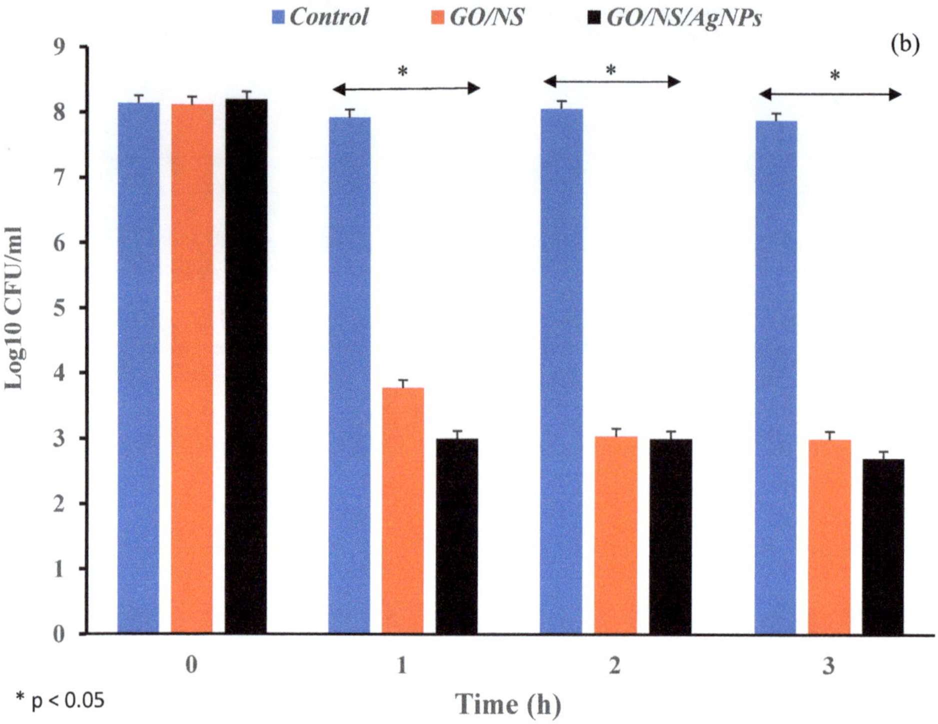

Figure 9. Bar graphs depicting the log reduction against (**a**) *E. coli* and (**b**) *B. subtilis* when exposed to GO/NS and GO/NS/AgNPs. The samples were compared with the control and analyzed at 0 h, 1 h, 2 h, and 3 h.

In Figure 9b, the log reduction of *B. subtilis* for GO/NS (4.3-log) was lower than that of GO/NS/AgNPs (5.2-log) after an hour. After 3 h, the log reduction of GO/NS/AgNPs against *B. subtilis* was higher than that of GO/NS. There was a significant difference in the *B. subtilis* concentration between the composites and control ($F(2,11) = 8.71$, $p = 0.02$).

When the effectiveness of both materials was compared against *E. coli* and *B. subtilis*, the results indicated that the inactivation of *B. subtilis* was significantly lower than the inactivation of *E. coli* ($F(5,23) = 8.34$, $p = 0.0006$). Bacteria tend to form spores to help them survive in harsh environmental conditions. It has been shown in previous studies that the spores of *B. subtilis* have a high resistance to radiation, heat, and chemicals [44]. The sturdiness of *B. subtilis* might suggest the difference in log reduction results between *B. subtilis* and *E. coli* when exposed to GO/NS and GO/NS/AgNPs. This will attract further future studies to explain the phenomenon. These results demonstrated the potential of GO/NS and GO/NS/AgNPs in sterilizing resistant bacterial strains.

From Figure 5, the results indicate that more than 99% of *E. coli* and *B. subtilis* populations were killed by GO/NSN and GO/NSN/AgNPs within one hour. Table 2 shows the bacterial log reduction difference between graphene oxide, graphene oxide composites, mesoporous silica, and silver composites. GO/NS and GO/NS/AgNPs both exhibited higher log reduction of Gram-negative and Gram-positive bacteria than other materials within 1 h. A similar result was reported in a study by Nguyen et al. where the material rGO-Ag achieved antibacterial effectiveness against Gram-negative and Gram-positive bacteria after 24 h. Their material took a longer time to achieve effectiveness, possibly due to the loss of functional groups on the surface of the graphene oxide [45]. In the Nguyen et al. study, it was observed that the graphene oxide lost the C=O and O-H carboxyl group after being reduced [45].

Table 2. Bacterial log reduction comparison of different materials from various studies.

Materials	Time Length	Log Reduction (Gram-Negative Bacteria)	Log Reduction (Gram-Positive Bacteria)	Reference
rGO-Ag [1]	24 h	5	5	[45]
rGO-nAg [1]	2–2.5 h	1.31	1	[46]
rGO [2]	2–2.5 h	0.4	0.4	[46]
nAg	2–2.5 h	0.4	0.4	[46]
AgNC-MSNs [3]	12 h	5.2	3.5	[47]
Ag/PVAGr [4]	1 h	5	~3	[48]
TA-GA [5]	4 h	Less than 0.5	4	[49]
GO [6]	3 h	0.5–1	NA	[50]
GO/CeO_2 [7]	3 h	6	NA	[50]
Silver coated PMMA [8] microsphere	24 h	4	4	[51]
GO/NS	1 h	~7	4.3	This work
GO/NS/AgNPs	1 h	~7	5.2	This work

[1] Reduced graphene oxide–silver nanoparticles, [2] reduced graphene oxide, [3] silver nanoclusters decorated mesoporous silica nanoparticles, [4] silver/polyvinyl alcohol/graphene, [5] tannic acid–graphene aerogel, [6] graphene oxide, [7] graphene oxide–cerium oxide nanoparticles, and [8] poly(methylmethacrylate).

Table 3 indicates the inhibition capabilities of GO/NS and GO/NS/AgNPs against *E. coli* and B. *Subtilis*. The inhibition effect of GO/NS/AgNPs was higher than GO/NS. Both composites showed a stronger inhibition effect against *B. Subtilis* than that against *E. coli*. It was demonstrated in the previous study that the silica nanoparticles significantly boosted the antibacterial property of silver nanoparticles [46]. These results highly support the capability of the composites in suppressing the growth of pathogens.

Table 3. Zone of inhibition diameter (mm) of GO/NS and GO/NS/AgNPs against *E. coli* and *B. subtilis*.

Species	Zone of Inhibiton Diameter (mm)	
	GO/NS	GO/NS/AgNPs
E. coli	70 ± 1.0	90 ± 1.0
B. subtilis	80 ± 1.0	100 ± 1.0

In a Bhargav et al. study, the inhibition zone diameters of antibiotics, including Cefixime, Cefotaxime, Gatifloxacin, and Levofloxacin, against Gram-negative bacteria were 20.76 mm, 25.04 mm, 25.10 mm, and 22.71 mm, respectively [52]. In another study, the amoxicillin exhibited inhibition zones of 35.50 mm and 39.40 mm against some Gram-positive bacteria such as *Streptococcus aureus* and *Bacillus subtilis* [53]. When comparing the results of this literature with Table 3, it showed a promising potential of GO/NS and GO/NS/AgNPs for various future biomedical applications including dental filling, wound treatment, and medical coating [54,55].

4. Conclusions

The novel GO/NS/AgNPs composite was successfully synthesized and compared with the GO/NS. The GO/NS/AgNPs contained 3% of silver nanoparticles attached to the nanosilica grown over graphene oxide. The composites in this study showed extraordinarily bacterial inactivation over time. Both GO/NS/AgNPs and GO/NS achieved more than 99% antibacterial efficiency against *E. coli* and *B. subtilis*. Through the zone of inhibition studies, it is highly suggested that GO/NS/AgNPs have a high potential to be applied as an effective antibacterial coating for medical equipment and other surfaces. The composite may become attractive for future biocompatibility studies to explore further applications in the medical field.

Supplementary Materials: The following supporting information can be downloaded at: https://www.mdpi.com/article/10.3390/nano12193341/s1, Figure S1: Examples of Colony Forming Unit (CFU) plate count images of (A) The control *E. coli* samples from 0 h to 3 h, (B) The *E. coli* +GO/NS from 0 h to 3 h, (C) The *E. coli* + GO/NS/AgNPs from 0 h to 3 h; Figure S2: Examples of Colony Forming Unit (CFU) plate count images of (A) The control *B. subtilis* samples from 0 h to 3 h, (B) The *B. Subtilis* + GO/NS from 0 h to 3 h, (C) The *B. subtilis* + GO/NS/AgNPs from 0 h to 3 h; Figure S3: Examples of Kirby-Bauer disk diffusion images of (A) GO/NS and GO/NS/AgNPs against *E. coli*, (B) GO/NS and GO/NS/AgNPs against *B. subtilis*.

Author Contributions: Q.Q.: data curation, formal analysis, writing—original draft. T.M.A.-F.: conceptualization, validation, formal analysis, investigation, resources, supervision, writing—review & editing. All authors have read and agreed to the published version of the manuscript.

Funding: This research received no external funding.

Institutional Review Board Statement: Not applicable.

Informed Consent Statement: Not applicable.

Data Availability Statement: The data presented in this study are available on request from the corresponding author.

Acknowledgments: Corresponding author acknowledges Lawrence J. Sacks professorship in chemistry.

Conflicts of Interest: The authors declare no conflict of interest.

References

1. Prestinaci, F.; Pezzotti, P.; Pantosti, A. Antimicrobial resistance: A global multifaceted phenomenon. *Pathog. Glob. Health* **2015**, *109*, 309–318. [CrossRef] [PubMed]
2. Concillo, S.; Sessa, L.; Petrone, A.M.; Porta, A.; Diana, R.; Diana, R.; Lannell, P.; Piotto, S. Structure Modification of an Active Azo-Compound as a Route to New Antimicrobial Compounds. *Molecules* **2017**, *22*, 875. [CrossRef] [PubMed]
3. Demir, B.; Broughton, R.M.; Qiao, M.; Huang, T.S.; Worley, S.D. N-Halamine Biocidal Materials with Superior Antimicrobial Efficacies for Wound Dressings. *Molecules* **2017**, *22*, 1582. [CrossRef]
4. Ardila, N.; Daigle, F.; Heuzey, M.C.; Ajji, A. Antibacterial Activity of Neat Chitosan Powder and Flakes. *Molecules* **2017**, *22*, 100. [CrossRef]
5. Lardani, L.; Derchi, G.; Marchio, V.; Carli, E. One-Year Clinical Performance of Activa™ Bioactive-Restorative Composite in Primary Molars. *Children* **2022**, *9*, 433. [CrossRef] [PubMed]
6. Franci, G.; Falanga, A.; Galdiero, S.; Palomba, L.; Rai, M.; Morelli, G.; Galdiero, M. Silver Nanoparticles as Potential Antibacterial Agents. *Molecules* **2015**, *20*, 8856–8874. [CrossRef] [PubMed]
7. Masri, A.; Anwar, A.; Ahmed, D.; Siddiqui, R.B.; Shah, M.S.; Khan, N.A. Silver Nanoparticle Conjugation-Enhanced Antibacterial Efficacy of Clinically Approved Drugs Cephradine and Vildagliptin. *Antibiotics* **2018**, *7*, 100. [CrossRef]
8. Kvitek, A.P.L.; Smekalova, M.; Vecerova, R.; Kolar, M.; Roderova, M.; Dycka, F.; Sebela, M.; Prucek, R.; Tomanec, O.; Zboril, R. Bacterial resistance to silver nanoparticles and how to overcome it. *Nature Nanotech.* **2018**, *13*, 65–71. [CrossRef]
9. McNeilly, O.; Mann, R.; Hamidian, M.; Gunawan, C. Emerging Concern for Silver Nanoparticle Resistance in Acinetobacter baumannii and Other Bacteria. *Front. Microbiol.* **2021**, *12*, 894. [CrossRef]
10. Tejamaya, M.; Romer, I.; Merrifield, R.C.; Lead, J.R. Stability of Citrate, PVP, and PEG Coated Silver Nanoparticles in Ecotoxicology Media. *Environ. Sci. Technol.* **2012**, *46*, 7011–7017. [CrossRef]
11. Dong, Y.; Wang, Q.; Wan, L.; Chi, Y. Carbon based dot capped silver nanoparticles for efficient surface-enhanced Raman scattering. *J. Mater. Chem. C.* **2016**, *4*, 7472–7477. [CrossRef]
12. Quach, Q.; Briehler, E.; Elzamzami, A.; Huff, C.; Long, J.M.; Abdel-Fattah, T.M. Catalytic Activity of Beta-Cyclodextrin-Gold Nanoparticles Network in Hydrogen Evolution Reaction. *Catalysts.* **2021**, *11*, 118. [CrossRef]
13. Huff, C.; Quach, Q.; Long, J.M.; Abdel-Fattah, T.M. Nanocomposite Catalyst Derived from Ultrafine Platinum Nanoparticles and Carbon Nanotubes for Hydrogen Generation. *ECS J. Solid State Sci. Technol.* **2020**, *9*, 101008. [CrossRef]

14. Biehler, E.; Quach, Q.; Huff, C.; Abdel-Fattah, T.M. Organo-Nanocups Assist the Formation of Ultra-Small Palladium Nanoparticle Catalysts for Hydrogen Evolution Reaction. *Materials.* **2022**, *15*, 2692. [CrossRef] [PubMed]
15. Abdel-Fattah, T.M.; Loftis, D.; Biomed, A.M.J. Nanosized Controlled Surface Pretreatment of Biometallic Alloy 316L Stainless Steel. *Nanotechnology* **2011**, *7*, 794–800. [CrossRef]
16. Mahapatro, A.; Negron, T.D.M.; Bonner, C.; Fattah, T.M.A. Nanolayers on Magnesium (Mg) Alloy for Metallic Bone Tissue Engineering Scaffolds. *J. Biomater. Tissue Eng.* **2013**, *3*, 196–204. [CrossRef]
17. Mahapatro, R.B.A.; Bonner, C.; Abdel-Fattah, T.M. In vitro stability study of organophosphonic self assembled monolayers (SAMs) on cobalt chromium (Co–Cr) alloy. *Mater. Sci. Eng. C* **2013**, *33*, 2050–2058. [CrossRef]
18. Kaya, S.; Cresswell, M.; Boccaccini, A.R. Mesoporous silica-based bioactive glasses for antibiotic-free antibacterial applications. *Mater. Sci. Eng. C* **2018**, *83*, 99–107. [CrossRef]
19. Rojas-Andrade, M.D.; Chata, G.; Rouholiman, D.; Liu, J.; Saltikov, C.; Chen, S. Antibacterial mechanisms of graphene-based composite nanomaterials. *Nanoscale* **2017**, *9*, 994–1006. [CrossRef]
20. Vallet-Regi, M.; Balas, F. Silica Materials for Medical Applications. *Open Biomed. Eng. J.* **2008**, *2*, 1–9. [CrossRef]
21. Campbell, E.; Hasan, M.T.; Pho, C.; Callaghan, K.; Akkaraju, G.R.; Naumov, A.V. Graphene Oxide as a Multifunctional Platform for Intracellular Delivery, Imaging, and Cancer Sensing. *Sci. Rep.* **2019**, *9*, 416. [CrossRef] [PubMed]
22. Liu, R.; Wang, X.; Ye, J.; Xue, X.; Zhang, F.; Zhang, H.; Hou, X.; Liu, X.; Zhang, Y. Enhanced antibacterial activity of silver-decorated sandwich-like mesoporous silica/reduced graphene oxide nanosheets through photothermal effect. *Nanotechnology* **2018**, *29*, 105704. [CrossRef]
23. Liu, S.; Zheng, T.H.; Hofmann, M.; Burcombe, E.; Wei, J.; Jiang, R.; Kong, J.; Chen, Y. Antibacterial Activity of Graphite, Graphite Oxide, Graphene Oxide, and Reduced Graphene Oxide: Membrane and Oxidative Stress. *ACS Nano.* **2011**, *5*, 6971–6980. [CrossRef]
24. Wang, Y.; Nor, Y.A.; Song, H.; Yang, Y.; Xu, C.; Yu, M.; Yu, C. Small-sized and large-pore dendritic mesoporous silica nanoparticles enhance antimicrobial enzyme delivery. J. *Mater. Chem. B* **2016**, *4*, 2646–2653. [CrossRef] [PubMed]
25. Kumar, P.; Huo, P.; Zhang, R.; Liu, B. Antibacterial Properties of Graphene-Based Nanomaterials. *Nanomaterials* **2019**, *9*, 737. [CrossRef] [PubMed]
26. Szunerits, S.; Boukherroub, R. Antibacterial activity of graphene-based materials. *J. Mater. Chem. B* **2016**, *4*, 6892–6912. [CrossRef] [PubMed]
27. Yang, K.; Chen, B.; Zhu, L. Graphene-coated materials using silica particles as a framework for highly efficient removal of aromatic pollutants in water. *Sci. Rep.* **2015**, *5*, 11641. [CrossRef]
28. Kou, L.; Gao, C. Making silicananoparticle-covered graphene oxide nanohybrids as general building blocks for large-area superhydrophilic coatings. *Nanoscale* **2011**, *3*, 519–528. [CrossRef]
29. Abdelkhalek, A.; El-Latif, M.A.; Ibrahim, H.; Hamad, H.; Showman, M. Controlled synthesis of graphene oxide/silica hybrid nanocomposites for removal of aromatic pollutants in water. *Sci. Rep.* **2022**, *12*, 7060. [CrossRef]
30. Bhawal, P.; Ganguly, S.; Chaki, T.K.; Das, N.C. Synthesis and characterization of graphene oxide filled ethylene methyl acrylate hybrid nanocomposites. *RSC Adv.* **2016**, *6*, 20781. [CrossRef]
31. Liu, X.; Wu, X.; Xing, Y.; Zhang, Y.; Zhang, X.; Pu, Q.; Wu, M.; Zhao, J.X. Reduced Graphene Oxide/Mesoporous Silica Nanocarriers for pH-Triggered Drug Release and Photothermal Therapy. *ACS Appl. Bio Mater.* **2020**, *3*, 2577–2587. [CrossRef]
32. Rajapaksha, P.; Cheeseman, S.; Hombsch, S.; Murdoch, B.J.; Gangadoo, S.; Blanch, E.W.; Truong, Y.; Cozzolino, D.; McConville, C.F.; Crawford, E.J.; et al. Antibacterial Properties of Graphene Oxide-Copper Oxide Nanoparticle Nanocomposites. *ACS Appl. Bio Mater.* **2019**, *2*, 5687–5696. [CrossRef] [PubMed]
33. Zhang, W.L.; Choi, H.J. Silica-Graphene Oxide Hybrid Composite Particles and Their Electroresponsive Characteristics. *Langmuir.* **2012**, *28*, 7055–7062. [CrossRef] [PubMed]
34. Mehmood, Y.; Khan, I.U.; Shahzad, Y.; Khalid, S.H.; Asghar, S.; Irfan, M.; Asif, M.; Khalid, I.; Yousaf, A.M.; Hussain, T. Facile synthesis of mesoporous silica nanoparticles using modified sol-gel method: Optimization and in vitro cytotoxicity studies. *Pak. J. Pharm. Sci.* **2019**, *32*, 1805–1812. [PubMed]
35. Meva, F.E.; Segnou, M.L.; Ebongue, C.O.; Ntoumba, A.A.; Kedi, P.B.E.; Deli, V.; Etoh, M.-A.; Mpondo, E.M. Spectroscopic synthetic optimizations monitoring of silver nanoparticles formation from Megaphrynium macrostachyum leaf extract. *Rev. Bras. Farmacogn.* **2016**, *26*, 640–646. [CrossRef]
36. Kim, D.; Jeong, S.; Moon, J. Synthesis of silver nanoparticles using the polyol process and the influence of precursor injection. *Nanotechnology* **2006**, *17*, 4019–4024. [CrossRef]
37. Ciplak, Z.; Yildiz, N.; Calimli, A. Investigation of Graphene/Ag Nanocomposites Synthesis Parameters for Two Different Synthesis Methods. *Fuller. Nanotub. Carbon Nanostructures* **2015**, *23*, 361–370. [CrossRef]
38. He, D.; Peng, Z.; Gong, W.; Luo, Y.; Zhao, P.; Kong, L. Mechanism of a green graphene oxide reduction with reusable potassium carbonate. *RSC Adv.* **2015**, *5*, 11966–11972. [CrossRef]
39. Nodeh, H.R.; Ibrahim, W.A.W.; Ali, I.; Sanagi, M.M. Development of magnetic graphene oxide adsorbent for the removal and preconcentration of As(III) and As(V) species from environmental water samples. *Environ. Sci. Pollut. Res.* **2016**, *23*, 9759–9773. [CrossRef]
40. Su, G.; Yang, C.; Zhu, J.J. Fabrication of Gold Nanorods with Tunable Longitudinal Surface Plasmon Resonance Peaks by Reductive Dopamine. *Langmuir* **2015**, *31*, 817–823. [CrossRef]

41. Perumbilavil, S.; Sankar, P.; Rose, T.P.; Philip, R. White light Z-scan measurements of ultrafast optical nonlinearity in reduced graphene oxide nanosheets in the 400–700 nm region. *Appl. Phys. Lett.* **2015**, *107*, 051104. [CrossRef]
42. Carboni, D.; Lasio, B.; Alzari, V.; Mariani, A.; Loche, D.; Casula, M.F.; Malfatti, L.; Innocenzi, P. Graphene-mediated surface enhanced Raman scattering in silica mesoporous nanocomposite films. *Phys. Chem. Chem. Phys.* **2014**, *16*, 25809–25818. [CrossRef] [PubMed]
43. Kora, A.J.; Arunachalam, J. Green Fabrication of Silver Nanoparticles by Gum Tragacanth (Astragalus gummifer): A Dual Functional Reductant and Stabilizer. *J. Nanomater.* **2012**, *2012*, 869765. [CrossRef]
44. Setlow, P. Spores of Bacillus subtilis: Their resistance to and killing by radiation, heat and chemicals. *J. Appl. Microbiol.* **2006**, *101*, 514–525. [CrossRef] [PubMed]
45. Nguyen, V.H.; Kim, B.K.; Jo, Y.L.; Shim, J.J. Preparation and antibacterial activity of silver nanoparticles-decorated graphene composites. *J. Supercrit. Fluids.* **2012**, *72*, 28–35. [CrossRef]
46. Prasad, K.; Lekshmi, G.S.; Ostrikov, K.; Lussini, V.; Blinco, J.; Mohandas, M.; Vasilev, K.; Bottle, S.; Bazaka, K.; Ostrikov, K. Synergic bactericidal effects of reduced graphene oxide and silver nanoparticles against Gram-positive and Gram-negative bacteria. *Sci. Rep.* **2017**, *7*, 1591. [CrossRef]
47. Liu, J.; Li, S.; Fang, Y.; Zhu, Z. Boosting antibacterial activity with mesoporous silica nanoparticles supported silver nanoclusters. *J. Colloid. Interface Sci.* **2019**, *555*, 470–479. [CrossRef]
48. Abdubabbus, M.M.; Jevremovic, I.; Jankovic, A.; Peric-Grujjic, A.; Matic, I.; Vukasinovic-Sekulic, M.; Hui, D.; Rhee, K.Y.; Miskovic-Stankovic, V. Biological activity of electrochemically synthesized silver doped polyvinyl alcohol/graphene composite hydrogel discs for biomedical applications. *Compos. B Eng.* **2016**, *104*, 26–34. [CrossRef]
49. Luo, J.; Lai, J.; Zhang, N.; Liu, Y.; Liu, R.; Liu, X. Tannic Acid Induced Self-Assembly of Three-Dimensional Graphene with Good Adsorption and Antibacterial Properties. *ACS Sustain. Chem. Eng.* **2016**, *4*, 1404–1413. [CrossRef]
50. Montes-Duarte, G.G.; Tostado-Blazquez, G.; Castro, K.L.S.; Araujo, J.R.; Achete, C.A.; Sanchez-Salas, J.L.; Campos-Delgado, J. Key parameters to enhance the antibacterial effect of graphene oxide in solution. *RSC Adv.* **2021**, *11*, 6509–6516. [CrossRef]
51. Dutta, D.; Goswami, S.; Dubey, R.; Dwivedi, S.K.; Puzari, A. Antimicrobial activity of silver-coated hollow poly(methylmethacrylate) microspheres for water decontamination. *Environ. Sci. Eur.* **2021**, *33*, 22. [CrossRef]
52. Bhargav, H.S.; Shastri, S.D.; Poornav, S.P.; Darshan, K.M.; Nayak, M.M. Measurement of the Zone of Inhibition of an Antibiotic. *IEEE* **2016**, *5*, 409–414. [CrossRef]
53. Chamidah, A.; Hardoko, H.; Prihanto, A.A. Antibacterial activities of β-glucan (laminaran) against gram-negative and gram-positive bacteria. *AIP Conf. Proc.* **2017**, *1884*, 020011. [CrossRef]
54. Arkowski, J.; Obremska, M.; Kedzierski, K.; Slawuta, A.; Wawrzynska, M. Applications for graphene and its derivatives in medical devices: Current knowledge and future applications. *Adv. Clin. Exp. Med.* **2020**, *12*, 1497–1504. [CrossRef]
55. Pagano, S.; Costanzi, G.L.E.; Balloni, S.; Bruscoli, S.; Flamini, S.; Coniglio, M.; Valenti, C.; Cianetti, S.; Marinucci, L. Morpho-functional effects of different universal dental adhesives on human gingival fibroblasts: An in vitro study. *Odontology* **2021**, *109*, 524–539. [CrossRef]

nanomaterials

MDPI

Article

Mechanical Properties and Antibacterial Effect on Mono-Strain of *Streptococcus mutans* of Orthodontic Cements Reinforced with Chlorhexidine-Modified Nanotubes

Elias Nahum Salmerón-Valdés [1,*], Ana Cecilia Cruz-Mondragón [1], Víctor Hugo Toral-Rizo [1], Leticia Verónica Jiménez-Rojas [2], Rodrigo Correa-Prado [3], Edith Lara-Carrillo [1], Adriana Alejandra Morales-Valenzuela [1], Rogelio José Scougall-Vilchis [1], Alejandra Itzel López-Flores [1], Lia Hoz-Rodriguez [4] and Ulises Velásquez-Enríquez [1]

1 Center for Research and Advanced Studies in Dentistry, Faculty of Dentistry, School of Dentistry, Autonomous University of Mexico State, Toluca 50130, Mexico
2 Infectious Diseases Research Unit of the Mexico Children's Hospital Federico Gómez, Mexico City 06720, Mexico
3 Center for Applied Physics and Advanced Technology, National Autonomous University of Mexico, A.P. 1-1010, Queretaro 76000, Mexico
4 Periodontal Biology Laboratory, School of Dentistry, National Autonomous University of Mexico, Mexico City 04510, Mexico
* Correspondence: salmeron81@hotmail.es; Tel.: +52-7224190463

Citation: Salmerón-Valdés, E.N.; Cruz-Mondragón, A.C.; Toral-Rizo, V.H.; Jiménez-Rojas, L.V.; Correa-Prado, R.; Lara-Carrillo, E.; Morales-Valenzuela, A.A.; Scougall-Vilchis, R.J.; López-Flores, A.I.; Hoz-Rodriguez, L.; et al. Mechanical Properties and Antibacterial Effect on Mono-Strain of *Streptococcus mutans* of Orthodontic Cements Reinforced with Chlorhexidine-Modified Nanotubes. *Nanomaterials* **2022**, *12*, 2891. https://doi.org/10.3390/nano12172891

Academic Editors: Nebojša Pantelić and Goran Kaluđerović

Received: 5 August 2022
Accepted: 13 August 2022
Published: 23 August 2022

Abstract: Recently, several studies have introduced nanotechnology into the area of dental materials with the aim of improving their properties. The objective of this study is to determine the antibacterial and mechanical properties of type I glass ionomers reinforced with halloysite nanotubes modified with 2% chlorhexidine at concentrations of 5% and 10% relative to the total weight of the powder used to construct each sample. Regarding antibacterial effect, 200 samples were established and distributed into four experimental groups and six control groups (4 +ve and 2 −ve), with 20 samples each. The mechanical properties were evaluated in 270 samples, assessing microhardness (30 samples), compressive strength (120 samples), and setting time (120 samples). The groups were characterized by scanning electron microscopy and Fourier transform infrared spectroscopy, and the antibacterial activity of the ionomers was evaluated on *Streptococcus mutans* for 24 h. The control and positive control groups showed no antibacterial effect, while the experimental group with 5% concentration showed a zone of growth inhibition between 11.35 mm and 11.45 mm, and the group with 10% concentration showed a zone of growth inhibition between 12.50 mm and 13.20 mm. Statistical differences were observed between the experimental groups with 5% and 10% nanotubes. Regarding the mechanical properties, microhardness, and setting time, no statistical difference was found when compared with control groups, while compressive strength showed higher significant values, with ionomers modified with 10% concentration of nanotubes resulting in better compressive strength values. The incorporation of nanotubes at concentrations of 5% and 10% effectively inhibited the presence of *S. mutans*, particularly when the dose–response relationship was taken into account, with the advantage of maintaining and improving their mechanical properties.

Keywords: glass ionomer cements; chlorhexidine; nanotubes; microhardness; compressive strength

1. Introduction

Conventional glass ionomer cements (GICs) are a restorative material and the first choice of cement in dentistry. Specifically, type I ionomers are optimal for the adhesion of orthodontic restorations or bands used in conventional orthopedic treatments. Due to their excellent properties including biocompatibility, a desirable thermal expansion coefficient, and good adherence to enamel and dentin, type I ionomers provide extraordinary clinical benefits. However, the accumulation of dentobacterial plaque around orthodontic bands, along with microfiltration, facilitates the passage of oral fluids and bacteria to the dental tissue, frequently generating lesions by demineralization or white spot lesions, caries, and periodontal disorders, causing a risk for patients receiving orthodontic treatment. Recently, hybrid materials have been developed to improve their properties, including resin-modified glass ionomers (RMGIs), which have been analyzed in various studies and are mainly characterized by anticariogenic activity, an ability to remineralize dentin, and resistance to fracture [1–7]. However, to date, it has not been possible to increase the antibacterial capacity of these materials, to reduce the number of lesions by demineralization or the recurrence of caries [6–11].

Research has been conducted to evaluate the antibacterial capacity of type I glass ionomers with respect to cariogenic microorganisms [8]. Bacteria that have shown greater proliferation in patients with fixed appliances, mainly orthodontic bands, are principally *Streptococcus mutans (S. mutans)* and *Porphyromonas gingivalis*. Specifically, *S. mutans* is the microorganism most frequently involved in the development of carious lesions [3,7].

Recently, nanotechnology has been introduced in the area of dental materials with the aim of improving various properties of these materials. Various antimicrobial components, such as triclosan, fluoride, chlorhexidine (CX), and xylitol, have been incorporated into different dental materials to improve their antibacterial activity [12–14]. However, previous studies have been reported that direct incorporation of these antimicrobial components altered the mechanical properties of dental materials [15]. Halloysite nanotubes (HNs) are clay nanostructures with high levels of mechanical strength, thermal stability, and biocompatibility. Their main advantage over other nanocarriers is their low cost. Due to their internal tubular structure, they can be loaded with different drugs for slow release through nanopores located at their ends, prolonging the time of action [13,16–19]. Some studies have indicated that drugs released from HNs can last 30–100 times longer than the drug alone [17].

HNs can improve the beneficial properties of dental materials [14,20], without altering their mechanical properties. The present study performed different tests to characterize and evaluate the antibacterial effect on *Streptococcus mutans*, and the mechanical properties (microhardness, compressive strength, and setting time) of conventional and hybrid type I glass ionomers modified with and without HNs loaded with CX. The first hypothesis of the study was that the incorporation of HNs with CX into glass ionomer would confer antibacterial effects on these materials. The second hypothesis was that a higher quantity of HNs incorporated into glass ionomers would increase the antibacterial effects conferred on these materials. The third hypothesis was that the incorporation of HNs into ionomers would not negatively affect their mechanical properties. The null hypothesis would be accepted if an absence of an antibacterial effect were observed in experimental groups, or if the dose-response of modified HNs in glass ionomers showed no changes in antibacterial effect, or if the mechanical properties of experimental groups were altered negatively compared with the control groups.

2. Materials and Methods

2.1. Modification of Halloysite Nanotubes with Chlorhexidine

One gram of HNs (Sigma–Aldrich, St. Louis, MO, USA) that had been previously dried in a HERAtherm drying oven (Thermo Fisher Scientific, Waltham, MA, USA) was weighed using an analytical balance (Shimadzu Scientific Instruments, Kyoto, Japan). A solution of 3-(trimethoxysilyl) propyl-methacrylate-98% (Sigma–Aldrich, St. Louis, MO, USA) diluted to 5% and 95% acetone (Sigma–Aldrich, St. Louis, MO, USA) was used for the immersion of the nanotubes for 24 h at 110 °C in a drying oven.

Subsequently, 1 g of silanized nanotubes was mixed with 10 mL of 2% CX (Consepsis, Ultradent Products, South Jordan, UT, USA), commonly used for disinfection in dentistry [21], and 10 mL of 95% pure ethanol and sonicated for 1 h. The CX-loaded nanotubes were then placed in a drying oven for 10 days at 30 °C to eliminate residual solvent [17].

2.2. *Incorporation of Modified Nanotubes into Glass Ionomers*

In this study, a conventional glass ionomer KC (Ketac Cem, 3M ESPE, Minnesota, USA) and a resin-modified glass ionomer FO (Fuji Ortho, GC CORPORATION, Tokyo, Japan) were used. Two hundred blocks, with a diameter of 3 mm and thickness of 1 mm, were fabricated in a Teflon matrix. The materials were handled according to the manufacturer's instructions. The materials were activated with light and polymerized by an LED device (Elipar, 3M ESPE, Saint Paul, MN, USA) for 40 s; a Demetron LED radiometer (Kavo Kerr, Charlotte, USA) was used to verify that the minimum intensity of light emitted was 400 mW/cm^2.

The amount of powder recommended by the manufacturer was weighed into 10 samples, using the spoon provided by the manufacturer, to determine the average weight. KC showed an average of 0.3631 g, while FO showed an average of 0.2543 g. Once these averages were obtained, 5% and 10% of the powder in each sample was replaced with HNs with and without loaded CX, to form the experimental and positive control groups, respectively. In previous studies, nanostructures including nanotubes have been incorporated to dental materials at percentages from 3 to 20% [12,14,22,23] Some studies have mentioned that a concentration of 10% is necessary to improve the mechanical properties [24]. The modified powder was mixed following the manufacturer's instructions.

Forty KC and FO ionomer blocks were used for the control group. As a positive control, four groups of ionomers were formed with HNs without loaded CX (80 blocks), which were distributed as follows: KC5HN (KC with 5% HNs), KC10HN (KC with 10% HNs), FO5HN (FO with 5% HNs), and FO10HN (FO with 10% HNs).

For the experimental group, 80 ionomer blocks with HNs loaded with CX were used, which were distributed in the following groups: KC5CX (KC with 5% HNs with CX), KC10CX (KC with 10% HNs with CX), FO5CX (FO with 5% HNs with CX), and FO10CX (FO with 10% HNs with CX). The distribution can be observed in Figure 1. A total of 200 circumferential blocks (5 mm × 1 mm) were fabricated to evaluate antibacterial effect.

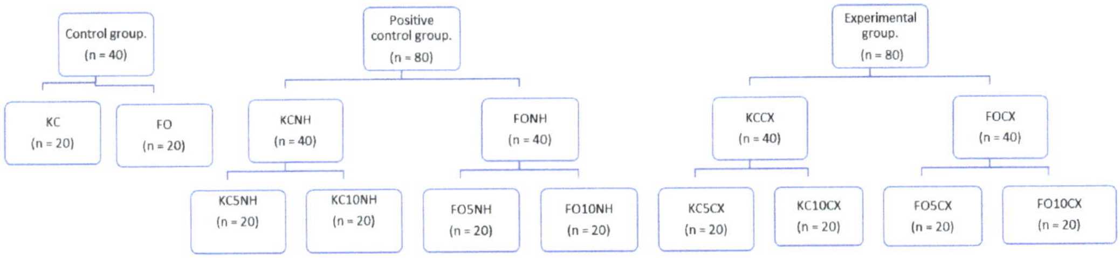

Figure 1. Sample distribution diagram for evaluation of antibacterial effect.

2.3. *Sample Characterization with Fourier Transform Infrared Spectroscopy (FTIR)*

Fourier transform infrared (FTIR) analysis or FTIR spectroscopy was employed to determine the presence of chlorhexidine in experimental groups, and to compare their chemical properties with the control group. The samples were analyzed on a 6700 FTIR spectrometer (Perkin Elmer, Waltham, MA, USA) by ATR (attenuated total reflectance) using a diamond/ZnSe crystal plate. Thirty-two scans were performed on each sample at spectral resolution of 5 cm^{-1} with an infrared spectrum range of 400 to 4000 cm^{-1}.

2.4. *Scanning Electron Microscopy (SEM)*

The specimens were mounted and observed by a cold field emission scanning electron microscope (Hitachi SU8230, Hitachi High-Technologies Corporation, Tokyo, Japan) at 1.0 keV equipped with a Bruker XFlash 6/60.

2.5. Microbiology Assay

The microbiological tests performed in this study were carried out according to the guidelines established in standard M100 of the Clinical and Laboratory Standards Institute (CLSI). *Streptococcus mutans (S. mutans)* 33688 (ATCC) was seeded in Petri dishes with Muller Hinton agar (MHA) supplemented with 5% sheep blood (BD Columbia II, Germany) using the cross-streaking technique, and incubated at 37 °C for 18 h. Five colonies were taken from the fresh culture and adjusted to the 0.5 turbidity standard of the McFarland nephelometer (1.5×10^8 CFU/mL) with 0.9% $NaCl_2$ solution for dilutions [25].

Subsequently, the MHA Petri dishes were inoculated with *S. mutans*, and the glass ionomer blocks corresponding to each group were added. The plates were placed in an incubator (Thermo Fisher Scientific, Massachusetts, USA) for 24 h at 37 °C in an anaerobic atmosphere with 5% CO_2. The entire procedure was performed in triplicate.

After the plates were removed from the incubator, they were examined to verify that the bacterial growth was uniform. the bacterial inhibition was evaluated by measuring the zones of growth inhibition in millimeters with a Vernier caliper, taking into account the diameter of the glass ionomer blocks.

2.6. Microhardness

The Vickers scale based on ISO 9917-1: 2007 standards was used to evaluate microhardness. The sample was conformed of 5 circumferential blocks (10 mm × 3 mm) for each Ketac and Fuji group (KC, KC5CX, KC10CX, FO, FO5CX and FO10CX). A total of 30 blocks with 25 indentations per block were used to evaluate microhardness. The blocks were placed on the microdurometer (SXHV-1000TA, Sinowen, Dongguan, China) and a force of 10 Newtons for 10 s was applied, using a diamond indenter certified by ISO 9001:2008.

2.7. Compression Strength

For evaluation of compressive strength, 120 rectangular blocks (4 mm × 3 mm × 3 mm) were fabricated, divided into six groups previously mentioned with 20 samples in each. The blocks were analyzed with a universal testing machine (Autograph AGS-X, Shimadzu Corporation, Tokyo, Japan); the flat tip was placed in the center of the sample, and the formula $CS = 2P/\pi dh$ was used to calculate the compressive strength, where CS represents the compressive strength, P is the load at the fracture, d is the width of the sample and h is the thickness of the sample. The results were obtained in MPa based on the ISO 9917:1991 standard.

2.8. Setting Time

Experimental and control pastes were placed into 120 rectangular molds (4 mm × 3 mm × 3 mm) divided as previously described. The setting time was measured according the ISO method for water-based dental cement (ISO 9917-1:2007) recording the time elapsed between the start of mixing and the moment where the needle (1.06 mm diameter and 400 g weight established in the indenter) did not mark the surface with a complete circular indentation.

The mechanical properties data obtained from the microbiological assay were analyzed using the statistical program IBM SPSS statistical software (Version 25, IBM Corporation, New York, NY, USA). Shapiro–Wilk, Kruskal–Wallis and the Mann–Whitney U test were performed to evaluate the inhibitory effect of the experimental and control groups. For the evaluation of microhardness, setting time, and compressive strength in the experimental and control groups, Shapiro–Wilk, one way ANOVA, and Tukey testing were used.

3. Results

3.1. IR Spectroscopy

The IR spectra (Figure 2) of halloysite showed absorption bands at 746 cm^{-1} corresponding to hydroxyl groups (OH) and at 908 cm^{-1} corresponding to Al–OH stretching; the vibrational bands at 1006 cm^{-1} and 1119 cm^{-1} were attributed to silicate groups (Si–O–Si and Si–O_2, respectively), and two bands at 3621 cm^{-1} and 3697 cm^{-1} were related to Si–O–Al groups.

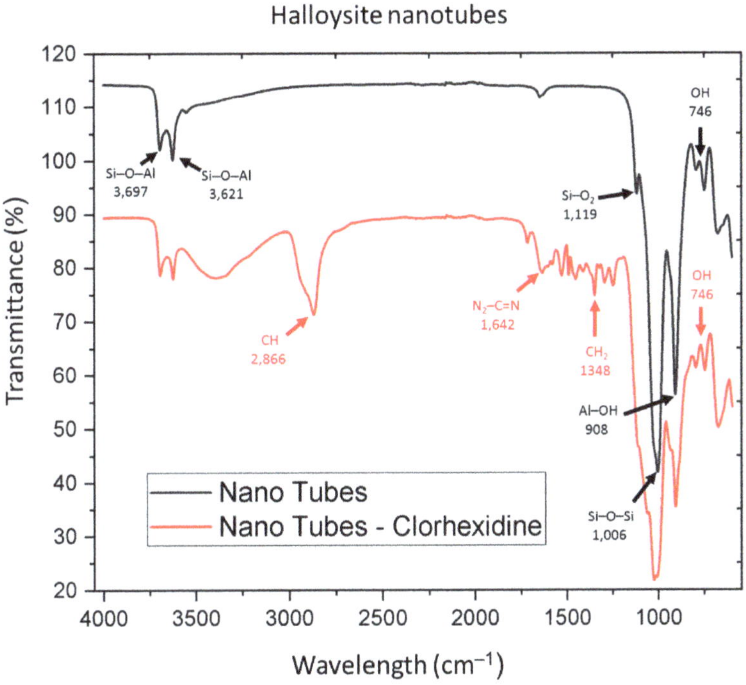

Figure 2. Spectroscopy of halloysite nanotubes and chlorhexidine nanotubes.

The presence of CX was corroborated by the N_2–C=N stretching band at approximately 1642 cm^{-1}, and vibrational bands at 1348 cm^{-1} and 2866 cm^{-1} related to CH_2 and CH methyl groups, respectively (Figure 2).

In the FO sample spectra (Figure 3), the vibrational bands at 1725 cm^{-1} and 1584 cm^{-1} were attributed to C=O polymeric acid carboxyl groups. The peak at 1538 cm^{-1} was related to the stretching of C=C double bonds, reflecting the increased interaction between the glass ionomer and the HNs at a concentration of 10%. The peaks at 2979 and 2883 cm^{-1} were attributed to CH_2 and CH_3 methyl groups, and silicon group peaks were observed at 1068 and 998 cm^{-1}, corresponding to Al–O–Si and Si–O–Si, respectively, corroborating the presence of HNs in the FO glass ionomer at concentrations of 5% and 10%.

Regarding the KC spectra (Figure 4), bands associated with aluminum polyacrylate C=O were found at 1570 cm^{-1} and 1454 cm^{-1}, the peak at 1395 cm^{-1} was attributed to methyl CH, and the peaks at 2887 cm^{-1} and 2981 cm^{-1} were found to correspond to CH_2 and CH_3, respectively. The SiO_2 group was clearly observed in the peaks at 1156 cm^{-1} and 1026 cm^{-1}, which may correspond to the increase in HNs in the KC group at 5%.

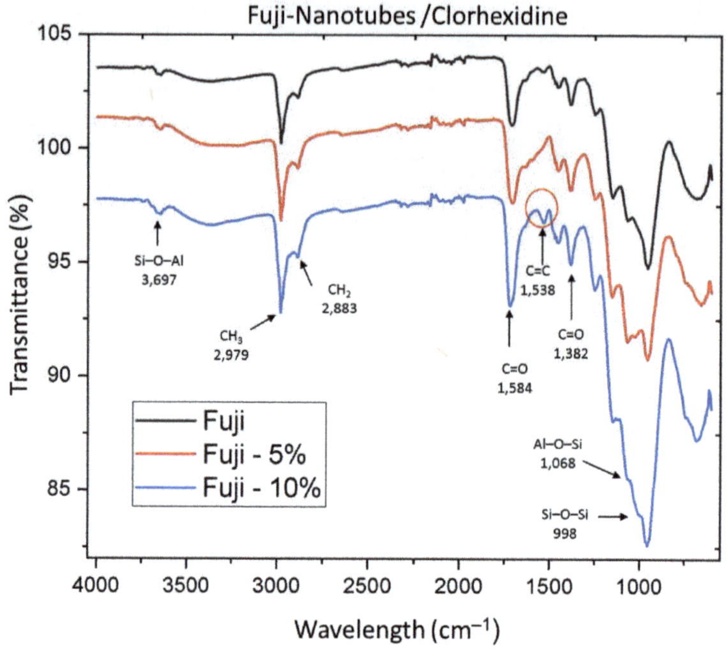

Figure 3. Fuji Ortho (FO) sample spectra.

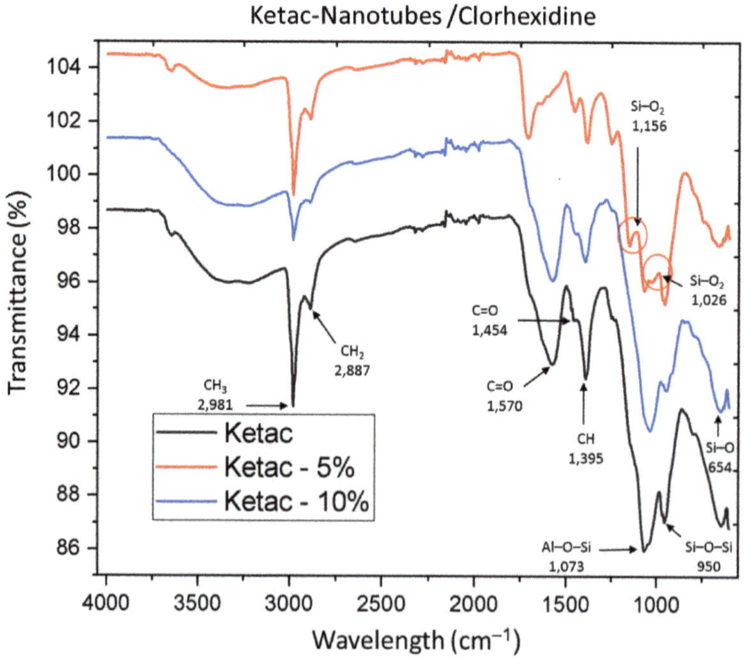

Figure 4. Fuji Ortho sample spectra.

3.2. SEM Results

The HN micrograph shows symmetrical, agglomerated, and disorganized nanotubes with an average size of 200–500 nm in length and a width of approximately 50 nm (Figure 5).

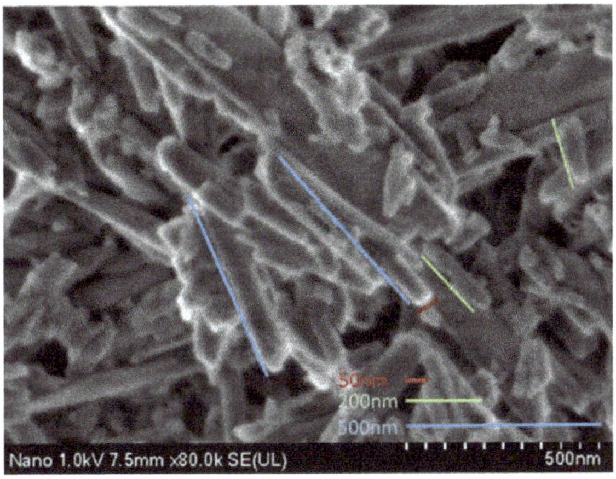

Figure 5. Halloysite nanotubes.

The image of the FO control group (Figure 6a) shows irregular areas with a rough surface, in addition to particles where the surface is smoother, with an approximate dimension of 5 to 10 μm.

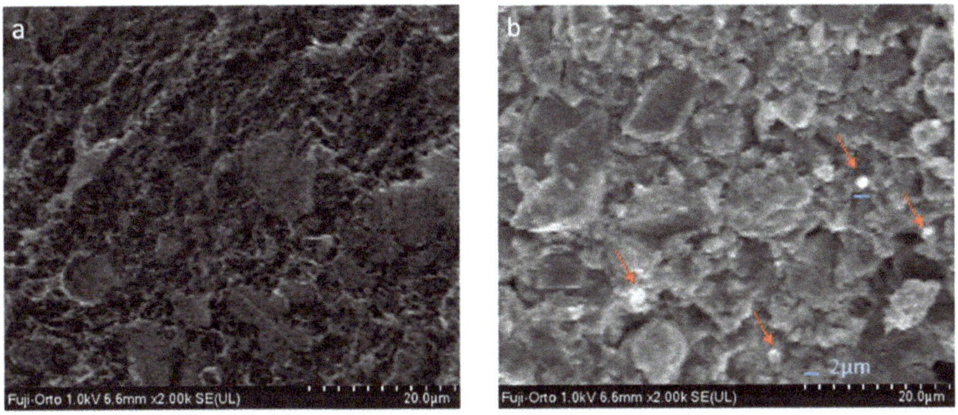

Figure 6. (**a**) Fuji Ortho group, (**b**) Fuji Ortho group with halloysite nanotubes.

In the micrograph of the FO group with HNs (Figure 6b), irregular, agglomerated particles are observed, as well as small spherical particles of approximately 1–2 μm.

The photomicrograph corresponding to the KC group (Figure 7a) shows an irregular surface with particles ranging from 5 to 10 μm.

In the image of the KC group with HNs (Figure 7b), agglomerated particles with an approximate size of 1 to 2 μm are visible.

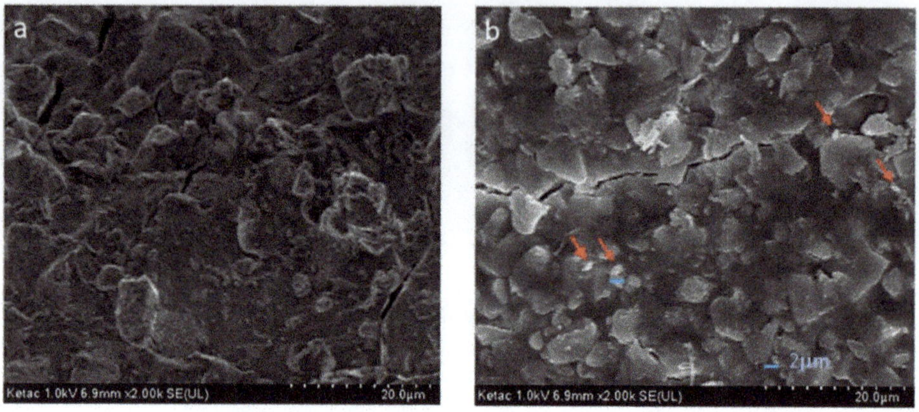

Figure 7. (**a**) Ketac Cem group, (**b**) Ketac Cem group with halloysite nanotubes.

3.3. Microbiology Assay

The control and positive control groups did not show inhibition of *S. mutans*. However, the four experimental groups showed an antibacterial effect on this microorganism, with a mean between 11.35 mm and 13.2 mm.

Figure 8 shows that the FO10CX group had a mean of 12.45 mm, one more millimeter of inhibitory effect than the FO5CX group, which had a mean of 11.45 mm. The difference between the KC5CX and FO5CX groups was only 0.10 mm, which indicated no significant change between the two ionomers. However, the KC10CX group had the greatest inhibitory effect, with a mean of 13.20 mm, a difference of 1.85 mm compared with the KC5CX group, which could be due to the higher percentage of incorporated nanotubes. After 72 h, bacterial growth was observed in all experimental groups.

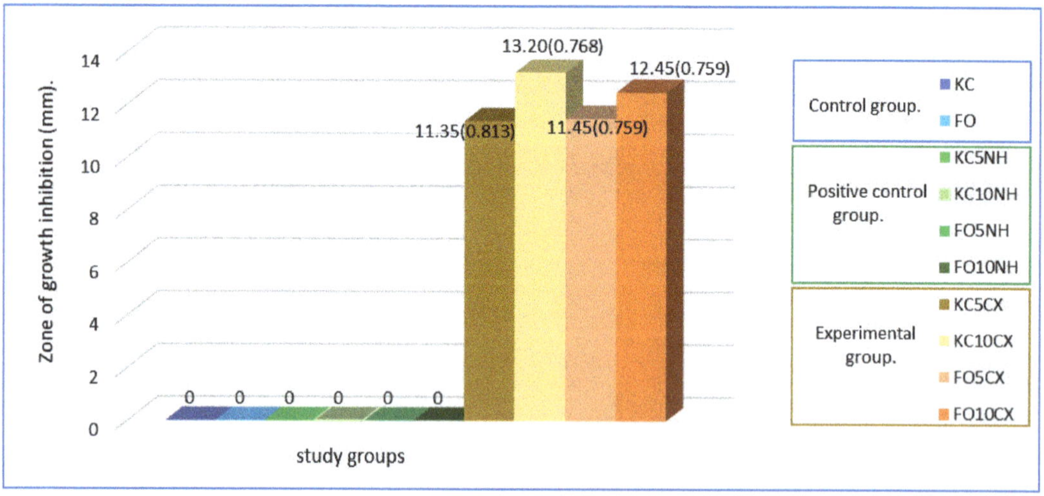

Figure 8. Inhibitory effect of the different groups analyzed in this study; mean (standard deviation); mm: millimeters.

The normality of the data was verified with the Shapiro–Wilk test. Regarding the inhibitory effect of the experimental and control groups, a Kruskal–Wallis test was performed for multiple comparisons, and the Mann–Whitney U test was used to analyze differences between two groups. On the other hand, microhardness, setting time, and compressive strength of the experimental and control groups showed a normal distribution; they were compared with one way ANOVA testing and Tukey testing to analyze differences between groups.

Statistically significant differences were observed between the experimental groups evaluated in this study, using the Kruskal–Wallis test with a value of $p = 0.001$ (Table 1).

Table 1. Comparison of inhibitory effect from experimental groups analyzed in this study.

Groups	Mean (SD)
Fuji ortho 5% of NH + CX. (FO5CX)	11.45 (0.759)
Fuji Ortho 10% of NH + CX. (FO10CX)	12.45 (0.759)
Ketac Cem 5% of NH + CX. (KC5CX)	11.35 (0.813)
Ketac Cem 10% of NH + CX. (KC10CX)	13.20 (0.768)
Total samples	80
Contrast Statistics	41.735
Degrees of freedom	3
p value Kruskal Wallis Test	0.001 *

SD: Standard deviation, *: significative differences $p \leq 0.05$.

A pairwise comparison was performed with the Mann–Whitney U test to determine the differences between the four experimental groups. Statistically significant differences were observed in all groups that contained a higher percentage of nanotubes (10%) compared with those that contained 5%, which suggests that greater inhibitory effect is obtained when the percentage of nanotubes is increased (Table 2). The greatest differences were between the FO5CX and KC10CX groups, and between the KC5CX and KC10CX groups, yielding $p = 0.001$. Finally, when the KC10CX group was compared with the FO10CX group, a minimal difference of 0.75 mm was observed, which indicates no significant difference between the groups loaded at 10% ($p = 0.223$).

Table 2. Comparison between experimental groups.

Groups	Contrast Statistics	Contrast Statistics Deviation	*p* Value
KC5CX-FO5CX	−1.725	−0.244	1.000
KC5CX-FO10CX	−24.025	−3.400	0.004 *
KC5CX-KC10CX	−38.760	−5.485	0.001 *
FO5CX-FO10CX	−22.300	−3.156	0.010 *
FO5CX-KC10CX	37.025	5.240	0.001 *
FO10CX-KC10CX	14.725	2.084	0.223

FO5CX: Fuji Ortho 5% of NH + CX; FO10CX: Fuji Ortho 10% of NH + CX; KC5CX: Ketac Cem 5% of NH + CX; KC10CX: Ketac Cem 10% of NH + CX; *: significative differences $p \leq 0.05$.

Descriptive results from the mechanical tests and setting times for control and experimental groups analyzed in this study can be observed in Table 3. In terms of microhardness, statistically significant differences were not observed when the Fuji control (FO) was compared with experimental groups FO5CX and FO10CX, with means of 68.83, 67.96, and 67.66 respectively, $p = 0.766$. In the same way, Ketac Cem control (KC) was compared with experimental groups KC5CX and KC10CX and no statistically significant differences were observed from the ANOVA test ($p = 0.056$), with means of 80.03, 77.87, and 77.66 respectively. The results for setting time were equal at 7.56 min for the Ketac Cem groups, and 9.57 min. for FO and FO5CX, while only FO10CX was different with 9.56 min. However, statistically significant differences were not observed.

Table 3. Comparison of mechanical properties from experimental and control groups.

Groups	VMHN Mean (SD)	ST Mean (SD)	CS Mean (SD)
KC (control group)	80.03 (4.56)	7.56 (0.024)	84.16 (0.92)
KC5CX	77.87 (3.63)	7.56 (0.017)	88.78 (1.12)
KC10CX	77.66 (2.99)	7.56 (0.019)	93.96 (1.66)
Total degrees of freedom	74	59	59
Sum of squares	1119.245	0.025	1056.608
Fisher's statistic	2.992	0.697	293.83
ANOVA test FO groups	0.056	0.502	0.001 *
FO (control group)	68.83 (5.26)	9.57 (0.011)	125.42 (1.79)
FO5CX	67.96 (5.85)	9.57 (0.014)	128.26 (2.26)
FO10CX	67.66 (6.50)	9.56 (0.018)	133.17 (2.13)
Total degrees of freedom	74	59	59
Sum of squares	2520.976	0.014	860.412
Fisher's statistic	0.267	1.043	71.39
ANOVA test FO groups	0.766	0.359	0.001 *

KC: ketac cem cement, FO: Fuji Orto cement, 5CX: 5% of Chlorhexidine-modified nanotubes, 10CX: 10% of Chlorhexidine-modified nanotubes VMHN: Vickers microhardness, CS: Compressive strength, ST: Setting time, SD: Standard deviation, *: $p \leq 0.05$ (significative differences).

On the other hand, values of compressive strength were observed to increase in all experimental groups (FO5CX, FO10CX, KC5CX, KC10CX), and statistically significant differences were indicated by ANOVA testing with a p value of 0.001 (Table 3). In Table 4, a Tukey test comparison shows statistically significative differences between all the study groups.

Table 4. Comparison of compressive strength between experimental and control groups.

Groups	Mean Difference	95% Confidence Intervals	p Value
Control-KC5CX	−4.62800	IL: −5.6022, SL: −3.6538	0.001 *
Control-KC10CX	−9.80900	IL: −10.7832, SL: −8.8348	0.001 *
KC5CX-KC10CX	−5.18100	IL: −6.1552, SL: −4.2068	0.001 *
Control-FO5CX	−2.84500	IL: −4.4242, SL: −1.2658	0.001 *
Control-FO10CX	−7.75100	IL: −9.3302, SL: −6.1718	0.001 *
FO5CX-FO10CX	−4.90600	IL: −6.4852, SL: −3.3268	0.001 *

* p value ≤ 0.05, KC: ketac cem cement, FO: Fuji Orto cement, 5CX: 5% of Chlorhexidine-modified nanotubes, 10CX: 10% of Chlorhexidine-modified nanotubes, inferior limit (IL) and superior limit (SL).

4. Discussion

Certain treatments require the use of orthodontic bands due to the stability these provide to appliances, despite the disadvantages of hindering oral hygiene, causing the accumulation of dentobacterial plaque and giving rise to white lesions [26,27].

An investigation by Tasios et al. [28] mentioned that 24% of teeth treated with orthodontics developed at least one white spot, with the maxillary and mandibular first molars being most affected. Therefore, different alternatives have been implemented with the objective of reducing the presence of these lesions. Among the main alternatives are methods to improve the bactericidal properties of dental materials. However, an appropriate material has not yet been found that can act as a bactericide in the mouth and is efficient as a cementing agent for orthodontic bands [29,30].

Previous studies have concluded that glass ionomers show bacterial inhibition due to the release of fluoride [31]. Several studies have mentioned that this antibacterial effect occurs at a minimum concentration of 5000 parts per million (ppm) [32]. Other studies report that glass ionomers alone are capable of releasing fluoride between 32.6 and 17.4 ppm [33]. Therefore, despite being efficient as restoratives, bases, or cementing agents, these dental materials are limited in their antibacterial effect. Several studies have developed different materials and compounds (nanoparticles of hidroxiapatite, fluorapatite and TiO_2, fiberglass, zirconia, amino acids, chloroxylenol, boric acid and thymol, triclosan, silver nanoparticles, etc.) with the objective of enhancing the mechanical and antibacterial properties of glass ionomer cements. However, those studies indicated that adding some secondary filler into the glass ionomer cements improved some properties and altered others [34]. In the present investigation, an enhancement of antibacterial and mechanical properties was obtained by using CX preloaded on HNs.

Recently, HNs have been incorporated into different dental materials with the aim of improving the materials' properties. Previous studies concluded that HNs are excellent nanocarriers for drugs, as well as fillings for restorations, because they very efficiently promote the physical and chemical effects of dental materials. However, these nanostructures do not present bacterial inhibition by themselves [18]. In the present investigation, the control groups for glass ionomers without HN loading corroborated the null capacity of bacterial inhibition.

Degrazia et al. [17] incorporated triclosan-loaded HNs into dental resins, and the results demonstrated their efficacy and potential antibacterial effects. In the present investigation, HNs were loaded with CX and subsequently incorporated into type I glass ionomers, and the antibacterial effects of these dental materials were evident in all experimental groups.

An inhibitory effect was observed in all the experimental groups analyzed in this study, showing zones of growth inhibition between 11 mm and 13 mm with loads of 5% and 10% CX respectively, which agrees with previous studies in which CX treatment of *Streptococcus mutans* resulted in slightly lower zones of growth inhibition, between 7 mm and 9 mm [35]. *Streptococcus mutans* (ATCC 33688) was solely used in the present study, due to the ability of the organism to inhabit and invade various areas of the oral cavity making it a prime perpetrator of tooth decay. Moreover, scientific investigators have validated the dominance of *S. mutans* on the depressions of the tooth surface, constituting 39% of streptococci in the oral ecosystem, and their production of glucan is directly proportional to the extent of biofilm formation [36].

Previous studies have stated that is necessary to chose an antimicrobial agent for addition to a restorative material that will provide effective antibacterial action without adversely affecting the material's mechanical properties [37]. Takahashi conducted a study in which CX was directly incorporated into a glass ionomer, and the results obtained were similar to those of the present work: notably, a greater antibacterial effect was observed when the CX concentration increased. In the present study, the antibacterial effect of ionomers with higher concentrations of nanotubes preloaded with CX increased the zone of growth inhibition. It is important to mention that in the study conducted by Takahashi, increasing the CX concentration affected physical properties [15]. According to our results, most of the mechanical properties analyzed in this study were not negatively affected and compressive strength (CS) was positively modified.

Microhardness (VHN) values of control groups analyzed in this study were similar to values observed in previous studies where conventional and resin-modified glass ionomer cements have been evaluated [38–40], and the microhardness of experimental groups was not found significantly altered. The setting time (ST) was consistent with that reported by the manufacturer in control and experimental groups, although some studies mention that incorporation of nanostructures (magnesium nanoparticles) to glass ionomer cements increased the ST in this material [37].

Regarding CS, previous studies found similar values to those observed in conventional and RMGI cements analyzed in our study [41–44]. On the other hand, in this study the experimental groups with 5% and 10% of Halloysite nanotubes preloaded with chlorhexidine showed an increase in CS, which can be considered a positive modification. These results are similar to reported in previous studies where mechanical properties were improved by adding different nanostructures [23]. Specifically, CS of glass ionomers was improved with the addition of magnesium oxide nanoparticles, although an increase in the setting time was observed, probably because the presence of magnesium ions may impede or interfere with the acid-base reaction [37]. Other studies added 5%, 10%, 15%, and 20% of ceramic powder to glass ionomer cements to improve the CS of the material, concluding that only at a concentration of 10% was it possible to improve the CS without compromising the ST [24]. In our study, concentrations of 5% and 10% halloysite nanotubes with chlorhexidine added to glass ionomer cements increased the CS without compromising the ST.

In a study by Pazourkova et al. [45] in 2019, the presence of CX was corroborated by IR spectroscopic analysis: methylene CH stretching at 2940 cm^{-1} and 2860 cm^{-1}, N_2–C=N– stretching at approximately 1646 cm^{-1}, and CH_2 groups at 1492 cm^{-1} were observed. The spectral results for CX in this study were similar, showing peaks for CH at 2866 cm^{-1}, N_2–C=N– at 1642 cm^{-1}, and CH_2 stretching at 1348 cm^{-1}.

In 2018, Zhang et al. [46] confirmed the presence of HNs by observing bands corresponding to the hydroxyl group stretching at 3697 cm^{-1} and 3624 cm^{-1}, and Si–O–Si stretching at approximately 1036 cm^{-1}, while the bands below 1000 cm^{-1} corresponded to the symmetrical stretching of Si–O or Al–O groups. Similar results were observed in the current investigation: –OH groups were identified at 3697 cm^{-1} and 3621 cm^{-1}, Si–O–Si was observed at 1006 cm^{-1}, and finally, Al–O stretching was observed at 908 cm^{-1}.

With respect to the analysis of glass ionomers through IR spectroscopy, previous studies reported a peak at 3354 cm^{-1}, corresponding to the stretching of OH groups in the ionomer liquid, a peak at 1705 cm^{-1} corresponding to C=O, and C=C stretching observed at 1640 cm^{-1} [47]. These results are similar to those observed in the present study, with the OH peak at 3697 cm^{-1}, C=O peak at 1725 cm^{-1}, and C=C peak at 1538 cm^{-1}; the latter group may be associated with the increased interaction of HNs with the glass ionomer at 10% loading.

Previous studies have reported aromatic CH stretching bands at 2914 and 2852 cm^{-1} that correspond to an ionomer. The above coincides with the CH groups observed in the present study, where aromatic CH stretching was observed at 2979 and 2883 cm^{-1} [48].

The presence of HNs in the glass ionomers was corroborated by the presence of bands at 1068 and 998 cm^{-1}, corresponding to aluminum (Al–O–Si) and silica (Si–O–Si) stretching, respectively, in the FO ionomer. For the KC ionomer, stretching was observed at 1073 and 950 cm^{-1}.

The null hypothesis was rejected because all experimental groups showed antibacterial effect, the higher concentration of nanotubes with chlorhexidine in glass ionomer cements showed an increase in antibacterial effect, whilst microhardness and setting time were not altered, and compressive strength was enhanced. Previous studies investigated the mechanical properties of dental adhesives modified with nanotubes [12,13], and others investigated the mechanical properties of resin-based materials modified with nanotubes [17], however, glass ionomers modified with halloysite nanotubes have not been evaluated, nor their antibacterial effect. Some studies evaluated the antibacterial effect of composites modified with clay nanotubes, without evaluating the mechanical properties (microhardness, compressive strength and setting time) [16,19]. Other studies evaluated antimicrobial activity on different dental materials using only one strain of *Streptococcus mutans* as the most common bacterial strain that causes dental caries [1,6,7,9,16,32]. In this study, the mechanical properties and antibacterial effect on *Streptococcus mutans* were evaluated in glass ionomer cements modified with halloysite nanotubes pre-loaded with chlorhexidine.

Limitations of the Study

The present study was limited through the evaluation of one bacterial strain (*Streptococcus mutans*). Evaluation of the effects on more bacterial strains would have been more suitable, other oral bacteria should be considered for proper microbiological analysis; nevertheless, the authors considered that *S. mutans* is one of the most significant causative of caries-related pathologies. Mechanical properties were limited to three parameters due to the high cost analysis of each.

5. Conclusions

Authors concluded that the addition of nanotubes preloaded with chlorhexidine into glass ionomers at concentrations of 5% and 10% showed a notable inhibitory effect on *S. mutans*, without altering the microhardness and setting time. Furthermore, compressive strength was experimentally enhanced using 10% nanotubes. These results suggest that halloysite nanotubes added to conventional and resin-modified glass ionomer cements could be novel method to counteract injuries caused by orthodontic bands, with the advantage of maintaining and improving mechanical properties.

Author Contributions: Conceptualization, A.C.C.-M. and E.N.S.-V.; Methodology, A.I.L.-F., R.J.S.-V. and R.C.-P.; Formal analysis, E.N.S.-V. and A.I.L.-F.; Writing-Original Draft, E.N.S.-V., A.C.C.-M.; Project administration E.N.S.-V.; Visualization, L.V.J.-R. and U.V.-E.; Supervision, U.V.-E.; Software, A.A.M.-V. and L.H.-R.; Validation, A.A.M.-V. and L.V.J.-R.; Data Curation, E.L.-C. and V.H.T.-R.; Resources E.L.-C. and L.H.-R.; Investigation, A.C.C.-M., R.J.S.-V. and V.H.T.-R.; Writing-Review & Editing, R.C.-P. and E.N.S.-V. All authors have read and agreed to the published version of the manuscript.

Funding: This research received no external funding.

Institutional Review Board Statement: Not applicable.

Informed Consent Statement: Not applicable.

Data Availability Statement: Not applicable.

Conflicts of Interest: The authors declare no conflict of interest.

References

1. Sun, L.; Yan, Z.; Duan, Y.; Zhang, J.; Liu, B. Improvement of the mechanical, tribological and antibacterial properties of glass ionomer cements by fluorinated graphene. *Dent. Mater.* **2018**, *34*, e115–e127. [CrossRef] [PubMed]
2. Heravi, F.; Bagheri, H.; Rangrazi, A. Evaluation of Microleakage of Orthodontic Bands Cemented With CPP-ACP-Modified Glass Ionomer Cement. *J. Adv. Oral Res.* **2019**, *10*, 128–131. [CrossRef]
3. Araújo, J.L.D.S.; Alvim, M.M.A.; Campos, M.J.D.S.; Apolônio, A.C.M.; Carvalho, F.G.; Lacerda-Santos, R. Analysis of Chlorhexidine Modified Cement in Orthodontic Patients: A Double-Blinded, Randomized, Controlled Trial. *Eur. J. Dent.* **2021**, *15*, 639–646. [CrossRef]
4. Zandi Karimi, A.; Rezabeigi, E.; Drew, R.A.L. Glass ionomer cements with enhanced mechanical and remineralizing properties containing 45S5 bioglass-ceramic particles. *J. Mech. Behav. Biomed. Mater.* **2019**, *97*, 396–405. [CrossRef]
5. Heravi, F.; Omidkhoda, M.; Koohestanian, N.; Hooshmand, T.; Bagheri, H.; Ghaffari, N. Retentive Strength of Orthodontic Bands Cemented with Amorphous Calcium Phosphate-Modified Glass Ionomer Cement: An In-Vitro Study. *Front. Dent. J. Dent. Tehran Univ. Med Sci.* **2017**, *14*, 13–20.
6. Tarasingh, P.; Sharada Reddy, J.; Suhasini, K.; Hemachandrika, I. Comparative evaluation of antimicrobial efficacy of resin-modified glass ionomers, compomers and giomers—An invitro study. *J. Clin. Diagn. Res.* **2015**, *9*, ZC85–ZC87. [CrossRef]
7. Zayed, M.M.; Hassan, R.E.; Riad, M.I. Evaluation of the antibacterial efficacy of different bioactive lining and pulp capping agents. *Tanta Dent. J.* **2015**, *12*, 132–139. [CrossRef]
8. Tüzüner, T.; Dimkov, A.; Nicholson, J.W. The effect of antimicrobial additives on the properties of dental glass-ionomer cements: A review. *Acta Biomater. Odontol. Scand.* **2019**, *5*, 9–21. [CrossRef]
9. Kurt, A.; Tüzüner, T.; Baygın, Ö. Antibacterial characteristics of glass ionomer cements containing antibacterial agents: An in vitro study. *Eur. Arch. Paediatr. Dent.* **2021**, *22*, 49–56. [CrossRef]
10. Tüzüner, T.; Ulusu, T. Effect of antibacterial agents on the surface hardness of a conventional glass-ionomer cement. *J. Appl. Oral Sci.* **2012**, *20*, 45–49. [CrossRef]
11. Soygun, K.; Soygun, A.; Dogan, M.C. The effects of chitosan addition to glass ionomer cement on microhardness and surface roughness. *J. Appl. Biomater. Funct. Mater.* **2021**, *19*, 2280800021989706. [CrossRef] [PubMed]
12. Kalagi, S.; Feitosa, S.A.; Münchow, E.A.; Martins, V.M.; Karczewski, A.E.; Cook, N.B.; Diefenderfer, K.; Eckert, G.J.; Geraldeli, S.; Bottino, M.C. Chlorhexidine-modified nanotubes and their effects on the polymerization and bonding performance of a dental adhesive. *Dent. Mater.* **2020**, *36*, 687–697. [CrossRef]
13. Feitosa, S.A.; Palasuk, J.; Geraldeli, S.; Windsor, L.J.; Bottino, M.C. Physicochemical and biological properties of novel chlorhexidine-loaded nanotube-modified dentin adhesive. *J. Biomed. Mater. Res. Part B Appl. Biomater.* **2019**, *107*, 868–875. [CrossRef] [PubMed]

14. Degrazia, F.W.; Genari, B.; Leitune, V.C.B.; Arthur, R.A.; Luxan, S.A.; Samuel, S.M.W.; Collares, F.M.; Sauro, S. Polymerisation, antibacterial and bioactivity properties of experimental orthodontic adhesives containing triclosan-loaded halloysite nanotubes. *J. Dent.* **2018**, *69*, 77–82. [CrossRef] [PubMed]
15. Takahashi, Y.; Imazato, S.; Kaneshiro, A.V.; Ebisu, S.; Frencken, J.E.; Tay, F.R. Antibacterial effects and physical properties of glass-ionomer cements containing chlorhexidine for the ART approach. *Dent. Mater.* **2006**, *22*, 647–652. [CrossRef] [PubMed]
16. Cunha, D.A.; Rodrigues, N.S.; Souza, L.C.; Lomonaco, D.; Rodrigues, F.P.; Degrazia, F.W.; Collares, F.M.; Sauro, S.; Saboia, V.P.A. Physicochemical and microbiological assessment of an experimental composite doped with triclosan-loaded halloysite nanotubes. *Materials* **2018**, *11*, 1080. [CrossRef] [PubMed]
17. Degrazia, F.W.; Leitune, V.C.B.; Takimi, A.S.; Collares, F.M.; Sauro, S. Physicochemical and bioactive properties of innovative resin-based materials containing functional halloysite-nanotubes fillers. *Dent. Mater.* **2016**, *32*, 1133–1143. [CrossRef]
18. Massaro, M.; Lazzara, G.; Milioto, S.; Noto, R.; Riela, S. Covalently modified halloysite clay nanotubes: Synthesis, properties, biological and medical applications. *J. Mater. Chem. B* **2017**, *5*, 2867–2882. [CrossRef]
19. Stavitskaya, A.; Batasheva, S.; Vinokurov, V.; Fakhrullina, G.; Sangarov, V.; Lvov, Y.; Fakhrullin, R. Antimicrobial applications of clay nanotube-based composites. *Nanomaterials* **2019**, *9*, 708. [CrossRef]
20. Lvov, Y.M.; DeVilliers, M.M.; Fakhrullin, R.F. The application of halloysite tubule nanoclay in drug delivery. *Expert Opin. Drug Deliv.* **2016**, *13*, 977–986. [CrossRef]
21. Suma, N.K. Effect of Dentin Disinfection with 2% Chlorhexidine Gluconate and 0.3% Iodine on Dentin Bond Strength: An in vitro Study. *Int. J. Clin. Pediatr. Dent.* **2017**, *10*, 223–228. [CrossRef] [PubMed]
22. Morales-Valenzuela, A.A.; Scougall-Vilchis, R.J.; Lara-Carrillo, E.; Garcia-Contreras, R.; Salmeron-Valdes, E.N.; Aguillón-Sol, L. Comparison of Fluoride release in conventional glass-Ionomer cements with a new mechanical mixing cement. *Oral Health Prev. Dent.* **2020**, *18*, 319–323. [CrossRef] [PubMed]
23. Kheur, M.; Kantharia, N.; Iakha, T.; Kheur, S.; Husain, N.A.H.; Özcan, M. Evaluation of mechanical and adhesion properties of glass ionomer cement incorporating nano-sized hydroxyapatite particles. *Odontology* **2020**, *108*, 66–73. [CrossRef] [PubMed]
24. Gupta, A.A.; Mulay, S.; Mahajan, P.; Raj, A.T. Assessing the effect of ceramic additives on the physical, rheological and mechanical properties of conventional glass ionomer luting cement—An in-vitro study. *Heliyon* **2019**, *5*, e02094. [CrossRef] [PubMed]
25. Patel, J.B. *Performance Standards for Antimicrobial Susceptibility Testing*; Clinical and Laboratory Standards Institute: Malvern, PA, USA, 2017; ISBN 1562388053.
26. Teubner, S.; Schmidlin, P.R.; Menghini, G.; Attin, T.; Baumgartner, S. The Impact of Orthodontic Bands on the Marginal Periodontium of Maxillary First Molars: A Retrospective Cross-Sectional Radiographic Analysis. *Open Dent. J.* **2018**, *12*, 312–321. [CrossRef] [PubMed]
27. Bourouni, S.; Dritsas, K.; Kloukos, D.; Wierichs, R.J. Efficacy of resin infiltration to mask post-orthodontic or non-post-orthodontic white spot lesions or fluorosis—A systematic review and meta-analysis. *Clin. Oral Investig.* **2021**, *25*, 4711–4719. [CrossRef]
28. Tasios, T.; Papageorgiou, S.N.; Papadopoulos, M.A.; Tsapas, A.; Haidich, A.B. Prevention of orthodontic enamel demineralization: A systematic review with meta-analyses. *Orthod. Craniofacial Res.* **2019**, *22*, 225–235. [CrossRef]
29. Kamber, R.; Meyer-Lückel, H.; Kloukos, D.; Tennert, C.; Wierichs, R.J. Efficacy of sealants and bonding materials during fixed orthodontic treatment to prevent enamel demineralization: A systematic review and meta-analysis. *Sci. Rep.* **2021**, *11*, 16556. [CrossRef]
30. Greene, L.E.; Bearn, D.R. Reducing white spot lesion incidence during fixed appliance therapy. *Dent. Update* **2013**, *40*, 487–492. [CrossRef]
31. Palenik, C.J.; Behnen, M.J.; Setcos, J.C.; Miller, C.H. Inhibition of microbial adherence and growth by various glass ionomers in vitro. *Dent. Mater.* **1992**, *8*, 16–20. [CrossRef]
32. Pradiptama, Y.; Purwanta, M.; Notopuro, H. Antibacterial Effects of Fluoride in Streptococcus mutans Growth in Vitro. *Biomol. Health Sci. J.* **2019**, *2*, 1. [CrossRef]
33. Prabhakar, A.R.; Balehosur, D.V.; Basappa, N. Comparative evaluation of shear bond strength and fluoride release of conventional glass Ionomer with 1% ethanolic extract of propolis incorporated glass Ionomer cement—Invitro study. *J. Clin. Diagn. Res.* **2016**, *10*, ZC88–ZC91. [CrossRef] [PubMed]
34. Ching, H.S.; Luddin, N.; Kannan, T.P.; Ab Rahman, I.; Abdul Ghani, N.R.N. Modification of glass ionomer cements on their physical-mechanical and antimicrobial properties. *J. Esthet. Restor. Dent.* **2018**, *30*, 557–571. [CrossRef]
35. Boaro, L.C.C.; Campos, L.M.; Varca, G.H.C.; dos Santos, T.M.R.; Marques, P.A.; Sugii, M.M.; Saldanha, N.R.; Cogo-Müller, K.; Brandt, W.C.; Braga, R.R.; et al. Antibacterial resin-based composite containing chlorhexidine for dental applications. *Dent. Mater.* **2019**, *35*, 909–918. [CrossRef] [PubMed]
36. Ranganathan, V.; Akhila, C. Streptococcus mutans: Has it become prime perpetrator for oral manifestations? *J. Microbiol. Exp.* **2019**, *7*, 207–213. [CrossRef]
37. Noori, A.J.; Kareem, F.A. Setting time, mechanical and adhesive properties of magnesium oxide nanoparticles modified glass-ionomer cement. *J. Mater. Res. Technol.* **2020**, *9*, 1809–1818. [CrossRef]
38. Farret, M.M.; de Lima, E.M.; Mota, E.G.; Mitsuo, H.S.O.; Maguilnik, G.; Scheid, P.A. Assessment of the mechanical properties of glass ionomer cements for orthodontic cementation. *Dental Press J. Orthod.* **2012**, *17*, 154–159. [CrossRef]
39. Moshaverinia, M.; Navas, A.; Jahedmanesh, N.; Shah, K.C.; Moshaverinia, A.; Ansari, S. Comparative evaluation of the physical properties of a reinforced glass ionomer dental restorative material. *J. Prosthet. Dent.* **2019**, *122*, 154–159. [CrossRef]

40. Sharafeddin, F.; Jowkar, Z.; Bahrani, S. Comparison between the effect of adding microhydroxyapatite and chitosan on surface roughness and Microhardness of resin modified and conventional glass ionomer cements. *J. Clin. Exp. Dent.* **2021**, *13*, 737–744. [CrossRef]
41. Wajong, K.H.; Damiyanti, M.; Irawan, B. The effects of shelf life on the compressive strength of resin-modified glass ionomer cement. *J. Phys. Conf. Ser.* **2017**, *884*, 12101. [CrossRef]
42. Aguiar, D.A.; Ritter, D.E.; Rocha, R.; Locks, A.; Borgatto, A.F. Evaluation of mechanical properties of five cements for orthodontic band cementation. *Braz. Oral Res.* **2013**, *27*, 136–141. [CrossRef] [PubMed]
43. Mallmann, A.; Oliveira Ataíde, J.C.; Amoeda, R.; Rocha, P.V.; Jacques, L.B. Compressive strength of glass ionomer cements using different specimen dimensions. *Braz. Oral Res.* **2007**, *21*, 204–208. [CrossRef] [PubMed]
44. Piwowarczyk, A.; Ottl, P.; Lauer, H.C. Laboratory strength of glass ionomer and zinc phosphate cements. *J. Prosthodont.* **2001**, *10*, 140–147. [CrossRef] [PubMed]
45. Pazourková, L.; Reli, M.; Hundáková, M.; Pazdziora, E.; Predoi, D.; Martynková, G.S.; Lafdi, K. Study of the structure and antimicrobial activity of Ca-deficient ceramics on chlorhexidine nanoclay substrate. *Materials* **2019**, *12*, 2996. [CrossRef]
46. Zhang, H.; Cheng, C.; Song, H.; Bai, L.; Cheng, Y.; Ba, X.; Wu, Y. A facile one-step grafting of polyphosphonium onto halloysite nanotubes initiated by Ce(iv). *Chem. Commun.* **2019**, *55*, 1040–1043. [CrossRef]
47. Fareed, M.A.; Stamboulis, A. Effect of nanoclay dispersion on the properties of a commercial glass ionomer cement. *Int. J. Biomater.* **2014**, *2014*, 685389. [CrossRef]
48. Mukhopadhyay, S.; Sahu, P.; Bhajiwala, H.; Mohanty, S.; Gupta, V.; Bhowmick, A.K. Synthesis, characterization and properties of self-healable ionomeric carboxylated styrene–butadiene polymer. *J. Mater. Sci.* **2019**, *54*, 14986–14999. [CrossRef]

nanomaterials MDPI

Article

β-Glucan-Functionalized Nanoparticles Down-Modulate the Proinflammatory Response of Mononuclear Phagocytes Challenged with *Candida albicans*

Tânia Lima [1,2,3,*], Stefán B. Gunnarsson [4,5], Elisabete Coelho [6], Dmitry V. Evtuguin [7], Alexandra Correia [1,2,3], Manuel A. Coimbra [6], Tommy Cedervall [4,5] and Manuel Vilanova [1,2,3]

1 I3S-Instituto de Investigação e Inovação em Saúde, Universidade do Porto, 4200-135 Porto, Portugal; alexandra.correia@ibmc.up.pt (A.C.); vilanova@icbas.up.pt (M.V.)
2 IBMC-Instituto de Biologia Molecular e Celular, Universidade do Porto, 4200-135 Porto, Portugal
3 ICBAS-Instituto de Ciências Biomédicas de Abel Salazar, Universidade do Porto, 4050-313 Porto, Portugal
4 Department of Biochemistry and Structural Biology, Lund University, 221 00 Lund, Sweden; sbragig@gmail.com (S.B.G.); tommy.cedervall@biochemistry.lu.se (T.C.)
5 NanoLund, Center for Nanoscience, Lund University, 221 00 Lund, Sweden
6 LAQV-REQUIMTE, Departamento de Química, Universidade de Aveiro, 3810-193 Aveiro, Portugal; ecoelho@ua.pt (E.C.); mac@ua.pt (M.A.C.)
7 CICECO, Department of Chemistry, University of Aveiro, 3810-193 Aveiro, Portugal; dmitrye@ua.pt
* Correspondence: tania.lima@i3s.up.pt

Citation: Lima, T.; Gunnarsson, S.B.; Coelho, E.; Evtuguin, D.V.; Correia, A.; Coimbra, M.A.; Cedervall, T.; Vilanova, M. β-Glucan-Functionalized Nanoparticles Down-Modulate the Proinflammatory Response of Mononuclear Phagocytes Challenged with *Candida albicans*. *Nanomaterials* **2022**, *12*, 2475. https://doi.org/10.3390/nano12142475

Academic Editors: Goran Kaluđerović and Nebojša Pantelić

Received: 16 June 2022
Accepted: 10 July 2022
Published: 19 July 2022

Abstract: Systemic fungal infections are associated with significant morbidity and mortality, and *Candida albicans* is the most common causative agent. Recognition of yeast cells by immune cell surface receptors can trigger phagocytosis of fungal pathogens and a pro-inflammatory response that may contribute to fungal elimination. Nevertheless, the elicited inflammatory response may be deleterious to the host by causing excessive tissue damage. We developed a nanoparticle-based approach to modulate the host deleterious inflammatory consequences of fungal infection by using β1,3-glucan-functionalized polystyrene (β-Glc-PS) nanoparticles. β-Glc-PS nanoparticles decreased the levels of the proinflammatory cytokines TNF-α, IL-6, IL-1β and IL-12p40 detected in in vitro culture supernatants of bone marrow-derived dendritic cells and macrophage challenged with *C. albicans* cells. Moreover, β-Glc-PS nanoparticles impaired the production of reactive oxygen species by bone marrow-derived dendritic cells incubated with *C. albicans*. This immunomodulatory effect was dependent on the nanoparticle size. Overall, β-Glc-PS nanoparticles reduced the proinflammatory response elicited by fungal cells in mononuclear phagocytes, setting the basis for a targeted therapy aimed at protecting the host by lowering the inflammatory cost of infection.

Keywords: *Candida albicans*; β-glucan; nanoparticles; infection; inflammation

1. Introduction

In recent decades, the incidence of human fungal infections has increased, especially in immunocompromised and hospitalized individuals. The fungal infection associated morbidity and mortality are significant, and it is clear that these infections have emerged as important public health problems [1,2]. *Candida albicans* is one of the most frequently recovered human pathogen from fungal infections [3,4]. *C. albicans* cell wall is a dynamic and complex structure essential to almost every aspect of pathogenicity. The cell wall inner layer act as the cell skeleton and is composed of chitin and β-glucans. The outermost layer of the *C. albicans* cell wall is composed of mannoproteins bound to the β-glucan/chitin inner layer. The recognition of *C. albicans* cell wall and subsequent host response depends on the chemical composition and linkages of the *C. albicans* cell wall polysaccharides [5–9]. Immune responses elicited during *C. albicans* infections are related to the host–pathogen interaction. Nevertheless, the inflammatory response elicited by fungal recognition may also be deleterious to the host by causing excessive tissue damage [10]. For instance, an exacerbated inflammatory response is often observed during neutrophil recovery in acute leukemia patients with systemic candidiasis [11]. These observations raise important questions concerning host evolution towards maintaining a balance between control of fungal burden and excessive inflammation [10,12]. A better understanding of these mechanisms may offer new insights into the pathophysiology of these infections, as well as open new possibilities for targeted anti-microbial therapy. In recent decades, β-glucans have been used in many ways to improve health conditions in various situations [13]. Administration of these polysaccharides proved to enhance efficacy of monoclonal antibodies used for cancer immunotherapy, induce faster regeneration of physically damaged tissues, reduce toxicity of bacterial endotoxin, reduce gut inflammation, enhance antibody production against mucosal antigens, and enhance inflammation defense [13–18]. Taken together, these works have demonstrated that in a context of excessive inflammatory response, administration of β-glucans can depress these deleterious responses, suggesting that β-glucans may be used as an anti-inflammatory treatment, namely, in gut inflammation and sepsis.

The emergent application of nanotechnology is revolutionizing several aspects of modern medicine, including diagnostics and therapeutics [19,20]. Nanoparticles (NP) present a great immunomodulatory potential, as they can lead to activation or suppression of immune function [21]. Some experiments have shown that different NP trigger inflammatory responses in different ways [22–24]. Furthermore, the NP properties play a central role in determining the diffusion rate, the absorption of biomolecules, uptake by immune cells, biodistribution, and NP retention time. These parameters can be used to control biological functions and optimize the therapeutic applications. Here, we used bioengineered NP, functionalized with β1,3-glucans, the major component of *C. albicans* cell wall, in order to modulate the host immune recognition of *C. albicans* and reduce the inflammatory consequences of a fungal infection.

2. Materials and Methods

2.1. NP and β-glucans

Whole glucan particles (WGP) from *Saccharomyces cerevisiae*, composed mainly of β1,3-glucans ((β1→3)-Glc), were purchased from InvivoGen (San Diego, CA, USA). Carboxyl-terminated polystyrene (COOH-PS) NP (200, 80 and 26 nm) were purchased from Bangs Laboratories, Inc. (Fishers, IN, USA). All the polystyrene (PS) NP were used right after 24 h of dialysis (molecular weight cut-off (MWCO): 3500 kDa) against deionized (DI) water.

2.2. Chemical Characterization of Soluble β-Glucans

WGP (5 mg) were dissolved in 0.1 M $NaNO_3$ solution and analyzed by Size Exclusion Chromatography (SEC) using a PL-GPC 110 chromatograph (Polymer Laboratories, Venice, CA, USA) with a pre-column PL aquagel-OH Guard 8 µm and two PL aquagel-OH MIXED 8 µm D 300 mm × 7.5 mm columns at 36 °C. A 0.1 M $NaNO_3$ eluent solution was pumped at a flow rate of 0.9 mL/min. Sugar composition of soluble WGP was analyzed by gas chromatography-flame ionization detection (GC-FID) (Perkin Elmer-Clarus 400, PerkinElmer, Waltham, MA, USA) and quantified using 2-deoxyglucose as internal standard [25], as previously described [26]. The alditol acetates were evaluated in a GC-FID with a capillary column DB-225 (30 m length, 0.25 mm inner diameter and 0.15 µm film thickness). The oven temperature was: 200 °C to 220 °C at a rate of 40 °C/min (7 min), increasing to 230 °C at a rate of 20 °C/min (1 min). The temperature of the injector was 220 °C and the detector was at 230 °C. Hydrogen was used as a carrier gas at a flow rate of 1.7 mL/min [26,27].

The glycosidic-linkage composition was determined by gas chromatography-mass spectrometry (GC-MS) of partially methylated alditol acetates (PMAA) using a previously described methodology [26]. The PMAA were analyzed by GC-MS using a Shimadzu GCMS-QP2010, with a DB-1 (J&W Scientific, Folsom, CA, USA) capillary column (30 m length, 0.25 mm internal diameter, and 0.10 µm film thickness). Samples were injected in "split" mode using an injector temperature of 220 °C. The temperature program used was: 50 °C with a linear increase of 8 °C/min until 140 °C, followed by a linear increase of 0.5 °C/min until 150 °C and finally by a linear increase of 40 °C/min until 250 °C. The helium carrier gas had a flow rate of 1.84 mL/min and a column head pressure of 124.1 kPa. The GC was connected to a QP2010 Ultra (Shimadzu Corporation, Kyoto, Japan) mass quadrupole selective detector operating with an electron impact mode at 70 eV and scanning the range *m/z* 40–500, 1 s cycle, in a full scan mode acquisition.

2.3. Surface Functionalization of PS NP with Soluble (β1→3)-Glc

Soluble WGP (5 mg) was dissolved in phosphate buffer (10 mM sodium phosphate, 150 mM NaCl, pH 7.5). Subsequently, sodium periodate (0.1 M) was added, and the solution was incubated for 30 min. The resultant oxidizing glucan solution was then mixed with excess of carbohydrazide (100:1) and sodium cyanoborohydride (10 mM). Dialyzed COOH-PS (1 mg/mL) was mixed with previous glucan solution and 1-ethyl-3(3-dimethylaminopropyl)carbodiimide (EDC) (1 mg) during 2 h. β-Glucan-conjugated nanoparticles (β-Glc-PS) were desalted on a PD-10 column to remove excess reactants.

2.4. Size and ζ-Potential Analysis of β-Glc-PS

The hydrodynamic size of NP was determined by dynamic light scattering (DLS) on a DynaPro Plate Reader II (Wyatt Technology Corporation, Santa Barbara, CA, USA), at 37 °C, with 10 acquisitions per sample. Differential centrifugal sedimentation (DCS) was performed on DC24000 disc centrifuge (CPS Instruments, Prairieville, LA, USA), with a linear 8–24% sucrose gradient at 23,677 rotations per minute (RPM). The ζ-potential of the particles was evaluated by DLS at 25 °C with a Malvern ZetaSizer Nano ZS particle analyzer (Malvern Panalytical, Malvern, UK). Malvern Dispersion Technology Software (DTS; Malvern Panalytical, Malvern, UK) with monomodal mode data processing was used to determine average ζ-potential (mV) and error values. The analysis was conducted between 1 h and 24 h after incubation at 25 °C using 1 mg/mL, 0.1 mg/mL, and 0.01 mg/mL of NP in phosphate buffer. Each condition was set in triplicate and the measurements were analyzed with the Dynamics 7.1.7 (Wyatt Technology Corporation, Santa Barbara, CA, USA).

2.5. Dot Blot Analysis and Glucan Quantification of β-Glc-PS NP

For the dot blot immunoassay, 2 µL of each sample was spotted onto nitrocellulose membrane and then allowed to air dry. Each nitrocellulose membrane was incubated in blocking solution (Tris Buffer Saline-Tween (TBS-T) + 2% low fat dry milk) for 1 h at room temperature (RT). The membranes were then washed three times 10 min in TBS-T and incubated with anti-(β1→3)-Glc (1:800, Biosupplies, Yagoona, Australia) for 2 h, washed, and further incubated with polyclonal rabbit anti-mouse IgG/AP (1:500, Dako, Glostrup, Denmark) and, subsequently, with substrate solution (BCIP/NBT, Sigma-Aldrich, St. Louis, MO, USA) until spots were visible.

The amount of (β1→3)-Glc on the surface of the NP was analyzed using a phenol-sulphuric acid assay [28].

2.6. FTIR Spectroscopy

For Fourier transformed infrared (FTIR) analysis, 2 mL of each NP suspension (26, 80, 200 nm β-Glc-PS NP and COOH-PS NP) was scanned using PerkinElmer Frontier FTIR (Universal Attenuated Total Reflection (ATR) Sampling Accessory; PerkinElmer, Waltham, MA, USA) with a resolution of 4 cm^{-1} in the ATR sampling mode. For each sample, 32 scans were acquired and averaged, Energy = 371 and OPD = 0.2 and recorded in the 4000–400 cm^{-1} spectral range.

2.7. NP Aggregation under Physiological Conditions

PS NP were diluted with either water, PBS, RPMI plus 10% fetal bovine serum (FBS) or mouse serum (MS). The analyses were conducted between 0 h and 24 h after incubation. The hydrodynamic size of NP was measured by DLS on DynaPro Plate Reader II (Wyatt Technology Corporation, Santa Barbara, CA, USA), at 37 °C, with 10 acquisitions per sample. Before the analysis of samples, the refractive index and viscosity were adjusted according to solvent used.

2.8. Mouse Serum Protein Corona Analysis

NP suspensions (0.5 mg/mL) were incubated with MS for 1 h, 12 h and 24 h. The samples were centrifuged (15 min, 14,000 rpm, 20 °C) to pellet the particle–protein complexes. The pellet was resuspended in PBS, transferred to a new vial, and centrifuged again to pellet the particle–protein complexes. This procedure was repeated three times. After the third washing step, the proteins were eluted from the particles by adding SDS-sample buffer (dH_2O, 0.5 M Tris-HCl, Glycerol, 10% SDS, 2-mercaptoethanol, 1% bromophenol blue) to the pellet and boiling the solution. Corona proteins were separated through 10% SDS/PAGE 1D gels. Each gel run included one lane of a molecular weight ladder standard, PageRuler Prestained Protein Ladder (Thermo Scientific, Waltham, MA, USA).

2.9. Yeast Cell and Culture Conditions

C. albicans SC5314 (ATCC MYA-2876, ATCC, Manassas, VA, USA) cells were maintained frozen (30% glycerol at −80 °C). Viable cells were obtained from cultures performed on solid yeast extract-peptone-dextrose (YPD) medium (1% yeast extract, 2% peptone, 2% glucose, 2% agar). Isolates were then grown in liquid YPD medium (37 °C, 140 rpm) in a shaking incubator to late exponential growth (14–16 h), recovered by centrifugation and washed twice in sterile phosphate-buffered saline (PBS). To prepare heat-killed (HK) cells, *C. albicans* yeast cells were grown as described before and incubated at 70 °C for 30 min in sterile PBS.

2.10. Mice

BALB/c mice were bred under specific-pathogen-free conditions at the Animal Facility of the Instituto de Investigação e Inovação em Saúde (i3S), Porto, Portugal. Procedures involving mice were performed according to the European Convention for the Protection of Vertebrate Animals used for Experimental and Other Scientific Purposes (ETS 123), directive 2010/63/EU, of the European Parliament and of the council of 22 September 2010 on the protection of the animals used for scientific purposes, and Portuguese rules (DL 113/2013). Experiments were approved by the institutional board responsible for animal welfare (ORBEA) at i3S and authorization to perform the experiments was issued by the competent national authority (Direção Geral de Alimentação e Veterinária) with the reference number 014036/2019-07-24.

2.11. In Vitro Differentiation of BMM and BMDC

To generate bone marrow-derived macrophages (BMM) and dendritic cells (BMDC), bone marrow cells were collected from femurs and tibias of BALB/c mice by flushing with cold Roswell Park Memorial Institute (RPMI) 1640 (Sigma-Aldrich, St. Louis, MO, USA). The collected cells (1 × 106 cells/mL) were seeded onto six-well plates and incubated at 37 °C, 5% CO_2 in RPMI supplemented with 10% fetal bovine serum (FBS) (Biowest, Nuaillé, France), penicillin (100 IU/mL)/streptomycin (100 mg/mL) (Sigma-Aldrich, St. Louis, MO, USA), L-glutamine (2 mM) (Sigma-Aldrich, St. Louis, MO, USA), 1% (*v*/*v*) 4-(2-hydroxyethyl)-1piperazineethanesulfonic acid (HEPES) buffer (Sigma-Aldrich, St. Louis, MO, USA), 5 μM 2-mercaptoethanol (Sigma-Aldrich, St. Louis, MO, USA) and 20% (*v*/*v*) J558-cell supernatant, containing murine granulocyte-macrophage colony-stimulating factor (GM-CSF), or 10% (*v*/*v*) L-cell conditioned medium (LCCM), to differentiate BMM and BMDC, respectively. The media were renewed every two days for BMDC cultures and at day 3 and 5 for BMM. On day 7, non-adherent and loosely adherent BMDC were harvested by gentle washing with PBS and BMM were carefully recovered using a cell scraper. The proportions of differentiated dendritic cells (DC) and macrophages were assessed by flow cytometry upon staining with anti-CD11c FITC-conjugated (clone HL3), anti-F4/80 PE-Cy5.5-conjugated (clone BM8), anti-MHC class II PE-conjugated (clone 2G9) and anti-CD86 PE-Cy7-conjugated (clone GL1).

2.12. *β-Glc-PS Cytotoxicity Analysis*

BMM and BMDC were plated in 96-well tissue culture plates (2 × 105 cells/well) and incubated at 37 °C in a humidified atmosphere in 5% CO_2. Three different concentrations of NP were evaluated. Cell viability was assessed using the 3-[4,5-dimethylthiazol-2-yl]-2,5-diphenyltetrazolium bromide (MTT) assay. Enzymatic activity was quantified after solubilization of MTT formazan with dimethyl sulfoxide (DMSO):ethanol (1:1) solution. Absorbance was then measured at 570 nm. Untreated cells were used as viability control (100%). Normal, apoptotic, and necrotic cells were determined using an Annexin V-FITC/propidium iodide (PI) assay kit (eBioscience, San Diego, CA, USA) according to manufacturer's instructions. The flow cytometry analysis was performed within 15 min in a BD FACSCanto^{TM}II (BD Biosciences, Franklin Lakes, NJ, USA), using the FlowJo software (version 10.0.7; FlowJo, Ashland, OR, USA).

2.13. *Phagocytosis Quantification Assay*

The phagocytosis of *C. albicans* by BMDC and BMM was analyzed by flow cytometry using a previously described methodology [29]. BMDC and BMM suspensions were incubated with Sytox Green-labeled *C. albicans* HK at a MOI (Multiplicity of infection) of 1:5 (1 BMDC/BMM to 5 yeasts) and treated with 10 μg/mL NP for 30 min, at 37 °C and 5% CO_2. Cells were analyzed by flow cytometry in a BD FACSCanto^{TM} II, using the FlowJo software (version 10.0.7; FlowJo, Ashland, OR, USA). The percentage of yeasts interacting with immune cells was calculated from dot plot analysis of Sytox Green vs. PI fluorescence intensities.

2.14. *Evaluation of ROS Production*

Reactive oxygen species (ROS) production was evaluated using the Superoxide Detection Kit for flow cytometry (Enzo Life Sciences, Farmingdale, NY, USA). *C. albicans* HK yeast cells were incubated with BMM or BMDC (MOI of 1 BMDC/BMM: 5 yeasts) and treated at same time with 10 μg/mL NP or 10 μg/mL WGP. As a positive control, BMM and BMDC were treated with 100 nM PMA (Sigma-Aldrich, St. Louis, MO, USA). Plates were incubated for 15 min at 37 °C and 5% CO_2. Upon incubation, cells were prepared according to manufacturer's instructions. Samples were immediately analyzed in a BD FACSCanto II and acquired data were analyzed by using FlowJo software (version 10.0.7; FlowJo, Ashland, OR, USA).

2.15. *Cytokine Quantification by Sandwich ELISA*

BMDC were incubated for 24 h with yeast cells at a MOI of 1 BMDC: 5 Yeasts. Cultures were treated with 10 μg/mL of NP or 10 μg/mL of WGP.

Cytokine levels in cell culture supernatants and serum samples were evaluated by sandwich enzyme-linked immunosorbent assay (ELISA) using commercial kits, according to the manufacturer's instructions (Mouse IL-6, IL-12p70 ELISA Ready-SET-Go!®, eBioscience, San Diego, CA, USA; Mouse TNF-α and IL-1β MAX Standard, Biolegend, San Diego, CA, USA).

2.16. *Statistical Analysis*

All experiments were performed at least in triplicate ($n \geq 3$). Data are reported as means ± SD and were analyzed by one-way ANOVA and Tukey's post hoc test using

GraphPad Prism (Version 7.0; GraphPad Software, San Diego, CA, USA). Statistically significant results were defined as follows: * $p < 0.05$, ** $p < 0.01$, *** $p < 0.001$ and **** $p < 0.0001$.

3. Results and Discussion

3.1. Characterization of Soluble Whole Glucan Particle (WGP)

Soluble WGP was used to functionalize carboxyl-terminated polystyrene nanoparticles (COOH-PS NP) sized 26, 80 and 200 nm. The WGP was characterized according to their molecular weight (MW) distribution, neutral sugars, and glycosidic-linkage analyses (Figure S1, Tables 1 and 2).

Table 1. Characterization of soluble WGP. Neutral sugar analysis of soluble WGP. Carbohydrate composition of (β1,3)-Glucan (WGP) was determined by GC-FID and quantified using 2-deoxyglucose as internal standard. Mean results are presented in molecular percentage (mol %) and mass concentration (mg/g). Each condition was set in triplicate.

	WGP	
	mol %	**C (mg/g)**
Glucose	93.35	607.91
Mannose	0.58	3.54
Arabinose	0.07	0.33

Table 2. Characterization of soluble WGP. Glycosidic linkage composition (molecular percentage) of soluble WGP. The glycosidic-linkage composition of soluble WGP was determined by GC-MS of partially methylated alditol acetates. Man—Mannose; Glc—Glucose; GlcNAc—*N*—Acetylglucosamine Each condition was set in triplicate.

	WGP (mol %)
t-Man	0.80
2,3-Man	0.21
Total Man	**1.01**
t-Glc	20.60
3-Glc	53.53
4-Glc	1.88
6-Glc	12.57
3,4-Glc	1.00
2,3-Glc	3.70
3,6-Glc	3.81
2,3,4-Glc	1.28
2,3,6-Glc	0.18
Total Glc	**98.55**
4-GlcNAc	1.4

Neutral sugar analysis (Table 1) showed that soluble WGP was composed of 93.5% glucose. As expected, since soluble WGP is obtained from *S. cerevisiae* cell wall extractions, minor contaminations with other cell wall sugars, like mannose and arabinose, were also detected. The used soluble WGP was composed mostly of (1,3)-linked glucose (53.5%) followed by 12.6% of (1,6)-linked glucose (Table 2) distributed into two populations with a MW of 50 and 280 kDa (Figure S1), where the highest MW corresponds mainly to (β1→3)-Glc and the lowest MW corresponds mainly to (β1→3)-Glc and (β1→6)-Glc (Table S1 and Figure S2).

3.2. Production and Characterization of β-Glc-PS

Directional conjugation of WGP onto COOH-PS was achieved through carbonyl-reactive chemistry (Figure 1). Three different NP sizes (200, 80 and 26 nm) were chosen as the size of NP may influence WGP delivery to and recognition by cells of the immune system. Protein binding is dependent on NP size as shown for unconjugated COOH-PS

NP in mouse and human serum [30,31], which together with the size itself can affect biodistribution and secretion pathways [32,33].

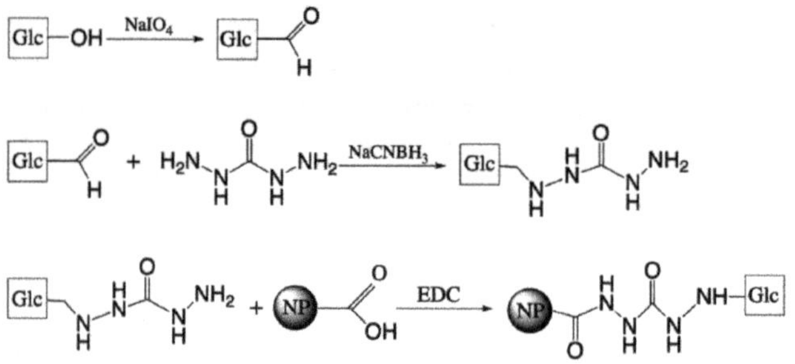

Figure 1. Schematic representation of the chemical conjugation of soluble WGP onto COOH-PS NP surface.

Soluble WGP were oxidized with periodate in order to create aldehyde residues between C2−C3 and C3-C4 bonds of the non-reducing end or (β1→6)-Glc residues [34]. This means that the polymer of lower molecular weight was more prone to be modified than the one with the highest molecular weight which, due to its richness in (β1→3)-Glc residues, resistant to periodate oxidation, and lower content in terminal residues. These oxidized β-glucans were coupled to carbohydrazide and this bond was further stabilized by reduction with sodium cyanoborohydride. β-glucan-hydrazide molecules were incubated with COOH-PS NP in the presence of EDC for a complete conjugation [35].

The size of conjugated NP was determined by DLS and DCS (Table 3 and Figure S3). All NP showed increased radius after β-glucan conjugation. Moreover, all β-Glc-PS NP presented less negative surface charge than naked COOH-PS NP counterparts (Table 3). This charge shift was strongly related with the presence of (β1→3)-Glc onto PS NP surface and the consequent reduction of surface carboxyl groups.

Table 3. Physical properties of polystyrene NP before (COOH-PS) and after conjugation with (β1,3)-glucans (β-Glc-PS). The average hydrodynamic size of COOH-PS and β-Glc-PS NP was evaluated through dynamic light scattering (DLS) at a concentration of 0.5 mg mL^{-1}. PDI values were determined by DLS using DynaPro Plate Reader II. Zeta-potential analysis of COOH-PS and β-Glc-PS was performed using Malvern Zetasizer Nano ZS particle analyser. Surface area valeues were obtained from manufacturer. Values represent mean ± SD from three independent experiments.

Manufacturer Nominal Size (nm)	Surface Conjugation	Diameter (nm)	PDI	ζ-Potential (mV)	Surface Area ($\mu m^2\ g^{-1}$)
200	COOH-PS	189.9 ± 3.3	0.02 ± 0.01	−35.2 ± 2.0	2.9×10^{13}
	β-Glc-PS	199.3 ± 3.6	0.03 ± 0.02	−12.2 ± 0.4	
80	COOH-PS	87.1 ± 0.7	0.09 ± 0.02	−35.8 ± 1.4	7.1×10^{13}
	β-Glc-PS	92.0 ± 0.9	0.08 ± 0.01	−12.9 ± 1.8	
26	COOH-PS	26.1 ± 1.3	0.36 ± 0.01	−27.3 ± 1.3	2.2×10^{14}
	β-Glc-PS	47.5 ± 0.1	0.57 ± 0.02	−8.6 ± 0.6	

The amount of β-glucans on the surface of NP was determined using the phenol-sulphuric acid assay (Table 4) [28]. The average β-glucan content increased as the NP size increased. These quantifications were in accordance with dot blot analysis using (β1→3)-Glc-specific mAb (Figure 2). Although 200 nm NP presented lower surface area per mass and less possible conjugation sites available (1.0×10^{20} COOH groups/mL) when compared with 80 nm (1.3×10^{20} COOH groups/mL) and 26 nm (3.4×10^{20} COOH groups/mL), the 200 nm NP bound more (β1→3)-Glc than the other NP used. Size-related

(β1→3)-Glc conjugation efficacy could be associated with NP size and curvature effect. As previously reported, lower NP curvature facilitates ligand–ligand interactions and hence a higher packing density [36,37].

Table 4. β-glucan content on 0.5 mg/mL NP surface after conjugation was quantified according to Dubois et al. Values represent means ± SD.

β-Glc-PS NP	Glucan Concentration (μg/mL)
200 nm	4.44 ± 1.67
80 nm	2.97 ± 1.10
26 nm	1.41 ± 0.32

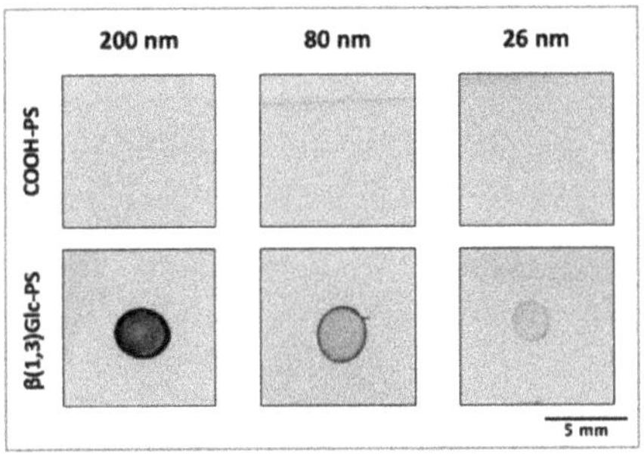

Figure 2. Analysis of (β1,3)-glucan onto NP surface after conjugation. Dot Blot analysis of COOH-PS and β-Glc-PS. After conjugation protocol, 0.5 mg/mL of β-Glc-PS and COOH-PS NP were spotted (5 μL) on nitrocellulose membranes and incubated with (β1,3)-glucan-specific mAb.

The successful covalent conjugation of (β1→3)-Glc onto PS NP surface was also confirmed using Fourier transform infrared (FTIR) spectroscopy (Figure 3). The β-Glc-PS spectra showed a band at ~1650 cm^{-1} (amine I) and 1550 cm^{-1} (amine II), indicating the successful conjugation of β-glucans onto PS surface [38–41]. Moreover, the band at the fingerprint region of 950–1200 cm^{-1}, due to the stretching vibration associated with the C-O-C and C-O-H groups of polysaccharides [27], also confirming the existence of β-glucans on the NP surface [42].

The formation of protein corona modifies the physical and chemical NP properties, namely their hydrodynamics size and surface charge. Moreover, the opsonization of NP by protein corona formation has a central role in the recognition of NP by immune cells. To understand the corona profile of NP in a physiological environment, NP were incubated with MS and the protein corona formation was analyzed by SDS-page (Figure S4). The protein binding after incubation with MS was compared between the three sizes of NP. In previous work, we demonstrated a size-dependent protein corona formation around COOH-PS NP after MS incubation. However, after the NP surface conjugation with soluble β-glucans, no significant differences in protein corona were observed between the three β-Glc-PS NP sizes.

Aggregation of NP was also evaluated after incubation with MS and RPMI + 10% FBS (Figure S5). The 26 nm β-Glc-PS NP showed higher size variations after incubation with RPMI + 10% FBS and MS when compared with 80 nm and 200 nm NP. The aggregation is not obviously associated with differences in NP protein corona. However, the structure of absorbed proteins could be different among NP sizes affecting aggregation and consequently affecting biodistribution, immune recognition, and NP uptake [43].

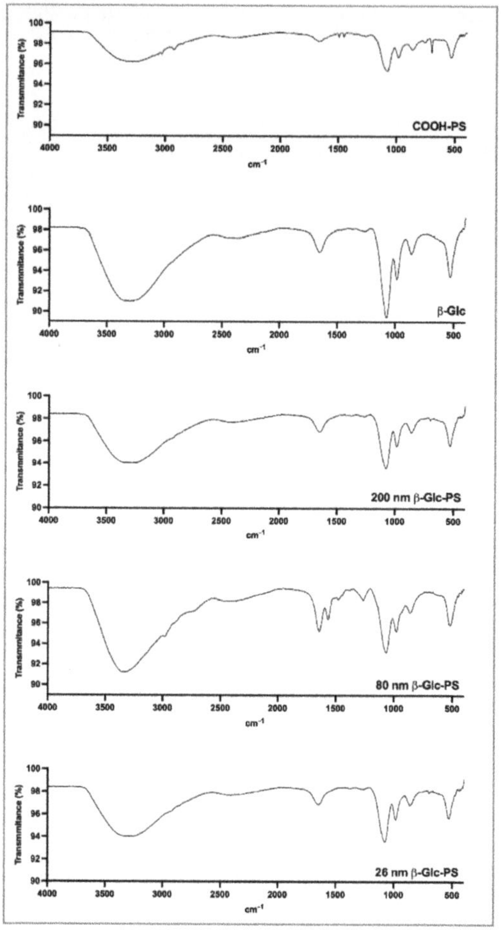

Figure 3. Fourier transformed infrared (FTIR) spectra of COOH-PS, soluble β-Glc and β-Glc-PS NP. Sample suspensions were scanned using PerkinElmer Frontier FTIR (Universal ATR Sampling Assessory) with a resolution of 4 cm^{-1} in ATR sampling mode. Each condition was set in triplicate.

3.3. Biocompatibility of β-Glc-PS with BMDC and BMM

Biocompatibility of β-Glc-PS was assessed with MTT assay and Annexin V/Propidium Iodide (PI) staining using BMDC and BMM (Figure 4a,b). A slight decrease in cell viability was detected when using 26 nm NP while no significant effects were detected using 80 and 200 nm NP. This decrease in cell viability could be related with the high aggregation observed for 26 nm NP when incubated in serum-containing media which might influence its cell interaction and induce cytotoxicity (Figure S5). For the same mass of NP, smaller NP present a higher surface area and thus more available surface to interact with macromolecules, in addition to higher aggregation as already mentioned. Particle size and available surface area may thus contribute to the higher cell toxicity of the 26 nm NP [44]. The protein corona composition does not seem likely to have had a role in toxicity since after conjugation with β-glucans, as all the different sized NP showed similar protein corona profile (Figure S4). Although we cannot exclude that aggregation could influence the NP physical properties and decrease the surface area available, the complexity of 26 nm NP aggregates and the high polydispersity and heterogenicity can still result in cell toxicity since they cannot be assumed to behave like single and stable larger-size NP.

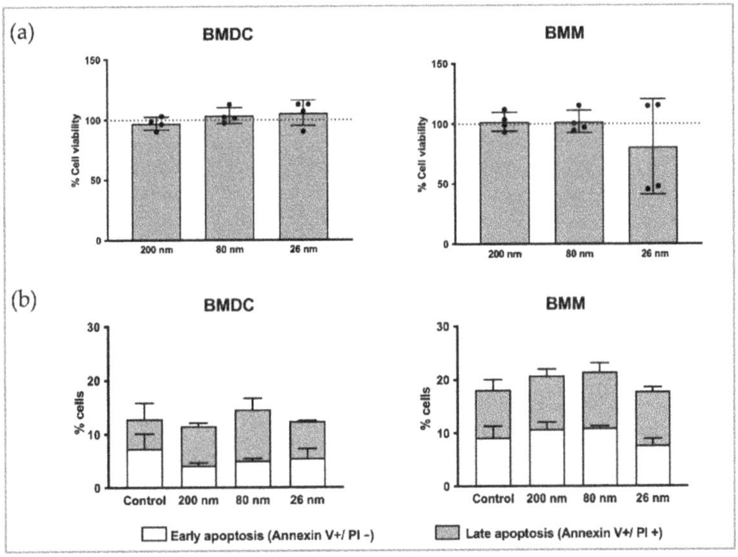

Figure 4. Cell viability of BMDC and BMM treated with β-Glc-PS. BMDC (left) and BMM (right) were stimulated with three different-sized β-Glc-PS for 24 h at a concentration of 10 µg/mL and analyzed in a (**a**) MTT assay and by (**b**) FITC-annexin V staining, using flow cytometry. Samples were acquired on FACSCantoII (BD Biosciences) and data analyzed with FlowJo software (Version 10.0.7). Each black dot corresponds to the mean value of three technical replicates of an independent experiment.

3.4. *Effect of β-Glc-PS on Phagocyte Function*

Phagocytosis plays a critical role in innate immunity, promoting the removal and killing of pathogens, and triggering the adaptive immune response [45].

Phagocytosis assays were performed in vitro using BMDC and BMDC incubated with *C. albicans* and treated with different sized β-Glc-PS. Naked NP (COOH-PS) were used as controls to assess a possible effect of the used polymer. Treatment with 200 or 80 nm β-Glc-PS induced a reduction of yeast phagocytosis, observed in both BMDC and BMM (Figure 5a, gating strategy shown in Figure S6). No reduction in yeast phagocytosis was observed using 26 nm NP, which could indicate a size-dependent effect. Also, no effect was observed on phagocytic function using COOH-PS.

Production of reactive oxygen metabolites is considered a major antifungal mechanism in phagocytes [46,47]. Therefore, we studied the ability of β-Glc-PS to modulate ROS production (Figure 5, Figure S7 and Table S2). As shown in Figure 5b, lower levels of ROS were produced in *C. albicans*-challenged BMDC and BMM treated with β-Glc-PS when compared with non-treated cells. This could be related to the observed reduction of β-glucan-mediated *C. albicans* phagocytosis (Figure 5a). These results indicate that β-Glc-PS may interfere with effector mechanisms used by phagocytic cells in the course of *C. albicans* infections. The immobilization of β-Glc onto NP surface showed to be more effective in reducing ROS production than did free soluble β-Glc treatment. No significant ROS production was detected in non-infected cells treated with β-Glc-PS alone, which reinforces the absence of a nanoparticle cytotoxic effect [48]. COOH-PS had no effect on ROS production by BMM and BMDC after *C. albicans* stimulation (Figure S7).

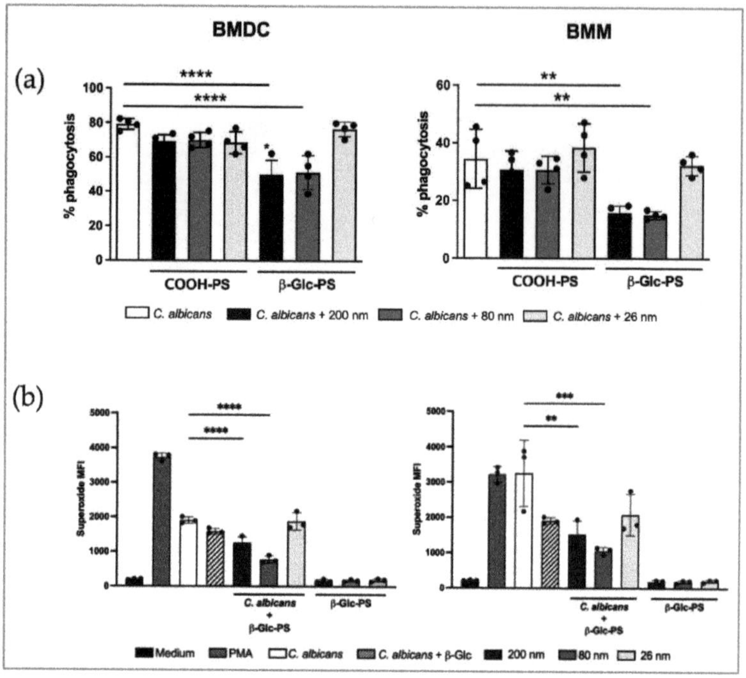

Figure 5. Analysis of β-Glc-PS treatment on phagocyte function. (**a**) Quantification of yeast phagocytosis by BMDC and BMM was assessed by flow cytometry. Immune cells were incubated for 30 min with heat-killed *C. albicans* SC5314 (MOI 1 BMDC/BMM: 5 yeasts) labeled with Sytox Green plus 80 nm β-Glc-PS (10 μg/mL) and stained with Propidium Iodide prior to analysis. (**b**) Production of ROS was assessed by using Superoxide Detection kit. BMM and BMDC incubated for 15 min with heat-killed *C. albicans* SC5314 at a MOI 1:5, in the presence of 200, 80 or 26 nm β-Glc-PS NP (10 μg/mL) and analyzed by flow cytometry. Samples were acquired on FACSCantoII (BD Biosciences) and data analyzed with FlowJo software (Version 10.0.7). Bars correspond to means ± SD. Each condition was set in triplicate. Each black dot corresponds to the mean value of the three technical replicates of an independent experiment. One-way ANOVA with Tukey's post Hoc test, (** $p < 0.01$; *** $p < 0.001$; **** $p < 0.0001$). The complete statistical analysis data is presented in the Supplementary Material (Table S2).

3.5. *β-Glc-PS NP Reduced the Proinflammatory Response of Dendritic Cells Induced by C. albicans*

Following recognition of pathogens by Pathogen Recognition Receptors, distinct signaling pathways are induced. These pathways act through a cross-regulation mechanism, which results in the production of proinflammatory cytokines [49]. The induction of proinflammatory cytokines is an important component of antifungal host defense [50]. Thus, the levels of TNF-α, IL-1β, IL-6, and IL-12p70 were quantified in the culture supernatants of *C. albicans*-challenged BMDC treated and untreated with β-Glc-PS. All the cultures treated with β-Glc-PS showed a marked reduction of TNF-α, IL-1β, IL-6, IL-12p70 levels (Figure 6, Figure S8 and Table S3). β-Glc-PS anti-inflammatory effect showed size-dependency. The 26 nm β-Glc-PS NP was the least effective size. No significant reduction in cytokines levels was observed in BMM and BMDC cultures challenged with *C. albicans* and treated with COOH-PS (Figure S8).

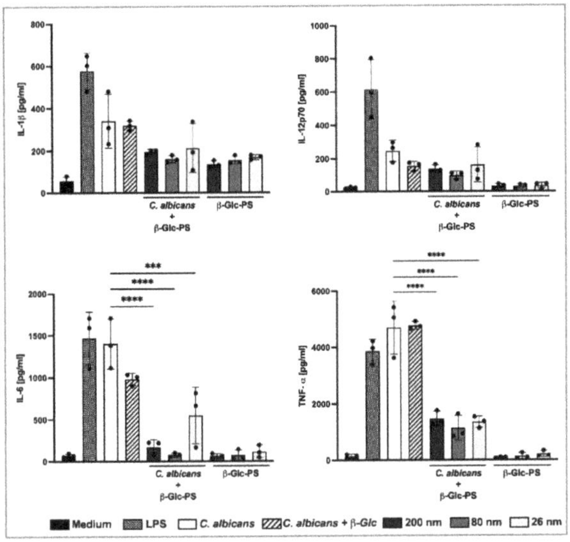

Figure 6. Quantification of cytokines in supernatants of BMDC cultures. The levels of TNF-α, IL-1β, IL-6 and IL-12p70 secreted by BALB/c mice BMDC were quantified by sandwich ELISA. Immune cells were incubated with live *C. albicans* SC5314 at a MOI of 1:5 (1 DC: 5 yeast cells) for 24 h. The 80 nm β-Glc-PS (10 μg/mL) were added 1 h prior to infection. Each condition was set in triplicate. Each black dot corresponds to the mean value of the three technical replicates of an independent experiment. Bars correspond to means ± one SD. One-way ANOVA with Tukey's post hoc test. (*** $p < 0.001$; **** $p < 0.0001$). The complete statistical analysis data is presented in the Supplementary Material (Table S3).

The β-Glc-PS NP treatment showed size dependency. In fact, 80 nm β-Glc-PS was the most biologically effective size. Smaller NP presents a higher surface area per mass, and thus more available surface to interact with cells and cellular components, in addition to higher aggregation and less stability. This may therefore have contributed to the less effective results obtained with 26 nm β-Glc-PS NP (Figure S4) [44]. The curvature could potentially affect the density of conjugated WGP as well as the angles between the WGP chains affecting the binding to Dectin-1 receptor [35,36]. Moreover, the diffusion rate, which increases with decreasing particle size, is also an important factor for ligand receptor recognition [51]. The immobilization of (β1→3)-Glc onto NP surface enhances the anti-inflammatory effect of these polysaccharides when compared with treatment with free soluble (β1→3)-Glc.

4. Conclusions

C. albicans asymptomatically colonizes more than 30% of individuals in a population at any given time. However, when the host immune system of the carrier individuals is weakened, this fungus can cause both mucosal and systemic infections. Some risk factors, such as neutropenia, central venous catheters, or systemic antibiotic exposure, may predispose individuals to invasive and even life-threatening systemic candidiasis. Invasive *C. albicans* infections can result in an uncontrolled hyper-inflammatory response, leading to severe host damage. With this study we showed that β-glucan-functionalized nanoparticles could down-modulate the proinflammatory response of host immune cells induced by *C. albicans*, in a size-dependent manner. This NP-based approach showed a promising therapeutic potential, extending the described potential of (β1→3)-Glc in host inflammation control.

Supplementary Materials: The following supporting information can be downloaded at: https://www.mdpi.com/article/10.3390/nano12142475/s1, Figure S1: Characterization of soluble WGP. Chromatogram obtained by gel permeation chromatography (GPC); Figure S2. Light Scattering analysis of WGP SEC separation. Table S1. Glycosidic linkage composition of WGP fractions after SEC separation. Figure S3: Physical properties of NP after conjugation with β-(1,3)-glucans; Figure S4: Protein corona of β-Glc-PS after mouse serum incubation; Figure S5: Hydrodynamics radius of 200 nm, 80 nm and 26 nm in biological conditions; Figure S6: Representative FACS gating strategy of phagocytosis assay; Figure S7: Effect of COOH-PS NP treatment in BMDC and BMM ROS production; Table S2: Statistical analysis of BMDC and BMM ROS production; Figure S8: Effect of COOH-PS NP treatment in BMDC pro-inflammatory cytokine production. Table S3: Statistical analysis of BMDC proinflammatory cytokine production.

Author Contributions: Conceptualization, Methodology, Investigation, Formal analysis, Data Curation, Writing—Original Draft, Visualization, T.L.; Conceptualization, Methodology, Investigation, S.B.G.; Conceptualization, Investigation, A.C.; Methodology, E.C. and D.V.E.; Conceptualization, Methodology, Resources, Supervision, M.A.C.; Conceptualization, Methodology, Resources, Writing—Review and Editing, Supervision, Project administration, T.C.; Conceptualization, Methodology, Resources, Validation, Writing—Review and Editing, Supervision, Project administration, Funding acquisition, M.V. All authors have read and agreed to the published version of the manuscript.

Funding: This work received financial support from PT national funds (FCT/MCTES, Fundação para a Ciência e Tecnologia and Ministério da Ciência, Tecnologia e Ensino Superior). LAQV/REQUIMTE (UIDB/50006/2020, UIDP/50006/2020), through national founds and, where applicable, co-funded by the FEDER, within the PT2020 Partnership Agreement and Compete 2020. HEALTH-UNORTE: Setting-up biobanks and regenerative medicine strategies to boost research in cardiovascular, musculoskeletal, neurological, oncological, immunological and infectious diseases (NORTE-01-0145-FEDER-000039), supported by Norte Portugal Regional Operational Programme (NORTE 2020), under the PORTUGAL 2020 Partnership Agreement, through the European Regional Development Fund (ERDF). T.L. was supported by FCT PhD grant (PD/BD/128393/2017) from Biotech-Health International Doctoral programme. A.C. was supported by FCT Individual CEEC (CEEC IND/01514/2017). E.C. was supported by FCT through program DL 57/2016-Norma transitória (CDL-CTTRI-88-ARH/2018-REF.049-88-ARH/2018).

Institutional Review Board Statement: The animal study protocol was approved by the institutional board responsible for animal welfare (ORBEA) of Instituto de Investigação e Inovação em Saúde (i3S), Porto, Portugal (2019-5/2019-06-11) and authorization to perform the experiments was issued by the competent national authority (Direção Geral de Alimentação e Veterinária) with the reference number 014036/2019-07-24.

Informed Consent Statement: Not applicable.

Data Availability Statement: The datasets generated and/or analyzed during the current study are available from the corresponding author on reasonable request.

Acknowledgments: The authors gratefully acknowledge the support of the i3S Scientific Platforms, namely Translational Cytometry and Biointerfaces and Nanotechnology platforms.

Conflicts of Interest: The authors declare that they have no known competing financial interest or personal relationships that could have appeared to influence the work reported in this paper.

References

1. Tam, J.M.; Mansour, M.K.; Khan, N.S.; Yoder, N.C.; Vyas, J.M. Use of fungal derived polysaccharide-conjugated particles to probe Dectin-1 responses in innate immunity. *Integr. Biol.* **2012**, *4*, 220–227. [CrossRef] [PubMed]
2. Pfaller, M.A.; Diekema, D.J. Epidemiology of Invasive Candidiasis: A Persistent Public Health Problem. *Clin. Microbiol. Rev.* **2007**, *20*, 133–163. [CrossRef] [PubMed]
3. Kim, J.; Sudbery, P. Candida albicans, a major human fungal pathogen. *J. Microbiol.* **2011**, *49*, 171–177. [CrossRef] [PubMed]
4. Pappas, P.G.; Lionakis, M.S.; Arendrup, M.C.; Ostrosky-Zeichner, L.; Kullberg, B.J. Invasive candidiasis. *Nat. Rev. Dis. Primers* **2018**, *4*, 18026. [CrossRef] [PubMed]
5. Gozalbo, D.; Roig, P.; Villamon, E.; Gil, M. Candida and Candidiasis: The Cell Wall as a Potential Molecular Target for Antifungal Therapy. *Curr. Drug Target-Infect. Disord.* **2004**, *4*, 117–135. [CrossRef]

6. Ruiz-Herrera, J.; Elorza, M.V.; Valentin, E.; Sentandreu, R. Molecular organization of the cell wall of Candida albicans and its relation to pathogenicity. *FEMS Yeast Res.* **2006**, *6*, 14–29. [CrossRef]
7. Hall, R.A. Dressed to impress: Impact of environmental adaptation on the Candida albicans cell wall. *Mol. Microbiol.* **2015**, *97*, 7–17. [CrossRef]
8. Nakagawa, Y.; Kanbe, T.; Mizuguchi, I. Disruption of the Human Pathogenic Yeast*Candida albicans*Catalase Gene Decreases Survival in Mouse-Model Infection and Elevates Susceptibility to Higher Temperature and to Detergents. *Microbiol. Immunol.* **2003**, *47*, 395–403. [CrossRef]
9. Bain, J.M.; Louw, J.; Lewis, L.E.; Okai, B.; Walls, C.A.; Ballou, E.R.; Walker, L.A.; Reid, D.; Munro, C.A.; Brown, A.J.P.; et al. Candida albicans hypha formation and mannan masking of beta-glucan inhibit macrophage phagosome maturation. *mBio* **2014**, *5*, e01874. [CrossRef] [PubMed]
10. Duggan, S.; Leonhardt, I.; Hunniger, K.; Kurzai, O. Host response to Candida albicans bloodstream infection and sepsis. *Virulence* **2015**, *6*, 316–326.
11. Tran, V.G.; Kim, H.J.; Kim, J.; Kang, S.W.; Moon, U.J.; Cho, H.R.; Kwon, B. IL-33 Enhances Host Tolerance to Candida albicans Kidney Infections through Induction of IL-13 Production by CD4+ T Cells. *J. Immunol.* **2015**, *194*, 4871–4879. [CrossRef] [PubMed]
12. Mortaz, E.; Alipoor, S.D.; Adcock, I.M.; Mumby, S.; Koenderman, L. Update on Neutrophil Function in Severe Inflammation. *Front. Immunol.* **2018**, *9*, 2171. [CrossRef] [PubMed]
13. Brown, G.D.; Gordon, S. Fungal beta-glucans and mammalian immunity. *Immunity* **2003**, *19*, 311–315. [CrossRef]
14. Aaarsæther, E.J.R.M.; Engstad, R.E.; Busund, R. Cardioprotective effect of pre-treatment with beta-glucan in coronary artery bypass grafting. *Scand. Cardiovasc. J.* **2006**, *40*, 298304. [CrossRef] [PubMed]
15. Zykova, S.N.B.K.; Vorokhobina, N.V.; Kuznetsova, A.V.; Engstad, R.Z.T. Macrophage stimulating agent soluble yeast b-1,3/1,6-glucan as a topical treatment of diabetic foot and leg ulcers: A randomized, double blind, placebo-controlled phase II study. *J. Diabetes Investig.* **2014**, *5*, 392–399. [CrossRef]
16. Sandvik, A.; Wang, Y.Y.; Morton, H.C.; Aasen, A.O.; Wang, J.E.; Johansen, F.E. Oral and systemic administration of beta-glucan protects against lipopolysaccharide-induced shock and organ injury in rats. *Clin. Exp. Immunol.* **2007**, *148*, 168–177. [CrossRef]
17. Johansen, F.E.; Sandvik, A.; Engstad, R.E. Methods of Treating or Preventing Inflammatory Diseases of the Intestinal Tract. International Publication WO/2009/063221, 22 May 2009.
18. Zhang, M.; Kim, J.A.; Huang, A.Y. Optimizing Tumor Microenvironment for Cancer Immunotherapy: Beta-Glucan-Based Nanoparticles. *Front. Immunol.* **2018**, *9*, 341. [CrossRef]
19. Mankes, R.F.; Silver, C.D. Quantitative study of controlled substance bedside wasting, disposal and evaluation of potential ecologic effects. *Sci. Total Environ.* **2013**, *444*, 298–310. [CrossRef]
20. Onoue, S.; Yamada, S.; Chan, K. Nanodrugs: Pharmacokinetics and safety. *Int. J. Nanomed.* **2014**, *9*, 1025–1037. [CrossRef]
21. Foroozandeh, P.; Aziz, A.A. Insight into Cellular Uptake and Intracellular Trafficking of Nanoparticles. *Nanoscale Res. Lett.* **2018**, *13*, 339. [CrossRef]
22. Shvedova, A.A.; Fabisiak, J.; Kisin, E.R.; Murray, A.R.; Roberts, J.R.; Tyurina, Y.; Antonini, J.M.; Feng, W.H.; Kommineni, C.; Reynolds, J.; et al. Sequential Exposure to Carbon Nanotubes and Bacteria Enhances Pulmonary Inflammation and Infectivity. *Am. J. Respir. Cell Mol. Biol.* **2008**, *38*, 579–590. [CrossRef] [PubMed]
23. Shvedova, A.A.; Kisin, E.R.; Porter, D.; Schulte, P.; Kagan, V.E.; Fadeel, B.; Castranova, V. Mechanisms of pulmonary toxicity and medical applications of carbon nanotubes: Two faces of Janus? *Pharmacol. Ther.* **2009**, *121*, 192–204. [CrossRef] [PubMed]
24. Sanfins, E.; Augustsson, C.; Dahlbäck, B.; Linse, S.; Cedervall, T. Size-Dependent Effects of Nanoparticles on Enzymes in the Blood Coagulation Cascade. *Nano Lett.* **2014**, *14*, 4736–4744. [CrossRef] [PubMed]
25. Blakeney, A.B.; Harris, P.J.; Henry, R.J.; Stone, B.A. A simple and rapid preparation of alditol acetates for monosaccharide analysis. *Carbohydr. Res.* **1983**, *113*, 291–299. [CrossRef]
26. Bastos, R.; Coelho, E.; Coimbra, M.A. Modifications of Saccharomyces pastorianus cell wall polysaccharides with brewing process. *Carbohydr. Polym.* **2015**, *124*, 322–330. [CrossRef]
27. Pinto, M.; Coelho, E.; Nunes, A.; Brandao, T.; Coimbra, M.A. Valuation of brewers spent yeast polysaccharides: A structural characterization approach. *Carbohydr. Polym.* **2015**, *116*, 215–222. [CrossRef]
28. DuBois, M.; Gilles, K.A.; Hamilton, J.K.; Rebers, P.A.; Smith, F. Colorimetric method for determination of sugars and related substances. *Anal. Chem.* **1956**, *28*, 350–356. [CrossRef]
29. Carneiro, C.; Vaz, C.; Carvalho-Pereira, J.; Pais, C.; Sampaio, P. A new method for yeast phagocytosis analysis by flow cytometry. *J. Microbiol. Methods* **2014**, *101*, 56–62. [CrossRef]
30. Lima, T.; Bernfur, K.; Vilanova, M.; Cedervall, T. Understanding the Lipid and Protein Corona Formation on Different Sized Polymeric Nanoparticles. *Sci. Rep.* **2020**, *10*, 1129. [CrossRef]
31. Lundqvist, M.; Stigler, J.; Elia, G.; Lynch, I.; Cedervall, T.; Dawson, K.A. Nanoparticle size and surface properties determine the protein corona with possible implications for biological impacts. *Proc. Natl. Acad. Sci. USA* **2008**, *105*, 14265–14270. [CrossRef]
32. Li, B.; Lane, L.A. Probing the biological obstacles of nanomedicine with gold nanoparticles. *WIREs Nanomed. Nanobiotechnol.* **2018**, *11*, e1542. [CrossRef] [PubMed]
33. Hoshyar, N.; Gray, S.; Han, H.; Bao, G. The effect of nanoparticle size on in vivo pharmacokinetics and cellular interaction. *Nanomedicine* **2016**, *11*, 673–692. [CrossRef] [PubMed]

34. Kristiansen, K.A.; Potthast, A.; Christensen, B.E. Periodate oxidation of polysaccharides for modification of chemical and physical properties. *Carbohydr. Res.* **2010**, *345*, 1264–1271. [CrossRef] [PubMed]
35. Hermanson, G.T. *Microparticles and Nanoparticles, in Bioconjugate Techniques*; Elsevier: Amsterdam, The Netherlands, 2013; pp. 571–573.
36. Solveyra, E.G.; Szleifer, I. What is the role of curvature on the properties of nanomaterials for biomedical applications? Wiley interdisciplinary reviews. *Nanomed. Nanobiotechnol.* **2016**, *8*, 334–354. [CrossRef] [PubMed]
37. Panja, S.; Saha, B.; Ghosh, S.K.; Chattopadhyay, S. Synthesis of Novel Four Armed PE-PCL Grafted Superparamagnetic and Biocompatible Nanoparticles. *Langmuir* **2013**, *29*, 12530–12540. [CrossRef]
38. Villarreal, E.; Li, G.G.; Zhang, Q.; Fu, X.; Wang, H. Nanoscale Surface Curvature Effects on Ligand-Nanoparticle Interactions: A Plasmon-Enhanced Spectroscopic Study of Thiolated Ligand Adsorption, Desorption, and Exchange on Gold Nanoparticles. *Nano Lett.* **2017**, *17*, 4443–4452. [CrossRef]
39. Maruyama, T.; Katoh, S.; Nakajima, M.; Nabetani, H.; Abbott, T.P.; Shono, A.; Satoh, K. FT-IR analysis of BSA fouled on ultrafiltration and microfiltration membranes. *J. Membr. Sci.* **2001**, *192*, 201–207. [CrossRef]
40. Chittur, K.K. FTIR/ATR for protein adsorption to biomaterial surfaces. *Biomaterials* **1998**, *19*, 357–369. [CrossRef]
41. Teng, Y.; Pong, P.W.T. Fe_2O_3-SiO_2 Core-Shell Nanoparticles Conjugated with Bovine Serum Albumin. *IEEE Trans. Magn.* **2017**, *53*, 1–6. [CrossRef]
42. Irshad, A.; Sarwar, N.; Sadia, H.; Riaz, M.; Sharif, S.; Shahid, M.; Khan, J.A. Silver nano-particles: Synthesis and characterization by using glucans extracted from Pleurotus ostreatus. *Appl. Nanosci.* **2019**, *10*, 3205–3214. [CrossRef]
43. Lundqvist, M.; Sethson, I.; Jonsson, B.-H. Protein Adsorption onto Silica Nanoparticles: Conformational Changes Depend on the Particles' Curvature and the Protein Stability. *Langmuir* **2004**, *20*, 10639–10647. [CrossRef] [PubMed]
44. Huang, Y.W.; Cambre, M.; Lee, H.J. The Toxicity of Nanoparticles Depends on Multiple Molecular and Physicochemical Mechanisms. *Int. J. Mol. Sci.* **2017**, *18*, 2702. [CrossRef] [PubMed]
45. Herre, J.; Marshall, A.S.; Caron, E.; Edwards, A.D.; Williams, D.L.; Schweighoffer, E.; Tybulewicz, V.; e Sousa, C.R.; Gordon, S.; Brown, G.D. Dectin-1 uses novel mechanisms for yeast phagocytosis in macrophages. *Blood* **2004**, *104*, 4038–4045. [CrossRef] [PubMed]
46. Cheng, S.C.; Joosten, L.A.; Kullberg, B.J.; Netea, M.G. Interplay between Candida albicans and the mammalian innate host defense. *Infect. Immun.* **2012**, *80*, 1304–1313. [CrossRef]
47. Gantner, B.N.; Simmons, R.M.; Underhill, D.M. Dectin-1 mediates macrophage recognition of Candida albicans yeast but not filaments. *EMBO J.* **2005**, *24*, 1277–1286. [CrossRef]
48. Fu, P.P.; Xia, Q.; Hwang, H.-M.; Ray, P.C.; Yu, H. Mechanisms of nanotoxicity: Generation of reactive oxygen species. *J. Food Drug Anal.* **2014**, *22*, 64–75. [CrossRef]
49. Mogensen, T.H. Pathogen Recognition and Inflammatory Signaling in Innate Immune Defenses. *Clin. Microbiol. Rev.* **2009**, *22*, 240–273. [CrossRef]
50. Ifrim, D.C.; Bain, J.M.; Reid, D.M.; Oosting, M.; Verschueren, I.; Gow, N.A.R.; van Krieken, J.H.; Brown, G.D.; Kullberg, B.-J.; Joosten, L.A.B.; et al. Role of Dectin-2 for Host Defense against Systemic Infection with Candida glabrata. *Infect. Immun.* **2014**, *82*, 1064–1073. [CrossRef]
51. Dell'Orco, D.; Lundqvist, M.; Cedervall, T.; Linse, S. Delivery success rate of engineered nanoparticles in the presence of the protein corona: A systems-level screening. *Nanomedicine* **2012**, *8*, 1271–1281. [CrossRef]

nanomaterials

MDPI

Article

Graphene Oxide Nanoplatforms to Enhance Cisplatin-Based Drug Delivery in Anticancer Therapy

Elena Giusto [1,†], Ludmila Žárská [2,†], Darren Fergal Beirne [3], Arianna Rossi [1,4], Giada Bassi [1,5], Andrea Ruffini [1], Monica Montesi [1], Diego Montagner [3,*], Vaclav Ranc [2,6,*] and Silvia Panseri [1,*]

1 Institute of Science and Technology for Ceramics–National Research Council (CNR), 48018 Faenza (RA), Italy; elena.giusto@aol.com (E.G.); arianna.rossi@istec.cnr.it (A.R.); giada.bassi@istec.cnr.it (G.B.); andrea.ruffini@istec.cnr.it (A.R.); monica.montesi@istec.cnr.it (M.M.)
2 Regional Centre of Advanced Technologies and Materials, Czech Advanced Technology and Research Institute, Palacký University Olomouc, 78371 Olomouc, Czech Republic; ludmila.zarska@centrum.cz
3 Department of Chemistry, Maynooth University, Maynooth, Ireland; darren.beirne.2017@mumail.com
4 Department of Chemical, Biological, Pharmaceutical and Environmental Sciences, University of Studies of Messina, 98100 Messina (ME), Italy
5 Department of Neuroscience, Imaging and Clinical Sciences, University of Studies G. d'Annunzio Chieti-Pescara, 66100 Chieti (CH), Italy
6 Institute of Molecular and Translation Medicine, Faculty of Medicine and Dentistry, Palacký University in Olomouc, Hnevotinska 5, 77900 Olomouc, Czech Republic
* Correspondence: diego.montagner@mu.ie (D.M.); vaclav.ranc@upol.cz (V.R.); silvia.panseri@istec.cnr.it (S.P.)
† These authors contributed equally to this work.

Abstract: Chemotherapeutics such as platinum-based drugs are commonly used to treat several cancer types, but unfortunately, their use is limited by several side effects, such as high degradation of the drug before entering the cells, off-target organ toxicity and development of drug resistance. An interesting strategy to overcome such limitations is the development of nanocarriers that could enhance cellular accumulation in target cells in addition to decreasing associated drug toxicity in normal cells. Here, we aim to prepare and characterize a graphene-oxide-based 2D nanoplatform functionalised using highly branched, eight-arm polyethylene-glycol, which, owing to its high number of available functional groups, offers considerable loading capacity over its linear modalities and represents a highly potent nanodelivery platform as a versatile system in cancer therapy. The obtained results show that the GO@PEG carrier allows for the use of lower amounts of Pt drug compared to a Pt-free complex while achieving similar effects. The nanoplatform accomplishes very good cellular proliferation inhibition in osteosarcoma, which is strictly related to increased cellular uptake. This enhanced cellular internalization is also observed in glioblastoma, although it is less pronounced due to differences in metabolism compared to osteosarcoma. The proposed GO@PEG nanoplatform is also promising for the inhibition of migration, especially in highly invasive breast carcinoma (i.e., MDA-MB-231 cell line), neutralizing the metastatic process. The GO@PEG nanoplatform thus represents an interesting tool in cancer treatment that can be specifically tailored to target different cancers.

Keywords: nanomaterials; platinum-based drug; graphene oxide; breast cancer; osteosarcoma; glioblastoma; nanomedicine; drug delivery systems

Citation: Giusto, E.; Žárská, L.; Beirne, D.F.; Rossi, A.; Bassi, G.; Ruffini, A.; Montesi, M.; Montagner, D.; Ranc, V.; Panseri, S. Graphene Oxide Nanoplatforms to Enhance Cisplatin-Based Drug Delivery in Anticancer Therapy. *Nanomaterials* **2022**, *12*, 2372. https://doi.org/10.3390/nano12142372

Academic Editors: Goran Kaluđerović and Nebojša Pantelić

Received: 21 June 2022
Accepted: 9 July 2022
Published: 11 July 2022

1. Introduction

Cancer is one of the leading causes of death worldwide [1,2], and it is frequently caused by many factors, including genetic variables, physical and chemical insults and lifestyle habits, such as poor diet or smoking [3,4]. It has been widely studied based on its epidemiology and mechanism, with the main therapies adopted to date based on various surgeries, chemotherapy and radiotherapy, as well as relatively new strategies, such as immune or gene therapies. Following surgical removal, chemotherapy is the most common treatment strategy, as it is based on the use of specific molecules targeting the high cancer cell proliferation rate, interfering with their metabolism [5]. Among chemotherapeutic agents, platinum(II)-based compounds are the most successful, and three compounds, i.e., cisplatin, carboplatin and oxaliplatin, are used worldwide to cure several cancer types (Figure 1) [6,7].

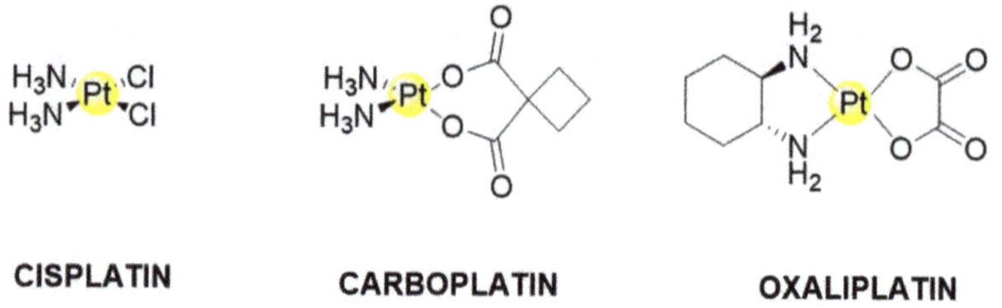

Figure 1. Pt-based drugs. Cisplatin, carboplatin and oxaliplatin.

The main target of platinum-based drugs is nuclear DNA; upon penetration of the cell membrane, cisplatin infiltrates the hydrolysis pathways forming cationic aqua species that reach and covalently bind to the nuclear DNA, causing its damage, as well as the arrest of the cancer cell cycle in the G2/M transition phase, leading to apoptosis [7]. Although platinum-based chemotherapeutics are the most frequently used anticancer drugs, they still have several side effects. Platinum is administered intravenously, abd the targeted cancer cells are reached the via blood stream, where the drug is usually bound to plasma proteins, such as albumin, and degraded in a high percentage before entering the target cells [7,8]. Systemic administration potentially leads to higher toxicity for off-target organs, eventually causing adverse side effects, such as nephrotoxicity and ototoxicity [9,10]. Moreover, tumour cell resistance to platinum has adverse and serious consequences for the fate of patients [11]. Ab interesting strategy to overcome these limitations is the development of nanocarriers that could enhance cellular accumulation, decreasing the associated toxicity [10–12]. Graphene oxide (GO)-based nanoplatforms have interesting physicochemical and surface properties, making them potentially attractive for medical applications as nanomaterials for cancer-targeted drug delivery [13,14]. However, water-dispersible GO often aggregates under physiological conditions in the presence of salts due to the charge screening effect [15]. In addition, depending on the concentration used, GO can per se induce cytotoxicity due to oxidative stress [16,17]. To target these problems, GO readily undergoes bipartite surface modification thanks to carboxyl groups abundantly present on its surface that allow for functionalisation with many biomolecules and drugs [18]. In order to reduce the level of cytotoxicity while promoting its cellular uptake, the use of bio-mimicking molecules, such as polyethylene glycol (PEG), has been recently studied [18], and it has been shown that the addition of PEG increases stability, improves solubility and reduces aggregation, prolonging the circulation of GO in the bloodstream [19,20].

In the present study, we focused on the preparation and characterization of a GO-based nanoplatform functionalised using highly branched PEG modality, namely 8-arm PEG, to enhance the efficacy and loading capacity of Pt-based drugs, thereby achieving a highly performant and stable nanodelivery system. Physicochemical characterization was carried out to investigate the features of the functionalised nanomaterial in terms of its dimensions, platinum drug-loading capacity and long-term stability. Its biological activity, such as cellular uptake, viability, morphology and migration, was then evaluated in seven human tumour cell lines selected as representative of three cancers with high incidence and morbidity worldwide, i.e., breast cancer: MDA-MB-231 and MDA-MB-486 cell lines; glioblastoma: U87 and U118 cell lines; and osteosarcoma: MG63, U2-OS and SAOS-2 cell lines [12,21,22].

2. Materials and Methods

2.1. Synthesis of Pt-Based Drug 1 (Pt-Free)

Compound **1** was synthesized as previously reported with few modifications (Figure 2) [23]. Oxoplatin, c,c,t-[$Pt(NH_3)_2Cl_2(OH)_2$] (0.1 g, 0.3 mmol) was suspended in 10 mL of DMSO. Succinic anhydride (0.028 g, 0.28 mmol) was added to the mixture, and the suspension was stirred at 45 °C overnight. The obtained solution was filtered to remove the small amount of unreacted oxoplatin and lyophilized overnight. The residue was washed with cold acetone, cold methanol and diethyl ether and dried under a vacuum (0.058 g, yield 62%). ^{1}H-NMR (DMSO-d_6) 6.50 (br.tr, 6H), 2.95–1.93 (m, 4H). Elemental analyses calc for $C_4H_{12}Cl_2N_2O_5Pt$: C, 11.07; H, 2.79; N, 6.45. Found: C, 11.35; H, 2.51; N, 6.28.

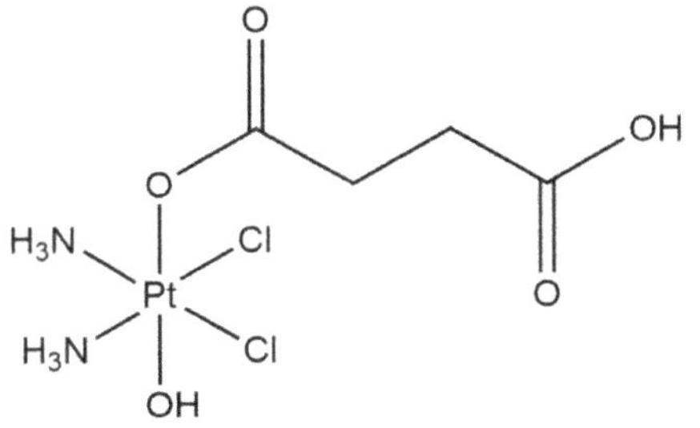

Figure 2. Chemical structure of Compound **1** (Pt-free).

2.2. GO Flake Size Optimization

Commercially available graphene oxide (GO, Sigma Aldrich, Saint Louis, MO, USA) was used as a starting material. The flakes size adjustment and size selection were based on a combination of two previously published protocols [24,25]. Briefly, the GO stock solution (4 mg/mL) was diluted to a concentration of 400 µg/mL in PBS buffer. The diluted GO solution was further sonicated in an ultrasonic bath (Sonorex Digitec DT 103 H, Bandelin, Berlin, ermany) at 70 °C for 6 h. The sample was then agitated for 18 h with a Heidolph (Schwabach, Germany) Unimax 1010 shaker (500 RPM, 65 °C) and sonicated again for 6 h in the ultrasonic bath at 70 °C. Large- flakes were removed by centrifugation (Benchtop 4–16 K, 21191 RCF, 5 min), and the supernatant containing GO dispersion was used in all further experiments.

2.3. PEGylation of GO

An amount of 25 mg 8-arm polyethylene glycol-amine (10 kDa, Sigma-Aldrich, St. Louis, MI, USA) was added to the 5 mL of GO dispersion prepared in the previous step and sonicated for 10 min. Subsequently, 40 µL of 5 mg/mL N-(3-Dimethylaminopropyl)-N′-ethylcarbodiimide hydrochloride (EDC, Sigma-Aldrich) was added to the mixture dropwise. Next, a second cycle of agitation and sonification was performed for 18 h (500 RPM, 65 °C) and 6 h (70 °C). The infrared spectra were obtained on a Nicolet iS5 FTIR spectrometer (Fisher Scientific, Waltham, MA, USA) in ATR mode using a ZnSe crystal.

2.4. Pt Loading on GO@PEG Nanoplaftorms

Loading of compound **1** onto GO@PEG was carried out in two consecutive steps. A stock solution containing 5 mg of compound **1** and 22.1 mg EDC dissolved in 1 mL deionized H_2O was resuspended by a brief sonication to form a homogeneous, clear solution, which was agitated (500 RPM) at room temperature for 1 h. A volume of 100 µL of 5 mg/mL compound **1** stock solution was added to 1 mL GO@PEG and agitated for 24 h (500 RPM, 23 °C). Unbound compound **1** was removed by centrifugation at 21,191 RCF for 10 min. The total amount of the anchored compound **1** was determined by atomic absorption spectroscopy (AAS) performed in triplicate ($N = 3$). The resulting pellet GO@PEG-Pt was gently resuspended in 1 mL PBS and stored at 4 °C until use.

2.5. Characterization of GO-Based Nanoplatforms

2.5.1. Determination of GO Amount and Size Distribution

Raman spectra of GO@PEG nanoplatforms were obtained using a Witec Alpha 300 R+ Raman spectroscopic system (Witec, Ulm, Germany) followed by excitation operating at 532 nm. The power of the laser on the sample was 5 mW. In total, 30 microscans were averaged to obtain one spectrum. In total, six spectra from six random locations over the flake were averaged to obtain the data shown in the Section 3. The concentration and size distribution of the prepared dispersion were determined by means of atomic force microscopy (AFM). The concentration was 1.5×10^9 GO flakes/mL, with a median of 266 nm. An atomic force microscope (AFM, Ntegra spectra, NT-MDT, Moscow, Russia) was used to analyse the height and size profile of the GO flakes in the stock solution and in the supernatant solution, as well as GO@PEG (Supplementary Materials, Figure S1). Based on previous AFM observations by many groups, the thickness of single-layer graphene has been experimentally demonstrated to be approximately 1.1 nm [26–28]. This microscopic method was also used to determine the amount of GO flakes in 1 mL of supernatant solution and, subsequently, to analyse the size distribution of GO in the supernatant solution.

A volume of 5 μL of the sample was pipetted onto a mica substrate with a radius of 0.5 cm. AFM images of 50 × 50 μm samples were taken in semi-contact mode with an ACTA-SS-10 tip at a scan speed of 0.3 Hz. Subsequently, the captured images were edited in the Gwyddion program, and the number of GO flakes per image was analysed using ImageJ software. The final amount of GO in the supernatant solution was then calculated. The evaluation results in ImageJ were also used for size distribution ($N = 2142$).

Scanning electron microscopy (SEM) images taken using a Hitachi SU6600 scanning electron microscope (Hitachi, Tokyo, Japan). For analysis of GO in stock solution, a small drop of material dispersion in water was placed on a carbon tape and dried at room temperature. An accelerating voltage of 7 kV was used for imaging. A small drop of GO@PEG dispersion in water was placed on a copper grid with carbon film and dried at room temperature. The sample was analysed with an accelerating voltage of 5 kV.

2.5.2. Pt Loading on GO-PEG Nanoplatforms

The loading efficiency of compound **1** on GO@PEG was determined from the supernatant obtained by centrifugation in the final step of GO@PEG-Pt preparation (Section 2.4) using atomic absorption spectroscopy (AAS). The loading ratio of Pt was analysed by AAS and calculated according to a previously published definition of loading efficiency [28]. LE% is defined as (concentration of the drug loaded on GO/the initial concentration of the drug) × 100:

$$\text{LE} = \frac{Concentration\ of\ Pt\ loaded}{concentration\ of\ Pt\ initially} \times 100\%$$

2.5.3. GO-PEG Pt Stability Testing

GO@PEG Pt solution in the dialysis membrane was incubated in a flask with PBS at pH 7.4 for 1, 2, 3, 5, 14 and 21 days under two different temperature conditions (4 °C and 23 °C). The concentration of released Pt from the GO@PEG nanoplatform to PBS was also measured by means of AAS. Measurements were performed in triplicate for each time point ($N = 3$).

2.6. In Vitro Biological Study

Compound **1** (Pt-free) was dissolved in DMSO at 1 mg/mLm diluted in the cultured medium at different concentrations (i.e., 15 μM, 30 μM and 60 μM) and used as the Pt-free group. Based on the quantification of compound **1** loaded on the nanoplatforms, the bioactivity of GO@PEG-Pt was tested at the same concentration of Pt-free that corresponded to 1.0 μg/mL, 2.0 μg/mL and 4.0 μg/mL of GO@PEG nanoplatforms loaded with 15 μM, 30 μM and 60 μM of Pt-free, respectively. The GO@PEG nanoplatforms alone were also tested at the same concentration (1.0 μg/mL, 2.0 μg/mL and 4.0 μg/mL) to verify the cytotoxicity of the nanoplatform itself. Cells alone were used as a control group.

2.6.1. Cell Culture

The following cell lines were used: three human osteosarcoma cell lines: MG63 (ATCC CRL1427), SAOS-2 (ATCC HTB-85) and U2-OS (ATCC HTB-96); two human adenocarcinoma cell lines isolated from breast cancers: MDA-MB-231 (ATCC HTB26) and MDA-MB-468 (ATCC HTB231); and two human glioblastoma cell lines: U118 (ATCC HTB15) and U87 (ATCC HTB14). MG63 cells were cultured in a growth medium composed of DMEM F12 GlutaMAX™ modified medium (Gibco, Waltham, MA, USA), 10% foetal bovine serum (FBS) (Gibco) and 1% penicillin/streptomycin (Pen/Strep) (100 U/mL-100 μg/mL, Gibco); SAOS-2 cells were cultured using McCoy's 5 A (modified) medium (Gibco) supplemented with 15% FBS and 1% Pen/Strep; and U2-OS cells were cultured using McCoy's 5 A (modified) medium supplemented with 10% FBS and 1% Pen/Strep. MDA MB 321 and 468 cells were cultured in growth media using RPMI 1640 (Gibco), 10% FBS and 1% Pen/Strep; U87 cells were grown in complete medium composed of MEM-α nucleosides no-ascorbic-acid medium (Gibco), 10% FBS and 1% Pen/Strep; and U118 cells were cultured using DMEM high-glucose pyruvate medium (Gibco), 10% FBS and 1% Pen/Strep.

Cells were grown at 37 °C in a 5% CO_2 atmosphere under controlled humidity conditions, detached from culture flasks by trypsinization and then centrifuged. The cell number and viability were assessed by trypan blue dye exclusion test. All cell-handling procedures were performed in a sterile laminar flow hood.

2.6.2. Cell Viability

The MTT assay is a colorimetric protocol used to quantitatively assess cell viability. 3-(4,5-dimethylthiazol-2—yl)2,5-diphenyltetrazolium bromide (MTT) can be reduced to formazan crystals by metabolically active cells. Briefly, 1.6×10^4 cells/cm^2 were plated in a 96-well plate. After 24 h, Pt-free, GO@PEG-Pt and GO@PEG at the above-reported concentrations were added to the culture medium and left for 72 h. MTT reagent was resuspended in 1X phosphate-buffered saline (PBS) at a 5 mg/mL final concentration and added to cell culture media in a 1:10 ratio. After 2 h incubation at 37 °C, the solution was discarded, and formazan crystals were dissolved by adding DMSO and shaking for 15 min. The absorbance of three biological replicates (n = 3) for each condition was read at 570 nm using a Multiskan FC microplate photometer (Thermo Scientific). The results are represented in graphs with % with respect to cells only.

2.6.3. Cell Morphology

Cell morphology of 1.6×10^4 cells/cm^2 was analysed 72 h after incubation with Pt-free and GO@PEG-Pt at 15 μM, 30 μM and 60 μM concentrations and GO@PEG at the corresponding nanoplatform concentrations. Briefly, cells were washed with PBS 1X and fixed using 4% paraformaldehyde (Merck) for 15 min at room temperature (RT). Membrane permeabilization was performed in PBS 1X with 0.1% (v/v) Triton X-100 (Sigma Aldrich, Saint Louis, MO, USA) for 5 min at RT. A PBS 1X wash was performed, and then F-actin filaments were highlighted with rhodamine phalloidin (Actin Red 555 Ready Probes™ Reagent, Invitrogen, Waltham, MA, USA), following the manufacturer's indications for 30 min at RT to visualize the cytoskeletal conformation. DAPI (600 nM) counterstaining was performed for identification of cell nuclei, following the manufacturer's instructions. Images were captured with an inverted Ti-E fluorescence microscope (Nikon, Tokyo, Japan). A single biological replicate was performed for each condition (n = 1).

2.6.4. Quantification of Cellular Uptake of Pt

Inductively coupled plasma optical emission spectrometry (ICP-OES, Agilent Technologies 5100 ICP-OES, Santa Clara, CA, USA) was used to evaluate the platinum drug **1** cell internalization (Pt-free and GO@PEG-Pt). Briefly, 1.6×10^4 cell/cm^2 were seeded in a 6-well plate for each cell line. After 24 h, Pt-free and GO@PEG-Pt were added at a concentration of 30 µM and incubated for 4 and 24 h, respectively. At each time point, cells were washed with PBS 1X, trypsinised and scraped, reaching a final volume of 400 µL for each sample. Cells were counted using a trypan blue dye exclusion test. ICP-OES was used for quantitative determination of Pt ions. Briefly, the samples were dissolved with 500 µL nitric acid (65 wt.%) and 2.1 mL of Milli-Q water, followed by sonication for 30 min in an ultrasonicated bath. Cells alone were prepared similarly. The analytical wavelength of Pt was 265.945 nm. Pt per cell was quantified, and biological analysis was performed in triplicate for each condition (n = 3).

2.6.5. Migration Assay

Cell migration ability was analysed by applying an optimized model of the scratch assay [29]. All the cell lines were seeded in a 24-well plate at a density of 50×10^3 cells/cm^2. After 24 h, the cell monolayer was scraped in a straight line to create a "scratch" with a p200 pipet tip; then, cells were washed with the same cell medium supplemented with only 2% of FBS to remove cell debris, and Pt-free and GO@PEG-Pt were added at a 30 µM concentration. A first image of the scratch was acquired at time 0, then after 24, 48 and 72 h by an inverted Ti-E Fluorescent microscope. For each acquired image, six measures of scratch width were obtained and analysed quantitatively using ImageJ software. In addition, at times 0 and 72 h, cells were fixed with 4% w/v paraformaldehyde (PFA), cell nuclei were highlighted by DAPI staining and images were acquired using an inverted Ti-e fluorescent microscope. A biological duplicate was performed for each condition (N = 2).

2.6.6. Statistical Analyses

Results of MTT assays are reported as percentage (%) with respect to cells only ± standard error of the mean (SEM), and data were analysed by two-way analysis of variance (two-way ANOVA) and Tukey's multiple comparisons test. ICP-OES results are elaborated as picograms of Pt per cell, reported in graphs as mean ± SEM and analysed by two-way ANOVA and Sidak's multiple comparisons test. Scratch assay results are graphically represented as distance covered (µm) by cells over time towards the centre of the performed scratch, expressed as mean ± SEM plotted on the graphs and analysed by two-way ANOVA and Tukey's multiple comparisons test. Statistical analysis was performed with GraphPad Prism software (version 8.0.1).

3. Results and Discussion

In this work, we focused on the development of GO-based nanoplatforms with high performance in delivering platinum-based drugs in order to overcome the current limitations of clinically used chemotherapeutics, including carboplatin, oxaliplatin and cisplatin, which, once in the human body, are rapidly degraded, with considerable toxicity to off-target organs [23,30].

Among various cancer types, we selected three tumours based on their incidence (breast cancer) and aggressiveness (osteosarcoma and glioblastoma). Breast cancer is the leading cause of death in women, with an incidence of 2.3 million cases worldwide in 2020 [11]. Breast cancer often metastasizes in the liver, lungs, brain and, in 70% of cases, bones [31]. Osteosarcoma is the most common bone cancer affecting young patients, with a poor response to chemotherapy, with a negative impact on patients' life expectancy [12]. Glioblastoma is a very aggressive type of cancer of the central nervous system generated from the glial cells, with a poor prognosis of life expectancy, with 5-year survival rates of 5 % without any significant improvements in recent decades [21]. A distinctive feature of this work is the wide in vitro testing performed in seven cancer cell lines of selected tumours, where cell viability and morphology were first evaluated with three different concentrations of GO@PEG-Pt. The most promising group was investigated further to evaluate the cellular uptake of GO@PEG-Pt, as it is well-known that Pt becomes activated only once it enters the cells. In addition, our in vitro study focused on another extremely important aspect of developing an anticancer strategy: the possibility of inhibiting cancer cell migration in order to reduce the infiltration of the tumour surrounding parenchyma, which often metastasize throughout the body.

3.1. Characterization of GO-Based Nanoplatforms

The mean flake size of the graphene oxide particles found in the GO stock solution was considerably decreased by a combination of protocols previously described by Ma and Chen [24,25] (see Section 2). The treated GO and GO functionalised using PEG were consequently characterized by AFM (Figure 3A,B). Statistical analysis of the obtained image data confirmed the successful preparation of pristine GO flakes with a mean lateral size of 130 nm for more than 85% of the identified flakes (n = 2142) (Figure 3C). A full histogram of the flake size distribution is shown in Figure S2 (Supplementary Materials). Modification of the surface with PEG resulted in a change in the height of GO flakes; in the stock solution, the typical height was approximately 1.1 nm (1–2 layers, Figure S1B), and after surface modification, the height increased by approximately nine times to 9 nm (corresponding to fewer than 20 layers; Figure 3B). The Raman spectrum in Figure 3D shows dominant D and G bands characteristic of sp2 and sp3 hybridization of carbon containing groups in the nanomaterial. The D/G ratio was 0.75, which indicates several defects in the lattice caused by the present functional groups. GO morphology was characterized by the means of scanning electron microscopy (SEM); the resulting micrograph is shown in Figure 3E.

Despite their hydrophilic nature, the prepared GO flakes rapidly aggregated in the presence of salts and serum components, as observed by Wang and Loutfy [32–34]. Therefore, the branched eight-arm PEG-NH_2 polymer was covalently conjugated to the as-prepared GO to increase its stability under physiological conditions. The successful PEGylation of GO flakes was confirmed by data acquired using infrared spectroscopy (Figure 3F), with prominent bands characteristic of PEG observed. Furthermore, PEG-coated GO flakes were imaged by SEM (Figure 3G). The NH_2 groups of the GO@PEG were covalently linked to the carboxylic part of Pt-based drug 1, as shown in Figure 2. The resulting GO@PEG-Pt showed high temperature-independent stability in PBS buffer (pH 7.4) solution (Figure 3H). After 24 h, 35% of Pt was released from the GO@PEG nanoplatform before reaching a plateau. A loading efficiency (LE) of 64% was determined for the GO@PEG nanoplatform; this higher LE based on GO is a major advantage over conventional nanocarriers, such as liposomes [35] and solid lipid nanocarriers [36]. The use of eight-arm PEG in our study allowed us to achieve a significantly higher LE of cisplatin (LE = 64%) compared to previous studies, wherein linear PEG (LE = 4.5%) [37] and six-arm PEG (LE = 11%) [38] were used.

In general, GO-based nanoplatforms enable the delivery of higher concentrations of drugs to the tumour region. In the case of drug loading for liposomes and solid lipid nanoparticles, the drug needs to be dissolved or added within the matrix. However, due to the limited solubility of hydrophobic drugs in these matrices, the loading capacities are generally lower than those observed for GO. There is also a significant loss of drug during the synthesis of liposomes and solid lipids [32]. For example, Zhou et al. developed a drug delivery platform based on microsomes composed of poly(2-methacryloyloxyethyl phosphorylcholine)-b-poly(methacrylic acid) copolymer. This system was loaded with cis-diamminedichloroplatinum and used for treatment of osteosarcoma cells. The reported drug loading content was 13.7% [33]. Son K.D. et al. described a drug delivery platform based on calcium-phosphate nanocomposites and evaluated its performance in the delivery of cisplatin, caffeic acid and chlorogenic acid for treatment of osteosarcoma. The drug loading content was between 1% (caffeic acid) and 1.7% (cisplatin) [34]. Li et al. described cisplatin-loaded poly(L-glutamic acid)-g-methoxy poly(ethylene glycol) nanoparticles with an average size of around 43 nm for treatment of osteosarcoma [35].

However, as previously mentioned, the use of GO-based nanoplatforms has its limitations. PEGylation, for which both linear and branched PEG can be used, is an effective strategy to overcome these shortcomings. However, linear PEGylation has several limitations, such as lack of targeting ability due to collapse in the bloodstream and low efficiency in drug loading [39–42]. Compared to linear PEGs, branched PEGs have more modifiable end groups, which can be used to prepare multifunctional systems with different molecules, including anticancer drugs, fluorescent molecules and targeted ligands. Branched PEGs have better targeting ability, environmental responsiveness and blood circulation, improving drug solubility and bioavailability [43].

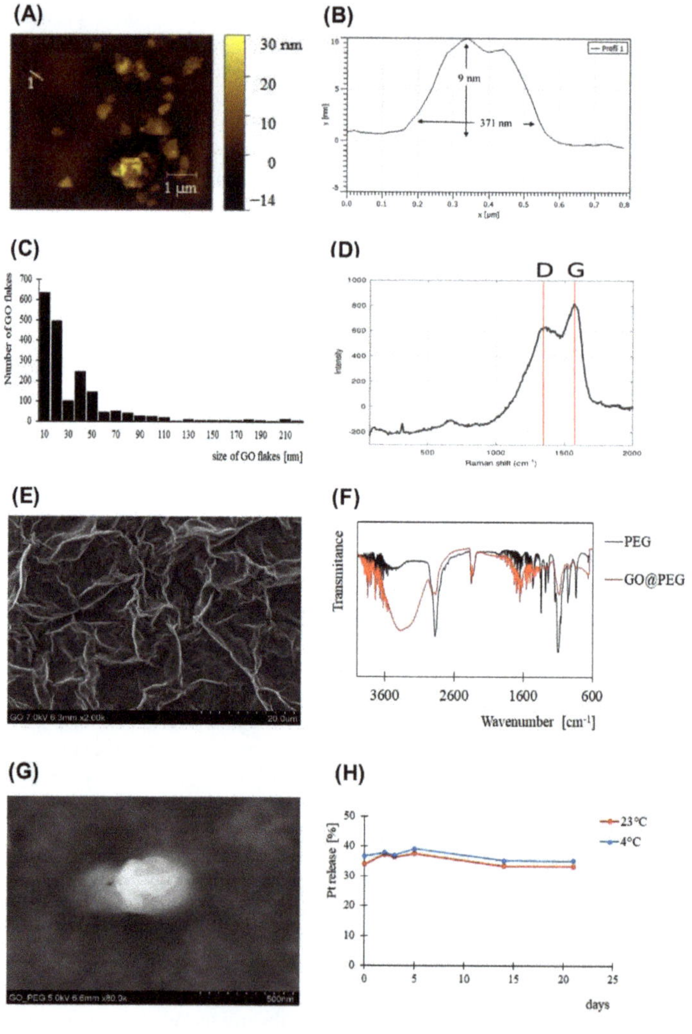

Figure 3. Characterization of GO-based nanoplatforms. (**A**) AFM image of PEGylated GO flakes. (**B**) Height profile was determined for marked GO flakes. (**C**) Size distribution of GO flakes in supernatant and typical D and G bands of GO are shown by the Raman spectrum (**D**). (**E**) Scanning electron microscopy images (SEM) of graphene oxide in stock solution and the presence of PEG in the sample with GO are demonstrated by IR spectra (**F**). (**H**) Successfully PEGylated GO flakes are displayed by SEM (**G**). Release kinetics of Pt-loaded GO@PEG at 4 °C and 23 °C (n = 3).

3.2. *Synthesis of the Platinum-Based Prodrug*

Complex **1** is a Pt(IV) compound based on a cisplatin scaffold with succinic acid in an axial position that allow for the conjugation to GO functionalized with PEG-NH_2 groups via coupling with EDC (see Section 2 for details). Pt(IV) species are considered prodrugs because they are intracellularly activated by reduction. It is well known that Pt(IV) species are reduced in the cellular environment (by reducing agents such as glutathione or ascorbic acid), releasing the active Pt(II) scaffold (in this case cisplatin). Using this strategy, cisplatin covalently bound to GO is released upon cellular internalization [44].

3.3. *In Vitro Evaluation of GO@PEG-Pt Bioactivity*

The cytotoxicity of the unloaded GO@PEG nanoplatform was investigated at three different concentrations (i.e., 1.0 μg/mL, 2.0 μg/mL and 4.0 μg/mL) corresponding to the concentrations of GO@PEG nanoplatforms loaded with 15 μM, 30 μM and 60 μM of Pt-free, respectively (see Section 2). Unloaded GO@PEG did not show any significant toxicity, given about 100% of viable cells (Figures 4–6 and S3–S9), without negatively affecting the cell morphology, visualized by actin filament staining, (Figures 4–6 and S3–S9) in all the seven cell lines tested. These results confirm that the GO@PEG-based nanoplatform is a promising nanodelivery system, in accordance with previous works showing the absence of toxicity strictly related to GO@PEG, even in vivo with higher concentrations [45,46]; therefore, GO@PEG nanoplatforms have the potential to be administered in vivo with no side effects.

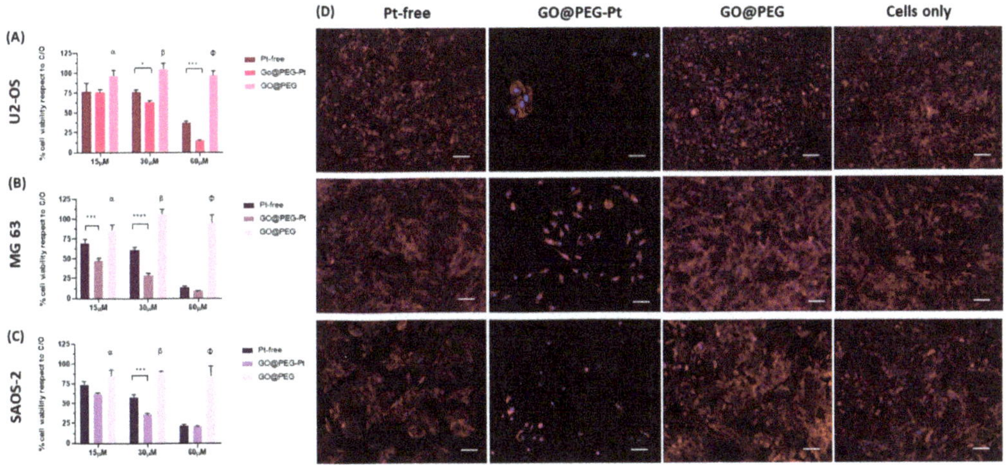

Figure 4. Cell viability and morphological analysis in osteosarcoma cell lines. An MTT assay was performed after 72 h of cell culture. The data show the percentage of viable cells compared to cells alone as the control, and the mean ± standard error of the mean is presented. The graphs show the viability of the U2-OS cell line (**A**), the MG63 cell line (**B**), the SAOS-2 cell line (**C**) and the respective morphological images of the cells cultured for 72 h in the presence of the 30 mM concentration. Phalloidin red stains for actin filaments, and DAPI blue stains for cell nuclei (**D**). Scale bars: 100 μm. (* p-value ≤ 0.05; *** p-value ≤ 0.001; **** p-value ≤ 0.0001). Significant differences between GO@PEG and the other compounds are reported in the graph as follows: a: GO@PEG vs. Pt-free; p-value ≤ 0.01 and GO@PEG vs. GO@PEG-Pt p-value ≤ 0.001 in U2-OS; GO@PEG vs. Pt-free p-value ≤ 0.01 and GO@PEG vs GO@PEG-Pt p-value ≤ 0.0001 in MG63; GO@PEG vs. Pt-free p-value ≤ 0.05 and GO@PEG vs GO@PEG-Pt p-value ≤ 0.001 in SAOS-2. b and F: GO@PEG vs Pt-free and GO@PEG vs. GO@PEG-Pt both p-value ≤ 0.0001 in all cell lines.

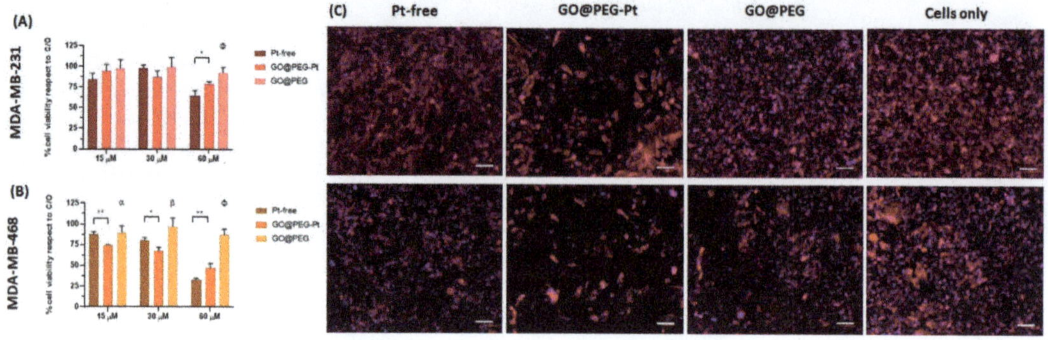

Figure 5. Cell viability and morphological analysis in glioblastoma cell lines. An MTT assay was performed after 72 h of cell culture. The data show the percentage of viable cells compared to cells alone as the control, and the mean ± standard error of the mean is presented. The graphs show the viability of the U87 cell line (**A**), the U118 cell line (**B**) and the respective morphological images of the cells cultured for 72 h in the presence of the 30 mM concentration. Phalloidin red stains for actin filaments, and DAPI blue stains for cell nuclei (**C**). Scale bars: 100 μm. (* p-value ≤ 0.05; ** p-value ≤ 0.01). Significant differences between GO@PEG and the other compounds are reported in the graph as follows, a: GO@PEG vs. Pt-free p-value ≤ 0.01 in U87; GO@PEG vs. GO@PEG-Pt p-value ≤ 0.05 in U118. b: GO@PEG vs. Pt-free p-value ≤ 0.0001 and GO@PEG vs. GO@PEG-Pt p-value ≤ 0.01 in U87, GO@PEG vs. Pt-free p-value ≤ 0.01 and GO@PEG vs. GO@PEG-Pt p-value ≤ 0.0001 in U118. F: GO@PEG vs. Pt-free p-value ≤ 0.0001 and GO@PEG vs. GO@PEG-Pt p-value ≤ 0.001 in U87, GO@PEG vs. Pt-free p-value ≤ 0.0001 and GO@PEG vs. GO@PEG-Pt p-value ≤ 0.0001 in U118.

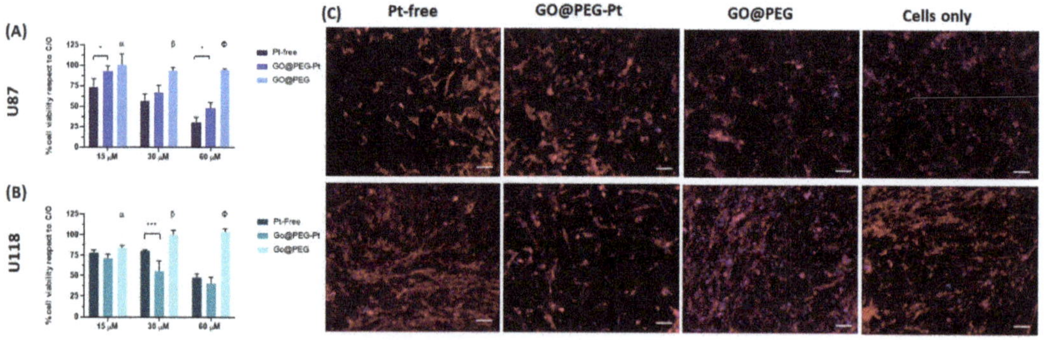

Figure 6. Cell viability and morphological analysis in breast adenocarcinoma cell lines. An MTT assay was performed after 72 h of cell culture. The data show the percentage of viable cells compared to cells alone as the control, and the mean ± standard error of the mean is presented. The graphs show the viability of the MDA-MB-231 cell line (**A**), the MDA-MB 468 cell line (**B**) and the respective morphological images of the cells cultured for 72 h in the presence of the 30 mM concentration. Phalloidin red stains for actin filaments, and DAPI blue stains for cell nuclei (**C**). Scale bars: 100 μm. (* p-value ≤ 0.05, *** p-value ≤ 0.001). Significant differences between GO@PEG and the other compounds are reported in the graph as follows, a: GO@PEG vs. GO@PEG-Pt p-value ≤ 0.05 in MDA-MB 468. b: GO@PEG vs. Pt-free p-value ≤ 0.01 and GO@PEG vs. GO@PEG-Pt p-value ≤ 0.0001 in MDA-MB 468. F: GO@PEG vs. Pt-free p-value ≤ 0.001 in MDA-MB 231, GO@PEG vs. Pt-free and GO@PEG vs. GO@PEG-Pt both p-value ≤ 0.0001 in MDA-MB 468.

The biological bioactivity of GO@PEG-Pt was then evaluated to validate the nanoplatform as a highly performant nanodelivery system. We investigated the cell viability, morphology, nanoplatform uptake and cell migration in different cancer cell lines compared to Pt-free at three different concentrations (15 μM, 30 μM and 60 μM) and cells alone as a control group (i.e., untreated cells). The data show an evident dose-dependent reduction in the cell metabolic activity in both cases (GO@PEG-Pt and Pt-free) compared to the cells alone in all tested cancer cell lines (Figures 4–6). These overall results demonstrate that Pt maintained its action when loaded on GO@PEG nanoplatforms.

A detailed review of the data indicates that osteosarcoma is the most affected tumour, with cell viability reduced to 90% compared to cells alone (Figure 4A–C), confirming the literature on osteosarcoma chemotherapy, which reports that cisplatin is a key and widely used drug [47,48]. The most promising result is related to the significant decrease in cell viability in the GO@PEG-Pt group compared to the Pt-free group, which validates the GO@PEG nanoplatforms as promising Pt vehicles for osteosarcoma treatment at a concentration of 30 μM [49,50] (Figure 4A–C).

In the MG63 cell line, this effect is also clearly visible (p-value ≤ 0.001) at the lowest tested Pt concentration (i.e., 15 μM) and, although less pronounced, was also detected in the U118 glioblastoma and the MDA-MB-468 breast cancer cell lines (Figures 5A,B and 6A,B).

A qualitative cell morphology analysis was performed, confirming the cell viability data. First, the cell density is lower in the GO@PEG-Pt group compared to the Pt-free group in the osteosarcoma cell lines, breast cancer cell lines and glioblastoma U118 cell line (Figures 4D, 5C, 6C and S3–S7 Supplementary Materials). MG63 and SAOS-2 cell lines are most negatively affected by the presence of GO@PEG-Pt, with a round and smaller cell morphology shape and with actin filaments aggregated at the cell's edges (Figure 4D). This analysis confirmed the absence of cytotoxicity of the GO@PEG nanoplatform in all the tested cells (Figures 4D, 5C and 6C). In fact, the cell morphology reflected the healthy state of the cells in the GO@PEG group, showing a high cell number and typical spindle-shaped cells without differences with the untreated cells-only group (Figures 4D, 5C and 6C).

To better elucidate whether the higher cell mortality was strictly related to enhanced Pt uptake driven by the GO@PEG nanoplatform, a cellular uptake analysis was performed and evaluated with ICP-OES, quantifying the Pt amount per cell after 4 and 24 h at 30 μM, selected as the most promising concentration. The GO@PEG nanoplatform enhanced Pt internalization in all the cell lines tested after 24 h, with the greatest differences compared to Pt-free in osteosarcoma and glioblastoma cell lines (MG63 p-value ≤ 0.0001; SAOS-2 and U87 p-value ≤ 0.05) (Figures 7A–C, 8A,B and 9A,B). However, even after 4 h, a visible trend of the increase in Pt uptake was observed. An evident discrepancy between Pt-free and GO@PEG was observed for osteosarcoma cell lines (p-value ≤ 0.0001 and p-value ≤ 0.001 in MG63 and U2-OS cells, respectively), glioblastoma cell lines and the MDA-MB-468 breast cancer cell line (Figures 7A–C, 8A,B and 9A,B).

These results are very promising and support our initial hypothesis with respect to the use of the GO@PEG nanoplatform as a highly performant nanodelivery system for platinum-based drugs. This strategy was demonstrated to be very efficient in reducing the amount of Pt needed in cancer therapy and, consequently, in diminishing the well-known side effects related to Pt-based drugs.

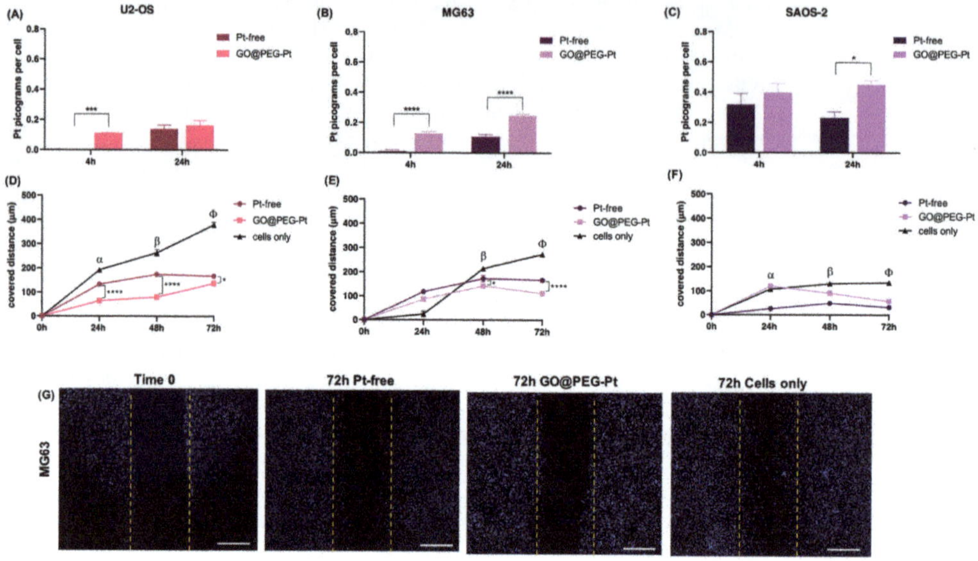

Figure 7. Human osteosarcoma cell lines: ICP-OES on U2-OS (**A**), MG63 (**B**) and SAOS-2 (**C**). Scratch test on U2-OS (**D**), MG63 (**E**) and SAOS-2 (**F**). α: Pt-free vs. cells only and GO@PEG-Pt vs. cells only, both p values ≤ 0.0001 in U2-OS; Pt-free vs. cells only. p value ≤ 0.0001 in SAOS-2. β: Pt-free vs. cells only and GO@PEG-Pt vs. cells only, both p values ≤ 0.0001 in U2-OS; Pt-free vs. cells only, p value ≤ 0.01, and GO@PEG-Pt vs. cells only, p value ≤ 0.0001 in MG63; Pt-free vs. cells only and GO@PEG-Pt vs. cells only, both p values ≤ 0.0001 in SAOS-2. φ: Pt-free vs. cells only and GO@PEG-Pt vs. cells only both, p value ≤ 0.0001 in U2-OS, MG63 and SAOS-2. Significant differences between Pt-free and GO@PEG-Pt are reported in the graph as follows: * p-value ≤ 0.05, *** p-value ≤ 0.001, **** p-value ≤ 0.0001. Representative DAPI staining of scratch test on MG63 cells (**G**) cells. Scale bars = 500 μm. Cell nuclei are indicated in blue.

A further investigation was performed to verify whether the proposed GO@PEG nanoplatform delivery system could also play a key role in the inhibition of the cell migration/invasiveness. One clinically distinctive trait of several tumours is the extensive infiltration by cancer cells of the tumour surrounding parenchyma, which often metastasize throughout the body [51]. An in vitro scratch assay was used as an easy, low-cost and well-developed method to measure cell migration of all seven cell lines in contact with GO@PEG-Pt compared to Pt-free at 30 μM for 72 h (Figures 7D–G, 8C–E and 9C–E). The distance covered by cells towards the centre of the scratch was quantified as an indicator of migrating movements, and overall, the results showed that GO@PEG-Pt significantly inhibits the migration of all tested cell lines compared to the control group, with the exception of U87 glioblastoma cells, for which a significant inhibition of cell migration was observed in the presence of the nanoplatform only after 48 h (Figures 7D–F, 8C,D and 9C,D). It is important to note that GO@PEG-Pt was able to significantly reduce the migration of MDA-MB-231 breast cancer cells compared to Pt-free (p value ≤ 0.0001 at 48 and 72 h; Figure 9C,E). In particular, at 72 h, a distance of only 141 μm was covered by MDA-MB-231 cells in the presence of the loaded nanoplatform compared to the more than doubled migration distance showed by the cells cultured with Pt-free (i.e., a distance of 300 μm covered). These are very interesting and promising results, as although the MDA-MB-231 cell line is not particularly sensitive to the GO@PEG system in terms of cell viability inhibition, as verified by MTT test, the proposed GO-based nanoplatforms could be used as an effective therapy to reduce tumour metastatic invasion, which is the primary cause of patient mortality during breast cancer progression [52]. MDA-MB-231 is a highly invasive breast carcinoma cell line compared to MDA-MB-468 and is commonly used to model late-stage breast cancer. It is invasive in vitro and, when implanted orthotopically, spontaneously metastasizes to lymph nodes [53,54]. A slight difference in migration was also observed in U118 glioblastoma cells, as well as MG63 and U2-OS osteosarcoma cell lines, when cultured with GO@PEG-Pt compared to Pt-free (p-value ≤ 0.0001 at 48 h and p-value ≤ 0.01 at 72 h for U118; p-value ≤ 0.05 and p-value ≤ 0.0001 at 48 and 72 h, respectively, for MG63; p-value ≤ 0.0001 and p-value ≤ 0.05 at 24 and 48 h, and 72 h, respectively, for U2-OS) (Figures 7D,E,G and 8D,E). U118, MG63 and U2-OS cells were able to cover about 360 μm, 112 μm and 137 μm, respectively, when cultured with GO@PEG-Pt compared to Pt-free (about 411 μm and 167 μm for U118, MG63 and U2-OS, respectively). However, the invasive nature of glioblastoma cells is shown by the scratch test, as well as the malignancy of U118 and U87 strictly related to the high proliferative ability of cells, which makes them difficult to treat, as confirmed by our cell viability results [55]. An increasing number of scientific reports are still discussing the critical relationship between the proliferation and migration of glioblastoma cells in relation to platinum-based nanoplatforms, depending on genetically and morphological different cell lines and various platinum sources affecting these two cellular properties. In this case, the results suggest a increased effect of the GO@PEG-Pt nanoplatform on the migration of U118 cells compared to U87. Moreover, the slight but significant inhibition of MG63 and U2-OS migration confirmed the aggressiveness and metastatic potential of osteosarcoma cells relative to conventional chemotherapeutic drugs, including high-dose platinum-based drugs [56–58]. On these bases, our results are very exciting overall, suggesting an improvement of Pt drug action in the migration inhibition of osteosarcoma, especially in MG63, U2-OS and glioblastoma cells when loaded on the GO-based nanoplatform. We hypothesize that this behaviour could be attributed to the cell metabolism GO, which decreases the electron transfer chain activity, limiting ATP production and compromising the assembly of actin filaments fundamental to cell migration and tumour invasiveness [59,60].

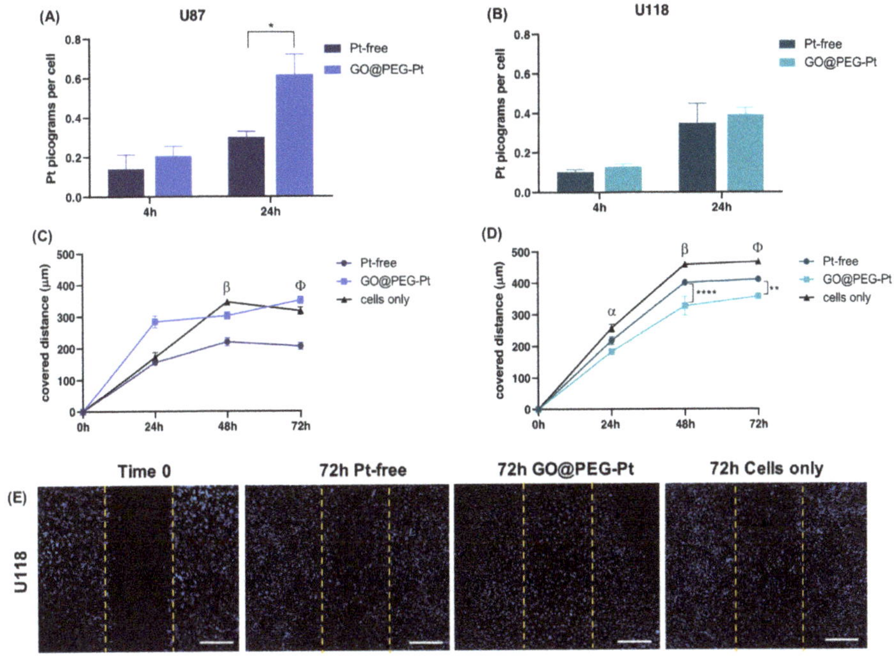

Figure 8. Human glioblastoma cell lines. ICP-OES on U87 (**A**) and U118 (**B**). Scratch test on U87 (**C**) and U118 (**D**); α: Pt-free vs. cells alone, p value ≤ 0.05, and GO@PEG-Pt vs. cells alone,

p value $\leq$ 0.0001 for U87. β: Pt-free vs. cells alone, *p* value $\leq$ 0.0001, and GO@PEG-Pt vs. cells alone, *p* value $\leq$ 0.05 in U87; Pt-free vs. cells alone, *p* value $\leq$ 0.001, and GO@PEG-Pt vs. cells alone, *p* value $\leq$ 0.0001 in U118. φ: Pt-free vs. cells alone, *p* value $\leq$ 0.0001 in U87; Pt-free vs. cells alone, *p* value $\leq$ 0.001, and GO@PEG-Pt vs. cells alone, *p* value $\leq$ 0.0001 in U118. Significant differences between Pt-free and GO@PEG-Pt are reported in the graph as follows: * *p*-value $\leq$ 0.05, ** *p*-value $\leq$ 0.01, **** *p*-value $\leq$ 0.0001. Representative DAPI staining of scratch test on U118 cells (**E**). Scale bars = 500 μm. Cell nuclei are indicated in blue.

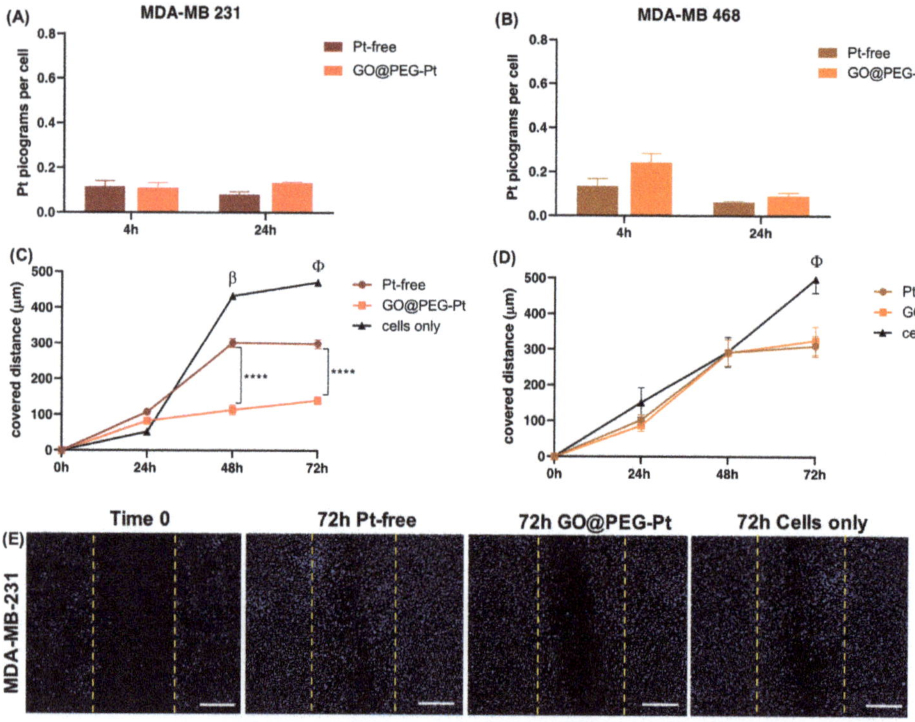

Figure 9. Human breast adenocarcinoma cell lines. ICP-OES on MDA-MB-231 (**A**) and MDA-MB-468 (**B**). Scratch test on MDA-MB-231 (**C**) and MDA-MB-468 (**D**); β: Pt-free vs. cells alone and GO@PEG-Pt vs. cells alone, both *p* values $\leq$ 0.0001 in MDA-MB 231. φ: Pt-free vs. cells alone and GO@PEG-Pt vs. cells alone, both *p* values $\leq$ 0.0001 in MDA-MB 231; Pt-free vs. cells alone, *p* value $\leq$ 0.0001, and GO@PEG-Pt vs. cells alone, *p* value $\leq$ 0.001 in MDA-MB 468. Significant differences between Pt-free and GO@PEG-Pt are reported in the graph as follows: **** *p*-value $\leq$ 0.0001. Representative DAPI staining of scratch test on MDA-MB-231 cells (**E**). Scale bars = 500 μm. Cell nuclei are indicated in blue.

4. Conclusions

Several nanomaterials, such as liposomes, polymeric nanoparticles, metal or carbon nanostructures, are useful tools to selectively target tumour cells, implementing the drug's pharmacokinetics, increasing the anticancer effect and diminishing the toxic effect at the same time. Their large surface area makes carbon-based nanomaterials excellent drug carrier candidates. In this study, we validated a GO-based nanoplatform modified using eight-arm PEG to improve functionalisation with a Pt(IV) anticancer drug based on a cisplatin scaffold, obtaining a highly performant nanodelivery system as a versatile system in cancer therapy. The obtained data demonstrate that the use of a GO@PEG carrier permits the use of less Pt drugs, achieving a very good cellular proliferation inhibition in osteosarcoma strictly related to a higher cellular uptake. This enhanced cellular internalization is also observed in glioblastoma; however, due to a different cell metabolism, the Pt drug bioactivity, once inside the cells, is less pronounced, although it is a good starting point to drive more selective drugs (e.g., triazene analogue of dacarbazine). The proposed GO@PEG nanoplatform is also promising for the inhibition of migration, especially in highly invasive breast carcinoma (i.e., MDA-MB-231 cell line), neutralizing the metastatic process. In conclusion the GO@PEG nanoplatform represents a promising tool in nanomedicine, in particular for cancer treatment, due to its drug type and loading capacity, which can be specifically tailored to target different cancers.

Supplementary Materials: The following supporting information can be downloaded at: https://www.mdpi.com/article/10.3390/nano12142372/s1, Figure S1: AFM analysis; Figure S2: Size distribution of GO flakes. Figure S3: U2-OS cells' morphology evaluation at 72 h; Figure S4.: MG63 cells' morphology evaluation at 72 h; Figure S5. SAOS-2 cells' morphology evaluation at 72 h; Figure S6. U87 cells' morphology evaluation at 72 h; Figure S7. U118 cells' morphology evaluation at 72 h; Figure S8. MDA-MB-231 cells' morphology evaluation at 72 h; Figure S9. MDA-MB-468 cells' morphology evaluation at 72 h.

Author Contributions: Conceptualization: M.M., S.P., D.M. and V.R.; investigation: E.G., L.Ž., D.F.B., A.R. (Arianna Rossi), A.R. (Andrea Ruffini) and G.B.; data analysis: E.G., L.Ž., A.R. (Arianna Rossi), G.B. and A.R. (Andrea Ruffini); writing—original draft preparation: E.G., L.Ž., A.R. (Arianna Rossi), G.B., D.M. and S.P.; writing—review and editing: D.M., S.P. and V.R.; funding acquisition: S.P., D.M. and V.R. All authors have read and agreed to the published version of the manuscript.

Funding: This work was funded by the European Project Horizon 2020 NANO4TARMED (H2020-WIDESPREAD-2020-5; GA number 952063) and by the Ministry of Health of the Czech Republic (grant no. NV19-04-00281).

Data Availability Statement: Not applicable.

Acknowledgments: D.F.B. kindly acknowledges the Irish Research Council for funding this research via a Government of Ireland Postgraduate Scholarship (GOIPG/2020/55).

Conflicts of Interest: The authors declare no conflict of interest.

References

1. De Martel, C.; Georges, D.; Bray, F.; Ferlay, J.; Clifford, G.M. Global burden of cancer attributable to infections in 2018: A worldwide incidence analysis. *Lancet Glob Health* **2020**, *8*, e180–e190. [CrossRef]
2. World Health Organization—Cancer. 2021. Available online: https://www.who.int/news-room/fact-sheets/detail/cancer (accessed on 3 February 2022).
3. Golemis, E.A.; Scheet, P.; Beck, T.N.; Scolnick, E.M.; Hunter, D.J.; Hawk, E.; Hopkins, N. Molecular mechanisms of the preventable causes of cancer in the United States. *Genes Dev.* **2018**, *32*, 868–902. [CrossRef]
4. Hulvat, M.C. Cancer Incidence and Trends. *Surg. Clin. North. Am.* **2020**, *100*, 469–481. [CrossRef]
5. Schirrmacher, V. From chemotherapy to biological therapy: A review of novel concepts to reduce the side effects of systemic cancer treatment (Review). *Int. J. Oncol.* **2019**, *54*, 407–419. [CrossRef]
6. Johnstone, T.C.; Suntharalingam, K.; Lippard, S.J. The Next Generation of Platinum Drugs: Targeted Pt(II) Agents, Nanoparticle Delivery, and Pt(IV) Prodrugs. *Chem. Rev.* **2016**, *116*, 3436–3486. [CrossRef]
7. Gmeiner, W.H.; Ghosh, S. Nanotechnology for cancer treatment. *Nanotechnol. Rev.* **2014**, *3*, 111–122. [CrossRef]
8. Zhang, Y.; Li, M.; Gao, X.; Chen, Y.; Liu, T. Nanotechnology in cancer diagnosis: Progress, challenges and opportunities. *J. Hematol. Oncol.* **2019**, *12*, 137. [CrossRef]
9. Bersini, S.; Jeon, J.S.; Dubini, G.; Arrigoni, C.; Chung, S.; Charest, J.L.; Moretti, M.; Kamm, R.D. A microfluidic 3D invitro model for specificity of breast cancer metastasis to bone. *Biomaterials* **2014**, *35*, 2454–2461. [CrossRef]
10. Brook, N.; Brook, E.; Dharmarajan, A.; Dass, C.R.; Chan, A. Breast cancer bone metastases: Pathogenesis and therapeutic targets. *Int. J. Biochem. Cell Biol.* **2018**, *96*, 63–78. [CrossRef]
11. Lei, S.; Zheng, R.; Zhang, S.; Wang, S.; Chen, R.; Sun, K.; Zeng, H.; Zhou, J.; Wei, W. Global patterns of breast cancer incidence and mortality: A population-based cancer registry data analysis from 2000 to 2020. *Cancer Commun.* **2021**, *41*, 1183–1194. [CrossRef]
12. Ottaviani, G.J.N. The epidemiology of osteosarcoma. *Cancer Treat. Res.* **2009**, *152*, 3–13. [CrossRef] [PubMed]
13. Zhang, H.; Peng, C.; Yang, J.; Lv, M.; Liu, R.; He, D.; Fan, C.; Huang, Q. Uniform ultrasmall graphene oxide nanosheets with low cytotoxicity and high cellular uptake. *ACS Appl. Mater. Interfaces* **2013**, *5*, 1761–1767. [CrossRef] [PubMed]

14. Gonçalves, G.; Vila, M.; Portolés, M.T.; Vallet-Regi, M.; Gracio, J.; Marques, P.A.A.P. Nano-graphene oxide: A potential multifunctional platform for cancer therapy. *Adv. Healthc. Mater.* **2013**, *2*, 1072–1090. [CrossRef]
15. Yang, K.; Feng, L.; Shi, X.; Liu, Z. Nano-graphene in biomedicine: Theranostic applications. *Chem. Soc. Rev.* **2013**, *42*, 530–547. [CrossRef] [PubMed]
16. Sima, L.E.; Chiritoiu, G.; Negut, I.; Grumezescu, V.; Orobeti, S.; Munteanu, C.V.A.; Sima, F.; Axente, E. Functionalized Graphene Oxide Thin Films for Anti-tumor Drug Delivery to Melanoma Cells. *Front. Chem.* **2020**, *8*, 184. [CrossRef] [PubMed]
17. Wang, P.; Wang, X.; Tang, Q.; Chen, H.; Zhang, Q.; Jiang, H.; Wang, Z. Functionalized graphene oxide against U251 glioma cells and its molecular mechanism. *Mater. Sci. Eng. C* **2020**, *116*, 111187. [CrossRef] [PubMed]
18. D'souza, A.A.; Shegokar, R. Polyethylene glycol (PEG): A versatile polymer for pharmaceutical applications. *Expert Opin. Drug. Deliv.* **2016**, *13*, 1257–1275. [CrossRef]
19. Orecchioni, M.; Cabizza, R.; Bianco, A.; Delogu, L.G. Graphene as cancer theranostic tool: Progress and future challenges. *Theranostics* **2015**, *5*, 710–723. [CrossRef]
20. Zhao, X.; Si, J.; Huang, D.; Li, K.; Xin, Y.; Sui, M. Application of star poly(ethylene glycol) derivatives in drug delivery and controlled release. *J. Control. Release* **2020**, *323*, 565–577. [CrossRef]
21. Rajaratnam, V.; Islam, M.M.; Yang, M.; Slaby, R.; Ramirez, H.M.; Mirza, S.P. Glioblastoma: Pathogenesis and current status of chemotherapy and other novel treatments. *Cancers* **2020**, *12*, 937. [CrossRef]
22. Liang, Y.; Zhang, H.; Song, X.; Yang, Q. Metastatic heterogeneity of breast cancer: Molecular mechanism and potential therapeutic targets. *Semin. Cancer Biol.* **2020**, *60*, 14–27. [CrossRef] [PubMed]
23. Dhar, S.; Daniel, W.L.; Giljohann, D.A.; Mirkin, C.A.; Lippard, S.J. Polyvalent oligonucleotide gold nanoparticle conjugates as delivery vehicles for platinum(IV) warheads. *J. Am. Chem. Soc.* **2009**, *131*, 14652–14653. [CrossRef]
24. Ma, X.; Tao, H.; Yang, K.; Feng, L.; Cheng, L.; Shi, X.; Li, Y.; Guo, L.; Liu, Z. A functionalized graphene oxide-iron oxide nanocomposite for magnetically targeted drug delivery, photothermal therapy, and magnetic resonance imaging. *Nano Res.* **2012**, *5*, 199–212. [CrossRef]
25. Chen, J.; Liu, H.; Zhao, C.; Qin, G.; Xi, G.; Li, T.; Wang, X.; Chen, T. One-step reduction and PEGylation of graphene oxide for photothermally controlled drug delivery. *Biomaterials* **2014**, *35*, 4986–4995. [CrossRef] [PubMed]
26. Stankovich, S.; Dikin, D.A.; Piner, R.D.; Kohlhaas, K.A.; Kleinhammes, A.; Jia, Y.; Wu, Y.; Nguyen, S.T.; Ruoff, R.S. Synthesis of graphene-based nanosheets via chemical reduction of exfoliated graphite oxide. *Carbon* **2007**, *45*, 1558–1565. [CrossRef]
27. Schniepp, H.C.; Li, J.L.; McAllister, M.J.; Sai, H.; Herrera-Alonso, M.; Adamson, D.H.; Prud'Homme, R.K.; Car, R.; Saville, D.A.; Aksay, I.A. Functionalized single graphene sheets derived from splitting graphite oxide. *J. Phys. Chem. B* **2006**, *110*, 8535–8539. [CrossRef]
28. Jung, I.; Pelton, M.; Piner, R.; Dikin, D.A.; Stankovich, S.; Watcharotone, S.; Hausner, A.M.; Ruoff, R.S. Simple approach for high-contrast optical imaging and characterization of graphene-based sheets. *Nano Lett.* **2007**, *7*, 3569–3575. [CrossRef]
29. Liang, C.C.; Park, A.Y.; Guan, J.L. In vitro scratch assay: A convenient and inexpensive method for analysis of cell migration in vitro. *Nat. Protoc.* **2007**, *2*, 329–333. [CrossRef]
30. Sazonova, E.V.; Kopeina, G.S.; Imyanitov, E.N.; Zhivotovsky, B. Platinum drugs and taxanes: Can we overcome resistance? *Cell Death Discov.* **2021**, *7*, 155. [CrossRef]
31. Sowder, M.E.; Johnson, R.W. Bone as a Preferential Site for Metastasis. *JBMR Plus* **2019**, *3*, e10126. [CrossRef]
32. Wang, J.; Tian, Q.; Ding, F.; Yu, Y.; Wu, F. CRGDyK-modified camretastain A4-loaded graphene oxide nanosheets for targeted anticancer drug delivery. *RSC Adv.* **2015**, *5*, 40258–40268. [CrossRef]
33. Zhou, H.; Wang, G.; Lu, Y.; Pan, Z. Bio-inspired cisplatin annocarriers for osteosarcoma treatment. *Biomater. Sci.* **2016**, *4*, 1212–1218. [CrossRef] [PubMed]
34. Son, K.D.; Kim, Y.J. Anticancer activity of drug-loaded calcium phosphate nanocomposites against human osteosarcoma. *Biomater. Res.* **2017**, *21*, 13. [CrossRef] [PubMed]
35. Li, Y.F.; Yu, H.Y.; Hai, S.; Liu, J.G. Cisplatin-loaded poly(L-glutamic acid)-g-methoxy poly(ethylene glycol) nanoparticles as a potential chemotherapeutic agent against osteosarcoma. *Chin. J. Polym. Sci.* **2015**, *33*, 763–771. [CrossRef]
36. Loutfy, S.A.; Salaheldin, T.A.; Ramadan, M.A.; Farroh, K.Y.; Abdallah, Z.F.; Eloahed, T.Y.A. Synthesis, characterization and cytotoxic evaluation of graphene oxide nanosheets: In vitro liver cancer model. *Asian Pacific J. Cancer Prev.* **2017**, *18*, 955–961. [CrossRef]
37. Zhao, J.; Wang, X.; Sun, T.; Li, L. Crystal phase transition and properties of titanium oxide nanotube arrays prepared by anodization. *J. Alloys Compd.* **2007**, *434–435*, 792–795. [CrossRef]
38. Li, Y.; Liu, R.; Yang, J.; Ma, G.; Zhang, Z.; Zhang, X. Dual sensitive and temporally controlled camptothecin prodrug liposomes codelivery of siRNA for high efficiency tumor therapy. *Biomaterials* **2014**, *35*, 9731–9745. [CrossRef]
39. Jang, D.J.; Moon, C.; Oh, E. Improved tumor targeting and antitumor activity of camptothecin loaded solid lipid nanoparticles by preinjection of blank solid lipid nanoparticles. *Biomed. Pharmacother.* **2016**, *80*, 162–172. [CrossRef]
40. Charmi, J.; Nosrati, H.; Amjad, J.M.; Mohammadkhani, R.; Danafar, H. Polyethylene glycol (PEG) decorated graphene oxide nanosheets for controlled release curcumin delivery. *Heliyon* **2019**, *5*, e01466. [CrossRef]
41. Muñoz, R.; Singh, D.P.; Kumar, R.; Matsuda, A. Graphene oxide for drug delivery and cancer therapy. In *Nanostructured Polymer Composites for Biomedical Applications*; Nanostructured Polymer Composites for Biomedical Applications Micro and Nano Technologies; Elsevier: Amsterdam, The Netherlands, 2019; pp. 447–488. [CrossRef]

42. Li, W.; Zhan, P.; De Clercq, E.; Lou, H.; Liu, X. Current drug research on PEGylation with small molecular agents. *Prog. Polym. Sci.* **2013**, *38*, 421–444. [CrossRef]
43. Hassanzadeh, P.; Atyabi, F.; Dinarvand, R. Linkers: The key elements for the creation of efficient nanotherapeutics. *J. Control. Release* **2018**, *270*, 260–267. [CrossRef] [PubMed]
44. Kenny, R.G.; Marmion, C.J. Toward Multi-Targeted Platinum and Ruthenium Drugs—A New Paradigm in Cancer Drug Treatment Regimens? *Chem. Rev.* **2019**, *119*, 1058–1137. [CrossRef]
45. Du, L.; Wu, S.; Li, Y.; Zhao, X.; Ju, X.; Wang, Y. Cytotoxicity of PEGylated graphene oxide on lymphoma cells. *Biomed. Mater. Eng.* **2014**, *24*, 2135–2141. [CrossRef] [PubMed]
46. Xu, M.; Zhu, J.; Wang, F.; Xiong, Y.; Wu, Y.; Wang, Q.; Weng, J.; Zhang, Z.; Chen, W.; Liu, S. Improved In Vitro and In Vivo Biocompatibility of Graphene Oxide through Surface Modification: Poly(Acrylic Acid)-Functionalization is Superior to PEGylation. *ACS Nano* **2016**, *10*, 3267–3281. [CrossRef] [PubMed]
47. Yamamoto, N.; Tsuchiya, H. Hemotherapy for osteosarcoma Where does it come from What is it Where is it going. *Expert Opin. Pharm.* **2013**, *14*, 2183–2193. [CrossRef] [PubMed]
48. Zhang, Y.; Yang, J.; Zhao, N.; Wang, C.; Kamar, S.; Zhou, Y.; He, Z.; Yang, J.; Sun, B.; Shi, X.; et al. Progress in the chemotherapeutic treatment of osteosarcoma. *Oncol Lett.* **2018**, *16*, 6228–6237. [CrossRef]
49. Robl, B.; Botter, S.M.; Pellegrini, G.; Neklyudova, O.; Fuchs, B. Evaluation of intraarterial and intravenous cisplatin chemotherapy in the treatment of metastatic osteosarcoma using an orthotopic xenograft mouse model. *J. Exp. Clin. Cancer Res.* **2016**, *35*, 113. [CrossRef]
50. Winkler, K.; Bielack, S.; Delling, G.; Salzer-Kuntschik, M.; Kotz, R.; Greenshaw, C.; Jürgens, H.; Ritter, J.; Kusnierz-Glaz, C.; Erttmann, R. Effect of Intraarterial Versus Intravenous Cisplatin in Addition to Systemic Doxorubicin, High-Dose Methotrexate, and Ifosfamide on Histologic Tumor Response in Osteosarcoma (Study COSS-8s). *Cancer* **1990**, *66*, 1703–1710. [CrossRef]
51. Mortezaee, K. Organ tropism in solid tumor metastasis: An updated review. *Future Oncol.* **2021**, *17*, 194. [CrossRef]
52. Phromnoi, K.; Yodkeeree, S.; Anuchapreeda, S.; Limtrakul, P. Inhibition of MMP-3 activity and invasion of the MDA-MB-231 human invasive breast carcinoma cell line by bioflavonoids. *Acta Pharmacol. Sin.* **2009**, *30*, 1169–1176. [CrossRef]
53. Welsh, J.E. *Animal Models for Studying Prevention and Treatment of Breast Cancer*; Elsevier: Amsterdam, The Netherlands, 2013. [CrossRef]
54. Liu, Y.L.; Chou, C.K.; Kim, M.; Vasisht, R.; Kuo, Y.-A.; Ang, P.; Liu, C.; Perillo, E.P.; Chen, Y.-I.; Blocher, K.; et al. Assessing metastatic potential of breast cancer cells based on EGFR dynamics. *Sci. Rep.* **2019**, *9*, 3395. [CrossRef] [PubMed]
55. Kutwin, M.; Sawosz, E.; Jaworski, S.; Wierzbicki, M.; Strojny, B.; Grodzik, M.; Chwalibog, A. Assessment of the proliferation status of glioblastoma cell and tumour tissue after nanoplatinum treatment. *PLoS ONE* **2017**, *12*, e0178277. [CrossRef]
56. Luetke, A.; Meyers, P.A.; Lewis, I.; Juergens, H. Osteosarcoma treatment—Where do we stand? A state of the art review. *Cancer Treat. Rev.* **2014**, *40*, 523–532. [CrossRef]
57. Du, G.Y.; He, S.W.; Zhang, L.; Sun, C.X.; Mi, L.D.; Sun, Z.G. Hesperidin exhibits in vitro and in vivo antitumor effects in human osteosarcoma MG-63 cells and xenograft mice models via inhibition of cell migration and invasion, cell cycle arrest and induction of mitochondrial-mediated apoptosis. *Oncol Lett.* **2018**, *16*, 6299–6306. [CrossRef]
58. Wedekind, M.F.; Wagner, L.M.; Cripe, T.P. Immunotherapy for osteosarcoma: Where do we go from here? *Pediatr. Blood Cancer* **2018**, *65*, 315–327. [CrossRef] [PubMed]
59. Zhou, H.; Zhang, B.; Zheng, J.; Yu, M.; Zhou, T.; Zhao, K.; Jia, Y.; Gao, X.; Chen, C.; Wei, T. The inhibition of migration and invasion of cancer cells by graphene via the impairment of mitochondrial respiration. *Biomaterials* **2014**, *35*, 1597–1607. [CrossRef] [PubMed]
60. Zhang, B.; Wei, P.; Zhou, Z.; Wei, T. Interactions of graphene with mammalian cells: Molecular mechanisms and biomedical insights. *Adv. Drug. Deliv. Rev.* **2016**, *105*, 145–162. [CrossRef]

nanomaterials

MDPI

Article

Controlled Delivery of an Anti-Inflammatory Toxin to Macrophages by Mutagenesis and Nanoparticle Modification

Ayaka Harada [1,†], Hiroyasu Tsutsuki [2,†], Tianli Zhang [2], Kinnosuke Yahiro [3], Tomohiro Sawa [2] and Takuro Niidome [1,*]

[1] Faculty of Advanced Science and Technology, Kumamoto University, 2-39-1 Kurokami, Chuo-ku, Kumamoto 860-8555, Japan; 201d8604@st.kumamoto-u.ac.jp

[2] Department of Microbiology, Graduate School of Medical Sciences, Kumamoto University, 1-1-1 Honjo, Chuo-ku, Kumamoto 860-8556, Japan; tsutsuki@kumamoto-u.ac.jp (H.T.); zhangtianli220@hotmail.com (T.Z.); sawat@kumamoto-u.ac.jp (T.S.)

[3] Department of Microbiology and Infection Control Sciences, Kyoto Pharmaceutical University, 5 Misasagi-Nakauchi-cho, Yamashina-ku, Kyoto 607-8414, Japan; kin816tas@mb.kyoto-phu.ac.jp

* Correspondence: niidome@kumamoto-u.ac.jp

† These authors contributed equally to this work.

Abstract: Advances in drug delivery systems (DDSs) have enabled the specific delivery of drugs to target cells. Subtilase cytotoxin (SubAB) produced by certain enterohemorrhagic *Escherichia coli* strains induces endoplasmic reticulum (ER) stress and suppresses nitric oxide generation in macrophages. We previously reported that modification of SubAB with poly(D,L-lactide-co-glycolic) acid (PLGA) nanoparticles (SubAB-PLGA NPs) increased intracellular uptake of SubAB and had an anti-inflammatory effect on macrophages. However, specific delivery of SubAB to macrophages could not be achieved because its effects on other cell types were not negligible. Therefore, to suppress non-specific SubAB binding, we used low-binding mutant $SubAB_{S35A}$ (S35A) in which the 35th serine of the B subunit was mutated to alanine. In a macrophage cell line, PLGA NPs modified with S35A (S35A-PLGA NPs) induced ER stress and had anti-inflammatory effects similar to WT-PLGA NPs. However, in an epithelial cell line, S35A-PLGA NPs induced lower ER stress than WT-PLGA NPs. These results suggest that S35A is selectively delivered to macrophages rather than epithelial cells by modification with PLGA NPs and exerts anti-inflammatory effects. Our findings provide a useful technique for protein delivery to macrophages and encourage medical applications of DDSs for the treatment of inflammatory diseases.

Keywords: PLGA nanoparticles; anti-inflammatory; controlled drug delivery

Citation: Harada, A.; Tsutsuki, H.; Zhang, T.; Yahiro, K.; Sawa, T.; Niidome, T. Controlled Delivery of an Anti-Inflammatory Toxin to Macrophages by Mutagenesis and Nanoparticle Modification. *Nanomaterials* **2022**, *12*, 2161. https://doi.org/10.3390/nano12132161

Academic Editors: Goran Kaluđerović and Nebojša Pantelić

Received: 19 May 2022
Accepted: 20 June 2022
Published: 23 June 2022

1. Introduction

A drug delivery system (DDS) is an engineered technology for targeted delivery and/or controlled release of therapeutic agents, which plays an important role in advances in pharmaceutical and medical fields. The goal of a DDS is to achieve better therapeutic effects without any side effects. Therefore, the development of attractive drug carriers with both targeting of lesions and controlled release is eagerly awaited [1]. Synthetic nanoparticles (NPs) as drug carriers are a potential clinical therapeutic tool against cancer and inflammation [2]. NPs have been developed on the basis of various materials including liposomes [3], chitosan [4], dextran [5], silica [6], metals such as silver and gold [7], and biodegradable poly(D,L-lactide-co-glycolic) acid (PLGA) [8]. Recently, Jin et al. reported the chirality-controlled protein NPs driven by molecular interactions for cancer therapy. The developed NPs were effectively taken up into HCT116 cells and showed significant antitumor activity [9]. Other polymer materials including hydrogels, micro/nanofibers, and inorganic hybrid NPs have also been developed and are expected to be next-generation DDSs. [10–12] Moreover, drugs are delivered to not only specific tissues and cells, but also to specific cellular organelles, such as nuclei, mitochondria, and lysosomes, using a DDS [13–15].

The endoplasmic reticulum (ER) is a crucial organelle for protein quality control (e.g., protein synthesis, folding, maturation, and stabilization) [16]. Many intracellular proteins are folded in the ER by various chaperones and folding enzymes [17]. Subsequently, the proteins are subjected to post-translational modifications, including glycosylation and disulfide bond formation by glycosylation enzymes and oxidoreductase, respectively, to generate mature forms [18]. Genetic and environmental damage, such as nutrient depletion, and oxidative stress, causes the accumulation of unfolded proteins, resulting in ER stress. This stress activates the unfolded protein response (UPR) via ER stress sensor proteins [19]. Although ER stress is associated with many pathological conditions, recent studies have revealed that adaptive UPR signaling protects cells from cytotoxicity and excessive inflammation in a context-dependent manner [20]. Therefore, controlling the UPR may provide new prospects for future treatments of ER-related diseases and inflammation. On the basis of the above concept, some researchers have reported targeting the ER using sulfonyl ligands and peptides [21,22].

Subtilase cytotoxin (SubAB), a member of the AB_5 toxin family, was identified in enterohemorrhagic *Escherichia coli* O113:H21 that was associated with an outbreak in Australia [23–25]. SubAB consists of an enzymatically active A subunit (SubA) that cleaves host ER chaperon binding immunoglobulin protein (BiP) and a pentameric B subunit (SubB) that binds to sialoglycan-modified cell surface receptors and mediates uptake into target cells [26,27]. SubAB translocates to the ER via the COPI-dependent Golgi retrograde trafficking pathway [28,29]. After translocation into the ER, SubAB cleaves BiP and results in ER stress-induced cytotoxicity through activation of ER stress sensors protein kinase R-like endoplasmic reticulum kinase (PERK), inositol-requiring enzyme 1α (IRE1α), and activating transcription factor 6 (ATF6) that mediate a unique signaling pathway including eukaryotic initiation factor 2α (eIF2α) phosphorylation and C/EBP homologous protein (CHOP) expression [30,31]. Several studies have demonstrated that SubAB induces cytotoxicity, stress granule formation [31], inhibition of autophagy [32] and inflammasome [33] in vitro, and lethal severe hemorrhagic inflammation and enteropathogenic bacterial infection in mice [34]. Additionally, SubAB inhibits lipopolysaccharide (LPS)-induced nitric oxide (NO) production in macrophages through suppression of nuclear factor kappa-B (NF-κB) activation and subsequent inducible NO synthase (iNOS) expression [35]. SubAB may have therapeutic potential for immunosuppression in macrophages because of inhibiting activation of NF-κB, the master regulator of LPS-induced pro-inflammatory gene expression.

Recently, we reported that modification of NPs with PLGA improves the efficiency of wildtype SubAB (WT) uptake into macrophages and the subsequent anti-inflammatory effect [36]. However, macrophage-specific delivery of WT was difficult because the B subunit of WT has a potent cell adhesion property without cell specificity. Therefore, in this study, we attempted macrophage-specific delivery of the toxin using a mutant SubAB with a low-affinity cell adhesion property engineered through site-directed mutagenesis in the B subunit.

2. Materials and Methods

2.1. Materials

Poly(D,L-lactide-co-glycolic) acid (PLGA, 85:15, MW: 190,000–240,000) and fluorescein isothiocyanate isomer I were purchased from Sigma-Aldrich (St. Louis, MO, USA). N-hydroxysuccinimide (NHS), 1-ethyl-3-(3-dimethylaminopropyl) carbodiimide hydrochloride (water-soluble carbodiimide, WSC), and N-(5-amino-1-carboxypentyl) iminodiacetic acid (AB-NTA) were from DOJINDO (Kumamoto, Japan). A purified mouse anti-BiP/GRP78 antibody (Cat#610979) was from BD Transduction Laboratories™ (Franklin Lakes, NJ, USA). Oleylamine and an anti-β-actin mouse monoclonal antibody were purchased from FUJIFILM Wako Pure Chemical (Osaka, Japan). Anti-mouse immunoglobulin G (IgG), horseradish peroxidase (HRP)-linked (#7076) and anti-rabbit IgG, HRP-linked (#7074), anti-CHOP (#2895), anti-poly(ADP-ribose) polymerases (PARP) (#9542), anti-cleaved PARP (#5625), anti-cleaved cysteine aspartate-specific protease (caspase) 7 (#9491), anti-cleaved caspase 3 (#9661), anti-phospho-eIF2α (#3398), and anti-total eIF2α (#5324) antibodies were obtained from Cell Signaling Technology (Danvers, MA, USA). An anti-iNOS mouse monoclonal antibody [37] and anti-SubAB anti-serum [38] were prepared as reported previously.

2.2. Purification of His-Tagged Toxins

Oligo histidine-tagged (6× His-tagged) recombinant SubAB (His-tagged SubAB; WT), catalytically inactive mutant His-tagged SubA$_{S272A}$B (S272A), and His-tagged A subunit (SubA) were purified by affinity chromatography using Ni-NTA agarose (Qiagen, Hilden, Germany) as reported previously [39].

2.3. Site-Directed Mutagenesis of B Subunit of SubAB

To replace a serine residue (S35) with alanine in the B subunit of SubAB, we performed PCR-based site-directed mutagenesis on a SubAB expression plasmid (pET23b-SubAB) using a QuikChange site-directed mutagenesis kit (Agilent Technologies, Santa Clara, CA, USA), in accordance with the manufacturer's instructions. Primers used for mutagenesis were as follows: 5′-GGCATGTTTGCAGGCGTTGTTATTACCC-3′ and 5′-CAACGCCTGCAAACATGCCATCCCGGGC-3′ (mutated bases are underlined). The sequence was confirmed by an ABI PRISM 377 DNA sequencer (Applied Biosystems, Foster City, CA, USA). The his-tagged SubAB$_{S35A}$ (S35A) mutant was purified using the same method described above.

2.4. HiLyte Fluor™ 555 (HF555) Labeling of Toxins

To investigate SubAB uptake by RAW264.7 and HeLa cells, WT and mutant toxins (S272A, SubA, and S35A) were labeled with HiLyte Fluor™ 555 Labeling Kit-NH_2 (Dojindo, Kumamoto, Japan) in accordance with the manufacturer's instructions. Briefly, to replace the solvent in which the toxins were dissolved, 200 μg toxins were applied to the ultrafiltration spin column and centrifuged at 8000× *g* for 10 min. Toxins were dissolved in reaction buffer, mixed with the succinimidyl ester (SE) reactive form of Hilyte Fluor™ 555 (HF555-SE) (dissolved in DMSO), and incubated at 37 °C for 30 min. WS buffer was added to the mixtures in the spin column and then washed by centrifugation at 8000× *g* for 10 min to remove the unbound HF555-SE. The HF555-labeled toxins (HF555-toxins) were dissolved in 200 μL PBS by pipetting and stored at 4 °C.

2.5. Cell Culture

Murine macrophage cell lines J774.1 and RAW264.7 cells, and human cervical cancer cell line HeLa were cultured in Dulbecco's modified Eagle's medium (FUJIFILM Wako) supplemented with 10% heat-inactivated fetal bovine serum (MP Biomedicals, Santa Ana, CA, USA) and 1% penicillin-streptomycin (Nacalai Tesque, Kyoto, Japan) in a 5% CO_2 humidified incubator at 37 °C.

2.6. Fluorescence Microscopy

RAW264.7 and HeLa cells were seeded on a glass-bottom dish at 5×10^4 cells/dish and 1×10^4 cells/dish, respectively. After overnight culture, 100 μL HF555-toxins (HF555-WT, HF555-S272A, HF555-SubA, and HF555-S35A) and 1 μL Cellstain® DAPI solution (DOJINDO) were added to the cells. After incubation for 1 h at 37 °C, cells were observed under a BZ-X800 All-in-one Fluorescence Microscope (Keyence, Osaka, Japan).

2.7. Preparation of PLGA NPs

PLGA NPs were synthesized in accordance with our previous study [36]. Briefly, PLGA and stearic acid at a 1:37 ratio were prepared in chloroform (oil phase). To prepare PLGA NPs, the oil phase was added dropwise to the ultrapure water under probe sonication and evaporation solvent for 3 h.

2.8. Preparation of FITC-Conjugated Oleylamine

FITC-conjugated oleylamine was synthesized by a reaction between the amino group of oleylamine (OA)-fluorescein isothiocyanate isomer I (FITC) and the isothiocyanate group of FITC. FITC (2 mg) was dissolved in 2 mL N,N-dimethylformamide (DMF). Then, 30 mL OA was dissolved in 50 mL DMF. The FITC and OA solution was then mixed for 48 h at 50 °C. After the reaction, FITC-OA was collected by column chromatography.

2.9. Preparation of FITC-Labeled PLGA NPs

FITC-labeled PLGA NPs were prepared by an emulsion/evaporation method. Briefly, PLGA, stearic acid, and FITC-OA at a 1:7:30 ratio were prepared in chloroform (oil phase). To prepare PLGA NPs, the oil phase was added dropwise to the ultrapure water under probe sonication and evaporation solvent for 3 h.

2.10. Characterization of PLGA NPs

The size distribution and zeta potential of NPs were measured by a Zetasizer Nano ZS (Malvern Instruments Ltd., Worcestershire, UK). The shape and size of NPs were observed by transmission electron microscopy (TEM, JEM-1400Plus; JEOL, Tokyo, Japan). For TEM observation, 100 μL PLGA NPs was mixed with 100 μL of 0.01% phosphotungstic acid for 5 min and then 20 μL of the solution was placed on parafilm. A carbon-film-coated TEM grid (ELS-C10, Okenshoji Co., Ltd., Tokyo, Japan) was placed inside the droplet. After incubation for 2 min, the grid was dried in a vacuum at room temperature overnight.

2.11. Surface Modification of PLGA NPs with SubAB Toxins

Modification of the PLGA NP surface with SubAB was performed in accordance with a previous report [30]. In brief, 100 μL PLGA NPs dispersed in water was centrifuged at 12,000× *g* for 5 min. The precipitate was redispersed in 100 μL of a solution containing 10 μL of 0.9 mg/mL NHS and 1.5 mg/mL WSC in 20 mM phosphate buffer (pH 8.2). The mixture was shaken for 20 min at 37 °C, followed by washing with 20 mM phosphate buffer (pH 8.2). Then, 5 μL of 2 mg/mL AB-NTA was added to the NHS-activated PLGA NP solution and the mixture was shaken for 2 h at 37 °C. After washing the mixture, the resultant NTA-modified PLGA NPs were reacted with 5 μL of a 2 mg/mL $NiCl_2$ aqueous solution and shaken at 37 °C for 30 min. The mixture was washed with 20 mM phosphate buffer (pH 8.2) and redispersed in 95 μL of 20 mM phosphate buffer (pH 8.2). Then, 5 μL of 347 μg/mL His-tagged SubAB was added. The mixture was shaken for 1 h at room temperature and washed with 20 mM phosphate buffer (pH 8.2). Then, the resultant His-tagged SubAB-conjugated PLGA NPs (WT-PLGA NPs) were stored at 4 °C. Mutant $SubAB_{S35A}$-conjugated PLGA NPs (S35A-PLGA NPs) and SubA-conjugated PLGA NPs (A-PLGA NPs) were prepared in a similar manner.

2.12. Evaluation of pH-Dependent Release of His-Tagged SubAB Toxins

One-hundred microliters of SubAB toxins-PLGA NPs (WT-PLGA NPs, S35A-PLGA NPs, and A-PLGA NPs) were centrifuged (12,000× *g*, 5 min) to remove the supernatant. The NPs were redispersed in 100 μL ultrapure water. The solution was centrifuged again (12000× *g*, 5 min) to remove the supernatant. The NPs were then redispersed in 30 μL of 8 mM phosphate buffer (pH 7.4) or acetate buffer (pH 5.0) and shaken at 37 °C. After 0, 1, 10, and 30 min, the samples were centrifuged (12,000× g, 5 min) to collect the supernatants. Released His-tagged toxins in the supernatants were analyzed by Western blotting using anti-SubAB anti-serum.

2.13. Intracellular Uptake of SubAB Toxin-PLGA NPs

Intracellular uptake of SubAB toxins-PLGA NPs in RAW264.7 and HeLa cells was evaluated by fluorescence microscopy and a fluorescence microplate reader. RAW264.7 and HeLa cells were seeded at 5×10^4 and 1×10^4 cells/dish in glass-bottom dishes at 5×10^4 and 1×10^4 cells/well in a 96-well plate, respectively, and incubated overnight at 37 °C.

Cells were incubated for 20 h with FITC-labeled PLGA NPs modified with HF555-labeled SubAB toxins. The distributions of HF555-labeled SubAB toxins and FITC-labeled PLGA NPs were observed under the BZ-X800 All-in-one Fluorescence Microscope. The fluorescence intensity of FITC-labeled PLGA NPs in each well was measured and quantified using a TECAN Infinite F200 Pro fluorescence microplate reader (Tecan Group Ltd., Männedorf, Switzerland) at an excitation wavelength of 485 nm and emission wavelength of 535 nm.

2.14. Western Blotting

J774.1 and HeLa cells were seeded at 5×10^4 and 1×10^4 cells/well, respectively, and cultured overnight at 37 °C. The cells were treated with or without LPS (100 ng/mL) in the presence of 5 μg/mL SubAB toxins and 20 μg/mL SubAB toxins-PLGA NPs, and then incubated for 3, 8, or 20 h at 37 °C. Cells were lysed in SDS sample buffer (62.5 mM Tris-HCl, pH 6.8, 2% SDS, 6% glycerol, 0.005% bromophenol blue, and 2.5% 2-mercaptoethanol) and then boiled for 3 min. Proteins were separated by SDS-PAGE, and transferred onto a polyvinylidene difluoride membrane (Merck Millipore, Darmstadt, Germany) at 100 V for 1 h. The membranes were blocked with 5% dry non-fat milk in TBS-T (20 mM Tris-HCl, pH 7.5, 137 mM NaCl, and 0.1% Tween 20) for 1 h and then incubated for 1 h at room temperature or overnight at 4 °C with the indicated primary antibody. The membranes were washed with TBS-T, followed by incubation with an HRP-conjugated secondary antibody at room temperature for 1 h. Protein bands were detected using Immobilon Western Chemiluminescent HRP Substrate (Merck Millipore) and a luminescent image analyzer, the ChemiDoc™ XRS system (Bio-Rad, Hercules, CA, USA).

2.15. MTT Assay

Cell viability was assessed by the MTT assay in accordance with our previous study [40]. To evaluate the toxicity of SubAB toxin-PLGA NPs, J774.1 and HeLa cells were seeded at 5×10^4 and 1×10^4 cells/well, respectively, in a 96-well plate and treated for 24 h with or without 5 μg/mL PLGA NPs, WT-PLGA NPs, A-PLGA NPs, or S35A-PLGA NPs. Culture supernatants were then replaced with a culture medium containing 0.75 mg/mL MTT. After 1 h of incubation, a stop solution (isopropanol containing 0.4% HCl and 10% Triton X-100) was added to each well. Relative cell viability was analyzed by measuring absorbance at 570 nm using the microplate reader. Results are expressed as the mean cell viability ± standard deviation (S.D.) of triplicate cultures.

2.16. Griess Assay

Quantification of nitrite ions produced by macrophages was examined by the Griess assay. J774.1 cells seeded in a 96-well plate were treated with or without 5 μg/mL PLGA NPs, WT-PLGA NPs, A-PLGA NPs, or S35A-PLGA NPs. After treatment for 20 h, 50 μL of culture supernatant was mixed with 25 μL Griess Reagent I (1% sulfanilamide in 5% HCl) and incubated for 5 min. Then, 25 μL Griess Reagent II [0.1% N-(1-naphthyl)-ethylenediamine] was added, followed by incubation at room temperature for 10 min. Absorbance at 570 nm was measured by the microplate reader.

2.17. Statistical Analysis

All data are expressed as means ± S.D. Data for each experiment were obtained from at least three independent experiments. Statistical analyses were performed using Student's *t*-test with the level of significance set at $p < 0.05$.

3. Results and Discussion

3.1. Cell Recognition-Inactivated SubAB$_{S35A}$ (S35A) Mutant Had Decreased Uptake by Cells

Wildtype SubAB (WT) binds to cell surface receptors and enters through lipid rafts and actin-dependent macropinocytosis-like mechanisms in HeLa cells [26,41,42]. The 35th serine residue (S35) of the B subunit (corresponds to S12 of the secreted B subunit without the N-terminal signal peptide) is required for binding to N-glycolylneuraminic acid-terminated glycans of receptors [23,41]. To evaluate cellular uptake of mutant SubAB toxins including the catalytically inactive mutant $SubA_{S272A}B$ (S272A), site-directed B subunit mutant $SubAB_{S35A}$ (S35A), and B subunit-deficient mutant SubA, we treated cells with fluorescent dye-labeled toxins [HF555-wildtype SubAB (WT), S272A, S35A, and SubA (A)] and observed the distribution by fluorescence microscopy. Figure 1 shows fluorescent signals of HF555 around nuclei in both RAW264.7 and HeLa cells treated with HF555-WT and HF555-S272A, suggesting that both WT and mutant S272A, which have intact B subunits, were internalized and localized in the ER (Figure 1a,b). However, intracellular accumulation of HF555 fluorescence was not observed in cells treated with HF555-A and HF555-S35A. In agreement with a previous report [23,43], our results indicated that intracellular uptake of SubAB was dramatically decreased by deficiency of the B subunit or mutation in serine at position 35. *Yersinia pestis* and *Salmonella enterica* serovar Typhi encode homologs of SubB, namely YpeB and PltB [44,45]. Mutations in the serine residue corresponding to S35 of SubB negate the binding activity, demonstrating the importance of the serine residue at this position for sialoglycan-modified cell surface receptors. However, mutation in the active site of the A subunit did not affect cellular uptake and accumulation around the nucleus.

SubAB is known to specifically cleave the endoplasmic reticulum chaperone BiP. Figure 1c shows the Western blot images of BiP cleavage by incubation of RAW264.7, HeLa, and HeLa cell lysates with the indicated toxins. WT cleaved BiP in both RAW 264.7 and HeLa cells, and in cell lysates, whereas SubA (A) and $SubAB_{S35A}$ (S35A) cleaved BiP only in cell lysates (Figure S1a–c). These results indicate that both A and S35A have an enzymatic activity that cleaves BiP but cannot enter cells because of reduced cell binding. Therefore, the accumulation of HF555-SubAB toxins around the nucleus indicates ER accumulation that depends on the binding activity of the B subunit. However, S272A translocates to the ER but does not cleave BiP in macrophages, HeLa cells, or cell lysates because of mutation in the active site.

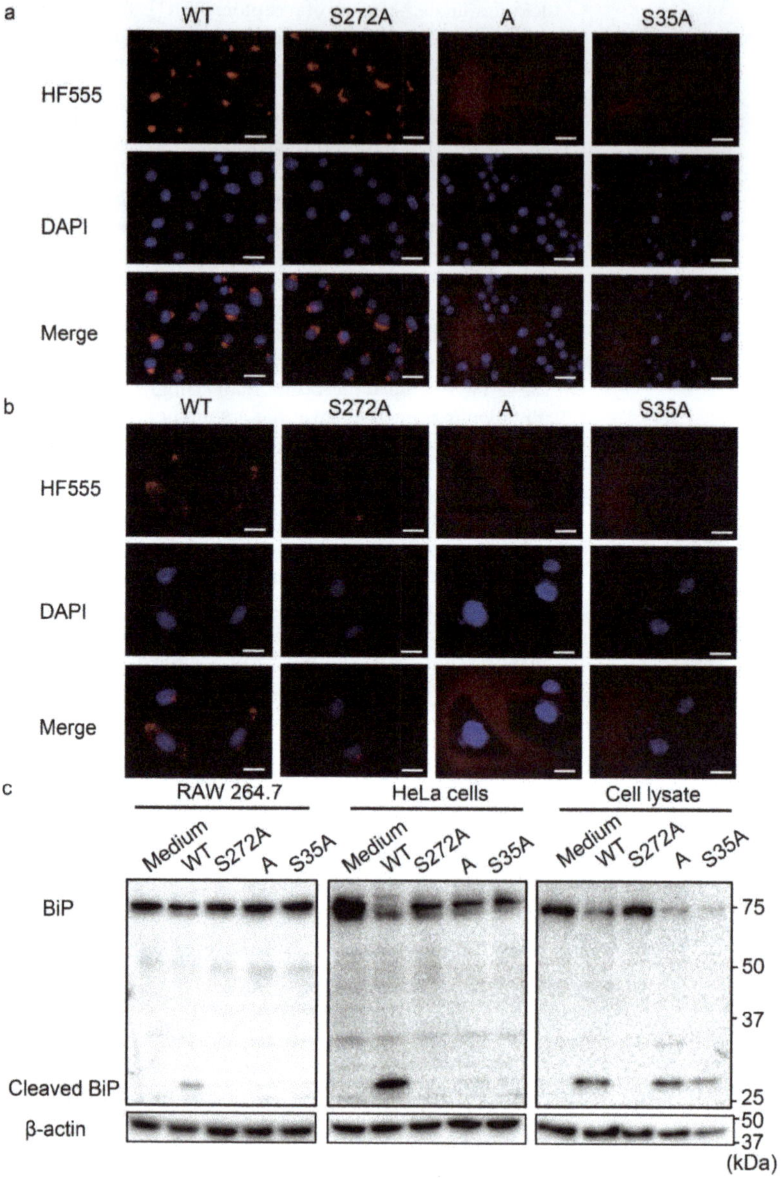

Figure 1. Evaluation of cellular uptake and BiP cleavage activity of various SubAB toxins. (**a**,**b**) RAW264.7 cells (**a**) and HeLa cells (**b**) were treated with 5 μg/mL Hylite555 (HF555)-labeled toxins and then incubated for 1 h at 37 °C. After adding a DAPI solution, cells were observed by fluorescence microscopy. WT: HF555-labeled wildtype SubAB; S272A: HF555-labeled $SubA_{S272A}B$; A: HF555-labeled SubA; S35A: HF555-labeled $SubAB_{S35A}$. Scale bars: 10 μm. (**c**) RAW264.7 and HeLa cells and a HeLa cell lysate were treated for 1 h with 5 μg/mL toxins. BiP cleavage in cultured cells and the cell lysate were evaluated by Western blotting. WT: wildtype SubAB; S272A: $SubA_{S272A}B$; A: SubA; S35A: $SubAB_{S35A}$. β-Actin served as a loading control.

3.2. Preparation and Characterization of PLGA NPs Modified with Various SubAB Toxins

Poly(D,L-lactide-co-glycolic) acid (PLGA) is a synthetic copolymer composed of lactic and glycolic acids. PLGA has amphipathic properties because of the lactic acid-derived hydrophobic crystalline characteristic and glycolic acid-derived hydrophilic amorphous characteristic. It is degraded by cellular hydrolysis and metabolized by the tricarboxylic acid cycle, and subsequently eliminated from the body as carbon dioxide and water [46]. PLGA NPs are one of the most characterized biomaterials available for the development of drug delivery systems in terms of design and performance as a biodegradable micro/nano-device [47,48]. To expose carboxyl groups outside of the particle, we prepared PLGA NPs by the single oil-in-water emulsion method in presence of stearic acids (Figure 2a). TEM observation showed that the PLGA NPs were spherical and dispersed individually (Figure 2b). The PLGA NPs had an average particle size of 260.7 ± 14.7 nm and a zeta potential of −32.8 ± 0.32 mV (Table 1 and Figure S2a,b). The observed size of the PLGA NPs on the TEM grids was smaller than the diameters measured by DLS, which is related to the different states of the particles during these measurements. In DLS measurement, the NPs were dispersed in an aqueous solution and therefore swollen [49,50]. Meanwhile, in TEM observation, the NPs were observed in dry conditions. Thus, the NPs may be shrunk during the dry process, resulting in a smaller size than that of DLS measurement [51]. Furthermore, DLS measurement is a technique to characterize particle size from the decay of light scattering as a result of Brownian motion. In the case of a high polydispersity index, light scattering intensity is strengthened by large particles, meaning that the size of characterized particle was larger than that of the intrinsic particle [52]. The negative zeta potential indicated that the carboxyl groups of stearic acid were embedded in the NP surface. The carboxyl groups of stearic acid were modified by His-tagged WT, A, or S35A through a Ni^{2+} chelate complex (Figure 2c). The poly-histidine-tagged (His-tagged) recombinant protein binds strongly to transition metal chelates such as the Ni(II) nitrilotriacetate (Ni-NTA) complex at pH 8.0. Protonation of the imidazole nitrogen atom in the histidine residue (pKa_3 = 6.04) and the coordination bond between histidine and transition metal ions including Ni^{2+} are disrupted by a pH reduction to 5.0 [53,54]. In a previous study, macrophage-selective delivery was not achieved by WT-PLGA NPs. WT-PLGA has also delivered WT to epithelial cells. Thus, to target macrophages, we modified PLGA NPs through the Ni-NTA complex at pH 8.0 with the His-tagged A and S35A, which has a low cell-binding activity. The particle size and zeta potential of WT-, A-, and S35A-PLGA NPs were examined by DLS. As shown in Table 1 and Figure S2c–h, the average size of the toxin-modified PLGA NPs was approximately 300 nm, and the zeta potential had increased positively. Theoretically calculated isoelectric point (pI) values of SubAB subunits obtained by the Compute pI/Mw tool ExPASy (https://web.expasy.org/computepi/) (accessed on 14 February 2022) were 9.28 (A subunit) and 8.54 (B subunit) in water. These results suggested that PLGA NPs were successfully modified with toxins via the stearic-acid-based Ni-NTA complex.

Table 1. Particle size, polydispersity index, and zeta potential of PLGA NPs, WT-PLGA NPs, A-PLGA NPs, and S35A-PLGA NPs in water.

	Size (nm)	Polydispersity Index	Zeta Potential (mV)
PLGA NPs	260.7 ± 14.7	0.215 ± 0.012	−32.8 ± 0.32
WT-PLGA NPs	303.4 ± 4.9	0.367 ± 0.030	7.09 ± 1.32
A-PLGA NPs	302.5 ± 5.6	0.355 ±0.008	8.89 ± 0.73
S35A-PLGA NPs	302.7 ± 2.3	0.362 ±0.035	3.19 ± 0.15

Data are expressed as means ± S.D. (n = 3).

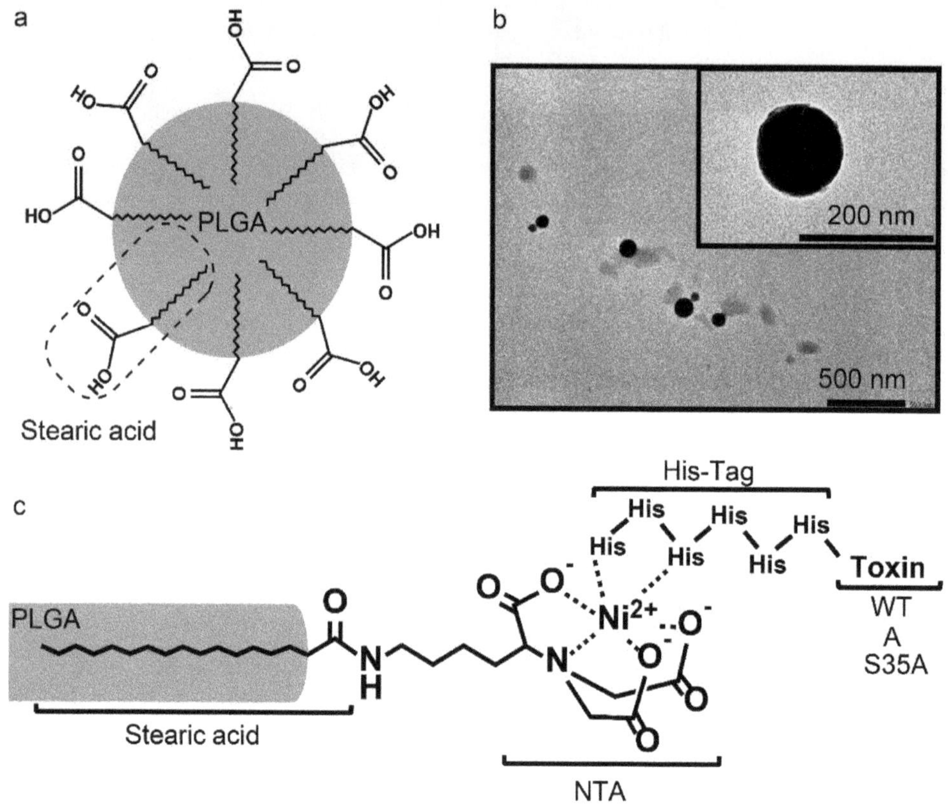

Figure 2. Preparation of PLGA NPs and modification with SubAB. (**a**) Schematic diagram of PLGA NPs in which stearic acids are anchored on the surface. (**b**) Transmission electron microscopy image of PLGA NPs. (**c**) Schematic diagram of toxin-PLGA NPs in which toxins are anchored on the surface.

To evaluate pH-responsive dissociation of His-tagged SubAB toxins and SubA from PLGA NPs, we treated WT-, A-, and S35A-PLGA NPs with buffer solutions of different pH values as indicated in Figure S3. Most His-tagged toxins had dissociated within 1 min at pH 5.0, but not at pH 7.4 (Figure S3a–d). Recently, functional vesicular systems consisting of non-ionic surfactants and cholesterol such as niosome have attracted much attention in the field of drug delivery because of their advantages, such as being capable of entrapping both hydrophilic and hydrophobic drugs in their aqueous inner core and lipid bilayer, high stability, biocompatible, biodegradable [55,56]. Niosome is a well-designed bilayer membrane carrier, but it takes several hours to release the drug [57]. On the other hand, the release of toxins from PLGA NPs is considerably faster and is expected to have a rapid drug effect in an acidic environment. Consistent with a previous study [36], these data suggested that the His-tagged toxins would be released from PLGA NPs in a pH-dependent manner when taken up by cells and exposed to the acidic environment of endosomes and lysosomes. This study evaluated the stability of NPs in water (Table 1 and Figure S2) and the release of toxins under acid conditions (Figure S3). However, in a biological environment containing blood or plasma, the stability of NPs may change and uncontrolled toxin release may occur. Therefore, further studies under various biological conditions are needed to better explore its stability and toxin release.

3.3. WT- and S35A-PLGA NPs Induce ER Stress and Inhibit iNOS Expression in Macrophages

SubAB cleaves BiP to produce 44 and 28 kDa polypeptides [27]. The cleavage of BiP induces an ER stress response, leading to expression of CHOP, cleavage of PARP, and activation of caspases [58–60]. To investigate the effect of toxin-modified PLGA NPs on J774.1 cells, the cells were incubated for 3, 8, and 20 h with free or toxin-modified PLGA NPs, followed by detection of BiP cleavage, PARP cleavage, and CHOP expression.

After 3 h of incubation, free WT cleaved BiP, but not free S35A, A, and PLGA NPs alone. However, WT and S35A-modified PLGA NPs induced BiP cleavage, but not A-PLGA NPs (Figure 3a,b). PARP was cleaved from 116 kDa to 85 kDa as apoptosis progressed. Evaluation of PARP cleavage at 8 h revealed more fragments in cells treated with WT, WT-PLGA NPs, and S35A-PLGA NPs compared with other treatments (Figure 3c,d). Consistent with the results of BiP and PARP cleavage, CHOP expression at 8 and 20 h was higher after WT, WT-PLGA NP, and S35A-PLGA NP treatments compared with other treatments. S35A-PLGA NPs induced a lower level of CHOP expression compared with WT at 8 h, but the expression was equal to WT at 20 h (Figure 3c,e,f and Figure S4a). In any time period, A did not induce PARP cleavage or CHOP expression even when modified to PLGA NPs. Therefore, these results suggest that S35A-PLGA NPs affect macrophages, although it is slightly less than WT and WT-PLGA NPs.

Previously, we reported that SubAB-mediated ER stress inhibits LPS-induced iNOS expression by suppressing nuclear translocation of NF-κB [35]. iNOS is an inflammatory marker, which produces NO from L-arginine as a substrate [61]. The produced NO is oxidized to nitrite as a stable metabolite [62]. At 20 h of treatment, iNOS and nitrite generation were quantified by Western blotting and the Griess assay, respectively. S35A-PLGA NPs inhibited iNOS expression and nitrite production in addition to free WT and WT-PLGA NPs (Figure 3g–i). Additionally, cell viability assessed by MTT assays after treatment for 24 h revealed that WT, WT-PLGA NPs, and S35A-PLGA NPs showed cytotoxicity (Figure S4b).

These results revealed that free S35A did not induce ER stress and suppressed inflammation because it was not taken up by cells. The PLGA NP modification induced S35A uptake by J774.1 cells, followed by ER stress due to BiP cleavage, and suppressed iNOS expression and nitrite production in J774.1 cells. However, A-PLGA NPs did not induce ER stress or BiP cleavage, suggesting that the B subunit of SubAB is important not only for cell adhesion but also for delivery of the A subunit to the ER. In general, the ER retention signal sequence, such as KDEL and KKXX-like motif sequences, play a crucial role in the localization of transmembrane proteins in the ER [63], and the B subunit has the KKNS sequence at amino acids 128–131. Although S35A has a mutation in serine at position 35 of the B subunit to alanine, this mutation does not affect the sequence of KKNS. Therefore, WT and S35A transferred to the ER via KKNS sequences after being taken up by macrophages.

3.4. S35A-PLGA NPs Do Not Induce ER Stress or Cytotoxicity in HeLa Cells

As shown in Figure 3, S35A-PLGA NPs showed anti-inflammatory effects in macrophages. To develop macrophage-specific delivery, it was necessary to prove that S35A-PLGA NPs did not enter non-phagocytic cells such as epithelial cells and do not induce ER stress. Therefore, we used HeLa cells to investigate whether S35A-PLGA NPs induced ER stress in non-phagocytic cells. Caspase is activated by the ER stress response and is a useful marker of cell death due to apoptosis [64]. Phosphorylation of eIF2α at Ser51 is induced by various environmental stresses such as ER and oxidative stresses or amino acid starvation. There are four types of protein kinases known to phosphorylate eIF2α in mammals, including PERK, general control nonderepressible-2 (GCN2), heme-regulated inhibitory (HRI), and protein kinase R (PKR) [65]. PERK activation induces eIF2α phosphorylation, caspase 3 and 7 cleavage, CHOP expression, and PARP cleavage [66–69]. We next evaluated eIF2α phosphorylation, caspase and PARP cleavage, and CHOP expression to assess the cytotoxicity of S35A-PLGA NPs

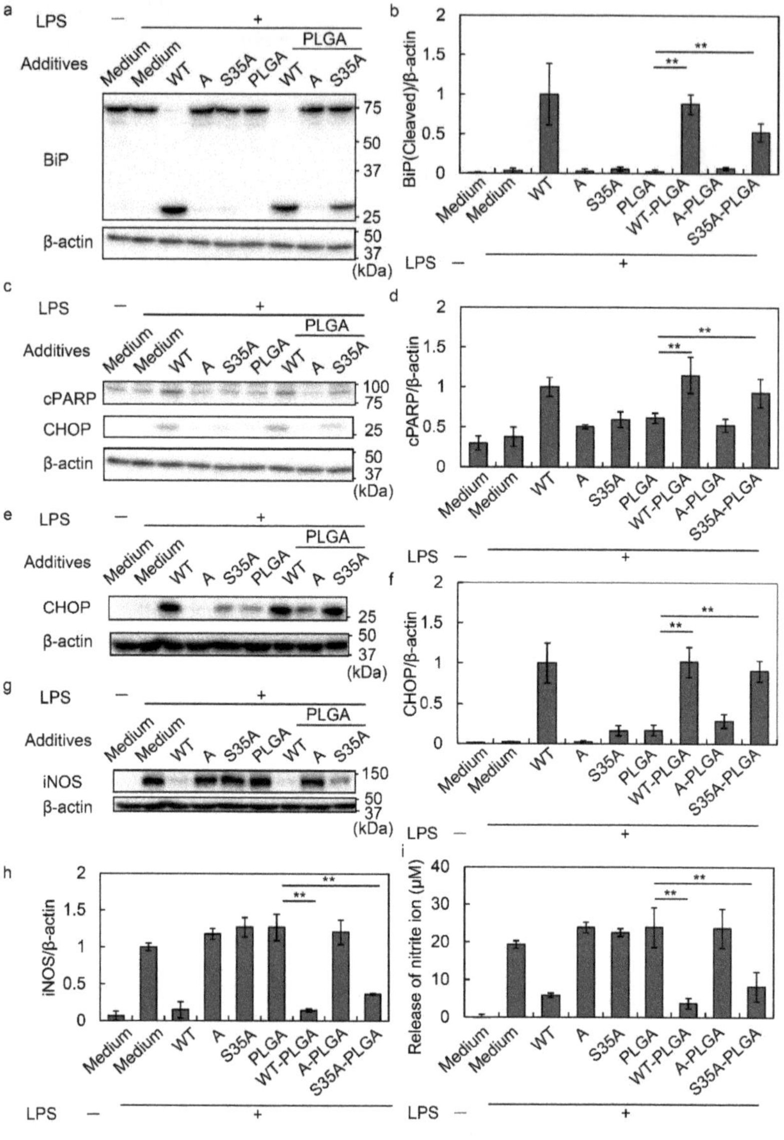

Figure 3. Induction of ER stress and anti-inflammatory effect by SubAB-PLGA NPs on macrophages. (**a–h**) J774.1 cells were treated with or without LPS (100 ng/mL) in the presence or absence of 5 μg/mL toxins and 20 μg/mL PLGA NPs for 3, 8, and 20 h. After incubation, the cells were lysed in SDS sample buffer and subjected to Western blotting. (**a**) Evaluation of BiP cleavage at 3 h by Western blot analysis. (**b**) Quantification of BiP cleavage by densitometry. (**c**) Evaluation of cleaved PARP (cPARP) and CHOP expression at 8 h by Western blot analysis. (**d**) Quantification of PARP cleavage by densitometry. (**e**) Evaluation of CHOP expression at 20 h by Western blot analysis. (**f**) Quantification of CHOP expression by densitometry. (**g**) Evaluation of iNOS expression at 20 h by Western blot analysis. (**h**) Quantification of iNOS expression by densitometry. (**i**) Quantification of nitrite production by the Griess assay. Data are expressed as means ± S.D. (n = 3). Statistical analyses were performed using Student's t-test. ** $p < 0.01$.

We evaluated the BiP cleavage, eIF2α phosphorylation, and cleaved caspase 7 at 3 h in HeLa cells (Figure 4a,b). WT and WT-PLGA NPs cleaved BiP, which correlated with caspase 7 activation and eIF2α phosphorylation. Interestingly, BiP cleavage was not detected in S35A-PLGA NP-treated HeLa cells (Figure 4a,b and Figure S5a,b). Caspase 7 activation and eIF2α phosphorylation in S35A-PLGA NP-treated cells were lower than those in WT or WT-PLGA NP-treated cells. At 8 h, PARP cleavage, as well as caspase 3 and 7 activation, were detected in WT or WT-PLGA NP-treated HeLa cells (Figure 4c,d and Figure S5c,d), but not in cells treated with S35A-PLGA NPs. CHOP expression induced by S35A-PLGA NPs was hardly detected compared with WT or WT-PLGA NP treatments, suggesting that ER stress did not occur in HeLa cells (Figure 4e,f). Consistent with these results, WT and WT-PLGA, but not S35A-PLGA NPs, caused cell death after 24 h of treatment (Figure S5e). ER stress mediated by activation of the PERK-eIF2α pathway is important as the main cell death mechanism induced by SubAB [70]. SubAB inhibits protein synthesis through eIF2α phosphorylation and caspase 3/7 activation and fragmentation, followed by apoptosis [71,72]. Notably, S35A-PLGA NPs showed lower activation of these factors than WT-PLGA NPs in HeLa cells. By summarizing the results from Figure 3, Figure 4, Figures S4 and S5, our data suggested that S35A-PLGA NPs induce ER stress and cytotoxicity in macrophages through activation of these cell death mechanisms, but not in endothelial cells.

3.5. *S35A-PLGA NPs Specifically Enter Macrophages, but Not Epithelial Cells*

To examine the cellular uptake of PLGA NPs by RAW264.7 and HeLa cells, we synthesized FITC-labeled PLGA NPs and then modified them with WT or S35A. The intracellular localization of these fluorescent PLGA NPs was evaluated by fluorescence microscopy and a fluorescence microplate reader. Fluorescence images revealed that FITC-labeled PLGA NPs were taken up by macrophages regardless of the toxin type (Figure 5a). Additionally, evaluation of PLGA NP uptake by the fluorescence microplate reader revealed no significant difference in the uptake by RAW264.7 cells of any of the NPs (Figure 5b). However, WT-PLGA NPs were taken up and HF555-WT was detected in HeLa cells, but not PLGA NPs alone or S35A-PLGA NPs (Figure 5c). The fluorescence intensity read by the microplate reader showed that S35A-PLGA NPs did not enter HeLa cells (Figure 5d). These data suggested that S35A-PLGA specifically targets macrophages. In addition to macrophages, S35A-PLGA NPs have the potential to target neutrophils and dendritic cells that have a phagocytic activity.

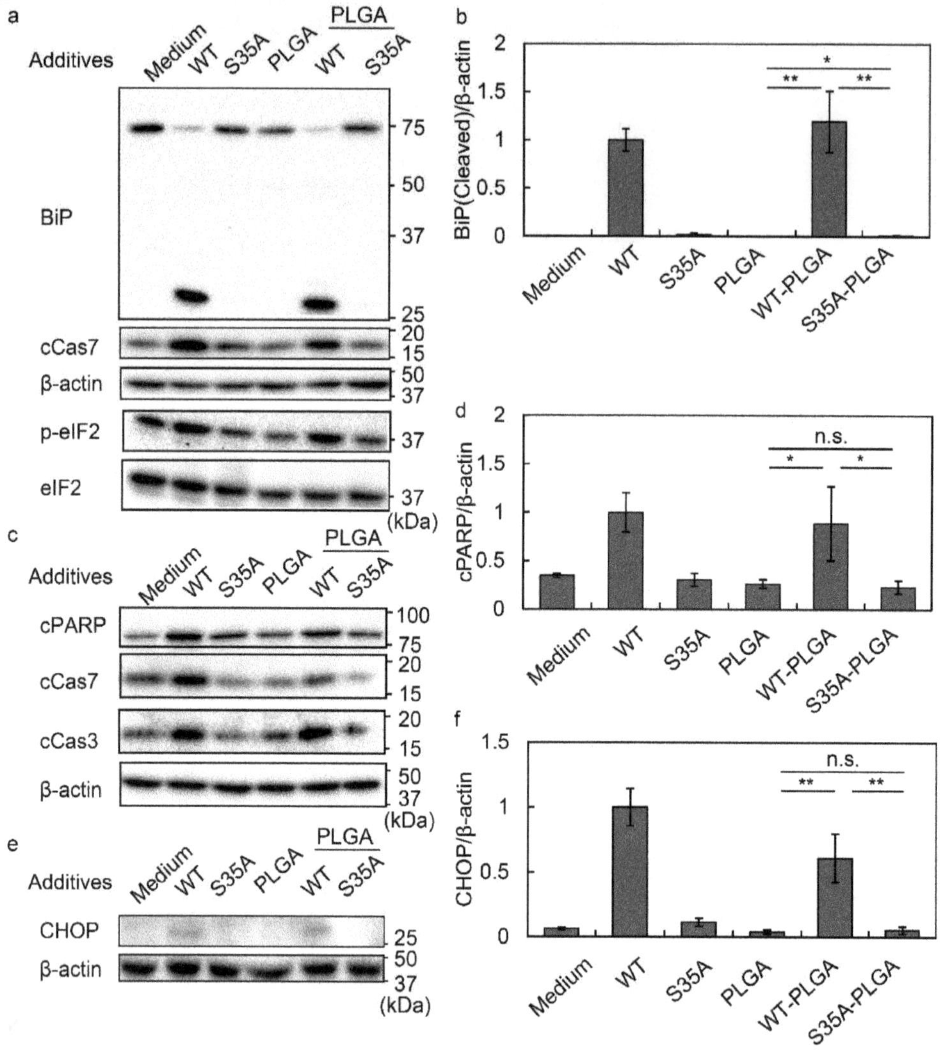

Figure 4. Induction of ER stress and cytotoxicity by SubAB-PLGA NPs in HeLa cells. (**a**–**f**) HeLa cells were treated with or without 5 μg/mL toxins and 20 μg/mL toxin-PLGA NPs for 3, 8, and 20 h. After treatment, the cells were lysed in SDS sample buffer and subjected to Western blotting. cCas7, cleaved caspase 7; cCas3, cleaved caspase 3; cPARP, cleaved PARP. (**a**) Evaluation of BiP cleavage, caspase 7 activation, and eIF2 phosphorylation at 3 h. (**b**) Quantification of BiP cleavage by densitometry. (**c**) Evaluation of PARP cleavage, and caspase 3 and 7 activation at 8 h. (**d**) Quantification of PARP cleavage by densitometry. (**e**) Evaluation of CHOP expression at 20 h. (**f**) Quantification of CHOP expression by densitometry. Data are expressed as means ± S.D. ($n = 3$). Statistical analyses were performed using Student's *t*-test. n.s., not significant. * $p < 0.05$, ** $p < 0.01$.

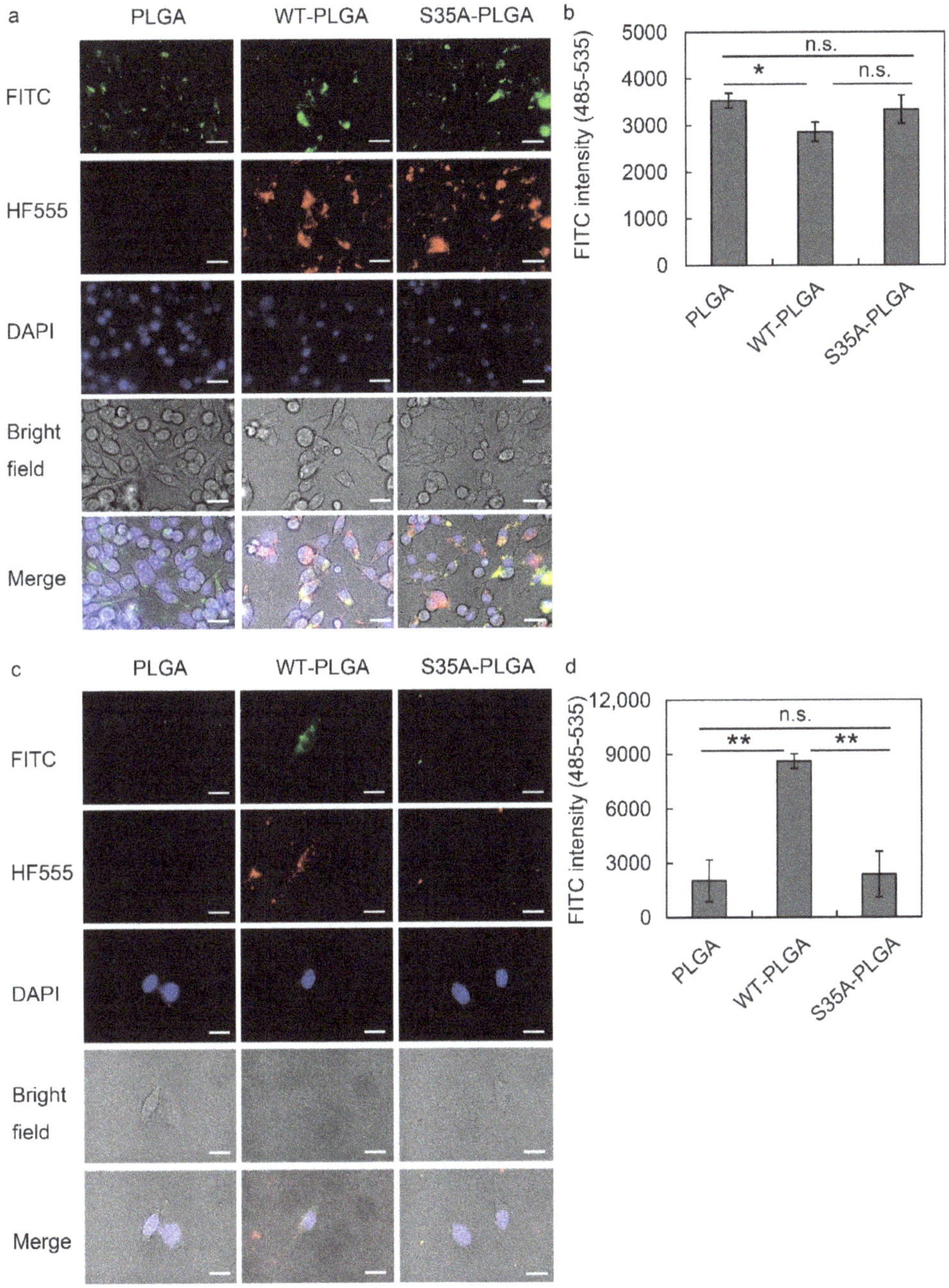

Figure 5. Uptake of SubAB-PLGA NPs by RAW264.7 and HeLa cells. RAW264.7 (**a**,**b**) and HeLa (**c**,**d**) cells were treated with 5 µg/mL FITC-labeled PLGA NPs modified with or without HF555-labeled WT or S35A. After incubation for 20 h, intracellular localization of fluorescent dye-labeled PLGA NPs and toxins was analyzed by fluorescence microscopy (**a**,**c**) and a fluorescence microplate reader (**b**,**d**). Scale bars: 10 µm. Data are expressed as means ± S.D. ($n = 3$). Statistical analyses were performed using Student's *t*-test. n.s., not significant. * $p < 0.05$, ** $p < 0.01$.

WT-PLGA NPs exert a higher effect on macrophages than free SubAB at low concentrations [36]. It has been considered that this may be due to uptake of PLAGA NPs via phagocytosis caused by the entry of SubAB through other pathways or promotion of the normal pathway, including endocytosis or a macropinocytosis-like mechanism. However, WT-PLGA NPs showed cytotoxicity in HeLa cells (Figure S5e), which are non-phagocytic epithelial cells. SubAB enters HeLa cells via an actin/lipid-raft-dependent macropinocytosis-like uptake pathway [26]. The B subunit binds to sialoglycan which mainly has N-glycolylneuraminic acid as a sialic acid [43]. S35A is a synthetic mutant in which the 35th serine (S35) of the B subunit is replaced with alanine, but the BiP cleavage ability of the A subunit and ER retention signal sequence of the B subunit are active. Although the entry of S35A into cells was reduced, if taken up, it was transported to the ER, cleaved BiP (Figure 1c), and induced ER stress in macrophages (Figure 3). These data indicated that S35 is responsible for the binding activity of the B subunit. By focusing on this property, we conceived delivery using the phagocytosis of PLGA NPs by macrophages. In fact, S35A was not incorporated as a free toxin, but S35A-PLGA NPs were taken up by macrophages (Figure 5a), leading to ER stress and anti-inflammatory effects (Figure 3). As expected, S35A-PLGA NPs did not enter HeLa cells (Figure 5b) and produced neither ER stress nor cytotoxicity. Interestingly, we achieved specific delivery of S35A to the macrophage ER by targeting the uptake with PLGA NP. Previous KO cell screening using the CRISPR/Cas9 gRNA library has demonstrated that KDEL receptors 1 and 2 are important host factors for SubAB toxicity [41]. Cell death is attenuated in KDELR1 and 2 KO cells, suggesting that the ER retention KKNS sequence in the C-terminal of the SubAB B subunit contributes to the localization of the ER and the cytotoxicity of SubAB. S35A reached the ER via this sequence, but SubA was unable to suppress macrophage inflammation even when modified with PLGA NPs. As SubA does not have an ER retention KKNS sequence, these results suggest that SubA cannot reach the ER even if this subunit enters the cell with PLGA NPs. Our observation raises the possibility that PLGA NPs modified with SubB containing an ER retention signal sequence could deliver a reagent of interest to the ER of phagocytic cells. By modifying proteins or peptides with organelle-targeting sequences, our system may deliver substances to various organelles such as nuclei and mitochondria. The ER and Golgi pathways play important roles in the post-translational modification and maturation of synthesized proteins. In particular, the ER is an essential organelle for redox regulation of disulfide bond formation and sugar chain modification, and enzymes that mediate these maturation processes are mainly abundant in the ER. In the future, it is expected that this technology will be applied to DDSs that target organelles, leading to the development of treatment strategies for many diseases such as infectious diseases and cancer as well as inflammation.

4. Conclusions

In this study, we established a macrophage-specific delivery system using site-directed mutagenesis of bacterial toxins and PLGA NP modification. $SubAB_{S35A}$ (S35A), which has a mutated serine residue (S35) of the B subunit to reduce cell-binding activity [23,43], was not taken up by cells without the PLGA modification. However, S35A-PLGA NPs were efficiently taken up by J774.1 cells and suppressed inflammation. S35A-PLGA NPs were hardly taken up by HeLa cells and did not show toxicity, suggesting that S35A PLGA NPs do not have a non-specific effect on epithelial cells. Additionally, PLGA NPs conjugated with His-tagged S35A via the Ni-NTA complex resulted in rapid dissociation of S35A from PLGA NPs at a low pH, which was similar to the results observed in our previous study. This suggests that S35A is efficiently released from NPs by intracellular acidification during cellular trafficking after phagocytosis. Our well-designed DDS has achieved macrophage-specific delivery of anti-inflammatory toxins and it is exerting their effects in the ER. S35A-PLGA NPs are nano-sized materials that are biodegradable and expected to be an anti-inflammatory nanomaterial with low side effects. However, further confirmation of the stability of PLGA NPs and the release of toxins from NPs under biological conditions is necessary in future studies. Toxins from various microorganisms are bioactive and act specifically on the target molecule in the host. Many toxins have useful functions, such as anti-inflammatory, anti-cancer, and immune-potentiating effects, to support bacterial survival and transmission. Therefore, such bioactive toxins could be applied as new therapeutic agents without side effects if their binding activity and cellular uptake are regulated by modification with NP. These results are also expected to be applied to organelle-specific delivery of drugs and therapeutic strategies for autoimmune diseases, infection, and cancer. This study has been limited to experiments using cultured cells, especially mouse macrophages. In the future, we would like to establish a therapeutic treatment for mouse inflammation models and an experimental system using human NO-producing cells and a co-culture system consisting of macrophages and epithelial cells, and evaluate the potential of anti-inflammatory NPs for clinical application.

Supplementary Materials: The following supporting information can be downloaded at: https://www.mdpi.com/article/10.3390/nano12132161/s1, Figure S1: Densitometry of BiP cleavage in (a) J774.1 cells, (b) HeLa cells, and (c) a HeLa cell lysate; Figure S2: Characterization of PLGA NPs; Figure S3: pH-responsive dissociation of His-tagged WT, A, and S35A from PLGANPs; Figure S4: Evaluation of ER stress and cytotoxicity in J774.1 cells; Figure S5: Evaluation of ER stress and apoptosis induction in HeLa cells.

Author Contributions: Conceptualization, A.H. and H.T.; methodology, H.T., T.Z. and K.Y.; formal analysis, A.H.; writing—original draft preparation, A.H.; writing—review and editing, H.T., K.Y. and T.N.; supervision, H.T.; project administration, T.S. and T.N.; funding acquisition, H.T., K.Y., T.S. and T.N. All authors have read and agreed to the published version of the manuscript.

Funding: This study was supported in part by Grants-in-Aid for Scientific Research [(B) and (C)] from the Ministry of Education, Science, Sports, and Technology (MEXT), Japan, to H.T. (20K08823) and T.S. (21H02071), a grant from the Takeda Science Foundation to H.T., a grant from the Japan Agency for Medical Research and Development (AMED) to K.Y. (22fk0108611h0902) and a grant from the Japan Science and Technology Agency (JST), Core Research for Evolutionary Science and Technology (CREST) to T.N. (JPMJCR18H5).

Data Availability Statement: The data presented in this study are available on request from the corresponding author.

Acknowledgments: We thank Mitchell Arico from Edanz (https://jp.edanz.com/ac) (accessed on 20 April 2022) for editing a draft of this manuscript.

Conflicts of Interest: The authors declare no conflict of interest.

References

1. Vargason, A.M.; Anselmo, A.C.; Mitragotri, S. The evolution of commercial drug delivery technologies. *Nat. Biomed. Eng.* **2021**, *5*, 951–967. [CrossRef] [PubMed]
2. Hu, G.; Guo, M.; Xu, J.; Wu, F.; Fan, J.; Huang, Q.; Yang, G.; Lv, Z.; Wang, X.; Jin, Y. Nanoparticles Targeting Macrophages as Potential Clinical Therapeutic Agents Against Cancer and Inflammation. *Front. Immunol.* **2019**, *10*, 1998. [CrossRef] [PubMed]
3. Karchemski, F.; Zucker, D.; Barenholz, Y.; Regev, O. Carbon nanotubes-liposomes conjugate as a platform for drug delivery into cells. *J. Control. Release* **2012**, *160*, 339–345. [CrossRef] [PubMed]
4. Ding, Y.; Lv, B.; Zheng, J.; Lu, C.; Liu, J.; Lei, Y.; Yang, M.; Wang, Y.; Li, Z.; Yang, Y.; et al. RBC-hitchhiking chitosan nanoparticles loading methylprednisolone for lung-targeting delivery. *J. Control. Release* **2022**, *341*, 702–715. [CrossRef]
5. Heo, R.; You, D.G.; Um, W.; Choi, K.Y.; Jeon, S.; Park, J.S.; Choi, Y.; Kwon, S.; Kim, K.; Kwon, I.C.; et al. Dextran sulfate nanoparticles as a theranostic nanomedicine for rheumatoid arthritis. *Biomaterials* **2017**, *131*, 15–26. [CrossRef]
6. Yang, Y.; Chen, F.; Xu, N.; Yao, Q.; Wang, R.; Xie, X.; Zhang, F.; He, Y.; Shao, D.; Dong, W.F.; et al. Red-light-triggered self-destructive mesoporous silica nanoparticles for cascade-amplifying chemo-photodynamic therapy favoring antitumor immune responses. *Biomaterials* **2022**, *281*, 121368. [CrossRef]
7. Chang, Y.; Cho, B.; Lee, E.; Kim, J.; Yoo, J.; Sung, J.S.; Kwon, Y.; Kim, J. Electromagnetized gold nanoparticles improve neurogenesis and cognition in the aged brain. *Biomaterials* **2021**, *278*, 121157.
8. Cruz, L.J.; van Dijk, T.; Vepris, O.; Li, T.; Schomann, T.; Baldazzi, F.; Kurita, R.; Nakamura, Y.; Grosveld, F.; Philipsen, S.; et al. PLGA-Nanoparticles for Intracellular Delivery of the CRISPR-Complex to Elevate Fetal Globin Expression in Erythroid Cells. *Biomaterials* **2021**, *268*, 120580. [CrossRef]

9. Yan, J.; Yao, Y.; Yan, S.; Gao, R.; Lu, W.; He, W. Chiral Protein supraparticles for tumor suppression and synergistic immunotherapy: An enabling strategy for bioactive supramolecular chirality construction. *Nano Lett.* **2020**, *20*, 5844–5852. [CrossRef]
10. Lai, W.F. Non-conjugated polymers with intrinsic luminescence for drug delivery. *J. Drug Deliv. Sci. Technol.* **2020**, *59*, 101916. [CrossRef]
11. Lai, W.F.; Huang, E.; Lui, K.H. Alginate-based complex fibers with the janus morphology for controlled release of co-delivered drugs. *Asian J. Pharm. Sci.* **2021**, *16*, 77–85. [CrossRef] [PubMed]
12. Arkaban, H.; Barani, M.; Akbarizadeh, M.R.; Pal Singh Chauhan, N.; Jadoun, S.; Dehghani Soltani, M.; Zarrintaj, P. polyacrylic acid nanoplatforms: Antimicrobial, tissue engineering, and cancer theranostic applications. *Polymers* **2022**, *14*, 1259. [CrossRef] [PubMed]
13. Han, G.; You, C.C.; Kim, B.J.; Turingan, R.S.; Forbes, N.S.; Martin, C.T.; Rotello, V.M. Light-regulated release of DNA and its delivery to nuclei by means of photolabile gold nanoparticles. *Angew. Chem. Int. Ed. Engl.* **2006**, *45*, 3165–3169. [CrossRef] [PubMed]
14. Yu, H.; Jin, F.; Liu, D.; Shu, G.; Wang, X.; Qi, J.; Sun, M.; Yang, P.; Jiang, S.; Ying, X.; et al. ROS-responsive nano-drug delivery system combining mitochondria-targeting ceria nanoparticles with atorvastatin for acute kidney injury. *Theranostics* **2020**, *10*, 2342–2357. [CrossRef] [PubMed]
15. Piao, S.; Amaravadi, R.K. Targeting the lysosome in cancer. *Ann. N. Y. Acad. Sci.* **2016**, *1371*, 45–54. [CrossRef] [PubMed]
16. Schwarz, D.S.; Blower, M.D. The endoplasmic reticulum: Structure, function and response to cellular signaling. *Cell Mol. Life Sci.* **2016**, *73*, 79–94. [CrossRef]
17. Gidalevitz, T.; Stevens, F.; Argon, Y. Orchestration of secretory protein folding by ER chaperones. *Biochim. Biophys. Acta* **2013**, *1833*, 2410–2424. [CrossRef]
18. Fuson, K.L.; Zheng, M.; Craxton, M.; Pataer, A.; Ramesh, R.; Chada, S.; Sutton, R.B. Structural mapping of post-translational modifications in human interleukin-24: Role of N-linked glycosylation and disulfide bonds in secretion and activity. *J. Biol. Chem.* **2009**, *284*, 30526–30533. [CrossRef]
19. Jäger, R.; Bertrand, M.J.; Gorman, A.M.; Vandenabeele, P.; Samali, A. The unfolded protein response at the crossroads of cellular life and death during endoplasmic reticulum stress. *Biol. Cell* **2012**, *104*, 259–270. [CrossRef]
20. Salminen, A.; Kaarniranta, K.; Kauppinen, A. ER stress activates immunosuppressive network: Implications for aging and Alzheimer's disease. *J. Mol. Med.* **2020**, *98*, 633–650. [CrossRef]
21. Zhou, Y.; Cheung, Y.K.; Ma, C.; Zhao, S.; Gao, D.; Lo, P.C.; Fong, W.P.; Wong, K.S.; Ng, D.K.P. Endoplasmic Reticulum-Localized Two-Photon-Absorbing Boron Dipyrromethenes as Advanced Photosensitizers for Photodynamic Therapy. *J. Med. Chem.* **2018**, *61*, 3952–3961. [CrossRef] [PubMed]
22. Shi, L.; Gao, X.; Yuan, W.; Xu, L.; Deng, H.; Wu, C.; Yang, J.; Jin, X.; Zhang, C.; Zhu, X. Endoplasmic Reticulum-Targeted Fluorescent Nanodot with Large Stokes Shift for Vesicular Transport Monitoring and Long-Term Bioimaging. *Small* **2018**, *14*, e1800223. [CrossRef] [PubMed]
23. Le Nours, J.; Paton, A.W.; Byres, E.; Troy, S.; Herdman, B.P.; Johnson, M.D.; Paton, J.C.; Rossjohn, J.; Beddoe, T. Structural basis of subtilase cytotoxin SubAB assembly. *J. Biol. Chem.* **2013**, *288*, 27505–27516. [CrossRef] [PubMed]
24. Beddoe, T.; Paton, A.W.; Le Nours, J.; Rossjohn, J.; Paton, J.C. Structure, biological functions and applications of the AB5 toxins. *Trends Biochem. Sci.* **2010**, *35*, 411–418. [CrossRef] [PubMed]
25. Paton, A.W.; Srimanote, P.; Talbot, U.M.; Wang, H.; Paton, J.C. A new family of potent AB(5) cytotoxins produced by Shiga toxigenic Escherichia coli. *J. Exp. Med.* **2004**, *200*, 35–46. [CrossRef]
26. Nagasawa, S.; Ogura, K.; Tsutsuki, H.; Saitoh, H.; Moss, J.; Iwase, H.; Noda, M.; Yahiro, K. Uptake of Shiga-toxigenic Escherichia coli SubAB by HeLa cells requires an actin- and lipid raft-dependent pathway. *Cell Microbiol.* **2014**, *16*, 1582–1601. [CrossRef]
27. Paton, A.W.; Beddoe, T.; Thorpe, C.M.; Whisstock, J.C.; Wilce, M.C.; Rossjohn, J.; Talbot, U.M.; Paton, J.C. AB5 subtilase cytotoxin inactivates the endoplasmic reticulum chaperone BiP. *Nature* **2006**, *443*, 548–552. [CrossRef]
28. Chong, D.C.; Paton, J.C.; Thorpe, C.M.; Paton, A.W. Clathrin-dependent trafficking of subtilase cytotoxin, a novel AB5 toxin that targets the endoplasmic reticulum chaperone BiP. *Cell Microbiol.* **2008**, *10*, 795–806. [CrossRef]
29. Smith, R.D.; Willett, R.; Kudlyk, T.; Pokrovskaya, I.; Paton, A.W.; Paton, J.C.; Lupashin, V.V. The COG complex, Rab6 and COPI define a novel Golgi retrograde trafficking pathway that is exploited by SubAB toxin. *Traffic* **2009**, *10*, 1502–1517. [CrossRef]
30. Wolfson, J.J.; May, K.L.; Thorpe, C.M.; Jandhyala, D.M.; Paton, J.C.; Paton, A.W. Subtilase cytotoxin activates PERK, IRE1 and ATF6 endoplasmic reticulum stress-signalling pathways. *Cell Microbiol.* **2008**, *10*, 1775–1786. [CrossRef]
31. Tsutsuki, H.; Yahiro, K.; Ogura, K.; Ichimura, K.; Iyoda, S.; Ohnishi, M.; Nagasawa, S.; Seto, K.; Moss, J.; Noda, M. Subtilase cytotoxin produced by locus of enterocyte effacement-negative Shiga-toxigenic Escherichia coli induces stress granule formation. *Cell Microbiol.* **2016**, *18*, 1024–1040. [CrossRef] [PubMed]
32. Yahiro, K.; Tsutsuki, H.; Ogura, K.; Nagasawa, S.; Moss, J.; Noda, M. A negative regulator of autophagy, controls SubAB-mediated apoptosis and autophagy. *Infect. Immun.* **2014**, *82*, 4899–4908. [CrossRef] [PubMed]
33. Tsutsuki, H.; Zhang, T.; Yahiro, K.; Ono, K.; Fujiwara, Y.; Iyoda, S.; Wei, F.Y.; Monde, K.; Seto, K.; Ohnishi, M.; et al. Subtilase cytotoxin from Shiga-toxigenic *Escherichia coli* impairs the inflammasome and exacerbates enteropathogenic bacterial infection. *iScience* **2022**, *25*, 104050. [CrossRef] [PubMed]
34. Furukawa, T.; Yahiro, K.; Tsuji, A.B.; Terasaki, Y.; Morinaga, N.; Miyazaki, M.; Fukuda, Y.; Saga, T.; Moss, J.; Noda, M. Fatal hemorrhage induced by subtilase cytotoxin from Shiga-toxigenic Escherichia coli. *Microb. Pathog.* **2011**, *50*, 159–167. [CrossRef]

35. Tsutsuki, H.; Yahiro, K.; Suzuki, K.; Suto, A.; Ogura, K.; Nagasawa, S.; Ihara, H.; Shimizu, T.; Nakajima, H.; Moss, J.; et al. Subtilase cytotoxin enhances Escherichia coli survival in macrophages by suppression of nitric oxide production through the inhibition of NF-kappaB activation. *Infect. Immun.* **2012**, *80*, 3939–3951. [CrossRef]
36. Harada, A.; Tsutsuki, H.; Zhang, T.; Lee, R.; Yahiro, K.; Sawa, T.; Niidome, T. Preparation of Biodegradable PLGA-Nanoparticles Used for pH-Sensitive Intracellular Delivery of an Anti-inflammatory Bacterial Toxin to Macrophages. *Chem. Pharm. Bull.* **2020**, *68*, 363–368. [CrossRef]
37. Nakamura, Y.; Kitagawa, T.; Ihara, H.; Kozaki, S.; Moriyama, M.; Kannan, Y. Potentiation by high potassium of lipopolysaccharide-induced nitric oxide production from cultured astrocytes. *Neurochem. Int.* **2006**, *48*, 43–49. [CrossRef]
38. Yahiro, K.; Morinaga, N.; Satoh, M.; Matsuura, G.; Tomonaga, T.; Nomura, F.; Moss, J.; Noda, M. Identification and characterization of receptors for vacuolating activity of subtilase cytotoxin. *Mol. Microbiol.* **2006**, *62*, 480–490. [CrossRef]
39. Morinaga, N.; Yahiro, K.; Matsuura, G.; Watanabe, M.; Nomura, F.; Moss, J.; Noda, M. Two distinct cytotoxic activities of subtilase cytotoxin produced by shiga-toxigenic Escherichia coli. *Infect. Immun.* **2007**, *75*, 488–496. [CrossRef]
40. Tsutsuki, H.; Zhang, T.; Harada, A.; Rahman, A.; Ono, K.; Yahiro, K.; Niidome, T.; Sawa, T. Involvement of protein disulfide isomerase in subtilase cytotoxin-induced cell death in HeLa cells. *Biochem. Biophys. Res. Commun.* **2020**, *525*, 1068–1073. [CrossRef]
41. Yamaji, T.; Hanamatsu, H.; Sekizuka, T.; Kuroda, M.; Iwasaki, N.; Ohnishi, M.; Furukawa, J.I.; Yahiro, K.; Hanada, K. A CRISPR Screen Using Subtilase Cytotoxin Identifies SLC39A9 as a Glycan-Regulating Factor. *iScience* **2019**, *15*, 407–420. [CrossRef] [PubMed]
42. Yahiro, K.; Satoh, M.; Morinaga, N.; Tsutsuki, H.; Ogura, K.; Nagasawa, S.; Nomura, F.; Moss, J.; Noda, M. Identification of subtilase cytotoxin (SubAB) receptors whose signaling, in association with SubAB-induced BiP cleavage, is responsible for apoptosis in HeLa cells. *Infect. Immun.* **2011**, *79*, 617–627. [CrossRef] [PubMed]
43. Byres, E.; Paton, A.W.; Paton, J.C.; Lofling, J.C.; Smith, D.F.; Wilce, M.C.; Talbot, U.M.; Chong, D.C.; Yu, H.; Huang, S.; et al. Incorporation of a non-human glycan mediates human susceptibility to a bacterial toxin. *Nature* **2008**, *456*, 648–652. [CrossRef] [PubMed]
44. Deng, L.; Song, J.; Gao, X.; Wang, J.; Yu, H.; Chen, X.; Varki, N.; Naito-Matsui, Y.; Galan, J.E.; Varki, A. Host adaptation of a bacterial toxin from the human pathogen Salmonella Typhi. *Cell* **2014**, *159*, 1290–1299. [CrossRef]
45. Khan, N.; Sasmal, A.; Khedri, Z.; Secrest, P.; Verhagen, A.; Srivastava, S.; Varki, N.; Chen, X.; Yu, H.; Beddoe, T.; et al. Sialoglycan binding patterns of bacterial AB5 toxin B subunits correlate with host range and toxicity, indicating evolution independent of A subunits. *J. Biol. Chem.* **2022**, *298*, 101900. [CrossRef]
46. Silva, A.T.C.R.; Cardoso, B.C.O.; e Silva, M.E.S.R.; Freitas, R.F.S.; Sousa, R.G. Synthesis, Characterization, and study of PLGA copolymer in vitro degradation. *J. Biomater. Nanobiotechnol.* **2015**, *6*, 8–19. [CrossRef]
47. Yuan, Z.; Wei, P.; Huang, Y.; Zhang, W.; Chen, F.; Zhang, X.; Mao, J.; Chen, D.; Cai, Q.; Yang, X. Injectable PLGA microspheres with tunable magnesium ion release for promoting bone regeneration. *Acta Biomater.* **2019**, *85*, 294–309. [CrossRef]
48. Kim, J.H.; Park, J.S.; Yang, H.N.; Woo, D.G.; Jeon, S.Y.; Do, H.J.; Lim, H.Y.; Kim, J.M.; Park, K.H. The use of biodegradable PLGA nanoparticles to mediate SOX9 gene delivery in human mesenchymal stem cells (hMSCs) and induce chondrogenesis. *Biomaterials* **2011**, *32*, 268–278. [CrossRef]
49. Blasi, P.; D'Souza, S.S.; Selmin, F.; DeLuca, P.P. Plasticizing effect of water on poly(lactide-co-glycolide). *J. Control. Release* **2005**, *108*, 1–9. [CrossRef]
50. Rapier, C.E.; Shea, K.J.; Lee, A.P. Investigating PLGA microparticle swelling behavior reveals an interplay of expansive intermolecular forces. *Sci. Rep.* **2021**, *11*, 14512. [CrossRef]
51. Courant, T.; Roullin, V.G.; Cadiou, C.; Delavoie, F.; Molinari, M.; Andry, M.C.; Chuburu, F. Development and physic chemical characterization of copper complexes-loaded PLGA nanoparticles. *Int. J. Pharm.* **2009**, *379*, 226–234. [CrossRef] [PubMed]
52. Garms, B.C.; Poli, H.; Baggley, D.; Han, F.Y.; Whittaker, A.K.; Anitha, A.; Grøndahl, L. Evaluating the effect of synthesis, isolation, and characterisation variables on reported particle size and dispersity of drug loaded PLGA nanoparticles. *Mater. Adv.* **2021**, *2*, 5657–5671. [CrossRef]
53. Liu, X.; Gillespie, M.; Ozel, A.D.; Dikici, E.; Daunert, S.; Bachas, L.G. Electrochemical properties and temperature dependence of a recombinant laccase from Thermus thermophilus. *Anal. Bioanal. Chem.* **2011**, *399*, 361–366. [CrossRef] [PubMed]
54. Patel, J.D.; O'Carra, R.; Jones, J.; Woodward, J.G.; Mumper, R.J. Preparation and characterization of nickel nanoparticles for binding to his-tag proteins and antigens. *Pharm. Res.* **2007**, *24*, 343–352. [CrossRef] [PubMed]
55. Khan, D.H.; Bashir, S.; Khan, M.I.; Figueiredo, P.; Santos, H.A.; Peltonen, L. Formulation optimization and in vitro characterization of rifampicin and ceftriaxone dual drug loaded niosomes with high energy probe sonication technique. *J. Drug Deliv. Sci. Technol.* **2020**, *58*, 101763. [CrossRef]
56. Barani, M.; Sargazi, S.; Hajinezhad, M.R.; Rahdar, A.; Sabir, F.; Pardakhty, A.; Zargari, F.; Anwer, M.K.; Aboudzadeh, M.A. Preparation of pH-responsive vesicular deferasirox: Evidence from in silico, in vitro, and in vivo evaluations. *ACS Omega* **2021**, *6*, 24218–24232. [CrossRef] [PubMed]
57. Barani, M.; Hajinezhad, M.R.; Sargazi, S.; Rahdar, A.; Shahraki, S.; Lohrasbi-Nejad, A.; Baino, F. In vitro and in vivo anticancer effect of pH-responsive paclitaxel-loaded niosomes. *J. Mater. Sci.-Mater. Med.* **2021**, *32*, 147. [CrossRef] [PubMed]
58. Chen, X.; Zhong, J.; Dong, D.; Liu, G.; Yang, P. Endoplasmic Reticulum Stress-Induced CHOP Inhibits PGC-1alpha and Causes Mitochondrial Dysfunction in Diabetic Embryopathy. *Toxicol. Sci.* **2017**, *158*, 275–285. [CrossRef]

59. Song, J.; Zhang, Q.; Wang, S.; Yang, F.; Chen, Z.; Dong, Q.; Ji, Q.; Yuan, X.; Ren, D. Cleavage of caspase-12 at Asp94, mediated by endoplasmic reticulum stress (ERS), contributes to stretch-induced apoptosis of myoblasts. *J. Cell Physiol.* **2018**, *233*, 9473–9487. [CrossRef]
60. Cornelis, R.; Hahne, S.; Taddeo, A.; Petkau, G.; Malko, D.; Durek, P.; Thiem, M.; Heiberger, L.; Peter, L.; Mohr, E.; et al. Stromal Cell-Contact Dependent PI3K and APRIL Induced NF-kappaB Signaling Prevent Mitochondrial- and ER Stress Induced Death of Memory Plasma Cells. *Cell Rep.* **2020**, *32*, 107982. [CrossRef]
61. Rubiolo, J.A.; Lence, E.; Gonzalez-Bello, C.; Roel, M.; Gil-Longo, J.; Campos-Toimil, M.; Ternon, E.; Thomas, O.P.; Gonzalez-Cantalapiedra, A.; Lopez-Alonso, H.; et al. Crambescin C1 Acts as A Possible Substrate of iNOS and eNOS Increasing Nitric Oxide Production and Inducing In Vivo Hypotensive Effect. *Front. Pharm.* **2021**, *12*, 694639. [CrossRef] [PubMed]
62. Danilov, A.I.; Andersson, M.; Bavand, N.; Wiklund, N.P.; Olsson, T.; Brundin, L. Nitric oxide metabolite determinations reveal continuous inflammation in multiple sclerosis. *J. Neuroimmunol.* **2003**, *136*, 112–118. [CrossRef]
63. Stornaiuolo, M.; Lotti, L.V.; Borgese, N.; Torrisi, M.R.; Mottola, G.; Martire, G.; Bonatti, S. KDEL and KKXX retrieval signals appended to the same reporter protein determine different trafficking between endoplasmic reticulum, intermediate compartment, and Golgi complex. *Mol. Biol. Cell* **2003**, *14*, 889–902. [CrossRef] [PubMed]
64. Fuchs, Y.; Steller, H. Programmed cell death in animal development and disease. *Cell* **2011**, *147*, 742–758. [CrossRef] [PubMed]
65. Zhou, D.; Palam, L.R.; Jiang, L.; Narasimhan, J.; Staschke, K.A.; Wek, R.C. Phosphorylation of eIF2 directs ATF5 translational control in response to diverse stress conditions. *J. Biol. Chem.* **2008**, *283*, 7064–7073. [CrossRef]
66. Schroder, M. Endoplasmic reticulum stress responses. *Cell Mol. Life Sci.* **2008**, *65*, 862–894. [CrossRef]
67. Kim, K.W.; Moretti, L.; Mitchell, L.R.; Jung, D.K.; Lu, B. Endoplasmic reticulum stress mediates radiation-induced autophagy by perk-eIF2alpha in caspase-3/7-deficient cells. *Oncogene* **2010**, *29*, 3241–3251. [CrossRef]
68. Chan, S.W.; Egan, P.A. Hepatitis C virus envelope proteins regulate CHOP via induction of the unfolded protein response. *FASEB J.* **2005**, *19*, 1510–1512. [CrossRef]
69. Pan, M.Y.; Shen, Y.C.; Lu, C.H.; Yang, S.Y.; Ho, T.F.; Peng, Y.T.; Chang, C.C. Prodigiosin activates endoplasmic reticulum stress cell death pathway in human breast carcinoma cell lines. *Toxicol. Appl. Pharm.* **2012**, *265*, 325–334. [CrossRef]
70. Zhao, Y.; Tian, T.; Huang, T.; Nakajima, S.; Saito, Y.; Takahashi, S.; Yao, J.; Paton, A.W.; Paton, J.C.; Kitamura, M. Subtilase cytotoxin activates MAP kinases through PERK and IRE1 branches of the unfolded protein response. *Toxicol. Sci.* **2011**, *120*, 79–86. [CrossRef]
71. Morinaga, N.; Yahiro, K.; Matsuura, G.; Moss, J.; Noda, M. Subtilase cytotoxin, produced by Shiga-toxigenic Escherichia coli, transiently inhibits protein synthesis of Vero cells via degradation of BiP and induces cell cycle arrest at G1 by downregulation of cyclin D1. *Cell Microbiol.* **2008**, *10*, 921–929. [CrossRef] [PubMed]
72. Matsuura, G.; Morinaga, N.; Yahiro, K.; Komine, R.; Moss, J.; Yoshida, H.; Noda, M. Novel subtilase cytotoxin produced by Shiga-toxigenic Escherichia coli induces apoptosis in vero cells via mitochondrial membrane damage. *Infect. Immun.* **2009**, *77*, 2919–2924. [CrossRef] [PubMed]

nanomaterials

MDPI

Article

Strong and Elastic Hydrogels from Dual-Crosslinked Composites Composed of Glycol Chitosan and Amino-Functionalized Bioactive Glass Nanoparticles

Qing Min [1,†], Congcong Wang [2,†], Yuchen Zhang [1], Danlei Tian [2], Ying Wan [2,*] and Jiliang Wu [1,*]

1 School of Pharmacy, Hubei University of Science and Technology, Xianning 437100, China; baimin0628@hbust.edu.cn (Q.M.); zhangych@hbust.edu.cn (Y.Z.)

2 College of Life Science and Technology, Huazhong University of Science and Technology, Wuhan 430074, China; congcongwang@hust.edu.cn (C.W.); tiandanlei@hust.edu.cn (D.T.)

* Correspondence: ying_wan@hust.edu.cn (Y.W.); jlwu@hbust.edu.cn (J.W.)

† These authors contributed equally to this work.

Abstract: Mesoporous bioactive glass (BG) nanoparticles (NPs) with a high specific surface area were prepared. The surfaces of BG NPs were further modified using an amino-containing compound or synthesized precursors to produce three kinds of amino-functionalized bioactive glass (ABG) NPs via devised synthetic routes. The achieved ABG NPs possessed various spacer lengths with free amino groups anchored at the end of the spacer. These ABG NPs were then combined with glycol chitosan (GCH) to construct single- or dual-crosslinked ABG/GCH composite hydrogels using genipin (GN) alone as a single crosslinker or a combination of GN and poly(ethylene glycol) diglycidyl ether (PEGDE) as dual crosslinkers. The spacer length of ABG NPs was found to impose significant effects on the strength and elasticity of GN-crosslinked ABG/GCH hydrogels. After being dually crosslinked with GN and PEGDE, the elastic modulus of some dual-crosslinked ABG/GCH hydrogels reached around 6.9 kPa or higher with their yielding strains larger than 60%, indicative of their strong and elastic features. The optimally achieved ABG/GCH hydrogels were injectable with tunable gelation time, and also able to support the growth of seeded MC3T3-E1 cells and specific matrix deposition. These results suggest that the dual-crosslinked ABG/GCH hydrogels have the potential for some applications in tissue engineering.

Keywords: dual-crosslinked hydrogel; glycol chitosan; bioactive glass nanoparticles; amino functionalization; strength and elasticity

Citation: Min, Q.; Wang, C.; Zhang, Y.; Tian, D.; Wan, Y.; Wu, J. Strong and Elastic Hydrogels from Dual-Crosslinked Composites Composed of Glycol Chitosan and Amino-Functionalized Bioactive Glass Nanoparticles. *Nanomaterials* **2022**, *12*, 1874. https://doi.org/10.3390/nano12111874

Academic Editors: Goran Kaluđerović and Nebojsa Pantelic

Received: 28 April 2022
Accepted: 19 May 2022
Published: 30 May 2022

1. Introduction

Injectable hydrogels have drawn a lot of attention in tissue engineering applications over the last decade due to their several advantages, such as a minimally invasive injection procedure, the in situ formation of self-supporting objects and convenient filling of complex tissue defects with arbitrary shapes [1,2]. In addition to these, they can serve as injectable carriers for delivering cells, therapeutic drugs and bioactive molecules because they have highly interconnected porous structures with a large amount of water retention as well as good permeability, which is particularly conducive to the easy transport of nutrients and metabolites, and, in turn, beneficial for the cell-involved tissue repair and reconstruction [2–4]. Nowadays, many kinds of naturally sourced polymers, majorly including collagen, gelatin, silk fibroin, chitosan (CH), dextran, alginate and hyaluronic acid, have been extensively used in the form of hydrogels since they usually show good biocompatibility, easy biodegradation and better biomedical performance than other types of hydrogels made from synthetic polymers [1–7].

Among the mentioned natural polymers, CH is recognized as a versatile biomaterial and has been broadly investigated for hydrogel applications due to its meritorious properties, including biodegradability, nontoxicity, nonantigenicity, anti-microbial activity, adherence, cell affinity and a broad range of abilities for chemical modification through its amino at the C-2 sites or hydroxyl groups at the C-6 sites [8,9]. In particular, CH molecules have a chemical structure highly similar to glycosaminoglycans (GAGs) that commonly exist in the extracellular matrix (ECM) of different types of human tissues. These advantages of CH have made it a preferable biomaterial for a large variety of applications in tissue repair and regeneration [8,10]. To date, many types of CH-based hydrogels have been developed via various physical or chemical methods and they have been widely utilized for the repair and regeneration of injured skin, nerve, cartilage and bone [8,9,11]. Despite the wide-ranging usability, the hydrogels based on innate CH often show poor mechanical properties such as fragility and insufficient strength, limiting their applications to some extent.

In the case of bone repair and regeneration, the applicable hydrogels need to have the ability to build a strong and elastic microenvironment for housing cells because the mechanical properties of the hydrogels applied can markedly modulate the growth behavior of the cells encapsulated in or migrating from the periphery of injured tissue and, concomitantly, exert a strong effect on the remodeling of the neonatal tissues. Nowadays, it is generally recognized that a mechanically poor hydrogel would usually result in inferior outcomes if it is used in the repair and regeneration of some skeletal tissues, such as bone or articular cartilage [12,13]. Accordingly, how to endow a hydrogel with high strength and good elasticity as well as sufficient safety is always an important issue that needs to be addressed if it is intended for use in bone repair.

A large number of studies have suggested that a dual or multiple network polymer hydrogel could be substantially strengthened in terms of its three-dimensional (3D) stability, mechanical performance and degradation tolerance through interactions such as the mutual restriction of networks, chain entanglement and intermolecular interactions when compared to a single network gel [14–16]. Hence, it can be envisaged that a strong and elastic CH-based composite hydrogel with the potential for use in bone repair could be developed, provided that some other suitable components are employed while the resulting composite hydrogel is effectively crosslinked using suitable crosslinkers. In addition to the employment of strong and elastic hydrogels for housing cells in bone repair, the incorporation of certain osteogenic ingredients into the gels is one of the practicable strategies to promote bone repair [3,17]. Various kinds of inorganic biomaterials such as hydroxyapatite, beta-tricalcium phosphate and bioactive glass (BG) have been widely investigated for use in bone repair. Of them, BG nanoparticles (NPs) have two specific advantages in bone repair: (1) they can firmly bond to bone tissue in the defect and transform into hydroxyapatite-like layers under the action of physiological fluids [3,17,18], and (2) their bioactive dissolution products such as Si and Ca ions have the capability for osteogenic stimulation [17,18]. Nevertheless, it is generally difficult to build a strong and elastic composite hydrogel by simply mixing CH and BG NPs while crosslinking the resulting mixture because BG NPs only contain hydroxyl groups on their surface and these hydroxyl groups are difficult to use for crosslinking with CH. Hence, new strategies for effectively crosslinking BG NPs with CH need to be explored for constructing strong and elastic BG/CH composite hydrogels if the resulting gels are intended for use in bone repair.

With these considerations in mind, in this study, an effort was made to develop a new type of injectable composite hydrogel using glycol chitosan (GCH), a water-soluble derivative of CH, and amino-functionalized bioactive glass (ABG) NPs while employing genipin (GN) as a single crosslinker or a combination of GN and poly(ethylene glycol) diglycidyl ether (PEGDE) as dual crosslinkers. Mesoporous BG NPs with a high specific surface area were first prepared, and their surfaces were modified using an amino-containing compound or synthesized precursors to produce three kinds of ABG NPs via newly designed synthetic routes. These ABG NPs were endowed with different spacers with varying spacer lengths and the free amino groups were anchored at the end of the spacer. ABG NPs were then combined with GCH to construct single or dual-crosslinked ABG/GCH composite hydrogels. GN, a natural product that is derived from the fruits of the *Gardenia jasminoides* plant, was selected as an amino-associated crosslinker due to its much higher safety when compared to many other kinds of crosslinkers suited for crosslinking amino groups [8,19,20]. PEGDE was employed as another crosslinker considering that its epoxy groups have reactivity toward amino and hydroxyl groups while showing much faster reaction rates compared to GN. Some optimally constructed ABG/GCH gels were found to exhibit mechanically strong and elastic characteristics with well-defined injectability and tunable gelation time. They were able to support the growth of seeded MC3T3-E1 cells and specific matrix deposition. All these results suggest that the presently developed ABG/GCH composite gels could find a new avenue for potential application in bone tissue engineering.

2. Materials and Methods

2.1. Materials

GCH (degree of polymerization $\geq$ 400), GN and PEGDE (M_n of PEG: 2000) were purchased from Sigma-Aldrich (Shanghai, China). Cetyltrimethylammonium bromide (CTAB), tetraethyl orthosilicate (TEOS), 3-(aminopropyl)triethoxysilane (APTES), diethylenetriamine (DETA), 3-chloropropyldimethylmethoxysilane (CPDMMS) and pentaethylenehexamine (PEHA) were supplied by Aladdin Inc. (Shanghai, China). Other chemicals were of analytical grade and purchased from Sinopharm (Shanghai, China).

2.2. Synthesis of Bioglass Nanoparticles

BG NPs were synthesized following a reported method with some modification [21]. In a typical process, 0.5 g of CTAB was dissolved in a mixed solution composed of 70 mL of deionized water, 0.8 mL of aqueous ammonia, 5 mL of ethyl ether, and 10 mL of ethanol, and the resulting mixture was vigorously stirred for 30 min. To this mixture, 2.5 mL of TEOS was added with additional stirring for 30 min, followed by the addition of 0.47 g of calcium nitrate ($Ca(NO_3)_2{\cdot}4H_2O$). The mixture was allowed to react with stirring for 6 h at room temperature. The resulting sediment was collected by centrifugation at 10,000 rpm for 5 min, washed with ethanol three times first, and then with deionized water repeatedly. The obtained BG NPs were dried at 70 °C for 24 h. The ground powder was heated up to 550 °C at a rate of 2 °C/min to remove organic residues. CaO/SiO_2 molar ratio for the prepared BG NPs was measured to be around 0.14 from their energy dispersive spectra.

2.3. Amino Functionalization of Bioglass Nanoparticles

Three kinds of ABG NPs were prepared via different synthetic routes. BG NPs were directly reacted with APTES in dry toluene to produce a kind of ABG NP (denoted ABG-1 NPs) using a method similar to that described in the literature [22]. Two other kinds of ABG NPs were prepared as follows. In brief, a given amount of DETA was dissolved in 5 mL of anhydrous ethanol, in presence of CPDMMS and triethylamine, with stirring for 1 h to produce a precursor. After that, a suspension of BG NPs in ethanol (0.2 wt%) was added to the system containing the precursor with stirring at reflux for 24 h. The product was recovered by centrifugation, washed with anhydrous ethanol repeatedly and dried at 120 °C to achieve the second kind of ABG NP (denoted ABG-2 NPs). The third kind of ABG NP, denoted ABG-3 NPs, was prepared using the same method except that DETA was replaced by PEHA. A schematic illustration for the synthesis of BG NPs and three kinds of ABG NPs with varied spacer lengths is presented in Figure 1. The feed ratios of BG NPs to the amino-containing compound or other compounds were optimized using the orthogonal design method to ensure that three kinds of ABG NPs had similar amino amounts but varied spacer lengths. The amino content of these ABG NPs was detected by the ninhydrin assay [23].

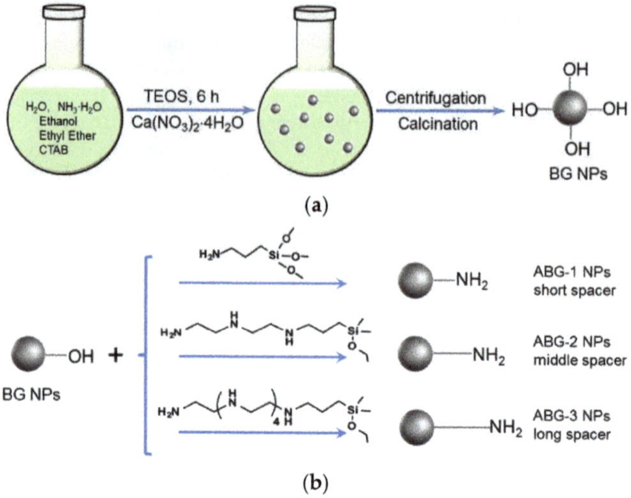

Figure 1. Scheme for synthesis of BG NPs (**a**) and different kinds of ABG NPs (**b**) with varied spacer lengths.

2.4. Characterization

NPs were viewed with a transmission electron microscope (TEM, Tecnai, FEI, Hillsboro, OR, USA) to identify their morphology, size and dispersity. A dynamic light scattering instrument (Nano-ZS90, Malvern, Worcestershire, UK) was used to detect the hydrodynamic size and zeta (ζ) potential of NPs. Energy dispersive X-ray (EDX) spectra of NPs were detected during scanning electron microscopy (SEM, Quanta, FEI, Eindhoven, The Netherlands). For the measurements of isotherms and pore-size distributions, NPs were first dried in a vacuum oven at 100 °C for 12 h before loading into the sample chamber of the surface area and pore size analyzer (ASAP 2020 Plus, Micromeritics, Norcross, GA, USA). After being degassed at 120 °C for 24 h, the volume of nitrogen adsorption–desorption was measured at different pressures. The specific surface areas of NPs were determined using the BET method, and their pore size and volume were calculated with the BJH method.

2.5. Preparation of Hydrogels

GCH solutions with varying concentrations were prepared by dissolving GCH in deionized water. Three kinds of ABG NPs were dispersed in deionized water to produce their respective suspensions. A series of composite solutions with formulated compositions were prepared by mixing the selected GCH solution, ABG suspension and GN solution together. The prepared composite solutions were then introduced into different vials, and incubated at 37 °C for 12 h to produce the single-crosslinked gels.

In the case of dual-crosslinked ABG/GCH gel preparation, PEGDE was added to the above prepared composite solutions to serve as another crosslinker. The obtained composite solutions were also introduced into different vials and the vials were incubated at 37 °C for varied periods. Gelation time for dual-crosslinked gels was determined using a tube-inverting method. In a typical process, one of the preparatory composite solutions was introduced into a vial, and the vial was placed in an ice/water bath with stirring for 5 min before incubation. Once the incubation of the vial in the water bath (37 °C) began, the fluidity of the solution in the vial was regularly checked, and gelation time was recorded starting from the time point for the vial incubation and ending at the moment when the solution stopped flowing.

2.6. Rheological Measurements

A rheometer (Kinexus Pro KNX2100, Southborough, MA, USA) was used for rheological measurements. Frequency sweep spectra for elastic modulus (G') and viscous modulus (G'') of gel samples were detected in a frequency range between 0.1 and 100 Hz at 37 °C and a constant strain of 1%. Concerning strain sweep spectra, G' and G'' of gels were detected by setting the temperature at 37 °C and frequency at 1 Hz, respectively. Shear viscosity measurement was conducted in a shear rate range between 0.1 and 100 s^{-1} at 25 °C using liquid samples.

2.7. Cell Culture

An osteoblast-like cell line (MC3T3-E1) was procured from the Type Culture Collection of the Chinese Academy of Sciences and used to evaluate the gel potential in supporting the cell growth. The purchased cells were expanded by culturing them in the α-MEM medium that contains fetal bovine serum (10%), penicillin (1%) and streptomycin (1%) in a 5% CO_2 humidified atmosphere at 37 °C with medium changes every two days. The harvested cells were resuspended in PBS and used for subsequent experiments.

Cell proliferation was assessed by measuring the DNA content in cell–gel constructs. In a typical procedure, the selected composite solutions were sterilized by introducing them into different glass dishes to form their respective thin layers, and these dishes were irradiated by UV light at 4 °C for a required period of time. Afterward, an aliquot of the sterilized solution was placed in a sterile petri dish and mixed with a given volume of medium containing MC3T3-E1 cells to produce a mixture. The cell-containing mixture was loaded into 24-well culture dishes at a designated volume of 200 μL per well and incubated at 37 °C for gelling. The formed gels were then cultured with a complete medium for 7 days with medium changes every 2 days. At predetermined time intervals, gel samples were withdrawn, washed with PBS and crushed into powder in liquid nitrogen. The powder samples were then cultured with a solution containing proteinase K at 55 °C for 48 h to digest proteins in the powder samples. The supernatant matching with each sample was collected by centrifugation, and then subjected to DNA content determination using a Quant-iT PicoGreen dsDNA kit (Invitrogen) following the manufacturer's instructions. Two-dimensional cell culture was used as a control (named 2D-control) [24].

The selected cell–gel constructs were detected using a live/dead staining method to examine the viability of the seeded cells. Typically, the above prepared cell-containing mixture was loaded into confocal dishes and incubated at 37 °C for gelling. After 3 and 7 days of incubation, the cell–gel constructs were washed with PBS and then cultured with a serum-free medium containing calcein acetoxymethyl ester and propidium iodide in the dark for cell staining. After washing with PBS, the gels were immediately imaged using a confocal microscope (LSM 510 META, Zeiss, Shanghai, China).

The activity of alkaline phosphatase (ALP) of the seeded cells and the amount of cell-synthesized type-I collagen were quantitatively measured, respectively. The above prepared cell–gel constructs were cultured in the complete medium for various durations up to 14 days, and at prescribed time points, they were washed with PBS, crushed and lysed in the lysis buffer at 4 °C. The collected supernatants were then assayed with an ALP kit (Beyotime, Shanghai, China) and a collagen type I ELISA kit (Biological, Salem, MA, USA), respectively. The amount of total protein content in cell–gel constructs was also assayed with a bicinchoninic acid protein kit (Beyotime, China).

2.8. Statistical Analysis

Data were shown as means ± standard deviation. Statistical difference between two groups was determined using Student's t-test, and multiple comparisons were made using one-way analysis of variance. A p-value < 0.05 was considered statistically significant.

3. Results and Discussion

3.1. Characterization of Bioglass Nanoparticles

BG NPs with a high specific surface area were first synthesized under optimized conditions, intending to achieve higher surface amino substitution in subsequent chemical modification. The inserted TEM image in Figure 2a shows that BG NPs were approximately spherical with highly porous morphology. Their size varied from around 140 to 400 nm and exhibited a nearly symmetrical size distribution. The recorded N_2 adsorption–desorption isotherm for BG NPs (Figure 2b) signifies that they had a typical hysteresis loop with the inception turning point of around 0.57 (p/p_0), indicative of the presence of mesoporous pores inside these BG NPs [25]. The pore-size distribution in Figure 2c exhibits that most of the pores in BG NPs were about 5 nm in size, and a small percentage of pores had a size notably larger than 5 nm. Oxygen, silicon and calcium elements were detected from the EDX spectrum of BG NPs (Figure 2d). Based on the data shown in Figure 2d, the presently synthesized BG NPs were estimated to contain 88 mol% SiO_2 and 12 mol% CaO, respectively. Several sets of BG NPs were measured to determine their specific surface area, pore volume, pore size, ζ-potential and hydrodynamic size; relevant parameters are provided in Table 1.

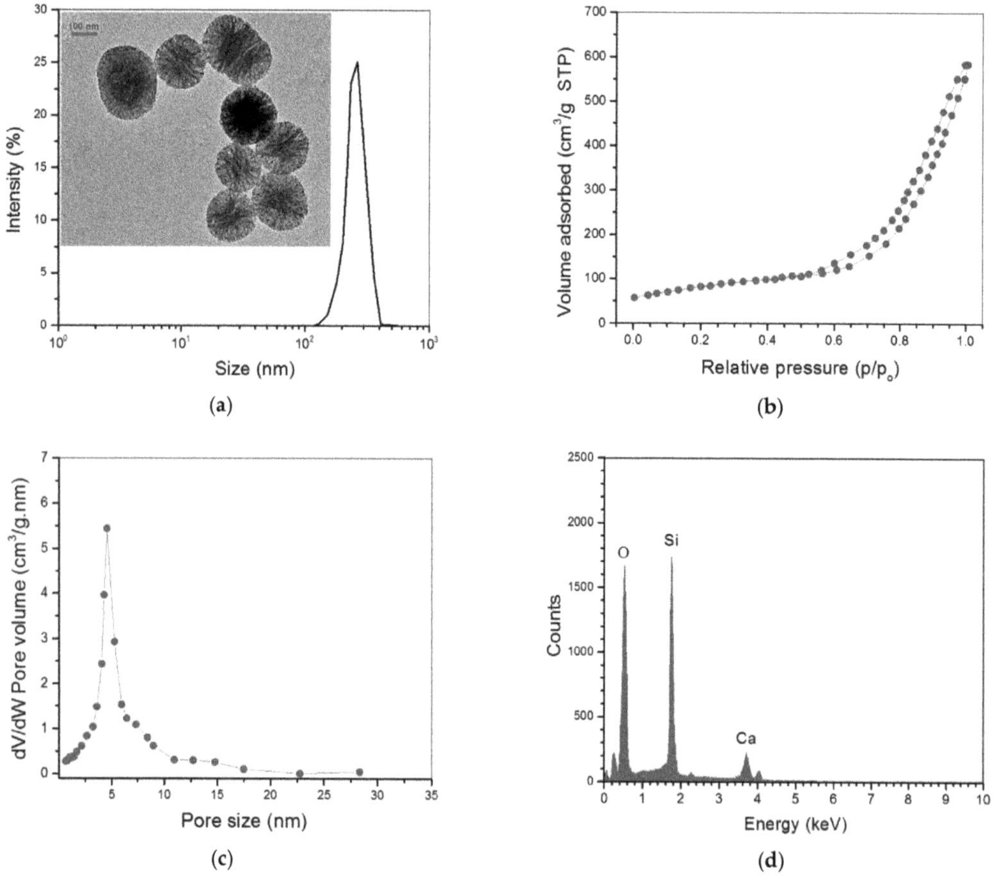

Figure 2. Size distribution with inserted TEM micrograph (**a**), nitrogen adsorption–desorption isotherm (**b**), pore-size distribution (**c**) and EDX spectrum (**d**) for BG NPs.

Table 1. Parameters for different kinds of nanoparticles.

Sample Name	Surface Area (m^2/g)	Pore Volume (mL/g)	Pore Size (nm)	ζ-Potential (mV)	Particle Size (nm)	Content of Amino Groups (mmol/g)
BG	794.4 ± 63.2	1.03 ± 0.09	4.94 ± 0.21	−12.8 ± 0.81	251.6 ± 10.8	–
ABG-1	462.1 ± 31.7	0.81 ± 0.06	4.06 ± 0.17	27.3 ± 1.27	279.9 ± 17.5	0.426 ± 0.031
ABG-2	445.4 ± 27.4	0.74 ± 0.07	3.74 ± 0.13	26.7 ± 1.32	298.1 ± 20.7	0.407 ± 0.046
ABG-3	417.8 ± 23.1	0.68 ± 0.05	3.38 ± 0.15	26.4 ± 1.49	314.4 ± 21.1	0.386 ± 0.042

Three kinds of ABG NPs were produced through different synthetic routes. ABG-1 NPs were prepared via a condensation reaction between ethoxy groups in APTES and hydroxyl groups on the surface of BG NPs. Thus, free amino groups of APTES were exposed on the surface of ABG-1 NPs and connected to Si atoms by the spacer containing four chain units, as shown in Figure 1. With respect to ABG-2 and ABG-3 NPs, they were prepared by reacting BG NPs with different precursors. One of the precursors was synthesized via the reaction between CPDMMS and DETA, and the other was obtained by reacting CPDMMS with PEHA. Based on the presently designed synthetic routes, ABG-2 and ABG-3 NPs had notably longer spacer lengths in comparison to ABG-1 NPs. The spacer for ABG-2 NPs consisted of 10 chain units from the free terminal amino group to the surface Si atom of ABG-2 NPs, and, correspondingly, the spacer for ABG-3 NPs contained 19 chain units, as illustrated in Figure 1. Theoretically, it seems that the precursor with longer spacers can be synthesized following the current method. Our tentative preliminary experiments showed that if a diamino-terminal linear molecule with a longer chain length was used to synthesize the intended precursor, the obtained product would likely contain more by-products and these by-products were difficult to remove. Therefore, the spacer length for these ABG NPs was controlled at 19 chain units or lower.

Several representative TEM micrographs for three kinds of ABG NPs are shown in Figure 3. These ABG NPs were still approximately spherical with porous morphologies and well-defined dispersity; their size distribution intervals became wider but the distribution curves showed better symmetry in comparison to BG NPs. Many sets of specimens for each kind of ABG NP were subject to the assigned measurements, and obtained parameters for them are also summarized in Table 1. It can be noticed that these ABG NPs had a smaller specific surface area, pore volume and pore size but a larger particle size, and, in particular, positive ζ-potential when compared to BG NPs. As described in the experimental section, ABG NPs were produced by modifying BG NPs with an amino-containing compound or two kinds of precursors. Since BG NPs are highly porous with a hydroxyl group-exposed surface, the compound or precursors employed would thus react with the hydroxyl groups on the surface of pores inside the BG NPs in addition to their reaction with the hydroxyl groups on the surface of BG NPs, which would thus lead to decreases in the specific surface area, pore volume and pore size for ABG NPs. The three kinds of ABG NPs had similar ζ-potentials, indirectly suggesting that the amount of amino groups exposed on their surfaces would be similar. Quantitative measurements further confirmed that their surface contained about 0.4 mmol/g amino groups without significant difference (Table 1).

It is known that the hydroxyl group-exposed surface of BG NPs generally makes their ζ-potential negative. After modifying the BG NPs with the selected compound and synthesized precursors with free amino groups situated at the end of their spacers, these free amino groups will thus be exposed on the surface of ABG NPs, and, meanwhile, these amino groups are connected to the NPs by different spacers with varying lengths (Figure 1). As a result, ABG NPs attain positive ζ-potential and an enlarged hydrodynamic size compared to BG NPs.

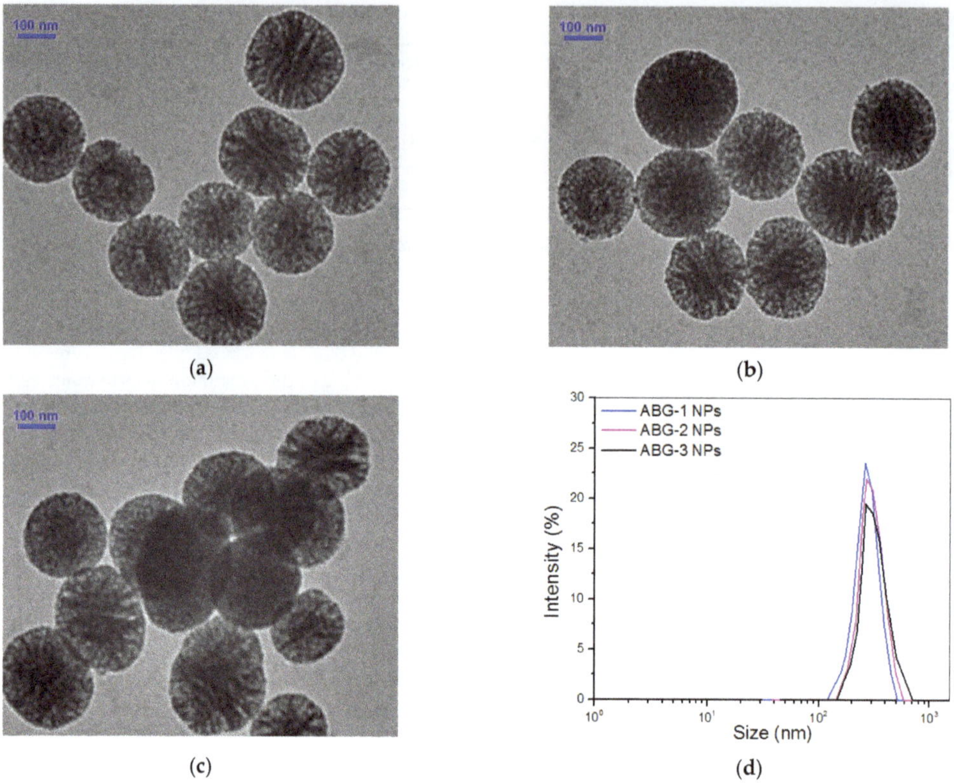

Figure 3. TEM micrographs for ABG-1 (**a**), ABG-2 (**b**) and ABG-3 (**c**) NPs as well as their size distributions (**d**).

3.2. Single-Crosslinked Hydrogels

Three kinds of ABG NPs were combined with GCH to prepare single-crosslinked ABG/GCH composite gels using GN only as the crosslinker. The compositional proportions and frequency sweep spectra for the resulting gels are provided in Table 2 and Figure 4, respectively. The applied amount of GN was already preset to a level of no more than 0.2% for effectively crosslinking these gels while endowing them with sufficient safety. The frequency sweep spectrum of the GL-1 gel in Figure 4a exhibits that the G′ value in the linear viscoelastic region (LVR) was around 800 Pa, meaning that the GL-1 gel is mechanically weak [26]. GL-2, GL-3 and G-4 gels that were incorporated with varied amounts of ABG-1 NPs showed much higher G′ compared to the GL-1 gel, and the G′ values of these gels in the LVR remarkably increased with an increasing amount of ABG-1 NPs. Although an amount of ABG-1 NPs higher than 2 (w/v)% can also be incorporated into these gels, the resulting composite solutions were found to be inconducive for injectable applications due to their increased viscosity. Accordingly, the amount of ABG-1 NPs incorporated was controlled at 2 (w/v)% or lower. As shown in Table 2, GL-4, GL-5 and G-6 gels were built by combining GCH with different kinds of ABG NPs having various spacer lengths, and their frequency sweep spectra exhibit that these gels had their G′ values in the LVR much higher than those of others. To quantitatively compare the gels enumerated in Table 2, many sets of gel samples were measured to determine their G′ and G″ at 1 Hz, and the relevant results are depicted in Figure 4c,d. In principle, the magnitude of G′ and the G′/G″ ratio can be used together to assess the strength of the hydrogel [26,27]. In general, a mechanically strong hydrogel has large G′, and, conjointly, its G′ is one order or even two orders of magnitude greater than its G″ [26,28,29]. The bar graphs in Figure 4c,d explicate that GL-*i* (*i* = 2, 3, 4, 5 and 6) gels had their respective G′ values of about 1.4, 2.2, 2.6, 2.9 and 2.9 kPa, respectively, with matched G′/G″ ratios of around 19.3, 23.1, 20.7, 21.6 and 20.6, respectively. Among these gels, GL-3, GL-4, GL-5 and GL-6 gels can be considered to have mechanically strong characteristics because their G′ values are larger than 2 kPa and the corresponding G′/G″ ratio is higher than 20. It can also be noticed that GL-5 and GL-6 gels showed significantly larger G′ compared to GL-4 gel. Taking into consideration the difference in ABG NPs employed for producing GL-4, GL-5 and GL-6 gels (Tables 1 and 2), it can be concluded that the ABG NPs with a longer spacer length can serve as a better component for enhancing the strength of the resulting composite gels.

Table 2. Parameters for single-crosslinked hydrogels.

Sample Name	GCH (*w*/*v*%)	ABG-1 (*w*/*v*%)	ABG-2 (*w*/*v*%)	ABG-3 (*w*/*v*%)	GN (*w*/*v*%)
GL-1 (a)	2.5	–	–	–	0.2
GL-2	2.5	1	–	–	0.2
GL-3	2.5	1.5	–	–	0.2
GL-4	2.5	2.0	–	–	0.2
GL-5	2.5	–	2.0	–	0.2
GL-6	2.5	–	–	2.0	0.2

(a) This gel was used as control.

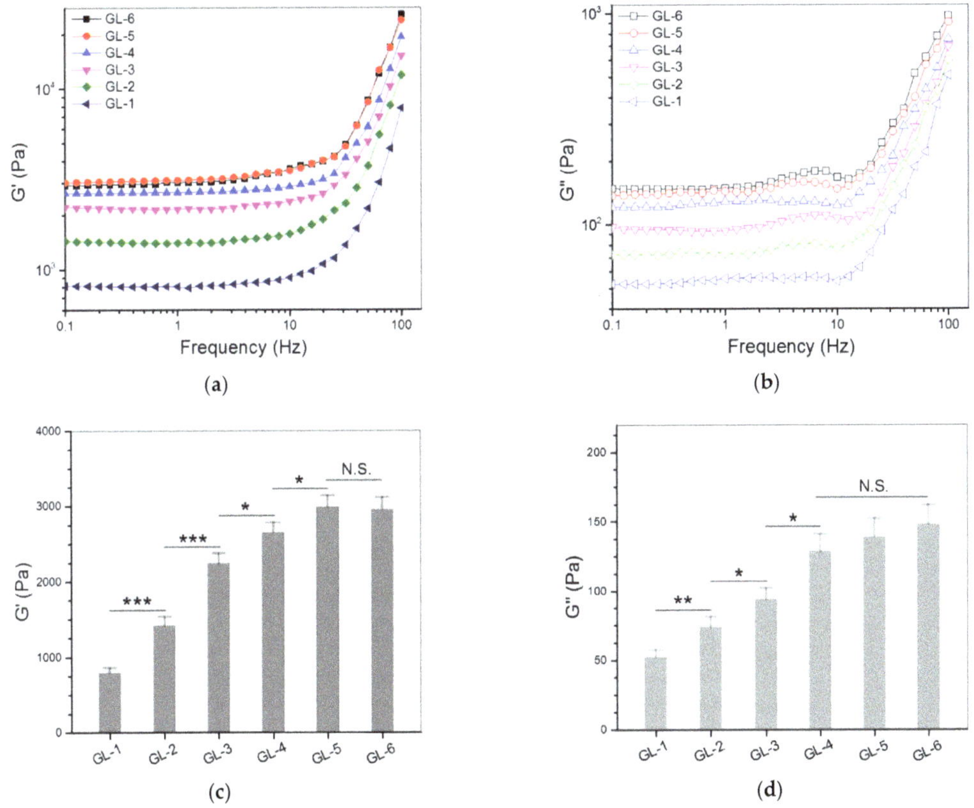

(**a**) (**b**) (**c**) (**d**)

Figure 4. Frequency-dependent variations in G′ (**a**) and G″ (**b**) as well as average values of G′ (**c**) and G″ (**d**) at 1 Hz for gels illustrated in Table 2 (*, $p < 0.05$; **, $p < 0.01$; ***, $p < 0.001$; N.S., not significant).

The strain dependency of G′ and G″ is often used as an indicator to assess the elasticity of hydrogels [26,28]. In most cases, the strain sweep spectrum of a hydrogel has a crossover point at which G′ is equal to G″, commonly called the yielding strain. Starting from the yielding strain, the increasing strain of the gel will not be in phase with the applied stress, and G′ of the gel will drop down rapidly, which connotes the occurrence of disruption of the 3D network in the gel [26,28]. Figure 5 presents several plots of G′ and G″ versus the strain for different gels illustrated in Table 2. GL-1, GL-2, GL-3 and GL-4 gels were seen to have yielding strains of around 20% or slightly higher without significant differences, suggesting that they are inelastic. GL-5 and GL-6 gels showed significantly larger yielding strains than other gels, manifesting that these gels gain certain elasticity.

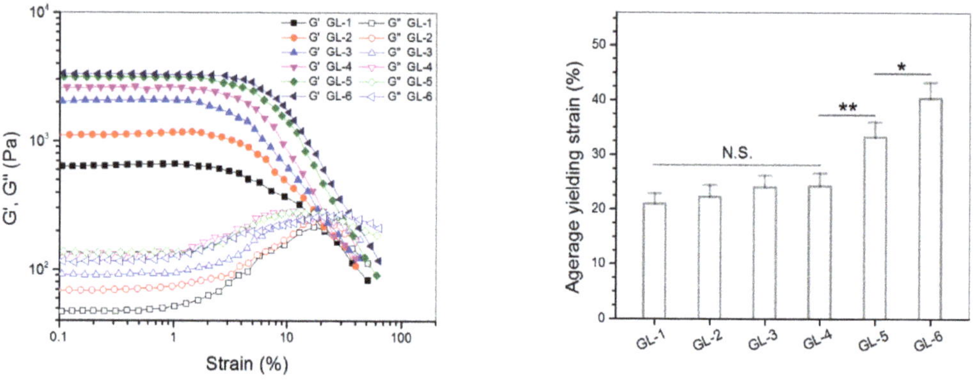

(**a**) (**b**)

Figure 5. Strain sweep spectra (**a**) and average yielding strain (**b**) for gels illustrated in Table 2 (*, $p < 0.05$; **, $p < 0.01$; N.S., not significant).

As shown in Table 2, GL-4, GL-5 and GL-6 gels were built in the same proportions, but they differed from each other in the ABG NPs incorporated. The bar graphs in Figure 5b propose that the ABG NPs with a longer spacer length can endow the resulting composite gels with significantly improved elasticity.

3.3. *Dual-Crosslinked Hydrogels*

In general, an injectable gel needs a suitable gelation time so that it has good fluidity before injection and, on the other hand, is able to solidify into a self-supporting object within a rational time interval after injection [30]. It is known that GN-involved crosslinking reactions are time-consuming and the gelation time for GN-crosslinked CH hydrogels is usually long [31,32]. In the present instance, the prepared solutions with their formulations shown in Table 2 were estimated to take more than 2 h at 37 °C to become nonflowing, and thereafter, they also needed a long time to form well-solidified gels. The long gelation time for the GN-only crosslinked gels makes them unsuitable for certain in situ gelling applications where fast gelatinization is needed. In addition, the stronger single-crosslinked gels, namely GL-5 and GL-6, only have their G′ values close to 3kPa, which still requires a significant improvement in strength if they are intended for use in bone repair. Considering that GL-6 gel is similar to GL-5 gel in G′ value (Figure 4c) but shows a significantly larger yielding strain (Figure 5b) than GL-5 gel, the GL-6 gel was thus selected for the preparation and subsequent investigation of dual-crosslinked gels.

PEGDE was employed as another crosslinker due to the high reactivity of the diepoxy groups in PEGDE toward amino and hydroxyl groups [33,34], which could yield multiple networks in the resulting gels when used together with GN and, in turn, impart the heightened strength and enlarged elasticity to the resulting gels. To ensure the adequate safety of composite gels, in the present study, the PEGDE dosage was controlled to a level of 0.03 *w*/*v*% or lower. The compositional proportions, gelation time and some rheological measurement results for the dual-crosslinked gels are provided in Table 3 and Figure 6, respectively. Data in Table 3 reveal that the gelation time for these gels was measured to be from a few minutes to about 10 min with significant dependence on the amount of PEGDE applied, signifying that the gelatinization of the composite solutions was accelerated by PEGDE-involved crosslinking and the gelation rate of the gels can also be tuned by the amount of PEGDE applied.

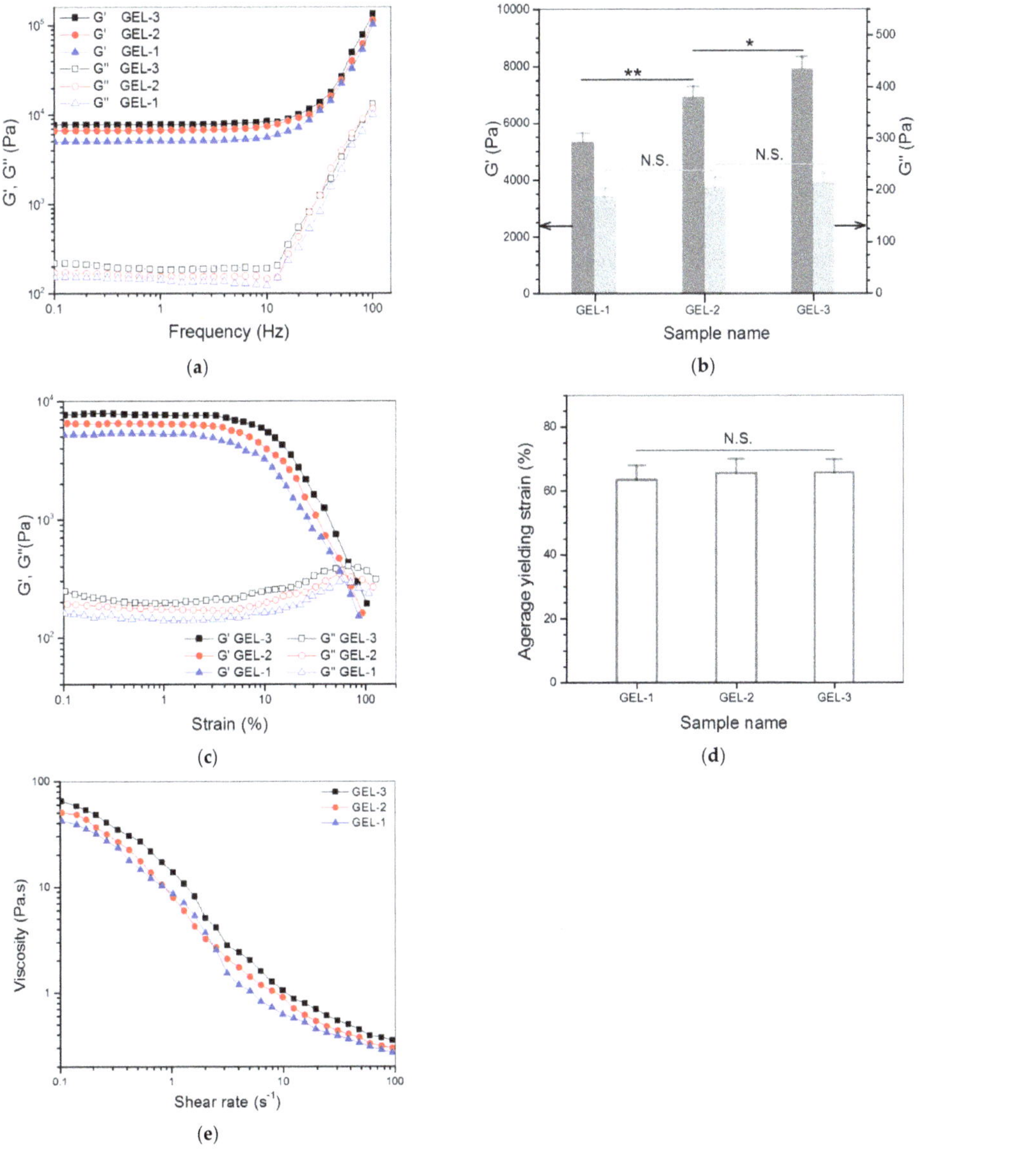

Figure 6. Frequency-dependent functions (**a**) of G′ and G″, average values (**b**) of G′ and G″ at 1 Hz, variations (**c**) in G′ and G″ versus strain, average yielding strain (**d**) and shear rate dependency (**e**) of viscosity (C, 25 °C) for gels illustrated in Table 3 (*, $p < 0.05$; **, $p < 0.01$; N.S., not significant).

Table 3. Parameters for dual-crosslinked hydrogels.

Sample Name	GCH (*w*/*v*%)	ABG-3 (*w*/*v*%)	GN (*w*/*v*%)	PEGDE (*w*/*v*%)	Gelation Time at 37 °C (s) (a)	Degree of Crosslinking ($\times 10^{-5}$ mol/cm^3) (b)	ϕ
GEL-1	2.5	2.0	0.2	0.01	630 ± 24.4	0.902 (±0.063)	0.0128
GEL-2	2.5	2.0	0.2	0.02	285 ± 17.3	1.122 (±0.051) *	0.0147
GEL-3	2.5	2.0	0.2	0.03	207.5 ± 20.6	1.23 (±0.047) #	0.0161

(a) The gelation time was determined by inverting the vials every 30 s. (b) *, $p < 0.05$ compared to GEL-1; #, $p < 0.05$ compared to GEL-2.

Figure 6a,b explicate the frequency sweep spectra of G′ and G″ for GEL-1, GEL-2 and GEL-3 gels, and their average values of G′ and G″ at 1 Hz, respectively. These gels showed their G′ of around 5.3, 6.9 and 7.8 kPa, respectively; much greater than that of their counterpart, GE-6 gel. In particular, the G′ value of GEL-2 and GEL-3 gels was more than twice that of GE-6 gel. In addition, the G′/G″ ratios of GEL-2 and GEL-3 gels reached about 33.7 and 37.1, respectively, which were considerably larger than that for GE-6 gel. These results demonstrate that GEL-2 and GEL-3 gels behave as strong gels given their large G′ and high G′/G″ ratios. Figure 6c shows the functions of G′ and G″ versus strain for GEL-1, GEL-2 and GEL-3 gels, and the measured average yielding strains for them are plotted in Figure 6d. The curves and bar graphs symbolize that GEL-1, GEL-2 and GEL-3 gels had similar yielding strains without significant difference, and their yielding strain was notably higher than that of their counterpart, GE-6 gel, providing the evidence that these gels have been further improved in elasticity when compared to GE-6 gel.

It is known that GN is a kind of amino-specific crosslinker [19]. In the single-crosslinked ABG/GCH composite gels, GN can react with amino groups that only belong to GCH chains to build a GCH alone constructed network, and at the same time, it is also able to crosslink amino groups that respectively belong to GCH and ABG NPs to build another network consisting of ABG NPs and GCH. Such built networks would thus enable ABG/GCH composite gels to have certain strength and elasticity with significant dependence on the spacer length of the ABG NPs employed (Figures 4 and 5). Unlike GN, a small molecule crosslinker, PEGDE is a kind of chain-like oligomer with a molecular weight larger than 2 kDa (M_n for the PEG segment in PEGDE: 2kDa), and its epoxy groups are capable of reacting with amino or hydroxyl groups. Accordingly, in the case of dual-crosslinked ABG/GCH composite gels, in addition to the presence of the mentioned networks built by the GN-involved linkages, PEGDE would also contribute to the construction of some other networks. The diepoxy groups situated at the two ends of PEGDE molecules can not only react with amino groups belonging to either GCH molecules or ABG NPs, but also react with the hydroxyl groups located at the side chains in GCH molecules. As a consequence, PEGDE-correlated linkages would additionally result in the formation of at least two networks in the ABG/GCH composite gels. Moreover, the PEGDE-bridged linkages in these networks would have certain extensibility since the applied PEGDE molecules have a linear structure and the PEG segment in the PEGDE molecules will exist inside dual-crosslinked ABG/GCH composite gels in the form of random curls, and these PEG curls will unbend during the gel strain. It can also be inferred that the established multiple networks inside dual-crosslinked ABG/GCH composite gels would be randomly interpenetrated and entangled together. All these mentioned factors will synergistically impart the dual-crosslinked ABG/GCH composite gels with significantly enhanced strength and elasticity in comparison to the single-crosslinked composite gels, as evidenced by the results shown in Figure 6.

The degree of crosslinking is known to impose important effects on the structures and properties of hydrogel [26], and, in particular, it is closely correlated to the mechanical performance and swelling behavior of the gels [26,35,36]. The equilibrium swelling measurement and uniaxial compression test are commonly used approaches for estimating the degree of crosslinking of hydrogels [35,36], and the solid state nuclear magnetic resonance spectroscopy has also been shown to be effective for determining the degree of crosslinking of certain hydrogels [37,38]. Based on the Flory theory while assuming a Poisson ratio of 0.5 for the rubber-like hydrogel, the shear modulus (G) of an ideal rubber-like hydrogel can be expressed as follows with a good approximation [39]

$$G = \nu RT\phi^{1/3}\frac{\overline{r_0^2}}{\overline{r_f^2}} \quad (1)$$

where ν is the degree of crosslinking, R is the gas constant, T is the absolute temperature, $\overline{r_0^2}/\overline{r_f^2}$ is the front factor and ϕ is the volume fraction of the polymer in the swollen hydrogel. By assigning a value of unity to the front factor, Equation (1) can be simplified into the following formula [40]

$$|G^*| = \nu RT\phi^{1/3} \quad (2)$$

The degree of crosslinking for the gels illustrated in Table 3 was calculated using Equation (2) and such approximately obtained results are also provided in Table 3. It can be observed that there were significant differences in the degree of crosslinking among GEL-1, GEL-2 and GEL-3 gels, and the degree of crosslinking increased with the amount of PEGDE applied. By comparing the data in Table 3 with those shown in Figure 6b, it can be seen that the crosslinking degree of the gels changed in a manner similar to the variation trend of their G', signifying that the gels can be strengthened by raising their degree of crosslinking. This is understandable because PEGDE-bridged linkages will increase as the amount of PEGDE applied rises, which would thus lead to augmented network entanglement in the resulting gel, thereby enhancing its strength. Despite the differences in the degree of crosslinking among GEL-1, GEL-2 and GEL-3 gels, they had similar average yielding strains (Figure 6d), implying that the degree of crosslinking has an insignificant effect on their elasticity. One possible reason for the insignificant effect of the crosslinking degree may be ascribed to the following tentative interactions. In general, an increase in the gel strength will result in a corresponding rise in its deformation stress. In the present instance, for the gel having a higher degree of crosslinking, its enhanced resistant force against the deformation could be counteracted by its increased deformation stresses, resulting in a similar yield strain when compared to the gel having a relatively lower degree of crosslinking.

Taking into account the injection applicability of gels, GEL-1, GEL-2 and GEL-3 solutions were tested for their viscosity versus shear rate, and the results are elucidated in Figure 6e. In the lower shear rate range with an upper limit of around 10 s^{-1}, these composite solutions were viscous with similarity in their viscosity, and their viscosity became notably low once the applied shear rate was higher than 10 s^{-1}, indicating that they have shear-thinning features. Considering that the injection of composite solutions is usually conducted at ambient temperature, the results in Figure 6e account for their well-defined injectability.

3.4. Cell Growth and Analysis

The GEL-3 gel had its G' significantly higher than GEL-1 and GEL-2 gels, and it was similar to GL-6 gel with a difference in the applied crosslinkers. GL-6 and GEL-3 gels were thus selected for the in vitro evaluation. Figure 7 presents representative fluorescence images for the stained MC3T3-E1 cells that were seeded in GL-6 and GEL-3 gels and cultured for various periods up to 7 days. Images in both columns display that very few dead cells were detected from these gels after the culture of cell–gel constructs for 3 (left column) or 7 (right column) days, and the cell density in the images matching with 7-day culture was markedly higher than that corresponding to 3-day culture. These images confirm that the seeded MC3T3-E1 cells had high viability, and GL-6 and GEL-3 gels had similar capabilities to support the growth of seeded cells.

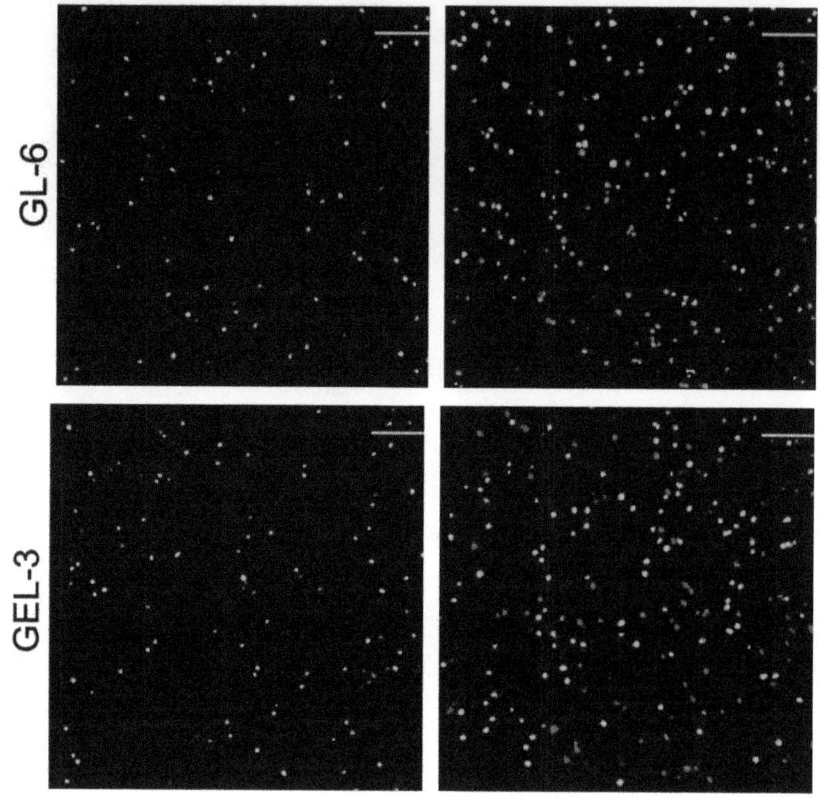

Figure 7. Confocal images for stained MC3T3-E1cells that were seeded GL-6 and GEL-3 gels (green: viable cells; red: dead cells; scale bar: 100 μm; culture time: 3 days (left column) and 7 days (right column)).

Figure 8a delineates the results for the cell proliferation in different gel matrices. It can be observed that the growth of seeded cells can be roughly divided into two phases: they grew slowly from days 1 to 3 and then grew faster from days 3 to 7. The slow cell growth in the first phase is due to the cell attachment as well as subsequent population recovery, and the fast cell growth in the second phase can be ascribed to the occurrence of cell proliferation. There were no significant differences in the detected DNA amounts among these groups during one-week culture, revealing that they have similar abilities to support the proliferation of seeded MC3T3-E1 cells. Considering that the GL-6 and GEL-3 gels are different in terms of the crosslinkers applied, the results shown in Figures 7 and 8a suggest that the applied amount of GN and the combined amount of GN and PEGDE are all within the safe dose range.

ALP activity is a commonly used indicator for evaluating the osteogenic development of cells in the early stage [41,42]. The cell–gel constructs were thus detected to assess the ALP activity of seeded cells, and relevant results are graphed in Figure 8b. The bar graphs exhibit that there was no significant difference in the ALP activity for the cells seeded in the GL-6 or GEL-3 gels after 7-day culture. After being extensively cultured for a period of up to 14 days, the ALP activity detected from these gels remarkably increased without significant difference. As shown in Tables 2 and 3, the GL-6 and GEL-3 gels differed from each other in the crosslinkers applied and, in turn, in their strength and elasticity. The results depicted in Figure 8b indicate that the mentioned differences between the GL-6 and GEL-3 gels impose an insignificant impact on the ALP activity of the cells seeded in these gels.

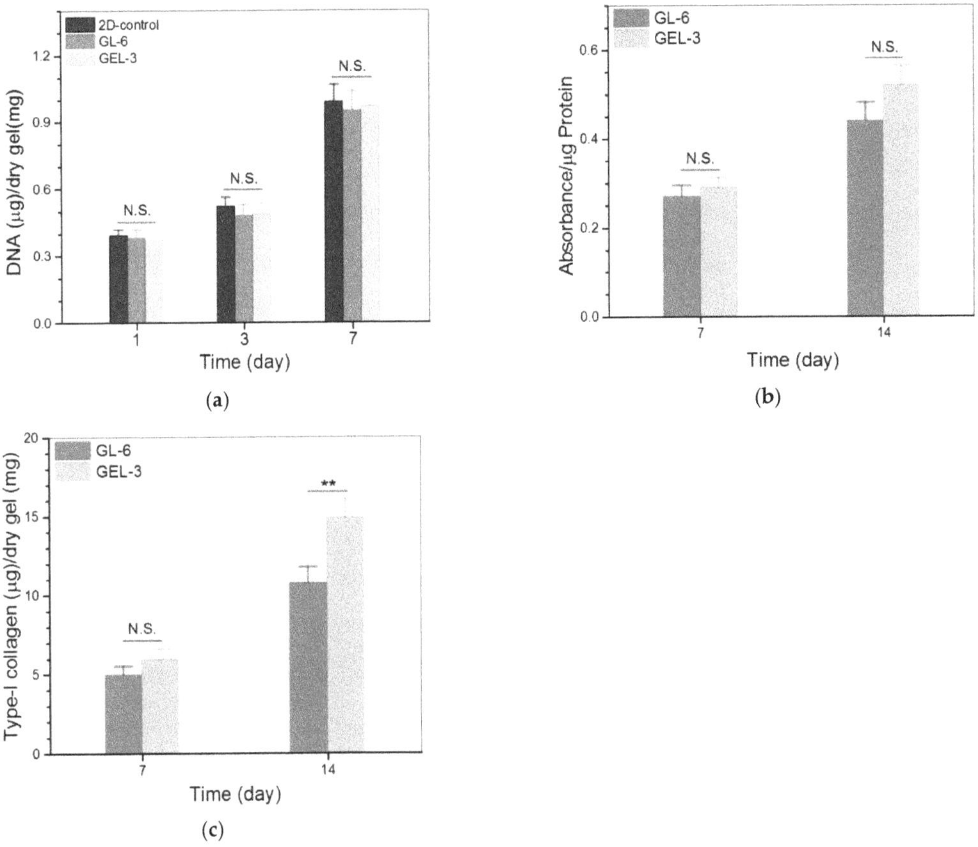

Figure 8. Proliferation (**a**), ALP activity (**b**) and type-I collagen deposition (**c**) of MC3T3-E1 cells growing inside gels (**, $p < 0.01$; N.S., not significant).

The BG NPs composed of CaO and SiO_2 are known to be able to release Si and Ca ions during their dissolution, and these two kinds of ions are demonstrated to have osteogenic activity [17,18]. In the present study, ABG-3 NPs were made from BG NPs by surface chemical modification; hence, they would also be able to release Si and Ca ions during the dissolution and, accordingly, endue the resulting GL-6 and GEL-3 gels with certain osteogenic activity, denoted by the increasing ALP levels. The insignificant difference in the ALP activity detected from the GL-6 and GEL-3 gels can be ascribed to the fact that the GL-6 and GEL-3 gels are incorporated with the same amount of ABG-3 NPs.

Type-I collagen synthesized by the osteoblast-like cells is another important indication correlated to the osteogenic development of the cells [43]. The amount of type-I collagen in the cell–gel constructs was thus measured to figure out the effect of the gels on the type-I collagen deposition, and the obtained data are plotted in Figure 8c. It is shown that the type-I collagen amounts deposited in the GL-6 and GEL-3 gels were similar without significant difference after 7-day culture. After being seeded in these gels, MC3T3-E1 cells need to go through the process of adhesion, restorative growth and the subsequent proliferation, and a 7-day growth period could be too short for them to secrete a large amount of type-I collagen, leading to the insignificant difference in their type-I collagen deposition. Type-I collagen deposition was noticed to increase at varied rates for GL-6 and

GEL-3 gels as the culture time advanced from day 7 to day 14, and the deposited amount of type-I collagen for GEL-3 gel was remarkably higher than that for GL-6 gel on day 14. Data presented in Table 1 and Figures 4–6 exhibit that GEL-3 gel differed from GL-6 gel in the ingredient of PEGDE as well as in the strength and elasticity. Hence, the significantly higher type-I collagen deposition in GEL-3 gel should be majorly attributed to its higher strength and elasticity when compared to GL-6 gel.

Both CH and BG NPs have been widely used in bone tissue engineering since the former can act as bone affinitive material to well support cell growth and ECM synthesis, and the latter has a demonstrated osteogenic activity [11,44–46]. Nevertheless, A CH-based hydrogel intended for use in bone repair still needs improvement to increase its strength and make it elastic enough because a strong and elastic microenvironment is required for the cells implanted or migrated from the host tissue to repair the injured bone tissue. In the present study, GCH, a water-soluble derivative of CH, and ABG NPs, a modified product of BG NPs, were employed as two major components for the preparation of composite hydrogels, and the optimally achieved dual-crosslinked ABG/GCH gels were demonstrated to be strong and elastic with the abilities to support the growth of seeded MC3T3-E1 cells and the synthesis of matrix components. These results suggest that these composite hydrogels have the potential for applications in bone tissue engineering.

4. Conclusions

Three kinds of mesoporous ABG NPs with similar amounts of surface amino groups but varied spacer lengths were successfully synthesized via newly devised synthetic routes. By combining ABG NPs with GCH while using GN as a single crosslinker, the constructed ABG/GCH gels had their strength and elasticity with certain dependence on the spacer length of ABG NPs. The ABG NPs with a longer spacer length would endow the resulting ABG/GCH gels with higher strength or larger elasticity when compared to those ABG NPs having a shorter spacer length. By employing GN and PEGDE as dual crosslinkers, the dual-crosslinked ABG/GCH gels would be remarkably improved in their strength and elasticity while showing tunable gelation time. Some optimally dual-crosslinked ABG/GCH gels were capable of supporting the growth of the seeded osteoblast-like cells as well as the matrix deposition. The higher strength and larger elasticity of the gels were found to be conducive to the synthesis of type-I collagen. Results demonstrate that this new type of dual-crosslinked hydrogel has potential for applications in bone repair.

Author Contributions: Q.M., Y.W. and J.W. conceived and designed the experiments; C.W., Y.Z. and D.T. performed the experiments; Q.M., Y.W. and J.W. wrote the paper. All authors have read and agreed to the published version of the manuscript.

Funding: This research was funded by the National Natural Science Foundation of China (Grant No. 81972065).

Institutional Review Board Statement: Not applicable.

Informed Consent Statement: Not applicable.

Data Availability Statement: The data presented in this study are available on request from the corresponding author.

Conflicts of Interest: The authors declare no conflict of interest.

References

1. Sharma, S.; Tiwari, S. A review on biomacromolecular hydrogel classification and its applications. *Int. J. Biol. Macromol.* **2020**, *163*, 737–747. [CrossRef]
2. Tan, H.; Marra, K.G. Injectable, biodegradable hydrogels for tissue engineering applications. *Materials* **2010**, *3*, 1746–1767. [CrossRef]
3. Couto, D.S.; Hong, Z.; Mano, J.F. Development of bioactive and biodegradable chitosan-based injectable systems containing bioactive glass nanoparticles. *Acta Biomater.* **2009**, *5*, 115–123. [CrossRef] [PubMed]

4. Kretlow, J.D.; Klouda, L.; Mikos, A.G. Injectable matrices and scaffolds for drug delivery in tissue engineering. *Adv. Drug Deliv. Rev.* **2007**, *59*, 263–273. [CrossRef] [PubMed]
5. Hunt, J.A.; Chen, R.; Veen, T.; Bryan, N. Hydrogels for tissue engineering and regenerative medicine. *J. Mater. Chem. B* **2014**, *2*, 5319–5338. [CrossRef]
6. Wu, J.; Liu, J.; Shi, Y.; Wan, Y. Rheological, mechanical and degradable properties of injectable chitosan/silk fibroin/hydroxyapatite/glycerophosphate hydrogels. *J. Mech. Behav. Biomed. Mater.* **2016**, *64*, 161–172. [CrossRef]
7. Lee, E.J.; Kang, E.; Kang, S.W.; Huh, K.M. Thermo-irreversible glycol chitosan/hyaluronic acid blend hydrogel for injectable tissue engineering. *Carbohydr. Polym.* **2020**, *244*, 116432. [CrossRef]
8. Muzzarelli, R.A. Genipin-crosslinked chitosan hydrogels as biomedical and pharmaceutical aids. *Carbohydr. Polym.* **2009**, *77*, 1–9. [CrossRef]
9. Bhattarai, N.; Gunn, J.; Zhang, M. Chitosan-based hydrogels for controlled, localized drug delivery. *Adv. Drug Deliv. Rev.* **2010**, *62*, 83–99. [CrossRef]
10. Majeti, N.V.; Kumar, R. A review of chitin and chitosan applications. *React. Funct. Polym.* **2000**, *46*, 1–27.
11. Muzzarelli, R.A. Chitins and chitosans for the repair of wounded skin, nerve, cartilage and bone. *Carbohydr. Polym.* **2009**, *77*, 167–182. [CrossRef]
12. Engler, A.J.; Sen, S.; Sweeney, H.L.; Discher, D.E. Matrix elasticity directs stem cell lineage specification. *Cell* **2006**, *126*, 677–689. [CrossRef]
13. Discher, D.E.; Janmey, P.; Wang, Y.L. Tissue cells feel and respond to the stiffness of their substrate. *Science* **2005**, *310*, 1139–1143. [CrossRef]
14. Chen, Q.; Chen, H.; Zhu, L.; Zheng, J. Fundamentals of double network hydrogels. *J. Mater. Chem. B* **2015**, *3*, 3654–3676. [CrossRef]
15. Xu, C.; Dai, G.; Hong, Y. Recent advances in high-strength and elastic hydrogels for 3D printing in biomedical applications. *Acta Biomater.* **2019**, *95*, 50–59. [CrossRef]
16. Li, J.; Suo, Z.; Vlassak, J.J. Stiff, strong, and tough hydrogels with good chemical stability. *J. Mater. Chem. B* **2014**, *2*, 6708–6713. [CrossRef]
17. El-Rashidy, A.A.; Roether, J.A.; Harhaus, L.; Kneser, U.; Boccaccini, A.R. Regenerating bone with bioactive glass scaffolds: A review of in vivo studies in bone defect models. *Acta Biomater.* **2017**, *62*, 1–28. [CrossRef]
18. Hoppe, A.; Guldal, N.S.; Boccaccini, A.R. A review of the biological response to ionic dissolution products from bioactive glasses and glass-ceramics. *Biomaterials* **2011**, *32*, 2757–2774. [CrossRef]
19. Park, J.E.; Lee, J.Y.; Kim, H.G.; Hahn, T.R.; Paik, Y.S. Isolation and characterization of water-soluble intermediates of blue pigments transformed from geniposide of Gardenia jasminoides. *J. Agric. Food Chem.* **2002**, *50*, 6511–6514. [CrossRef]
20. Sung, H.W.; Huang, R.N.; Huang, L.; Tsai, C.C. In vitro evaluation of cytotoxicity of a naturally occurring crosslinking reagent for biological tissue fixation. *J. Biomater. Sci. Polym. Ed.* **1999**, *10*, 63–68. [CrossRef]
21. Du, X.; He, J. Hierarchically mesoporous silica nanoparticles: Extraction, amino-functionalization, and their multipurpose potentials. *Langmuir* **2011**, *27*, 2972–2979. [CrossRef]
22. Walcarius, A.; Etienne, M.; Lebeau, B. Rate of access to the binding sites in organically modified silicates. 2. ordered mesoporous silicas grafted with amine or thiol groups. *Chem. Mater.* **2003**, *15*, 2161–2173. [CrossRef]
23. Mori, T.; Kubo, T.; Kaya, K.; Hosoya, K. Quantitative evaluations of surface-concentrated amino groups on monolithic-type solid supports prepared by copolymerization method. *Colloid Polym. Sci.* **2009**, *287*, 513–523. [CrossRef]
24. Lu, H.; Ko, Y.G.; Kawazoe, N.; Chen, G. Cartilage tissue engineering using funnel-like collagen sponges prepared with embossing ice particulate templates. *Biomaterials* **2010**, *31*, 5825–5835. [CrossRef]
25. Mahapatra, C.; Singh, R.K.; Kim, J.J.; Patel, K.D.; Perez, R.A.; Jang, J.H.; Kim, H.W. Osteopromoting reservoir of stem cells: Bioactive mesoporous nanocarrier/collagen gel through slow-releasing FGF18 and the activated BMP signaling. *ACS Appl. Mater. Interfaces* **2016**, *8*, 27573–27584. [CrossRef]
26. Clark, A.H.; Ross-Murphy, S.B. Structural and mechanical properties of biopolymer gels. *Adv. Polym. Sci.* **1987**, *83*, 57–192.
27. Lejardi, A.; Hernandez, R.; Criado, M.; Santos, J.I.; Etxeberria, A.; Sarasua, J.R.; Mijangos, C. Novel hydrogels of chitosan and poly(vinyl alcohol)-g-glycolic acid copolymer with enhanced rheological properties. *Carbohydr. Polym.* **2014**, *103*, 267–273. [CrossRef]
28. Kavanagh, G.M.; Ross-Murphy, S.B. Rheological characterization of polymer gels. *Prog. Polym. Sci.* **1998**, *23*, 533–562. [CrossRef]
29. Martinez-Ruvalcaba, A.; Chornet, E.; Rodrigue, D. Viscoelastic properties of dispersed chitosan/xanthan hydrogels. *Carbohydr. Polym.* **2007**, *67*, 586–595. [CrossRef]
30. Yang, J.A.; Yeom, J.; Hwang, B.W.; Hoffman, A.S.; Hahn, S.K. In situ-forming injectable hydrogels for regenerative medicine. *Prog. Polym. Sci.* **2014**, *39*, 1973–1986. [CrossRef]
31. Delmar, K.; Bianco-Peled, H. The dramatic effect of small pH changes on the properties of chitosan hydrogels crosslinked with genipin. *Carbohydr. Polym.* **2015**, *127*, 28–37. [CrossRef] [PubMed]
32. Bhattarai, N.; Ramay, H.R.; Gunn, J.; Matsen, F.A.; Zhang, M. PEG-grafted chitosan as an injectable thermosensitive hydrogel for sustained protein release. *J. Control. Release* **2005**, *103*, 609–624. [CrossRef] [PubMed]
33. Jóźwiak, T.; Filipkowska, U.; Szymczyk, P.; Rodziewicz, J.; Mielcarek, A. Effect of ionic and covalent crosslinking agents on properties of chitosan beads and sorption effectiveness of reactive black 5 dye. *React. Funct. Polym.* **2017**, *114*, 58–74. [CrossRef]

34. Tanabe, T.; Okitsu, N.; Yamauchi, K. Fabrication and characterization of chemically crosslinked keratin films. *Mater. Sci. Eng. C* **2004**, *24*, 441–446. [CrossRef]
35. Baker, J.P.; Blanch, H.V.; Prausnits, J.M. Swelling properties of carylamide-based ampholytic hydrogels: Comparison of experiment with theory. *Polymer* **1995**, *36*, 1061–1069. [CrossRef]
36. Senna, A.M.; Novack, K.M.; Botaro, V.R. Synthesis and characterization of hydrogels from cellulose acetate by esterification crosslinking with EDTA dianhydride. *Carbohydr. Polym.* **2014**, *114*, 260–268. [CrossRef]
37. Capitani, D.; Nobile, M.A.D.; Mensitieri, G.; Sannino, A.; Segre, A.L. ^{13}C solid-state NMR determination of cross-linking degree in superabsorbing cellulose-based networks. *Macromolecules* **2000**, *33*, 430–437. [CrossRef]
38. Lenzi, F.; Sannino, A.; Borriello, A.; Porro, F.; Capitani, D.; Mensitieri, G. Probing the degree of crosslinking of a cellulose based superabsorbing hydrogel through traditional and NMR techniques. *Polymer* **2003**, *44*, 1577–1588. [CrossRef]
39. Flory, P.J. *Principles of Polymer Chemistry*; Cornel University Press: Ithica, NY, USA, 1953.
40. Lionetto, F.; Sannino, A.; Mensitieri, G.; Maffezzoli, A. Evaluation of the degree of cross-linking of cellculose based superabsorbent hydroles: A comparison between different techniques. *Macromol. Symp.* **2003**, *200*, 199–207. [CrossRef]
41. Samavedi, S.; Whittington, A.R.; Goldstein, A.S. Calcium phosphate ceramics in bone tissue engineering: A review of properties and their influence on cell behavior. *Acta Biomater.* **2013**, *9*, 8037–8045. [CrossRef]
42. Koons, G.L.; Diba, M.; Mikos, A.G. Materials design for bone-tissue engineering. *Nature Rev. Mater.* **2020**, *5*, 584–603. [CrossRef]
43. Shi, M.; Zhou, Y.; Shao, J.; Chen, Z.; Song, B.; Chang, J.; Wu, C.; Xiao, Y. Stimulation of osteogenesis and angiogenesis of hBMSCs by delivering Si ions and functional drug from mesoporous silica nanospheres. *Acta Biomater.* **2015**, *21*, 178–189. [CrossRef]
44. LogithKumar, R.; KeshavNarayan, A.; Dhivya, S.; Chawla, A.; Saravanan, S.; Selvamurugan, N. A review of chitosan and its derivatives in bone tissue engineering. *Carbohydr. Polym.* **2016**, *151*, 172–188. [CrossRef]
45. Saravanan, S.; Vimalraj, S.; Thanikaivelan, P.; Banudevi, S.; Manivasagam, G. A review on injectable chitosan/beta glycerophosphate hydrogels for bone tissue regeneration. *Int. J. Biol. Macromol.* **2019**, *121*, 38–54. [CrossRef]
46. Moreira, C.D.F.; Carvalho, S.M.; Sousa, R.G.; Mansur, H.S.; Pereira, M.M. Nanostructured chitosan/gelatin/bioactive glass in situ forming hydrogel composites as a potential injectable matrix for bone tissue engineering. *Mater. Chem. Phys.* **2018**, *218*, 304–316. [CrossRef]

nanomaterials

Article

Fluorinated PEG-PEI Coated Magnetic Nanoparticles for siRNA Delivery and CXCR4 Knockdown

Yixiang Cao [1], Shiyin Zhang [2], Ming Ma [1,*] and Yu Zhang [1,*]

1 State Key Laboratory of Bioelectronics, Jiangsu Key Laboratory for Biomaterials and Devices, School of Biological Sciences and Medical Engineering, Southeast University, Nanjing 210096, China; 220195143@seu.edu.cn
2 Nanjing Nanoeast Biotech Co., Ltd., Nanjing 211000, China; syzhang@nanoeast.net
* Correspondence: maming@seu.edu.cn (M.M.); zhangyu@seu.edu.cn (Y.Z.)

Abstract: CXC chemokine receptor 4 (CXCR4) is a promising therapeutic target. Previous studies have shown that intracellular delivery of siRNA to knockdown CXCR4 expression in cancer cells is an effective therapeutic strategy. To prepare efficient magnetic nucleic acid carriers, it is now necessary to improve the endocytosis efficiency of PEGylated magnetic nanoparticles. In our work, Heptafluorobutyryl-polyethylene glycol-polyethyleneimine (FPP) was first prepared and then used to coat magnetic nanoparticles (MNPs) to obtain magnetic nanocarriers FPP@MNPs. The materials were characterized by ^{19}F-Nuclear Magnetic Resonance (NMR), transmission electron microscope (TEM), energy dispersive spectroscopy (EDS), and dynamic light scattering (DLS). The biosecurity of FPP@MNPs was confirmed by cell viability and apoptosis experiments. Cellular uptake of FPP@MNPs and siRNA transfection enhanced by external magnetic fields were detected by fluorescence microscopy, confocal laser microscopy, and flow cytometry. The results show that the cellular uptake efficiency of FPP@MNPs was significantly improved, and transfection efficiency reached more than 90%. The knockdown of CXCR4 on the 4 T1 cell membrane was confirmed by real-time polymerase chain reaction (RT-PCR) and flow cytometry. In conclusion, the fluorinated cationic polymer-coated magnetic nanoparticles FPP@MNPs can be loaded with siRNA to reduce CXCR4 expression as well as be expected to be efficient universal siRNA carriers.

Keywords: magnetic nanoparticles; fluorination; polyethyleneimine; polyethylene glycol; RNA interference; CXC chemokine receptor 4

Citation: Cao, Y.; Zhang, S.; Ma, M.; Zhang, Y. Fluorinated PEG-PEI Coated Magnetic Nanoparticles for siRNA Delivery and CXCR4 Knockdown. *Nanomaterials* **2022**, *12*, 1692. https://doi.org/10.3390/nano12101692

Academic Editors: Goran Kaluđerović and Nebojša Pantelić

Received: 15 April 2022
Accepted: 12 May 2022
Published: 16 May 2022

1. Introduction

The CXC chemokine receptor 4 (CXCR4) has become a very promising target for tumor therapies. The expression of chemokine receptors varies among cancer cells of different origins, but CXCR4 is widely expressed in human cancers [1–5].

The interaction between CXC chemokine ligand 12/chemokine stromal cell-derived factor-1 (CXCL12/SDF-1) and the corresponding receptor CXCR4 promotes tumor metastasis [6]. Furthermore, CXCR4 overexpression is associated with poor prognosis in many cancer subtypes. As is reported, CXCR4 is highly expressed in breast cancers, with the ligand CXCL12 showing the highest levels of expression in the organ where cancer metastases first occur [7]. Moreover, CXCL12 activating CXCR4 promotes the transfer of proto-cancer mitochondria between cells [8]. Activation of CXCR4 promotes trafficking and homing of acute myeloid leukemia cells to the bone marrow and spleen [9]. The CXCL12-CXCR4 axis is one of the most important factors in tumor metastasis, angiogenesis, and drug resistance. Therefore, it is of great significance to develop efficient methods to block CXCR4 in tumors.

Currently, studies of medicine against CXCR4 focus on inhibitors or antagonists, such as Plerixafor (AMD3100) [10], T140 peptide [11], and E5 peptide [12]. RNA interference (RNAi) refers to the highly conserved phenomenon of efficient and specific degradation of homologous mRNA induced by double-stranded RNA (dsRNA). The applications of RNAi are generally mediated by small interfering RNA (siRNA) or micro-RNA (miRNA) [13]. The main difference between siRNA and miRNA is that siRNA specifically degrades the target mRNA, while miRNA can regulate the degradation of various mRNA sequences [14–16]. Since RNAi can specifically reduce the expression of target genes, it has been widely used to explore the fields of gene function and disease treatment. RNAi mediated by siRNA would be an effective strategy for reducing or even silencing CXCR4 expression [17].

In general, the delivery of siRNA requires carriers, so the development of efficient siRNA carriers is an important part of RNAi technology. In many studies, cationic polymer-modified magnetic nanoparticles were used for nucleic acid delivery. As is reported, polymers such as Polyethyleneimine (PEI) [18], Poly-L-lysine (PLL) [19], and chitosan [20] coated magnetic nanoparticles were used as magnetic carriers. In particular, external magnetic fields have been widely demonstrated to increase the transfection capacity of magnetic nanoparticles loaded with drugs [21]. Therefore, in our work, we focus on the development of magnetic nanocarriers that can efficiently deliver siRNA into cells based on PEI-modified magnetic nanoparticles (PEI@MNP) and verify the transfection effect.

Polyethylene glycol (PEG) has the advantages of good aqueous solubility and high biocompatibility, which makes it suitable for biomedical applications [22]. Currently, PEG is widely used to prepare biomedical materials such as liposomes [23,24], nanoparticles [25–27], and hydrogels [28,29]. To improve the stability and safety of PEI@MNP in a physiological environment, polyethylene glycol (PEG) modification is necessary. As is reported, PEGylation of cationic polymeric coated nucleic acid carriers can improve stability and reduce toxicity [30]. Compared with PEI@MNP, PEG-PEI@MNP has a lower positive surface charge, significantly improved biocompatibility, and reduced levels of oxidative stress and cell membrane damage [31]. PEGylation of nucleic acid carriers can effectively prevent protein adsorption and can help these carriers escape the reticuloendothelial system (RES) while maintaining prolonged blood circulation [32].

However, PEGylation of the nanoparticles would inhibit cellular uptake, and the subsequent endosomal escape efficiency would be significantly reduced [33,34]. This phenomenon is known as the "PEG dilemma". Several approaches have been reported to overcome the "PEG dilemma". Using cleavable chemical bonds, such as hydrazide-hydrazone bonds [35], Schiff base bonds [36], thioketal bonds [37], and matrix metalloproteinase substrates [38] to connect PEG and carriers would be effective strategies to overcome this dilemma. In addition, the nanoparticles can also be modified with peptides (e.g., RGD peptide [39,40], TAT peptide [41]) to improve the efficiency of cellular uptake.

Fluoroalkyl chains are both hydrophobic and lipophobic [42]. Many studies show that fluorination is an effective strategy to improve the transfection efficiency of nucleic acid delivery vectors. As is reported, fluorinated nucleic acid carriers combine the features of liposomes and cationic polymers [43]. Fluorination enhances cellular uptake of dendrimer/DNA complexes and facilitates endosomal escape [44]. Furthermore, compared with bioreducible poly(amido amine)s (bPAA) fluorinated with trifluoroacetic anhydride or pentafluoropropionic anhydride, heptafluorobutyric anhydride-modified bPAA achieved better gene silencing effect [45]. As fluorination to overcome the "PEG dilemma" has not been reported in previous studies, we propose that fluorination may be an effective strategy to overcome this dilemma and improve the cellular uptake efficiency of PEGylated nanomaterials.

Herein, we developed a fluorinated PEG-PEI modified MNP for in vitro siRNA delivery. This fluorinated magnetic nanocarrier has low cytotoxicity and can effectively overcome the "PEG dilemma". At the same time, high transfection efficiency can be obtained under the action of external magnetic fields, and the siRNA delivery efficiency was validated in a variety of cell lines. Furthermore, fluorinated PEG-PEI@MNP were used to deliver CXCR4 siRNA(siCXCR4) into 4 T1 cells, which could effectively reduce the expression of CXCR4 on the cell membrane.

2. Materials and Methods

2.1. Materials

Ferric acetylacetonate (AR, Sinopharm Chemical Reagent Co., Ltd., SCRC, Shanghai, China), oleic acid (85%, Aladdin, Shanghai, China), benzyl ether (98%, Sigma-Aldrich, St. Louis, MS, USA), ethanol (AR, SCRC), n-hexane (97%, SCRC), dimercaptosuccinic acid (98%, TCI, Shanghai, China), acetone (AR, SCRC), triethylamine (AR, Aladdin, Shanghai, China), methanol (AR, SCRC), polyethyleneimine (Mw 25000, Sigma-Aldrich, St. Louis, MS, USA), heptafluorobutyric anhydride (97%, Macklin, Shanghai, China), amino-polyethylene glycol-carboxyl (Mw 2000, Xi'an Ruixi, Xi'an, China), methoxy-polyethylene glycol-polyethyleneimine (mPEG-PEI, Mw 2000-25000, Xi'an Ruixi Co., Ltd., Xi'an, China) FITC (Beyotime, Shanghai, China), CCK8 kit (KeyGen, Nanjing, China), Hoechst33342 (KeyGen, Nanjing, China), rhodamine-phalloidin (KeyGen, Nanjing, China), LysoTracker (KeyGen, Nanjing, China), Annexin V-FITC/PI kit (KeyGen, Nanjing, China), Anti-CXCR4 antibody (Abcam of Thermo Fisher Scientific Inc., Waltham, MA, USA), DMEM high glucose medium (KeyGen, Nanjing, China), fetal bovine serum (Gibco of Thermo Fisher Scientific Inc., Waltham, MA, USA), trypsin (Gibco of Thermo Fisher Scientific Inc., Waltham, MA, USA), siRNA(GenePharma Co., Ltd., Shanghai, China), 4 T1, A549, HeLa cell line was provided by Southeast University School of Medicine (Nanjing, China).

2.2. Methods

2.2.1. Synthesis of Fluorinated Polyethylene Glycol-Polyethyleneimine (F_7-PEG-PEI)

The synthesis steps of F_7-PEG-PEI (FPP) are shown in Figure 1.

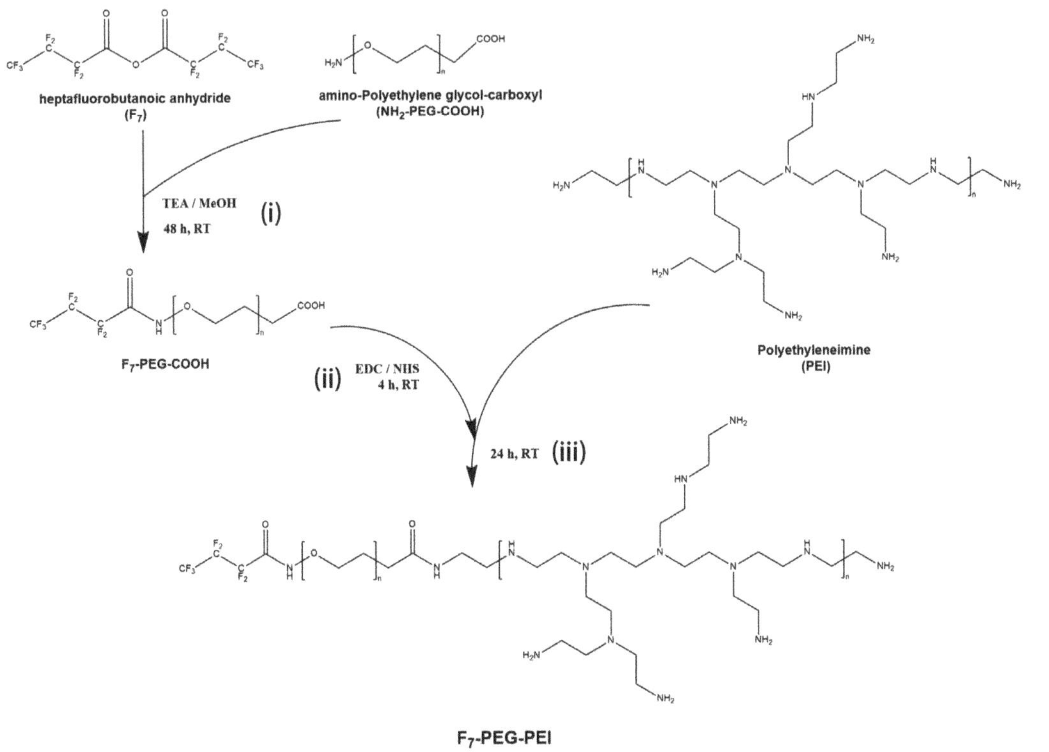

Figure 1. Synthetic process of F_7-PEG-PEI. TEA means triethylamine, MeOH means methanol, RT means room temperature.

Step (i):

(1) 200 μmol of heptafluorobutyric anhydride (F_7) and 100 μmol of amino-polyethylene glycol-carboxyl (NH_2-PEG-COOH) were dissolved in 10 mL methanol, catalyzed by 50 μL triethylamine, magnetic stirred at 240 rpm, and reacted at room temperature for 48 h.;
(2) Then, dialysis membrane (1200 MWCO) was used, and the product was dialyzed against ultrapure water for 3 days. After dialysis, a rotary evaporator was used to evaporate most of the water. Vacuum drying was performed for 48 h to obtain F_7-PEG-COOH.

Step (ii):

(3) 100 mg of F_7-PEG-COOH was dissolved in 50 mL pure water. Then 20 mg of 1-(3-dimethylaminopropyl)-3-ethylcarbodiimide hydrochloride (EDC) and 10 mg of N-hydroxysuccinimide (NHS) were added, and the reaction was stirred at room temperature for 2 h.

Step (iii):

(4) Then, 1200 mg of polyethyleneimine was added, and the reaction was stirred at room temperature for 48 h;
(5) Dialysis membrane (8000 MWCO) was used, and the product was dialyzed against pure water for 3 days;
(6) After dialysis, a rotary evaporator was used to evaporate most of the water. Vacuum drying was performed for 48 h to obtain F_7-PEG-PEI.

2.2.2. Synthesis of Fe_3O_4 Magnetic Nanoparticles

Firstly, 0.7 g (2 mmol) iron acetylacetonate ($Fe(acac)_3$), 20 mL of benzyl ether, and 30 mL of oleic acid were mixed. Then, nitrogen gas was introduced below the liquid level, and the mixture was condensed and refluxed without stirring while constantly bubbling nitrogen through the reaction mixture. Solution was heated from room temperature to 220 °C at a rate of 3.3 °C/min and kept at 220 °C for 60 min. Then, the temperature was raised to 290 °C at a rate of 3.3 °C/min and kept for 30 min. After the solution was naturally cooled to room temperature, absolute ethanol was added to the system to wash the products, followed by magnetic separation for 5 min to discard the supernatant, and repeat 3 times. The oleic acid-modified Fe_3O_4 magnetic nanoparticles (OA@MNPs) were prepared and were dissolved in 100 mL of n-hexane. The Fe concentration of OA@MNPs determined by UV-Vis spectrophotometry (Method S2) was 8.22 mg/mL suspension, and the yield was 74%.

The suspension of OA@MNPs (containing 800 mg of Fe) in 100 mL of hexane as described above was added to a solution of dimercaptoscuccinic acid (DMSA, 400 mg) in acetone (200 mL). The mixture was condensed and refluxed for 4 h in water bath at 60 °C under electric stirring at 500 rpm. After that, 200 mL of ultrapure water was added to extract the product. The product was magnetically separated and the supernatant was discarded. Then, 200 mL of pure water was added to wash the product. The product was magnetically separated and the supernatant was discarded, repeated 3 times. The products were additionally subjected to dialysis (48 h, MWCO 10000) and kept in aqueous suspension (100 mL of water) to obtain DMSA@MNPs. The Fe concentration was 5.11 mg/mL suspension, measured by UV-Vis spectrophotometry (Shimadzu Inc., Kyoto, Japan) (Method S2), with a yield of 64% based on Fe content.

2.2.3. Preparation of F_7-PEG-PEI Coated MNPs (FPP@MNPs)

The suspension of DMSA@MNPs (containing 5 mg Fe) and 20 mg of FPP were mixed in 40 mL of pure water, and the mixture was reacted under electric stirring at 500 rpm with 100 W ultrasonic vibration for 1 h. The solution was then purified by ultrafiltration (100 kDa NMWL) 6 times and kept in aqueous suspension (3 mL of water). The Fe concentration of FPP@MNPs was 1.17 mg/mL, measured by UV-Vis spectrophotometry, with a yield of 70%

based on Fe content. FPP@MNPs were characterized by magnetization, TEM, EDS, DLS, and zeta potential. mPEG-PEI coated MNPs were prepared by Method S1. Size distribution of mPEG-PEI@MNPs was shown in Figure S5.

2.2.4. Cell Viability and Apoptosis

(1) Cell viability was detected by CCK8 kit. HeLa cells were cultured in a 96-well plate, 10^4 cells per well.100 μL of culture medium (DMEM and 10% FBS) was added per well, and cells were cultured at 37 °C, 5% CO_2 for 16 h. After that, the culture medium was replaced, and FPP@MNPs were added at the final concentrations of 2, 4, 6, 8, 10, and 12 μg/mL. Cells were incubated for 24 or 48 h. Then 10 μL of CCK-8 solution was added to each well, and the OD value was detected at λ = 450 nm after incubation for 1 h.
The cell viability was calculated according to the following formula:
Cell viability (%) = [(As − Ab)/(Ac − Ab)] × 100;
As = experimental well absorbance (cells, medium, CCK-8 and FPP@MNPs);
Ab = absorbance of blank wells (medium and CCK-8);
Ac = control well absorbance (cells, medium and CCK-8).

(2) Cell apoptosis was detected by Annexin V-FITC/PI apoptosis kit. HeLa cells were cultured in a 6-well plate, 5 × 10^5 cells per well. In total, 2 mL of culture medium (DMEM and 10% FBS) was added per well, and cells were cultured at 37°C, 5% CO_2 for 16 h. Then FPP@MNPs were added at the final concentrations of 2.5 or 5 μg/mL. Cells were incubated for 24 or 48 h. After the cells were digested with 0.25% trypsin solution, Annexin V-FITC and PI were added and measured by flow cytometry Attune NxT (Invitrogen Inc., Carlsbad, CA, USA).

2.2.5. Observation of Cellular Uptake and Transfection by Laser Confocal Microscopy

(3) Cells were cultured in glass-bottom dishes with a diameter of 35 mm and a thickness of 0.17 mm, with 5 × 10^5 cells per dish, and the volume of the culture medium was 2 mL.
(4) Preparation of FITC-labeled(green) nanoparticles:
1 mg of F_7-PEG-PEI@MNPs (FPP@MNPs) or mPEG-PEI@MNPs (mPP@MNPs) and 50 μg FITC were mixed at room temperature and placed in the dark for 24 h. After that, the ultrafiltration purification (100 kDa NMWL) was repeated three times.
(5) Observation of cellular uptake:
FITC-labeled FPP@MNP or mPP@MNP (containing 10 μg Fe) was mixed with 2 mL DMEM medium containing LysoTracker (red) and Hoechst33342 (blue). Then the mixture was added to the cells, placed under a confocal microscope for observation.
(6) Observation of cell transfection:
The transfected cells were added to a 2.5% glutaraldehyde solution, fixed at 4 °C for 30 min, and then washed three times with PBS. F-actin was stained with rhodamine-phalloidin (red), and nuclei were stained with Hoechst33342. siRNA labeled with FAM (green).
(7) The excitation wavelength:
405 nm for blue fluorescence, 488 nm for green fluorescence, and 562 nm for red fluorescence.

2.2.6. Cell Transfection

Cells were cultured in 24-well plates, 1 × 10^5 cells per well, with 500 μL DMEM medium (10% FBS), incubated at 37 °C, 5% CO_2 until the cell density reaches 60%.

External magnetic field-enhanced transfection (MagTrans) used the following steps:

(1) The medium in each well was replaced with fresh 500 μL of DMEM medium (10% FBS);

(2) For each well, 40 pmol of siRNA was dissolved in 10 μL of DEPC water, and then mixed with 0.25, 0.5, 0.75, or 1.0 μg of FPP@MNP (calculated as Fe content). Incubated at room temperature for 10–15 min to obtain FPP@MNP/siRNA complexes;
(3) The above complexes were added to the medium in the well;
(4) The 24-well plate was placed on the magnetic plate (400 mT) and incubated at 37 °C for 10–20 min;
(5) The magnetic plate was removed, and the cells were cultured at 37 °C, 5% CO_2.

2.2.7. RT-PCR

RNA was extracted from the transfected cells using Trizol (Invitrogen). RNA concentration was detected by Nanodrop spectrophotometer (Thermo Fisher Scientific Inc., Waltham, MA, USA). Both reverse transcription and SYBR green fluorescence quantitative PCR were performed on LightCycler96 (Roche Inc., Basel, Switzerland) using TB Green Premix Ex Taq II kit (Takara Bio Inc., Kusatsu, Japan). The primer sequences are shown in the table below. The relative expression of CXCR4 was calculated with Actin as reference. The sequences of primer are shown in Table 1.

Table 1. Sequences of primer.

Target (Mouse)		Primer
Actin	F	CTCCTGAGCGCAAGTACTCT
	R	TACTCCTGCTTGCTGATCCAC
CXCR4	F	TTCATCTTTGCCGACGTCAG
	R	CGAGACCCACCATTATATGCT

2.2.8. Transfection Efficiency and CXCR4 Expression Measured by Flow Cytometry

Cells were digested with 0.25% trypsin solution, centrifuged at 1000 rpm, and washed three times with PBS. Cells were transfected with FAM-labeled RNA with excitation light of 488 nm wavelength and detection channel of 530 nm wavelength.

Cells transfected with siCXCR4 were fixed with 4% paraformaldehyde solution at 4 °C for 30 min. Then cells were centrifuged at 1000 rpm and washed for 3 times with PBS. In total, 5×10^5 cells were mixed with 0.5 μg Cy5-labeled Anti-CXCR4 antibody and then incubated at room temperature for 30 min. Then the cells were centrifuged at 1000 rpm and washed for 3 times with PBS. An excitation light of 637 nm wavelength and a detection channel of 695 nm wavelength were used.

2.2.9. Characterization

Elements' content measured by energy dispersive spectroscopy (EDS): 50 μL of F_7-PEG-COOH solution was dropped on a single-crystal silicon wafer (0.5 cm × 0.5 cm). After drying at room temperature, the silicon wafers were placed in scanning electron microscope (Ultra Plus, Zeiss Inc., Oberkochen, Germany) for characterization.

Nanoparticles observed by transmission electron microscope (TEM): 20 μL of OA@ MNPs, DMSA@MNPs, or FPP@MNPs were dropped on a 200-mesh copper grid. The copper grid was dried at room temperature and then placed in a TEM for observation. HRTEM imaging and SAED were performed on JEM-2100 (JEOL Ltd., Akishima, Japan) at an operating voltage of 200 kV. EDS mapping was performed on FEI Tecnai G2 F30(FEI Inc., Hillsboro, OR, USA) at an operating voltage of 300 kV.

Magnetization curves: The OA@MNPs (containing 20 mg Fe) were dried and put into vibrating sample magnetometer (Lakeshore 7407, Lakeshore Inc., Columbus, OH, USA) for measurement. Set the magnetic field range from −7500 G to 7500 G.

^{19}F-Nuclear Magnetic Resonance (NMR): 20 mg of F_7-PEG-PEI was dissolved in 0.5 mL of CD_3 OD and using Bruker 600M (Bruker Inc., Karlsruhe, Germany) for characterization, the operating frequency was 600 MHz.

Characterization of size and zeta potential: the size distribution and zeta potential of DMSA@MNPs or FPP@MNPs were measured by Zetasizer ZS90 (Malvern Panalytical

Ltd., Malvern, UK). The sample was diluted with pure water to a final concentration of 0.1 mg/mL Fe, and then 1 mL of the sample was added to the sample pool before testing. DLS distribution of mPEG-PEI@MNPs is shown in Figure S5.

50 particles in Figure 2b were randomly selected for TEM size measurement by ImageJ. Figure S1 shows the TEM size distribution. Figure S2 shows the filtered HRTEM image for measuring lattice spacing of OA@MNPs. Figure S3 shows the absorbance standard curve for measuring Fe concentration.

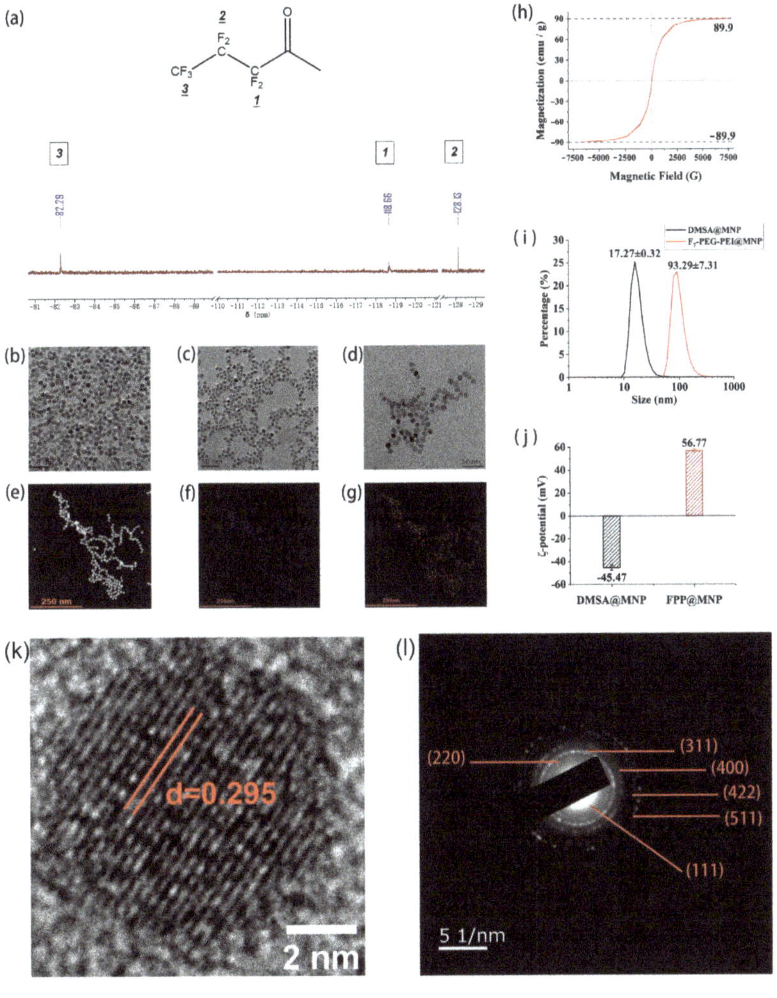

Figure 2. (**a**) [19] F-NMR spectrum of F_7-PEG-PEI (FPP). (**b**) TEM micrograph of oleic acid-modified MNPs. (**c**) TEM micrograph of DMSA@MNP. (**d**) TEM micrograph of FPP@MNP. (**e**) HAADF micrograph of FPP@MNP. (**f**) Distribution of F element of FPP@MNP. (**g**) Distribution of Fe element of FPP@MNP. (**h**) Magnetization curve of MNPs. (**i**) Hydrodynamic size distribution (intensity) of DMSA@MNP and FPP@MNP. (**j**) Zeta potential of DMSA@MNP and FPP@MNP. (**k**) HRTEM image of oleic acid-modified MNPs. (**l**) SAED image of oleic acid-modified MNPs.

2.3. Statistical Analysis

The data were expressed as the mean ± standard deviation (SD). Graphs were drawn by Origin Software (version 2022, OriginLab Co., Northampton, MA, USA), and the significances were calculated by Student's *t*-test.

The * represents $p < 0.05$, ** represents $p < 0.01$. ($n \geq 3$).

3. Results

3.1. Characterization of Fluorinated PEG-PEI Coated MNPs

The elemental composition of F_7-PEG-COOH was analyzed by SEM-EDS. The results are shown in Table S1. The atomic ratio of F, O, and C in F_7-PEG-COOH is F:O:C = 1:7.288:15.25, which is basically consistent with the theoretical calculation of F:O:C =1:6.92:13.70. The FPP was dissolved in CD_3 OD and characterized by ^{19} F-NMR, and the results are shown in Figure 2a. Compared with the published study [42], peak $\delta_1 = -118.66$ ppm is assigned to the fluorine atoms attached to the α carbon atom *(1)* of the heptafluorobutyryl. $\delta_2 = -128.13$ ppm is assigned to the fluorine atoms attached to the β carbon atom *(2)*. $\delta_3 = -82.29$ ppm is assigned to the fluorine atom attached to the γ carbon atom *(3)*. Hereby, it can be proved that the prepared product contains heptafluorobutyryl.

Magnetic nanoparticles (MNPs) were synthesized by pyrolysis of $Fe(acac)_3$ at high temperatures and stored as a suspension in hexane. The TEM image of oleic acid-modified Fe_3O_4 magnetic nanoparticles (OA@MNPs) prepared by high-temperature pyrolysis is shown in Figure 2b. The size of MNPs is 11.92 ± 0.94 nm; the distribution of size is shown in Figure S1. After drying the OA@MNPs, the magnetization measured by the vibrating sample magnetometer (VSM) is shown in Figure 2h. The magnetization curves approximately pass through the (0,0) point, indicating that when the external magnetic field is 0, the magnetization of the MNPs is about 0. In addition, the maximum value indicates that the saturation magnetization of the OA@MNPs is 89.9 emu/g Fe (the Fe content of OA@MNPs was 70.2%). Figure 2k shows the high-resolution TEM (HRTEM) image of OA@MNPs. The lattice spacing (d) of MNPs is 0.295 nm, which is matched with the (220) plane of the cubic structure of Fe_3O_4. Figure 2l shows the selected area electron diffraction (SAED) of OA@MNPs. The electron diffraction rings are assigned to (111) (220) (311) (400) (422) (511) planes of the cubic structure of Fe_3O_4, respectively. The measurement of lattice spacing by HRTEM images is presented in Figure S2. The lattice spacings measured by HRTEM and SAED are shown in Table S2.

Then, using dimercaptosuccinic acid (DMSA), OA@MNPs were converted into water-soluble DMSA@MNPs and stored as a suspension in water. The TEM image of DMSA@MNPs is shown in Figure 2c, indicating that the morphology of MNPs remained after DMSA modification. F_7-PEG-PEI coated MNPs (FPP@MNPs) were prepared by electrostatic adsorption. Figure 2d shows the TEM image of FPP@MNPs. It can be seen that the FPP-coated MNPs are dispersed instead of forming aggregates. The HAADF image of FPP@MNPs is shown in Figure 2e, and the distribution of F and Fe elements of FPP@MNPs is shown in Figure 2f,g. As can be seen, the distribution of F and Fe is highly correlated, indicating that FPP was combined with MNPs.

The hydrodynamic size and zeta potential of DMSA@MNPs and FPP@MNPs are shown in Figure 2i,j, which indicates that after F_7-PEG-PEI modification, the size increased from 17.27 ± 0.32 nm to 93.29 ± 7.31 nm, and the zeta potential increased from −45.57 mV to 56.77 mV. A previous study reported that PEI-coated magnetic nanoparticles exhibited increased particle size and positive zeta potential [46]. The increase in size and the change of zeta potential from negative to positive illustrate that FPP is adsorbed on the surface of MNPs. Compared with the TEM images of FPP@MNPs and the hydrodynamic size of DMSA@MNPs, the hydrodynamic size of FPP@MNPs is significantly increased, suggesting that FPP@MNPs may aggregate in water. This phenomenon may be caused by the fluorophilic character between the fluorine chains.

3.2. Cell Viability and Apoptosis Assays

Figure 3i shows the effect of FPP@MNPs on cell viability, which was characterized by the CCK8 kit. Respectively, 2, 4, 6, 8, 10, and 12 μg/mL of FPP@MNP (calculated by the mass of Fe) were added to HeLa cells and then cultured for 24 h or 48 h. Compared with the cells incubating for 24 h, the viability of HeLa cells after 48 h of incubation was slightly reduced. When the concentration was lower than 4 μg/mL, FPP@MNPs showed low cytotoxicity (viability $\geq$ 80%). When the concentration reached 6 μg/mL, the cell viability decreased significantly. Furthermore, FPP@MNPs were toxic to HeLa cells when the concentration was $\geq$ 8 μg/mL. Meanwhile, the results of FPP@MNPs-induced HeLa cell apoptosis are shown in Figure 3a–e, and the analysis of cell viability is shown in Figure 3ii. In Figure 3a–e, the FITC-/PI- represents cells with normal viability. Figure 3ii shows that the inhibition of cell viability by FPP@MNPs with a concentration lower than 5 μg/mL was relatively slight, while the incubation time was not longer than 48 h.

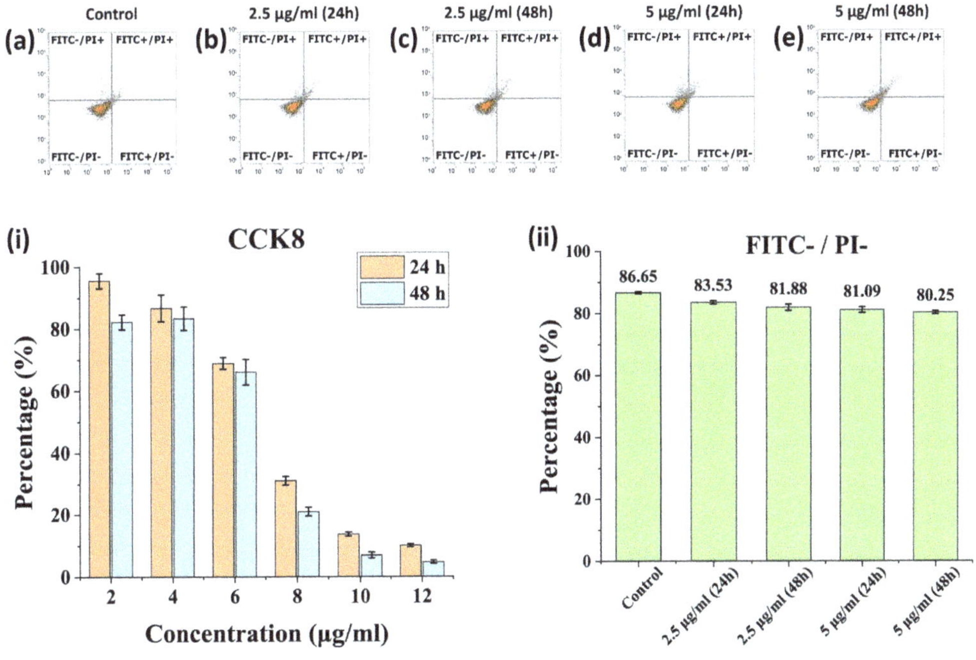

Figure 3. (**a–e**) Scatter diagram of HeLa cells apoptosis stained by Annexin V-FITC/PI measured by flow cytometry. (**i**) Relative cell viability measured by CCK8 kit. (**ii**) Analysis of cell viability measured by flow cytometry (**a–e**). The concentration is calculated by mass of Fe.

Combining the results of CCK8 and apoptosis assays, it can be shown that when the concentration of FPP@MNPs is less than 4 μg/mL, the toxicity to cells is acceptable, which is much higher than that in our subsequent studies.

3.3. Validation of Cellular Uptake and Transfection Efficacy

Figure 4 shows the uptake of FITC-labeled F_7-PEG-PEI@MNPs (FPP@MNPs) and mPEG-PEI@MNPs (mPP@MNPs) by HeLa cells. As shown in Figure 4a, 15 min after adding FPP@MNPs to HeLa cells, it was observed that particles (green fluorescence) were distributed around the cells and FPP@MNPs co-localized with lysosomes (red fluorescence). Moreover, from the beginning to 40 min, more co-localization can be observed. Prussian blue staining confirmed intracellular iron distribution (Figure S4).

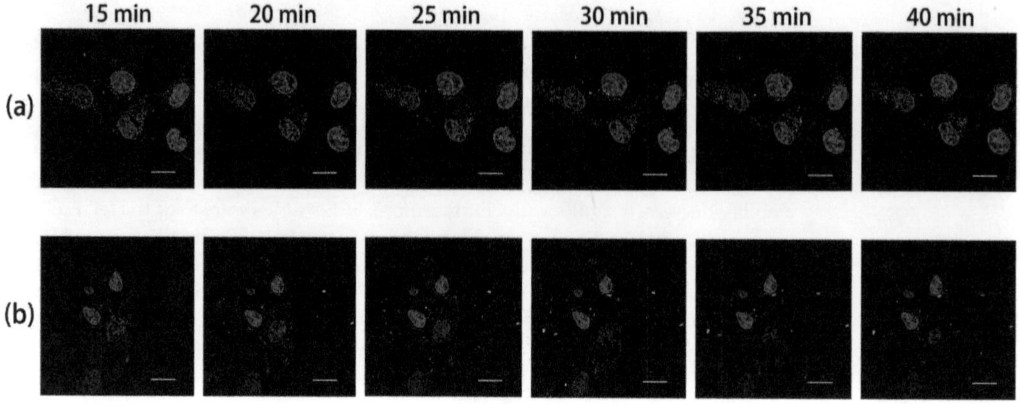

Figure 4. Uptake of (**a**) F_7-PEG-PEI@MNP; (**b**) mPEG-PEI@MNP by HeLa cells. Figures were captured by confocal laser microscopy. Blue is Hoechst 33342 stained nuclei, Red is LysoTracker stained lysosomes, and Green is FITC-labeled nanoparticles. The result of co-localization of red and green fluorescence is yellow. (Scale = 20 μm).

As is shown in Figure 4b, from 15 min to 40 min after adding mPP@MNPs to HeLa cells, the nanoparticles around the cells could be observed. The intensity of green fluorescence gradually increased, indicating mPP@MNPs were gradually adsorbed on the cell membrane. In contrast, there is no observable co-localization of mPP@MNPs (green fluorescence) with lysosomes (red fluorescence) from the beginning until 40 min. Comparing Figure 4a,b, it is obvious that the cellular uptake efficiency of FPP-coated MNPs is significantly higher than mPP-coated MNPs.

Afterward, we used FPP@MNPs to load FAM-labeled siRNA (green fluorescence) and verified the transfection efficiency on human HeLa, A549 cell line, and mouse 4 T1 cell line. Cells were transfected at different FPP@MNPs/siRNA ratios and with or without external magnetic field enhancement. Specifically, the dosage of siRNA was 40 pmol, and the FPP@MNPs were 0.25, 0.5, 0.75, or 1.0 μg (mass of Fe). The method of external magnetic field enhancement is to place the cell culture plate on the magnetic plate for 15 min.

Figure 5 shows the images taken by fluorescence microscopy. After 12 h, the results of each group show that the efficiency of siRNA delivery by FPP@MNPs can be significantly enhanced by an external magnetic field (mag+) compared to the transfections without magnetic field enhancement (mag−). For HeLa cells and A549 cells, 0.5 μg or 0.75 μg of FPP@MNPs combined with 40 pmol of siRNA-NC can achieve high transfection efficiency by the magnetic field enhancement. For 4 T1 cells, 0.75 μg of FPP@MNPs combined with 40 pmol of siRNA NCs can achieve high-efficiency transfection under the enhancement of the magnetic field. Overall, 0.75 μg FPP@MNPs with 40 pmol of siRNA are preferred.

As is shown in Figure 6, in order to further explore the distribution of siRNA(FAM-labeled, green fluorescence) in cells, the cells were scanned layer by layer and three-dimensionally reconstructed by laser confocal microscopy. HeLa, A549, and 4 T1 cells were transfected with FPP@MNPs under external magnetic field enhancement (MagTrans). The commercial nucleic acid vector Invitrogen Lipofectamine 3000 Reagent (Lipo3000) was used as the control. For each group of cells, it is evident that the amount of siRNA inside MagTrans treated cells is significantly higher than Lipo3000 transfected cells. For Lipo3000 transfected cells, the distribution of siRNA can be observed in almost every cell, but the amount of siRNA contained in a single cell is far less than that of cells treated by MagTrans.

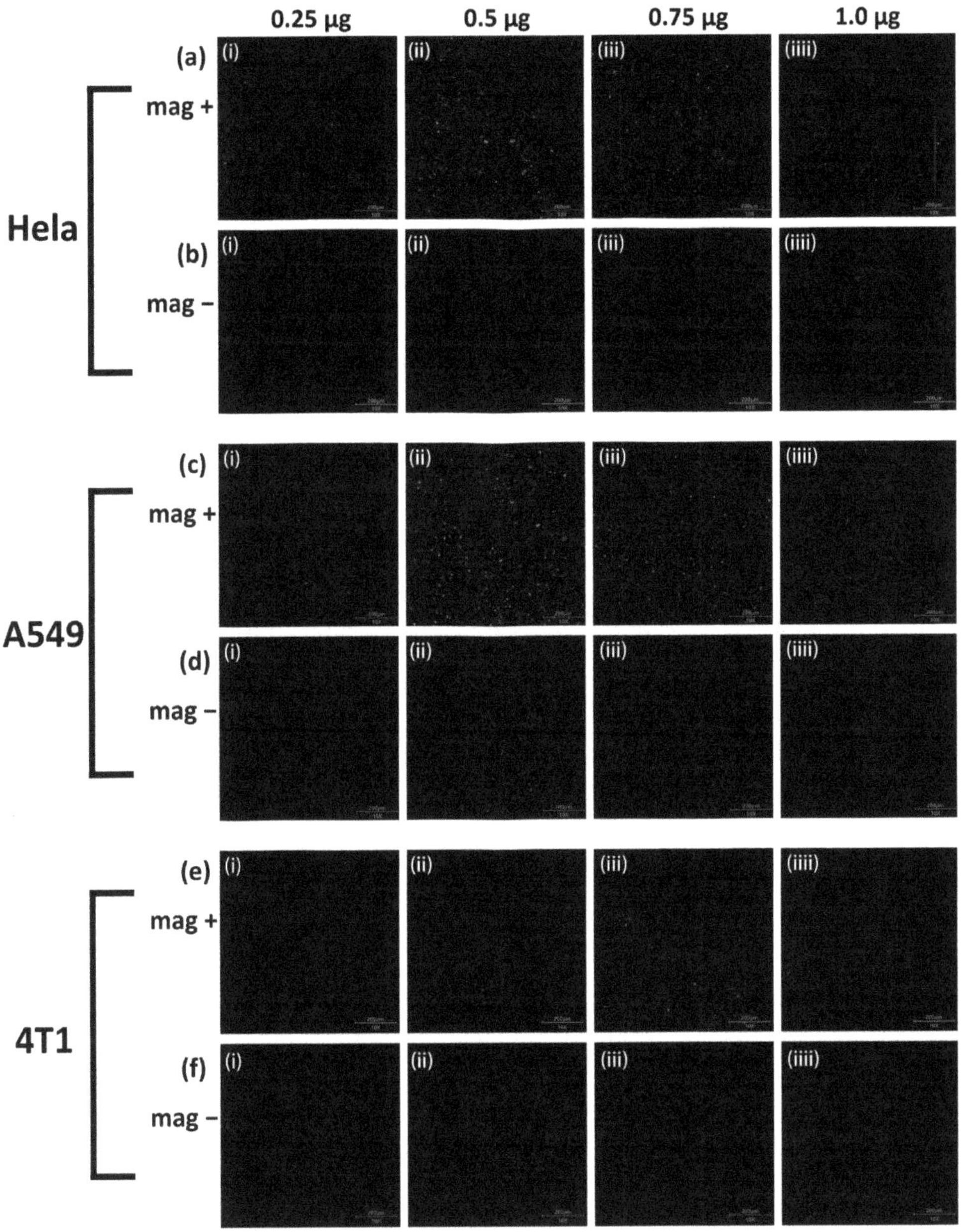

Figure 5. Fluorescence microscopic images of (**a**,**b**) HeLa, (**c**,**d**) A549, (**e**,**f**) 4 T1 cells transfected with FAM-siRNA. In total, 40 pmol FAM-siRNA mixed with 0.25 μg, 0.5 μg, 0.75 μg, or 1.0 μg (mass of Fe) F7-PEG-PEI@MNP, and then added to cells in 24-well plate. The effect of cell transfection enhanced by magnetic fields (mag+) or not (mag−). Excitation λ = 488 nm. (Scale = 200 μm).

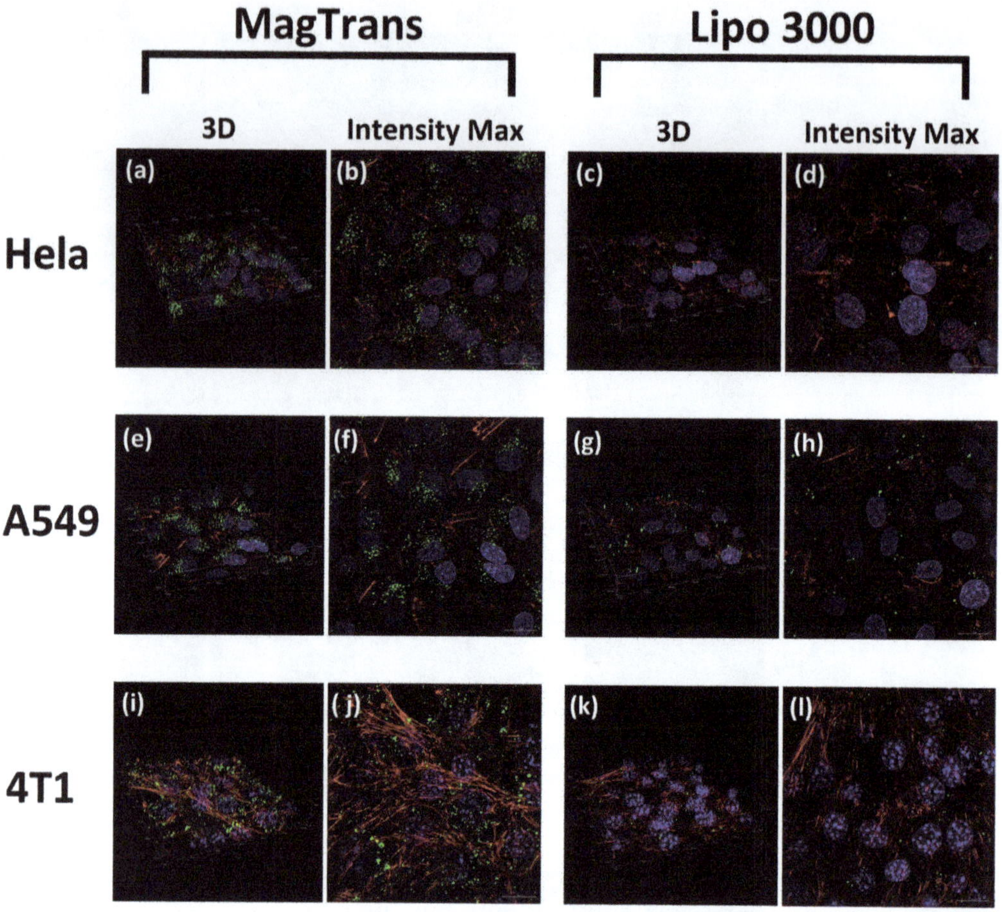

Figure 6. Results of FAM-siRNA transfection captured by confocal microscopy. Blue is Hoechst33342 stained nuclei, Green is FAM-labeled siRNA NC, and Red is Rhodamine-Phalloidin stained F-actin (**a–d**) HeLa, (**e–h**) A549, (**i–l**) 4 T1. Transfection method used were magnetic field-enhanced F_7-PEG-PEI@MNP transfection (MagTrans) or Lipo3000 transfection (Lipo 3000). Images are reconstructed in 3 D, and the images with the highest fluorescence intensity (Intensity Max) are given by NIS-Elements software (Nikon Co., Tokyo, Japan).

The transfection efficiency of HeLa cells was analyzed by flow cytometry, and the results are shown in Figure 7. The dots inside the rectangular gate indicate transfected cells. Figure 7A shows untreated HeLa cells as a negative control, and Figure 7B shows HeLa cells transfected with Lipo3000 as the positive control. The results of magnetic field enhanced FPP@MNPs transfection (mag+) are shown in Figure 7a–d, and the results of FPP@MNPs transfection without magnetic-field enhancement (mag−) are shown in Figure 7e–h. The statistical analysis of transfection efficiency and mean fluorescence intensity of each group is shown in Figure 7i,ii.

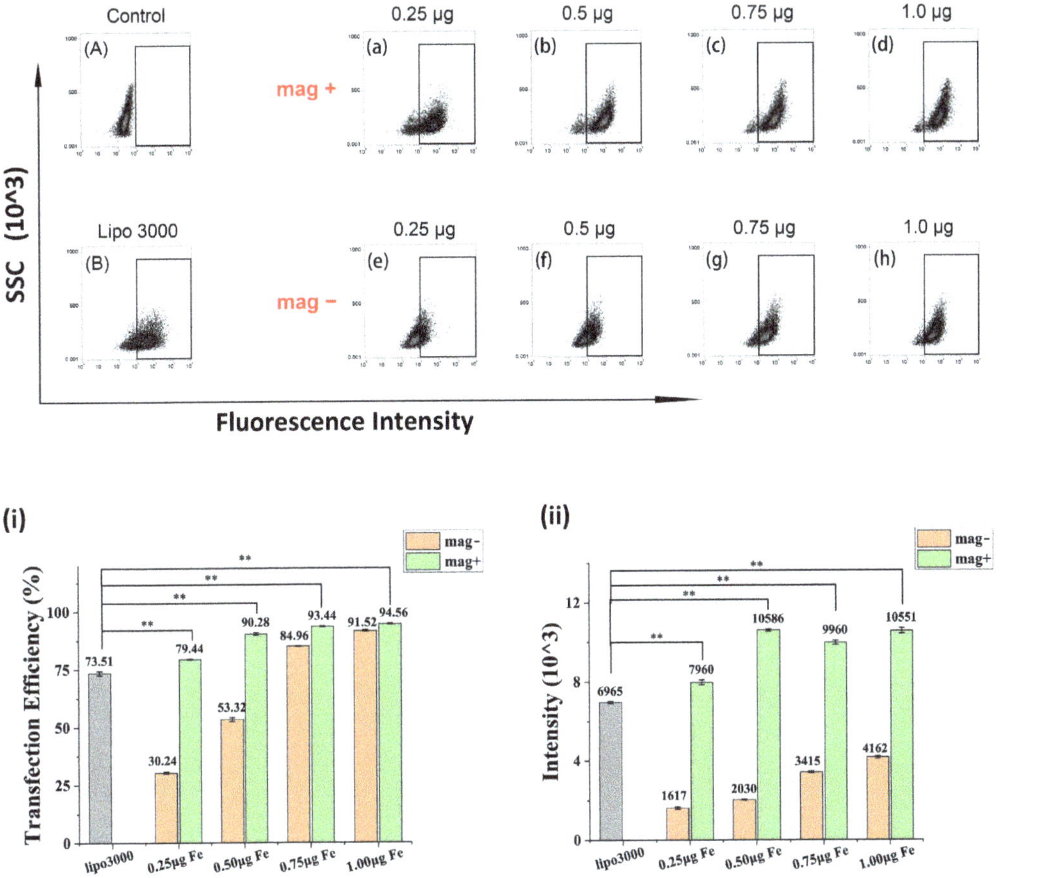

Figure 7. Transfection efficiency of HeLa cells measured by flow cytometry. (**A**) Control. (**B**) Lipo 3000 transfection. (**a**–**d**) 40 pmol FAM-siRNA mixed with 0.25 μg, 0.5 μg, 0.75 μg, or 1.0 μg F_7-PEG-PEI@MNP, transfection enhanced by magnetic fields (mag+). (**e**–**h**) 40 pmol FAM-siRNA mixed with F_7-PEG-PEI@MNP (mag-). (**i**) Transfection efficiency of Lipo3000, mag+ and mag-. (**ii**) Mean fluorescence intensity of Lipo3000, mag+ and mag− transfected cells (gated). ** $p < 0.01$.

In Figure 7i, comparing the Mag+ group with the Mag- group, the external magnetic field could effectively improve the transfection efficiency. In particular, the improvement was more pronounced between the groups that used fewer FPP@MNPs (0.25 μg or 0.5 μg, mass of Fe, the same below). Furthermore, the difference in transfection efficiency between the mag+ group and Lipo3000 is significant. The transfection efficiency of Lipo3000 was 73.51%, and the transfection efficiency of 0.25 μg FPP@MNP was 79.44%, which was slightly improved. For the transfection results of 0.5 μg, 0.75 μg, and 1.0 μg FPP@MNP, the efficiency was higher than 90%, which was greatly improved compared with Lipo3000.

Figure 7ii shows the statistical analysis of the Mean Fluorescence Intensity (MFI) of transfected cells (within the rectangular gate). Compared with the mag+ group and the mag - group, the magnetic field-enhanced transfection could not only improve the transfection efficiency but also increase the content of siRNA transfected into cells. Comparing the mag+ group with the Lipo3000 group, it can be seen that the MFI of the 0.25 μg mag+ was slightly higher than that of the Lipo3000. The MFI of 0.5 μg, 0.75 μg, and 1.0 μg FPP@MNP transfection was much higher than that of the Lipo3000, and the difference was significant.

3.4. Knockdown of CXCR4 Expression in 4 T1 Cells

Then, siRNA sequences (R1-R3) for silencing CXCR4 (siCXCR4) were designed and synthesized. The sequences of negative control (NC) and siCXCR4 are given in Table 2. For 4 T1 cells in the 24-well cell culture plate, 0.75 μg of FPP@MNP and 40 pmol of siRNA were mixed for transfection, which was enhanced with an external magnetic field. Cells were cultured for 12 h after transfection, and CXCR4 mRNA was detected by RT-PCR. The results are shown in Figure 8i. With the CXCR4 mRNA in the NC group as a reference, the relative expression levels of the R1, R2, and R3 groups were calculated after normalization. As can be seen in Figure 8i, the knockdown rate of R1 is not ideal and may even lead to increased CXCR4 expression. R2 and R3 can effectively reduce CXCR4 mRNA, and R2 achieves the best effect. The above differences are all significant.

Table 2. siRNA Sequences.

siRNA	Sequence
NC	(5′-3′) UUCUCCGAACGUGUCACGUTT (3′-5′) TTAAGAGGCUUGCACAGUGCA
R1	(5′-3′) CGAUCAGUGUGAGUAUAUATT (3′-5′) TTGCUAGUCACACUCAUAUAU
R2	(5′-3′) GUCCAUUUCAAUAGGAUCUTT (3′-5′) TTCGGAGUUCUAGGAAAGGUU
R3	(5′-3′) GCCUCAAGAUCCUUUCCAATT (3′-5′) TTCGGAGUUCUAGGAAAGGUU

The CXCR4 expression on the membrane of 4 T1 cells after transfection for 24 h or 48 h was detected by flow cytometry. The fluorescence intensity distribution of Cy5 labeled anti-CXCR4 antibody is shown in Figure 8a–f, and the analysis is shown in Figure 8ii. After 24 h, the expression of CXCR4 in the R2 and R3 groups decreased, while the expression of CXCR4 in the R1 group increased compared with the negative control. The difference between R1 and R2, R1 and R3 was extremely significant, but the difference between R2 and R3 was not significant. After 48 h, the expression of CXCR4 in the R1, R2, and R3 groups was decreased compared with the cells transfected for 24 h. However, the knockdown efficiency of R1 was still lower than that of R2 or R3, and the difference was significant. Besides, the difference between R2 and R3 is still not significant.

To sum up, the CXCR4 knockdown rate increased as the culture time of the transfected cells increased from 24 h to 48 h. Compared to the efficiencies of the three sequences (R1, R2, and R3), the effect of R1 was poor. Even R1 may cause an increase in CXCR4 expression for a short period of time and should therefore be excluded. At the protein level, the effect of R2 and R3 was not significantly different. However, at the mRNA level, the knockdown effect of R2 was extremely significant compared to R3.

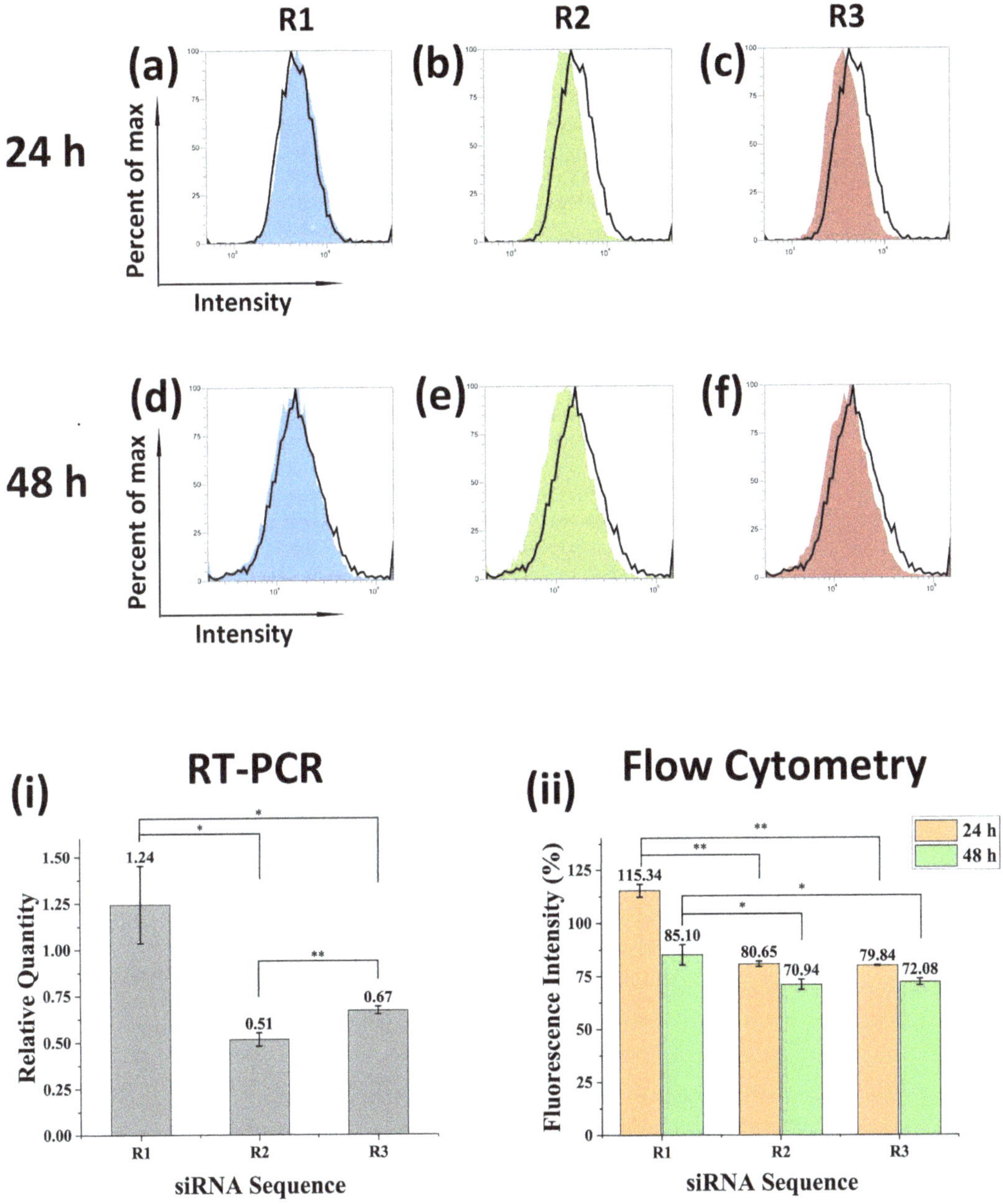

Figure 8. Results of CXCR4 expression and knockdown on 4 T1 cells. (**a–c**) 24 h after transfection, fluorescence intensity of Cy5. (**d–f**) 48 h after transfection, fluorescence intensity of Cy5. CXCR4 were detected by Cy5 labelled antibody, peaks with black lines represent negative control. (**i**) Analysis of CXCR4 mRNA measured by RT-PCR. (**ii**) Analysis of the relative intensity of CXCR4 expression on cell membranes measured by flow cytometry (**a–f**). * $p < 0.05$, ** $p < 0.01$.

4. Discussion

The preparation process of F_7-PEG-PEI (FPP) includes the preparation of F_7-PEG-COOH and the reaction of F_7-PEG-COOH with polyethyleneimine (PEI). The measurement of the three elements (C, O, and F) of F_7-PEG-COOH shows that the proportion of the F element is 4.23%. Theoretically, when a heptafluorobutyryl group is attached to a PEG_{2000} molecule, the proportion of the F element is 4.63%. The ratio of the actual value to the theoretical value was 4.23/4.63 = 91.36%, which indicated that most of the PEG molecules were modified with heptafluorobutyryl. The EDC/NHS system is widely used for activating carboxyl groups. The carboxyl group of F_7-PEG-COOH was activated by EDC/NHS and then reacted with PEI to synthesize FPP. The ^{19} F-NMR results of FPP proved the existence of heptafluorobutyryl.

Nowadays, MNPs are widely used in biomedical fields such as MRI, drug delivery, and hyperthermia [47]. Compared with the magnetic nanoparticles prepared by co-precipitation, the size distribution of the MNPs prepared by high-temperature pyrolysis is more uniform, and the saturation magnetization is higher [48]. Magnetite nanoparticles smaller than 25 nm in size are considered to be superparamagnetic, which is defined as the magnetic material that has no magnetization in the absence of an external magnetic field [21]. The TEM size of prepared Fe_3O_4 nanoparticles is 11.92 $\pm$ 0.94 nm, and the magnetization curve in Figure 2h proves that our prepared Fe_3O_4 nanoparticles are superparamagnetic. The increase in hydrodynamic size and the change of zeta potential demonstrated that FPP was coated on the surface of DMSA@MNPs, and the EDS mapping confirmed that the nanoparticles were fluorinated. In our work, the results of SAED are in good agreement with previous studies [49,50], which is consistent with the inverse spinel structure of Fe_3O_4. Besides, the lattice spacing of the (220) plane measured by HRTEM is 0.295 nm, which corroborates with the 0.298 nm referred to in a previous study [50].

As shown in the introduction part [42–45], it is shown in many studies that fluorination can improve the transfection efficiency of polymer nanoparticles. Different from previous studies, we modified PEG with heptafluorobutyric anhydride and then connected it with PEI. Using fluorinated PEG-PEI to coat MNPs can not only enhance cellular uptake by fluorination but also enhance transfection efficiency by a magnetic field. Compared with the commercial vector Lipo3000, our prepared FPP@MNPs can transfect more siRNA into cells. As can be seen from Figure 6, Lipo3000 has high transfection efficiency, but the amount of siRNA delivered into cells is not as high as that of FPP@MNPs (enhanced by magnetic field). These claims can also be confirmed by the results of flow cytometry in Figure 7.

Previous studies have established that CXCR4 is required for breast cancer cell proliferation or survival, and CXCR4 inhibitors will improve the treatment of primary and metastatic breast cancer [51,52]. The high expression of CXCR4 in 4 T1 breast cancer cells was widely demonstrated in much research [53–55]. From the previous studies, this mouse cell line is suitable for breast cancer research, including in vitro studies or the construction of mouse models for in vivo studies. We used the FPP@MNPs prepared in this study as siRNA carriers to transfect 4 T1 cells and knock down the expression of CXCR4 in vitro. The effect of FPP@MNPs as siRNA carriers has been confirmed, and our research on HeLa cells and A549 cells shows that FPP@MNPs can be extended to more applications.

5. Conclusions

Based on magnetic nanoparticles, fluorinated siRNA carriers F_7-PEG-PEI@MNPs were developed in this study. The carriers were less toxic to cells at conventional doses and allowed for efficient cellular uptake. Our findings suggest that fluorination is an effective strategy to overcome the "PEG dilemma". The results of transfection showed that our prepared FPP@MNPs could achieve an efficiency of more than 90%. In particular, the transfection efficiency can be further improved by an external magnetic field, even with a smaller amount of FPP@MNPs. In vitro experiments confirmed that FPP@MNPs could be used to knockdown CXCR4 expression in breast cancer cells. Furthermore, this vector was verified to be universal on different cells.

6. Patents

Heptafluorobutyryl-polyethylene glycol-polyethyleneimine (F_7-PEG-PEI) and nanoparticles prepared with F_7-PEG-PEI and the above technical solutions and applications have been submitted for patent examination in the People's Republic of China. All rights are hereby declared. Application Number [202210144467.8].

Supplementary Materials: The following supporting information can be downloaded at: https://www.mdpi.com/article/10.3390/nano12101692/s1, Table S1: Relative Content of C, O, F Elements in F7-PEG-COOH measured by EDS; Table S2: Lattice spacing measured by HRTEM and SAED; Figure S1: TEM size distribution of MNPs; Figure S2: Filtered HRTEM image for measuring lattice spacing of Fe_3O_4 NPs; Figure S3: Standard curve for measuring Fe concentration; Figure S4. Fe^{3+} distribution in Hela cells measured by Prussian blue Staining; Figure S5. DLS distribution of mPEG-PEI@MNPs; Method S1: Preparation of mPEG-PEI@MNPs (mPP@MNPs); Method S2. Measurement of Fe concentration by UV-Vis Spectrophotometry; Method S3. Prussian blue staining.

Author Contributions: Y.C. conceived the study and conducted the experiments. S.Z. provided technical support for transfection and participated in the design of this study. This manuscript was written by Y.C. and critically reviewed by M.M. and Y.Z. All authors have read and agreed to the published version of the manuscript.

Funding: This work was supported by the National Key Research and Development Program of China [No. 2017 YFA0205502]; National Natural Science Foundation of China [No. 82072067, 61821002]; Special Fund for Transformation of Scientific and Technological Achievements of Jiangsu Province [BA2020016] and the Fundamental Research Funds for the Central Universities.

Institutional Review Board Statement: Not applicable.

Informed Consent Statement: Not applicable.

Data Availability Statement: Not applicable.

Acknowledgments: We are particularly grateful to Nanoeast Biotech for providing magnetic plates for cell transfection and technical support in verifying the transfection effect. Cell lines were provided by Southeast University School of Medicine, and School of Biological Science and Medical Engineering.

Conflicts of Interest: The authors declare no conflict of interest.

References

1. Li, Z.; Wang, Y.; Shen, Y.; Qian, C.; Oupicky, D.; Sun, M. Targeting pulmonary tumor microenvironment with CXCR4-inhibiting nanocomplex to enhance anti–PD-L1 immunotherapy. *Sci. Adv.* **2020**, *6*, eaaz9240. [CrossRef] [PubMed]
2. Arya, M.; Patel, H.R.H.; McGurk, C.; Tatoud, R.; Klocker, H.; Masters, J.; Williamson, M. The Importance of the CXCL12-CXCR4 chemokine ligand-receptor interaction in prostate cancer metastasis. *J. Exp. Ther. Oncol.* **2004**, *4*, 291–303. [PubMed]
3. Liang, Z.; Yoon, Y.; Votaw, J.; Goodman, M.M.; Williams, L.; Shim, H. Silencing of CXCR4 blocks breast cancer metastasis. *Cancer Res.* **2005**, *65*, 967–971. [PubMed]
4. Han, M.; Lv, S.; Zhang, Y.; Yi, R.; Huang, B.; Fu, H.; Bian, R.; Li, X. The prognosis and clinicopathology of CXCR4 in gastric cancer patients: A meta-analysis. *Tumor Biol.* **2014**, *35*, 4589–4597. [CrossRef]
5. Passaro, D.; Irigoyen, M.; Catherinet, C.; Gachet, S.; Da Costa De Jesus, C.; Lasgi, C.; Tran Quang, C.; Ghysdael, J. CXCR4 is required for leukemia-initiating cell activity in T cell acute lymphoblastic leukemia. *Cancer Cell* **2015**, *27*, 769–779. [CrossRef] [PubMed]
6. Tseng, D.; Vasquez-Medrano, D.A.; Brown, J.M. Targeting SDF-1/CXCR4 to inhibit tumour vasculature for treatment of glioblastomas. *Br. J. Cancer* **2011**, *104*, 1805–1809. [CrossRef]
7. Müller, A.; Homey, B.; Soto, H.; Ge, N.; Catron, D.; Buchanan, M.E.; McClanahan, T.; Murphy, E.; Yuan, W.; Wagner, S.N.; et al. Involvement of chemokine receptors in breast cancer metastasis. *Nature* **2001**, *410*, 50–56. [CrossRef]
8. Giallongo, C.; Dulcamare, I.; Tibullo, D.; del Fabro, V.; Vicario, N.; Parrinello, N.; Romano, A.; Scandura, G.; Lazzarino, G.; Conticello, C.; et al. CXCL12/CXCR4 axis supports mitochondrial trafficking in tumor myeloma microenvironment. *Oncogenesis* **2022**, *11*, 6. [CrossRef]
9. Meng, J.; Ge, Y.; Xing, H.; Wei, H.; Xu, S.; Liu, J.; Yan, D.; Wen, T.; Wang, M.; Fang, X.; et al. Synthetic CXCR4 antagonistic peptide assembling with nanoscaled micelles combat acute myeloid leukemia. *Small* **2020**, *16*, 2001890. [CrossRef]

10. Donzella, G.A.; Schols, D.; Lin, S.W.; Esté, J.A.; Nagashima, K.A.; Maddon, P.J.; Allaway, G.P.; Sakmar, T.P.; Henson, G.; De Clercq, E.; et al. AMD3100, a small molecule inhibitor of HIV-1 Entry via the CXCR4 Co-Receptor. *Nat. Med.* **1998**, *4*, 72–77. [CrossRef]
11. Tamamura, H.; Hiramatsu, K.; Kusano, S.; Terakubo, S.; Yamamoto, N.; Trent, J.O.; Wang, Z.; Peiper, S.C.; Nakashima, H.; Otaka, A.; et al. Synthesis of Potent CXCR4 inhibitors possessing low cytotoxicity and improved biostability based on T140 derivatives. *Organic Biomol. Chem.* **2003**, *1*, 3656. [CrossRef] [PubMed]
12. Li, X.; Guo, H.; Yang, Y.; Meng, J.; Liu, J.; Wang, C.; Xu, H. A designed peptide targeting CXCR4 displays anti-acute myelocytic leukemia activity in vitro and in vivo. *Sci. Rep.* **2015**, *4*, 6610. [CrossRef] [PubMed]
13. Lam, J.K.W.; Chow, M.Y.T.; Zhang, Y.; Leung, S.W.S. SiRNA versus MiRNA as therapeutics for gene silencing. *Mol. Ther. Nucl. Acids* **2015**, *4*, e252. [CrossRef]
14. Elbashir, S.M.; Harborth, J.; Lendeckel, W.; Yalcin, A.; Weber, K.; Tuschl, T. Duplexes of 21-Nucleotide RNAs Mediate RNA interference in cultured mammalian cells. *Nature* **2001**, *411*, 494–498. [CrossRef] [PubMed]
15. Setten, R.L.; Rossi, J.J.; Han, S. The current state and future directions of RNAi-based therapeutics. *Nat. Rev. Drug Discov.* **2019**, *18*, 421–446. [CrossRef] [PubMed]
16. Lim, L.P.; Lau, N.C.; Garrett-Engele, P.; Grimson, A.; Schelter, J.M.; Castle, J.; Bartel, D.P.; Linsley, P.S.; Johnson, J.M. Microarray analysis shows that some MicroRNAs downregulate large numbers of target MRNAs. *Nature* **2005**, *433*, 769–773. [CrossRef] [PubMed]
17. Landry, B.; Gül-Uludağ, H.; Plianwong, S.; Kucharski, C.; Zak, Z.; Parmar, M.B.; Kutsch, O.; Jiang, H.; Brandwein, J.; Uludağ, H. Targeting CXCR4/SDF-1 axis by lipopolymer complexes of SiRNA in acute myeloid leukemia. *J. Control. Release* **2016**, *224*, 8–21. [CrossRef] [PubMed]
18. Mykhaylyk, O.; Vlaskou, D.; Tresilwised, N.; Pithayanukul, P.; Möller, W.; Plank, C. Magnetic nanoparticle formulations for DNA and SiRNA delivery. *J. Magn. Magn. Mater.* **2007**, *311*, 275–281. [CrossRef]
19. Liu, D.; Cheng, Y.; Cai, R.; Wang, B.W.; Cui, H.; Liu, M.; Zhang, B.; Mei, Q.; Zhou, S. The enhancement of SiPLK1 penetration across BBB and its anti glioblastoma activity in vivo by magnet and transferrin co-modified nanoparticle. *Nanomed. Nanotechnol. Biol. Med.* **2018**, *14*, 991–1003. [CrossRef]
20. Babu, A.; Wang, Q.; Muralidharan, R.; Shanker, M.; Munshi, A.; Ramesh, R. Chitosan coated polylactic acid nanoparticle-mediated combinatorial delivery of cisplatin and SiRNA/Plasmid DNA chemosensitizes cisplatin-resistant human ovarian cancer cells. *Mol. Pharm.* **2014**, *11*, 2720–2733. [CrossRef]
21. Estelrich, J.; Escribano, E.; Queralt, J.; Busquets, M. Iron oxide nanoparticles for magnetically-guided and magnetically-responsive drug delivery. *Int. J. Mol. Sci.* **2015**, *16*, 8070–8101. [CrossRef] [PubMed]
22. Hoang Thi, T.T.; Pilkington, E.H.; Nguyen, D.H.; Lee, J.S.; Park, K.D.; Truong, N.P. The importance of poly(ethylene glycol) alternatives for overcoming PEG immunogenicity in drug delivery and bioconjugation. *Polymers* **2020**, *12*, 298. [CrossRef] [PubMed]
23. Koren, E.; Apte, A.; Jani, A.; Torchilin, V.P. Multifunctional PEGylated 2C5-immunoliposomes containing PH-sensitive bonds and TAT peptide for enhanced tumor cell internalization and cytotoxicity. *J. Control. Release* **2012**, *160*, 264–273. [CrossRef] [PubMed]
24. Mohamed, M.; Abu Lila, A.S.; Shimizu, T.; Alaaeldin, E.; Hussein, A.; Sarhan, H.A.; Szebeni, J.; Ishida, T. PEGylated Liposomes: Immunological responses. *Sci. Technol. Adv. Mater.* **2019**, *20*, 710–724. [CrossRef]
25. Zhang, W.; Guo, Z.; Huang, D.; Liu, Z.; Guo, X.; Zhong, H. Synergistic effect of chemo-photothermal therapy using PEGylated graphene oxide. *Biomaterials* **2011**, *32*, 8555–8561. [CrossRef]
26. Chen, M.; Tang, S.; Guo, Z.; Wang, X.; Mo, S.; Huang, X.; Liu, G.; Zheng, N. Core-Shell Pd@Au nanoplates as theranostic agents for in-vivo photoacoustic imaging, CT imaging, and photothermal therapy. *Adv. Mater.* **2014**, *26*, 8210–8216. [CrossRef]
27. Wang, M.; Chang, M.; Chen, Q.; Wang, D.; Li, C.; Hou, Z.; Lin, J.; Jin, D.; Xing, B. Au2Pt-PEG-Ce6 nanoformulation with dual nanozyme activities for synergistic chemodynamic therapy/phototherapy. *Biomaterials* **2020**, *252*, 120093. [CrossRef]
28. Ciocci, M.; Cacciotti, I.; Seliktar, D.; Melino, S. Injectable silk fibroin hydrogels functionalized with microspheres as adult stem cells-carrier systems. *Int. J. Biol. Macromol.* **2018**, *108*, 960–971. [CrossRef]
29. Masood, N.; Ahmed, R.; Tariq, M.; Ahmed, Z.; Masoud, M.S.; Ali, I.; Asghar, R.; Andleeb, A.; Hasan, A. Silver nanoparticle impregnated chitosan-PEG hydrogel enhances wound healing in diabetes induced rabbits. *Int. J. Pharm.* **2019**, *559*, 23–36. [CrossRef]
30. Grun, M.K.; Suberi, A.; Shin, K.; Lee, T.; Gomerdinger, V.; Moscato, Z.M.; Piotrowski-Daspit, A.S.; Saltzman, W.M. PEGylation of poly(amine-co-ester) polyplexes for tunable gene delivery. *Biomaterials* **2021**, *272*, 120780. [CrossRef]
31. Hoskins, C.; Wang, L.; Cheng, W.P.; Cuschieri, A. Dilemmas in the reliable estimation of the in-vitro cell viability in magnetic nanoparticle engineering: Which tests and what protocols? *Nanoscale Res. Lett.* **2012**, *7*, 77. [CrossRef] [PubMed]
32. Sun, C.-Y.; Shen, S.; Xu, C.-F.; Li, H.-J.; Liu, Y.; Cao, Z.-T.; Yang, X.-Z.; Xia, J.-X.; Wang, J. Tumor Acidity-Sensitive Polymeric Vector for Active Targeted SiRNA Delivery. *J. Am. Chem. Soc.* **2015**, *137*, 15217–15224. [CrossRef] [PubMed]
33. Fang, Y.; Xue, J.; Gao, S.; Lu, A.; Yang, D.; Jiang, H.; He, Y.; Shi, K. Cleavable PEGylation: A strategy for overcoming the "PEG Dilemma" in efficient drug delivery. *Drug Deliv.* **2017**, *24*, 22–32. [CrossRef] [PubMed]
34. Chen, Y.; Zhang, M.; Jin, H.; Tang, Y.; Wang, H.; Xu, Q.; Li, Y.; Li, F.; Huang, Y. Intein-mediated site-specific synthesis of tumor-targeting protein delivery system: Turning PEG dilemma into prodrug-like feature. *Biomaterials* **2017**, *116*, 57–68. [CrossRef] [PubMed]

35. Kanamala, M.; Palmer, B.D.; Jamieson, S.M.; Wilson, W.R.; Wu, Z. Dual PH-sensitive liposomes with low PH-triggered sheddable PEG for enhanced tumor-targeted drug delivery. *Nanomedicine* **2019**, *14*, 1971–1989. [CrossRef] [PubMed]
36. Han, X.; Li, Y.; Xu, Y.; Zhao, X.; Zhang, Y.; Yang, X.; Wang, Y.; Zhao, R.; Anderson, G.J.; Zhao, Y.; et al. Reversal of pancreatic desmoplasia by re-educating stellate cells with a tumour microenvironment-activated nanosystem. *Nat. Commun.* **2018**, *9*, 3390. [CrossRef] [PubMed]
37. Zhu, Y.; Chen, C.; Cao, Z.; Shen, S.; Li, L.; Li, D.; Wang, J.; Yang, X. On-demand PEGylation and DePEGylation of PLA-Based nanocarriers via amphiphilic MPEG-*TK*-Ce6 for nanoenabled cancer chemotherapy. *Theranostics* **2019**, *9*, 8312–8320. [CrossRef]
38. Fan, G.; Fan, M.; Wang, Q.; Jiang, J.; Wan, Y.; Gong, T.; Zhang, Z.; Sun, X. Bio-inspired polymer envelopes around adenoviral vectors to reduce immunogenicity and improve in vivo kinetics. *Acta Biomater.* **2016**, *30*, 94–105. [CrossRef]
39. Li, G.; Song, Y.; Huang, Z.; Chen, K.; Chen, D.; Deng, Y. Novel, nano-sized, liposome-encapsulated polyamidoamine dendrimer derivatives facilitate tumour targeting by overcoming the polyethylene glycol dilemma and integrin saturation obstacle. *J. Drug Target.* **2017**, *25*, 734–746. [CrossRef]
40. Hou, M.; Wu, X.; Zhao, Z.; Deng, Q.; Chen, Y.; Yin, L. Endothelial cell-targeting, ROS-Ultrasensitive Drug/SiRNA co-delivery nanocomplexes mitigate early-stage neutrophil recruitment for the anti-inflammatory treatment of myocardial ischemia reperfusion injury. *Acta Biomater.* **2022**, *143*, 344–355. [CrossRef]
41. Shuai, Q.; Cai, Y.; Zhao, G.; Sun, X. Cell-penetrating peptide modified PEG-PLA micelles for efficient PTX Delivery. *Int. J. Mol. Sci.* **2020**, *21*, 1856. [CrossRef] [PubMed]
42. Lv, J.; Chang, H.; Wang, Y.; Wang, M.; Xiao, J.; Zhang, Q.; Cheng, Y. Fluorination on polyethylenimine allows efficient 2D and 3D cell culture gene delivery. *J. Mater. Chem. B* **2015**, *3*, 642–650. [CrossRef] [PubMed]
43. Wang, H.; Wang, Y.; Wang, Y.; Hu, J.; Li, T.; Liu, H.; Zhang, Q.; Cheng, Y. Self-assembled fluorodendrimers combine the features of lipid and polymeric vectors in gene delivery. *Angew. Chem. Int. Ed.* **2015**, *54*, 11647–11651. [CrossRef] [PubMed]
44. Wang, M.; Liu, H.; Li, L.; Cheng, Y. A fluorinated dendrimer achieves excellent gene transfection efficacy at extremely low nitrogen to phosphorus ratios. *Nat. Commun.* **2014**, *5*, 3053. [CrossRef] [PubMed]
45. Chen, G.; Wang, Y.; Ullah, A.; Huai, Y.; Xu, Y. The effects of fluoroalkyl chain length and density on sirna delivery of bioreducible poly(amido amine)s. *Eur. J. Pharm. Sci.* **2020**, *152*, 105433. [CrossRef] [PubMed]
46. Zhang, L.; Li, Y.; Yu, J.C.; Chen, Y.Y.; Chan, K.M. Assembly of polyethylenimine-functionalized iron oxide nanoparticles as agents for DNA transfection with magnetofection technique. *J. Mater. Chem. B* **2014**, *2*, 7936–7944. [CrossRef]
47. Obaidat, I.; Issa, B.; Haik, Y. Magnetic properties of magnetic nanoparticles for efficient hyperthermia. *Nanomaterials* **2015**, *5*, 63–89. [CrossRef]
48. Sun, S.; Zeng, H.; Robinson, D.B.; Raoux, S.; Rice, P.M.; Wang, S.X.; Li, G. Monodisperse MFe_2O_4 (M = Fe, Co, Mn) Nanoparticles. *J. Am. Chem. Soc.* **2004**, *126*, 273–279. [CrossRef]
49. Ghosh, R.; Pradhan, L.; Devi, Y.P.; Meena, S.S.; Tewari, R.; Kumar, A.; Sharma, S.; Gajbhiye, N.S.; Vatsa, R.K.; Pandey, B.N.; et al. Induction Heating Studies of Fe_3O_4 magnetic nanoparticles capped with oleic acid and polyethylene glycol for hyperthermia. *J. Mater. Chem.* **2011**, *21*, 13388. [CrossRef]
50. Wan, Q.; Xie, L.; Gao, L.; Wang, Z.; Nan, X.; Lei, H.; Long, X.; Chen, Z.-Y.; He, C.-Y.; Liu, G.; et al. Self-assembled magnetic theranostic nanoparticles for highly sensitive MRI of minicircle DNA delivery. *Nanoscale* **2013**, *5*, 744–752. [CrossRef]
51. Smith, M.C.P.; Luker, K.E.; Garbow, J.R.; Prior, J.L.; Jackson, E.; Piwnica-Worms, D.; Luker, G.D. CXCR4 regulates growth of both primary and metastatic breast cancer. *Cancer Res.* **2004**, *64*, 8604–8612. [CrossRef] [PubMed]
52. Wang, Z.; Ma, Y.; Yu, X.; Niu, Q.; Han, Z.; Wang, H.; Li, T.; Fu, D.; Achilefu, S.; Qian, Z.; et al. Targeting CXCR4-CXCL12 Axis for visualizing, predicting, and inhibiting breast cancer metastasis with theranostic AMD3100-Ag_2S Quantum Dot Probe. *Adv. Funct. Mater.* **2018**, *28*, 1800732. [CrossRef]
53. Zhang, F.; Gong, S.; Wu, J.; Li, H.; Oupicky, D.; Sun, M. CXCR4-targeted and redox responsive dextrin nanogel for metastatic breast cancer therapy. *Biomacromolecules* **2017**, *18*, 1793–1802. [CrossRef] [PubMed]
54. Li, H.; Zhang, X.; Wu, H.Y.; Sun, L.; Ma, Y.; Xu, J.; Lin, Q.; Zeng, D. ^{64}Cu-labeled ubiquitin for PET imaging of CXCR4 expression in mouse breast tumor. *ACS Omega* **2019**, *4*, 12432–12437. [CrossRef] [PubMed]
55. Mikaeili, A.; Erfani, M.; Goudarzi, M.; Sabzevari, O. Breast tumor targeting in mice bearing 4T1 tumor with labeled CXCR4 antagonist analogue. *Int. J. Peptide Res. Ther.* **2021**, *27*, 2449–2457. [CrossRef]

nanomaterials

Review

Nanoparticles for Coronavirus Control

Maryam Kianpour [1], Mohsen Akbarian [2,*] and Vladimir N. Uversky [3,4,*]

1 Institute of Biomedical Sciences, National Sun Yat-sen University, Kaohsiung 804, Taiwan; maryam.kianpour89@gmail.com
2 Department of Chemistry, National Cheng Kung University, Tainan 701, Taiwan
3 Department of Molecular Medicine and Health Byrd Alzheimer's Institute, Morsani College of Medicine, University of South Florida, Tampa, FL 33612, USA
4 Laboratory of New Methods in Biology, Institute for Biological Instrumentation of the Russian Academy of Sciences, Federal Research Center "Pushchino Scientific Center for Biological Research of the Russian Academy of Sciences", 142290 Pushchino, Moscow Region, Russia
* Correspondence: makbarian@gs.ncku.edu.tw (M.A.); vuversky@usf.edu (V.N.U.)

Abstract: More than 2 years have passed since the SARS-CoV-2 outbreak began, and many challenges that existed at the beginning of this pandemic have been solved. Some countries have been able to overcome this global challenge by relying on vaccines against the virus, and vaccination has begun in many countries. Many of the proposed vaccines have nanoparticles as carriers, and there are different nano-based diagnostic approaches for rapid detection of the virus. In this review article, we briefly examine the biology of SARS-CoV-2, including the structure of the virus and what makes it pathogenic, as well as describe biotechnological methods of vaccine production, and types of the available and published nano-based ideas for overcoming the virus pandemic. Among these issues, various physical and chemical properties of nanoparticles are discussed to evaluate the optimal conditions for the production of the nano-mediated vaccines. At the end, challenges facing the international community and biotechnological answers for future viral attacks are reviewed.

Keywords: coronaviruses; vaccines; nanoparticles; diagnostic nanoparticles

Citation: Kianpour, M.; Akbarian, M.; Uversky, V.N. Nanoparticles for Coronavirus Control. *Nanomaterials* **2022**, *12*, 1602. https://doi.org/10.3390/nano12091602

Academic Editors: Goran Kaluđerović and Nebojša Pantelić

Received: 23 March 2022
Accepted: 6 May 2022
Published: 9 May 2022

1. Introduction

Human life has always been in danger of various sources of harm. Millennia ago, our ancestors thought that all of their problems were limited to finding enough food and staying safe from predators. Later, with the advancement of science, more civilized humans were able to partially overcome these long-standing challenges but were unaware that new problems would arise [1]. During the entire history of humankind, we have used our knowledge to raise the quality of life. However, the mortality rates caused by numerous maladies, such as neurodegeneration, contagions (including various viral infections), cardiovascular disorders, diabetes, and different forms of cancer remain high. Although some of the current health issues, such as cancer, diabetes, cardiovascular diseases, and neurodegeneration, are associated with the modern lifestyle and can potentially be reduced by changing habits, exposure to dangerous microorganisms and viruses clearly cannot be easily ignored. Furthermore, with its ever-increasing connectivity among countries and continents, the modern world is clearly facing global risks of fast spread of dangerous infections. These days, many countries in the world are struggling with the second and third waves of the coronavirus (CoV) disease 2019 (COVID-19) caused by severe acute respiratory syndrome coronavirus-2 (SARS-CoV-2) infection. Every day, thousands of people around the world fall victim to this virus. Due to COVID-19, the economies of countries are worsening daily, many businesses and industries have gone bankrupt, and millions of people have lost their jobs [2–5]. In this sense, the global SARS-CoV-2 pandemic filled the 21st century with serious challenges.

In the last two decades, the human race has been repeatedly attacked by human coronaviruses (HCoVs). In 2002, the severe acute respiratory syndrome CoV (SARS-CoV) claimed 8096 lives, whereas 2494 people succumbed to the Middle East respiratory syndrome CoV (MERS-CoV) in 2012, and finally the uninvited guest of 2019–2022, SARS-CoV-2, is still collecting its toll. In comparison with the rather moderate incidences of SARS-CoV and MERS-CoV, the occurrence of the SARS-CoV-2 is significantly higher [6–9]. As of June 2021, 176,482,998 people were reported to be infected with SARS-CoV-2, of which 3,812,194 died [8]. According to the Worldometer (https://www.worldometers.info/coronavirus/), as of 23 April 2022, the number of SARS-CoV-2 infected individuals climbed to 509,048,132, and 6,241,704 COVID-19 patients passed away.

HCoVs, which were discovered in 1960, belong to the orthocoronavirinae subfamily belonging to the coronaviridae family. There are seven HCoVs, which, in addition to the aforementioned SARS-CoV, SARS-CoV-2, and MERS-CoV, include HCoV-229E, HCoV-NL63, HCoV-OC43, and HCoV-HKU1 types that usually cause mild-to-moderate upper-respiratory tract illnesses, with symptoms similar to the common cold. HCoV-229E and HCoV-NL63 are alpha CoVs, whereas SARS-CoV, SARS-CoV-2, MERS-CoV, and the lesser-known types of HCoV-OC43 and HCoV- HKU1 all belong to the beta CoV genus [10]. This genus in particular has a high potential for infecting humans. All CoVs are capsid-coated viruses, usually spherical, and contain a single-stranded RNA genome with a length of 32–37 kb [11–13].

With a deeper look at health problems and by studying modern ways to deal with them, we can boldly say that the obvious human hope to combat most of these diseases is the use of proteins and vaccines [14–16]. In the last three decades, with the advent of biotechnology and nanotechnology for the recombinant production of human proteins in different hosts, diverse medicinal proteins have found their way into the field of treatment for the management of many diseases. To date, nearly 400 types of these biologics have been introduced [17,18]. These proteins could potentially reach the market faster than chemically synthesized drugs due to better predictability of their behavior in the living environment, as well as their lower toxicity [19,20]. In this study, with the help of recent data, we first describe the structure of SARS-CoV-2. In the next section, we look at how the virus causes pathogenesis and the body's immune response to this infection. Useful information on the use of nanoparticles in the production of the coronavirus vaccines such as DNA vaccine, RNA vaccine, viral vector vaccine, and adenovirus-vector vaccine will also be provided. The use of monoclonal antibodies in passive immunization/adaptive immunity and the 2022 status of the virus vaccines will be summarized. Finally, the challenges in the field of vaccination, problems related to vaccines, and solutions to overcome them will be presented in the form of a list of suggestions. Also this paper will describe the utilization of various types of polymer nanoparticles, metal nanoparticles, and peptide nanoparticles for the detection and suppression of coronavirus infection. Properties of these nanoparticles, such as particle size, surface charge, particle shape, and hydrophobicity and hydrophilicity will also be examined in the types of immunological responses nanoparticles may generate. The use of nanoparticles in SARS-CoV-2 diagnostic methods and the strategies used will also be reviewed.

2. SARS-CoV-2 Structure and Infection Mechanism

Knowing the structural features and life cycle peculiarities of the virus will enable researchers to suggest solutions to deal with the virus outbreak. With their genomes approaching 30 kb in length, CoVs are among the largest known RNA viruses. SARS-CoV-2 is an enveloped positive-strand single-strand RNA virus (+ssRNA virus), whose genomic ssRNA is condensed by the nucleocapsid (N) proteins at the center of the viral particle. The size of the viral particle of SARS-CoV-2 can be up to 100 nm [21]. The outermost layer of the viral particle is made of a phospholipid membrane similar to mammalian cells, which contains three types of viral proteins. These proteins include membrane (M) protein in high abundance, coating proteins (envelop protein, E protein) in relatively low abundance, and most importantly, spike protein (S protein) [22,23]. S protein is a trimeric glycoprotein, a monomer that has a total length of 1273 amino acids. There is a 76% sequence similarity between the S proteins of SARS-CoV and SARS-CoV-2, and they are glycosylated at 21 to 35 sites, respectively.

The S protein consists of two main parts called S1 and S2 subunits. The S1 subunit has a segment that detects mammalian cellular receptors (receptor-binding domain, RBD) and is responsible for binding the viral particle to the host cell, while the S2 subunit has trans-membrane domains that hold the protein like a rod into the viral membrane [21,24]. Thanks to the structural studies of this protein, it was observed that as soon as S protein binds to receptors on the surface of the host cell, the open structure of this protein becomes closed. This structural switch ensures a strong attachment between the virus and the host cell.

M and E proteins play structural roles for viral particles [25–27]. Figure 1 illustrates the overall view of a CoV particle. Attachment to the surface of the host cell followed by cell entry is among the most important aspects of the life cycle [28]. A correct and in-depth understanding of this stage of viral infection can help in designing drugs to prevent the virus entrance into the host cell. Although this stage of the virus' life cycle has not yet been completely uncovered, significant progress has been made in this field. In this article, a detailed description of the stages of viral infection will be presented, and the mechanisms that have been proposed to fight the virus at each step will be discussed.

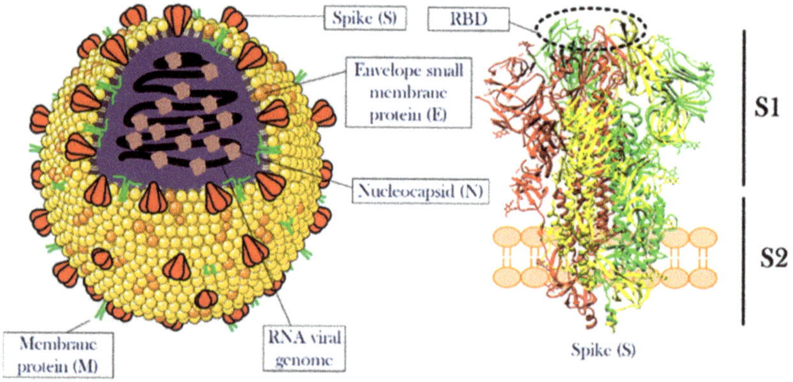

Figure 1. The overall structure of CoV and the spike protein. The left side illustrates a single particle of CoV. Three membrane proteins and RNA viral genome in a complex with nucleocapsid proteins were shown in the scheme. The right side shows the detailed structure of a spike protein. This protein, which is the most important functional protein during the attachment of the viral particle to the host cell, has two subunits named S1 and S2. The receptor-binding domain is located at the top of S1. For more information see the text. The structure of the spike protein was extracted from PDB 6ZGI [29]. RBD is receptor-binding domain.

To design a safe drug, one has to understand the effects of the target virus on the body. Figure 2 presents details of how the body responds to the virus and shows that the virus first enters the lungs through the respiratory system, and then the immune system in the alveoli would be used as the first level of defense against the virus.

To that end, immune cells that have entered through blood vessels into the alveoli, secrete a variety of cytokines in contact with the invading virus [30]. This response, by itself, triggers more immune cells that would generate more cytokines, resulting in a cytokine storm [31]. Among the most important cytokines secreted at this stage are the interleukins 6 and 1 (IL-6 and IL-1), as well as α-interferon (INF-α). The total secretion of cytokines causes the overproduction and exudation of fibrin in the alveoli, which, by disruption of cell junction, can eventually increase the flow of blood fluid through the capillaries to the lung chamber, resulting in pulmonary edema, focal hemorrhage, and pulmonary consolidation [32–35]. All of this occurs due to the body's intense inflammatory responses against the invading virus [36].

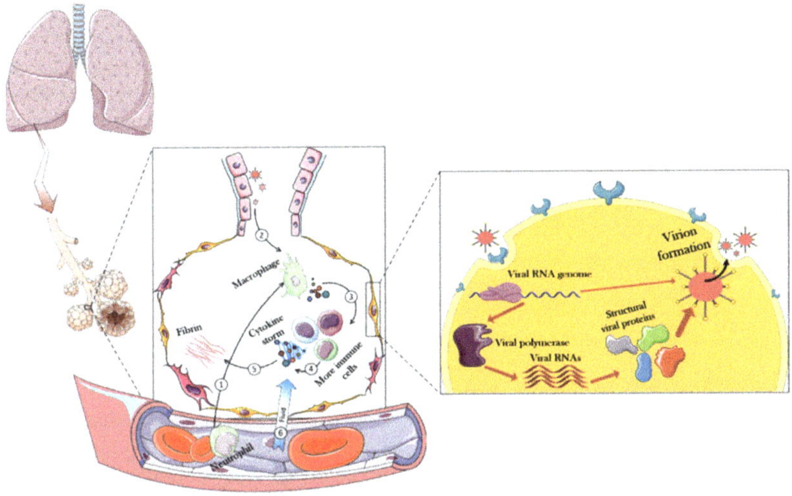

Figure 2. Immune responses and overall intracellular events triggered in the face of CoV-2. In normal conditions, immune cells enter the alveoli of the lungs through the blood. When CoV arrives in the alveolar compartment, these immune cells oppose them and eventually produce cytokines. Positively regulating, cytokines trigger more immune cells, eventually producing more cytokines. During these events, the fibrin of the alveoli increases, resulting in partial destruction and increased permeability of the alveoli. Consequently, fluid goes on the battlefield (alveoli) through the capillaries, causing destruction and edema of the infected lung. However, when a virus can compete with these challenges, it will be able to enter the host cell. In the host cell, the viral genome is generally released and the host equipment is used to replicate the viral particles. The following sections of the article provide more details in this regard.

Upon closer inspection of the infected cell, the first step in the virus entering the host cell is the interaction between the S protein (RBD of the S1 subunit) and the angiotensin-converting enzyme 2 (ACE2) receptor at the surface of the host cell [37–39]. The analog for this receptor in the MERS-CoV infection mechanism is dipeptidyl peptidase-4 (DPP4) receptor [40,41]. The angiotensin converting-enzyme is an important membrane protein that is abundantly expressed on the surface of cells of various human and animal tissues. The tissues with the highest expression levels of ACE2 are the lungs, gastrointestinal tract, blood vessels, kidneys, liver, and heart. The history of recognizing the importance of this protein dates back to three decades ago [42].

During these years, scientists speculated on many functions for this protein, including its crucial role in the renin–angiotensin–aldosterone system (RAAS) pathway regulating blood pressure, wound healing, and inflammation. Here, ACE2 modulates activities of angiotensin II (ANG II), a protein that increases blood pressure, body water and sodium content, and inflammation, as well as increases damage to blood vessel linings and various types of tissue injury, by converting ANG II to other molecules that counteract the effects of ANG II. Concerning the molecular mechanisms of CoV-2 infection, ACE2 serves as a major receptor controlling the main route of SARS-CoV-2 entry to the host cells [43].

The importance of this receptor is further emphasized by finding a close relationship between COVID-19 vulnerability and gastrointestinal symptoms due to high expression levels of the ACE2 receptor at the gastrointestinal epithelial cells [44,45]. However, some researchers have suggested that SARS-CoV-2 may also enter the host cell through interaction with another receptor, the CD147 protein [46], although this claim has been questioned in more recent studies [47]. This receptor is also expressed on the surface of many tissues and is responsible for changing the shape of the matrix in some phenomena, such as cancer, inflammation, and wound healing [48]. It is believed that the reason for the different responses of different patients to viral infection is a difference in the expression level of these receptors at the surface of their cells. With all this, as soon as the interaction between ACE2 and S protein has taken place, the viral particle enters the host cell by endocytosis, and the virus genome is released [49]. However, one should take a closer look at the roles of another protein in this process. The binding of S protein to ACE2 is not sufficient for cell entry, and the internalization of the virus particle to the host cell is activated by the specific cleavage of the ACE2-bound S protein by a transmembrane serine protease protein-2 (TMPSS2) [50,51]. As another serendipity of the human body, this protein has been closely linked to many health problems. For example, the crucial roles of TMPSS2 in the metastasis and progression of prostate cancer are well-known [52].

After entering the cell, due to the suitable conditions for viral protein translation, the host translational machinery starts producing viral RNA-dependent RNA polymerase, and this event marks the beginning of a cascade of intracellular processes in favor of the invading virus. After the viral polymerase is translated, a variety of viral RNA genes is produced, resulting in the expression of structural and functional proteins needed to form a complete viral particle inside the host cell [53,54]. Consequently, at the last stage of viral infection, whole viral particles are expelled by the explosion of the host cell or exocytosis, infecting the surrounding healthy cells. All of these pathways have been used to design strategies to combat the SARS-CoV-2 infection [55–57]. Although no promising and definitive approaches to combat CoVs infection and COVID-9 progression have been reported so far, in this article, we intend to introduce all the proposed strategies based on using the biological materials, or the combination of these materials with non-biological materials, to limit virus progression.

3. Our New Comrades-in-Arms: Nanomaterials and the Development of CoV-2 Vaccines

Nanomaterials have been used in various fields for decades. This group of materials in general, with their very large surface-to-volume ratios, the ability to place molecules with different properties on their surface (functionalization), and simplicity in their production process have played a major role in advancing human knowledge and increasing quality of life [58–60]. Nonetheless, although some scientists have discovered new usages for the substances, others pointed to their side effects. Thereby, this type of material should be considered a double-edged sword. In this section, different applications of nanomaterials in the inhibition of the virus life cycle are presented.

3.1. Different Classes of Nanomaterials against Coronavirus Disease

This ongoing pandemic, with all its bitterness, proved that the harmony between the sciences could be a great help in achieving human goals. For example, it was observed that the use of various nanoparticles along with molecular biology eventually introduced vaccines into the consumer markets, which turned the dark days of the epidemic into a beacon of hope. Scientists in the nanotechnology field have followed various strategies, each with a vision to generate means to fight the virus. Some nanoparticles have been used to directly fight the virus [61], while others have been used for the rapid detection of viruses in laboratory samples [62]. Additionally, some nanocarriers are used to deliver anti-SARS-Cov-2 drugs/vaccines [63]. In the subsequent sections of this study, the types of these nanoparticles will be discussed. In general, these nanoparticles are classified into groups based on their chemical nature, such as polymer nanoparticles, metal nanoparticles, and peptide nanoparticles. Each of these groups can be subdivided into more detailed subgroups on their own, which will be explained further.

To deal with global epidemics, careful review and utilization of all available tools/means are important. In this regard, the use of nanotechnology as a new field in medical sciences and its multifunctional structures can be a solution. Nanotechnology can be used for a variety of medical purposes, such as clinical diagnosis, pharmaceutical research, immune system activation, and the extraction of the biological materials. To defeat COVID-19, better understanding of the virus, better diagnosis of infection, its treatment and prevention are steps in which nanotechnology is expected to help [64].

Polymer nanoparticles have found their place in many industrial and medical fields. These substances, especially in regenerative medicine, have been able to give much hope to the patient community to recover from tissue degeneration diseases [65]. Having a high level of safety, biodegradability, simplicity of synthesis, and the ability to control their properties through different functionalizations initiated strong attention to nanoparticles in the field of anti-SARS-CoV-2 research [61]. Another positive point for this group of nanoparticles is that some of them have been approved by the Food and Drug Administration (FDA) and have been studied in great detail in other fields [66]. By selecting this group of nanoparticles, the distance to reach the final goal may not be as long as the one needed for the development of novel and completely unknown nanoparticles.

Certainly, the synthesis methods of these nanoparticles, which directly affect their properties, will also indirectly determine the effectiveness of nanoparticles in combating the new coronavirus. The sizes of nanoparticles, which are generally between 1 and 100 nanometers, guarantee a very high surface to volume ratio, as well as the ability to load significant amounts of drugs in small amounts of nanoparticles, which can help a lot in combating pathogens, including SARS-CoV-2 virus [67].

3.2. Antiviral Mechanism of Nanoparticles

The antiviral mechanisms of nanoparticles include inhibiting the binding of the virus to the target cell, preventing the virus from entering into the host cell, and attacking the growth and proliferation stage of the virus. Possible mechanisms of nanoparticles include direct and indirect inactivation of viruses. These mechanisms vary depending on the three-dimensional shape and type of nanoparticles [68,69]. Another mechanism proposed for the antiviral action of nanoparticles is the local field action of nanoparticles. In this way, the designed nanoparticles change the membrane potential at the surface of the host cell as soon as they are adsorbed on the cell surface. Following this, membrane potential change and the penetration of the virus into the host cell are affected and reduced [70].

Other studies have suggested that metal nanoparticles, such as those containing silver ions with oxidizing properties in infected host cells, can prevent the virus from spreading to the healthy cells [71]. According to an in silico study, iron nanoparticles have also been shown to form a stable complex with the CoV spike protein and prevent the virus from attaching to the host cell [72].

Table 1 lists some of the studies that used nanoparticles in fighting respiratory viral diseases [73].

Table 1. Previously used different classes of nanoparticles in respiratory viral diseases.

Compound	Virus	Antigen	Adjuvant	NP Size (Diameter, nm)	Outcome
Polyanhydride	RSV	G and F glycoproteins	-	200–800	The replication of virus was suppressed in infected mice
HPMA/NIPAM	RSV	F protein	TLR-7/8 agonist	12–25	By having significant antigenicity, TH1 isotype anti-RSV F antibodies was produce in the blood.
Chitosan	IF(H1N1)	IF(H1N1)	Heat shock proteins	200–250	After administration, the nanosystem produced antibody and induced T cell immunity.

Table 1. *Cont.*

Compound	Virus	Antigen	Adjuvant	NP Size (Diameter, nm)	Outcome
PLGA	BPI3V	BPI3V proteins	-	225.4	The infected pigs had low virus penetration (loading) in their lungs.
Gold	IF	Antigen M2e	CpG	12	Full protection of vaccinated mice against the virus by the increasing M2e-specific IgG in serum.
Q11 peptide	IF(H1N1)	Antigen M2e	-	15–100	Protection against homologous challenge of IF PR8 H1N1 and heterologous challenge of avian IF H7N9.
Viral-like particle	RSV	M1 protein of IF and RSV-F or -G	MPL and trehalose 6,6 dimycolate	10–1000	Induction the memory of T cell responses.

Abbreviations: RSV, respiratory syncytial virus; TLR, toll-like receptor; TH1, T helper type 1; PLGA, Poly(D,L-lactide-*co*-glycolide).

3.3. Properties of Nanoparticles for Efficient Vaccine Production

Vaccines have shown great potential for use in the prevention and treatment of infectious diseases. With the rapid development of biotechnology and materials science, nanomaterials have found an essential place in the formulation of new vaccines, as they can enhance the effect of antigens by acting as a release system and/or as an immune-boosting aid [74]. The analysis of the effects of nanoparticles on vaccine properties shows the improvement of the antigen stability and immunogenicity as well as a capability of targeted and controlled release of active substances [74]. However, there are still obstacles in this field due to the lack of fundamental knowledge on how nanoparticles act at the molecular scale, and what the biological effects of nanoparticles in living organisms are [75]. Nanoparticle-based vaccines are classified based on the function of the nanoparticles in them as a release system and immune response enhancer. Therefore, a fundamental understanding of the distribution of nanoparticles in the body and their fate will accelerate the logical design of new nanoparticles that will change the future of vaccines. Nanotechnology has provided the opportunity to design different nanoparticles in terms of composition, size, shape, and surface properties for various pharmaceutical applications [76].

Nanoparticles with the same size as cellular components can show biophysical function and biotherapy similar to their biological counterparts. There are several systematic studies which showed that the nanoparticles designed with polyethylene glycol (PEG) are able to delay the clearance of the drug from the body and thus make the systematic circulation of the drug in the body longer than in the free drug state [77]. This can eventually be useful for the accumulation of more drug at the site of treatment. In addition, nanoparticle delivery systems can have several salient features, including high drug loading capacity, controlled release rate, and reduced drug toxicity in the body [78]. As a result, nanoparticle-based approaches as release systems provide new opportunities to enhance innate immune activation and induce a strong immune response to the slightest toxicity [75]. The most important components of an effective vaccine include an antigen to activate the immune system, an enhancer of the immune response to stimulate the innate immune system, and a release system to ensure proper antigen delivery and targeting [79]. To achieve these goals, the design of nanoparticles focuses on the chemical composition, size, surface charge, and surface properties of the nanoparticles, as these are used to control the distribution of these particles in the environment, the release of antigen, the efficiency of immune stimulation, and the final immune response [80]. Emulsions, liposomes, and synthetic polymers are nanoparticles that serve as helpers for the proper release of immune response enhancers. Antigen-carrying nanoparticles are able to affect the immune response and significantly enhance the T-cell cytotoxic response against the antigen fused to the nanoparticles [81]. This is due to the specialized ability of some antigen-presenting cells (APCs), which can effectively absorb foreign particles, such as microparticles and bacteria [82]. This process is performed by detecting antigenic material to analyze and express foreign antigens to other cells in the immune system. However, there are limitations to the utilization of these approaches, such as the existence of the nonspecific uptake and immunosuppression activities of these compounds.

As mentioned, controlling the size, shape, and chemical properties of nanoparticles enables these tiny particles to have a controllable cell uptake coefficient. In order to provide organized information for efficient vaccine production, Figure 3 and the discussion below indicate the properties that nanoparticles must have to be considered [80].

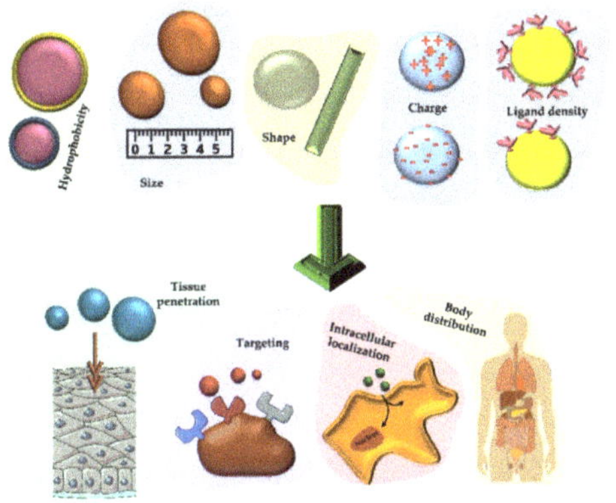

Figure 3. Various chemical and physical properties that can determine how nanoparticles will act as vaccine carriers. The size, hydrophobicity, charge, shape, and ligand density of the nanoparticles that carry the therapeutic factor will have a significant effect on their cellular uptake, distribution in the body, accumulation in the cell (especially phagocytic cells) and the rate of tissue penetration.

As shown in Figure 3, there are many criteria to consider when designing a nanoparticle-based vaccine. Nanoparticles with the size of less than ten nanometers are usually reported to be excreted by the kidneys, while larger particles are excreted by the liver [83]. Particles larger than 200 nm have also been shown to be able to be repelled by the spleen as long as they have good flexibility [84]. Therefore, in addition to the items listed in Figure 3, the flexibility of the nanomaterials is another important factor [85]. It has been seen that the shape of nanoparticles can also affect their cellular uptake and ultimately affect immune response they trigger [86]. Traditionally, the nanoparticles for formulations were considered spherical, but currently, different types of nanoparticles, such as disks, rods, prisms, and stars are being designed and studied [87]. It was reported that even the symmetry of nanomaterials is important for effective tissue distribution and cellular uptake, which can be due to the amount of reactions that occur at different levels [88].

It is also possible to generate positive and negative charges with different densities through chemical modifications applied to the surface of the nanoparticles. By applying this feature, the interaction of materials with targets is driven by electrostatic forces [80]. In many studies, the effects of different charges on immune responses were investigated. For example, positively charged hydrogel nanoparticles (modified with antigens) have been shown to stimulate antibody production, T-cell activation, and class II MHC expression [89]. The same effect was observed with hydrophobicity in mesoporous silica nanoparticles that affect the expression of CD3, 4 and 8 [90]. In terms of tissue penetration, it has also been observed that positively charged nanoparticles are capable of penetrating the skin 2–4 times more efficiently than their negatively charged counterparts [91].

In addition to the features reviewed in previous sections, the ligand density at the surface of nanoparticles also has a significant effect on the immunological response they generate. These differences in response may be due to the differences in cellular uptake. For example, it has been shown that the amino content of silica mesoporous particles significantly reduces the cytotoxicity of these particles, while PEGylation is effective in increasing the hydrophilicity degree and ultimately leads to an increase in the renal filtration rate [92,93].

3.4. Different Nanoparticle-Based Vaccines for CoVs

We focus here on the classification of nanoparticles that have been designed to fight CoVs. From a structural viewpoint, four types of nanoparticles have been introduced for this purpose (Figure 4). The first group of these nanoparticles originates from proteins that can self-assemble; e.g., viral proteins that can aggregate into virus-like particles or form protein micelles [94–96]. Similar to a virus particle, there is another group of nanoparticles that are liposomes along with the capsid proteins. Liposomes themselves can also be considered as nanoparticles capable of carrying therapeutic agents inside their body against CoV. Finally, exosomes are another group of nanoparticles that are very similar to viruses, except that these particles are usually produced by exocytosis from virus-infected cells [97,98].

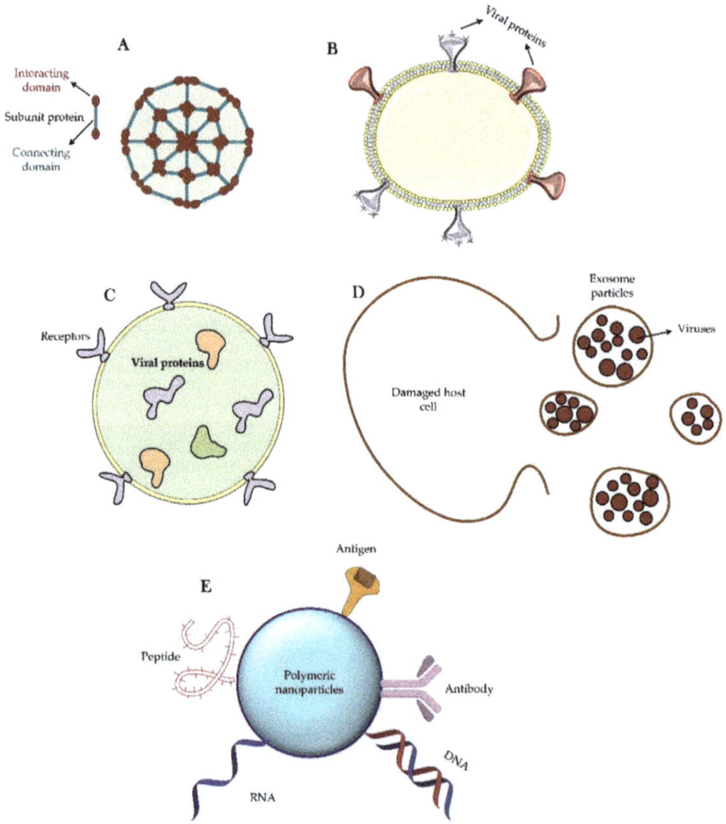

Figure 4. Different classes of nanoparticles are used as virus vaccines. (**A**) Self-assembling capsid protein nanoparticle. This type of nanoparticle is made up entirely of proteins that are able to self-aggregate. Sometimes two or three types of proteins are used to make this nanoparticle. (**B**) Virus-

like particle. Particle engineering has created the ability to design and synthesize a virus-like particle that is an assembly of a phospholipid and a set of viral proteins. (**C**) Liposome. Liposomes are free of any viral proteins on their surface. Sometimes they may have receptors for the correct targeting of the particle, but it should be noted that viral proteins are trapped inside the liposome structure and enter the immunological pathways into the host cell after the endocytosis of the particle. (**D**) Exosome particle. Once the host cell is infected with the virus, an exosome will emerge from the damaged cell that contain the newly synthesized viruses. These particles, after extraction and purification, can be suitable treatment options. (**E**) Corresponds many available nano-based polymeric materials that functionalized with different therapeutic agents such as DNA, RNA, antigens, peptides, and antibodies.

Today, perhaps the most important application of nanoparticles due to the global CoV-2 challenge is the use of these materials to load and transport viral antigens and viral DNA or RNA genomes [99,100]. In the meantime, physical and chemical interactions, such as adsorption, entrapment, and attachment have been used to load viral materials into the nanoparticles. For this important purpose, a variety of nanoparticles, such as nanopolymers [101], liposomes [102,103], and quantum dots [104,105] have been used [100]. Between the years 2014–2018, scientists used protein micelles consisting of the S protein of SARS-CoV-1 and MERS-CoV to fight these viruses [106,107]. Of these, some remain in the early clinical stages, but recently, this method has also been used to deal with the SARS-CoV-2 infection.

The vaccine-related virus-mimicking nanoparticles (NPs) such as self-assembled viral proteins and virus-like particles are in phase I clinical trials [97]. The advantages of using this group of nanoparticles include simplicity of their production, safety, high resistance in vitro, and virus-like body distribution. However, this type of nanoparticle also has some disadvantages, such as high production cost, the difficulty of industrialization, low stability in vivo, and occurrence of unwanted immune reactions [97]. Studies to suppress the progression of MERS-CoV [108,109] and SARS-CoV-2 [97] have been performed using virus-like particles of S proteins and RBD domains, respectively. Both of these approaches are in the early clinical phase.

Virus-like particles have very high immunogenicity and have recently been considered for their various applications in vaccination, targeted drug delivery, gene therapy, and immunotherapy. All four recombinant vaccines—Engerix, Cervarix, Recombivax HB, and Gardasil—on the market are based on highly pure virus-like particles (VLPs) [110]. However, there are several potential barriers in the development of virus-based vaccines from the research phase to the clinical phase. One of these problems is the lack of information on the folding and proper structure of these particles as compared with the parent infectious virus. Another problem with these particles is that the binding pattern of these particles is not the same as that of the parent infectious virus. Although these particles contain capsid proteins and can stimulate the body's immune system, they lack other viral components. Another problem is the complexity of the related clinical studies. However, human health has historically been more valuable than the problems and shortcomings that are considered for these vaccines [111]. Virus-like particles can be divided into two main categories based on the structure of parental viruses [112]: non-enveloped viral particles and enveloped viral particles. Non-enveloped virus-like particles typically consist of one or more pathogenic components that are self-assembled into the particles. These particles also do not contain any of the host components. This kind of VLP has been used to develop vaccines against pathogens such as HPV and RV [112]. Enveloped VLPs are relatively complex structures consisting of host cell membranes (as envelop) with target antigens on the outer surface. These types of particles provide a higher degree of flexibility for integrating most antigens from the same or different pathogens. The most prominent examples of enveloped viral-like particles are those engineered to express vaccine target antigens from influenza viruses, retroviruses, and hepatitis C virus (HCV) [113].

In the case of the liposomes, the most important challenge is the limited cargo capacity and the fast release of the cargo to the environment. However, these types of nanoparticles have several advantages, such as relative easiness of production, long-term physical stability, and high control over surface properties. These nanoparticles, in conjunction with S and R protein-encoding RNAs, have also been used to control the progression of CoVs such as SARS-CoV [114] and SARS-CoV-2 [115].

Two types of methods (mechanical and non-mechanical) are used to make liposomes and nanoliposomes. The mechanical methods include sonication, homogenization, extrusion, and microfluidization. The non-mechanical methods are reverse-phase evaporation, discharge of lipid–detergent micelle combination, freeze drying, solvent injection, thin layer dehydration, and the thermal method. The sonication method, the most common method in the production of liposomes, uses sound energy to create cavities and disperse particles [116]. The cavities created by the sound effect cause the gas bubbles in the liquid to expand and contract. As the waves increase, the bubbles begin to oscillate and eventually burst. As a result, small vesicles form. This method is used in the preparation of monolayer nanoliposomes. This method is also divided into three types of sonication with a probe, sonication in a water bath, and ultrasonication [117].

As far as exosomes are concerned, they have been used to combat the SARS-CoV virus [118]. In the study, 293 T cells were infected using plasmids containing the gene encoding the S protein. The exosomes obtained from these cells were then used as nanoparticle vaccines against the virus. The clinical research on the efficiency of this approach is in its early stages. The advantages of this method include high biocompatibility, but the high cost of production and the complex and long stages of nanoparticle development are disadvantageous. Scientists in the field have previously used this technology to design vaccines against infectious diseases, such as toxoplasmosis, HIV, influenza, etc. [109,119]. Given this evidence, it may make sense to consider using the same technique against the present virus (SARS-CoV-2).

In a recent study, a polyethylenimine nanopolymer was used to transport SARS-CoV-2 S protein antigens to stimulate immune responses. The use of these antigen-carrying nanoparticles induced immune and humoral responses in mice (production of IgG, IgA, g-interferon, and interleukin-2) [120]. Other researchers have used chitosan polymers to transport plasmids carrying the N protein gene. When these nano-carriers were delivered through the intranasal route, the production of IgG and IgA in mice increased [119]. In an interesting study by Kato et al., a nano-virus particle containing the N, S, and E proteins of MERS-CoV virus was artificially produced. Although no in vivo studies have been performed [109], the study was very important, since it demonstrated that one can fabricate artificial viral particles with the help of nanoparticles.

The most important challenge in using nanoparticle-based vaccines is their cytotoxicity and the need for adjuvants to increase the effectiveness of these vaccines [121]. In another study, using the RBD of MERS-CoV virus and ferritin nanoparticle aggregates, a 28–30 nm nano-scaffold was developed that could immunize mice against the SARS-CoV-2 for a limited time [122]. By and large, ref. [97] is useful for studying different types of nanoparticles in other aspects of dealing with viruses, such as use in hygiene and the prevention of virus pandemics.

3.5. Delivery Role of Nanoparticles: Focusing on the CoVs

Relying on human knowledge, which came from the use of DNA and RNA vaccines, the idea of using nanoparticles for vaccination was introduced in the field of COVID-19 treatment. The use of this strategy to deliver small interfering RNA (siRNA) vaccines has helped control the symptoms of several diseases, such as autoimmune and neurological diseases [123]. The fundamentals of the preparation of mRNA-1273 vaccine have been based on this idea. The vaccine comprises an RNA genome covered with a lipid-based nanoparticle envelop [124]. Nanoparticles can also be useful in the delivery of therapeutic antigens. Thanks to the development of science in this field, nanoparticles can be classified into two major types by considering whether the antigen is located inside (encapsulated antigen) or on the surface (surface-presented antigen) of the nanocarrier particle (Figure 5).

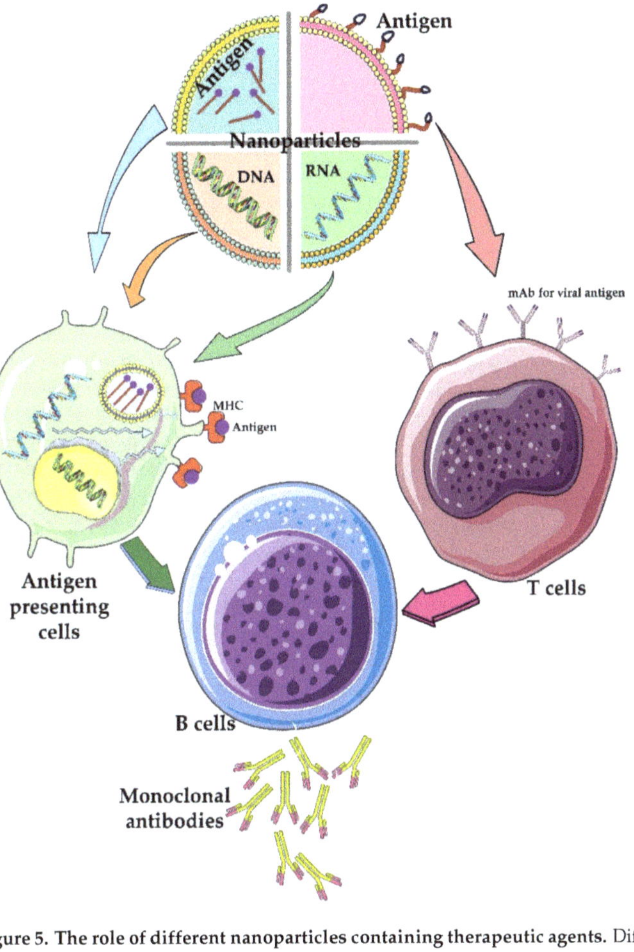

Figure 5. The role of different nanoparticles containing therapeutic agents. Different nanoparticles have been used to carry different components of viral particles such as genetic material and/or its antigens. The viral genetic materials are usually encapsulated or trapped inside the nanoparticles while the viral antigens are functionalized on the surface of the nanoparticles. Depending on which type of T-cell or antigen-presenting cells these engineered nanoparticles attach to, different immunological pathways are created in the body, which ultimately lead to the activation of B cells that produce monoclonal antibodies against the virus particle.

Different nanoparticles with different sources can be made depending on whether the therapeutic agent is to be placed on its surface or loaded into it. Nanoparticles containing the DNA genome can penetrate the host cell membrane and enter the cell and eventually arrive at the nucleus. The exogenous gene can enter transcriptional and eventually translational cycles to produce protein products that aid in the healing process. On the other hand, nanoparticles that carry antigens inside them can also enter antigen-presenting cells. After antigen processing by the cell, the antigen eventually would be present on the surface of these cells. These efficient antigens are ultimately able to invoke immune cells and activate immune responses. Alternatively, nanoparticles containing the RNA genome can enter antigen-presenting cells, enter directly into translocation cycles of the cell, and synthesize corresponded antigens. The disadvantage of such nanoparticles is their short half-life, but at the same time, they can produce relatively faster responses compared to DNA-containing nanoparticles. Another group of therapeutic nanomaterials is nanoparticles that carry antigens directly on their surface. These systems interact directly with immune cells. All of these nanoparticles will finally trigger the production of antibodies by immune cells.

RNA- and DNA-containing nanoparticles are generally placed inside nanoparticles due to the degradability of the nucleotides, but antigens can also be placed inside the nanoparticles or on the surface of these carriers [125,126]. The hope for the success of the genome-containing nanoparticles is very high. Among the group of nanoparticles that are being studied to oppose the CoV-2 are Moderna [127], Arcturus Therap [128], and CanSino [14]. There are also many cases of nanoparticles presenting a viral antigen on their surface [125]. The basis of the antigen placement at the nanoparticle surface is to mimic a viral particle. For example, by placing important antigens, such as the S protein, on the surface of a nanoparticle and injecting it into the body, an immune response can be activated against this particle without worrying about the generation of the new viral particles in the body.

3.6. Attacking the CoV-2 Life-Cycle with the Help of Nanoparticles

Luckily for us, various steps must be taken to finally create a new virus particle inside a host cell. As illustrated in Figure 6, which shows all the steps of the CoV-2 life cycle, scientists have been able to design various strategies to combat the virus by specifically attacking it at different stages (up to stage 5). To that end, according to the results obtained so far, it is clear that most success in the struggle between the therapeutic agents and the virus can be achieved before the extremely important viral enzyme, RNA-dependent RNA polymerase, starts to function. If the virus cycle cannot be controlled up to this stage, due to the speed of enzyme action and its specific activity and high turnover number of an enzyme, a significant number of viral RNAs will be formed in a short time, and the next steps will proceed at a worrying rate. After this stage, there is almost no hope for blocking the virus life cycle.

It makes sense that all efforts should be made to prevent the virus from entering human cells, but sometimes this is not the only recommended way. Up to now, two drugs have been suggested to inhibit this stage: Umifenovir and Camostat mesylate. The first drug, which is publicly licensed in China (People's Republic of China) and Russia, can prevent the virus from entering the body by binding to S proteins on the surface of the influenza viruses. In this regard, the drug has also been used as a candidate to prevent the entry of CoV-2 [129]. Camostat mesylate, with a similar mechanism, can prevent the entry of CoV-2 into the host cell [130]. Similarly, in 1994, a direct relationship between inhibition of the furin enzyme and attenuating penetration of influenza viruses to the host cell was proposed [131]. Several furin cleavage sites have been found in the S protein of SARS-CoV-2, which supports the theory of furin-dependent viral entry in the virus [132].

Given the above information, placing this protein on the surface of nanoparticles could serve as a good model for the viral binding inhibition studies. Of the parts of the virus life cycle, the entry of the viral genome into the host cell, led by S2 subunit, is an important stage [133]. As the virus nears the host cell membrane, two heptad repeat regions in S2 (HR1 and HR2) undergo structural changes to facilitate the integration of the virus and host cell membranes [134,135]. So far, many peptides have been made from these two protein regions that can mimic the behavior of these peptides and ultimately prevent the binding of the viruses to the host cells [136–138]. As an example, the synthetic peptide HR2 was injected into mice 5 h before the animals were infected with MERS-CoV, and it was observed that the viral infection in the lungs was significantly reduced as a result of such treatment [139].

However, the physical and chemical stabilities of a peptide (especially exogenous peptides) inside a cell are low [140]. Therefore, putting such therapeutic peptides in/on the nanocarrier is expected to provide more promising results. Alternatively, viral proteases can be considered good targets for attenuating the progression of viruses. It has been indicated that the *nsp3* and *nsp5* genes of CoVs encode papain-like cysteine protease (PLpro) and 3C-like serine protease (3CLpro), respectively [141]. In general, the function of these proteases plays a very important role in the process of viral genome transcription and virus duplication [142,143]. Therefore, targeting these proteases could be a key strategy to inhibit CoV-2 infection.

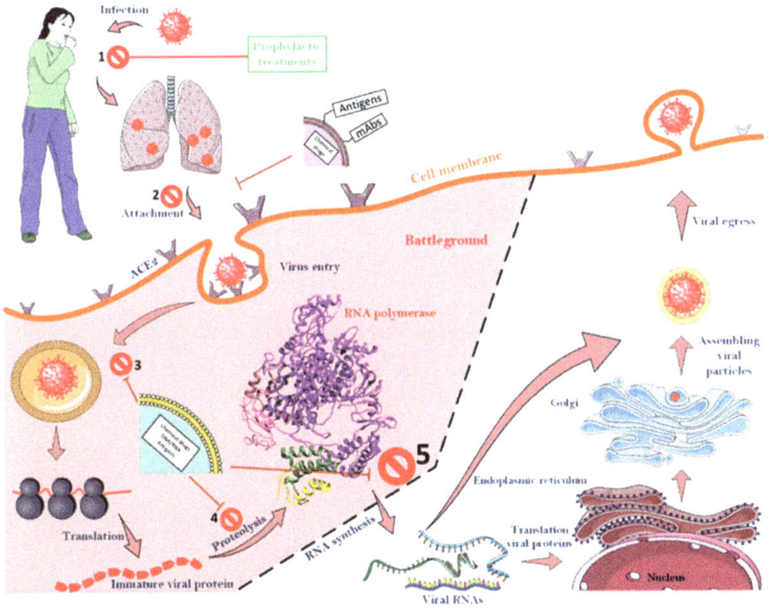

Figure 6. Different stages of a CoV-2 life cycle along with target points for fighting the virus. The CoVs can be fought almost from the beginning of the virus to the function of the RNA-dependent RNA polymerase. However, after the function of the enzyme, no specific strategy has been proposed to deal with the virus. As soon as the virus enters the body, a range of events will occur, though up to stage, 5 the virus can still be defeated. Nanoparticles that have monoclonal antibodies or antigens on their surface usually act before the virus enters the cell. The ACE2 receptor, which is among the most important cell surface receptors for the SARS-CoV-2, was a target of many studies for blocking virus cell entry. As soon as the virus binds to this receptor, the process of virus penetration into the host cell begins. Consequently, masking this agent on the surface of host cells can prevent the virus from entering the cell. Nanoparticles that encapsulated therapeutic agents will be able to fight the virus as it enters the host cell. See the text for more details.

A similar strategy has been used to combat HIV-1. Lopinavir (LPV) and ritonavir (RTV) are two important drugs that can target HIV-1 proteases. Fortunately, these drugs have been shown to target 3CLpro in both SARS-CoV-1 and MERS-CoV viruses and reduce the activity of these proteases [144,145]. The effect of these two important drugs in patients infected with SARS-CoV virus reduced viral load and viral-mediated mortality rate [146]. Also, the combined use of these two drugs in marmoset animal models has been shown to diminish viral loading and improve the general state of the body after infection [147]. However, due to the limited age of the tested animals and the different effects that the SARS-CoV virus had at various ages, the effects of these drugs could not be considered to be the same for all patients [148]. By looking at nanocarriers, their remarkable ability to cross membranes and high-performance delivery of such protease inhibitor drugs could be a promising strategy to overcome viral progression inside a host cell.

Unfortunately, not many mechanisms have been proposed to explain the inhibitory functions of nanoparticles at the cellular level, which indicates a general lack of knowledge and awareness in this area. In general, when talking about the virus outside the cell, it is possible to explore novel mechanisms by utilizing various approaches and designing new experiments. However, research opportunities are rather limited after the nanoparticles enter the cell space, and as a result, a rather restricted set of mechanisms have been explored in the many previously published studies. In the first step of the defense against SARS-CoV-2 virus, nanoparticles can create a protective "barrier" by blocking the entry of viruses utilizing designed functional groups and antibodies against viral antigens to interact with viruses outside the cell space; i.e., before their attachment to the host cell. Such nanoparticles, that are mostly nontoxic toward the host cells, disrupt the SARS-CoV-2 attachment of its receptor and thereby block the process of cell [149]. Another way to achieve this goal is by reducing the expression levels of coronavirus receptors (ACE2 and TMPRSS2) at the cell surface. Unfortunately, no cases were reported, where nanoparticles would be used to down-regulate these receptors to reduce the risk of host cell infection. Also, according to another study, nanoparticles themselves may have a destructive effect on the integrity of the viral structure, preventing the virus from entering the host cell, a result that was seen in the case of silver nanoparticles with a diameter of 2 to 15 nm [150]. However, among the metal nanoparticles studied so far, silver and copper nanoparticles are most frequently used in surface coatings to inhibit the spread of the virus in the environment, whereas gold nanoparticles have been used to combat the virus at the intracellular level [151].

One should keep in mind that since the patients usually receive high doses of the virus, the strategy based on using nanoparticles to prevent the virus from the entry to the host cells may not be able to create a very strong barrier for successful fight against viral infection. Once the virus enters the host cell, the first stage of its intracellular life is to extract the necessary information from the viral genome to synthesize the set of viral proteins. This stage could serve as another station, where nanoparticles should be able to control the function of viruses. At this level, nanoparticles coated with the antisense RNAs can be of great help [152]. Another option includes the utilization of lipid nanoparticles (LNP) as a delivery system for various highly effective small interfering RNA (siRNA) capable of targeting highly conserved regions of the SARS-CoV-2 virus [153]. One of the important considerations for this group of nanoparticles is their capability for rapid and efficient entrance into the host cells. It should also be borne in mind that the toxicity of these particles should be carefully checked to ensure that they do not cause problems entering the cell and coexisting with it for a certain period of time.

One of the attractive targets to attack the SARS-CoV-2 replication by many drugs and nanoparticles is the RNA-dependent RNA polymerase, which plays a central role in the viral infection cycle. At first glance, the fact that this target is an enzyme is very important. The enzyme with high catalytic activity is able to convert a large number of substrates into a product in a very short time (μs range, which is not precisely achieved for SARS-CoV-2 virus). In the case of SARS-CoV-2, RNA-dependent RNA polymerase catalyzes the synthesis of the viral genome from the triphosphate substrates of the host cell. Therefore, nanoparticle can be used to deliver drugs specifically affecting enzymatic activity of this important protein. One should also keep in mind that if viral polymerase cannot be e, the virus will be very difficult to control due to the high rate of synthesis of viral components [154].

The next step that can be affected and potentially controlled by nanoparticles is the efficiency of the viral protein expression. This step is usually conducted by the shared translation machineries of the host and the virus. Therefore, it can be modulated with less efficiency, and corresponding drugs could have more side effects. The final step at which nanoparticles can be expected to help is to prevent viral components from coming together and forming a new virus (packing and budding levels). However, modulation of this stage is not expected to be successful due to the involvement of cellular organelles, such as the

Golgi and the smooth endoplasmic reticulum. This could be the reason of why no articles have been published on this subject yet.

It is clear that various side effects of nanoparticles should be considered and carefully analyzed. For example, gold nanorods have been shown to inhibit the mitochondrial degradation of host cells by inhibiting caspase as soon as they enter a host cell infected with SARS-CoV-2 virus, and this intervention ultimately preserves the longevity of the infected cell [71]. A recent study showed that using siRNA/VIPER polyplexes it is possible to slow down the virus replication rate. The study showed promising results at the in vitro (air–liquid–interface) and ex vivo (human lung explant model) levels, and, most importantly, in vivo (lung epithelial cells). It was established that these siRNA/VIPER polyplexes not only were mostly non-toxic to the host cells, but were able to noticeably reduce viral replication [155]. Also, some studies, have found a close link between the pathogenicity induced by different viruses and the rate of autophagy in the host cells of infected people. Therefore, another way to fight the coronavirus is to use nanoparticles functionalized with autophagy inhibitors. Among these drugs, chloroquine and hydroxychloroquine have been able to inhibit autophagy by interfering with the connection of autophagosome to lysosome. The use of these drugs against viruses such as HIV, SARS-CoV, and Zika is also recommended [156].

4. Nano-Based Diagnostic Tests for COVID-19

Although a large fraction of the applications of nanoparticles has been in the field of drug delivery and development of SARS-CoV-2 vaccines, some researchers have also focused on the use of nanoparticles for diagnostic and treatment purposes. In the line of the diagnostic applications, the goals can be divided into two major categories: the use of biomolecular corona and colorimetric methods for corona detection. In the first case, research and efforts are hypothetical. Here, with the entry of nanoparticles into the complex and crowded living system (cell), a series of biomolecules such as proteins, polysaccharides/carbohydrates, and lipids will be in contact with the surface of nanoparticles and will form different patterns of bonding at different health conditions. Nanoparticles facing different proteins and biomolecular contents, such as those seen in patients infected with SARS-CoV-2, can form diverse microarrays of biomolecules (mainly proteins and peptides) in different states of health, and based on the analysis of the corresponding results, people with high risk to virus infection can be found [157]. Although numerous articles have already been published in the field of application of this approach in various diseases, no such research has been conducted yet for the SARS-CoV-2-related cases.

The second category of diagnostic research involves colorimetric (visible and/or non-visible) experiments with nanoparticles for the rapid and accurate detection of coronavirus in laboratory samples. In line with these considerations, gold nanoparticles are repeatedly used in the SARS-CoV-2 diagnosis, both in solution and in the form of nanochips. In a study using gold nanoparticles containing SARS-CoV-2 RNA antisense strands, a colorimetric system was developed for SARS-CoV-2 diagnostics without the need for complex instruments [158]. In this study, based on the interaction between suspended gold nanoparticles solution and viral RNA, after a period of incubation, the nano-system turns into larger aggregates, and finally, with the change in resonance of the surface of aggregated gold nanoparticles, diagnosis would be achievable [158]. In general, the physicochemical properties of nanoparticles enable them to be a tool that can produce a measurable signal once they interact with the virus (the interaction between the surface nanoparticles and part of the coronavirus such as surface proteins, viral DNA, and/or RNA). In another study that also used gold nanoparticles, the nanoparticle surface was functionalized using α, N-acetyl neuraminic acid, a specific sugar for binding to the SARS-CoV-2 S protein [159]. In this study, although virus-like particles containing surface S protein were used instead of the whole virus, and the results were promising. In detail, after the nanoparticle and virus complex is formed, it flows in the lateral flow immunoassay strip containing α,N-acetylneuraminic acid sugars. Followed by the formation of clots due to S protein re-interaction with the immobilized sugar on the immunoassay pathway, a signal to detect virus moiety will be produced. From the statistical viewpoint, it is stated that the detection sensitivity of this technique is high, and as little as 5 micrograms of S protein per milliliter can be detected [159].

Although more attention has been paid to the gold nanoparticles for the detection of SARS-CoV-2, some studies have looked at other groups of nanosystems as well. In one such study [160], poly (lactic-co-glycolic) (PLGA) nanoparticles conjugated to viral S protein were used. This system, which is very similar to the immunoassay sandwich, begins with the immobilization of antibodies against S protein in microplates. Then, S protein conjugated nanoparticles are incubated with the antibody, and finally, with the help of the peroxidation-like activity of copper nanoparticles and the external presence of hydrogen peroxide, 3.3%, 5,5′-tetramethylbenzidine immobilized in PLGA oxidases to change the color of the solution. This system is said to detect the S protein at the concentration of a femtogram per milliliter [160].

As aforementioned, the RNA content of SARS-CoV-2 has also been used for diagnostic purposes. In one of the related studies, europium-chelate nanoparticles (FNPs) conjugated to S9.6 antibodies (S9.6-FNPs) were used to detect the presence of SARS-CoV-2 using its RNA. S9.6 antibody is a protein that can bind to DNA and RNA hybrids. In this study, first, a DNA probe that could bind specifically to the viral RNA was used, and then S9.6-FNPs nanoparticles were used to bind to this hybrid. Based on the results of the analysis of samples taken from 734 patients, it was concluded that this test with its 99% specificity and 100% sensitivity could be a good option for the virus diagnosis [161]. Not only viral components but also host molecules released due to the viral entry into the body have been used to diagnose SARS-CoV-2 infection. In fact, diagnosis of immunoglobulin M (IgM) and immunoglobulin G (IgG) has been utilized as another diagnostic option using conjugated nanoparticles with anti-immunoglobulins [162,163].

Looking at the articles published so far, the strategies used in the diagnosis of SARS-CoV-2 by nanoparticles and/or fluorescently labeled nanoparticles can be summarized as shown in Figure 7 [164].

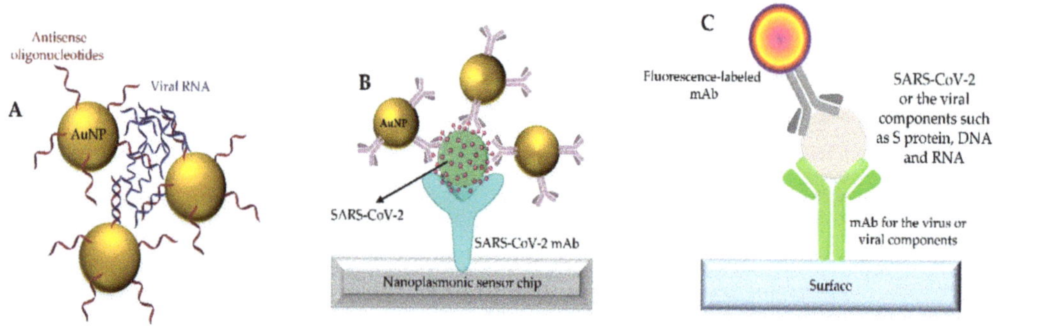

Figure 7. Different strategies used for the detection of SARS-CoV-2 by nanoparticles (**A**,**B**) and/or fluorescently labeled nanoparticles (**C**).

Although nanoparticles were successfully used for the rapid and accurate detection of SARS-CoV-2 in many studies, one of the challenges of these studies has been the high cost of the consumables required for conducting them. Most studies have used gold nanoparticles or nanoparticles labeled with fluorophores, which ultimately increases the cost of each test for people in the community. There is currently no method that is accurate, fast and cheap, but in this challenging era, the existing options are considered as the only available solutions.

Compared to the conventional SARS-CoV-2 detection methods, such as RT-PCR, nanoparticles can be more appropriate options in several respects. For example, by amino functionalizing (via 3-aminopropyl) the surfaces of magnetic nanoparticles, the RNA contents of SARS-CoV-2 can be co-precipitated in a rapid and efficient manner [165]. Likewise, quantum dots have proven their usefulness in many medical applications [166]. These particles are respectable options for studying the binding dynamics of the S protein to the cell surface receptor (ACE2). These particles can be easily detected as soon as they enter the host cell due to their very small size (between 1 to 10 nanometer) and photo-stability [167]. Quantum dots can also be used as probes in the study of inhibitors of binding between the S protein and its receptor [168]. Graphene, which has also been used in many diagnostic aspects [169], was studied in the field of SARS-CoV-2 diagnosis as well. In a corresponding study, S protein antibodies were attached to the surface of graphene using 1-pyrenebutyric acid N-hydroxysuccinimide ester linkers. The system was then used as a sensor to detect a protein with an accuracy of 1 femtogram of the S protein per milliliter. The diagnostic method was based on the field effect transistor (FET) [170].

Furthermore, different nanoparticles were also used in other platforms to fight the SARS-CoV-2. Some nanoparticles have been used as drug carriers against the virus, such as polymeric lipid nanoparticles and silica nanoparticles [171–173]. Silver nanoparticles [150] and nanoparticles capable of generating free radicals [174] have also been used in those studies for the killing the virus directly. In other experiments, specific nanoparticles, such as nano-exosomes having the viral contents, were used to neutralize the virus infection in the body [175].

5. Conclusions and Future Outlook

It would be very frightening to imagine the epidemic of another virus with SARS-CoV-2-like infectivity and Ebola-like mortality (around 90%) [176]. After taking this alarm seriously, many researchers put more focus on the investigation of various SARS-CoV-2 outbreak issues. In general, to combat any dangerous threats, the most important factors must first be carefully identified and then characterized. By the same token, to deal with the CoVs, exact structural knowledge of viral components and the molecular mechanisms of virus infection and stages of the life cycle must be understood. Some treatment strategies are purely preventative (prophylactic strategies), while others fight infection at different stages of the virus life cycle. Antigens and the viral genome can be used as functional vaccines [15,177–179]. While experimental observations from the use of vaccines have been promising, these therapeutic agents also have some challenges, such as short half-lives and unwanted immunological symptoms [124,148]. Recent studies indicate that the utilization of nanoparticles would be also able to help humans. The use of these new materials will enable researchers to meet the challenges of vaccines and raise hopes for the fight against the CoVs. Despite all the positive results and observations that have been made about the use of nanoparticles in the production of vaccines, due to the rush to make the SARS-CoV-2 vaccines, the toxicity of nanoparticle-based vaccines, their targeting accuracy, distribution in the body, and localization within the cell has not been properly assessed. Considering the need for a global vaccination and the fact that only some countries are able to produce vaccines, one of the problems that is expected to be solved in the future is the transfer of vaccine production technology to other countries or the investment of current companies in other companies to upgrade them to produce vaccines. In the process of controlling patients, such as COVID-19, vaccination will be effective when all sections of the population are vaccinated quickly for the sake of reducing the rates of the virus spread and mutation. On the other hand, looking at the origin of this virus, which is still in a state of ambiguity, the guess that it was transmitted from animal to human is not far from doubt. Therefore, the points of view on the lifestyle in the modern world should be re-examined to better control interactions between humans and animals so that in the future, we do not see the transmission of another disease from animals. More detailed studies on human food patterns should also be conducted. Last but not least, the training necessary to improve the scientific level of people in the community to accept vaccines should be increased, so that the doubts about vaccines are reduced, and those who are afraid of vaccination know the danger of this way of thinking for themselves and others.

Author Contributions: Conceptualization, M.A. and V.N.U.; methodology, M.K., M.A. and V.U; software, M.A.; validation, M.A. and V.N.U.; formal analysis, M.A. and V.N.U.; investigation, M.K., M.A. and V.N.U.; resources, M.A. and V.N.U.; data curation, M.A. and V.N.U.; writing—original draft preparation, M.K., M.A. and V.N.U.; writing—review and editing, M.A. and V.N.U.; visualization, M.A.; supervision, M.A. and V.N.U.; project administration, M.A. and V.N.U. All authors have read and agreed to the published version of the manuscript.

Funding: This research received no external funding.

Institutional Review Board Statement: Not applicable.

Informed Consent Statement: Not applicable.

Data Availability Statement: Not applicable.

Conflicts of Interest: The authors declare no conflict of interest.

References

1. Harari, Y.N. *Sapiens. A Brief History of Humankind/Yuval Noah Harari*; Vintage Books: London, UK, 2014.
2. Livingston, E.; Bucher, K. Coronavirus disease 2019 (COVID-19) in Italy. *JAMA* **2020**, *323*, 1335. [CrossRef] [PubMed]
3. Nicola, M.; Alsafi, Z.; Sohrabi, C.; Kerwan, A.; Al-Jabir, A.; Iosifidis, C.; Agha, M.; Agha, R. The socio-economic implications of the coronavirus pandemic (COVID-19): A review. *Int. J. Surg.* **2020**, *78*, 185. [CrossRef]
4. Bonaccorsi, G.; Pierri, F.; Cinelli, M.; Flori, A.; Galeazzi, A.; Porcelli, F.; Schmidt, A.L.; Valensise, C.M.; Scala, A.; Quattrociocchi, W. Economic and social consequences of human mobility restrictions under COVID-19. *Proc. Natl. Acad. Sci. USA* **2020**, *117*, 15530–15535. [CrossRef]
5. Padhan, R.; Prabheesh, K. The economics of COVID-19 pandemic: A survey. *Econ. Anal. Policy* **2021**, *70*, 220–237. [CrossRef] [PubMed]
6. World Health Organization. Summary of Probable SARS Cases with Onset of Illness from 1 November 2002 to 31 July 2003. Available online: http://www.who.int/csr/sars/country/table2004_04_21/en/index.html (accessed on 31 December 2003).
7. World Health Organization. Middle East Respiratory Syndrome Coronavirus (MERS-CoV). 2019. Available online: https://www.who.int/health-topics/middle-east-respiratory-syndrome-coronavirus-mers#tab=tab_1 (accessed on 31 December 2003).
8. European Centre for Disease Prevention and Control. Situation Update Worldwide, as of 10 May 2020. Available online: https://www.ecdc.europa.eu/en/geographical-distribution-2019-ncov-cases (accessed on 12 May 2020).
9. Fontanet, A.; Autran, B.; Lina, B.; Kieny, M.P.; Karim, S.S.A.; Sridhar, D. SARS-CoV-2 variants and ending the COVID-19 pandemic. *Lancet* **2021**, *397*, 952–954. [CrossRef]
10. Liu, D.X.; Liang, J.Q.; Fung, T.S. Human Coronavirus-229E, -OC43, -NL63, and -HKU1 (Coronaviridae). In *Encyclopedia of Virology*, 4th ed.; Bamford, D.H., Zuckerman, M., Eds.; Academic Press: Cambridge, MA, USA, 2021; pp. 428–440.
11. Raj, K.; Rohit, A.G.; Singh, S. Coronavirus as silent killer: Recent advancement to pathogenesis, therapeutic strategy and future perspectives. *VirusDisease* **2020**, *1*, 137–145. [CrossRef]
12. Shanmugaraj, B.; Siriwattananon, K.; Wangkanont, K.; Phoolcharoen, W. Perspectives on monoclonal antibody therapy as potential therapeutic intervention for Coronavirus disease-19 (COVID-19). *Asian Pac. J. Allergy Immunol.* **2020**, *38*, 10–18.
13. Rabaan, A.A.; Al-Ahmed, S.H.; Haque, S.; Sah, R.; Tiwari, R.; Malik, Y.S.; Dhama, K.; Yatoo, M.I.; Bonilla-Aldana, D.K.; Rodriguez-Morales, A.J. SARS-CoV-2, SARS-CoV, and MERS-CoV: A comparative overview. *Infez. Med.* **2020**, *28*, 174–184.
14. Le, T.T.; Andreadakis, Z.; Kumar, A.; Roman, R.G.; Tollefsen, S.; Saville, M.; Mayhew, S. The COVID-19 vaccine development landscape. *Nat. Rev. Drug Discov.* **2020**, *19*, 305–306. [CrossRef]
15. Graham, B.S. Rapid COVID-19 vaccine development. *Science* **2020**, *368*, 945–946. [CrossRef]
16. Abdool Karim, S.S.; de Oliveira, T. New SARS-CoV-2 Variants—Clinical, Public Health, and Vaccine Implications. *N. Engl. J. Med.* **2021**, *384*, 1866–1868. [CrossRef] [PubMed]
17. Dingermann, T. Recombinant therapeutic proteins: Production platforms and challenges. *Biotechnol. J. Healthc. Nutr. Technol.* **2008**, *3*, 90–97. [CrossRef] [PubMed]
18. Dimitrov, D.S. Therapeutic proteins. In *Therapeutic Proteins*; Springer: Berlin/Heidelberg, Germany, 2012; pp. 1–26.
19. Bruno, B.J.; Miller, G.D.; Lim, C.S. Basics and recent advances in peptide and protein drug delivery. *Ther. Deliv.* **2013**, *4*, 1443–1467. [CrossRef] [PubMed]
20. Banga, A.K. *Therapeutic Peptides and Proteins: Formulation, Processing, and Delivery Systems*; CRC Press: Boca Raton, FL, USA, 2015.
21. Li, F.; Li, W.; Farzan, M.; Harrison, S.C. Structure of SARS coronavirus spike receptor-binding domain complexed with receptor. *Science* **2005**, *309*, 1864–1868. [CrossRef]
22. Walls, A.C.; Park, Y.-J.; Tortorici, M.A.; Wall, A.; McGuire, A.T.; Veesler, D. Structure, function, and antigenicity of the SARS-CoV-2 spike glycoprotein. *Cell* **2020**, *181*, 281–292. [CrossRef]

23. Dai, W.; Zhang, B.; Jiang, X.-M.; Su, H.; Li, J.; Zhao, Y.; Xie, X.; Jin, Z.; Peng, J.; Liu, F. Structure-based design of antiviral drug candidates targeting the SARS-CoV-2 main protease. *Science* **2020**, *368*, 1331–1335. [CrossRef]
24. Xu, X.; Chen, P.; Wang, J.; Feng, J.; Zhou, H.; Li, X.; Zhong, W.; Hao, P. Evolution of the novel coronavirus from the ongoing Wuhan outbreak and modeling of its spike protein for risk of human transmission. *Sci. China Life Sci.* **2020**, *63*, 457–460. [CrossRef]
25. Xiong, X.; Qu, K.; Ciazynska, K.A.; Hosmillo, M.; Carter, A.P.; Ebrahimi, S.; Ke, Z.; Scheres, S.H.; Bergamaschi, L.; Grice, G.L. A thermostable, closed SARS-CoV-2 spike protein trimer. *Nat. Struct. Mol. Biol.* **2020**, *27*, 1–8. [CrossRef]
26. Gur, M.; Taka, E.; Yilmaz, S.Z.; Kilinc, C.; Aktas, U.; Golcuk, M. Conformational transition of SARS-CoV-2 spike glycoprotein between its closed and open states. *J. Chem. Phys.* **2020**, *153*, 075101. [CrossRef]
27. Lu, S.; Ye, Q.; Singh, D.; Cao, Y.; Diedrich, J.K.; Yates, J.R.; Villa, E.; Cleveland, D.W.; Corbett, K.D. The SARS-CoV-2 nucleocapsid phosphoprotein forms mutually exclusive condensates with RNA and the membrane-associated M protein. *Nat. Commun.* **2021**, *12*, 1–15. [CrossRef]
28. Seyedpour, S.; Khodaei, B.; Loghman, A.H.; Seyedpour, N.; Kisomi, M.F.; Balibegloo, M.; Nezamabadi, S.S.; Gholami, B.; Saghazadeh, A.; Rezaei, N. Targeted therapy strategies against SARS-CoV-2 cell entry mechanisms: A systematic review of in vitro and in vivo studies. *J. Cell. Physiol.* **2021**, *236*, 2364–2392. [CrossRef] [PubMed]
29. Wrobel, A.G.; Benton, D.J.; Xu, P.; Roustan, C.; Martin, S.R.; Rosenthal, P.B.; Skehel, J.J.; Gamblin, S.J. SARS-CoV-2 and bat RaTG13 spike glycoprotein structures inform on virus evolution and furin-cleavage effects. *Nat. Struct. Mol. Biol.* **2020**, *27*, 763–767. [CrossRef] [PubMed]
30. Sariol, A.; Perlman, S. Lessons for COVID-19 immunity from other coronavirus infections. *Immunity* **2020**, *3*, 248–263. [CrossRef]
31. Wong, R.S.; Wu, A.; To, K.; Lee, N.; Lam, C.W.; Wong, C.; Chan, P.K.; Ng, M.H.; Yu, L.; Hui, D.S. Haematological manifestations in patients with severe acute respiratory syndrome: Retrospective analysis. *BMJ* **2003**, *326*, 1358–1362. [CrossRef]
32. Wang, Y.-D.; Sin, W.-Y.F.; Xu, G.-B.; Yang, H.-H.; Wong, T.-y.; Pang, X.-W.; He, X.-Y.; Zhang, H.-G.; Ng, J.N.L.; Cheng, C.-S.S. T-cell epitopes in severe acute respiratory syndrome (SARS) coronavirus spike protein elicit a specific T-cell immune response in patients who recover from SARS. *J. Virol.* **2004**, *78*, 5612–5618. [CrossRef]
33. Buchholz, U.J.; Bukreyev, A.; Yang, L.; Lamirande, E.W.; Murphy, B.R.; Subbarao, K.; Collins, P.L. Contributions of the structural proteins of severe acute respiratory syndrome coronavirus to protective immunity. *Proc. Natl. Acad. Sci. USA* **2004**, *101*, 9804–9809. [CrossRef]
34. Zhou, P.; Yang, X.-L.; Wang, X.-G.; Hu, B.; Zhang, L.; Zhang, W.; Si, H.-R.; Zhu, Y.; Li, B.; Huang, C.-L. A pneumonia outbreak associated with a new coronavirus of probable bat origin. *Nature* **2020**, *579*, 270–273. [CrossRef]
35. Wang, C.; Xie, J.; Zhao, L.; Fei, X.; Zhang, H.; Tan, Y.; Nie, X.; Zhou, L.; Liu, Z.; Ren, Y. Alveolar macrophage dysfunction and cytokine storm in the pathogenesis of two severe COVID-19 patients. *EBioMedicine* **2020**, *57*, 102833. [CrossRef]
36. Moore, J.B.; June, C.H. Cytokine release syndrome in severe COVID-19. *Science* **2020**, *368*, 473–474. [CrossRef]
37. Boulant, S.; Stanifer, M.; Lozach, P.-Y. Dynamics of virus-receptor interactions in virus binding, signaling, and endocytosis. *Viruses* **2015**, *7*, 2794–2815. [CrossRef]
38. Yan, R.; Zhang, Y.; Li, Y.; Xia, L.; Guo, Y.; Zhou, Q. Structural basis for the recognition of SARS-CoV-2 by full-length human ACE2. *Science* **2020**, *367*, 1444–1448. [CrossRef] [PubMed]
39. Tandon, R.; Sharp, J.S.; Zhang, F.; Pomin, V.H.; Ashpole, N.M.; Mitra, D.; McCandless, M.G.; Jin, W.; Liu, H.; Sharma, P. Effective inhibition of SARS-CoV-2 entry by heparin and enoxaparin derivatives. *J. Virol.* **2021**, *95*, e01987-20. [CrossRef] [PubMed]
40. Solerte, S.B.; Di Sabatino, A.; Galli, M.; Fiorina, P. Dipeptidyl peptidase-4 (DPP4) inhibition in COVID-19. *Acta Diabetol.* **2020**, *1*, 779–783. [CrossRef] [PubMed]
41. Valencia, I.; Peiró, C.; Lorenzo, Ó.; Sánchez-Ferrer, C.F.; Eckel, J.; Romacho, T. DPP4 and ACE2 in diabetes and COVID-19: Therapeutic targets for cardiovascular complications? *Front. Pharmacol.* **2020**, *11*, 1161. [CrossRef]
42. Tikellis, C.; Bernardi, S.; Burns, W.C. Angiotensin-converting enzyme 2 is a key modulator of the renin–angiotensin system in cardiovascular and renal disease. *Curr. Opin. Nephrol. Hypertens.* **2011**, *20*, 62–68. [CrossRef]
43. Vitiello, A.; Pelliccia, C.; Ferrara, F. Drugs acting on the renin–angiotensin system and SARS-CoV-2. *Drug Discov. Today* **2021**, *26*, 870–874. [CrossRef]
44. Lamers, M.M.; Beumer, J.; van der Vaart, J.; Knoops, K.; Puschhof, J.; Breugem, T.I.; Ravelli, R.B.; van Schayck, J.P.; Mykytyn, A.Z.; Duimel, H.Q. SARS-CoV-2 productively infects human gut enterocytes. *Science* **2020**, *369*, 50–54. [CrossRef]
45. Hindson, J. COVID-19: Faecal–oral transmission? *Nat. Rev. Gastroenterol. Hepatol.* **2020**, *17*, 259. [CrossRef]
46. Wang, K.; Chen, W.; Zhou, Y.-S.; Lian, J.-Q.; Zhang, Z.; Du, P.; Gong, L.; Zhang, Y.; Cui, H.-Y.; Geng, J.-J. SARS-CoV-2 invades host cells via a novel route: CD147-spike protein. *bioRxiv* **2020**. [CrossRef]
47. Shilts, J.; Crozier, T.W.; Greenwood, E.J.; Lehner, P.J.; Wright, G.J. No evidence for basigin/CD147 as a direct SARS-CoV-2 spike binding receptor. *Sci. Rep.* **2021**, *11*, 1–10. [CrossRef]
48. Guindolet, D.; Gabison, E.E. Role of CD147 (EMMPRIN/basigin) in tissue remodeling. *Anat. Rec.* **2020**, *303*, 1584–1589. [CrossRef]
49. Song, Z.; Xu, Y.; Bao, L.; Zhang, L.; Yu, P.; Qu, Y.; Zhu, H.; Zhao, W.; Han, Y.; Qin, C. From SARS to MERS, thrusting coronaviruses into the spotlight. *Viruses* **2019**, *11*, 59. [CrossRef] [PubMed]
50. Hoffmann, M.; Kleine-Weber, H.; Schroeder, S.; Krüger, N.; Herrler, T.; Erichsen, S.; Schiergens, T.S.; Herrler, G.; Wu, N.-H.; Nitsche, A. SARS-CoV-2 cell entry depends on ACE2 and TMPRSS2 and is blocked by a clinically proven protease inhibitor. *Cell* **2020**, *181*, 271–280. [CrossRef] [PubMed]

51. Qiao, Y.; Wang, X.-M.; Mannan, R.; Pitchiaya, S.; Zhang, Y.; Wotring, J.W.; Xiao, L.; Robinson, D.R.; Wu, Y.-M.; Tien, J.C.-Y. Targeting transcriptional regulation of SARS-CoV-2 entry factors ACE2 and TMPRSS2. *Proc. Natl. Acad. Sci. USA* **2021**, *118*, e2021450118. [CrossRef]
52. Ko, C.-J.; Hsu, T.-W.; Wu, S.-R.; Lan, S.-W.; Hsiao, T.-F.; Lin, H.-Y.; Lin, H.-H.; Tu, H.-F.; Lee, C.-F.; Huang, C.-C. Inhibition of TMPRSS2 by HAI-2 reduces prostate cancer cell invasion and metastasis. *Oncogene* **2020**, *39*, 5950–5963. [CrossRef] [PubMed]
53. Wang, Q.; Wu, J.; Wang, H.; Gao, Y.; Liu, Q.; Mu, A.; Ji, W.; Yan, L.; Zhu, Y.; Zhu, C. Structural basis for RNA replication by the SARS-CoV-2 polymerase. *Cell* **2020**, *182*, 417–428.e13. [CrossRef] [PubMed]
54. Lung, J.; Lin, Y.S.; Yang, Y.H.; Chou, Y.L.; Shu, L.H.; Cheng, Y.C.; Liu, H.T.; Wu, C.Y. The potential chemical structure of anti-SARS-CoV-2 RNA-dependent RNA polymerase. *J. Med. Virol.* **2020**, *92*, 693–697. [CrossRef] [PubMed]
55. Chien, M.; Anderson, T.K.; Jockusch, S.; Tao, C.; Li, X.; Kumar, S.; Russo, J.J.; Kirchdoerfer, R.N.; Ju, J. Nucleotide analogues as inhibitors of SARS-CoV-2 polymerase, a key drug target for COVID-19. *J. Proteome Res.* **2020**, *19*, 4690–4697. [CrossRef] [PubMed]
56. Yin, W.; Mao, C.; Luan, X.; Shen, D.-D.; Shen, Q.; Su, H.; Wang, X.; Zhou, F.; Zhao, W.; Gao, M. Structural basis for inhibition of the RNA-dependent RNA polymerase from SARS-CoV-2 by remdesivir. *Science* **2020**, *368*, 1499–1504. [CrossRef]
57. Rohilla, S. Designing therapeutic strategies to combat severe acute respiratory syndrome coronavirus-2 disease: COVID-19. *Drug Dev. Res.* **2021**, *82*, 12–26. [CrossRef]
58. Jayakumar, R.; Menon, D.; Manzoor, K.; Nair, S.V.; Tamura, H. Biomedical applications of chitin and chitosan based nanomaterials—A short review. *Carbohydr. Polym.* **2010**, *82*, 227–232. [CrossRef]
59. Khot, L.R.; Sankaran, S.; Maja, J.M.; Ehsani, R.; Schuster, E.W. Applications of nanomaterials in agricultural production and crop protection: A review. *Crop Prot.* **2012**, *35*, 64–70. [CrossRef]
60. Lee, J.; Mahendra, S.; Alvarez, P.J. Nanomaterials in the construction industry: A review of their applications and environmental health and safety considerations. *ACS Nano* **2010**, *4*, 3580–3590. [CrossRef] [PubMed]
61. Joy, S.; Sunandana, D.; Sahasrabdi, A.; Prithu, B.; Krishnendu, A. Antiviral potential of nanoparticles for the treatment of Coronavirus infections. *J. Trace Elem. Med. Biol.* **2022**, *72*, 126977.
62. Wang, J.; Drelich, A.J.; Hopkins, C.M.; Mecozzi, S.; Li, L.; Kwon, G.; Hong, S. Gold nanoparticles in virus detection: Recent advances and potential considerations for SARS-CoV-2 testing development. *Wiley Interdiscip. Rev. Nanomed. Nanobiotechnol.* **2022**, *14*, e1754. [CrossRef]
63. Konrath, K.M.; Liaw, K.; Wu, Y.; Zhu, X.; Walker, S.N.; Xu, Z.; Schultheis, K.; Chokkalingam, N.; Chawla, H.; Du, J. Nucleic acid delivery of immune-focused SARS-CoV-2 nanoparticles drive rapid and potent immunogenicity capable of single-dose protection. *Cell Rep.* **2022**, *38*, 110318. [CrossRef]
64. Kumar, A.; Thakur, A. Use of Nanoparticles to Combat COVID-19. In *Handbook of Research on Green Synthesis and Applications of Nanomaterials*; IGI Global: Hershey, PA, USA, 2022; pp. 412–440.
65. Pina, S.; Reis, R.L.; Oliveira, J.M. Natural polymeric biomaterials for tissue engineering. In *Tissue Engineering Using Ceramics and Polymers*; Elsevier: Amsterdam, The Netherlands, 2022; pp. 75–110.
66. Bobo, D.; Robinson, K.J.; Islam, J.; Thurecht, K.J.; Corrie, S.R. Nanoparticle-based medicines: A review of FDA-approved materials and clinical trials to date. *Pharm. Res.* **2016**, *33*, 2373–2387. [CrossRef]
67. Murthy, N.T.V.; Paul, S.K.; Chauhan, H.; Singh, S. Polymeric Nanoparticles for Transdermal Delivery of Polyphenols. *Curr. Drug Deliv.* **2021**, *19*, 182–191. [CrossRef]
68. Synowiec, A.; Szczepański, A.; Barreto-Duran, E.; Lie, L.K.; Pyrc, K. Severe acute respiratory syndrome coronavirus 2 (SARS-CoV-2): A systemic infection. *Clin. Microbiol. Rev.* **2021**, *34*, e00133-20. [CrossRef]
69. Lara, Y.; Nguyen, T.; Marilena, L.; Alexander, M. Toxicological considerations of clinically applicable nanoparticles. *Nano Today* **2011**, *6*, 585–607.
70. Lysenko, V.; Lozovski, V.; Lokshyn, M.; Gomeniuk, Y.V.; Dorovskih, A.; Rusinchuk, N.; Pankivska, Y.; Povnitsa, O.; Zagorodnya, S.; Tertykh, V. Nanoparticles as antiviral agents against adenoviruses. *Adv. Nat. Sci. Nanosci. Nanotechnol.* **2018**, *9*, 025021. [CrossRef]
71. Du, T.; Zhang, J.; Li, C.; Song, T.; Li, P.; Liu, J.; Du, X.; Wang, S.J.B.C. Gold/silver hybrid nanoparticles with enduring inhibition of coronavirus multiplication through multisite mechanisms. *Bioconjugate Chem.* **2020**, *31*, 2553–2563. [CrossRef] [PubMed]
72. Abo-Zeid, Y.; Ismail, N.S.; McLean, G.R.; Hamdy, N.M. A molecular docking study repurposes FDA approved iron oxide nanoparticles to treat and control COVID-19 infection. *Eur. J. Pharm. Sci.* **2020**, *153*, 105465. [CrossRef] [PubMed]
73. Rashidzadeh, H.; Danafar, H.; Rahimi, H.; Mozafari, F.; Salehiabar, M.; Rahmati, M.A.; Rahamooz-Haghighi, S.; Mousazadeh, N.; Mohammadi, A.; Ertas, Y.N. Nanotechnology against the novel coronavirus (severe acute respiratory syndrome coronavirus 2): Diagnosis, treatment, therapy and future perspectives. *Nanomedicine* **2021**, *16*, 497–516. [CrossRef]
74. Smith, J.D.; Morton, L.D.; Ulery, B.D. Nanoparticles as synthetic vaccines. *Curr. Opin. Biotechnol.* **2015**, *34*, 217–224. [CrossRef]
75. Irvine, D.J.; Hanson, M.C.; Rakhra, K.; Tokatlian, T. Synthetic nanoparticles for vaccines and immunotherapy. *Chem. Rev.* **2015**, *115*, 11109–11146. [CrossRef]
76. Dhand, C.; Dwivedi, N.; Loh, X.J.; Ying, A.N.J.; Verma, N.K.; Beuerman, R.W.; Lakshminarayanan, R.; Ramakrishna, S.J.R.A. Methods and strategies for the synthesis of diverse nanoparticles and their applications: A comprehensive overview. *Rsc Adv.* **2015**, *5*, 105003–105037. [CrossRef]
77. Prencipe, G.; Tabakman, S.M.; Welsher, K.; Liu, Z.; Goodwin, A.P.; Zhang, L.; Henry, J.; Dai, H. PEG branched polymer for functionalization of nanomaterials with ultralong blood circulation. *J. Am. Chem. Soc.* **2009**, *131*, 4783–4787. [CrossRef]

78. Wilczewska, A.Z.; Niemirowicz, K.; Markiewicz, K.H.; Car, H. Nanoparticles as drug delivery systems. *Pharmacol. Rep.* **2012**, *64*, 1020–1037. [CrossRef]
79. Morens, D.M.; Taubenberger, J.K.; Fauci, A.S. Universal coronavirus vaccines—An urgent need. *N. Engl. J. Med.* **2022**, *386*, 297–299. [CrossRef]
80. Toy, R.; Roy, K.J.B. Engineering nanoparticles to overcome barriers to immunotherapy. *Bioeng. Transl. Med.* **2016**, *1*, 47–62. [CrossRef] [PubMed]
81. Nembrini, C.; Stano, A.; Dane, K.Y.; Ballester, M.; Van Der Vlies, A.J.; Marsland, B.J.; Swartz, M.A.; Hubbell, J.A. Nanoparticle conjugation of antigen enhances cytotoxic T-cell responses in pulmonary vaccination. *Proc. Natl. Acad. Sci. USA* **2011**, *108*, E989–E997. [CrossRef] [PubMed]
82. Yao, H.; Shen, N.; Ji, G.; Huang, J.; Sun, J.; Wang, G.; Tang, Z.; Chen, X. Cisplatin Nanoparticles Promote Intratumoral CD8+ T Cell Priming via Antigen Presentation and T Cell Receptor Crosstalk. *Nano Lett.* **2022**, *8*, 3328–3339. [CrossRef] [PubMed]
83. Parmar, K.; Patel, J.; Pathak, Y. Factors Affecting the Clearance and Biodistribution of Polymeric Nanoparticles. In *Pharmacokinetics and Pharmacodynamics of Nanoparticulate Drug Delivery Systems*; Springer: Berlin/Heidelberg, Germany, 2022; pp. 261–272.
84. Petros, R.A.; DeSimone, J.M. Strategies in the design of nanoparticles for therapeutic applications. *Nat. Rev. Drug Discov.* **2010**, *9*, 615–627. [CrossRef] [PubMed]
85. Anselmo, A.C.; Zhang, M.; Kumar, S.; Vogus, D.R.; Menegatti, S.; Helgeson, M.E.; Mitragotri, S. Elasticity of nanoparticles influences their blood circulation, phagocytosis, endocytosis, and targeting. *ACS Nano* **2015**, *9*, 3169–3177. [CrossRef]
86. Kumar, S.; Anselmo, A.C.; Banerjee, A.; Zakrewsky, M.; Mitragotri, S.J.J.o.C.R. Shape and size-dependent immune response to antigen-carrying nanoparticles. *J. Control. Release* **2015**, *220*, 141–148. [CrossRef]
87. Toy, R.; Peiris, P.M.; Ghaghada, K.B.; Karathanasis, E.J.N. Shaping cancer nanomedicine: The effect of particle shape on the in vivo journey of nanoparticles. *Nanomedicine* **2014**, *9*, 121–134. [CrossRef]
88. Agarwal, R.; Jurney, P.; Raythatha, M.; Singh, V.; Sreenivasan, S.V.; Shi, L.; Roy, K. Effect of shape, size, and aspect ratio on nanoparticle penetration and distribution inside solid tissues using 3D spheroid models. *Adv. Healthc. Mater.* **2015**, *4*, 2269–2280. [CrossRef]
89. Fromen, C.A.; Robbins, G.R.; Shen, T.W.; Kai, M.P.; Ting, J.P.; DeSimone, J.M. Controlled analysis of nanoparticle charge on mucosal and systemic antibody responses following pulmonary immunization. *Proc. Natl. Acad. Sci. USA* **2015**, *112*, 488–493. [CrossRef]
90. Shahbazi, M.-A.; Fernández, T.D.; Mäkilä, E.M.; Le Guével, X.; Mayorga, C.; Kaasalainen, M.H.; Salonen, J.J.; Hirvonen, J.T.; Santos, H.A. Surface chemistry dependent immunostimulative potential of porous silicon nanoplatforms. *Biomaterials* **2014**, *35*, 9224–9235. [CrossRef]
91. Fernandes, R.; Smyth, N.R.; Muskens, O.L.; Nitti, S.; Heuer-Jungemann, A.; Ardern-Jones, M.R.; Kanaras, A.G. Interactions of skin with gold nanoparticles of different surface charge, shape, and functionality. *Small* **2015**, *11*, 713–721. [CrossRef] [PubMed]
92. Yu, T.; Greish, K.; McGill, L.D.; Ray, A.; Ghandehari, H. Influence of geometry, porosity, and surface characteristics of silica nanoparticles on acute toxicity: Their vasculature effect and tolerance threshold. *ACS Nano* **2012**, *6*, 2289–2301. [CrossRef] [PubMed]
93. Liu, Y.; Hu, Y.; Huang, L. Influence of polyethylene glycol density and surface lipid on pharmacokinetics and biodistribution of lipid-calcium-phosphate nanoparticles. *Biomaterials* **2014**, *35*, 3027–3034. [CrossRef] [PubMed]
94. Scheerlinck, J.-P.Y.; Greenwood, D.L. Virus-sized vaccine delivery systems. *Drug Discov. Today* **2008**, *13*, 882–887. [CrossRef]
95. Liu, X.; Song, H.; Jiang, J.; Gao, X.; Yi, Y.; Shang, Y.; Li, J.; Li, D.; Zeng, Z.; Li, Y. Self-assembling SARS-CoV-2 nanoparticle vaccines targeting the S protein induces protective immunity in mice. *biorXiv* **2021**. [CrossRef]
96. Tang, Z.; Zhang, X.; Shu, Y.; Guo, M.; Zhang, H.; Tao, W. Insights from nanotechnology in COVID-19 treatment. *Nano Today* **2021**, *36*, 101019. [CrossRef]
97. Heinrich, M.A.; Martina, B.; Prakash, J. Nanomedicine strategies to target coronavirus. *Nano Today* **2020**, *35*, 100961. [CrossRef]
98. Machhi, J.; Shahjin, F.; Das, S.; Patel, M.; Abdelmoaty, M.M.; Cohen, J.D.; Singh, P.A.; Baldi, A.; Bajwa, N.; Kumar, R. A Role for Extracellular Vesicles in SARS-CoV-2 Therapeutics and Prevention. *J. Neuroimmune Pharmacol.* **2021**, *16*, 1–19. [CrossRef]
99. Neuman, B.W.; Adair, B.D.; Yoshioka, C.; Quispe, J.D.; Orca, G.; Kuhn, P.; Milligan, R.A.; Yeager, M.; Buchmeier, M.J. Supramolecular architecture of severe acute respiratory syndrome coronavirus revealed by electron cryomicroscopy. *J. Virol.* **2006**, *80*, 7918–7928. [CrossRef]
100. Zuercher, A.W.; Coffin, S.E.; Thurnheer, M.C.; Fundova, P.; Cebra, J.J. Nasal-associated lymphoid tissue is a mucosal inductive site for virus-specific humoral and cellular immune responses. *J. Immunol.* **2002**, *168*, 1796–1803. [CrossRef]
101. Charelli, L.E.; de Mattos, G.C.; de Jesus Sousa-Batista, A.; Pinto, J.C.; Balbino, T.A. Polymeric nanoparticles as therapeutic agents against coronavirus disease. *J. Nanopart. Res.* **2022**, *24*, 1–15. [CrossRef] [PubMed]
102. Satta, S.; Meng, Z.; Hernandez, R.; Cavallero, S.; Zhou, T.; Hsiai, T.K.; Zhou, C. An engineered nano-liposome-human ACE2 decoy neutralizes SARS-CoV-2 Spike protein-induced inflammation in both murine and human macrophages. *Theranostics* **2022**, *12*, 2639. [CrossRef] [PubMed]
103. Wang, J.; Yin, X.-G.; Wen, Y.; Lu, J.; Zhang, R.-Y.; Zhou, S.-H.; Liao, C.-M.; Wei, H.-W.; Guo, J. MPLA-Adjuvanted Liposomes Encapsulating s-Trimer or RBD or S1, But Not s-ECD, Elicit Robust Neutralization against SARS-CoV-2 and Variants of Concern. *J. Med. Chem.* **2022**, *65*, 3563–3574. [CrossRef] [PubMed]

104. Hatamluyi, B.; Rezayi, M.; Jamehdar, S.A.; Rizi, K.S.; Mojarrad, M.; Meshkat, Z.; Choobin, H.; Soleimanpour, S.; Boroushaki, M.T.J.B.; Bioelectronics, M.T. Sensitive and specific clinically diagnosis of SARS-CoV-2 employing a novel biosensor based on boron nitride quantum dots/flower-like gold nanostructures signal amplification. *Biosens. Bioelectron.* **2022**, *207*, 114209. [CrossRef] [PubMed]
105. Li, Y.; Ma, P.; Tao, Q.; Krause, H.-J.; Yang, S.; Ding, G.; Dong, H.; Xie, X.J.S.; Chemical, A.B. Magnetic graphene quantum dots facilitate closed-tube one-step detection of SARS-CoV-2 with ultra-low field NMR relaxometry. *Sens. Actuators B Chem.* **2021**, *337*, 129786. [CrossRef] [PubMed]
106. Jung, S.-Y.; Kang, K.W.; Lee, E.-Y.; Seo, D.-W.; Kim, H.-L.; Kim, H.; Kwon, T.; Park, H.-L.; Kim, H.; Lee, S.-M. Heterologous prime–boost vaccination with adenoviral vector and protein nanoparticles induces both Th1 and Th2 responses against Middle East Respiratory syndrome coronavirus. *Vaccine* **2018**, *36*, 3468–3476. [CrossRef]
107. Coleman, C.M.; Liu, Y.V.; Mu, H.; Taylor, J.K.; Massare, M.; Flyer, D.C.; Glenn, G.M.; Smith, G.E.; Frieman, M.B. Purified coronavirus spike protein nanoparticles induce coronavirus neutralizing antibodies in mice. *Vaccine* **2014**, *32*, 3169–3174. [CrossRef]
108. Wang, C.; Zheng, X.; Gai, W.; Wong, G.; Wang, H.; Jin, H.; Feng, N.; Zhao, Y.; Zhang, W.; Li, N. Novel chimeric virus-like particles vaccine displaying MERS-CoV receptor-binding domain induce specific humoral and cellular immune response in mice. *Antivir. Res.* **2017**, *140*, 55–61. [CrossRef]
109. Kato, T.; Takami, Y.; Deo, V.K.; Park, E.Y. Preparation of virus-like particle mimetic nanovesicles displaying the S protein of Middle East respiratory syndrome coronavirus using insect cells. *J. Biotechnol.* **2019**, *306*, 177–184. [CrossRef]
110. Hills, R.A.; Howarth, M. Virus-like particles against infectious disease and cancer: Guidance for the nano-architect. *Curr. Opin. Biotechnol.* **2022**, *73*, 346–354. [CrossRef]
111. Dai, S.; Wang, H.; Deng, F.J.J.o.I.S. Advances and challenges in enveloped virus-like particle (VLP)-based vaccines. *J. Immunol. Sci.* **2018**, *2*, 36–41.
112. Roy, P.; Noad, R.J.H.v. Virus-like particles as a vaccine delivery system: Myths and facts. *Hum. Vaccines* **2008**, *4*, 5–12. [CrossRef] [PubMed]
113. Mohsen, M.O.; Speiser, D.E.; Knuth, A.; Bachmann, M.F. Nanobiotechnology Virus-like particles for vaccination against cancer. *Wiley Interdiscip. Rev. Nanomed. Nanobiotechnol.* **2020**, *12*, e1579. [CrossRef] [PubMed]
114. Ohno, S.; Kohyama, S.; Taneichi, M.; Moriya, O.; Hayashi, H.; Oda, H.; Mori, M.; Kobayashi, A.; Akatsuka, T.; Uchida, T. Synthetic peptides coupled to the surface of liposomes effectively induce SARS coronavirus-specific cytotoxic T lymphocytes and viral clearance in HLA-A* 0201 transgenic mice. *Vaccine* **2009**, *27*, 3912–3920. [CrossRef] [PubMed]
115. Erasmus, J.H.; Khandhar, A.P.; O'Connor, M.A.; Walls, A.C.; Hemann, E.A.; Murapa, P.; Archer, J.; Leventhal, S.; Fuller, J.T.; Lewis, T.B. An alphavirus-derived replicon RNA vaccine induces SARS-CoV-2 neutralizing antibody and T cell responses in mice and nonhuman primates. *Sci. Transl. Med.* **2020**, *12*, eabc9396. [CrossRef] [PubMed]
116. Yousefi, M.; Ehsani, A.; Jafari, S.M. Lipid-based nano delivery of antimicrobials to control food-borne bacteria. *Adv. Colloid Interface Sci.* **2019**, *270*, 263–277. [CrossRef]
117. Mozafari, M.R. Nanoliposomes: Preparation and analysis. In *Liposomes*; Springer: Berlin/Heidelberg, Germany, 2010; pp. 29–50.
118. Kuate, S.; Cinatl, J.; Doerr, H.W.; Überla, K. Exosomal vaccines containing the S protein of the SARS coronavirus induce high levels of neutralizing antibodies. *Virology* **2007**, *362*, 26–37. [CrossRef]
119. Raghuwanshi, D.; Mishra, V.; Das, D.; Kaur, K.; Suresh, M.R. Dendritic cell targeted chitosan nanoparticles for nasal DNA immunization against SARS CoV nucleocapsid protein. *Mol. Pharm.* **2012**, *9*, 946–956. [CrossRef]
120. Shim, B.-S.; Park, S.-M.; Quan, J.-S.; Jere, D.; Chu, H.; Song, M.K.; Kim, D.W.; Jang, Y.-S.; Yang, M.-S.; Han, S.H. Intranasal immunization with plasmid DNA encoding spike protein of SARS-coronavirus/polyethylenimine nanoparticles elicits antigen-specific humoral and cellular immune responses. *BMC Immunol.* **2010**, *11*, 1–9. [CrossRef]
121. Badgujar, K.C.; Badgujar, V.C.; Badgujar, S.B. Vaccine development against coronavirus (2003 to present): An overview, recent advances, current scenario, opportunities and challenges. *Diabetes Metab. Syndr. Clin. Res. Rev.* **2020**, *14*, 1361–1376. [CrossRef]
122. Lainscek, D.; Fink, T.; Forstneric, V.; Hafner-Bratkovic, I.; Orehek, S.; Strmsek, Z.; Keber, M.M.; Pecan, P.; Esih, H.; Malensek, S. Immune response to vaccine candidates based on different types of nanoscaffolded RBD domain of the SARS-CoV-2 spike protein. *bioRxiv* **2020**. [CrossRef]
123. Yi Xue, H.; Guo, P.; Wen, W.-C.; Lun Wong, H. Lipid-based nanocarriers for RNA delivery. *Curr. Pharm. Des.* **2015**, *21*, 3140–3147. [CrossRef] [PubMed]
124. Corbett, K.S.; Flynn, B.; Foulds, K.E.; Francica, J.R.; Boyoglu-Barnum, S.; Werner, A.P.; Flach, B.; O'Connell, S.; Bock, K.W.; Minai, M. Evaluation of the mRNA-1273 vaccine against SARS-CoV-2 in nonhuman primates. *N. Engl. J. Med.* **2020**, *383*, 1544–1555. [CrossRef] [PubMed]
125. Al-Halifa, S.; Gauthier, L.; Arpin, D.; Bourgault, S.; Archambault, D. Nanoparticle-based vaccines against respiratory viruses. *Front. Immunol.* **2019**, *10*, 22. [CrossRef]
126. Chattopadhyay, S.; Chen, J.-Y.; Chen, H.-W.; Hu, C.-M.J. Nanoparticle vaccines adopting virus-like features for enhanced immune potentiation. *Nanotheranostics* **2017**, *1*, 244. [CrossRef]
127. Jackson, L.A.; Anderson, E.J.; Rouphael, N.G.; Roberts, P.C.; Makhene, M.; Coler, R.N.; McCullough, M.P.; Chappell, J.D.; Denison, M.R.; Stevens, L.J. An mRNA vaccine against SARS-CoV-2—Preliminary report. *N. Engl. J. Med.* **2020**. [CrossRef]

128. de Alwis, R.M.; Gan, E.S.; Chen, S.; Leong, Y.S.; Tan, H.C.; Zhang, S.L.; Yau, C.; Matsuda, D.; Allen, E.; Hartman, P. A Single Dose of Self-Transcribing and Replicating RNA Based SARS-CoV-2 Vaccine Produces Protective Adaptive Immunity in Mice. *bioRxiv* **2020**. [CrossRef]
129. Blaising, J.; Polyak, S.J.; Pécheur, E.-I. Arbidol as a broad-spectrum antiviral: An update. *Antivir. Res.* **2014**, *107*, 84–94. [CrossRef]
130. Uno, Y. Camostat mesilate therapy for COVID-19. *Intern. Emerg. Med.* **2020**, *15*, 1–2. [CrossRef]
131. Horimoto, T.; Nakayama, K.; Smeekens, S.P.; Kawaoka, Y. Proprotein-processing endoproteases PC6 and furin both activate hemagglutinin of virulent avian influenza viruses. *J. Virol.* **1994**, *68*, 6074–6078. [CrossRef]
132. Couture, F.; Kwiatkowska, A.; Dory, Y.L.; Day, R. Therapeutic uses of furin and its inhibitors: A patent review. *Expert Opin. Ther. Pat.* **2015**, *25*, 379–396. [PubMed]
133. Millet, J.K.; Whittaker, G.R. Physiological and molecular triggers for SARS-CoV membrane fusion and entry into host cells. *Virology* **2018**, *517*, 3–8. [PubMed]
134. Yuan, S.; Chu, H.; Chan, J.F.-W.; Ye, Z.-W.; Wen, L.; Yan, B.; Lai, P.-M.; Tee, K.-M.; Huang, J.; Chen, D. SREBP-dependent lipidomic reprogramming as a broad-spectrum antiviral target. *Nat. Commun.* **2019**, *10*, 1–15.
135. Walls, A.C.; Xiong, X.; Park, Y.-J.; Tortorici, M.A.; Snijder, J.; Quispe, J.; Cameroni, E.; Gopal, R.; Dai, M.; Lanzavecchia, A. Unexpected receptor functional mimicry elucidates activation of coronavirus fusion. *Cell* **2019**, *176*, 1026–1039.e15. [CrossRef]
136. Gao, J.; Lu, G.; Qi, J.; Li, Y.; Wu, Y.; Deng, Y.; Geng, H.; Li, H.; Wang, Q.; Xiao, H. Structure of the fusion core and inhibition of fusion by a heptad repeat peptide derived from the S protein of Middle East respiratory syndrome coronavirus. *J. Virol.* **2013**, *87*, 13134–13140.
137. Lu, L.; Liu, Q.; Zhu, Y.; Chan, K.-H.; Qin, L.; Li, Y.; Wang, Q.; Chan, J.F.-W.; Du, L.; Yu, F. Structure-based discovery of Middle East respiratory syndrome coronavirus fusion inhibitor. *Nat. Commun.* **2014**, *5*, 1–12.
138. Xia, S.; Yan, L.; Xu, W.; Agrawal, A.S.; Algaissi, A.; Tseng, C.-T.K.; Wang, Q.; Du, L.; Tan, W.; Wilson, I.A. A pan-coronavirus fusion inhibitor targeting the HR1 domain of human coronavirus spike. *Sci. Adv.* **2019**, *5*, eaav4580. [CrossRef]
139. Channappanavar, R.; Fehr, A.R.; Vijay, R.; Mack, M.; Zhao, J.; Meyerholz, D.K.; Perlman, S. Dysregulated type I interferon and inflammatory monocyte-macrophage responses cause lethal pneumonia in SARS-CoV-infected mice. *Cell Host Microbe* **2016**, *19*, 181–193. [CrossRef]
140. Akbarian, M.; Yousefi, R.; Farjadian, F.; Uversky, V.N. Insulin Fibrillation: Toward the Strategies for Attenuating the Process. *Chem. Commun.* **2020**, *56*, 11354–11373. [CrossRef]
141. Perlman, S.; Netland, J. Coronaviruses post-SARS: Update on replication and pathogenesis. *Nat. Rev. Microbiol.* **2009**, *7*, 439–450.
142. Ziebuhr, J.; Snijder, E.J.; Gorbalenya, A.E. Virus-encoded proteinases and proteolytic processing in the Nidovirales. *J. Gen. Virol.* **2000**, *81*, 853–879. [CrossRef] [PubMed]
143. Harcourt, B.H.; Jukneliene, D.; Kanjanahaluethai, A.; Bechill, J.; Severson, K.M.; Smith, C.M.; Rota, P.A.; Baker, S.C. Identification of severe acute respiratory syndrome coronavirus replicase products and characterization of papain-like protease activity. *J. Virol.* **2004**, *78*, 13600–13612. [CrossRef] [PubMed]
144. Wu, C.-Y.; Jan, J.-T.; Ma, S.-H.; Kuo, C.-J.; Juan, H.-F.; Cheng, Y.-S.E.; Hsu, H.-H.; Huang, H.-C.; Wu, D.; Brik, A. Small molecules targeting severe acute respiratory syndrome human coronavirus. *Proc. Natl. Acad. Sci. USA* **2004**, *101*, 10012–10017. [CrossRef] [PubMed]
145. De Wilde, A.H.; Jochmans, D.; Posthuma, C.C.; Zevenhoven-Dobbe, J.C.; Van Nieuwkoop, S.; Bestebroer, T.M.; Van Den Hoogen, B.G.; Neyts, J.; Snijder, E.J. Screening of an FDA-approved compound library identifies four small-molecule inhibitors of Middle East respiratory syndrome coronavirus replication in cell culture. *Antimicrob. Agents Chemother.* **2014**, *58*, 4875–4884. [CrossRef] [PubMed]
146. Chu, C.; Cheng, V.; Hung, I.; Wong, M.; Chan, K.; Chan, K.; Kao, R.; Poon, L.; Wong, C.; Guan, Y. Role of lopinavir/ritonavir in the treatment of SARS: Initial virological and clinical findings. *Thorax* **2004**, *59*, 252–256. [CrossRef]
147. Chan, J.F.-W.; Yao, Y.; Yeung, M.-L.; Deng, W.; Bao, L.; Jia, L.; Li, F.; Xiao, C.; Gao, H.; Yu, P. Treatment with lopinavir/ritonavir or interferon-β1b improves outcome of MERS-CoV infection in a nonhuman primate model of common marmoset. *J. Infect. Dis.* **2015**, *212*, 1904–1913. [CrossRef]
148. Tse, L.V.; Meganck, R.M.; Graham, R.L.; Baric, R.S. The current and future state of vaccines, antivirals and gene therapies against emerging coronaviruses. *Front. Microbiol.* **2020**, *11*, 658. [CrossRef]
149. Medhi, R.; Srinoi, P.; Ngo, N.; Tran, H.-V.; Lee, T.R. Nanoparticle-based strategies to combat COVID-19. *ACS Appl. Nano Materials* **2020**, *3*, 8557–8580. [CrossRef]
150. Jeremiah, S.S.; Miyakawa, K.; Morita, T.; Yamaoka, Y.; Ryo, A. Potent antiviral effect of silver nanoparticles on SARS-CoV-2. *Biochem. Biophys. Res. Commun.* **2020**, *533*, 195–200. [CrossRef]
151. Merkl, P.; Long, S.; McInerney, G.M.; Sotiriou, G.A. Antiviral activity of silver, copper oxide and zinc oxide nanoparticle coatings against SARS-CoV-2. *Nanomaterials* **2021**, *11*, 1312. [CrossRef]
152. Huber, H.F.; Jaberi-Douraki, M.; DeVader, S.; Aparicio-Lopez, C.; Nava-Chavez, J.; Xu, X.; Millagaha Gedara, N.I.; Gaudreault, N.N.; Delong, R.K. Targeting SARS-CoV-2 Variants with Nucleic Acid Therapeutic Nanoparticle Conjugates. *Pharmaceuticals* **2021**, *14*, 1012. [CrossRef] [PubMed]
153. Idris, A.; Davis, A.; Supramaniam, A.; Acharya, D.; Kelly, G.; Tayyar, Y.; West, N.; Zhang, P.; McMillan, C.L.D.; Soemardy, C.; et al. A SARS-CoV-2 targeted siRNA-nanoparticle therapy for COVID-19. *Mol. Ther.* **2021**, *29*, 2219–2226. [CrossRef] [PubMed]

154. Lu, G.; Zhang, X.; Zheng, W.; Sun, J.; Hua, L.; Xu, L.; Chu, X.-j.; Ding, S.; Xiong, W. Development of a simple in vitro assay to identify and evaluate nucleotide analogs against SARS-CoV-2 RNA-dependent RNA polymerase. *Antimicrob. Agents Chemother.* **2020**, *65*, e01508-20. [CrossRef] [PubMed]
155. Baldassi, D.; Ambike, S.; Feuerherd, M.; Cheng, C.-C.; Peeler, D.J.; Feldmann, D.P.; Porras-Gonzalez, D.L.; Wei, X.; Keller, L.-A.; Kneidinger, N.; et al. Inhibition of SARS-CoV-2 replication in the lung with siRNA/VIPER polyplexes. *J. Control. Release* **2022**, *345*, 661–674. [CrossRef]
156. García-Pérez, B.E.; González-Rojas, J.A.; Salazar, M.I.; Torres-Torres, C.; Castrejón-Jiménez, N.S. Taming the Autophagy as a Strategy for Treating COVID-19. *Cells* **2020**, *9*, 2679. [CrossRef]
157. Mahmoudi, M. Emerging biomolecular testing to assess the risk of mortality from COVID-19 infection. *Mol. Pharm.* **2020**, *18*, 476–482. [CrossRef]
158. Moitra, P.; Alafeef, M.; Dighe, K.; Frieman, M.B.; Pan, D. Selective naked-eye detection of SARS-CoV-2 mediated by N gene targeted antisense oligonucleotide capped plasmonic nanoparticles. *ACS Nano* **2020**, *14*, 7617–7627. [CrossRef]
159. Baker, A.N.; Richards, S.-J.; Guy, C.S.; Congdon, T.R.; Hasan, M.; Zwetsloot, A.J.; Gallo, A.; Lewandowski, J.R.; Stansfeld, P.J.; Straube, A.; et al. The SARS-COV-2 spike protein binds sialic acids and enables rapid detection in a lateral flow point of care diagnostic device. *ACS Cent. Sci.* **2020**, *6*, 2046–2052. [CrossRef]
160. Khoris, I.M.; Ganganboina, A.B.; Suzuki, T.; Park, E.Y.J.N. Self-assembled chromogen-loaded polymeric cocoon for respiratory virus detection. *Nanoscale* **2021**, *13*, 388–396. [CrossRef]
161. Wang, D.; He, S.; Wang, X.; Yan, Y.; Liu, J.; Wu, S.; Liu, S.; Lei, Y.; Chen, M.; Li, L.; et al. Rapid lateral flow immunoassay for the fluorescence detection of SARS-CoV-2 RNA. *Nat. Biomed. Eng.* **2020**, *4*, 1150–1158. [CrossRef]
162. Li, Z.; Yi, Y.; Luo, X.; Xiong, N.; Liu, Y.; Li, S.; Sun, R.; Wang, Y.; Hu, B.; Chen, W.; et al. Development and clinical application of a rapid IgM-IgG combined antibody test for SARS-CoV-2 infection diagnosis. *J. Med. Virol.* **2020**, *92*, 1518–1524. [CrossRef] [PubMed]
163. Huang, C.; Wen, T.; Shi, F.-J.; Zeng, X.-Y.; Jiao, Y.-J. Rapid detection of IgM antibodies against the SARS-CoV-2 virus via colloidal gold nanoparticle-based lateral-flow assay. *ACS Omega* **2020**, *5*, 12550–12556. [CrossRef] [PubMed]
164. Kevadiya, B.D.; Machhi, J.; Herskovitz, J.; Oleynikov, M.D.; Blomberg, W.R.; Bajwa, N.; Soni, D.; Das, S.; Hasan, M.; Patel, M.; et al. Diagnostics for SARS-CoV-2 infections. *Nat. Mater.* **2021**, *20*, 593–605. [CrossRef] [PubMed]
165. Chacón-Torres, J.C.; Reinoso, C.; Navas-León, D.G.; Briceño, S.; González, G. Optimized and scalable synthesis of magnetic nanoparticles for RNA extraction in response to developing countries' needs in the detection and control of SARS-CoV-2. *Sci. Rep.* **2020**, *10*, 1–10. [CrossRef] [PubMed]
166. Akbarian, M.; Gholinejad, M.; Mohammadi-Samani, S.; Farjadian, F.J. Theranostic mesoporous silica nanoparticles made of multi-nuclear gold or carbon quantum dots particles serving as pH responsive drug delivery system. *Microporous Mesoporous Mater.* **2022**, *329*, 111512. [CrossRef]
167. Hildebrandt, N.; Spillmann, C.M.; Algar, W.R.; Pons, T.; Stewart, M.H.; Oh, E.; Susumu, K.; Diaz, S.A.; Delehanty, J.B.; Medintz, I.L. Energy transfer with semiconductor quantum dot bioconjugates: A versatile platform for biosensing, energy harvesting, and other developing applications. *Chem. Rev.* **2017**, *117*, 536–711. [CrossRef]
168. Gorshkov, K.; Susumu, K.; Chen, J.; Xu, M.; Pradhan, M.; Zhu, W.; Hu, X.; Breger, J.C.; Wolak, M.; Oh, E. Quantum dot-conjugated SARS-CoV-2 spike pseudo-virions enable tracking of angiotensin converting enzyme 2 binding and endocytosis. *ACS Nano* **2020**, *14*, 12234–12247. [CrossRef]
169. Hoseini-Ghahfarokhi, M.; Mirkiani, S.; Mozaffari, N.; Sadatlu, M.A.A.; Ghasemi, A.; Abbaspour, S.; Akbarian, M.; Farjadain, F.; Karimi, M. Applications of graphene and graphene oxide in smart drug/gene delivery: Is the world still flat? *Int. J. Nanomed.* **2020**, *15*, 9469. [CrossRef]
170. Seo, G.; Lee, G.; Kim, M.J.; Baek, S.-H.; Choi, M.; Ku, K.B.; Lee, C.-S.; Jun, S.; Park, D.; Kim, H. Rapid detection of COVID-19 causative virus (SARS-CoV-2) in human nasopharyngeal swab specimens using field-effect transistor-based biosensor. *ACS Nano* **2020**, *14*, 5135–5142. [CrossRef]
171. Surnar, B.; Kamran, M.Z.; Shah, A.S.; Dhar, S. Clinically approved antiviral drug in an orally administrable nanoparticle for COVID-19. *ACS Pharmacol. Transl. Sci.* **2020**, *3*, 1371–1380. [CrossRef]
172. Kim, J.; Mukherjee, A.; Nelson, D.; Jozic, A.; Sahay, G. Rapid generation of circulating and mucosal decoy ACE2 using mRNA nanotherapeutics for the potential treatment of SARS-CoV-2. *bioRxiv* **2020**. [CrossRef]
173. Neufurth, M.; Wang, X.; Tolba, E.; Lieberwirth, I.; Wang, S.; Schröder, H.C.; Müller, W.E.G. The inorganic polymer, polyphosphate, blocks binding of SARS-CoV-2 spike protein to ACE2 receptor at physiological concentrations. *Biochem. Pharmacol.* **2020**, *182*, 114215. [CrossRef] [PubMed]
174. Khaiboullina, S.; Uppal, T.; Dhabarde, N.; Subramanian, V.R.; Verma, S.C. Inactivation of human coronavirus by titania nanoparticle coatings and UVC radiation: Throwing light on SARS-CoV-2. *Viruses* **2020**, *13*, 19. [CrossRef] [PubMed]
175. Lin, S.; Zhang, Q.; Li, S.; Zhang, T.; Wang, L.; Qin, X.; Zhang, M.; Shi, S.; Cai, X. Antioxidative and angiogenesis-promoting effects of tetrahedral framework nucleic acids in diabetic wound healing with activation of the Akt/Nrf2/HO-1 pathway. *ACS Appl. Mater. Interfaces* **2020**, *12*, 11397–11408. [CrossRef] [PubMed]
176. Kiley, M.P. Filoviridae: Marburg and ebola viruses. In *Laboratory Diagnosis of Infectious Diseases Principles and Practice*; Springer: Berlin/Heidelberg, Germany, 1988; pp. 595–601.

177. Kobinger, G.P.; Figueredo, J.M.; Rowe, T.; Zhi, Y.; Gao, G.; Sanmiguel, J.C.; Bell, P.; Wivel, N.A.; Zitzow, L.A.; Flieder, D.B. Adenovirus-based vaccine prevents pneumonia in ferrets challenged with the SARS coronavirus and stimulates robust immune responses in macaques. *Vaccine* **2007**, *25*, 5220–5231. [CrossRef] [PubMed]
178. Ura, T.; Okuda, K.; Shimada, M. Developments in viral vector-based vaccines. *Vaccines* **2014**, *2*, 624–641. [CrossRef] [PubMed]
179. Saif, L.J. Vaccines for COVID-19: Perspectives, prospects, and challenges based on candidate SARS, MERS, and animal coronavirus vaccines. *Euro. Med. J.* **2020**. [CrossRef]

nanomaterials

MDPI

Review

Nanotechnology-Assisted Cell Tracking

Alessia Peserico *, Chiara Di Berardino, Valentina Russo, Giulia Capacchietti, Oriana Di Giacinto, Angelo Canciello, Chiara Camerano Spelta Rapini and Barbara Barboni

Faculty of Bioscience and Technology for Food, Agriculture and Environment, University of Teramo, 64100 Teramo, Italy; cdiberardino@unite.it (C.D.B.); vrusso@unite.it (V.R.); gcapacchietti@unite.it (G.C.); odigiacinto@unite.it (O.D.G.); acanciello@unite.it (A.C.); chiara.cameranospelta@studenti.unite.it (C.C.S.R.); bbarboni@unite.it (B.B.)
* Correspondence: apeserico@unite.it

Abstract: The usefulness of nanoparticles (NPs) in the diagnostic and/or therapeutic sector is derived from their aptitude for navigating intra- and extracellular barriers successfully and to be spatiotemporally targeted. In this context, the optimization of NP delivery platforms is technologically related to the exploitation of the mechanisms involved in the NP–cell interaction. This review provides a detailed overview of the available technologies focusing on cell–NP interaction/detection by describing their applications in the fields of cancer and regenerative medicine. Specifically, a literature survey has been performed to analyze the key nanocarrier-impacting elements, such as NP typology and functionalization, the ability to tune cell interaction mechanisms under in vitro and in vivo conditions by framing, and at the same time, the imaging devices supporting NP delivery assessment, and consideration of their specificity and sensitivity. Although the large amount of literature information on the designs and applications of cell membrane-coated NPs has reached the extent at which it could be considered a mature branch of nanomedicine ready to be translated to the clinic, the technology applied to the biomimetic functionalization strategy of the design of NPs for directing cell labelling and intracellular retention appears less advanced. These approaches, if properly scaled up, will present diverse biomedical applications and make a positive impact on human health.

Keywords: nanoparticles; cell tracking; cell labelling; cell uptake; stem cell homing

Citation: Peserico, A.; Di Berardino, C.; Russo, V.; Capacchietti, G.; Di Giacinto, O.; Canciello, A.; Camerano Spelta Rapini, C.; Barboni, B. Nanotechnology-Assisted Cell Tracking. *Nanomaterials* **2022**, *12*, 1414. https://doi.org/10.3390/nano12091414

Academic Editors: Goran Kaluđerović and Nebojša Pantelić

Received: 27 March 2022
Accepted: 14 April 2022
Published: 20 April 2022

1. Introduction

Cell tracking refers to the investigation of living cells over a period allowing the visual depiction, characterization, and quantification of biological processes at the cellular and subcellular levels within intact living organisms. A fundamental property of any real-world object is that it extends in both space and time. This is particularly true and applicable for living organisms in which the study of subjects in three dimensions and the possibility to track it over time (denoted as 3D or 4D) assumes great significance.

Research community interest has grown exponentially in the last decades concerning the possibility of tracking objects (Figure 1A) [1–6] to automatically follow hundreds to thousands of cells. Imaging has provided the elective approach to enable cell tracking and it assumes the labeling of the target cell with objects, otherwise defined as contrast agents, with the aim of achieving contrast between the cells of interest and the other cells of the organism. Interestingly, nanoparticle (NP) systems have received great attention as the contrast agents to be used for cell labeling and followed by means of imaging techniques. Although the first attempts to automate the tracking of cells using NPs date back at least 20 years, the development of more advanced tracking methods really took off in the past decade (Figure 1B).

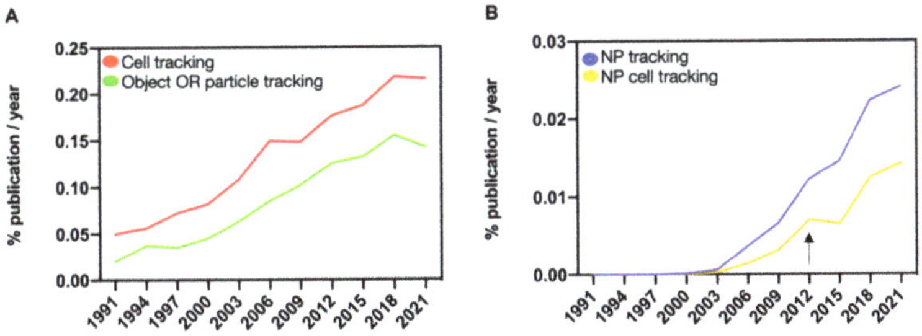

Figure 1. Scientific community interest in tracking related research. Percentage of publications in Web of Science (WoS) database as function of publication year for the indicated combinations of keywords with the field tag «Topic». "OR" was used as Boolean operator to combine the keywords "object" and "particle tracking" and to lunch the search in WoS database. The plot shows the exponentially increasing interest in cell, particle, or object tracking (**A**) as well as tracking with nano systems (**B**) in the biomedical and related literature. The impact of the intrinsic growth of the number of publications was weighed and corrected by plotting percentages.

Two main components have a big impact in defining the approaches for cell tracking: the NP systems and the imaging technologies adopted to assist in cell tracking. NP systems are characterized by particles or constituents with unique features making them attractive for tracking purposes: (1) a small size (range diameter between 1 and 100 nm) and (2) a surface to mass ratio, functional to adsorb and efficiently carry compounds including peptides, fluorescent probes, and drugs that can be used as tracers, therapeutic tools or cell uptake enhancers [7].

Due to their nanoscale dimensions, NPs can be easily transported across cell membranes and reach the crucial subcellular organelles [8]. Furthermore, a high surface area-over-volume ratio enhances their interaction with cellular components [9]. In addition, NPs offer attractive features since their structure and chemical properties can be modified to facilitate cellular incorporation and because they can carry a high payload of the relevant labels into cells, thus generating high contrast signals to be followed by imaging.

Aside from NPs, the selection of appropriate imaging modalities assumes a crucial role in tuning the effective cell tracking approaches. Many advances have been achieved by developing new optical imaging probes, which have been useful for setting up multimodal imaging systems, and by adopting sophisticated optics and optical systems able to overcome the resolution limitations imposed by the scattering and absorption properties of tissue [10].

Overall, NP systems and imaging techniques can be considered as two closely related components of any cell tracking approach, which, if well managed, have an important impact in the improvement of current imaging resolution and sensitivity. The cell-tracking imaging platform choice relies on the type of contrast agent used for cell labeling and vice versa. Currently, the most suitable platform categories for NP-based imaging are magnetic resonance imaging (MRI), X-ray (also known as X-ray computed tomography and abbreviated as CT), optical imaging (endoscopy and fluorescence near-infrared or bioluminescence-based imaging methodologies), ultrasound, radionuclide molecular imaging and photoacoustic imaging (PAI), a hybrid technology combining optical and ultrasound imaging [11].

Specifically, MRI and CT are the two most used and clinically approved techniques for in vivo cell tracking, whereas the other imaging approaches have been primarily investigated in the last decade in translational research protocols showing a great potential for their clinical application. Cell tracking applications can be found in different areas of biomedicine ranging from the imaging of tissue replacement cell therapy (regenerative medicine) to cancer diagnosis and treatment (cancer medicine).

In regenerative medicine, cell tracking is a key technology for the development and optimization of cell therapy for the replacement or renewal of damaged or diseased tissue using transplanted cells [12]. In the cancer context, it helps early-stage cancer diagnosis and therapeutic imaging of tumor foci by improving accuracy and sensitivity of the current clinical protocols [13].

Several NPs with unique magnetic and/or optical properties have been investigated preclinically for real-time, in vivo cell monitoring [11]. These NPs include inorganic (with metal, ceramic, or semiconductor elements), organic (with polymeric or lipidic elements) and hybrid/composite (with both inorganic and organic elements) NPs. Of note, since the current imaging technology of choice for clinical translation is MRI, NPs with magnetic properties have been also tested for clinical application. Consistently, 11 studies employing NPs with magnetic properties for intra-operative procedure monitoring and/or diagnosis purposes are currently in clinical trial phases (Supplementary Materials Table S1).

Different cell tracking approaches have been developed including the ability to track NPs enclosed in cell membranes, cell derivates (microvesicles) or actively and directly incorporated by the cells. The last, showing great potential for application in the field of cell tracking with imaging [14–16].

In the field of regenerative medicine, the tracking of stem cell transplantation procedures and regenerative performance through imaging have been investigated with success by employing direct cell labeling methods based on the introduction of the NPs working as a contrast agent into the cells to be transplanted prior to in vivo injection [17]. Stem cell tracking approaches enhancing the regenerative properties of stem cells to be transplanted have also been developed by exploiting the possibility of incorporating into these cells NPs bearing specific moieties with an immunomodulatory role [12] (Figure 2A).

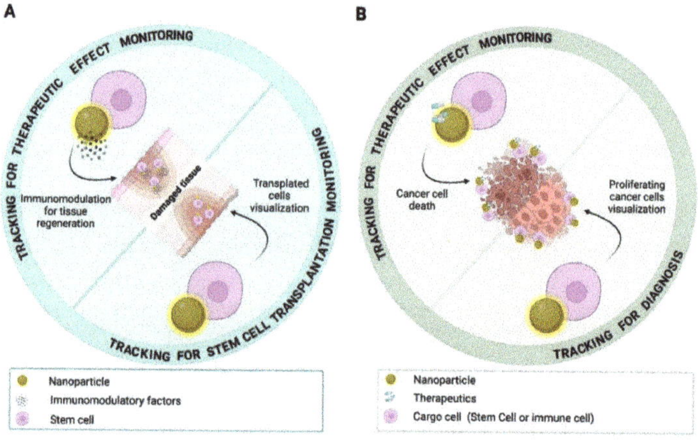

Figure 2. Direct cell tracking approaches suitable for imaging of cancer and cell transplantation for tissue regeneration, and related objectives. (**A**) Stem cells enclosing NPs can be used for monitoring transplantation events, as well as for directing therapeutic effects by releasing immunomodulatory factors into the tissue site requiring regeneration. (**B**) Stem cells or immune cells enclosing NPs can be adopted for homing in on tumors by allowing diagnosis and/or therapy. Figure created with BioRender. Available online: https://biorender.com/ (accessed on 25 March 2022).

Of note, this strategy emerged as also applicable in the field of cancer imaging, where the native attitude of stem cells to sense and home in on tumor foci as well as the ability of some immune cells (e.g., macrophages or T cells) to be recruited by tumor-derived inflammatory stimuli, were exploited by using these cells as vehicles for NPs and thus, as a tracer guiding the identification of cancerous districts [18–21]. Moreover, because of the availability of specific NP systems enclosing therapeutics within their structure, this approach was also found to be suitable for monitoring the therapeutic effects elicited by the NPs conveyed by the cargo cell or directly incorporated by the cancer cells to be targeted [22] (Figure 2B).

The application of cell tracking approaches has a big impact on the improvement of currently available diagnostic and therapeutic imaging protocols. In the context of regenerative medicine, NP-based cell tracking through imaging provides a safe way to monitor the viability, proliferation and differentiation of transplanted cells by enabling the guidance of cell replacement and testing the therapeutic success of the adopted transplantation procedure.

With respect to cancer medicine, NP-based cell tracking through imaging represents an invaluable tool to provide a rapid and sensitive detection of cancer foci, thus acquiring a relevant role in early tumor diagnosis [13,23], consequently yielding greater chances for cancer treatment, and extending life expectancy. Furthermore, it can be exploited to monitor the response to specific theranostics, allowing the assessment of disease state [24]. Interestingly, some recent evidence has shown a possible application in image-guided surgery, enabling the intra-operative detection of any residual tumor mass to be removed and/or therapeutically targeted [25–27]. Despite their great potential, there are currently some limitations regarding their application for in vivo cell tracking at the translational level yet to be overcome [28].

This review aims to provide an overview of the nanotechnologies and methodologies eligible for cell tracking which have been investigated in the last decade, the window period identified as critical for the development of NP-based tracking strategies, with emphasis on (1) the specific NP characteristics responsible for effects on cell labeling, consequently defining the effectiveness of cell tracking, and (2) the approaches and mechanisms of NP incorporation into cells.

Consequently, the next paragraphs will focus on the following key elements: the types of NPs currently adopted in cell imaging (Section 2); the NP functionalization strategies most used to improve cell targeting, cell uptake, therapeutic tracking, and imaging performance (Section 3); the approaches suitable for in vitro, in vivo and ex vivo tracking of NP delivery (Section 4); the uptake mechanisms of the NPs (Section 5) and the cellular systems mostly described for NP loading (Section 6).

All the above key aspects have been inferred from research evidence collected with a systematic literature review performed in Web of Science (WoS) database during the last decade which focused on the investigation of NP-based cell tracking strategies.

2. NP-Based Cell Tracking Derived Bibliographic Database

Literature data published in peer-reviewed international papers indexed in the advanced search of WoS [v.5.35] "Core collection" archive (https://apps.webofknowledge.com/WOS_AdvancedSearch (accessed on 3 July 2021) in the last 10 years were considered.

Specifically, three lists were created by searching for the following key words and filtering for original research articles:

1. Nanoparticles, cell tracking methods, fluorescent dyes;
2. Nanoparticles, cellular uptake, cell labelling, staining;
3. Tomography, magnetic resonance, nanoparticles, stem cell homing.

The words "AND" and "OR" were used as Boolean operators, "TS" as field tag.

The lists of papers originating from each key word search were then matched to create the final database of citations.

Overall, the adopted bibliographic approach allowed us to collect 151 original studies focusing on the investigation of strategies for direct cell labelling and tracking with NPs (Supplementary Materials Table S2). Ninety-eight of the total citations (64.5%) characterized the approaches to incorporate NPs into vehicle cells to be used to sense target cells or body districts after in vivo delivery into the body. In these studies, the aim was to find the most suitable way to improve uptake efficiency without affecting cell cargo wellness.

Fifty-four of the total citations (35.5%) investigated the effect of NP incorporation into cell models to be therapeutically targeted. In these studies, the aim was to find a protocol that could be effective in targeting desired cells, monitoring NP uptake and the cytotoxic effect of NP cell internalization (see Supplementary Materials Table S2).

Furthermore, based on the adopted cell models, the collected studies could find application in the field of cancer and tissue regeneration imaging. Specifically, 72 of the total citations were applied to cancer imaging, 57of the total citations to tissue regeneration imaging, whereas 23 of the total citations adopted models suitable in both fields (see Supplementary Materials Table S2).

3. NP Features and Cell Tracking Applications for Tissue Regeneration and Cancer Imaging

The literature survey allowed us to identify four distinctive classes of NPs as useful for developing suitable cell tracking platforms: inorganic, polymeric, hybrid and lipid based. The inorganic category resulted in the most represented NP typology in this survey, accounting for the 79.6% of the total collected citations, followed by polymeric (12.5%), hybrid (4.6%) and lipid based (3.3%), respectively (Figure 3). The prominent representation of the inorganic typology reflects the fact that these NPs are the ones best suited to cell tracking through the most characterized imaging platforms to date.

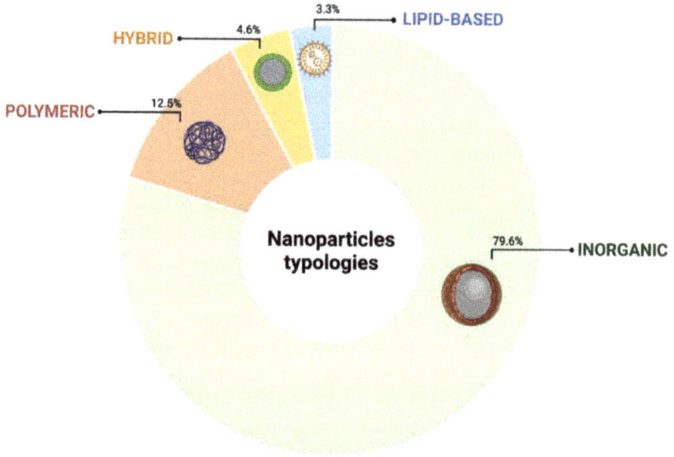

Figure 3. Classification of NPs based on the constitutive element. Classes and subclasses of nanoparticles based on their chemical properties (inorganic, polymeric, hybrid and lipid-based). Percentage (%) refers to the typology of NPs with respect to the total number of NPs identified in the literature survey. Figure created with BioRender. Available online: https://biorender.com/ (accessed on 25 March 2022).

3.1. *Inorganic*

Inorganic NPs are characterized by a typical core/shell structure which can assume different shapes: spherical, rods, wires, tubes, pyramids, and stars. The core size of these nanoparticles is approximately 3–6 nm and a coated system (shell) increases its size to 20–150 nm. The core can contain metals, other chemical elements or fluorescent dyes encapsulated in silica. The core defines the magnetic, electronic, fluorescence and optical properties of the NP [28]. The shell is usually made of metals or organic polymers that (I) protect the core from chemical interactions with the external environment, (II) serves as a substrate for conjugation with biomolecules (antibodies, peptides or oligonucleotides) and (III) preserves NP stability avoiding aggregation [29,30]. Due to their magnetic, radioactive, X-ray absorption or plasmonic properties, inorganic NPs are used for diagnostics and imaging purposes and most of them display good stability [31] and biocompatibility [32].

Four main subcategories of inorganic NPs applicable for cell tracking were identified by this survey, based on the nature of the core element:

(I) Metal-based NPs, accounting for 31.5% of the collected inorganic NPs, include silica, manganese, gold, silver, lanthanide, molybdenum, ruthenium, rubidium, gadolinium, and zinc elements.

(II) Metal oxide-based NPs, representing 64.5% of the collected inorganic NPs, enclose iron oxide, superparamagnetic iron oxide (SPIO), ultrasmall superparamagnetic iron oxide (USPIO), titanium oxide and cobalt iron oxide elements.

(III) Metal sulfide or phosphide-based NPs, accounting for 2.4% of the collected inorganic NPs, comprise quantum dots.

(IV) Mineral-based NPs, accounting for 2.4% of the collected inorganic NPs, take in hydroxyapatite and selenium elements.

3.2. Polymeric

Polymeric NPs are particles within the size range from 1 to 1000 nm and are constituted by a polymeric matrix core which can be loaded with bioactive molecules. They are categorized into two forms, spheres (the bioactive molecule is dispersed within a polymer matrix) and capsules (the bioactive compound is placed in the core of the particle covered by a layer of polymer). Polysaccharides and proteins are the commonly used materials for the fabrication of polymeric nanoparticles. The polymer provides biocompatibility and protection to the active molecules to the delivery site [33].

The most used polymers for the constitution of polymeric NPs collected by this survey were poly(ethylene glycol) (PEG; 33% of the collected polymeric NP), poly(lactide-co-glycolide) (PLGA; 24%) and polystyrene (PS; 13%). Although to a lesser extent, other polymers were also identified: poly(epsilon-caprolactone) (PCL), poly(lactide) (PLA), poly(2-methacryloyloxy) ethyl phosphorylcholine (PMPC), poly(glycerol-co-sebacate) (PGS), poly (diphenylamine)-poly(ethylene oxide) (PDPA-PEO) and poly[2-methoxy-5-(2-ethylhexyloxy)-1,4-phenylenevinylene] (MEH-PPV). Of note, these polymers have been used alone or in combination to design polymeric NPs for cell tracking.

3.3. Hybrid

Hybrid NPs are constructed from at least two different NPs, to overcome the limits of single-component nanoparticles, to improve properties, to achieve new properties not possible for single nanoparticles, and/or to achieve multiple functionalities for single nanoparticles. Size range can be variable depending on the NP typologies used for their fabrication [34].

3.4. Lipid-Based

Lipid-based NPs, defined as colloidal carriers for bioactive molecules, are assembled to constitute different structures: liposomes with size < 200 nm, solid lipid NP (SLN, solid lipids), and nanostructured lipid carriers (NLC, a combination of liquid and solid lipids) with sizes ranging from 10–1000 nm [35–38].

Liposomes are spherical vesicles with an aqueous core and bilayer lipid membrane. They have the capacity to encapsulate diverse bioactive compounds, which can be included into the aqueous core or at the bilayer interface [39].

SLNs are generally spherical in shape and consist of a solid lipid core stabilized by a surfactant. This construct can be used to deliver both hydrophilic and hydrophobic bioactive molecules [40].

Mixing solid lipids with small amounts of liquid lipids allows structural rearrangements of the matrix-generating NLC formulations to be produced, improving their properties in this respect, while maintaining the original benefits of SLNs. Of note, recently, cell-derived membrane lipidic vesicles with a small size range (40–100 nm) have been considered as alternative naturally derived lipid NPs versus the synthetic lipidic formulation [16].

With respect to the application of the different NP typologies in the field of cancer imaging or tissue regeneration imaging, the literature survey revealed that polymeric and lipid-based NPs are the most represented in the studies, pointing to the development of cancer imaging and therapeutic protocols; the hybrid category was applied for tissue regeneration imaging whereas use of the inorganic category was well documented for setting up both cancer and regenerative imaging protocols (see Supplementary Materials Table S1). Importantly, some of the collected studies, which will be further discussed below, moved a step forward in the characterization of the best NP platform for cell tracking by providing the adoption of in vitro and/or in vivo imaging procedures to be applicable in cancer or regenerative imaging.

3.5. NP-Based Cell Tracking for Imaging of Tissue Regeneration

In the context of tissue replacement cell-therapy imaging, protocols assuring the long-term monitoring of transplanted cells without affecting cell viability and/or properly modulating their state of differentiation depending on the tissue/organ to be targeted, have yet to be identified. Indeed, at present, clinical cell-tracking trials have only provided information on immediate cell delivery and short-term cell retention [41]. Overall, many key issues concerning labeling solutions able to augment the possibility of tracking cells enclosing NPs, avoiding label dilution during cell proliferation and improving the uptake and retention of NPs, as well as preserving the proper functions of the transplanted cell and their correct homing to the target site, remain to be addressed [42].

3.6. Labeling Solution Improving Cell Imaging

Over the last decade, direct cell labeling with inorganic NPs engineered with multiple tracers has received increased attention for improving the tracking and imaging performance of tissue and/or organ regeneration [39]. Indeed, the use of a combination of two or more chemical elements to be traced by means of different imaging techniques, known as the multi-labeling approach, represents a way to overcome the imaging detection limits of a specific technique. As some examples, inorganic gold NPs with red-emitting firefly luciferase (RfLuc)-based bioluminescence (BL) tags [43] and inorganic NPs with gold and gadolinium elements [44] have been developed for the noninvasive labeling and tracking of human mesenchymal stem cells (hMSCs) in a mouse model of pulmonary disease. Specifically, multimodal imaging with CT, MRI and BL was successfully adopted to characterize the in vivo migration, homing, functions, and survival of transplanted hMSCs, giving further insight for the treatment of pulmonary fibrosis.

Inorganic iron oxide NPs modified with melanin and Fe ions were successfully used for dual modal imaging through MRI and photoacoustic (PA) methods of engrafted bone marrow-derived mesenchymal stem cells in mouse models [45]. Multilayered inorganic NPs constituted by a combination of mesoporous silica, iron oxide and gold elements were successfully used to multilabel umbilical cord mesenchymal stem cells in ischemic mouse brain models [46]. Moreover, inorganic iron oxide labeled with a fluorophore have been used to trace in utero hematopoietic cell transplantation in a canine animal model, offering a potential approach for early intervention for the treatment of diseases before birth [47].

Interestingly, Geburek and colleagues experimented on an additional way for multilabeling by using adipose-derived mesenchymal stem cells traduced with a reporter gene for GFP and co-labelled with iron particles for tracking after the intralesional treatment of artificial equine tendon lesions [48]. A similar approach was used to efficiently track the homing of magnetically labeled adipose-derived stem cells [49] and human umbilical cord-derived mesenchymal stem cells [50] expressing GFP into mouse carotid artery and cutaneous injury sites, respectively. In contrast, Wu and colleagues suggested combining the reporter gene directly with the NPs. In this case, the adoption of inorganic iron oxide NPs complexed with polymers acting as a cargo for the luciferase system and or a fluorescence protein as reporter genes was shown to be effective for an MRI-visible gene delivery agent which could effectively label MSCs, providing the basis for bimodal bioluminescence and MRI tracking of transplantation to solve acute liver injury [51].

Hybrid NPs have also been adopted for generating a platform for multimodal imaging. Fluorinated polymeric NPs combined with fluorine as the MRI contrast agent and functionalized with a fluorophore (N-fluoresceinyl) maleimide) as optical contrast moiety were used for tracking MCS in vivo [52].

Advances in multimodal labeling and tracking of cell transplantations have also been made by functionalization of polymeric NPs. Zhang et al. identified how the fluorescent imaging of human mesenchymal stem cells that have endocytosed fluorescent polymeric NPs can be directly and clearly captured in the one-photon and two-photon modes, offering the possibility of the direct monitoring of stem cells with high resolution and quantifying NP uptake, encouraging future quantitative clinical assessment in imaging-guided cell therapies [53]. Similarly, the exploitation of the one-photon and two-photon imaging properties of polymeric NPs loaded with red phosphorescence dye of bis(2-methyldibenzo[f,h]quinoxaline) (acetylacetonate) iridium(III) (Ir(MDQ)2acac) provided a reliable approach for the efficient and sensitive tracking and monitoring of transplanted neural stem cells (NSCs) [54]. Multimodal tracking was also achieved by functionalizing NPs with fluorine, which served as contrast agent for MRI and as an effective platform to properly direct the transplanted MSC fate in a non-obese diabetic/severe combined immunodeficient mouse model [52], as well as for the tracking of immune and stem cells to be adopted for cellular therapies [55].

3.7. Labeling Solution Improving Cell Uptake and Homing

The labeling solutions aimed at improving cell uptake and homing to the desired site of the body have been investigated by functionalizing NPs with different biomaterials. Accordingly, the literature survey identified inorganic iron oxide NPs functionalized with several moieties including poly-lysine, dextran, zinc, protamine sulfate, heparin or poly-(l-lactide) as good platforms for direct label, home and track MSCs in preclinical models of wound healing [56], traumatic brain injury [57–61], after renal transplantation in mice [62], hepatic cirrhosis [63] and photothrombotic cerebral infarction [64] with MRI. Poly (dopamine)-coated iron oxide NPs were also used for noninvasive labeling, tracking, and targeted delivery of adipose tissue-derived stem cells in mouse models of liver diseases [65].

Efficient strategies for the traceable homing of therapeutic NPs were developed by using hybrid NPs composed of glucocorticoid betamethasone phosphate (BMP) and the fluorescent dye DY-647 (BMP-IOH-NPs). Uptake of these NPs into macrophages was proved to efficiently drive treatment of inflammation with simultaneous in vivo monitoring of NP delivery [66].

3.8. Labeling Solutions for Long-Term Imaging and Modulation of Stem Cell Differentiation State

Protocols assuring long-term monitoring and targeted modulation of transplanted stem cell differentiation have been investigated by functionalizing NPs with moieties such as bioactive factors and/or nucleic acids able to manage cell fate upon their in situ release. Consistently, the literature survey identifies studies in which inorganic mesoporous silica NPs functionalized with dexamethasone [67] or a specific miRNA (miRNA-26a-5p) [68] (p. 2) resulted in effective tracking of bone marrow stem-cell homing and enhancing osteogenic differentiation in preclinical mouse models. Furthermore, polymeric NPs bearing the stromal factor-1α (SDF-1α) as the bioactive molecule known to regulate MSC homing and localization and the antimir138, with a key role in the promotion of osteogenic differentiation proved to be efficient for driving cranial bone regeneration by means of MSCs in preclinical mouse models [69].

Other studies have noticed a strong influence of intrinsic physicochemical properties of some types of NPs in directing stem cell fate by modulating the differentiation and proliferation processes. Accordingly, inorganic gold NPs modified with 11-mercaptoundecanoic acid as well as inorganic upconverting NPs have shown the ability to modulate lineage differentiation of rat bone marrow MSCs [70,71].

3.9. NP-Based Cell Tracking for Imaging of Cancer

A NP-based cell tracking system represents a critical element, which if well managed, might allow the improvement of the resolution and sensitivity of current imaging techniques applied for targeted cancer diagnosis and/or therapy. Currently, the most widely used diagnostic imaging tools for the clinical detection of cancer, such as X-ray, magnetic resonance imaging (MRI), computed tomography (CT), endoscopy, and ultrasound, can only detect cancer when there is a visible change to the tissue. By that time, thousands of cancer cells may have proliferated and even metastasized. In addition, current imaging methods cannot distinguish benign lesions from malignant lesions [72]. Furthermore, cytology and histopathology examinations are usually adopted to characterize malignant cells and/or tissues primarily identified through imaging techniques such as CT or MRI and cannot be effectively and independently applied to detect cancer at an early stage [73].

The exploitation of the different physicochemical properties of NPs and the possibility of functionalizing NPs with a plethora of bioactive molecules have been demonstrated to be crucial in innovating the procedures for cancer imaging and to allow targeted therapeutic NP uptake.

3.10. Labeling Strategies to Improve Targeted Cell Uptake and Tracking

Improved targeted cancer cell tracking has been achieved by functionalizing several types of NPs with moieties which enhance tracking resolution.

In this context, inorganic silica NPs conjugated with innovative fluorescent nanomaterials bearing excellent photostability and with specific miRNAs and/or active molecules with a high affinity for cancer cells were successfully adopted to trace cancer foci. In detail, the synthesis of fluorescent dye-doped silica NPs carrying miRNA-21 or PEG peptide were demonstrated to be effective in NP delivery and imaging in breast [74] and liver [75] cancer cells. miRNA21 was found to be one of the most represented in breast cancer tumors [76], whereas PEG has been found to effectively recognize tumor-marker carcinoembryonic antigens (CEAs) [77]. Furthermore, inorganic iron oxide NPs carrying therapeutic siRNAs were efficiently used for tracking siRNA-based gene therapy targeting the BCL2 and BIRC5 of oral cancer and glioblastoma cells [78].

Another class of NPs, polymeric NPs conjugated with liposoluble fluorescent probes (fluorescein, coumarin 6 or DiR probes) as contrast agents and with oleanolic acid [79] or emodin and heparin sodium [80] as targeting moieties, were also used to track liver cancer cells to be therapeutically targeted. The multifunctionalization of polymeric NPs with coumarin 6 as labeling moieties and with either a therapeutic moiety and a peptide (peptide-22) with a special affinity for the low-density lipoprotein receptor (LDLR) was also successfully used to facilitate drug penetration of the blood–brain barrier and to improve the visualization of targeted cellular uptake, avoiding a non-selective brain drug accumulation in a preclinical mouse model of glioblastoma [81].

Furthermore, the tracking of polymeric NPs labeled with a near-infrared dye (NIR) and conjugated with either a chemotherapeutic drug paclitaxel (PTX) and transferrin as the targeting moieties allowed the biodistribution of tumor cells incorporating NPs and the antitumor effect to be studied in a preclinical mouse model of cancer overexpression in the transferrin receptor [82]. Conjugation of polymeric NPs with NIR fluorophore dye has also been proved efficient in mediating the long-term tracking of cancer cells in vivo [83].

Strategies for cancer imaging have also been developed by adopting other types of NPs. Hybrid NPs constituted by glyceryl monooleate-coated magnetic nanoparticles (MNPs) combined with glyceryl monooleate were found to act as a better labeling and efficient tracking agent without affecting the inherent properties of the MSCs to be used for tumor homing in preclinical rat models [84].

Lipid-based micelle NPs carrying a fluorescent-tagged antibody targeting the breast cancer cell antigen HER2 showed highly efficient internalization into the target cell, good tracking performance and a cytotoxic effect [85].

3.11. Labeling Strategies to Improve Imaging Sensitivity

Methods to augment the possibility of localizing the homing of stem cells or other cell types bearing NPs to tumors are being investigated with the goal of enhancing tracking and treatment efficacy.

As some examples, cationic magneto-liposomes were used to magnetically label human blood outgrowth endothelial cells (BOECs) and follow their homing to map cancer foci by magnetic resonance imaging (MRI) [86]. Importantly, these cells have shown extensive promise as gene-delivery vehicles for cancer cell treatment.

Furthermore, the two- and three-photon luminescence capabilities and strong optical absorption of an innovative class of gold NP structures, known as gold nanocages, enabled the quantitative tracking of labeled hMSCs using two imaging procedures: two-photon microscopy and photoacoustic microscopy. Importantly, this bi-modal tracking enabled the locating of cells that had migrated to glioblastoma regions formed by subcutaneous injection in nude mice in vivo [87].

In addition, luciferase-expressing human adipocyte-derived stem cells (ADSCs) were also used as a vehicle for Indium-111 radiolabeled inorganic iron oxide NPs to produce cells with tri-modal imaging capabilities. The ADSCs' homing and fate was monitored by using mouse models of breast cancer with bioluminescence imaging (BLI) as a measure of cell viability, magnetic resonance imaging (MRI) for cell localization and single photon emission computer tomography (SPECT) for cell quantification [88].

Dual-modal imaging has been investigated for improving the sensitivity of detection and accurate evaluation of benign and malignant lymph nodes by adopting hybrid NPs comprising superparamagnetic polymeric micelles conjugated with a fluorescent dye. These NPs showed efficient targeting and uptake by lymph node macrophages [89].

4. NP Functionalization Strategies for Cell Targeting, Cell Uptake, Therapeutic and Imaging Tracking

4.1. Key Factors for an Efficient and Targeted NP Uptake into Cells

Accurate cell tracking occurs if the NPs are correctly taken up by the cell to be targeted. The targetability and efficiency of the uptake are governed by the interactions between the NP surface groups and the plasma membrane antigens and/or receptors, which in turn depend on the density of the ligands and the antigens/receptors present on an NP and a cell, respectively [90]. NP functionalization with specific bioactive molecules (hereinafter referred as moieties) have been investigated to that purpose and those collected with the bibliographic survey were summarized in Table 1. Specifically, these moieties can elicit a targeted NP uptake by capturing specific cell biomarkers such as antigens/receptors (moieties with active action) or by enhancing NP permeation and retention (moieties with passive action) (Table 1).

4.2. Targeting and Uptake Moieties

Antibodies [91–94] with high specificity and affinity towards cancer cell surface molecules have been identified as active targeting and uptake moieties. Likewise, folic acid and riboflavin vitamins [95–100] have represented an attractive strategy for efficient NP uptake by cancer cells to be targeted, due to both overexpression of the folate and riboflavin receptors on the cancer cells and the rapid internalization of the receptor by receptor-mediated endocytosis [101,102].

Aptamers, consisting of small molecule DNA or RNA fragments that fold into well-defined 3D structures, have also been found to recognize specific receptors on the cell surface and mediate NP internalization [43,74,81,85,103–107]. The same is true for carbohydrates such as dextran and carbodextran [47,48,57,64,108–116], chitosan [69,117], glucose [118,119], beta cyclodextrin [120], and the glycoprotein transferrin [82] that have been mostly used for NP coating as a strategy to avoid immune response and, furthermore, to introduce active-targeting capabilities as well as increase cellular uptake [121].

Passive targeting and NP uptake have been efficiently achieved by polymers [56,60] and heparin [55,62] which, based on their biocompatibility, allow for enhanced NP permeation and retention.

Other types of moieties with passive action were quaternary ammonium cations [87], the polypeptide polylysine [49,58,63,64,122–126], and amino acids such as histidine [127,128], which, thanks to their positive charge, stabilize NPs and mediate the electrostatic interaction with the cell membrane, improving the endocytosis mechanisms. Similarly, some derivatives of vitamins, as an example TPGS compound [80], have been found to improve NP uptake by increase membrane permeability [129].

Table 1. NP surface targeting and uptake moieties.

Targeting and Uptake Moieties	Active or Passive Action	References
Antibodies, peptides	Active	[91–94]
Aminoacids	Passive	[127,128]
Aptamers	Active	[43,74,81,85,103–107]
Carbohydrates and glycoproteins (chitosan, beta cyclodextrin, dextran, transferrin, glucose)	Active	[47,48,57,64,69,82,108–121]
Vitamins (folate, riboflavin)	Active	[95–100]
Polymers (poly-L-lactide; hyaluronic acid, cholanic acid)	Passive	[56,60]
Polypeptide (Polylysine)	Passive	[49,58,63,64,122–126]
Heparin	Passive	[55,62]
D-α tocopheryl polyethylene glycol 1000 succinate (TPGS)	Passive	[80]
Quaternary ammonium cations	Passive	[87]

4.3. *Key Factors for an Efficient Imaging of NPs in Cells*

Identification of tissue alterations or lesion characterization as well as the diagnosis and treatment of disease through NP-based imaging techniques requires functionalization with contrast agents to be followed with imaging devices. Different molecules with optical properties such as fluorophore and bioluminescent dyes, isotopes or chemical elements with high molecular weight have been investigated as contrast agents. It is noteworthy that a considerable part of the reviewed studies suggested a combinatorial usage of contrast agents as an effective strategy for multimodal in vitro and/or in vivo tracking, as it could allow the limitations found with the use of a single-tracking approach to be overcome [130]. For this purpose, several protocols for the surface modification of inorganic NPs with specific elements have been developed. Because of this, the chemical element in the core of the NP represents one contrast moiety, and the molecule used for surface functionalization with an additional label. another. As an example, bi-modal tracking with MRI and PA has been performed by adopting NPs with two different metallic elements, such as gold and iron oxide [45,46]. A combination of fluorophores and metallic gold and iron oxide contrast agents was successfully used for optical and MRI tracking [47–50,87,88]. A review of the surface modification moieties for in vitro and/or in vivo tracking of the NPs identified by our bibliographic survey is included in Table 2.

Table 2. Moieties for labeling. NPs can be functionalized with labeling moieties allowing both in vitro and in vivo NP tracking delivery.

Moieties for Labeling	In Vitro	In Vivo
Fluorophores (fluorescein, rhodamine, squaraine, acridine, DY674, cianine, AxHD dye, SR-FLIVO dye, IR780, selenium, NIR dyes, becon probe, RuBpy, IRGD, I-BODIPY, Fluo4, PKH26)	[43,52–54,66,74,75,83,95,98,105,107,115, 120,123,131–176]	[43,52,54,61,66,83,95,96,105,115,123,157, 158,160–179]
Radionuclides (Indium-111; Gallium-68; Lutetium-177)	-	[55,88,180–186]

4.4. Key Factors for Tracking Therapeutic NPs through Imaging

Advances in the field of NP-based cell tracking have led to extraordinary progress, with several strategies being deployed to also track therapy delivery through imaging. Several therapeutic NPs that are capable of both self-reporting disease and/or tissue damage and delivering therapy have been developed to date and play a central role for the implementation of currently available diagnostic and therapeutic protocols [24]. As an example, therapy followed by imaging might be useful to test reactions in order to treat and identify patients in which therapy has an effect with the goal of providing personalized therapy for individual patients. Functionalization allowing the use of these NPs as therapeutic agents, referred to as therapeutic moieties consist of molecules managing the proliferative and inflammation status of the target cell.

Therapeutic moieties identified by the present literature survey include miRNA [68,69], siRNA [78,187], genes [164] and drugs or compounds with key roles in the modulation of cell proliferation and differentiation [66,67,79,80,104,146,166,188] (Table 3).

Some examples of the adoption of these moieties are listed as follows. The miRNA AntimiR-138 was shown to significantly promote the expression of osteogenesis-related genes and its conjugation to the NPs used to label the MSCs was found to efficiently enhance the osteogenic differentiation of transplanted MSCs and direct cranial bone regeneration in preclinical mouse models [69]. Similarly, Hosseinpour and colleagues adopted NPs functionalized with the miRNA -26a-5p to label MSCs to be transplanted for tissue regeneration purposes [68].

siRNAs targeting genes whose expression is enhanced in tumors, such as marker B-cell lymphoma-2 (BCL2) and baculoviral IAP repeat-containing 5 (BIRC5) as well as genes with regulatory roles in sustaining cancer cell proliferation and migration, were used for NP functionalization and were found to selectively blockade the cancer cell proliferation of oral [115] and glioblastoma [78] cancer cell models.

NP functionalization with a model gene, consisting of a plasmid containing the green fluorescent protein, enabled gene transfer to the hMSCs showing the potential use of NPs for gene therapy delivery [121].

Chemotherapy drugs such as doxorubicin and methotrexate as well as molecules with the ability to modulate cell proliferation and differentiation events such as dexamethasone, betamethasone, caspase inhibitors, ceramide, oleanolic acid, emodin, heparin, aspirin, curcumin and sulforaphane were also adopted as therapeutic moieties for NP functionalization and found to be effective in (1) blocking cancer cell proliferation in lymph node [183] glioma [188], ovarian [146], hepatocellular [189], liver [79,80], cervical [95], pancreatic ductal [190] tumor models; (2) exerting anti-inflammatory actions [66]; (3) directing stem cell differentiation [62]; and (4) targeting apoptotic events of cerebral ischemia [166].

Importantly, in all the mentioned studies the functionalization of NPs with tracking moieties also working as contrast agents enabled a solid strategy for monitoring the effect of NP-cell based therapy to be set up.

Table 3. NP surface moieties with therapeutic effects.

Therapeutic Moieties	References
miRNA	[68,69]
siRNA	[78,187]
Drugs (doxorubicin, dexamethasone, betamethasone, methotrexate, caspase inhibitors, ceramide, oleanoic acid, emodin, heparin, aspirin, curcumin and sulforaphane)	[66,67,79,95,103,132,147,167,189]
Gene	[164]

The surface modifications for NP delivery, diagnostic imaging and tracking of therapeutic effects are summarized in Figure 4.

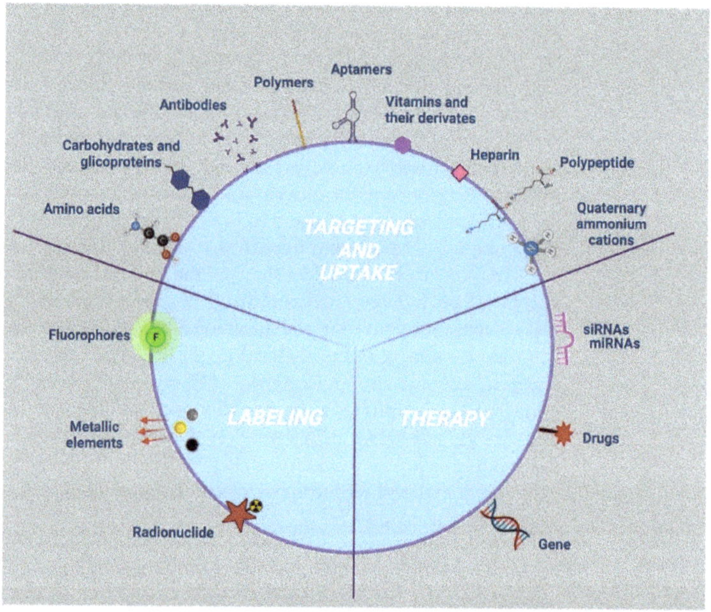

Figure 4. Multimodal NP surface modifications applied for enhancing targeting, uptake, tracking, and therapy. Identification of different elements useful to improve labeling, targeting, therapy and cellular uptake. Figure created with BioRender. Available online: https://biorender.com/ (accessed on 25 March 2022).

5. Approaches and Devices for In Vitro, In Vivo and Ex Vivo Tracking of NP Delivery and for Assessing NP Stability

5.1. In Vitro and In Vivo Tracking of NP Delivery

Based on the present literature review, fluorophore labeling moieties have been found to be suitable for both in vitro and in vivo tracking. Conversely, labeling with radionuclides was only applied for in vivo tracking.

In detail, optical imaging instruments such as confocal scanning, fluorescence microscopy and flow cytometry have been adopted for in vitro fluorophore tracking whereas the fluorescence imaging devices such as the NIR fluorescence IVIS Spectrum system, fluorescence molecular tomography imaging and single- and two-photon fluorescence imaging have been applied for in vivo fluorescence tracking. Conversely, radionuclide-labeled NP monitoring has been achieved in vivo by means of nuclear molecular imaging tools such as single photon emission computed tomography (SPECT) and positron emission tomography (PET).

The multimodality of imaging has also been investigated by incorporating labeled NPs with cells bearing ectopic expression of the luciferase gene [43,115,170] and provided a non-invasive and more effective imaging technology to track transplanted cells [43,170] and stem cell homing [115].

The multilabeling system represents an advantageous approach as it provides different molecules which can act as contrast agents to be distinctively used for in vitro and in vivo NP delivery characterization. Importantly, multilabeling possibilities rely on the definition of the NP material and type. Indeed, differently from polymeric and lipid-based NPs, the tracking of inorganic NPs also takes advantage of the intrinsic attitude of the constituent core element to work as a contrast agent. In that case, both the metallic core and the surface labeling moiety enable in vitro and in vivo tracking of NP delivery.

In vitro tracking of the core element can be achieved by means of transmission electron microscopy (TEM) and scanning electron microscopy (SEM), whereas NP surface-labeling moieties such as fluorophores and cell-labeling moieties such as luciferin can be followed in vitro by using confocal laser microscope and luminometer, respectively (Figure 5A and Supplementary Materials Table S3). On the other hand, in vivo tracking of the core element can be accomplished with MRI and tomography imaging such as computed tomography (CT), microcomputed tomography (μCT), photoacoustic tomography (PA) and X-ray fluorescence tomography (XFT) imaging.

It is noteworthy that, according to NP classification, our literature survey identified the MRI (52%), followed by CT (25%), optical (19%) and radionuclide (4%) imaging as the most used in vivo imaging devices (Figure 5B and Supplementary Materials Table S4).

Two different aspects bind the choice of the imaging technique: (I) the tissue to be targeted and (II) the sensitivity of the technique.

MRI represents the imaging method with the best soft tissue contrast, providing multiplanar capability without the use of ionizing radiation. It has emerged as the preferred modality for evaluating soft tissue masses and vascular structures, and it has been applied for checking soft tumors and inflammatory foci [191]. Iron oxide and gadolinium have been described as the most used contrast agents for MRI. While an MRI takes excellent pictures of soft tissue and blood vessels, a CT scan shows hard tissues much better, so it is often used to image bone structures [192]. Gold elements represent the elective contrast agent of CT. Many studies have shown that due to their strong X-ray absorption coefficient, good colloidal stability and biocompatibility, sustained contrast, shape and size controllability, surface modifiability, and few negative impacts on the marked cells, gold nanomaterials are ideal CT nano tracers for tracking [44]. As examples, with the use of this strategy, it was possible to demonstrate non-invasive cell tracking towards tumor sites [118], understand the fate of transplanted mesenchymal stem cells in certain diseases such as pulmonary fibrosis [43], and allow non-invasive and long-term CT monitoring of mesenchymal stem cells in tissue repair [123].

In contrast to MRI and CT, the two most used and clinically approved techniques for in vivo tracking, optical imaging is well suited for non-clinical use. However, it retains the potential for clinical application as it could exploit an enormous range of contrast agents such as molecules with fluorescent, near infrared or bioluminescent properties, that provide information about the structure and function of tissues ranging from single cells to entire organisms. An additional benefit of optical imaging that is often underexploited is its ability to acquire data at high speeds; a feature that enables it to not only observe static distributions of contrast, but to probe and characterize dynamic events related to physiology, disease progression and acute interventions in real time [193].

A few studies applied nuclear molecular imaging for NP tracking in vivo. Although this technique provides highly sensitive and quantitative imaging and shows good tissue penetration depth, it is limited by its low spatial resolution [194]. This was reflected in the small number of research articles collected by our bibliographic approach.

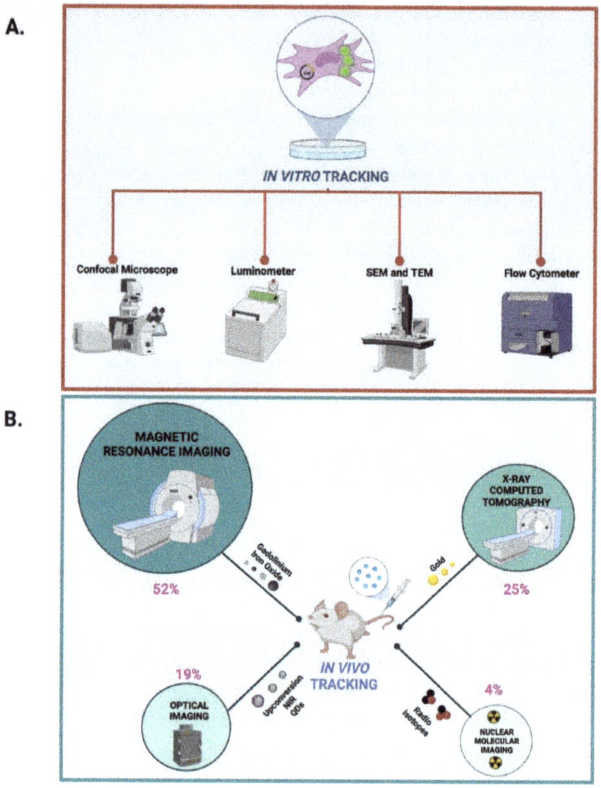

Figure 5. Methods and devices for in vivo and in vitro imaging of NPs. (**A**) In vitro tracking of NPs. (**B**) In vivo tracking of NPs. Percentage (%) refers to different types of imaging tools for NP tracking based on our literature survey. Figure created with BioRender. Available online: https://biorender.com/ (accessed on 25 March 2022).

Apart from these limitations, nuclear molecular imaging holds great potential for preclinical research and clinical applications as it might be a valuable tool granting long-term monitoring of cells, as well as the observation of cell clearance from various organs and homing at the tumor site.

NPs can be labeled with several radionuclides and have been proposed as innovative tools for imaging with single-photon emission computerized tomography (SPECT) and positron emission tomography (PET) and enhanced radioisotope therapy for several cancer types with limited toxicity [195].

Radionuclide-labeled NPs mostly characterized for diagnostic imaging and/or therapy purposes include technetium-99m (99mTc) [196], fluorine-18 (18F) [197], gallium-68 (68Ga) [198], indium-111 (111In) [199], and lutetium-177 (177Lu) [200].

The imaging and therapeutic applications of an isotope depend on its particle emission. Gamma (γ)-ray emitters such as 99mTC, 111In and positron emitters such as 18F, 68Ga are commonly used for diagnostic purposes using SPECT and PET imaging, respectively, and β- emitters such as 177Lu are mostly, but not exclusively, used for radionuclide therapy.

An appealing strategy that has proved beneficial for cell imaging with radionucleotides was its adoption in combination with other imaging tools such as those with magnetic and/or X-ray properties to combine the advantages of different imaging modalities and minimize the limitations [201]. Of note, multimodality imaging systems such as SPECT/MRI, PET/MRI, SPECT/CT, and PET/CT are currently under investigation. As some examples, the conjugation of 99mTc with silica [202], iron oxide [203] or gold [204] NPs was exploited for SPECT/MRI or SPECT/CT imaging of tumors in preclinical mouse models. Likewise, the conjugation of 18F with iron oxide NP was found effective for dual imaging with PET/MRI of stem cell biodistribution [205] as well as for monitoring anti-angiogenic therapeutic effects in breast cancer xenograft mouse models [206] and diagnosis of rectal cancer in clinics [207]. Importantly, since PET/MRI systems have become operational worldwide, several imaging protocols involving the use of 18F for whole body scanning in oncology are currently under standardization for patient management in clinics [208].

Radionuclide imaging for tumor diagnosis was also performed by adopting NPs conjugated with radiopharmaceutical formulations of the (γ)-ray emitter 111In (such as indium-111-oxine) allowing the monitoring of migration and homing of intravenously given stem cell-bearing NPs to the tumor location [55,88,180].

Advantageous strategies for simultaneous diagnostics and therapy are also being characterized. In this context, a promising strategy is the pairing of radionuclides of different elements [195]. As an example, 68Ga for PET imaging and 177Lu as β-emitter for a therapy couple have been investigated as a radiopharmaceutical tool to face prostate [181] colon [182], lymph node [183–185] and glioblastoma [186] cancers.

Overall, based on these premises, it would be helpful to find a compromise between invasiveness and diagnostic resolution to improve actual NP tracking methods.

5.2. Ex Vivo NP Tracking of NP Delivery

Specific ex vivo protocols have been developed to verify and quantify NP uptake on tissue and cell specimens prior to being applied for in vivo protocols. Several histochemical dyes can be adopted for that purpose (Table 4). Inorganic NP uptake has been reported to be obtained by using Prussian blue or alizarin red dyes as their ability to stain ferric ions and minerals, respectively. Conversely, Oil Red O dye, with its affinity to lipids, was successfully applied for staining lipidic NPs.

Table 4. Dyes used for ex vivo quantification of NP uptake. Dyes can react with elements characterizing the NPs or can be applied for NP coating.

Dye	Type	References
Histochemical staining	Prussian blue	[48,49,51,57–59,64,65,88,109, 114,116,117,124,127,138,154, 155,160,177,187,189,209–224]
	Oil Red O	[58,225]
	Alizarin red	[225]
Dyes for NP coating	Propidium iodide (PI)	[118]
	Nile red	[89]

Moreover, ex vivo checking, and quantification of cell uptake was performed by coating NPs with PI or Nile red dye. As example, NP coating with PI was successfully adopted to detect cells incorporating gold NPs through flow cytometry [118], whereas Nile red labeling of hybrid iron oxide/polymeric NPs was used to check in vitro uptake into macrophages [89].

5.3. NP Stability Assessment

Another key element influencing the performance of any NP-based cell tracking strategy is the stability of the NPs. Any NP formulation to be applied for tracking purposes needs to preserve a particular nanostructure property including aggregation, composition, shape, size, and surface chemistry.

Different environmental stresses such as extended storage, pH and mineral composition, thermal processing, freeze–thaw cycling, dehydration, mechanical stress and light exposure have been described as responsible for influencing NP stability [29].

The ability to measure the stability of the NP formulation prior to utilization for tracking purposes thus represents a key step to be pursued. Methods that provide quantitative metrics for measuring and modeling nanoparticle stability in terms of core composition, shape, size, and surface chemistry are distinguished by their physical- and chemical- [226,227] based approaches and are summarized in Table 5.

Table 5. Key parameters defining NP stability and strategies to determine stability preservation.

NP Stability	Definition	Approaches Used for Characterization of NP Stability	
		Physical	**Chemical**
Aggregation	Preservation of NPs upon collisions	Dynamic light scattering	Single particle inductively coupled plasma mass spectrometry UV–visible spectroscopy
Core Composition	Unchanged chemistry of the core during the use	X-ray diffraction	Single particle inductively coupled plasma mass spectrometry UV–visible spectroscopy Surface-enhanced Raman scattering X-ray photoelectron spectroscopy Energy dispersive X-ray
Shape	Preservation of NP architecture during the use	Transmission electron microscopy Scanning electron microscopy X-ray diffraction Atomic force microscopy	Single particle inductively coupled plasma-mass spectrometry UV–visible spectroscopy
Size	Preservation of NP dimension during use or storage	Dynamic light scattering Scanning electron microscopy Transmission electron microscopy Small-angle X-ray scattering Atomic force microscopy	Single particle inductively coupled plasma-mass spectrometry UV–visible spectroscopy
Surface chemistry	Preservation of the native surface functionality	Low energy ion scattering X-ray photoelectron spectroscopy	Single particle inductively coupled plasma-mass spectrometry UV–visible spectroscopy Surface-enhanced Raman scattering X-ray photoelectron spectroscopy Energy dispersive X-ray

6. NP Uptake Strategies and Mechanisms

The effects of the physicochemical properties of NPs, such as size, shape, charge and surface chemistry, are crucial for determining the cellular uptake mechanism and the NP function exerted upon the cells [228]. Importantly, knowledge of the underlying mechanisms involved in cellular NP incorporation is found to be relevant for the fate of the NPs and their cytotoxicity [90], as well as for safe and efficient therapeutic applications [229]. Our bibliographic survey identified the endocytosis mechanism as the most representative for the internalization of NPs (97%). Endocytosis can be accomplished through two main mechanisms: phagocytosis and pinocytosis. Phagocytosis is the preferred uptake for larger nanoparticles ($\geq$100 nm), while pinocytosis dominates the uptake of smaller nanoparticles ($\leq$100 nm) [230,231]. Of note, pinocytosis can be subdivided into clathrin-mediated and caveolae-dependent endocytosis, to date, considered as some of the most important mechanisms for NP uptake [229,230]. However, since the variation in endocytic mechanisms depends on cell and NP types, it is still challenging to generalize the current findings [229].

Another key aspect to be considered when testing cell uptake efficiency is the NP incubation time, which is influenced by the type and density of the cultured cells [232] and the shape, size and concentration of the NPs [233,234]. Our literature survey identified 24 h as the most validated incubation time for different types of cell target. However, it has been also proved that the addition of some compounds to NPs might shortening the time required for cell uptake: consistently, Mishra and colleagues showed that the use of protamine sulfate for inorganic iron oxide NP coating reduced the incubation time to 6 h in MSCs [59].

Cell density also influences NP uptake; the most suitable cell density reported in the reviewed studies refers to 10^4 cells/cm^2 with high variability defined by the type of cells adopted for the uptake assay. However, several uptake protocols are currently under investigation to increase the quantity of the recovered cells with the aim of having a bulk cell preparation to be used for in vivo NP delivery and tracking. Even in this case, tracking sensitivity reflects the resolution of the imaging technique applied; thus, it was not possible to identify a reference interval for cell number to be injected in in vivo mouse models. Generally, based on the reviewed studies, in vitro protocols pointed to the recovery of at least totally 10^6 cells. Likewise, NP concentration parameters were highly fluctuating and were strictly correlated to the NP architecture and size.

Alternative strategies identified with the presented literature review include the adoption of electroporation and transfection accounting, respectively, for about 2% and 1% of the total citations. Electroporation was reported to be useful in assisting the passage of NPs larger than 100 nm through the cell membrane [44] or to enhance the quantity of incorporated NPs [235].

Furthermore, the adoption of transfectant agents such as commercially available lipofectamine formulations has been proved to increase uptake efficiency. Accordingly, Wan et al. designed a new approach involving the surface modification of gold NPs with layers of silica and Transfectin 3000 (TS) to reduce cytotoxicity and improve the incorporation in bone marrow-derived stem cells (BMSCs) [236]. Uptake strategies identified in the literature survey are summarized in Figure 6.

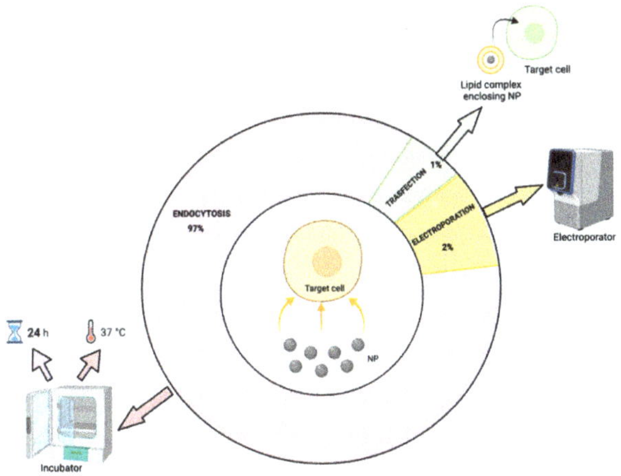

Figure 6. NP uptake mechanisms and strategies. Figure created with BioRender. Available online: https://biorender.com/ (accessed on 25 March 2022).

7. Cell Types Suitable for Safe NP Delivery

As previously mentioned, improved NP delivery can be achieved by using cells as the vehicle for NPs. Based on our literature survey, different types of cells have been investigated in the last decade as potential carriers for an efficient NP release.

Stem cells, with their intrinsic ability to home in on cancer or injury sites, represent the most suitable cell type source allowing NP delivery, with a 60% of the reviewed studies investigating their potential for biomedical application for cancer and regenerative medicine purposes. Conversely, a smaller percentage of studies (31%), focusing on the development of clinical protocols for precision cancer medicine, adopted cancer cells as the target of NP delivery. Only a few original research articles, pointing to the elucidation of inflammatory homing mechanisms, used other cell sources such as macrophages (7%) or T cells (2%) (Figure 7).

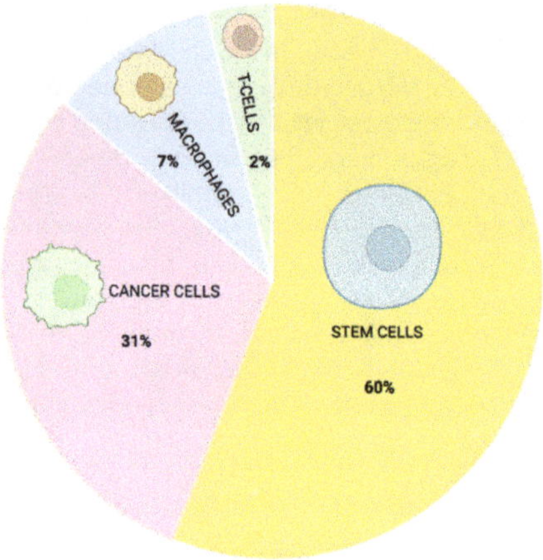

Figure 7. Cell sources adopted for efficient and safe NP delivery. Cell type subcategories and their relative percentage of utilization identified by the literature survey. Figure created with BioRender. Available online: https://biorender.com/ (accessed on 25 March 2022).

7.1. Stem Cells

Stem cell homing mechanisms are currently being investigated to provide further insights for their potential use as vehicles for contrast agents to be detected with currently available imaging techniques to improve the diagnosis and therapy of many diseases [237].

Different subtypes of stem cells were identified by reviewing our bibliographic dataset. In detail, mesenchymal stem cells (MSCs) accounted for 80% of the total citations using stem cells, 6% involved the use of adipose tissue-derived stem cells (ADSCs), and 14% instead involved the use of other different subtypes such as blastema cells [160], neuronal stem cells (NSC) [54], bone marrow dendritic cells (BMDCs) [55,219] and mouse embryonic stem cells (mESCs) [170].

Furthermore, stem cells derived from different species have been identified. Of note, many studies involved MSCs of human origin (84%), followed by those derived from rat (14%) and monkey (2%) species.

7.2. Cancer Cells

The most used cancer cell model identified with the present literature review is the HeLa cell line, with 44% of the total citations exploiting cancer cells for NP delivery purposes. HeLa was the first human cell line established in culture and has since become the most widely used human cell line in cancer research.

The choice of the cancer cell model relies on the type of cancer to be targeted. Thirteen percent of the studies focused their attention on the investigation of nanotechnology supporting hepatocellular carcinoma resolution by means of hepatocarcinoma-derived cell lines such as hepatoma cell line Hepa1–6 [91], HepG2 cells and HCa-F [80,100,136] and hepatocarcinoma cell line SMMC-7721 [111].

Cancer cell lines derived from glioma or glioblastoma, such as human GBM cell lines U-87 and U251 [78,180], U118 glioma [238] and F98 rat glioma [239] cells were represented by 13% of the total citations describing the application of cancer cell lines.

Similarly, 10% of studies aimed to identify therapeutic approaches against breast cancer with the MDA-MB-231 human breast cancer cell line [179,215,221] and breast cancer 4T1 cell line [162].

Ten percent of studies have investigated efficient protocols to cure lung cancer by means of lung cancer cell models such as A529, A549, NCI H441 and A2780 [134,141,152,169].

To a lesser extent, studies have been carried out with cell lines derived from pancreatic cancer such as Panc-1 and MIA Paca-2 cells [190] and human pancreatic cancer cell line BxPc-3 [94] and also derived from gastric cancer such as human gastric cancer cells (MGC80-3) [132,138], both of these cell types account for 5% of the total citations concerning the use of cancer cell line models for NP delivery.

7.3. *Macrophages*

Macrophage-bearing NPs have been adopted mostly for tracking. Macrophages were used to evaluate uptake of nanoparticles using imaging techniques [147,240] to investigate the metabolic conversion and elimination of nanoparticles [218] and since these macrophages are able to migrate to sites of inflammation [112], they have been labeled with NPs and their distribution assessed by clinically relevant imaging techniques [89,109,116,222]. Although little is known about NP delivery efficiency of macrophages, this cell source might represent a promising tool for inflammation-targeting purposes in the fields of both regenerative medicine and cancer diseases [241].

7.4. *T Cells*

T cells have been shown to kill malignant cells in vitro and in vivo, therefore, several studies have labeled these cells with gold NPs or SPIONs to follow their fate on CT or MRI. For example, T cells were transduced to express a melanoma-specific T cell, and then computed tomography analysis was performed, once labeled with gold NPs. CT efficiently detected the accumulation of these transduced and labeled T cells at the tumor site, leading to tumor regression [118]. Another study preferred to label T cells with ultrasmall SPIOs to follow their fate through MRI [242]. Overall, although T cells encapsulating NPs have been poorly characterized to date, their use in nanomedicine deserves attention for improving the efficacy of T-cell based immunotherapy [172].

8. Conclusions and Outlook for Effective NP-Based Cell Tracking

Issues determining the effectiveness of NP-assisted cell tracking such as NP design, material properties and functionalization with specific targeting, labeling (tracking) and/or uptake-enhancing moieties have been identified and almost exhaustively characterized during the last decade. Nevertheless, an in-depth analysis of the features relating to NPs and the interaction dynamics with the target body area, which in turn is influenced by the cell to be targeted, the NP loading time and concentration, still needs to be achieved.

The nexus of NPs with the biology is currently an area of intense study in the field of NP-based cell tracking as the success of NP incorporation into the cell and consequently the success of direct cell labeling in terms of non-invasive long-term targeted monitoring through imaging depends on the type of biological response triggered by NP uptake.

The investigation of mechanisms determining the efficacy of the active incorporation of NPs by the cell holds a prominent role in NP-assisted cell tracking as it keeps the potential clinical traceability for a targeted, sensitive, and long monitoring of cells with the current clinically approved imaging devices improving the performance of the currently used iodinated, barium or gadolinium-based contrast agents [14].

Much effort has been made to better characterize the biological cues of NP–cell dialogue with the aim of improving uptake and retention mechanisms to be exploited for direct cell tracking; despite its simple and straightforward nature, this approach has several drawbacks which deserve further investigation to improve their effective clinical use.

Specifically, the biggest limitation of any direct labeling method is the failure in the assessment of cell viability and division damping retention and the right delivery of NPs. Firstly, if the cells die, the NPs may be retained at the target site, being taken up by macrophages consequently leading to erroneous detection of cells. The NPs may also dissociate from the cells through efflux and exocytosis, leading to false quantification and monitoring of free labels instead of the cells of interest. Secondly, cell division dilutes the number of NPs, affecting the sensitivity of the detection of daughter cells and limiting the long-term tracking of labeled cells [243]. Overall, these aspects can be managed by deepening the characterization of cell-fate mechanisms and how they are modulated by NP incorporation.

Considering these premises, the definition of the impact of NP physicochemical properties on cell internalization mechanisms, together with the characterization of biological cell response might pave the way for effective NP-delivery solutions for cell tracking, finally providing future perspectives for a bench to bedside translation (Figure 8).

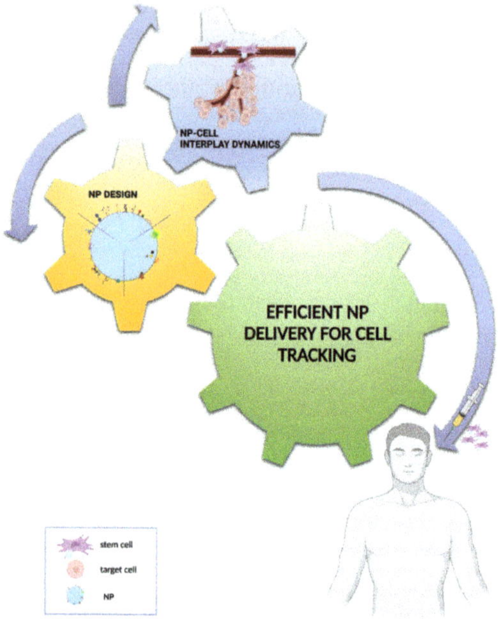

Figure 8. Essential ingredients for effective NP–cell tracking. NP design, cellular models and interaction mechanisms between NPs and cargo stem cells and/or target cells are key elements influencing NP-based cell tracking and delivery performances. Figure created with BioRender. Available online: https://biorender.com/ (accessed on 25 March 2022).

Supplementary Materials: The following supporting information can be downloaded at: https://www.mdpi.com/article/10.3390/nano12091414/s1, Table S1: Clinical trial studies investigating the use of magnetic NPs for cell tracking; Table S2: Database of citations exploring direct cell tracking approaches through imaging; Table S3: General operational principles of the technologies used for in vitro cell tracking; Table S4: General operational principles of the technologies used for in vivo cell tracking. References [244–247] are cited in the supplementary materials.

Author Contributions: Conceptualization, B.B.; methodology, A.P.; validation, B.B.; formal analysis, A.P., C.D.B., G.C., O.D.G. and C.C.S.R.; data curation, A.P., C.D.B., G.C. and C.C.S.R.; writing—original draft preparation, A.P. and C.D.B.; writing—review and editing, B.B., A.P., C.D.B., V.R., G.C., A.C. and C.C.S.R.; supervision, B.B. and V.R.; project administration, B.B.; funding acquisition, B.B. All authors have read and agreed to the published version of the manuscript.

Funding: This research was funded by FONDAZIONE TERCAS, Project Title: "New frontiers in stem cell homing" 2021.

Institutional Review Board Statement: Not applicable.

Informed Consent Statement: Not applicable.

Data Availability Statement: Not applicable.

Conflicts of Interest: The authors declare no conflict of interest.

References

1. Zimmer, C.; Zhang, B.; Dufour, A.; Thebaud, A.; Berlemont, S.; Meas-Yedid, V.; Marin, J.-C.O. On the digital trail of mobile cells. *IEEE Signal Process. Mag.* **2006**, *23*, 54–62. [CrossRef]
2. Meijering, E.; Smal, I.; Danuser, G. Tracking in molecular bioimaging. *IEEE Signal Process. Mag.* **2006**, *23*, 46–53. [CrossRef]
3. Meijering, E.; Dzyubachyk, O.; Smal, I.; van Cappellen, W.A. Tracking in cell and developmental biology. *Semin. Cell Dev. Biol.* **2009**, *20*, 894–902. [CrossRef] [PubMed]
4. Dorn, J.F.; Danuser, G.; Yang, G. Computational processing and analysis of dynamic fluorescence image data. *Methods Cell Biol.* **2008**, *85*, 497–538. [CrossRef] [PubMed]
5. Jaqaman, K.; Danuser, G. Computational image analysis of cellular dynamics: A case study based on particle tracking. *Cold Spring Harb. Protoc.* **2009**, *2009*, pdb.top65. [CrossRef] [PubMed]
6. Rohr, K.; Godinez, W.J.; Harder, N.; Wörz, S.; Mattes, J.; Tvaruskó, W.; Eils, R. Tracking and quantitative analysis of dynamic movements of cells and particles. *Cold Spring Harb. Protoc.* **2010**, *2010*, pdb.top80. [CrossRef]
7. Ray, S.S.; Bandyopadhyay, J. Nanotechnology-enabled biomedical engineering: Current trends, future scopes, and perspectives. *Nanotechnol. Rev.* **2021**, *10*, 728–743. [CrossRef]
8. McNamara, K.; Tofail, S.A.M. Nanoparticles in biomedical applications. *Adv. Phys. X* **2017**, *2*, 54–88. [CrossRef]
9. Bhirde, A.; Xie, J.; Swierczewska, M.; Chen, X. Nanoparticles for cell labeling. *Nanoscale* **2011**, *3*, 142–153. [CrossRef]
10. Arranz, A.; Ripoll, J. Advances in optical imaging for pharmacological studies. *Front. Pharmacol.* **2015**, *6*, 189. [CrossRef]
11. Ni, J.-S.; Li, Y.; Yue, W.; Liu, B.; Li, K. Nanoparticle-based Cell Trackers for Biomedical Applications. *Theranostics* **2020**, *10*, 1923–1947. [CrossRef] [PubMed]
12. Van Rijt, S.; Habibovic, P. Enhancing regenerative approaches with nanoparticles. *J. R. Soc. Interface* **2017**, *14*, 20170093. [CrossRef] [PubMed]
13. Zhang, Y.; Li, M.; Gao, X.; Chen, Y.; Liu, T. Nanotechnology in cancer diagnosis: Progress, challenges and opportunities. *J. Hematol. Oncol.* **2019**, *12*, 137. [CrossRef] [PubMed]
14. Bernsen, M.R.; Guenoun, J.; van Tiel, S.T.; Krestin, G.P. Nanoparticles and clinically applicable cell tracking. *Br. J. Radiol.* **2015**, *88*, 20150375. [CrossRef] [PubMed]
15. Zhang, M.; Cheng, S.; Jin, Y.; Zhang, N.; Wang, Y. Membrane engineering of cell membrane biomimetic nanoparticles for nanoscale therapeutics. *Clin. Transl. Med.* **2021**, *11*, e292. [CrossRef]
16. Thomas, S.C.; Kim, J.-W.; Pauletti, G.M.; Hassett, D.J.; Kotagiri, N. Exosomes: Biological Pharmaceutical Nanovectors for Theranostics. *Front. Bioeng. Biotechnol.* **2022**, *9*, 808614. [CrossRef]
17. Accomasso, L.; Gallina, C.; Turinetto, V.; Giachino, C. Stem Cell Tracking with Nanoparticles for Regenerative Medicine Purposes: An Overview. *Stem Cells Int.* **2016**, *2016*, 7920358. [CrossRef]
18. Huang, X.; Zhang, F.; Wang, H.; Niu, G.; Choi, K.Y.; Swierczewska, M.; Zhang, G.; Gao, H.; Wang, Z.; Zhu, L.; et al. Mesenchymal stem cell-based cell engineering with multifunctional mesoporous silica nanoparticles for tumor delivery. *Biomaterials* **2013**, *34*, 1772–1780. [CrossRef]
19. Perrin, J.; Capitao, M.; Mougin-Degraef, M.; Guérard, F.; Faivre-Chauvet, A.; Rbah-Vidal, L.; Gaschet, J.; Guilloux, Y.; Kraeber-Bodéré, F.; Chérel, M.; et al. Cell Tracking in Cancer Immunotherapy. *Front. Med.* **2020**, *7*, 34. [CrossRef]
20. Mu, Q.; Wang, H.; Zhang, M. Nanoparticles for imaging and treatment of metastatic breast cancer. *Expert Opin. Drug Deliv.* **2017**, *14*, 123–136. [CrossRef]
21. Ruiz-Garcia, H.; Alvarado-Estrada, K.; Krishnan, S.; Quinones-Hinojosa, A.; Trifiletti, D.M. Nanoparticles for Stem Cell Therapy Bioengineering in Glioma. *Front. Bioeng. Biotechnol.* **2020**, *8*, 558375. [CrossRef] [PubMed]
22. Tiet, P.; Berlin, J.M. Exploiting homing abilities of cell carriers: Targeted delivery of nanoparticles for cancer therapy. *Biochem. Pharmacol.* **2017**, *145*, 18–26. [CrossRef] [PubMed]
23. Mitchell, M.J.; Billingsley, M.M.; Haley, R.M.; Wechsler, M.E.; Peppas, N.A.; Langer, R. Engineering precision nanoparticles for drug delivery. *Nat. Rev. Drug Discov.* **2021**, *20*, 101–124. [CrossRef] [PubMed]
24. Zavaleta, C.; Ho, D.; Chung, E.J. Theranostic Nanoparticles for Tracking and Monitoring Disease State. *SLAS Technol.* **2018**, *23*, 281–293. [CrossRef]
25. Mason, E.E.; Mattingly, E.; Herb, K.; Śliwiak, M.; Franconi, S.; Cooley, C.Z.; Slanetz, P.J.; Wald, L.L. Concept for using magnetic particle imaging for intraoperative margin analysis in breast-conserving surgery. *Sci. Rep.* **2021**, *11*, 13456. [CrossRef]
26. Wei, Q.; Arami, H.; Santos, H.A.; Zhang, H.; Li, Y.; He, J.; Zhong, D.; Ling, D.; Zhou, M. Intraoperative Assessment and Photothermal Ablation of the Tumor Margins Using Gold Nanoparticles. *Adv. Sci.* **2021**, *8*, 2002788. [CrossRef]
27. Onishi, T.; Mihara, K.; Matsuda, S.; Sakamoto, S.; Kuwahata, A.; Sekino, M.; Kusakabe, M.; Handa, H.; Kitagawa, Y. Application of Magnetic Nanoparticles for Rapid Detection and In Situ Diagnosis in Clinical Oncology. *Cancers* **2022**, *14*, 364. [CrossRef]
28. Najahi-Missaoui, W.; Arnold, R.D.; Cummings, B.S. Safe Nanoparticles: Are We There Yet? *Int. J. Mol. Sci.* **2020**, *22*, 385. [CrossRef]

29. Phan, H.T.; Haes, A.J. What Does Nanoparticle Stability Mean? *J. Phys. Chem. C Nanomater. Interfaces* **2019**, *123*, 16495–16507. [CrossRef]
30. Núñez, C.; Estévez, S.V.; Del Pilar Chantada, M. Inorganic nanoparticles in diagnosis and treatment of breast cancer. *J. Biol. Inorg. Chem. JBIC Publ. Soc. Biol. Inorg. Chem.* **2018**, *23*, 331–345. [CrossRef]
31. Soenen, S.J.; Parak, W.J.; Rejman, J.; Manshian, B. (Intra)cellular stability of inorganic nanoparticles: Effects on cytotoxicity, particle functionality, and biomedical applications. *Chem. Rev.* **2015**, *115*, 2109–2135. [CrossRef] [PubMed]
32. Jiao, M.; Zhang, P.; Meng, J.; Li, Y.; Liu, C.; Luo, X.; Gao, M. Recent advancements in biocompatible inorganic nanoparticles towards biomedical applications. *Biomater. Sci.* **2018**, *6*, 726–745. [CrossRef] [PubMed]
33. Zielińska, A.; Carreiró, F.; Oliveira, A.M.; Neves, A.; Pires, B.; Venkatesh, D.N.; Durazzo, A.; Lucarini, M.; Eder, P.; Silva, A.M.; et al. Polymeric Nanoparticles: Production, Characterization, Toxicology and Ecotoxicology. *Molecules* **2020**, *25*, 3731. [CrossRef]
34. Ma, D. Hybrid Nanoparticles. In *Noble Metal-Metal Oxide Hybrid Nanoparticles*; Elsevier: Amsterdam, The Netherlands, 2019; pp. 3–6, ISBN 978-0-12-814134-2.
35. García-Pinel, B.; Porras-Alcalá, C.; Ortega-Rodríguez, A.; Sarabia, F.; Prados, J.; Melguizo, C.; López-Romero, J.M. Lipid-Based Nanoparticles: Application and Recent Advances in Cancer Treatment. *Nanomaterials* **2019**, *9*, 638. [CrossRef] [PubMed]
36. Singh, P.; Bodycomb, J.; Travers, B.; Tatarkiewicz, K.; Travers, S.; Matyas, G.R.; Beck, Z. Particle size analyses of polydisperse liposome formulations with a novel multispectral advanced nanoparticle tracking technology. *Int. J. Pharm.* **2019**, *566*, 680–686. [CrossRef] [PubMed]
37. Vitorino, C.; Carvalho, F.A.; Almeida, A.J.; Sousa, J.J.; Pais, A.A.C.C. The size of solid lipid nanoparticles: An interpretation from experimental design. *Colloids Surf. B Biointerfaces* **2011**, *84*, 117–130. [CrossRef]
38. Khosa, A.; Reddi, S.; Saha, R.N. Nanostructured lipid carriers for site-specific drug delivery. *Biomed. Pharmacother. Biomed. Pharmacother.* **2018**, *103*, 598–613. [CrossRef]
39. Fan, Y.; Marioli, M.; Zhang, K. Analytical characterization of liposomes and other lipid nanoparticles for drug delivery. *J. Pharm. Biomed. Anal.* **2021**, *192*, 113642. [CrossRef]
40. Mehnert, W.; Mäder, K. Solid lipid nanoparticles. *Adv. Drug Deliv. Rev.* **2012**, *64*, 83–101. [CrossRef]
41. Bulte, J.W.M.; Daldrup-Link, H.E. Clinical Tracking of Cell Transfer and Cell Transplantation: Trials and Tribulations. *Radiology* **2018**, *289*, 604–615. [CrossRef]
42. Sun, Y.; Lu, Y.; Yin, L.; Liu, Z. The Roles of Nanoparticles in Stem Cell-Based Therapy for Cardiovascular Disease. *Front. Bioeng. Biotechnol.* **2020**, *8*, 947. [CrossRef] [PubMed]
43. Bao, H.; Xia, Y.; Yu, C.; Ning, X.; Liu, X.; Fu, H.; Chen, Z.; Huang, J.; Zhang, Z. CT/Bioluminescence Dual-Modal Imaging Tracking of Mesenchymal Stem Cells in Pulmonary Fibrosis. *Small* **2019**, *15*, e1904314. [CrossRef] [PubMed]
44. Huang, J.; Huang, J.H.; Bao, H.; Ning, X.; Yu, C.; Chen, Z.; Chao, J.; Zhang, Z. CT/MR Dual-Modality Imaging Tracking of Mesenchymal Stem Cells Labeled with a Au/GdNC@SiO_2 Nanotracer in Pulmonary Fibrosis. *ACS Appl. Bio Mater.* **2020**, *3*, 2489–2498. [CrossRef] [PubMed]
45. Zhang, H.; Wang, Z.-J.; Wang, L.-J.; Li, T.-T.; He, S.; Li, L.-P.; Li, X.-Y.; Liu, S.-J.; Li, J.-D.; Li, S.-J.; et al. A dual-mode nanoparticle based on natural biomaterials for photoacoustic and magnetic resonance imaging of bone mesenchymal stem cells in vivo. *RSC Adv.* **2019**, *9*, 35003–35010. [CrossRef]
46. Chen, P.-J.; Kang, Y.-D.; Lin, C.-H.; Chen, S.-Y.; Hsieh, C.-H.; Chen, Y.-Y.; Chiang, C.-W.; Lee, W.; Hsu, C.-Y.; Liao, L.-D.; et al. Multitheragnostic Multi-GNRs Crystal-Seeded Magnetic Nanoseaurchin for Enhanced In Vivo Mesenchymal-Stem-Cell Homing, Multimodal Imaging, and Stroke Therapy. *Adv. Mater.* **2015**, *27*, 6488–6495. [CrossRef]
47. Vaags, A.K.; Gartley, C.J.; Halling, K.B.; Dobson, H.; Zheng, Y.; Foltz, W.D.; Dick, A.J.; Kruth, S.A.; Hough, M.R. Migration of cells from the yolk sac to hematopoietic tissues after in utero transplantation of early and mid gestation canine fetuses. *Transplantation* **2011**, *91*, 723–730. [CrossRef]
48. Geburek, F.; Mundle, K.; Conrad, S.; Hellige, M.; Walliser, U.; van Schie, H.T.M.; van Weeren, R.; Skutella, T.; Stadler, P.M. Tracking of autologous adipose tissue-derived mesenchymal stromal cells with in vivo magnetic resonance imaging and histology after intralesional treatment of artificial equine tendon lesions—A pilot study. *Stem Cell Res. Ther.* **2016**, *7*, 21. [CrossRef]
49. Qin, J.-B.; Li, K.-A.; Li, X.-X.; Xie, Q.-S.; Lin, J.-Y.; Ye, K.-C.; Jiang, M.-E.; Zhang, G.-X.; Lu, X.-W. Long-term MRI tracking of dual-labeled adipose-derived stem cells homing into mouse carotid artery injury. *Int. J. Nanomed.* **2012**, *7*, 5191–5203. [CrossRef]
50. Meng, Y.; Shi, C.; Hu, B.; Gong, J.; Zhong, X.; Lin, X.; Zhang, X.; Liu, J.; Liu, C.; Xu, H. External magnetic field promotes homing of magnetized stem cells following subcutaneous injection. *BMC Cell Biol.* **2017**, *18*, 24. [CrossRef]
51. Wu, C.; Li, J.; Pang, P.; Liu, J.; Zhu, K.; Li, D.; Cheng, D.; Chen, J.; Shuai, X.; Shan, H. Polymeric vector-mediated gene transfection of MSCs for dual bioluminescent and MRI tracking in vivo. *Biomaterials* **2014**, *35*, 8249–8260. [CrossRef]
52. Moonshi, S.S.; Zhang, C.; Peng, H.; Puttick, S.; Rose, S.; Fisk, N.M.; Bhakoo, K.; Stringer, B.W.; Qiao, G.G.; Gurr, P.A.; et al. A unique 19F MRI agent for the tracking of non phagocytic cells in vivo. *Nanoscale* **2018**, *10*, 8226–8239. [CrossRef] [PubMed]
53. Zhang, Q.; Nie, J.; Xu, H.; Qiu, Y.; Li, X.; Gu, W.; Tang, G.; Luo, J. Fluorescent microspheres for one-photon and two-photon imaging of mesenchymal stem cells. *J. Mater. Chem. B* **2017**, *5*, 7809–7818. [CrossRef] [PubMed]
54. Li, D.; Yan, X.; Hu, Y.; Liu, Y.; Guo, R.; Liao, M.; Shao, B.; Tang, Q.; Guo, X.; Chai, R.; et al. Two-Photon Image Tracking of Neural Stem Cells via Iridium Complexes Encapsulated in Polymeric Nanospheres. *ACS Biomater. Sci. Eng.* **2019**, *5*, 1561–1568. [CrossRef] [PubMed]

55. Sehl, O.C.; Gevaert, J.J.; Melo, K.P.; Knier, N.N.; Foster, P.J. A Perspective on Cell Tracking with Magnetic Particle Imaging. *Tomography* **2020**, *6*, 315–324. [CrossRef]
56. Vernikouskaya, I.; Fekete, N.; Bannwarth, M.; Erle, A.; Rojewski, M.; Landfester, K.; Schmidtke-Schrezenmeier, G.; Schrezenmeier, H.; Rasche, V. Iron-loaded PLLA nanoparticles as highly efficient intracellular markers for visualization of mesenchymal stromal cells by MRI. *Contrast Media Mol. Imaging* **2014**, *9*, 109–121. [CrossRef]
57. Shahror, R.A.; Ali, A.A.A.; Wu, C.-C.; Chiang, Y.-H.; Chen, K.-Y. Enhanced Homing of Mesenchymal Stem Cells Overexpressing Fibroblast Growth Factor 21 to Injury Site in a Mouse Model of Traumatic Brain Injury. *Int. J. Mol. Sci.* **2019**, *20*, 2624. [CrossRef]
58. Mishra, S.K.; Khushu, S.; Singh, A.K.; Gangenahalli, G. Homing and Tracking of Iron Oxide Labelled Mesenchymal Stem Cells After Infusion in Traumatic Brain Injury Mice: A Longitudinal In Vivo MRI Study. *Stem Cell Rev. Rep.* **2018**, *14*, 888–900. [CrossRef]
59. Mishra, S.K.; Khushu, S.; Gangenahalli, G. Biological effects of iron oxide-protamine sulfate complex on mesenchymal stem cells and its relaxometry based labeling optimization for cellular MRI. *Exp. Cell Res.* **2017**, *351*, 59–67. [CrossRef]
60. Huang, X.; Zhang, F.; Wang, Y.; Sun, X.; Choi, K.Y.; Liu, D.; Choi, J.; Shin, T.-H.; Cheon, J.; Niu, G.; et al. Design considerations of iron-based nanoclusters for noninvasive tracking of mesenchymal stem cell homing. *ACS Nano* **2014**, *8*, 4403–4414. [CrossRef]
61. Shen, W.-B.; Plachez, C.; Tsymbalyuk, O.; Tsymbalyuk, N.; Xu, S.; Smith, A.M.; Michel, S.L.J.; Yarnell, D.; Mullins, R.; Gullapalli, R.P.; et al. Cell-Based Therapy in TBI: Magnetic Retention of Neural Stem Cells In Vivo. *Cell Transplant.* **2016**, *25*, 1085–1099. [CrossRef]
62. Lee, J.; Jung, M.J.; Hwang, Y.H.; Lee, Y.J.; Lee, S.; Lee, D.Y.; Shin, H. Heparin-coated superparamagnetic iron oxide for in vivo MR imaging of human MSCs. *Biomaterials* **2012**, *33*, 4861–4871. [CrossRef] [PubMed]
63. Noorwali, A.; Faidah, M.; Ahmed, N.; Bima, A. Tracking iron oxide labelled mesenchymal stem cells(MSCs) using magnetic resonance imaging (MRI) in a rat model of hepatic cirrhosis. *Bioinformation* **2019**, *15*, 1–10. [CrossRef] [PubMed]
64. Ha, B.C.; Jung, J.; Kwak, B.K. Susceptibility-weighted imaging for stem cell visualization in a rat photothrombotic cerebral infarction model. *Acta Radiol.* **2015**, *56*, 219–227. [CrossRef] [PubMed]
65. Liao, N.; Wu, M.; Pan, F.; Lin, J.; Li, Z.; Zhang, D.; Wang, Y.; Zheng, Y.; Peng, J.; Liu, X.; et al. Poly (dopamine) coated superparamagnetic iron oxide nanocluster for noninvasive labeling, tracking, and targeted delivery of adipose tissue-derived stem cells. *Sci. Rep.* **2016**, *6*, 18746. [CrossRef]
66. Napp, J.; Markus, M.A.; Heck, J.G.; Dullin, C.; Möbius, W.; Gorpas, D.; Feldmann, C.; Alves, F. Therapeutic Fluorescent Hybrid Nanoparticles for Traceable Delivery of Glucocorticoids to Inflammatory Sites. *Theranostics* **2018**, *8*, 6367–6383. [CrossRef]
67. Ren, H.; Chen, S.; Jin, Y.; Zhang, C.; Yang, X.; Ge, K.; Liang, X.-J.; Li, Z.; Zhang, J. A traceable and bone-targeted nanoassembly based on defect-related luminescent mesoporous silica for enhanced osteogenic differentiation. *J. Mater. Chem. B* **2017**, *5*, 1585–1593. [CrossRef]
68. Hosseinpour, S.; Cao, Y.; Liu, J.; Xu, C.; Walsh, L.J. Efficient transfection and long-term stability of rno-miRNA-26a-5p for osteogenic differentiation by large pore sized mesoporous silica nanoparticles. *J. Mater. Chem. B* **2021**, *9*, 2275–2284. [CrossRef]
69. Wu, G.; Feng, C.; Quan, J.; Wang, Z.; Wei, W.; Zang, S.; Kang, S.; Hui, G.; Chen, X.; Wang, Q. In situ controlled release of stromal cell-derived factor-1α and antimiR-138 for on-demand cranial bone regeneration. *Carbohydr. Polym.* **2018**, *182*, 215–224. [CrossRef]
70. Ren, N.; Liang, N.; Yu, X.; Wang, A.; Xie, J.; Sun, C. Ligand-free upconversion nanoparticles for cell labeling and their effects on stem cell differentiation. *Nanotechnology* **2020**, *31*, 145101. [CrossRef]
71. Yuan, L.; Qi, X.; Qin, G.; Liu, Q.; Zhang, F.; Song, Y.; Deng, J. Effects of gold nanostructures on differentiation of mesenchymal stem cells. *Colloids Surf. B Biointerfaces* **2019**, *184*, 110494. [CrossRef]
72. Choi, Y.-E.; Kwak, J.-W.; Park, J.W. Nanotechnology for early cancer detection. *Sensors* **2010**, *10*, 428–455. [CrossRef] [PubMed]
73. Chinen, A.B.; Guan, C.M.; Ferrer, J.R.; Barnaby, S.N.; Merkel, T.J.; Mirkin, C.A. Nanoparticle Probes for the Detection of Cancer Biomarkers, Cells, and Tissues by Fluorescence. *Chem. Rev.* **2015**, *115*, 10530–10574. [CrossRef] [PubMed]
74. Li, H.; Mu, Y.; Qian, S.; Lu, J.; Wan, Y.; Fu, G.; Liu, S. Synthesis of fluorescent dye-doped silica nanoparticles for target-cell-specific delivery and intracellular microRNA imaging. *Analyst* **2015**, *140*, 567–573. [CrossRef]
75. Chen, M.-Y.; Chen, Z.-Z.; Wang, W.; Zhu, L.; Tang, H.-W.; Pang, D.-W. Preparation of RuBpy-doped Silica Fluorescent Nanoprobes and Their Applications to the Recognition of Liver Cancer Cells. *Chin. J. Anal. Chem.* **2014**, *42*, 326–331. [CrossRef]
76. Gao, J.; Zhang, Q.; Xu, J.; Guo, L.; Li, X. Clinical significance of serum miR-21 in breast cancer compared with CA153 and CEA. *Chin. J. Cancer Res.* **2013**, *25*, 743–748. [CrossRef] [PubMed]
77. Pedley, R.B.; Boden, J.A.; Boden, R.; Begent, R.H.; Turner, A.; Haines, A.M.; King, D.J. The potential for enhanced tumour localisation by poly(ethylene glycol) modification of anti-CEA antibody. *Br. J. Cancer* **1994**, *70*, 1126–1130. [CrossRef]
78. Wang, R.; Degirmenci, V.; Xin, H.; Li, Y.; Wang, L.; Chen, J.; Hu, X.; Zhang, D. PEI-Coated Fe_3O_4 Nanoparticles Enable Efficient Delivery of Therapeutic siRNA Targeting REST into Glioblastoma Cells. *Int. J. Mol. Sci.* **2018**, *19*, 2230. [CrossRef]
79. Gao, M.; Xu, H.; Bao, X.; Zhang, C.; Guan, X.; Liu, H.; Lv, L.; Deng, S.; Gao, D.; Wang, C.; et al. Oleanolic acid-loaded PLGA-TPGS nanoparticles combined with heparin sodium-loaded PLGA-TPGS nanoparticles for enhancing chemotherapy to liver cancer. *Life Sci.* **2016**, *165*, 63–74. [CrossRef] [PubMed]
80. Liu, H.; Xu, H.; Zhang, C.; Gao, M.; Gao, X.; Ma, C.; Lv, L.; Gao, D.; Deng, S.; Wang, C.; et al. Emodin-Loaded PLGA-TPGS Nanoparticles Combined with Heparin Sodium-Loaded PLGA-TPGS Nanoparticles to Enhance Chemotherapeutic Efficacy Against Liver Cancer. *Pharm. Res.* **2016**, *33*, 2828–2843. [CrossRef]
81. Zhang, B.; Sun, X.; Mei, H.; Wang, Y.; Liao, Z.; Chen, J.; Zhang, Q.; Hu, Y.; Pang, Z.; Jiang, X. LDLR-mediated peptide-22-conjugated nanoparticles for dual-targeting therapy of brain glioma. *Biomaterials* **2013**, *34*, 9171–9182. [CrossRef]

82. Yue, J.; Liu, S.; Wang, R.; Hu, X.; Xie, Z.; Huang, Y.; Jing, X. Transferrin-conjugated micelles: Enhanced accumulation and antitumor effect for transferrin-receptor-overexpressing cancer models. *Mol. Pharm.* **2012**, *9*, 1919–1931. [CrossRef] [PubMed]
83. Xiong, L.; Guo, Y.; Zhang, Y.; Cao, F. Highly luminescent and photostable near-infrared fluorescent polymer dots for long-term tumor cell tracking in vivo. *J. Mater. Chem. B* **2016**, *4*, 202–206. [CrossRef] [PubMed]
84. Singh, A.; Jain, S.; Senapati, S.; Verma, R.S.; Sahoo, S.K. Magnetic Nanoparticles Labeled Mesenchymal Stem Cells: A Pragmatic Solution toward Targeted Cancer Theranostics. *Adv. Healthc. Mater.* **2015**, *4*, 2078–2089. [CrossRef] [PubMed]
85. Shen, Y.; Zhang, J.; Hao, W.; Wang, T.; Liu, J.; Xie, Y.; Xu, S.; Liu, H. Copolymer micelles function as pH-responsive nanocarriers to enhance the cytotoxicity of a HER2 aptamer in HER2-positive breast cancer cells. *Int. J. Nanomed.* **2018**, *13*, 537–553. [CrossRef] [PubMed]
86. Soenen, S.J.; De Meyer, S.F.; Dresselaers, T.; Vande Velde, G.; Pareyn, I.M.; Braeckmans, K.; De Cuyper, M.; Himmelreich, U.; Vanhoorelbeke, K.I. MRI assessment of blood outgrowth endothelial cell homing using cationic magnetoliposomes. *Biomaterials* **2011**, *32*, 4140–4150. [CrossRef]
87. Zhang, Y.S.; Wang, Y.; Wang, L.; Wang, Y.; Cai, X.; Zhang, C.; Wang, L.V.; Xia, Y. Labeling human mesenchymal stem cells with gold nanocages for in vitro and in vivo tracking by two-photon microscopy and photoacoustic microscopy. *Theranostics* **2013**, *3*, 532–543. [CrossRef]
88. Zaw Thin, M.; Allan, H.; Bofinger, R.; Kostelec, T.D.; Guillaume, S.; Connell, J.J.; Patrick, P.S.; Hailes, H.C.; Tabor, A.B.; Lythgoe, M.F.; et al. Multi-modal imaging probe for assessing the efficiency of stem cell delivery to orthotopic breast tumours. *Nanoscale* **2020**, *12*, 16570–16585. [CrossRef]
89. Li, W.-J.; Wang, Y.; Liu, Y.; Wu, T.; Cai, W.-L.; Shuai, X.-T.; Hong, G.-B. Preliminary Study of MR and Fluorescence Dual-mode Imaging: Combined Macrophage-Targeted and Superparamagnetic Polymeric Micelles. *Int. J. Med. Sci.* **2018**, *15*, 129–141. [CrossRef]
90. Foroozandeh, P.; Aziz, A.A. Insight into Cellular Uptake and Intracellular Trafficking of Nanoparticles. *Nanoscale Res. Lett.* **2018**, *13*, 339. [CrossRef]
91. Ma, X.-H.; Wang, S.; Liu, S.-Y.; Chen, K.; Wu, Z.-Y.; Li, D.-F.; Mi, Y.-T.; Hu, L.-B.; Chen, Z.-W.; Zhao, X.-M. Development and in vitro study of a bi-specific magnetic resonance imaging molecular probe for hepatocellular carcinoma. *World J. Gastroenterol.* **2019**, *25*, 3030–3043. [CrossRef]
92. Zhang, W.; Qiao, L.; Wang, X.; Senthilkumar, R.; Wang, F.; Chen, B. Inducing cell cycle arrest and apoptosis by dimercaptosuccinic acid modified Fe_3O_4 magnetic nanoparticles combined with nontoxic concentration of bortezomib and gambogic acid in RPMI-8226 cells. *Int. J. Nanomed.* **2015**, *10*, 3275–3289. [CrossRef]
93. Al Faraj, A.; Shaik, A.S.; Al Sayed, B.; Halwani, R.; Al Jammaz, I. Specific targeting and noninvasive imaging of breast cancer stem cells using single-walled carbon nanotubes as novel multimodality nanoprobes. *Nanomedicine* **2016**, *11*, 31–46. [CrossRef] [PubMed]
94. Han, Y.; An, Y.; Jia, G.; Wang, X.; He, C.; Ding, Y.; Tang, Q. Facile assembly of upconversion nanoparticle-based micelles for active targeted dual-mode imaging in pancreatic cancer. *J. Nanobiotechnol.* **2018**, *16*, 7. [CrossRef]
95. Dumoga, S.; Rai, Y.; Bhatt, A.N.; Tiwari, A.K.; Singh, S.; Mishra, A.K.; Kakkar, D. Block Copolymer Based Nanoparticles for Theranostic Intervention of Cervical Cancer: Synthesis, Pharmacokinetics, and in Vitro/in Vivo Evaluation in HeLa Xenograft Models. *ACS Appl. Mater. Interfaces* **2017**, *9*, 22195–22211. [CrossRef] [PubMed]
96. Jayapaul, J.; Arns, S.; Bunker, M.; Weiler, M.; Rutherford, S.; Comba, P.; Kiessling, F. In vivo evaluation of riboflavin receptor targeted fluorescent USPIO in mice with prostate cancer xenografts. *Nano Res.* **2016**, *9*, 1319–1333. [CrossRef] [PubMed]
97. Pan, L.; He, M.; Ma, J.; Tang, W.; Gao, G.; He, R.; Su, H.; Cui, D. Phase and size controllable synthesis of NaYbF4 nanocrystals in oleic acid/ionic liquid two-phase system for targeted fluorescent imaging of gastric cancer. *Theranostics* **2013**, *3*, 210–222. [CrossRef]
98. Barar, J.; Kafil, V.; Majd, M.H.; Barzegari, A.; Khani, S.; Johari-Ahar, M.; Asgari, D.; Coukos, G.; Cokous, G.; Omidi, Y. Multifunctional mitoxantrone-conjugated magnetic nanosystem for targeted therapy of folate receptor-overexpressing malignant cells. *J. Nanobiotechnol.* **2015**, *13*, 26. [CrossRef]
99. Jayapaul, J.; Hodenius, M.; Arns, S.; Lederle, W.; Lammers, T.; Comba, P.; Kiessling, F.; Gaetjens, J. FMN-coated fluorescent iron oxide nanoparticles for RCP-mediated targeting and labeling of metabolically active cancer and endothelial cells. *Biomaterials* **2011**, *32*, 5863–5871. [CrossRef]
100. Chang, J.-Y.; Wang, G.-Q.; Cheng, C.-Y.; Lin, W.-X.; Hsu, J.-C. Strategies for photoluminescence enhancement of AgInS2 quantum dots and their application as bioimaging probes. *J. Mater. Chem.* **2012**, *22*, 10609. [CrossRef]
101. Narmani, A.; Rezvani, M.; Farhood, B.; Darkhor, P.; Mohammadnejad, J.; Amini, B.; Refahi, S.; Abdi Goushbolagh, N. Folic acid functionalized nanoparticles as pharmaceutical carriers in drug delivery systems. *Drug Dev. Res.* **2019**, *80*, 404–424. [CrossRef]
102. Darguzyte, M.; Drude, N.; Lammers, T.; Kiessling, F. Riboflavin-Targeted Drug Delivery. *Cancers* **2020**, *12*, 295. [CrossRef] [PubMed]
103. Jiang, D.; Gao, X.; Kang, T.; Feng, X.; Yao, J.; Yang, M.; Jing, Y.; Zhu, Q.; Feng, J.; Chen, J. Actively targeting D-α-tocopheryl polyethylene glycol 1000 succinate-poly(lactic acid) nanoparticles as vesicles for chemo-photodynamic combination therapy of doxorubicin-resistant breast cancer. *Nanoscale* **2016**, *8*, 3100–3118. [CrossRef] [PubMed]
104. Morita, Y.; Sakurai, R.; Wakimoto, T.; Kobayashi, K.; Xu, B.; Toku, Y.; Song, G.; Luo, Q.; Ju, Y. tLyP-1-conjugated core-shell nanoparticles, Fe3O4NPs@mSiO2, for tumor-targeted drug delivery. *Appl. Surf. Sci.* **2019**, *474*, 17–24. [CrossRef]

105. Li, H.; Wang, P.; Deng, Y.; Zeng, M.; Tang, Y.; Zhu, W.-H.; Cheng, Y. Combination of active targeting, enzyme-triggered release and fluorescent dye into gold nanoclusters for endomicroscopy-guided photothermal/photodynamic therapy to pancreatic ductal adenocarcinoma. *Biomaterials* **2017**, *139*, 30–38. [CrossRef]
106. Yoon, S.; Rossi, J.J. Aptamers: Uptake mechanisms and intracellular applications. *Adv. Drug Deliv. Rev.* **2018**, *134*, 22–35. [CrossRef]
107. Figueiredo, P.; Sipponen, M.H.; Lintinen, K.; Correia, A.; Kiriazis, A.; Yli-Kauhaluoma, J.; Österberg, M.; George, A.; Hirvonen, J.; Kostiainen, M.A.; et al. Preparation and Characterization of Dentin Phosphophoryn-Derived Peptide-Functionalized Lignin Nanoparticles for Enhanced Cellular Uptake. *Small* **2019**, *15*, e1901427. [CrossRef]
108. Hansen, L.; Hansen, A.B.; Mathiasen, A.B.; Ng, M.; Bhakoo, K.; Ekblond, A.; Kastrup, J.; Friis, T. Ultrastructural characterization of mesenchymal stromal cells labeled with ultrasmall superparamagnetic iron-oxide nanoparticles for clinical tracking studies. *Scand. J. Clin. Lab. Investig.* **2014**, *74*, 437–446. [CrossRef]
109. Tong, H.-I.; Kang, W.; Shi, Y.; Zhou, G.; Lu, Y. Physiological function and inflamed-brain migration of mouse monocyte-derived macrophages following cellular uptake of superparamagnetic iron oxide nanoparticles-Implication of macrophage-based drug delivery into the central nervous system. *Int. J. Pharm.* **2016**, *505*, 271–282. [CrossRef]
110. Fidler, F.; Steinke, M.; Kraupner, A.; Gruttner, C.; Hiller, K.-H.; Briel, A.; Westphal, F.; Walles, H.; Jakob, P.M. Stem Cell Vitality Assessment Using Magnetic Particle Spectroscopy. *IEEE Trans. Magn.* **2015**, *51*, 5100704. [CrossRef]
111. Xu, D.; Wu, F.; Chen, Y.; Wei, L.; Yuan, W. pH-sensitive degradable nanoparticles for highly efficient intracellular delivery of exogenous protein. *Int. J. Nanomed.* **2013**, *8*, 3405–3414. [CrossRef]
112. Saito, S.; Tsugeno, M.; Koto, D.; Mori, Y.; Yoshioka, Y.; Nohara, S.; Murase, K. Impact of surface coating and particle size on the uptake of small and ultrasmall superparamagnetic iron oxide nanoparticles by macrophages. *Int. J. Nanomed.* **2012**, *7*, 5415–5421. [CrossRef]
113. Nejadnik, H.; Henning, T.D.; Castaneda, R.T.; Boddington, S.; Taubert, S.; Jha, P.; Tavri, S.; Golovko, D.; Ackerman, L.; Meier, R.; et al. Somatic differentiation and MR imaging of magnetically labeled human embryonic stem cells. *Cell Transplant.* **2012**, *21*, 2555–2567. [CrossRef] [PubMed]
114. Mo, R.; Yang, J.; Wu, E.X.; Lin, S. Instant magnetic labeling of tumor cells by ultrasound in vitro. *J. Magn. Magn. Mater.* **2011**, *323*, 2287–2294. [CrossRef]
115. Menon, L.G.; Pratt, J.; Yang, H.W.; Black, P.M.; Sorensen, G.A.; Carroll, R.S. Imaging of human mesenchymal stromal cells: Homing to human brain tumors. *J. Neurooncol.* **2012**, *107*, 257–267. [CrossRef] [PubMed]
116. Cai, Q.-Y.; Lee, H.; Kim, E.-J.; Moon, H.; Chang, K.; Rho, J.; Hong, K.S. Magnetic resonance imaging of superparamagnetic iron oxide-labeled macrophage infiltrates in acute-phase renal ischemia-reperfusion mouse model. *Nanomed. Nanotechnol. Biol. Med.* **2012**, *8*, 365–373. [CrossRef] [PubMed]
117. Tong, M.; Xiong, F.; Shi, Y.; Luo, S.; Liu, Z.; Wu, Z.; Wang, Z. In vitro study of SPIO-labeled human pancreatic cancer cell line BxPC-3. *Contrast Media Mol. Imaging* **2013**, *8*, 101–107. [CrossRef] [PubMed]
118. Meir, R.; Shamalov, K.; Betzer, O.; Motiei, M.; Horovitz-Fried, M.; Yehuda, R.; Popovtzer, A.; Popovtzer, R.; Cohen, C.J. Nanomedicine for Cancer Immunotherapy: Tracking Cancer-Specific T-Cells in Vivo with Gold Nanoparticles and CT Imaging. *ACS Nano* **2015**, *9*, 6363–6372. [CrossRef]
119. Meir, R.; Betzer, O.; Motiei, M.; Kronfeld, N.; Brodie, C.; Popovtzer, R. Design principles for noninvasive, longitudinal and quantitative cell tracking with nanoparticle-based CT imaging. *Nanomed. Nanotechnol. Biol. Med.* **2017**, *13*, 421–429. [CrossRef]
120. Datz, S.; Illes, B.; Gößl, D.; Schirnding, C.V.; Engelke, H.; Bein, T. Biocompatible crosslinked β-cyclodextrin nanoparticles as multifunctional carriers for cellular delivery. *Nanoscale* **2018**, *10*, 16284–16292. [CrossRef]
121. Kröger, A.P.P.; Komil, M.I.; Hamelmann, N.M.; Juan, A.; Stenzel, M.H.; Paulusse, J.M.J. Glucose Single-Chain Polymer Nanoparticles for Cellular Targeting. *ACS Macro Lett.* **2019**, *8*, 95–101. [CrossRef]
122. Faidah, M.; Noorwali, A.; Atta, H.; Ahmed, N.; Habib, H.; Damiati, L.; Filimban, N.; Al-Qriqri, M.; Mahfouz, S.; Khabaz, M.N. Mesenchymal stem cell therapy of hepatocellular carcinoma in rats: Detection of cell homing and tumor mass by magnetic resonance imaging using iron oxide nanoparticles. *Adv. Clin. Exp. Med.* **2017**, *26*, 1171–1178. [CrossRef] [PubMed]
123. Ning, X.; Bao, H.; Liu, X.; Fu, H.; Wang, W.; Huang, J.; Zhang, Z. Long-term in vivo CT tracking of mesenchymal stem cells labeled with Au@BSA@PLL nanotracers. *Nanoscale* **2019**, *11*, 20932–20941. [CrossRef] [PubMed]
124. Wang, X.; Wei, F.; Liu, A.; Wang, L.; Wang, J.-C.; Ren, L.; Liu, W.; Tu, Q.; Li, L.; Wang, J. Cancer stem cell labeling using poly(L-lysine)-modified iron oxide nanoparticles. *Biomaterials* **2012**, *33*, 3719–3732. [CrossRef] [PubMed]
125. Jiang, J.; Chen, Y.; Zhu, Y.; Yao, X.; Qi, J. Efficient in vitro labeling of human prostate cancer cells with superparamagnetic iron oxide nanoparticles. *Cancer Biother. Radiopharm.* **2011**, *26*, 461–467. [CrossRef]
126. Wang, S.; Fang, J.; Zhang, T.; Wang, B.; Chen, J.; Li, X.; Zhang, S.; Zhang, W. Magnetic resonance imaging targeting of intracranial glioma xenografts by Resovist-labeled endothelial progenitor cells. *J. Neurooncol.* **2011**, *105*, 67–75. [CrossRef]
127. Shelat, R.; Bhatt, L.K.; Khanna, A.; Chandra, S. A comprehensive toxicity evaluation of novel amino acid-modified magnetic ferrofluids for magnetic resonance imaging. *Amino Acids* **2019**, *51*, 929–943. [CrossRef]
128. Han, Z.; Liu, S.; Pei, Y.; Ding, Z.; Li, Y.; Wang, X.; Zhan, D.; Xia, S.; Driedonks, T.; Witwer, K.W.; et al. Highly efficient magnetic labelling allows MRI tracking of the homing of stem cell-derived extracellular vesicles following systemic delivery. *J. Extracell. Vesicles* **2021**, *10*, e12054. [CrossRef]

129. Yu, L.; Bridgers, A.; Polli, J.; Vickers, A.; Long, S.; Roy, A.; Winnike, R.; Coffin, M. Vitamin E-TPGS increases absorption flux of an HIV protease inhibitor by enhancing its solubility and permeability. *Pharm. Res.* **1999**, *16*, 1812–1817. [CrossRef]
130. Burke, B.P.; Cawthorne, C.; Archibald, S.J. Multimodal nanoparticle imaging agents: Design and applications. *Philos. Transact. A Math. Phys. Eng. Sci.* **2017**, *375*, 20170261. [CrossRef]
131. Sun, M.; Sun, B.; Liu, Y.; Shen, Q.-D.; Jiang, S. Dual-Color Fluorescence Imaging of Magnetic Nanoparticles in Live Cancer Cells Using Conjugated Polymer Probes. *Sci. Rep.* **2016**, *6*, 22368. [CrossRef]
132. Wu, Y.; Tang, W.; Wang, P.; Liu, C.; Yuan, Y.; Qian, J. Cytotoxicity and Cellular Uptake of Amorphous Silica Nanoparticles in Human Cancer Cells. *Part. Part. Syst. Charact.* **2015**, *32*, 779–787. [CrossRef]
133. Shahabi, S.; Treccani, L.; Dringen, R.; Rezwan, K. Dual fluorophore doped silica nanoparticles for cellular localization studies in multiple stained cells. *Acta Biomater.* **2015**, *14*, 208–216. [CrossRef] [PubMed]
134. Efeoglu, E.; Keating, M.; McIntyre, J.; Casey, A.; Byrne, H.J. Determination of nanoparticle localisation within subcellular organelles in vitro using Raman spectroscopy. *Anal. Methods* **2015**, *7*, 10000–10017. [CrossRef]
135. Mumin, A.M.; Barrett, J.W.; Dekaban, G.A.; Zhang, J. Dendritic cell internalization of foam-structured fluorescent mesoporous silica nanoparticles. *J. Colloid Interface Sci.* **2011**, *353*, 156–162. [CrossRef] [PubMed]
136. Tang, W.; Yuan, Y.; Liu, C.; Wu, Y.; Lu, X.; Qian, J. Differential cytotoxicity and particle action of hydroxyapatite nanoparticles in human cancer cells. *Nanomedicine* **2014**, *9*, 397–412. [CrossRef] [PubMed]
137. Roy, D.; Mukhuty, A.; Fouzder, C.; Bar, N.; Chowdhury, S.; Kundu, R.; Chowdhury, P. Multi-emissive biocompatible silicon quantum dots: Synthesis, characterization, intracellular imaging and improvement of two fold drug efficacy. *Dyes Pigment.* **2021**, *186*, 109004. [CrossRef]
138. Lindemann, A.; Lüdtke-Buzug, K.; Fräderich, B.M.; Gräfe, K.; Pries, R.; Wollenberg, B. Biological impact of superparamagnetic iron oxide nanoparticles for magnetic particle imaging of head and neck cancer cells. *Int. J. Nanomed.* **2014**, *9*, 5025–5040. [CrossRef]
139. McCormick, S.C.; Stillman, N.; Hockley, M.; Perriman, A.W.; Hauert, S. Measuring Nanoparticle Penetration Through Bio-Mimetic Gels. *Int. J. Nanomed.* **2021**, *16*, 2585–2595. [CrossRef]
140. Lee, S.H.; Park, D.J.; Yun, W.S.; Park, J.-E.; Choi, J.S.; Key, J.; Seo, Y.J. Endocytic trafficking of polymeric clustered superparamagnetic iron oxide nanoparticles in mesenchymal stem cells. *J. Control. Release* **2020**, *326*, 408–418. [CrossRef]
141. Smyth, P.; Gibson, T.J.; Irvine, G.; Black, G.; Lavery, D.; Semsarilar, M.; Scott, C.J.; Themistou, E. pH-Responsive benzaldehyde-functionalized PEG-based polymeric nanoparticles for drug delivery: Effect of preparation method on morphology, dye encapsulation and attachment. *Eur. Polym. J.* **2020**, *124*, 109471. [CrossRef]
142. Luo, Y.; Liu, F.; Li, E.; Fang, Y.; Zhao, G.; Dai, X.; Li, J.; Wang, B.; Xu, M.; Liao, B.; et al. FRET-based fluorescent nanoprobe platform for sorting of active microorganisms by functional properties. *Biosens. Bioelectron.* **2020**, *148*, 111832. [CrossRef] [PubMed]
143. Qiu, K.; Du, Y.; Liu, J.; Guan, J.-L.; Chao, H.; Diao, J. Super-resolution observation of lysosomal dynamics with fluorescent gold nanoparticles. *Theranostics* **2020**, *10*, 6072–6081. [CrossRef] [PubMed]
144. Mortimer, G.M.; Jack, K.S.; Musumeci, A.W.; Martin, D.J.; Minchin, R.F. Stable non-covalent labeling of layered silicate nanoparticles for biological imaging. *Mater. Sci. Eng. C Mater. Biol. Appl.* **2016**, *61*, 674–680. [CrossRef] [PubMed]
145. Khalid, A.; Tran, P.A.; Norello, R.; Simpson, D.A.; O'Connor, A.J.; Tomljenovic-Hanic, S. Intrinsic fluorescence of selenium nanoparticles for cellular imaging applications. *Nanoscale* **2016**, *8*, 3376–3385. [CrossRef]
146. Chernenko, T.; Buyukozturk, F.; Miljkovic, M.; Carrier, R.; Diem, M.; Amiji, M. Label-Free Raman Microspectral Analysis for Comparison of Cellular Uptake and Distribution between Non-Targeted and EGFR-Targeted Biodegradable Polymeric Nanoparticles. *Drug Deliv. Transl. Res.* **2013**, *3*, 575–586. [CrossRef]
147. Zane, A.; McCracken, C.; Knight, D.A.; Young, T.; Lutton, A.D.; Olesik, J.W.; Waldman, W.J.; Dutta, P.K. Uptake of bright fluorophore core-silica shell nanoparticles by biological systems. *Int. J. Nanomed.* **2015**, *10*, 1547–1567. [CrossRef]
148. Lee, C.-M.; Lee, T.K.; Kim, D.-I.; Kim, Y.-R.; Kim, M.-K.; Jeong, H.-J.; Sohn, M.-H.; Lim, S.T. Optical imaging of absorption and distribution of RITC-SiO_2 nanoparticles after oral administration. *Int. J. Nanomed.* **2014**, *9* (Suppl. 2), 243–250. [CrossRef]
149. Kotsuchibashi, Y.; Zhang, Y.; Ahmed, M.; Ebara, M.; Aoyagi, T.; Narain, R. Fabrication of FITC-doped silica nanoparticles and study of their cellular uptake in the presence of lectins. *J. Biomed. Mater. Res. A* **2013**, *101*, 2090–2096. [CrossRef]
150. Zuber, A.; Purdey, M.; Schartner, E.; Forbes, C.; van der Hoek, B.; Giles, D.; Abell, A.; Monro, T.; Ebendorff-Heidepriem, H. Detection of gold nanoparticles with different sizes using absorption and fluorescence based method. *Sens. Actuators B Chem.* **2016**, *227*, 117–127. [CrossRef]
151. Madsen, J.; Canton, I.; Warren, N.J.; Themistou, E.; Blanazs, A.; Ustbas, B.; Tian, X.; Pearson, R.; Battaglia, G.; Lewis, A.L.; et al. Nile Blue-based nanosized pH sensors for simultaneous far-red and near-infrared live bioimaging. *J. Am. Chem. Soc.* **2013**, *135*, 14863–14870. [CrossRef]
152. Kasper, J.; Hermanns, M.I.; Bantz, C.; Koshkina, O.; Lang, T.; Maskos, M.; Pohl, C.; Unger, R.E.; Kirkpatrick, C.J. Interactions of silica nanoparticles with lung epithelial cells and the association to flotillins. *Arch. Toxicol.* **2013**, *87*, 1053–1065. [CrossRef] [PubMed]
153. Yang, C.-Y.; Hsiao, J.-K.; Tai, M.-F.; Chen, S.-T.; Cheng, H.-Y.; Wang, J.-L.; Liu, H.-M. Direct labeling of hMSC with SPIO: The long-term influence on toxicity, chondrogenic differentiation capacity, and intracellular distribution. *Mol. Imaging Biol.* **2011**, *13*, 443–451. [CrossRef] [PubMed]

54. Liu, W.-M.; Xue, Y.-N.; He, W.-T.; Zhuo, R.-X.; Huang, S.-W. Dendrimer modified magnetic iron oxide nanoparticle/DNA/PEI ternary complexes: A novel strategy for magnetofection. *J. Control. Release* **2011**, *152* (Suppl. 1), e159–e160. [CrossRef] [PubMed]
55. Dabrowska, S.; Del Fattore, A.; Karnas, E.; Frontczak-Baniewicz, M.; Kozlowska, H.; Muraca, M.; Janowski, M.; Lukomska, B. Imaging of extracellular vesicles derived from human bone marrow mesenchymal stem cells using fluorescent and magnetic labels. *Int. J. Nanomed.* **2018**, *13*, 1653–1664. [CrossRef]
56. Pužar Dominkuš, P.; Stenovec, M.; Sitar, S.; Lasič, E.; Zorec, R.; Plemenitaš, A.; Žagar, E.; Kreft, M.; Lenassi, M. PKH26 labeling of extracellular vesicles: Characterization and cellular internalization of contaminating PKH26 nanoparticles. *Biochim. Biophys. Acta Biomembr.* **2018**, *1860*, 1350–1361. [CrossRef]
57. Brown, K.; Thurn, T.; Xin, L.; Liu, W.; Bazak, R.; Chen, S.; Lai, B.; Vogt, S.; Jacobsen, C.; Paunesku, T.; et al. Intracellular in situ labeling of TiO_2 nanoparticles for fluorescence microscopy detection. *Nano Res.* **2018**, *11*, 464–476. [CrossRef]
58. Saladino, G.M.; Vogt, C.; Li, Y.; Shaker, K.; Brodin, B.; Svenda, M.; Hertz, H.M.; Toprak, M.S. Optical and X-ray Fluorescent Nanoparticles for Dual Mode Bioimaging. *ACS Nano* **2021**, *15*, 5077–5085. [CrossRef]
59. Sweeney, S.K.; Manzar, G.S.; Zavazava, N.; Assouline, J.G. Tracking embryonic hematopoietic stem cells to the bone marrow: Nanoparticle options to evaluate transplantation efficiency. *Stem Cell Res. Ther.* **2018**, *9*, 204. [CrossRef]
60. Lauridsen, H.; Foldager, C.B.; Hansen, L.; Pedersen, M. Non-invasive cell tracking of SPIO labeled cells in an intrinsic regenerative environment: The axolotl limb. *Exp. Ther. Med.* **2018**, *15*, 3311–3319. [CrossRef]
61. Hsu, F.-T.; Sun, R.; Hsieh, C.-L. Cellular Magnetic Resonance Imaging with Superparamagnetic Iron Oxide: Methods and Applications in Cancer. *SPIN* **2019**, *9*, 1940007. [CrossRef]
62. Xu, H.; Cheng, L.; Wang, C.; Ma, X.; Li, Y.; Liu, Z. Polymer encapsulated upconversion nanoparticle/iron oxide nanocomposites for multimodal imaging and magnetic targeted drug delivery. *Biomaterials* **2011**, *32*, 9364–9373. [CrossRef] [PubMed]
63. Andreiuk, B.; Reisch, A.; Lindecker, M.; Follain, G.; Peyriéras, N.; Goetz, J.G.; Klymchenko, A.S. Fluorescent Polymer Nanoparticles for Cell Barcoding In Vitro and In Vivo. *Small* **2017**, *13*, 1701582. [CrossRef] [PubMed]
64. Park, J.S.; Park, W.; Park, S.; Larson, A.C.; Kim, D.-H.; Park, K.-H. Multimodal Magnetic Nanoclusters for Gene Delivery, Directed Migration, and Tracking of Stem Cells. *Adv. Funct. Mater.* **2017**, *27*, 1700396. [CrossRef]
65. Wang, L.; Xu, K.; Hou, X.; Han, Y.; Liu, S.; Wiraja, C.; Yang, C.; Yang, J.; Wang, M.; Dong, X.; et al. Fluorescent Poly(glycerol-co-sebacate) Acrylate Nanoparticles for Stem Cell Labeling and Longitudinal Tracking. *ACS Appl. Mater. Interfaces* **2017**, *9*, 9528–9538. [CrossRef] [PubMed]
66. Saito, A.; Mekawy, M.M.; Sumiyoshi, A.; Riera, J.J.; Shimizu, H.; Kawashima, R.; Tominaga, T. Noninvasive targeting delivery and in vivo magnetic resonance tracking method for live apoptotic cells in cerebral ischemia with functional Fe2O3 magnetic nanoparticles. *J. Nanobiotechnol.* **2016**, *14*, 19. [CrossRef] [PubMed]
67. Domey, J.; Bergemann, C.; Bremer-Streck, S.; Krumbein, I.; Reichenbach, J.R.; Teichgräber, U.; Hilger, I. Long-term prevalence of NIRF-labeled magnetic nanoparticles for the diagnostic and intraoperative imaging of inflammation. *Nanotoxicology* **2016**, *10*, 20–31. [CrossRef] [PubMed]
68. Kim, S.M.; Jeong, C.H.; Woo, J.S.; Ryu, C.H.; Lee, J.-H.; Jeun, S.-S. In vivo near-infrared imaging for the tracking of systemically delivered mesenchymal stem cells: Tropism for brain tumors and biodistribution. *Int. J. Nanomed.* **2016**, *11*, 13–23. [CrossRef] [PubMed]
69. Kim, J.S.; Kim, Y.-H.; Kim, J.H.; Kang, K.W.; Tae, E.L.; Youn, H.; Kim, D.; Kim, S.-K.; Kwon, J.-T.; Cho, M.-H.; et al. Development and in vivo imaging of a PET/MRI nanoprobe with enhanced NIR fluorescence by dye encapsulation. *Nanomedicine* **2012**, *7*, 219–229. [CrossRef]
70. Chehade, M.; Srivastava, A.K.; Bulte, J.W.M. Co-Registration of Bioluminescence Tomography, Computed Tomography, and Magnetic Resonance Imaging for Multimodal In Vivo Stem Cell Tracking. *Tomography* **2016**, *2*, 159–165. [CrossRef]
71. Qi, S.; Zhang, P.; Ma, M.; Yao, M.; Wu, J.; Mäkilä, E.; Salonen, J.; Ruskoaho, H.; Xu, Y.; Santos, H.A.; et al. Cellular Internalization-Induced Aggregation of Porous Silicon Nanoparticles for Ultrasound Imaging and Protein-Mediated Protection of Stem Cells. *Small* **2019**, *15*, e1804332. [CrossRef]
72. Chen, M.; Betzer, O.; Fan, Y.; Gao, Y.; Shen, M.; Sadan, T.; Popovtzer, R.; Shi, X. Multifunctional Dendrimer-Entrapped Gold Nanoparticles for Labeling and Tracking T Cells Via Dual-Modal Computed Tomography and Fluorescence Imaging. *Biomacromolecules* **2020**, *21*, 1587–1595. [CrossRef] [PubMed]
73. Chen, Y.-C.; Wen, S.; Shang, S.-A.; Cui, Y.; Luo, B.; Teng, G.-J. Magnetic resonance and near-infrared imaging using a novel dual-modality nano-probe for dendritic cell tracking in vivo. *Cytotherapy* **2014**, *16*, 699–710. [CrossRef] [PubMed]
74. Schmidtke-Schrezenmeier, G.; Urban, M.; Musyanovych, A.; Mailänder, V.; Rojewski, M.; Fekete, N.; Menard, C.; Deak, E.; Tarte, K.; Rasche, V.; et al. Labeling of mesenchymal stromal cells with iron oxide-poly(L-lactide) nanoparticles for magnetic resonance imaging: Uptake, persistence, effects on cellular function and magnetic resonance imaging properties. *Cytotherapy* **2011**, *13*, 962–975. [CrossRef] [PubMed]
75. Ren, Z.; Wang, J.; Zou, C.; Guan, Y.; Zhang, Y.A. Labeling of cynomolgus monkey bone marrow-derived mesenchymal stem cells for cell tracking by multimodality imaging. *Sci. China Life Sci.* **2011**, *54*, 981–987. [CrossRef] [PubMed]
76. Namestnikova, D.; Gubskiy, I.; Kholodenko, I.; Melnikov, P.; Sukhinich, K.; Gabashvili, A.; Vishnevskiy, D.; Soloveva, A.; Abakumov, M.; Vakhrushev, I.; et al. Methodological aspects of MRI of transplanted superparamagnetic iron oxide-labeled mesenchymal stem cells in live rat brain. *PLoS ONE* **2017**, *12*, e0186717. [CrossRef] [PubMed]

177. Li, W.; Chen, R.; Lv, J.; Wang, H.; Liu, Y.; Peng, Y.; Qian, Z.; Fu, G.; Nie, L. In Vivo Photoacoustic Imaging of Brain Injury and Rehabilitation by High-Efficient Near-Infrared Dye Labeled Mesenchymal Stem Cells with Enhanced Brain Barrier Permeability. *Adv. Sci.* **2018**, *5*, 1700277. [CrossRef]
178. Ma, T.; Zheng, J.; Zhang, T.; Xing, D. Ratiometric photoacoustic nanoprobes for monitoring and imaging of hydrogen sulfide in vivo. *Nanoscale* **2018**, *10*, 13462–13470. [CrossRef]
179. Abbasi, A.Z.; Prasad, P.; Cai, P.; He, C.; Foltz, W.D.; Amini, M.A.; Gordijo, C.R.; Rauth, A.M.; Wu, X.Y. Manganese oxide and docetaxel co-loaded fluorescent polymer nanoparticles for dual modal imaging and chemotherapy of breast cancer. *J. Control. Release* **2015**, *209*, 186–196. [CrossRef]
180. Varma, N.R.S.; Shankar, A.; Iskander, A.; Janic, B.; Borin, T.F.; Ali, M.M.; Arbab, A.S. Differential biodistribution of intravenously administered endothelial progenitor and cytotoxic T-cells in rat bearing orthotopic human glioma. *BMC Med. Imaging* **2013**, *13*, 17. [CrossRef]
181. Moon, S.-H.; Yang, B.Y.; Kim, Y.J.; Hong, M.K.; Lee, Y.-S.; Lee, D.S.; Chung, J.-K.; Jeong, J.M. Development of a complementary PET/MR dual-modal imaging probe for targeting prostate-specific membrane antigen (PSMA). *Nanomed. Nanotechnol. Biol. Med.* **2016**, *12*, 871–879. [CrossRef]
182. Kim, S.; Chae, M.K.; Yim, M.S.; Jeong, I.H.; Cho, J.; Lee, C.; Ryu, E.K. Hybrid PET/MR imaging of tumors using an oleanolic acid-conjugated nanoparticle. *Biomaterials* **2013**, *34*, 8114–8121. [CrossRef] [PubMed]
183. Evertsson, M.; Kjellman, P.; Cinthio, M.; Andersson, R.; Tran, T.A.; In't Zandt, R.; Grafström, G.; Toftevall, H.; Fredriksson, S.; Ingvar, C.; et al. Combined Magnetomotive ultrasound, PET/CT, and MR imaging of 68Ga-labelled superparamagnetic iron oxide nanoparticles in rat sentinel lymph nodes in vivo. *Sci. Rep.* **2017**, *7*, 4824. [CrossRef] [PubMed]
184. Madru, R.; Tran, T.A.; Axelsson, J.; Ingvar, C.; Bibic, A.; Ståhlberg, F.; Knutsson, L.; Strand, S.-E. (68)Ga-labeled superparamagnetic iron oxide nanoparticles (SPIONs) for multi-modality PET/MR/Cherenkov luminescence imaging of sentinel lymph nodes. *Am. J. Nucl. Med. Mol. Imaging* **2013**, *4*, 60–69.
185. Yang, B.Y.; Moon, S.-H.; Seelam, S.R.; Jeon, M.J.; Lee, Y.-S.; Lee, D.S.; Chung, J.-K.; Kim, Y.I.; Jeong, J.M. Development of a multimodal imaging probe by encapsulating iron oxide nanoparticles with functionalized amphiphiles for lymph node imaging. *Nanomedicine* **2015**, *10*, 1899–1910. [CrossRef]
186. Malinge, J.; Géraudie, B.; Savel, P.; Nataf, V.; Prignon, A.; Provost, C.; Zhang, Y.; Ou, P.; Kerrou, K.; Talbot, J.-N.; et al. Liposomes for PET and MR Imaging and for Dual Targeting (Magnetic Field/Glucose Moiety): Synthesis, Properties, and in Vivo Studies. *Mol. Pharm.* **2017**, *14*, 406–414. [CrossRef] [PubMed]
187. Jin, L.; Wang, Q.; Chen, J.; Wang, Z.; Xin, H.; Zhang, D. Efficient Delivery of Therapeutic siRNA by Fe_3O_4 Magnetic Nanoparticles into Oral Cancer Cells. *Pharmaceutics* **2019**, *11*, 615. [CrossRef]
188. Wu, M.; Zhang, H.; Tie, C.; Yan, C.; Deng, Z.; Wan, Q.; Liu, X.; Yan, F.; Zheng, H. MR imaging tracking of inflammation-activatable engineered neutrophils for targeted therapy of surgically treated glioma. *Nat. Commun.* **2018**, *9*, 4777. [CrossRef]
189. Zhao, J.; Vykoukal, J.; Abdelsalam, M.; Recio-Boiles, A.; Huang, Q.; Qiao, Y.; Singhana, B.; Wallace, M.; Avritscher, R.; Melancon, M.P. Stem cell-mediated delivery of SPIO-loaded gold nanoparticles for the theranosis of liver injury and hepatocellular carcinoma. *Nanotechnology* **2014**, *25*, 405101. [CrossRef]
190. Thakkar, A.; Desai, P.; Chenreddy, S.; Modi, J.; Thio, A.; Khamas, W.; Ann, D.; Wang, J.; Prabhu, S. Novel nano-drug combination therapeutic regimen demonstrates significant efficacy in the transgenic mouse model of pancreatic ductal adenocarcinoma. *Am. J. Cancer Res.* **2018**, *8*, 2005–2019.
191. Afonso, D.; Mascarenhas, V. Imaging techniques for the diagnosis of soft tissue tumors. *Rep. Med. Imaging* **2015**, *8*, 63–70. [CrossRef]
192. Ladd, L.M.; Roth, T.D. Computed Tomography and Magnetic Resonance Imaging of Bone Tumors. *Semin. Roentgenol.* **2017**, *52*, 209–226. [CrossRef] [PubMed]
193. Hillman, E.M.C.; Amoozegar, C.B.; Wang, T.; McCaslin, A.F.H.; Bouchard, M.B.; Mansfield, J.; Levenson, R.M. In vivo optical imaging and dynamic contrast methods for biomedical research. *Philos. Trans. R. Soc. Math. Phys. Eng. Sci.* **2011**, *369*, 4620–4643. [CrossRef] [PubMed]
194. Ljungberg, M.; Pretorius, P.H. SPECT/CT: An update on technological developments and clinical applications. *Br. J. Radiol.* **2018**, *91*, 20160402. [CrossRef] [PubMed]
195. Barca, C.; Griessinger, C.M.; Faust, A.; Depke, D.; Essler, M.; Windhorst, A.D.; Devoogdt, N.; Brindle, K.M.; Schäfers, M.; Zinnhardt, B.; et al. Expanding Theranostic Radiopharmaceuticals for Tumor Diagnosis and Therapy. *Pharmaceuticals* **2021**, *15*, 13. [CrossRef] [PubMed]
196. Mushtaq, S.; Bibi, A.; Park, J.E.; Jeon, J. Recent Progress in Technetium-99m-Labeled Nanoparticles for Molecular Imaging and Cancer Therapy. *Nanomaterials* **2021**, *11*, 3022. [CrossRef]
197. Coenen, H.H.; Ermert, J. Expanding PET-applications in life sciences with positron-emitters beyond fluorine-18. *Nucl. Med. Biol.* **2021**, *92*, 241–269. [CrossRef]
198. Fernández-Barahona, I.; Muñoz-Hernando, M.; Pellico, J.; Ruiz-Cabello, J.; Herranz, F. Molecular Imaging with 68Ga Radio-Nanomaterials: Shedding Light on Nanoparticles. *Appl. Sci.* **2018**, *8*, 1098. [CrossRef]
199. Psimadas, D.; Georgoulias, P.; Valotassiou, V.; Loudos, G. Molecular nanomedicine towards cancer: ^{111}In-labeled nanoparticles. *J. Pharm. Sci.* **2012**, *101*, 2271–2280. [CrossRef]

200. Chakravarty, R.; Chakraborty, S. A review of advances in the last decade on targeted cancer therapy using 177Lu: Focusing on 177Lu produced by the direct neutron activation route. *Am. J. Nucl. Med. Mol. Imaging* **2021**, *11*, 443–475.
201. Lamb, J.; Holland, J.P. Advanced Methods for Radiolabeling Multimodality Nanomedicines for SPECT/MRI and PET/MRI. *J. Nucl. Med.* **2018**, *59*, 382–389. [CrossRef]
202. Gao, H.; Liu, X.; Tang, W.; Niu, D.; Zhou, B.; Zhang, H.; Liu, W.; Gu, B.; Zhou, X.; Zheng, Y.; et al. 99mTc-conjugated manganese-based mesoporous silica nanoparticles for SPECT, pH-responsive MRI and anti-cancer drug delivery. *Nanoscale* **2016**, *8*, 19573–19580. [CrossRef] [PubMed]
203. Xue, S.; Zhang, C.; Yang, Y.; Zhang, L.; Cheng, D.; Zhang, J.; Shi, H.; Zhang, Y. 99mTc-Labeled Iron Oxide Nanoparticles for Dual-Contrast (T1/T2) Magnetic Resonance and Dual-Modality Imaging of Tumor Angiogenesis. *J. Biomed. Nanotechnol.* **2015**, *11*, 1027–1037. [CrossRef] [PubMed]
204. Li, X.; Xiong, Z.; Xu, X.; Luo, Y.; Peng, C.; Shen, M.; Shi, X. (99m)Tc-Labeled Multifunctional Low-Generation Dendrimer-Entrapped Gold Nanoparticles for Targeted SPECT/CT Dual-Mode Imaging of Tumors. *ACS Appl. Mater. Interfaces* **2016**, *8*, 19883–19891. [CrossRef] [PubMed]
205. Belderbos, S.; González-Gómez, M.A.; Cleeren, F.; Wouters, J.; Piñeiro, Y.; Deroose, C.M.; Coosemans, A.; Gsell, W.; Bormans, G.; Rivas, J.; et al. Simultaneous in vivo PET/MRI using fluorine-18 labeled Fe_3O_4@$Al(OH)_3$ nanoparticles: Comparison of nanoparticle and nanoparticle-labeled stem cell distribution. *EJNMMI Res.* **2020**, *10*, 73. [CrossRef] [PubMed]
206. Wang, Y.; Liu, H.; Yao, D.; Li, J.; Yang, S.; Zhang, C.; Chen, W.; Wang, D. 18F-labeled magnetic nanoparticles for monitoring anti-angiogenic therapeutic effects in breast cancer xenografts. *J. Nanobiotechnol.* **2019**, *17*, 105. [CrossRef] [PubMed]
207. Crimì, F.; Valeggia, S.; Baffoni, L.; Stramare, R.; Lacognata, C.; Spolverato, G.; Albertoni, L.; Spimpolo, A.; Evangelista, L.; Zucchetta, P.; et al. [18F]FDG PET/MRI in rectal cancer. *Ann. Nucl. Med.* **2021**, *35*, 281–290. [CrossRef]
208. Umutlu, L.; Beyer, T.; Grueneisen, J.S.; Rischpler, C.; Quick, H.H.; Veit-Haibach, P.; Eiber, M.; Purz, S.; Antoch, G.; Gatidis, S.; et al. Whole-Body [18F]-FDG-PET/MRI for Oncology: A Consensus Recommendation. *ROFO Fortschr. Geb. Rontgenstr. Nuklearmed.* **2019**, *191*, 289–297. [CrossRef]
209. Hsu, F.-T.; Wei, Z.-H.; Hsuan, Y.C.-Y.; Lin, W.; Su, Y.-C.; Liao, C.-H.; Hsieh, C.-L. MRI tracking of polyethylene glycol-coated superparamagnetic iron oxide-labelled placenta-derived mesenchymal stem cells toward glioblastoma stem-like cells in a mouse model. *Artif. Cells Nanomed. Biotechnol.* **2018**, *46*, S448–S459. [CrossRef]
210. Qiao, Y.; Gumin, J.; MacLellan, C.J.; Gao, F.; Bouchard, R.; Lang, F.F.; Stafford, R.J.; Melancon, M.P. Magnetic resonance and photoacoustic imaging of brain tumor mediated by mesenchymal stem cell labeled with multifunctional nanoparticle introduced via carotid artery injection. *Nanotechnology* **2018**, *29*, 165101. [CrossRef]
211. Duan, L.; Zuo, J.; Zhang, F.; Li, B.; Xu, Z.; Zhang, H.; Yang, B.; Song, W.; Jiang, J. Magnetic Targeting of HU-MSCs in the Treatment of Glucocorticoid-Associated Osteonecrosis of the Femoral Head Through Akt/Bcl2/Bad/Caspase-3 Pathway. *Int. J. Nanomed.* **2020**, *15*, 3605–3620. [CrossRef]
212. Papadimitriou, N.; Thorfve, A.; Brantsing, C.; Junevik, K.; Baranto, A.; Barreto Henriksson, H. Cell viability and chondrogenic differentiation capability of human mesenchymal stem cells after iron labeling with iron sucrose. *Stem Cells Dev.* **2014**, *23*, 2568–2580. [CrossRef] [PubMed]
213. Kim, S.J.; Lewis, B.; Steiner, M.-S.; Bissa, U.V.; Dose, C.; Frank, J.A. Superparamagnetic iron oxide nanoparticles for direct labeling of stem cells and in vivo MRI tracking. *Contrast Media Mol. Imaging* **2016**, *11*, 55–64. [CrossRef] [PubMed]
214. Hachani, R.; Birchall, M.A.; Lowdell, M.W.; Kasparis, G.; Tung, L.D.; Manshian, B.B.; Soenen, S.J.; Gsell, W.; Himmelreich, U.; Gharagouzloo, C.A.; et al. Assessing cell-nanoparticle interactions by high content imaging of biocompatible iron oxide nanoparticles as potential contrast agents for magnetic resonance imaging. *Sci. Rep.* **2017**, *7*, 7850. [CrossRef] [PubMed]
215. Aşık, E.; Aslan, T.N.; Güray, N.T.; Volkan, M. Cellular uptake and apoptotic potential of rhenium labeled magnetic protein cages in MDA-MB-231 cells. *Environ. Toxicol. Pharmacol.* **2018**, *63*, 127–134. [CrossRef] [PubMed]
216. Salingova, B.; Simara, P.; Matula, P.; Zajickova, L.; Synek, P.; Jasek, O.; Veverkova, L.; Sedlackova, M.; Nichtova, Z.; Koutna, I. The Effect of Uncoated SPIONs on hiPSC-Differentiated Endothelial Cells. *Int. J. Mol. Sci.* **2019**, *20*, 3536. [CrossRef] [PubMed]
217. Ali, A.A.A.; Shahror, R.A.; Chen, K.-Y. Efficient Labeling Of Mesenchymal Stem Cells For High Sensitivity Long-Term MRI Monitoring In Live Mice Brains. *Int. J. Nanomed.* **2020**, *15*, 97–114. [CrossRef]
218. Zhang, L.; Xiao, S.; Kang, X.; Sun, T.; Zhou, C.; Xu, Z.; Du, M.; Zhang, Y.; Wang, G.; Liu, Y.; et al. Metabolic Conversion and Removal of Manganese Ferrite Nanoparticles in RAW264.7 Cells and Induced Alteration of Metal Transporter Gene Expression. *Int. J. Nanomed.* **2021**, *16*, 1709–1724. [CrossRef]
219. Su, H.; Mou, Y.; An, Y.; Han, W.; Huang, X.; Xia, G.; Ni, Y.; Zhang, Y.; Ma, J.; Hu, Q. The migration of synthetic magnetic nanoparticle labeled dendritic cells into lymph nodes with optical imaging. *Int. J. Nanomed.* **2013**, *8*, 3737–3744. [CrossRef]
220. Chang, Y.-K.; Liu, Y.-P.; Ho, J.H.; Hsu, S.-C.; Lee, O.K. Amine-surface-modified superparamagnetic iron oxide nanoparticles interfere with differentiation of human mesenchymal stem cells. *J. Orthop. Res.* **2012**, *30*, 1499–1506. [CrossRef]
221. Zhu, X.-M.; Wang, Y.-X.J.; Leung, K.C.-F.; Lee, S.-F.; Zhao, F.; Wang, D.-W.; Lai, J.M.Y.; Wan, C.; Cheng, C.H.K.; Ahuja, A.T. Enhanced cellular uptake of aminosilane-coated superparamagnetic iron oxide nanoparticles in mammalian cell lines. *Int. J. Nanomed.* **2012**, *7*, 953–964. [CrossRef]
222. Huang, C.; Neoh, K.G.; Wang, L.; Kang, E.-T.; Shuter, B. Surface functionalization of superparamagnetic nanoparticles for the development of highly efficient magnetic resonance probe for macrophages. *Contrast Media Mol. Imaging* **2011**, *6*, 298–307. [CrossRef] [PubMed]

223. Mou, Y.; Chen, B.; Zhang, Y.; Hou, Y.; Xie, H.; Xia, G.; Tang, M.; Huang, X.; Ni, Y.; Hu, Q. Influence of synthetic superparamagnetic iron oxide on dendritic cells. *Int. J. Nanomed.* **2011**, *6*, 1779–1786. [CrossRef]
224. Al Faraj, A.; Luciani, N.; Kolosnjaj-Tabi, J.; Mattar, E.; Clement, O.; Wilhelm, C.; Gazeau, F. Real-time high-resolution magnetic resonance tracking of macrophage subpopulations in a murine inflammation model: A pilot study with a commercially available cryogenic probe. *Contrast Media Mol. Imaging* **2013**, *8*, 193–203. [CrossRef] [PubMed]
225. Talukdar, Y.; Rashkow, J.; Lalwani, G.; Kanakia, S.; Sitharaman, B. The effects of graphene nanostructures on mesenchymal stem cells. *Biomaterials* **2014**, *35*, 4863–4877. [CrossRef]
226. Barbillon, G. Latest Novelties on Plasmonic and Non-Plasmonic Nanomaterials for SERS Sensing. *Nanomaterials* **2020**, *10*, 1200. [CrossRef]
227. Donahue, N.D.; Francek, E.R.; Kiyotake, E.; Thomas, E.E.; Yang, W.; Wang, L.; Detamore, M.S.; Wilhelm, S. Assessing nanoparticle colloidal stability with single-particle inductively coupled plasma mass spectrometry (SP-ICP-MS). *Anal. Bioanal. Chem.* **2020**, *412*, 5205–5216. [CrossRef]
228. Herd, H.; Daum, N.; Jones, A.T.; Huwer, H.; Ghandehari, H.; Lehr, C.-M. Nanoparticle geometry and surface orientation influence mode of cellular uptake. *ACS Nano* **2013**, *7*, 1961–1973. [CrossRef]
229. Sousa de Almeida, M.; Susnik, E.; Drasler, B.; Taladriz-Blanco, P.; Petri-Fink, A.; Rothen-Rutishauser, B. Understanding nanoparticle endocytosis to improve targeting strategies in nanomedicine. *Chem. Soc. Rev.* **2021**, *50*, 5397–5434. [CrossRef]
230. Zhao, J.; Stenzel, M.H. Entry of nanoparticles into cells: The importance of nanoparticle properties. *Polym. Chem.* **2018**, *9*, 259–272. [CrossRef]
231. Kafshgari, M.H.; Harding, F.J.; Voelcker, N.H. Insights into cellular uptake of nanoparticles. *Curr. Drug Deliv.* **2015**, *12*, 63–77. [CrossRef]
232. Behzadi, S.; Serpooshan, V.; Tao, W.; Hamaly, M.A.; Alkawareek, M.Y.; Dreaden, E.C.; Brown, D.; Alkilany, A.M.; Farokhzad, O.C.; Mahmoudi, M. Cellular uptake of nanoparticles: Journey inside the cell. *Chem. Soc. Rev.* **2017**, *46*, 4218–4244. [CrossRef] [PubMed]
233. Kinnear, C.; Moore, T.L.; Rodriguez-Lorenzo, L.; Rothen-Rutishauser, B.; Petri-Fink, A. Form Follows Function: Nanoparticle Shape and Its Implications for Nanomedicine. *Chem. Rev.* **2017**, *117*, 11476–11521. [CrossRef] [PubMed]
234. Moore, T.L.; Urban, D.A.; Rodriguez-Lorenzo, L.; Milosevic, A.; Crippa, F.; Spuch-Calvar, M.; Balog, S.; Rothen-Rutishauser, B.; Lattuada, M.; Petri-Fink, A. Nanoparticle administration method in cell culture alters particle-cell interaction. *Sci. Rep.* **2019**, *9*, 900. [CrossRef] [PubMed]
235. Zhang, Y.; Zhang, H.; Li, B.; Zhang, H.; Tan, B.; Deng, Z. Cell-assembled (Gd-DOTA)i-triphenylphosphonium (TPP) nanoclusters as a T2 contrast agent reveal in vivo fates of stem cell transplants. *Nano Res.* **2018**, *11*, 1625–1641. [CrossRef]
236. Wan, D.; Chen, D.; Li, K.; Qu, Y.; Sun, K.; Tao, K.; Dai, K.; Ai, S. Gold Nanoparticles as a Potential Cellular Probe for Tracking of Stem Cells in Bone Regeneration Using Dual-Energy Computed Tomography. *ACS Appl. Mater. Interfaces* **2016**, *8*, 32241–32249. [CrossRef]
237. Liesveld, J.L.; Sharma, N.; Aljitawi, O.S. Stem cell homing: From physiology to therapeutics. *Stem Cells* **2020**, *38*, 1241–1253. [CrossRef]
238. Schlorf, T.; Meincke, M.; Kossel, E.; Glüer, C.-C.; Jansen, O.; Mentlein, R. Biological properties of iron oxide nanoparticles for cellular and molecular magnetic resonance imaging. *Int. J. Mol. Sci.* **2010**, *12*, 12–23. [CrossRef]
239. Astolfo, A.; Schültke, E.; Menk, R.H.; Kirch, R.D.; Juurlink, B.H.J.; Hall, C.; Harsan, L.-A.; Stebel, M.; Barbetta, D.; Tromba, G.; et al. In vivo visualization of gold-loaded cells in mice using x-ray computed tomography. *Nanomed. Nanotechnol. Biol. Med.* **2013**, *9*, 284–292. [CrossRef]
240. Reifschneider, O.; Vennemann, A.; Buzanich, G.; Radtke, M.; Reinholz, U.; Riesemeier, H.; Hogeback, J.; Köppen, C.; Großgarten, M.; Sperling, M.; et al. Revealing Silver Nanoparticle Uptake by Macrophages Using SR-μXRF and LA-ICP-MS. *Chem. Res. Toxicol.* **2020**, *33*, 1250–1255. [CrossRef]
241. Hu, G.; Guo, M.; Xu, J.; Wu, F.; Fan, J.; Huang, Q.; Yang, G.; Lv, Z.; Wang, X.; Jin, Y. Nanoparticles Targeting Macrophages as Potential Clinical Therapeutic Agents Against Cancer and Inflammation. *Front. Immunol.* **2019**, *10*, 1998. [CrossRef]
242. Siegers, G.M.; Ribot, E.J.; Keating, A.; Foster, P.J. Extensive expansion of primary human gamma delta T cells generates cytotoxic effector memory cells that can be labeled with Feraheme for cellular MRI. *Cancer Immunol. Immunother. CII* **2013**, *62*, 571–583. [CrossRef] [PubMed]
243. Kim, M.H.; Lee, Y.J.; Kang, J.H. Stem Cell Monitoring with a Direct or Indirect Labeling Method. *Nucl. Med. Mol. Imaging* **2016**, *50*, 275–283. [CrossRef] [PubMed]
244. Elliott, A.D. Confocal Microscopy: Principles and Modern Practices. *Curr. Protoc. Cytom.* **2020**, *92*, 1. [CrossRef] [PubMed]
245. Berthold, F.; Tarkkanen, V. Luminometer development in the last four decades: Recollections of two entrepreneurs. *Luminescence* **2013**, *28*, 1–6. [CrossRef]
246. Stokes Debbie, J. Principles and Practice of Variable Pressure Environmental Scanning Electron Microscopy (VP-ESEM). In *Chichester*; John Wiley & Sons: Hoboken, NJ, USA, 2008; ISBN 978-0470758748.
247. Manohar, S.M.; Shah, P.; Nair, A. Flow cytometry: Principles, applications and recent advances. *Bioanalysis* **2021**, *13*, 181–198. [CrossRef]

nanomaterials

MDPI

Article

A Visual Discrimination of Existing States of Virus Capsid Protein by a Giant Molybdate Cluster

Yarong Xue [1], Mingfen Wei [1], Dingyi Fu [2], Yuqing Wu [1,*], Bo Sun [3], Xianghui Yu [3] and Lixin Wu [1,*]

[1] State Key Laboratory of Supramolecular Structure and Materials, Institute of Theoretical Chemistry, Jilin University, No.2699 Qianjin Street, Changchun 130012, China; xueyr17@mails.jlu.edu.cn (Y.X.); weimf19@mails.jlu.edu.cn (M.W.)
[2] School of Pharmacy, Nantong University, No. 19 Qixiu Road, Nantong 226001, China; fudingyi@ntu.edu.cn
[3] State Engineering Laboratory of AIDS Vaccine, Jilin University, No.2699 Qianjin Street, Changchun 130012, China; bo_sun@jlu.edu.cn (B.S.); xianghui@jlu.edu.cn (X.Y.)
* Correspondence: yqwu@jlu.edu.cn (Y.W.); wulx@jlu.edu.cn (L.W.)

Abstract: We report a unique phenomenon, the opposite color response of a giant polyoxometalate, $(NH_4)_{42}[Mo_{132}O_{372}(CHCOO)_{30}](H_2O)_{72}$ ($[Mo_{132}]$), to the existing states of human papillomavirus (HPV) major capsid protein, L1-pentamer (L1-p), and virus-like particles (VLPs). The color responses originate from the different assembly forms between $[Mo_{132}]$ and the capsid protein. The latter were inspected and separated by using CsCl gradient centrifugation, and validated in detail by sodium dodecyl sulfate-polyacrylamide gel-electrophoresis (SDS-PAGE), dynamic light scattering (DLS), and transmission electron microscopy (TEM) imaging. Furthermore, the intrinsic mechanisms were investigated in-depth by using XPS-based semi-quantitative analysis and well-designed peptides, revealing the critical points of L1 that determine the charge–transfer ratio between Mo(V) to Mo(VI), and consequently, the levels of $[Mo_{132}]$ hypochromic in different assemblies. Such a unique phenomenon is significant as it supplies a colorimetry approach to distinguish the existing states of the HPV capsid protein and would be significant in the quality assay of the HPV vaccine and existing states of other viruses in the future.

Keywords: colorimetric discrimination; giant polyoxometalate; HPV capsid protein; hypochromic of molybdate clusters; opposite color response

Citation: Xue, Y.; Wei, M.; Fu, D.; Wu, Y.; Sun, B.; Yu, X.; Wu, L. A Visual Discrimination of Existing States of Virus Capsid Protein by a Giant Molybdate Cluster. *Nanomaterials* 2022, *12*, 736. https://doi.org/10.3390/nano12050736

Academic Editors: Goran Kaluđerović and Nebojša Pantelić

Received: 14 January 2022
Accepted: 16 February 2022
Published: 22 February 2022

1. Introduction

Human papillomavirus (HPVs) causes commonly transmitted infections that occur in humans [1,2]. Some types of HPV lead to severe diseases including a series of verrucas and cancers [3]. To effectively prevent the occurrence of infections, vaccines that specifically target the virus-induced diseases have been developed recently, and many efforts have focused on building a security shielding system [4] Virus-like particles (VLPs), formed from the self-assembly of pentamer subunits comprising major capsid protein L1 without the participation of DNA, have been demonstrated to be an essential resource of vaccines because of the similar surface structure and antigenic epitopes to those of actual viruses. As a kind of prophylactic vaccine, VLPs were shown to be effective for protecting several subtypes of HPV from infections [5,6]. On one hand, these assembled particulate structures are also resilient to most environmental stresses and are promising for eliciting an efficient and potent immune response [7]. On the other hand, during the development of prophylactic vaccines against high-risk types of HPVs, VLPs can also be adopted directly in mimicking more closely to the neutralized epitopes, morphology, and keeping the size of native ones [8]. However, the production and usage of spherical capsids as vaccines relies significantly on the VLP's integrity, and any disassembly may result in losing effectiveness in producing antibodies. Therefore, the quick and convenient detection for the existing state of VLPs or their disassembly is crucial. Since the normal subunits of VLPs are the pentamer of capsid protein L1(L1-p), the sensitive determination of the species can demonstrate the incomplete state of VLPs [9]. Spectroscopic, dynamic light scattering, and electron microscopic measurements are commonly used characterization techniques [10,11], but they are not suitable for in situ quality assays during the preparation of vaccines. Therefore, it is imperative to develop a rapid and straightforward method for differentiating complete capsids and dissembled pentamers via a quick and direct procedure.

Polyoxometalates (POMs) are a kind of negatively charged inorganic clusters comprising early transition metal oxides [12]. The uniform shape and nanoscale size of POMs provides useful functional properties that can be useful in several emerging technologies including medicine [13,14]. Typically, the reduced coordination metal ions allow the deep-color clusters to have extended light absorption in the visible region. In the presence of oxidants, the reduced clusters exhibit reversible oxidation properties, shifting absorption toward another state [15]. Among the POM family, the giant molybdate clusters realized by Müller's group exhibited an elegant architecture and morphology with larger sizes of more than 2.5 nm [16]. As giant clusters, which have been extensively studied including, for example, self-assemblies into nanoscale capsules and forming co-assemblies together with cationic organic components via electrostatic interactions [17], and their functionalization was either performed on biomedical applications [18]. We have reported the electrostatic interaction of the giant clusters with cationic amphiphiles for synergistic self-assembly, the behavior with biomolecules in solutions in adhesion, anti-bacterial properties as well as bio-imaging [19,20]. However, up until now, there have been no reports concerning the use of these POMs and other inorganic clusters on the detection of the existing state of virus capsids, either in the assembly or disassembly state, neither the ionic binding position with capsid proteins.

The Mo(V) have been shown to be essential components in giant molybdate POMs, making them show colors in solid and solution due to their absorption, corresponding to the intervalence charge transfer (IVCT) between Mo(V) and Mo(VI) [21,22]. Thus, the valence state change of giant clusters always yields a color change, and as a result, Mo(V) can be oxidized into Mo(VI), accompanied by a color fading of brick-red in the presence of oxidants. Furthermore, the color changes are reversible because the Mo(VI) at the highest oxidized state are reduced while the cluster structure is maintained, just like the oxidation at a reduced state [23,24]. These properties of POMs and their charged surface provide an excellent opportunity to detect the existing state of the proteins comprising capsids through the electrostatic interaction with free N-terminals bearing redox features. Thus, the states of some proteins that can be oxidized and reduced can be characterized through the color fading of giant POMs after the redox process. In this context, we herein selected a giant brown Keplerate cluster, $(NH_4)_{42}[Mo_{132}O_{372}(CHCOO)_{30}](H_2O)_{72}$, abbreviated as $[Mo_{132}]$, which contains 72 Mo(VI) and 60 Mo(V), to sensitize the status of L1 proteins and the stability of the capsid [25]. Because L1 proteins at the states of assembly and disassembly show different binding modes with the $[Mo_{132}]$ cluster through exposing and shielding amino residues, we successfully realized the vital identification for the capsid proteins through simple color changes of $[Mo_{132}]$ (Scheme 1). The present method demonstrated a practical approach to examine the quality of the prophylactic HPV vaccines. Of course, the obtained results are also significant in widening bio-applications such as qualitative screening of HPV infections and tracking the differentiation process of HPVs in the human body by detecting L1 proteins.

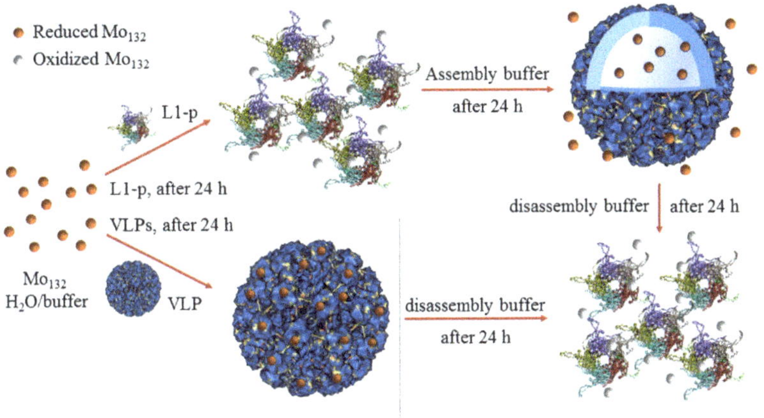

Scheme 1. Schematic illustration of the binding modes and different color responses of [Mo_{132}] to HPV L1-p and VLPs, respectively.

2. Results and Discussion

2.1. Hypochromic Response of [Mo_{132}] to L1-p

The [Mo_{132}] (2.5μM) in buffer A shows a typical absorption band at 456 nm similar to the cluster in aqueous solution (Figure 1A), which can be assigned to the intervalence charge transfer (IVCT) between Mo(V) and Mo(VI) centers bridging by O atoms [21,22]. Since the absorption bands of the L1 protein at both states of pentamer and VLP appeared at wavelengths less than 300 nm, the changes in absorption spectra of the [Mo_{132}] mixture with L1-p in the visible region was attributed to a source from the inorganic cluster. A gradual hypochromic behavior of [Mo_{132}] in buffer A solution with time exposure in air was observed, and nearly 20% absorbance was lost after 4 h, when a turning point occurred. After that, no further change occurred for the absorption with time over one day. The hypochromic properties of [Mo_{132}] can be attributed to the oxidation of partial Mo(V) in the giant cluster by oxygen in air, according to the opposite observation during its formation [24]. As further confirmed in the characterization, this color change in the aerobic environment does not affect the structural integrity and the detection for proteins because of the large number of reduced Mo(V) contained in one cluster.

In comparison to [Mo_{132}] in buffer A, a much stronger hypochromic effect of the cluster occurred in the same solution along with the addition of L1-p. A linear decrease in absorbance was observed, and more than 80% of the original absorbance disappeared until cultured to a turning point at 12 h. Further color fading proceeded continuously, and more than 90% of absorbance vanished after 24 h (Figure 1B). Interestingly, when [Mo_{132}] was added into the solution containing VLPs, only a little color degradation occurred over time (Figure 1C), and over 95% absorbance of the inorganic cluster was maintained at the time scale. The job plots of the [Mo_{132}] cluster at the states in the presence of protein L1-p or VLPs show the apparent differences in the absorbance at the visible region versus the time (Figure 1D).

[Mo_{132}] has been demonstrated to form blackberry-like hollow spherical self-assembled shapes in several tens of nanometers in aqueous solution [26] while maintaining a mono-dispersed state for a couple of days. As shown in Figure 2A, the DLS curve of the cluster alone showed a hydrodynamic diameter of 2.7 nm, similar to the size (2.9 nm) calculated from the crystal structure, illustrating the mono-dispersion and structural completeness [27,28]. The L1-p in buffer A had a size of about 12.0 nm (Figure 2B), being very close to the well-dispersed state of the pentamer subunit reported previously [29,30]. When the inorganic cluster was added into a solution containing L1-p, larger aggregates with a diameter of about 73 nm (Figure 2C) formed immediately. This size differs from each component or typical VLPs, implying the quick interaction between $[Mo_{132}]$ and the protein. The incubation for 24 h or longer did not lead to precipitation, but maintained a stabilized aggregate with a slight increase in size to 82 nm (Figure 2D). As already shown, the negatively charged POMs bind with some peptides containing basic residues through electrostatic interaction and hydrogen bonding [31,32]. The non-specific interaction can also induce the combination of the $[Mo_{132}]$ cluster bearing 42 negative charges with L1-p, causing the formation of larger assemblies or aggregates.

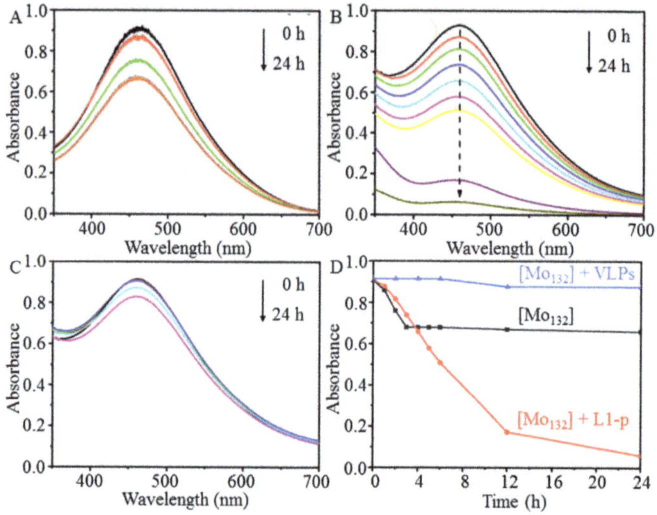

Figure 1. Time-dependent UV–Vis absorption spectra of $[Mo_{132}]$ (2.5 μM) in (**A**) buffer A, in the presence of (**B**) L1-p (10.0 μM) and (**C**) VLPs (10.0 μM), and (**D**) plots of the corresponding absorbance at 456 nm in (**A**–**C**) versus time for 24 h.

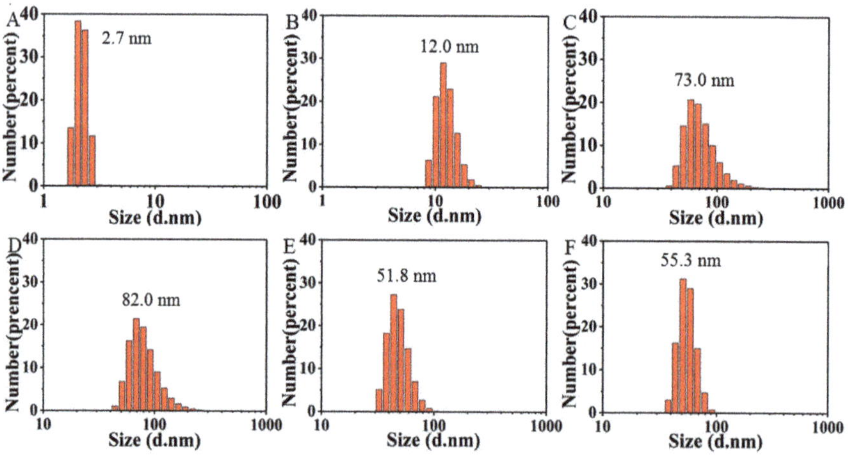

Figure 2. DLS histograms of (**A**) $[Mo_{132}]$ and (**B**) L1-p subunit in buffer A; the mixture of $[Mo_{132}]$ and L1-p (**C**) right after, and (**D**) after 24 h of incubation in buffer A; (**E**) the mixture of $[Mo_{132}]$ and L1-p after 24 h dialysis in assembly buffer; and (**F**) that of $[Mo_{132}]$ and the as-assembled VLPs in assembly buffer, respectively.

2.2. Visual Response of [Mo_{132}] to the Isolated L1-p and VLPs

The absorption spectral changes of [Mo_{132}] to L1-p and its VLP assembly (Figure S1) were examined in parallel with visual inspection through color observation. As shown in Figure 3, [Mo_{132}] is sensitive in detecting the existing states of protein L1. In a normal process, the L1 monomer is very unstable and exists typically in a pentamer state, which further self-assembles into an empty capsid, VLP, under high ionic concentration and low pH, spontaneously [27,28]. To identify the possible influence of the [Mo_{132}] cluster on the assembly process, the L1-p in a large amount of assembly buffer is monitored to allow for the assembly of VLP in the presence of [Mo_{132}], following a published standard process for other POM [27]. After encountering a dialysis procedure, the DLS assay revealed the formation of characteristic assemblies accompanied by a size change from 12 nm at the beginning to an average of 51.8 nm after 24 h of incubation (Figure 2E), in perfect agreement with the full-sized scale (50–55 nm) of VLPs comprising of the HPV16 L1 protein [29,30]. This result indicates that the presence of [Mo_{132}] does not affect the self-assembly of L1-p into VLP. However, after this assembly process, the dynamic small size attributed to [Mo_{132}] was no longer observed, even after 24 h of incubation (Figure 2F), implying that most of the clusters were trapped either inside or on the outside surfaces of the formed VLPs, or were eliminated through dialysis. The existence of [Mo_{132}] did not affect the further assembly of the L1-p as a subunit. Importantly, accompanying the self-assembly of L1-p, the faded color of [Mo_{132}] recovered because the solution changed from colorless back to brown, the original color of the [Mo_{132}] in solution (Figure 3).

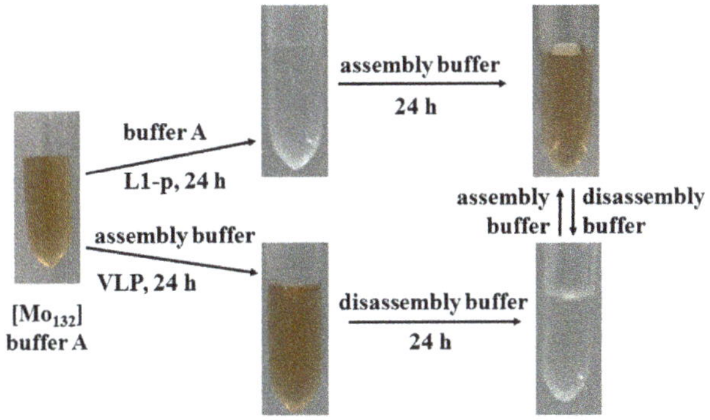

Figure 3. Color responses of [Mo_{132}] in buffer A in the absence and presence of either L1-pentamer or the VLPs, respectively.

2.3. Prevention of [Mo_{132}] Hypochromic by VLPs

More interestingly, the addition of [Mo_{132}] into an assembly buffer containing VLPs did not result in an obvious hypochromic after 24 h of incubation in the air; in contrast to L1-p, it showed much less color fading compared to a single cluster in buffer A, indicating the strong inhibition of VLPs on the color bleaching of [Mo_{132}]. Meanwhile, the DLS histogram (Figure 2F) confirmed a well-kept full-size of VLPs in 24 h. To identify whether [Mo_{132}] binds to VLPs or just remains isolated in solution, the ultracentrifugation of various components in gradient CsCl solution, were performed in tubes based on the principle of size and density dependence on the position to the rotating center, where components with the larger size and higher density will be located closer to the bottom.

After the CsCl gradient ultracentrifugation, the photograph of VLP alone in tube #0 showed a single blue band because of the concentrated protein, ascribed to the position of complete VLPs, at zone F2 (Figure 4A) in the middle of the centrifugation tube. In the case of VLP mixing with [Mo_{132}] in tube #1, besides a narrower blue belt at the F2#1, a wideband emerges at the lower position F3#1, suggesting the formation of aggregates with larger size differing from VLP or L1-p. Because the observed color of the belt near F3#1 was far from that of VLP at F2#1 while the sole VLP did not show any significant hint at the F3#0 position, we suggest that the [Mo_{132}] traps VLPs to form larger-sized aggregates. As a result, the inhibition to the hypochromic effect of [Mo_{132}] can be explained to be derived from protecting the VLPs from external oxidation. On the other hand, the photograph of L1-p mixing with [Mo_{132}] in tube #2 presented a weak belt with a pale color at the position F2#2, similar to that of VLP at F2#0, but there was no obvious belt emerging at zone F3#2. The result implies that part of L1-p self-assembled into VLPs automatically in the presence of inorganic clusters, but almost no formation of the proposed aggregates comprising the formed VLPs, and [Mo_{132}] was shown as the observation at zone F3#1. When L1-p was mixed with [Mo_{132}] in the assembly buffer, as seen in tube #3, both narrow blue belts at zone F2#3 and wide yellowish-brown belts emerged simultaneously at F3#3. The identical phenomenon of F3#3 to F3#1 demonstrates the formation of VLP in the presence of an inorganic cluster and the induced aggregation. Looking at these results together, we can draw the following conclusions: (1) the color degradation and the possible interaction with clusters did not affect the assembly of L1-p; and (2) a cluster induced aggregation of VLPs occurred, which is consistent with the DLS measurements (Figure 2C,D).

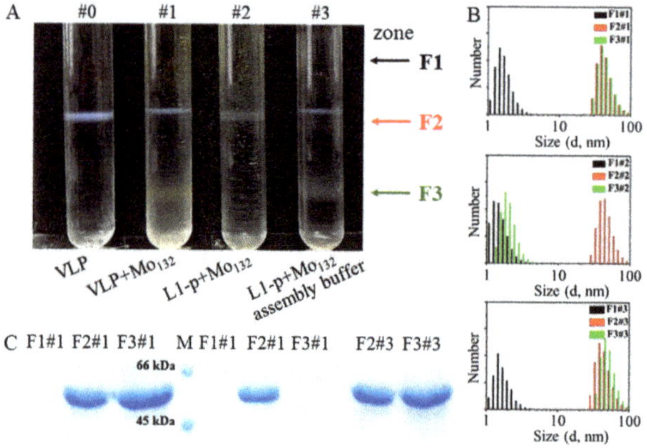

Figure 4. (**A**) The photographs after CsCl gradient ultracentrifugation in tubes for VLPs (tube #0), VLPs@[Mo_{132}] in assembly buffer (tube #1); the mixture of L1-p and [Mo_{132}] before (tube #2) and after (tube #3) monitoring in assembly buffer for 24 h, respectively. (**B**) DLS histograms of VLPs@[Mo_{132}] (upper), the mixture of L1-p and [Mo_{132}] before (middle) and after monitoring in assembly buffer (bottom), extracted from the position of zone F1 (black), F2 (red), and F3 (green), respectively. (**C**) SDS-PAGE of the corresponding component extracted from the position of F1#1, F2#1 and #3, and F3#1 and #3, respectively. The channel of M is the marker from the standard protein.

The belts at different zones in tubes after CsCl gradient ultracentrifugation were taken out, and the corresponding protein was assayed via the SDS-PAGE technique (Figure 4C). Taking two known proteins with molecular weights of 44 and 66 kDa as markers, all products extracted from F2 zones of tube#1–#3 point to a molecular weight of ~53 kDa, which is perfectly consistent with the L1-monomer. Besides emerging in the zone F2 in all tubes, the L1 protein also appeared at zone F3 in tubes #1 and #3, supporting the assignment for VLP aggregation with [Mo_{132}] at two zones. Meanwhile, the fact that no smaller proteins were observed in the F1 zones of tubes #2 and #3 shows that all the L1-p have already self-assembled into VLPs, especially in tube#2. Coincidently, the DLS histograms of the belts taken from tube #1 showed distributions of particle size that were consistent with [Mo_{132}] at zone F1#1, and the particle sizes corresponding to that of the VLPs at zones F2#1 and F3#1 were determined (Figure 4C, upper). As the cluster was proposed to be trapped in VLP aggregates, no isolated [Mo_{132}] was found at zone F3#1 due to the strong interaction between two components. For the belts extracted from tube #2 (Figure 4C, middle), besides identifying the particle that can be ascribed to [Mo_{132}] at F1#2, we also observed the particle size attributed to VLPs at F2#2. Although a small size close to [Mo_{132}] appeared at F3#2, no larger particles corresponding to the induced VLPs were observed excluding the possibility of it being there for [Mo_{132}]. Again, the DLS histogram of tube #3 (Figure 4C, bottom) shows the full spectra of size distributions similar to [Mo_{132}] at zone F1#3, VLPs at zone F2#3, and the cluster-triggered VLPs aggregation at zone F3#3, which were identical to the case observed in tube #1.

To further verify the morphology and completeness of the formed VLPs taken from F2#1 and F2#3, TEM images of [Mo_{132}] in buffer A, the mixture of [Mo_{132}] with L1-p before and after assembly monitoring, and the mixture of [Mo_{132}] and the as-prepared VLP were acquired. Because of the tiny size and mono-dispersion, the inorganic cluster (Figure 5A) could be well discerned from that of L1-p or VLPs in solution (Figure 5C,D). In the mixture of L1-p and [Mo_{132}], whether assembly monitoring was performed or not, we consistently obtained spherical particles (Figure 5B,C) that can be ascribed to the formation of VLPs as they matched the size of VLP cavity [29,30]. The amplified images (inset of Figure 5B,C) show small particles attributed to the inorganic clusters located inside the inner wall. Therefore, Figure 5B provides additional evidence for the cluster-triggered assembly of VLPs from L1-p, even without assembly monitoring. In addition, partial inorganic clusters were observed encapsulated inside the VLPs during the triggered co-assembly with L1-p, indicating a strong interaction between [Mo_{132}] and L1-p. The observation of VLPs can further support this analysis after mixing with [Mo_{132}] (Figure 5D), where the inorganic clusters are mainly located at the outside surface of VLPs being very different from that of the co-assembly of the two components.

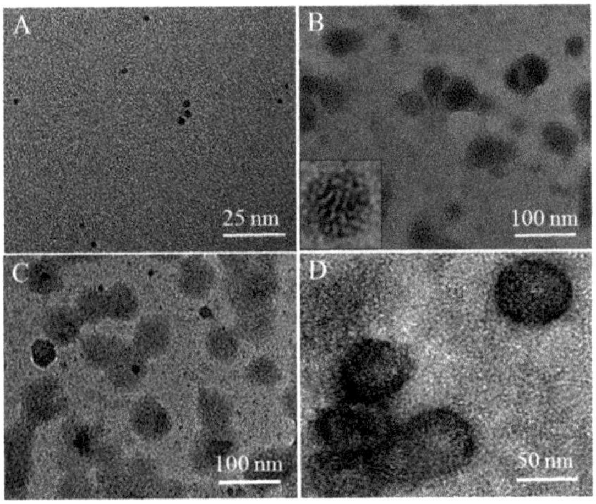

Figure 5. TEM images of (**A**) [Mo_{132}] in buffer A; (**B**) the mixture of [Mo_{132}] and L1-p in buffer A; (**C**) after 24 h dialysis in assembly buffer, and (**D**) the mixture of [Mo_{132}] and the as-prepared VLPs in assembly buffer, respectively. All images were acquired directly on samples without the phosphotungstate stain.

To obtain the essential profile, the TEM images for the samples extracted from zones F2#0, F2#1, and F2#3 were then acquired with negative staining (Figure 6). The micrographs show that the topography of VLPs and the particle size become more uniform and precise. Besides the transparent empty shell (Figure 6A), the mono-disperse particles with a statistical diameter of about 55 nm (Figure 6B,C) were in good agreement with the reported full-size of VLPs [29,30]. From the foregoing, we concluded that [Mo_{132}] could quickly bind with L1-p to form irregular aggregates, finally leading to the formation of VLP containing [Mo_{132}] inside. However, once [Mo_{132}] is mixed with the as-prepared VLPs, it assembles into [Mo_{132}]@VLP, where the cluster particles bind to the surface of VLPs (Figure 6C). As the molecular weight of [Mo_{132}] is much less than that of VLP, the sedimentation coefficient of [Mo_{132}]@VLPs is close to that of the VLPs, thus the encapsulation of a few [Mo_{132}] clusters does not significantly alter the surface properties of VLPs. Therefore, it is rational to explain that [Mo_{132}]@VLPs display a close level to VLPs in tubes after the CsCl gradient ultracentrifugation (Figure 4A). The two adjoining lancet belts at the position of F2#2 suggest an appreciable difference between VLPs and those encapsulated with [Mo_{132}]. Moreover, whether stained or not, the larger aggregates of VLPs induced by [Mo_{132}] were clearly discernible in Figures 5 and 6, which supports the observation of proteins in F3#1 and F3#3 (Figure 4B).

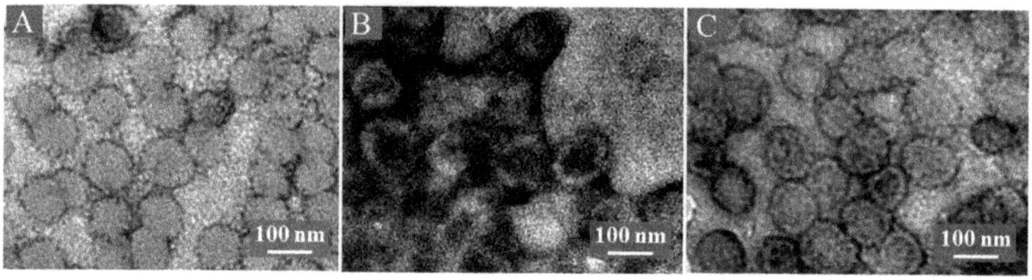

Figure 6. TEM images of (**A**) VLPs; (**B**) the mixture of [Mo_{132}] and L1-p after 24 h dialysis in assembly buffer; (**C**) the mixture of [Mo_{132}] and the as-prepared VLPs in assembly buffer, respectively. All of were extracted from the F2 fraction in tubes after CsCl gradient ultracentrifugation and stained with 2% phosphotungstate for 2 min before measurement.

2.4. Colorimetry Response Mechanism of Mo_{132} to L1-p and VLP

Redox nature of [Mo_{132}]. Coordination atoms such as Mo and W of POMs at the highest oxidation state are known to be photochemically reduced in the presence of a reductant, and the reduced POMs showed oxidation properties such as peroxidase for some organic and bio-molecules [33,34]. [Mo_{132}] has 60 reduced Mo(V) atoms and 72 oxidized Mo(VI) atoms, allowing the cluster to be both reduced and oxidized in suitable conditions. As the intervalent charge transfer absorption of Mo(V) to Mo(VI) emerged in the visible region, the [Mo_{132}] normally appeared as yellowish-brown. However, when Mo(V) atoms are oxidized, color degeneration will occur. For example, its incubation with a weak reductant L-ascorbic acid (Vc) for 24 h does not change the [Mo_{132}] cluster's color through reduction (Figure S2A). However, instead of the phenomenon displayed in Figure 1A, hypochromic properties are achieved in aerobic conditions when using Vc as a sacrifice against the air oxidation. To confirm that the hypochromic properties originate from the oxidation of [Mo_{132}] and to accelerate the process, 365 nm irradiation was then performed in parallel with and without Vc in buffer A (Figure S2B). After 12 h, 20% of the absorbance was diminished for the latter case; however, almost no change was observed for the former.

Interaction of capsid protein for colorimetry change of [Mo_{132}]. The non-covalent interactions of POMs with several types of biomolecules have been investigated extensively over the past years [35–37]. Based on the structural features of POMs, it is evident that the negatively charged [Mo_{132}] mainly provide electrostatic interactions with a variety of cationic species of protein. However, the observed hypochromic effect here should be essentially conducted by the external oxidation sourced from the protein rather than the buffer solution or the aerobic environment, since the color degradation does not fully occur in such a short time. Several peptides and proteins with specific surface charges have been used to examine the hypochromic effect of [Mo_{132}] (Figure S3). Under neutral conditions, the negative peptide of pTau-aac (pI = 4.5) and protein of BSA (pI = 4.6) do not induce a significant absorption change of [Mo_{132}]; however, the positively charged ones of dTau30 (pI = 10.4) and lysozyme (pI = 11.0) drive a significant decrease in absorption, indicating the vital role of positive partners to the [Mo_{132}] hypochromic effect. We can infer that the exposed basic residues in L1 accelerate the oxidation of [Mo_{132}].

The cryo-electron microscopy and image analysis on capsid proteins [37] revealed that the L1-p bind together to form VLPs via a long segment of the C-terminal (Figure S4A). The images revealed that the sequence after Asp401 at the C-terminal extended from one pentamer to the adjacent one to strengthen the VLPs structure. This segment was dominated by cationic residues such as arginine or lysine (Figure S4B), which are firmly prone to bind with the negatively charged [Mo_{132}]. After L1-p assembles into VLP, however, this segment is embedded in the wall of the VLP sphere, and the charged environment varies widely (Figure S4C). To further confirm the binding site of [Mo_{132}] with capsid protein, the sequence of two peptides, pep1–401 and pep401–495, were constructed and expressed separately, through identical approaches as the full-length L1.

After carrying out DNA sequence assays (Figure S5) and protein purification, each peptide was mixed with [Mo_{132}] in buffer A. This procedure yielded different phenomena. A quick decrease in [Mo_{132}] absorption at 456 nm was shown when mixed with pep401–495 (Figure 7A). The band disappeared completely, and the solution became pale within 90 min. The time-dependent plot of absorption (Figure 7C) revealed that the peptide drives the hypochromic effect of the cluster much faster than that of L1-p (Figure 1A), confirming a decisive role of pep401–495 in this process. In contrast, mixing pep1–401 with [Mo_{132}] only led to a very slight decrease in absorption centered at 456 nm (Figure 7B), indicating its feeble contribution in L1 to the [Mo_{132}] hypochromic effect. The plots of absorption intensity vs. time (Figure 7C) clearly illustrate the responsive differences between pep401–495, pep1–401, and L1-p to [Mo_{132}], further demonstrating that the binding with basic residues of pep401–495 causes the enhanced hypochromic effect by L1. Thus, it can be speculated that during the assembly of L1-p to VLP, the stronger binding affinity between L1-p subunits could force [Mo_{132}] to be released from the positive sites of L1-p and electrostatically attach to other positions of VLPs. Considering the positive areas at the interior surface of VLPs [38], the negatively charged [Mo_{132}] are easy to adsorb on the inner surface of VLP, which shield [Mo_{132}] from oxidation and consequently protect it from the hypochromic effect. Moreover, the observed larger aggregates of VLPs induced by [Mo_{132}] (Figures 5 and 6) suggest an additional protection of [Mo_{132}].

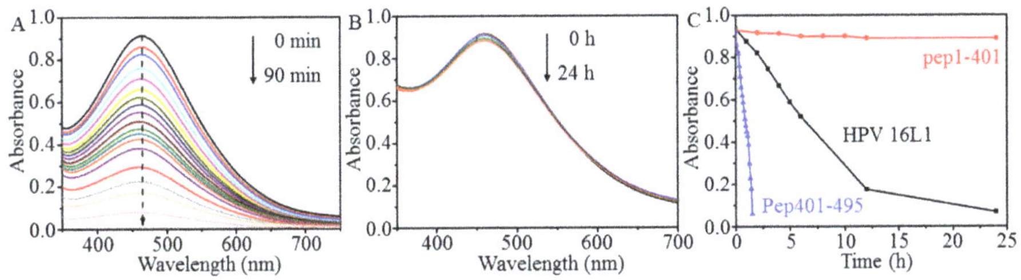

Figure 7. Time-dependent UV–Vis absorption spectra of [Mo_{132}] (2.5 μM) in buffer A in the presence of (**A**) pep401–495 (10.0 μM) and (**B**) pep1–401(10.0 μM), respectively. (**C**) The plot of the corresponding intensity changed at 456 nm in (**A**,**B**), and the mixture of [Mo_{132}] and L1-p in buffer A, respectively, vs. time for comparison.

Redox nature of [Mo_{132}] in the presence of L1-p and VLP. X-ray photoelectron spectroscopy (XPS) was used to analyze the redox state of pristine [Mo_{132}] (Figure 8A) and in the presence of L1-p (Figure 8B) or VLP (Figure 8C) under aerobic conditions. The characteristic Mo_{3d} doublet, composed of the $3d_{5/2}$ and $3d_{3/2}$ levels resulting from spin-orbit coupling, was observed in the spectra. Suitable fits of the data points, corresponding to two possible 3d doublets of Mo in different oxidation states of Mo(VI) and Mo(V), were achieved using two pairs of Lorentzian–Gaussian functions. The peaks centered at 232.4 and 235.5 eV were assigned to Mo(V), while those at 233.6 and 236.6 eV were attributed to Mo(VI), respectively [39,40]. Semi-quantitative calculations showed that the characteristic ratio of peak area for Mo(VI) and Mo(V) was 1.45:1 for [Mo_{132}] (Table 1), which was slightly over the value of 1.2 calculated from the ratio of 72 Mo(VI) to 60 Mo(V). The reason for this deviation can be deduced from the aerobic oxidation of partial Mo(V) atoms (approximate 10) in [Mo_{132}] solution during sample preparation.

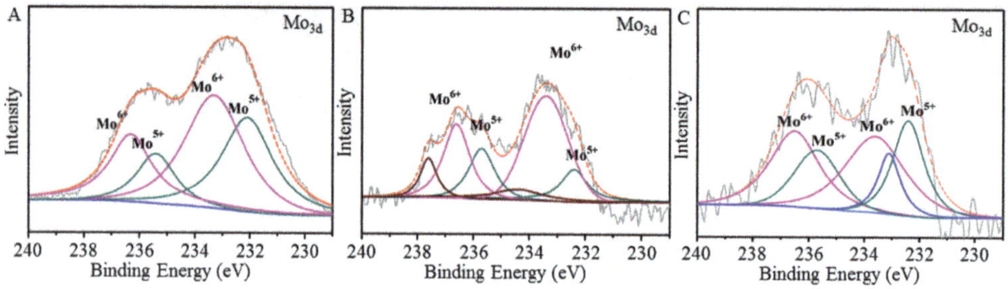

Figure 8. XPS of (**A**) [Mo_{132}] in buffer A; (**B**) [Mo_{132}] and L1-p in buffer A; (**C**) [Mo_{132}] and VLPs in the assembly buffer after incubation at the aerobic condition for 24 h.

Table 1. Integrated area of the simulated peak for Mo(V) and Mo(VI) from the XPS results in Figure 8A–C, respectively, and their ratio.

Samples	[Mo_{132}]	[Mo_{132}] + L1-p	[Mo_{132}] + VLPs
Mo(V)	115,359.60 + 43,910.59	15,696.10 + 18,827.48	17,349.15 + 21,444.59
Mo(VI)	147,757.50 + 82,513.98	58,321.11 + 28,475.33	26,737.09 + 32,049.94
Ratio of Mo(VI) to Mo(V)	1.45:1	2.5:1	1.5:1

The fitting curves of [Mo_{132}] in mixing with L1-p (Figure 8B) were vastly different from [Mo_{132}] alone. Besides the binding energy pairs attributed to Mo(V) and Mo(VI), the third coupled binding energy bands were observed. The latter were attributed to the intermediate of Mo during the transition. Furthermore, the quantitative calculation revealed that the ratio of peak area for Mo(VI) and Mo(V) increased to 2.5:1 (Table 1), indicating that more Mo(V) atoms had been oxidized. Larger amounts of Mo(V) (approximate 31) in [Mo_{132}] were oxidized into Mo(VI) after binding with L1-p, which is solid evidence for the enhanced hypochromic effect in the mixture.

Accompanying the assembly of L1-p, the XPS results of [Mo_{132}] almost returned back to the state of [Mo_{132}] alone in solution (Figure 8C). Although the intermediate component of Mo atoms still appeared in the fitting model, the calculation revealed a reduction in band ratio for Mo(VI) to Mo(V) of approximately 1.45:1 (Table 1). Seventeen of the oxidized Mo(VI) were reduced back to Mo(V). As a result, the L1-p presented here is similar to a sensitizer for [Mo_{132}] hypochromic properties. The electrostatic interaction between [Mo_{132}] and pep401–495 connecting the positive residue and [Mo_{132}] directly allows for the oxidation of Mo(V) to Mo(VI) more easily. Consequently, the intervalence charge transfer between Mo(V) and Mo(VI) centers was vastly weakened, and the color finally disappeared with time. As illustrated, such a process is accompanied by the binding with free basic residues in protein (Scheme 2) and the elimination of them in L1-p after assembly into VLP (Figure S4C).

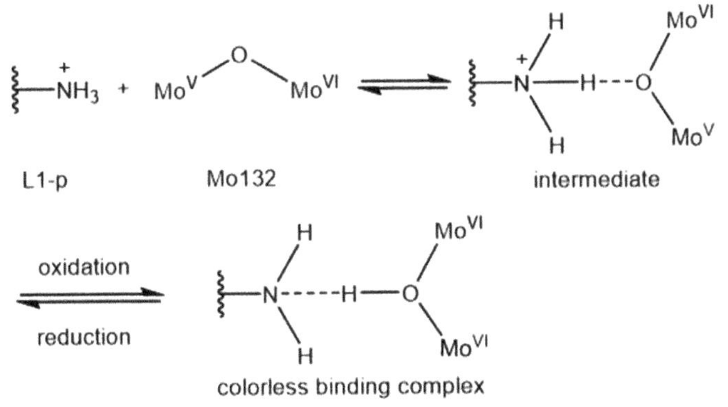

Scheme 2. Schematic drawing of the chromic mechanism of [Mo_{132}] induced by protein.

3. Conclusions and Perspectives

We report the opposite color response of a giant polyoxometalate, $(NH_4)_{42}[Mo_{132}O_{372}(CHCOO)_{30}]$ $(H_2O)_{72}$ ([Mo_{132}]), to the existing states of human papillomavirus (HPV) major capsid protein, L1-pentamer (L1-p) and virus-like particles (VLPs), originating from the assembly between [Mo_{132}] and capsid protein. Assembly with L1-p resulted in the improved hypochromic of [Mo_{132}] while assembly with the as-assembled VLPs led to an obvious protection of [Mo_{132}] from the hypochromic effect, compared to the single cluster in solution. Furthermore, both the size and morphology of the assemblies were characterized by using CsCl gradient centrifugation, SDS-PAGE, DLS, and TEM imaging. Remarkably, the in-depth mechanism studies determined by XPS and two well-designed peptides from L1 disclose that the electrostatic interaction between [Mo_{132}] and pep401–495 induces the change of Mo(V) into Mo (VI), which facilitates the hypochromic effect of [Mo_{132}]; however, the assembly from L1-p to VLPs reduced the binding possibility of [Mo_{132}] to pep401–495 and the transition from Mo(V) to Mo (VI), which protects [Mo_{132}] from the hypochromic effect. Therefore, the present study not only reports a unique phenomenon of contrasting color responses of [Mo_{132}] to HPV capsid protein, 16 L1-p and VLPs (which affords an easily performed colorimetry approach to evaluate the states of the HPV capsid protein), but also extends the potential of the molybdate polyoxometalate family toward new applications in medical science and could possibly be extended to other kinds of POMs.

4. Experimental Section

4.1. Reagents and Materials

Tryptone was purchased from OXOID Ltd. (Basingstoke, UK). Isopropyl-β-D-thiogalactoside (IPTG) and kanamycin were acquired from Tianjia Tech, China. Ethylene diamine tetraacetic acid (EDTA), acetic acid, ethanol, and methanol were obtained from BCIGC Ltd., China. Coomassie blue G-250 and Tween 80 were produced from Ding-Guo Ltd. (Beijing, China). 3-(*N*-morpholinyl) propanesulfonic acids (MOPs), and NaCl used for buffer preparation were purchased from Aladdin, China. D,L-dithiothreitol (DTT) and Tris(hydroxymethyl)aminomethane (Tris) are products from the Coolaber Company in China. The purity of all chemicals was higher than 99.9% and used as received.

4.2. Expression, Purification of HPV 16 L1 Protein and Peptides

The coding sequences and preparation of major capsid protein (L1) for HPV 16 followed the procedures described in a previous work where pET-30a vector and BL21 Star^{TM} (DE3) were employed for better expression [40]. Briefly, the recombinant *Escherichia coli* strain BL21 Star^{TM} (DE3) was cultured at 37 °C with stirring at 220 rpm, where the bacteria were induced by 0.1 mM IPTG. When OD_{600} reached 0.6–0.8, the cells were further cultivated at 25 °C for another 17 h, and then they were harvested by centrifugation. The cell pellets were re-suspended in buffer A (50 mM MOPs, 250 mM NaCl, 10 mM DTT, pH 7.0) with a concentration of 1.0 g in 10 mL. After being lysed by sonication and separated by centrifugation, a supernatant rich with L1 protein was obtained and further stabilized for another 30 min with increased DTT (20 mM) in buffer A. Then, purification was performed again by using a cationic exchange column filled with POROS® XS (xk16/10), where the supernatant was introduced into the column after being equilibrated with buffer A. A solution of 1.2 M NaCl finally eluted the target protein L1, and its final concentration was determined using the BCA method. The purity of L1 was finally identified by using 12% (wt/vol) SDS-PAGE. The obtained HPV 16 L1 existed essentially in the form of VLPs, which was further dialyzed against disassembly buffer (50 mM tris, 100 mM NaCl, 2 mM EDTA, 20 mM DTT, 0.01% Tween 80, pH 8.0) at 4 °C for 24 h (being changed every 4 h) to disassemble VLPs into L1-p for co-assembly monitoring.

The coding sequences of HPV 16 pep1–401 and pep401–495 were the same as the corresponding segment of HPV 16 L1 [40]. The pET-30a vector containing the respect gene was transformed into *Escherichia coli* strain BL21 Star™ (DE3) and cultured at 37 °C. The experimental conditions and method used for their expression and the following purifications were the same as that of the HPV 16 L1 protein described above.

4.3. Preparation of $(NH_4)_{42}[Mo_{132}O_{372}(CH_3COO)_{30}]\cdot 72H_2O$

The synthesis and structural characterization of the $[Mo_{132}]$ cluster followed procedures similar to those reported elsewhere [18]. Briefly, 0.8 g $N_2H_4\cdot H_2SO_4$ was added into 250 mL of H_2O containing 5.6 g of $(NH_4)[Mo_7O_{24}]\cdot 4H_2O$ and 12.5 g of CH_3COONH_4. After stirring for 10 min, 83 mL of 50% CH_3COOH (v/v) was added to replace the involved $-SO_4{}^{2-}$, further stabilizing and giving the product a crystalline precipitate. After desiccation in air, the product was characterized by X-ray diffraction, IR, and elemental analysis. The 100 μM of stock solution of $[Mo_{132}]$ was then prepared in deionized water, which was further diluted to the desired concentration when ready for use.

4.4. Assembly/Disassembly Monitoring of Mo_{132} and HPV 16 L1

Four typical samples were prepared, respectively, for the response assay of $[Mo_{132}]$ to the HPV capsid protein: (I) the stock solution of $[Mo_{132}]$ was mixed with L1-p and stirred at 4 °C for 24 h in buffer A to obtain full binding, where the concentration of inorganic clusters is 2.5 μM and that of proteins is 10 μM; (II) the mixture of (I) was subsequently dialyzed in a 1.0 L assembly buffer (10 mM phosphate, 500 mM NaCl, 0.03% Tween 80, pH 5.4) at 4 °C for 24 h to induce the assembly of L1-p into VLPs in the presence of $[Mo_{132}]$; (III) 10 μM of L1-p was dialyzed in 1.0 L assembly buffer to obtain full-sized empty VLPs, which was used either as a control or further mixed with $[Mo_{132}]$ for post-assembly monitoring as the number (IV) sample, after another 24 h stirring at 4 °C.

In a parallel experiment, sample (II) was further dialyzed against 1.0 L of disassembly buffer (50 mM Tris, 100 mM NaCl, 2 mM EDTA, 20 mM DTT, 0.01% Tween 80) at 4 °C for 24 h to record the disassembly behavior of VLPs into L1-p in the presence of $[Mo_{132}]$. The concentration of $[Mo_{132}]$ used in all of the above experiments was the same at 2.5 μM unless mentioned otherwise, while that for HPV 16 L1 was 10 μM, depending on the L1 monomer. In addition, the procedures used to monitor the conversion between L1-p and VLPs were the same as reported previously [27,28].

4.5. Cesium Chloride Gradient Centrifugation

First, we prepared the discontinuous densities of CsCl aqueous solution at 1.50, 1.35, and 1.25 g/mL, respectively. Then, the solutions in a volume of 0.5, 3.0, and 3.5 mL were slowly added into ultracentrifuge tubes in order, respectively. After further adding the mixture solution of the L1-p protein and Mo_{132} with equivalent volume to the upper part separately, the ultracentrifuge tubes were placed in the SW40 rotor and centrifuged at 220,000× g for 3.5 h at 4 °C. Finally, the interest bands and the counterparts were collected using a needle and dialyzed in PBS buffer for 24 h to remove the attached CsCl. The products were identified by 12% SDS-PAGE, dynamic light scattering (DLS), and transmission electron microscopy (TEM).

4.6. Instruments

UV–Vis absorption spectra were recorded on a Shimadzu (Kyoto, Japan) RF-5301PC spectrometer for the assembly and/or disassembly of HPV 16 L1 protein or peptides. A 1.0 mL target solution was added in a quartz cuvette, and the spectra in the 300–700 nm range were collected for each sample. The particle size was assayed by DLS. Briefly, after pre-filtration performed by placing the sample in a PCS1115 cuvette, the particle sizes were tested on a Malvern Zetasizer Nano-ZS 90 (Malvern, England) at 25 °C. All data were repeated for each sample in three parallel sets. The procedures in the sample preparation for TEM imaging were the same as those described previously [21,22]. The samples without staining were first spotted on copper grids coated with carbon and formvar. After drying in the air for 2 min, the measurements were carried out on a H-7650 transmission electron microscope (Hitachi, Tokyo, Japan) under an accelerating voltage of 80 kV. The images of samples with a stain of 2% phosphotungstate were collected under an accelerating voltage of 120 kV. The X-ray photoelectron spectroscopy (XPS) measurements were acquired using Thermo ESCALAB 250, where an Al K_α line (1486.6 eV) was employed as a monochromic X-ray source and the binding energy of C1s (284.6 eV) was used for correction.

Supplementary Materials: The following are available online at https://www.mdpi.com/article/10.3390/nano12050736/s1, Figure S1: (A) Time-dependent UV-vis absorption spectra of $[Mo_{132}]$@VLPs in disassembly buffer, to induce the disassembly of VLPs into L1-p again; (B) The plots of corresponding intensity in (A), which show obvious hypochromicity after 3 h incubation. Finally, more that 90% color was diminished again; Figure S2: Time-dependent UV-vis absorption spectra of $[Mo_{132}]$ in buffer A (2.5 μM) (A) in the presence of Vc (10.0 μM) under white light; (B) in the absence and presence of Vc (10.0 μM) under the irradiation of white light or 365 nm, respectively; Figure S3: Time-dependent UV-vis absorption spectra of $[Mo_{132}]$ in buffer A (2.5 μM) in the presence of the negative (A) peptide, pTau-aac (10.0 μM, pI = 4.5), (B) protein, BSA (10.0 μM, pI = 4.6); and positive (C) peptide, dTau30 (10.0 μM, pI = 10.4), (D) protein, lysozyme (10.0 μM, pI = 11.0), respectively; Figure S4: Illustration of 3D structural relationship between one HPV16 L1-p and its neighbor. The model is obtained from a Cryo-EM reconstruction structure (PDB ID, 3J6R).[S1] (A) Top view on L1-p; (B) the enlarge part in a box of (A), highlighting the involved arginine and lysine, respectively; (C) Side view of L1-p to show more clearly the burier of pep401–495 segment by the neighboring L1-p in VLP; Figure S5: Agarose gel electrophoresis to assay the DNA sequence of two peptides derived from HPV 16L1. (A) Enzyme digestion of T-easy-peptide 401–495. The gene of peptide 401–495 was 285 bp, corresponding to the site between 250 to 500 bp in marker. (B) Enzyme digestion of T-easy-peptide 1–401. The gene of peptide 1–401 is 1203 bp, corresponding to the site between 1000 to 1500 bp in marker, Table S1: Integrated area of the simulated peak for Mo(V) and Mo(VI) from the XPS results in Figure 8A,B, and C, respectively, and the ratio of them.

Author Contributions: Conceptualization, Y.W. and Y.X.; methodology, Y.W. and Y.X.; validation, B.S. and D.F.; formal analysis, Y.W. and Y.X.; investigation, Y.X. and D.F.; $[Mo_{132}]$ resources, M.W.; data curation, Y.X.; writing—original draft preparation, Y.X. and Y.W.; writing—review and editing, Y.W. and L.W.; visualization, X.Y.; supervision, Y.W.; project administration, Y.W. and L.W.; funding acquisition, Y.W. All authors have read and agreed to the published version of the manuscript.

Funding: We greatly appreciate the financial support from NSFC (Nos. 21875085 and 21373101) and the Open Project from the State Key Laboratory of Supramolecular Structure and Materials, Jilin University.

Conflicts of Interest: The authors declare that they have no competing interests.

References

1. Frega, A.; Stentella, P.; Ioris, A.D.; Piazze, J.J.; Fambrini, M.; Marchionni, M.; Cosmi, E.V. Young women, cervical intraepithelial neoplasia and human papillomavirus: Risk factors for persistence and recurrence. *Cancer Lett.* **2003**, *196*, 127–134. [CrossRef]
2. Colón-López, V.; Ortiz, A.P.; Palefsky, J. Burden of human papillomavirus infection and related comorbidities in men: Implications for research, disease prevention and health promotion among hispanic men. *P. R. Health Sci. J.* **2010**, *29*, 232–240.
3. Dunjic, M.; Stanisic, S.; Krstic, D.; Stanisic, M.; Dunjic, M. Integrative approach to diagnosis of genital human papillomaviruses (HPV) infection of female. *Acupunct. Electro-Ther. Res.* **2014**, *39*, 229–239. [CrossRef] [PubMed]
4. Pogoda, C.S.; Roden, R.B.S.; Garcea, R.L. Immunizing against anogenital cancer: HPV vaccines. *PLoS Pathog.* **2016**, *12*, e1005587. [CrossRef] [PubMed]
5. Harper, D.M.; Franco, E.L.; Wheeler, C.; Ferris, D.G.; Jenkins, D.; Schuind, A.; Zahaf, T.; Innis, B.; Naud, P.; De Carvalho, N.S.; et al. GlaxoSmithKline HPV Vaccine Study Group. Efficacy of a Bivalent L1 Virus-Like Particle Vaccine in Prevention of Infection with Human Papillomavirus Types 16 and 18 in Young Women: A Randomised Controlled Trial. *Lancet* **2004**, *364*, 1757–1765. [CrossRef]
6. Einstein, M.H. Acquired Immune Response to Oncogenic Human Papillomavirus Associated with Prophylactic Cervical Cancer Vaccines. *Cancer Immunol. Immunother.* **2008**, *57*, 443–451. [CrossRef]
7. Lua, L.H.L.; Connors, N.K.; Sainsbury, F.; Chuan, Y.P.; Wibowo, N.; Middelberg, A.P.J. Bioengineering virus-like particles as vaccines. *Biotechnol. Bioeng.* **2014**, *111*, 425–440. [CrossRef]
8. McCarthy, M.P.; White, W.I.; Palmer-Hill, F.; Koenig, S.; Suzich, J.A. Quantitative disassembly and reassembly of human papillomavirus type 11 virus-like particles in vitro. *J. Virol.* **1998**, *72*, 32–41. [CrossRef]
9. Zdanowicz, M.; Chroboczek, J. Virus-like particles as drug delivery vectors. *Acta Biochim. Pol.* **2016**, *63*, 469–473. [CrossRef]
10. Huber, B.; Schellenbacher, C.; Shafti-Keramat, S.; Jindra, C.; Christensen, N.; Kirnbauer, R. Chimeric L2-Based Virus-Like Particle (VLP) Vaccines Targeting Cutaneous Human Papillomaviruses (HPV). *PLoS ONE* **2017**, *12*, e0169533. [CrossRef]
11. Mach, H.; Volkin, D.B.; Troutman, R.D.; Wang, B.; Luo, Z.; Jansen, K.U.; Shi, L. Disassembly and reassembly of yeast-derived recombinant human papillomavirus virus-like particles (HPV VLPs). *J. Pharm. Sci.* **2006**, *95*, 2195–2206. [CrossRef] [PubMed]
12. Molina Sánchez, P. *Polyoxometalate Self-Assembly: From Molecules to Hybrid Materials*; University of Glasgow: Glasgow, UK, 2013.
13. Wu, L.; Liang, J. Polyoxometalates and Their Complexes Toward Biological Application. In *Supramolecular Chemistry of Biomimetic Systems*; Li, J., Ed.; Springer Nature Singapore Pte Ltd.: Singapore, 2017; pp. 311–354.
14. Bijelic, A.; Aureliano, M.; Rompel, A. Polyoxometalates as potential next-generation metallodrugs in the combat against cancer. *Angew. Chem. Int. Ed.* **2019**, *58*, 2980–2999. [CrossRef] [PubMed]
15. Liu, T.; Diemann, E.; Li, H.; Dress, A.W.M.; Müller, A. Self-assembly in aqueous solution of wheel-shaped Mo_{154} oxide clusters into vesicles. *Nature* **2003**, *426*, 59–62. [CrossRef] [PubMed]
16. Müller, A.; Rehder, D. Molecular Metal Oxides in Protein Cages/Cavities. Coordination Chemistry in Protein Cages. In *Coordination Chemistry in Protein Cages: Principles, Design, and Applications*; Ueno, T., Watanabe, Y., Eds.; John Wiley & Sons, Inc.: Hoboken, NJ, USA, 2013; pp. 25–42.
17. Ren, H.; Shehzad, F.K.; Zhou, Y.; Zhang, L.; Iqbal, A.; Long, Y. Incorporation of Keplerate-type Mo–O based macroanions into Zn_2Al-LDH results in the formation of all-inorganic composite films with remarkable third-order optical nonlinearity. *Dalton Trans.* **2018**, *47*, 6184–6188. [CrossRef] [PubMed]
18. Zhang, S.; Chen, H.; Zhang, G.; Kong, X.; Yin, S.; Li, B.; Wu, L. An ultra-small thermosensitive nanocomposite with a Mo_{154}-core as a comprehensive platform for NIR-triggered photothermal-chemotherapy. *J. Mater. Chem. B* **2018**, *6*, 241–248. [CrossRef] [PubMed]
19. Li, B.; Li, W.; Li, H.; Wu, L. Ionic Complexes of Metal Oxide Clusters for Versatile Self-Assemblies. *Acc. Chem. Res.* **2017**, *50*, 1391–1399. [CrossRef]
20. Li, J.; Chen, Z.; Zhou, M.; Jing, J.; Li, W.; Wang, Y.; Wu, L.; Wang, L.; Wang, Y.; Lee, M. Polyoxometalate-Driven Self-Assembly of Short Peptides into Multivalent Nanofibers with Enhanced Antibacterial Activity. *Angew. Chem. Int. Ed.* **2016**, *55*, 2592–2595. [CrossRef]
21. Müller, A.; Serain, C. Soluble molybdenum blues-"des Pudels Kern". *Acc. Chem. Res.* **2000**, *33*, 2–10. [CrossRef]
22. Müller, A.; Krickemeyer, E.; Meyer, J.; Bögge, H.; Peters, F.; Plass, W.; Diemann, E.; Dillinger, S.; Nonnenbruch, F.; Randerath, M.; et al. $[Mo_{154}(NO)_{14}O_{420}(OH)_{28}(H_2O)_{70}]^{(25\pm5)-}$: A Water-Soluble Big Wheel with More than 700 Atoms and a Relative Molecular Mass of about 24000. *Angew. Chem. Int. Ed. Engl.* **1995**, *34*, 2122–2124. [CrossRef]
23. Müller, A.; Krickemeyer, E.; Bögge, H.; Schmidtmann, M.; Peters, F. Organizational Forms of Matter: An Inorganic Super Fullerene and Keplerate Based on Molybdenum Oxide. *Angew. Chem. Int. Ed.* **1998**, *37*, 3359–3363. [CrossRef]
24. Müller, A.; Sarkar, S.; Shah, S.Q.N.; Bögge, H.; Schmidtmann, M.; Kögerler, P.; Hauptfleisch, B.; Trautwein, A.X.; Schünemann, V. Archimedean Synthesis and Magic Numbers: "Sizing" Giant Molybdenum-Oxide-Based Molecular Spheres of the Keplerate Type. *Angew. Chem. Int. Ed.* **1999**, *38*, 3238–3241. [CrossRef]
25. Guo, J.; Liu, W.; Dong, X.; Lei, H.; Li, X.M.; Sun, B.; Yin, Y.H. Expression and purification of human papillomavirus type 52 virus-like particles in Escherichia coli. *Biotechnology* **2019**, *2*, 127–132.
26. Liu, T.; Langston, M.L.K.; Li, D.; Pigga, J.M.; Pichon, C.; Todea, A.M.; Müller, A. Self-recognition among different polyprotic macroions during assembly processes in dilute solution. *Science* **2011**, *331*, 1590–1592. [CrossRef] [PubMed]

27. Fu, D.-Y.; Zhang, S.; Qu, Z.; Yu, X.; Wu, Y.; Wu, L. Hybrid Assembly toward Enhanced Thermal Stability of Virus-like Particles and Antibacterial Activity of Polyoxometalates. *ACS Appl. Mater. Interfaces* **2018**, *10*, 6137–6145. [CrossRef]
28. Jin, S.; Zheng, D.-D.; Sun, B.; Yu, X.; Zha, X.; Liu, Y.; Wu, S.; Wu, Y. Controlled Hybrid-Assembly of HPV16/18 L1 Bi VLPs in Vitro. ACS Appl. Mater. Interfaces. *ACS Appl. Mater. Interfaces* **2016**, *8*, 34244–34251. [CrossRef]
29. Zheng, D.-D.; Fu, D.-Y.; Wu, Y.; Sun, Y.-L.; Tan, L.-L.; Zhou, T.; Ma, S.-Q.; Zha, X.; Yang, Y.-W. Efficient inhibition of human papillomavirus 16 L1 pentamer formation by a carboxylatopillarene and a p-sulfonatocalixarene. *Chem. Commun.* **2014**, *50*, 3201–3203. [CrossRef]
30. Chen, X.S.; Garcea, R.L.; Goldberg, I.; Casini, G.; Harrison, S.C. Structure of Small Virus-like Particles Assembled from the L1 Protein of Human Papillomavirus 16. *Mol. Cell* **2000**, *5*, 557–567. [CrossRef]
31. Zhang, T.; Li, H.-W.; Wu, Y.; Wang, Y.; Wu, L. Self-Assembly of an Europium-Containing Polyoxometalate and the Arginine/Lysine-Rich Peptides from Human Papillomavirus Capsid Protein L1 in Forming Luminescence-Enhanced Hybrid Nanospheres. *J. Phys. Chem. C* **2015**, *119*, 8321–8328. [CrossRef]
32. Zhang, T.; Li, H.-W.; Wu, Y.; Wang, Y.; Wu, L. The Two-Step Assemblies of Basic-Amino-Acid-Rich Peptide with a Highly Charged Polyoxometalate. *Chem.-Eur. J.* **2015**, *21*, 9028–9033. [CrossRef]
33. Bernardini, G.; Wedd, A.G.; Zhao, C.; Bond, A.M. Photochemical oxidation of water and reduction of polyoxometalate anions at interfaces of water with ionic liquids or diethylether. *Proc. Natl. Acad. Sci. USA* **2012**, *109*, 11552–11557. [CrossRef]
34. Zhang, J.; Bond, A.M.; MacFarlane, D.R.; Forsyth, S.A.; Pringle, J.M.; Mariotti, A.W.A.; Glowinski, A.F.; Wedd, A.G. Voltammetric Studies on the Reduction of Polyoxometalate Anions in Ionic Liquids. *Inorg. Chem.* **2005**, *44*, 5123–5132. [CrossRef] [PubMed]
35. Bijelic, A.; Rompel, A. The use of polyoxometalates in protein crystallography—An attempt to widen a well-known bottleneck. *Coordin. Chem. Rev.* **2015**, *299*, 22–38. [CrossRef] [PubMed]
36. Bijelic, A.; Rompel, A. Ten Good Reasons for the Use of the Tellurium-Centered Anderson–Evans Polyoxotungstate in Protein Crystallography. *Acc. Chem. Res.* **2017**, *50*, 1441–1448. [CrossRef] [PubMed]
37. Cardone, G.; Moyer, A.L.; Cheng, N.; Thompson, C.D.; Dvoretzky, I.; Lowy, D.R.; Schiller, J.T.; Steven, A.C.; Buck, C.B.; Trus, B.L. Maturation of the Human Papillomavirus 16 Capsid. *mBio* **2014**, *5*, e01104–e01114. [CrossRef] [PubMed]
38. Li, M.; Cripe, T.P.; Estes, P.A.; Lyon, M.K.; Rose, R.C.; Garcea, R.L. Expression of the human papillomavirus type 11 capsid protein in escherichia coli: Characterization of protein domains involved in DNA binding and capsid assembly. *J. Virol.* **1997**, *71*, 2988–2995. [CrossRef] [PubMed]
39. Newton, G.N.; Cameron, J.M.; Wales, D.J. Shining a light on the photo-sensitisation of organic-inorganic hybrid polyoxometalates. *Dalton Trans.* **2018**, *47*, 5120–5136.
40. Shi, Z.; Zhou, Y.; Zhang, L.; Yang, D.; Mu, C.; Ren, H.; Shehzada, F.K.; Li, J. Fabrication and optical nonlinearities of composite films derived from the water-soluble Keplerate type polyoxometalate and chloroform-soluble porphyrin. *Dalton Trans.* **2015**, *44*, 4102–4107. [CrossRef]

MDPI
St. Alban-Anlage 66
4052 Basel
Switzerland
Tel. +41 61 683 77 34
Fax +41 61 302 89 18
www.mdpi.com

Nanomaterials Editorial Office
E-mail: nanomaterials@mdpi.com
www.mdpi.com/journal/nanomaterials

www.ingramcontent.com/pod-product-compliance
Lightning Source LLC
LaVergne TN
LVHW070231100526
838202LV00015B/2118